THE FOSSIL RECORD

West ⟷ East

- First men (in Plioc)
- First manlike primates
- First apes
- Grass spreads widely
- First elephants
- First horses
- Extinction of dinosaurs, giant marine reptiles, flying reptiles, and ammonites
- First primates
- Angiosperms spread widely
- First snakes
- First sequoias
- First birds
- First turtles and lizards
- First dinosaurs and mammals
- Last giant amphibians
- Extinction of trilobites, fusulinids, many corals, crinoids, and other invertebrates
- First mammal-like reptiles
- First conifers, ferns, and ginkgoes; first reptiles
- First flying insects
- First fusulinids
- Extinction of graptolites
- First seed plants
- First land-living vertebrates
- First sharks
- First forests and insect-like arthropods
- First ammonites
- First land vegetation and air-breathing arthropods
- First bone-bearing animals
- First bryozoans
- First corals
- First graptolites
- First gastropods
- First pelecypods and cepholopods
- Appearance of abundant fossils of many invertebrate phyla: arthropods, brachiopods, sponges, and echinoderms
- Fossils of coelenterates, annelids and "worms." Possible arthropods, molluscs and conodont animals.

CASCADIAN OROGENY
Pacific border

ROCKY MOUNTAIN (LARAMIDE) OROGENY
Cordilleran region of Mexico, United State, and Canada

NEVADAN OROGENY
Western Great Basin, Sierra Nevada

COLUMBIAN OROGENY
Western Canada

Continent moving relatively westward

LATE PALEOZOIC (APPALACHIAN) OROGENIES

Modern Atlantic Ocean begins to open

ALLEGHENY OROGENY
(Mississippian-Permian)
Middle and southern Appalachians

SONOMA OROGENY
Western Great Basin

ANCESTRAL ROCKIES
(Mid-Pennsylvanian)
Colorado, New Mexico, Utah

OUACHITA-MARATHON OROGENY
(Mississippian-Pennsylvania)
Texas, Oklahoma, Arkansas

ANTLER OROGENY
Western Great Basin

ACADIAN OROGENY
Southeastern Canada, New England, Piedmont

ELLESMERIAN OROGENY
Northern Canada

Proto-Atlantic Ocean closing

TACONIC OROGENY
Northern Appalachians, Piedmont

Proto-Atlantic Ocean between North America and Europe

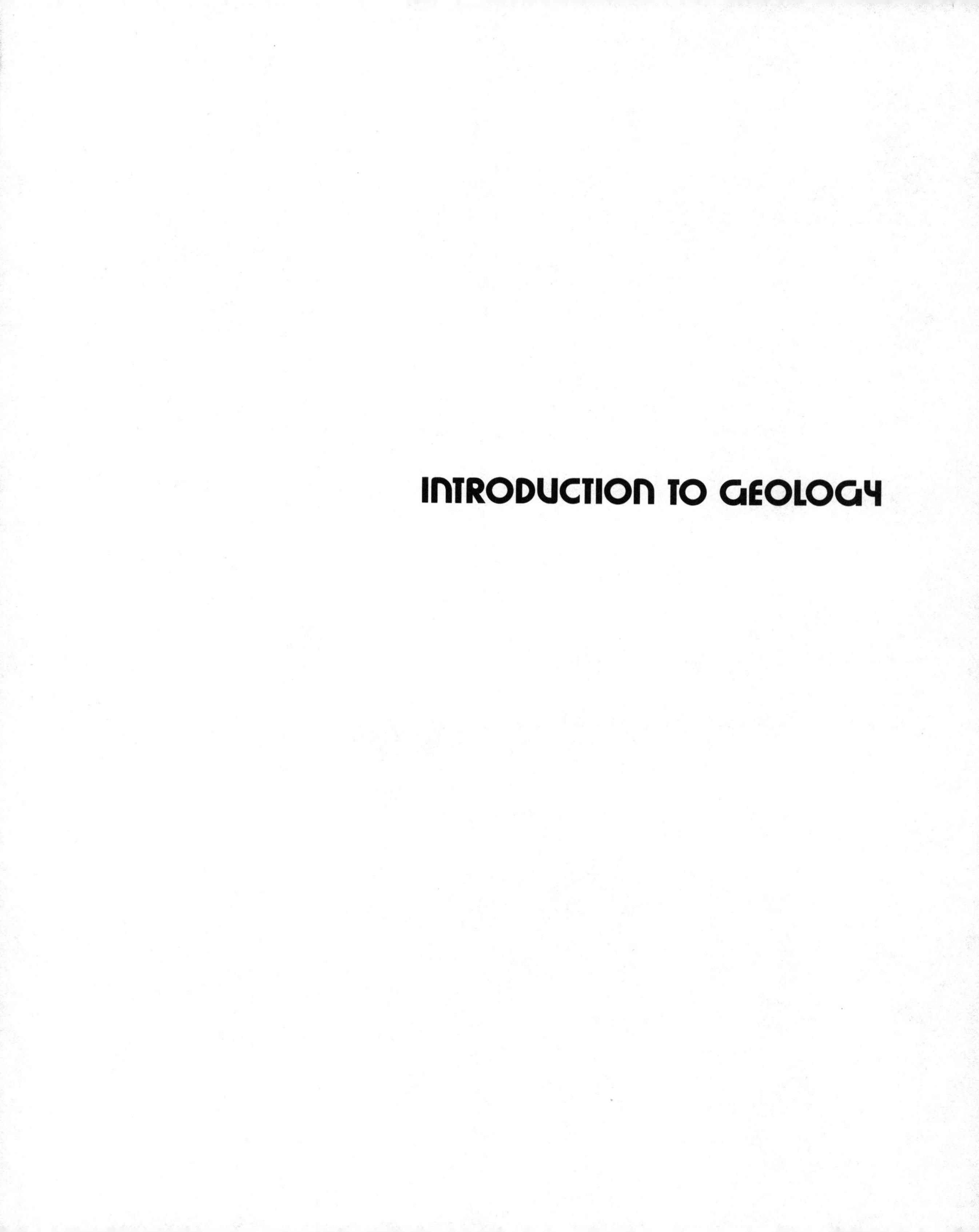
INTRODUCTION TO GEOLOGY

INTRODUCTION TO

PRENTICE-HALL, INC. Englewood Cliffs, New Jersey 07632

GEOLOGY PHYSICAL AND HISTORICAL

WILLIAM LEE STOKES
Professor, Department of Geology and Geophysics
University of Utah

SHELDON JUDSON
Professor, Department of Geological and Geophysical Sciences
Princeton University

M. DANE PICARD
Professor, Department of Geology and Geophysics
University of Utah

Library of Congress Cataloging in Publication Data

Stokes, William Lee, (date)
Introduction to geology, physical and historical.

Includes bibliographies and index.
1. Geology. I. Judson, Sheldon, joint author.
II. Picard, M. Dane, (date) joint author.
III. Title.
QE26.2.S75 551 77-21570
ISBN 0-13-484352-5

INTRODUCTION TO GEOLOGY

Physical and Historical, second edition

William Lee Stokes, Sheldon Judson, and M. Dane Picard

Printed in the United States of America

10 9 8 7 6 5

Prentice-Hall International, Inc., *London*
Prentice-Hall of Australia Pty. Limited, *Sydney*
Prentice-Hall of Canada, Ltd., *Toronto*
Prentice-Hall of India Private Limited, *New Delhi*
Prentice-Hall of Japan, Inc., *Tokyo*
Prentice-Hall of Southeast Asia Pte. Ltd., *Singapore*
Whitehall Books Limited, *Wellington, New Zealand*

In memory of Field, Hess, Howell, Jepsen, and MacClintock
and in appreciation
of Buddington, Dorf, Van Houten,
and our other professors of Geology at Princeton
who taught that the Earth is a unified whole.

Contents

17

The Early Paleozoic Periods 409

The Late Paleozoic Periods 434

The Mesozoic Era 465

Preface

Ten years have passed since this book was first published. Geology has changed greatly in that decade and what a student now needs to know and wants to know has also changed greatly. This is, therefore, an auspicious time for a second edition.

Geology stands in the midst of a revolution in science, a revolution ignited in the deep ocean basins as geologists grappled with continental drift and the origin of the deep oceans. Continental drift is now accepted by nearly all geologists. The unifying theory of crustal plate tectonics, which evolved from the early studies of continental drift and the deep ocean basins, is concerned with the dynamics of our planet and relates continents, ocean basins, earthquake belts, and volcanism in a single system of slow-moving material below the earth's crust. Continents have split, drifted apart, collided, and been joined together in new mosaics. The continents have grown; the oceans have continually been created and consumed as the earth's crust has moved. The revolution continues.

In addition to a new chapter on continental drift and crustal plate tectonics, we have included an account of the current planetary exploration and the discoveries of that exploration. Man's strongest longing is to not be alone and, although life has not yet been discovered on other planets, their exploration furthers our compelling need to understand the evolution of the solar system.

All of us are probably more concerned, however, with our environment and geological hazards than ever before in the history of the world. This book has therefore been reoriented to reflect such concerns and to help the student understand them. A long chapter near the end of the book on energy and mineral resources is also especially important in this time of daily attention to present and impending shortages and their cost in currency and misery throughout the world.

Finally, this book was written for students who are trying to understand their relationship to the earth on which they live. We have attempted to bring the student the essential present knowledge of the earth and its past, to show how this knowledge has been gained, and to focus attention on major unsolved problems that still confront geologists.

The number of persons who have helped us is large indeed. Credits for individual photographs are given in the text. Earle F. McBride, William P. Nash, Robert B. Smith, and Peter H. Roth critically read individual chapters and contributed many useful suggestions that were followed. Typing of various drafts was done by Donna Colton, Ginny Picard, and Betty Stokes. The photographic group at Medical Illustrations and Photography, College of Medicine, University of Utah, helped Picard with many of his photographs.

We are also indebted to the project editor, Logan Campbell, the designer, Lee Cohen, and to Phyllis Springmeyer, the production editor, for their considerable assistance along the way. No book is written and published in isolation, and we are very appreciative of the help that we've received.

W.L.S. / S.J. / M.D.P.

INTRODUCTION TO GEOLOGY

The Astronomical Background

The greatest technological achievement of man to date was to leave the Earth and reach the Moon. One of the results of this exploit was a better view of the Earth. Many photographs were returned including spectacular views of the entire globe. To the ordinary citizen this view of his home may have had more significance than the moon walks or moon rocks. It is no coincidence that the comprehensive views of Earth from space came when we also began to see clearly the nature and limitations of our earthbound mineral and energy resources.

Viewed from space, the Earth displays great alternating expanses of land and water partly hidden by the semitransparent atmosphere with ever-changing patterns of water-bearing clouds. Comparable space photographs have been made of the Moon, Mars, Mercury, and Venus. Venus is shrouded in

1

Earth as seen from the Moon. (Photo by NASA.)

thick unbroken mists; the Moon and Mercury appear to be frozen, static, and lifeless; and Mars shows considerable surface activity and water but also appears to be lifeless. Other sensational photographs have been returned from spacecrafts passing near Jupiter and Saturn. These show giant gaseous globes that contrast strongly with the smaller rocky planets such as Earth.

How can the members of the solar family be so diverse? And why, among the varied solid bodies, is Earth alone a suitable habitat for advanced forms of life? Only a few decades ago these questions could not even have been asked in a meaningful way. Until recently it was believed that bizarre but intelligent beings, more or less like earth people, existed on all planets and even on the Moon. Now we know that this belief is merely fanciful if not wishful thinking. We are utterly alone in the solar system. But this leads to another basic question: Are we alone in the universe? This problem may be beyond the power of science to solve simply because we cannot hope to travel far enough to find the answer.

A comparison of the planets

Who can lay claim to the Moon and planets? Many possible answers come to mind, but scientifically speaking, the exploration and explanation of those bodies that are earthlike or terrestrial belong to geology. The name *astrogeology* has considerable usage for the branch of knowledge concerned with the geology of the planets and other rocklike bodies of the solar system. A more inclusive term is *planetology,* which naturally includes the giant gaseous or icy planets: Jupiter, Saturn, Uranus, and Neptune.

Basic data about the various members of the solar family are given here in tabular form (Table 1-1); and the relative sizes of the planets are shown in Figure 1-1.

The inner planets (including the Moon) are much alike. All are stratified internally with thin shells of lighter material on the outside and a central heavier core. All are rich in iron and silica; common constituents would be ferrous oxide, FeO; olivine $(Mg,Fe)_2SiO_4$; and augite, $Ca(Mg,Fe,Al)(Al,Si)_2O_6$. Volcanic

Table 1-1 The Solar System

	AVERAGE DISTANCE FROM THE SUN	LENGTH OF THE YEAR	ROTATION	EQUATORIAL DIAMETER	MASS	DENSITY	NUMBER OF SATELLITES
	Millions of Kilometers	*Earth Units*	*Earth Units*	*Kilometers*	*Earth = 1*	*Water = 1*	
Sun	–	–	25–35 days	1,390,000	343.000	1.42	–
Moon	–	365.26 days	27.3 days	3,475	0.012	3.36	–
Mercury	58	87.97 days	58.64 days	4,830	0.10	5.4	0
Venus	108	224.70 days	243 days	12,108	0.82	5.27	0
Earth	147.5	365.26 days	23.9 hours	12,750	1.00	5.52	1
Mars	229	686.98 days	24.6 hours	6,800	0.11	3.9	2
Jupiter	780	11.86 years	9.93 hours	143,000	317.8	1.34	13
Saturn	1431	29.46 years	10.23 hours	121,000	95.1	0.69	11
Uranus	2880	84.02 years	12.3 hours	47,000	14.5	1.7	5
Neptune	4510	164.79 years	15.8 hours	45,000	17.2	2.3	2
Pluto	5950	248.4 years	6.39 days	6,000 (?)	0.8 (?)	3.0 (?)	0

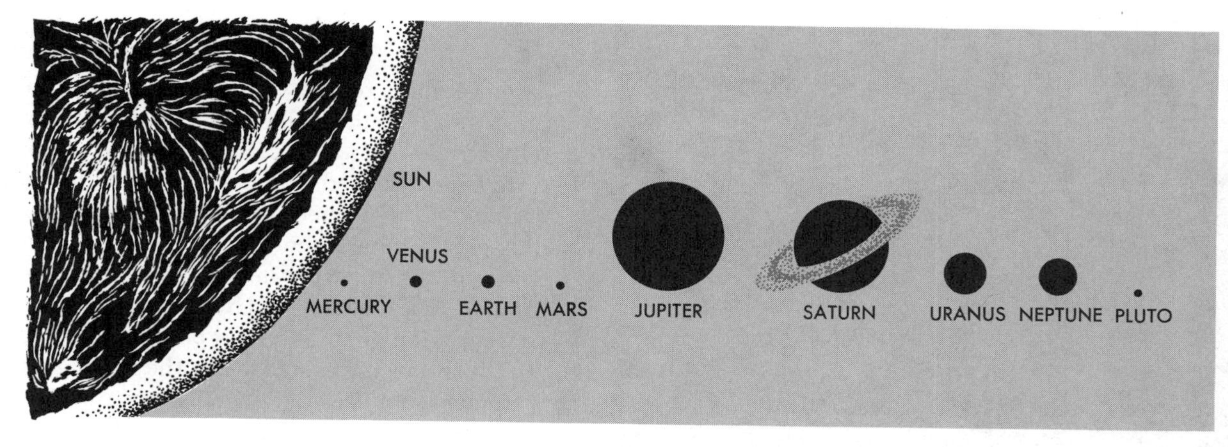

FIGURE 1-1 *The relative sizes of the planets in relation to a portion of the sun.*

Table 1-2 Facts and Figures Concerning the Earth

SIZE	
Polar radius	3,950 miles (6,357 km)
Equatorial radius	3,964 miles (6,378 km)
Mean radius	3,956 miles (6,371 km)
Circumference around the poles, approx.	24,857 miles (40,009 km)
Circumference around the equator, approx.	24,900 miles (40,079 km)
Ellipticity $\left(\frac{\text{equatorial radius} - \text{polar radius}}{\text{equatorial radius}}\right) = \frac{1}{283}$	

SHAPE

Oblate spheroid—a sphere flattened slightly at the poles. In nontechnical terms, the Earth is slightly pear-shaped, with the narrow end in the Arctic and the broader base in the Antarctic. The equator is slightly egg-shaped and not circular. The Earth has four high points arranged roughly like the corners of a pyramid.

VOLUMES	
Volume of total earth	260 billion cubic miles (1.08×10^{12} km^3)
Volume of water	330 million cubic miles ($1,370 \times 10^6$ km^3)
Volume of crust	2 billion cubic miles ($6,210 \times 10^6$ km^3)
Volume of mantle	216 billion cubic miles ($898,000 \times 10^6$ km^3)
Volume of core	41 billion cubic miles ($175,500 \times 10^6$ km^3)

SURFACE AND AREAS	
Total surface area, approx.	198 million square miles (510×10^6 km^2)
Land area (29.22% of total)	57.5 million square miles (149×10^6 km^2)
Water area (70.78% of total)	139.4 million square miles (361×10^6 km^2)
Oceans and seas less continental shelves	128.4 million square miles (322.6×10^6 km^2)
Land area plus continental shelves	68.5 million square miles (177.4×10^6 km^2)

DENSITIES (Water = 1)	
Density of entire earth	5.52
Density of crust, approx.	2.85
Density of mantle, approx.	4.53
Density of core, approx.	10.70

RELIEF	
Greatest height of land, Mt. Everest	29,028 feet (8,848 m)
Greatest known depth of ocean, Marianas Trench	36,198 feet (11,033 m)
Average height of land	2,757 feet (840 m)
Average depth of ocean	12,460 feet (3,808 m)

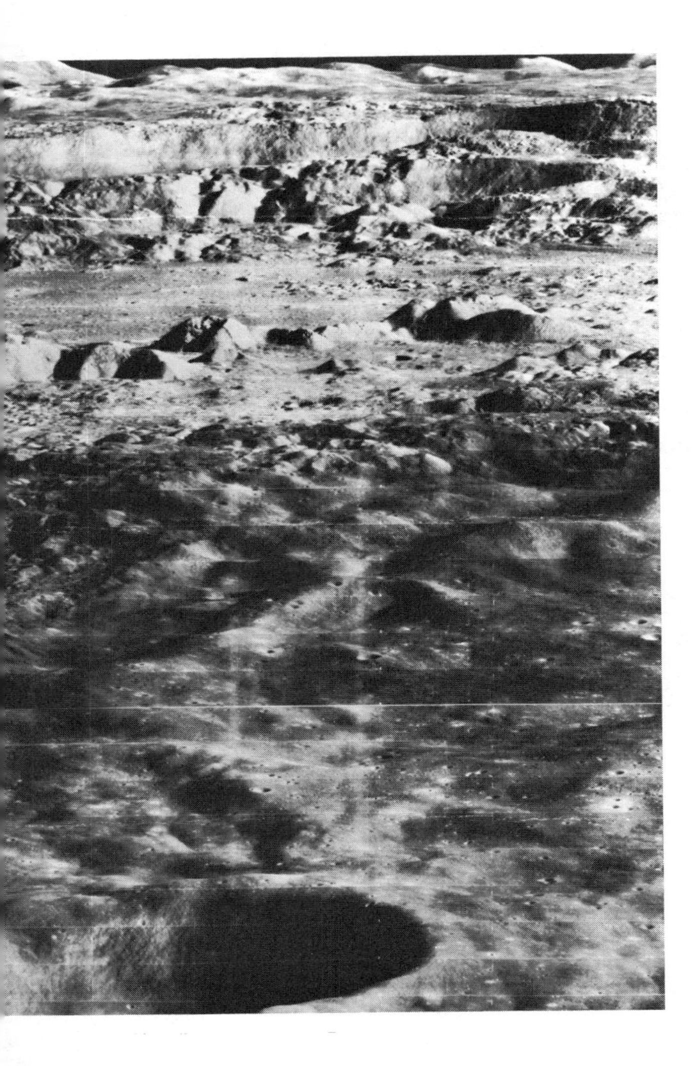

large asteroids or comets, took place early in the history of the solar system, about four billion years ago. Since that time few large craters have formed, and the rate of infall has declined. One mysterious fact with regard to the great bombardment epoch has also emerged: the impact scars are not evenly distributed. Mars, Moon, and Mercury all show a concentration of large craters on one side. For the Moon this is the side facing the Earth. It is tempting to think that this strange distribution is somehow related to the fact that Earth too is not symmetrical. It has a water hemisphere and a land hemisphere, a distinction that was much stronger at some time in the past than it now is.

FIGURE 1-2 *Low-level view of the crater Copernicus, showing numerous minor craters and the hummocky topography created on the lunar surface by successive meteoritic impacts over long periods of time. (Photo by NASA.)*

structures, particularly lava flows, are found on each of these bodies indicating melting at depth, but the evidences of volcanism differ greatly from planet to planet. In addition, Mars, Mercury, and the Moon show numerous superimposed craters of varied sizes. There are fairly good evidences that Venus too is marked by craters and that the Earth did not escape the general bombardment.

Crater formation has not been evenly distributed in time, there is positive evidence that the greatest collisions, involving very

The Earth

A brief tabular comparison of the planets, including Earth, was given in Table 1-1. This information is chiefly astronomical. Other specific facts and figures pertaining to the Earth are presented in Table 1-2. The data are chiefly geologic and will serve to introduce the factual material of succeeding chapters and also to draw attention to problems of the Earth's origin that will be briefly considered later in this chapter.

The Moon

The six moon missions of the Apollo project provided 475 kilograms of actual specimens and many kilometers of tape with other data. From this information a brief summary of the geological features can be given. The topography of the Moon consists of rough unorganized mountains and wide flat lava plains (*maria*). The whole surface is marked by innumerable craters of all sizes, which are impact scars created by fragments from space (see Figure 1-2). The highland mountains consist of shattered rocks rich in the mineral *plagioclase,* and the plains are lava rich in *titanium.* Analogous rocks are rare or nonexistent on earth. The lunar crust is 50 to 100 km thick, and it is assumed that a relatively dense, possibly molten core about 500 km in diameter occupies the center. The outer shell of brittle rock (*lithosphere*) is about 1,000 km thick and is quite rigid in contrast to the mobile outer shell of Earth. Six different views of the Moon are presented in Figure 1-3.

The Moon is thought to have formed initially about 4.6 billion years ago in the near vicinity of the Earth. Why the two bodies are dissimilar in many ways is not well understood. The outer 200 or so kilometers of the Moon were melted at an early time, probably aided by heat generated by abundant infalling fragments. At this stage there was a tendency for the relatively lighter crystals of plagioclase to rise toward the surface, and so the crust was formed.

FIGURE 1-3 *(a) The Moon from Earth. This fine photograph was taken through the 36-in. refractor telescope at Lick Observatory. The large crater, Ptolemaeus, is at the top; Alphonsus lies next below. The margin of the great lava field or mare, Oceanus Procellarum, is on the left. (Photo by NASA.)*

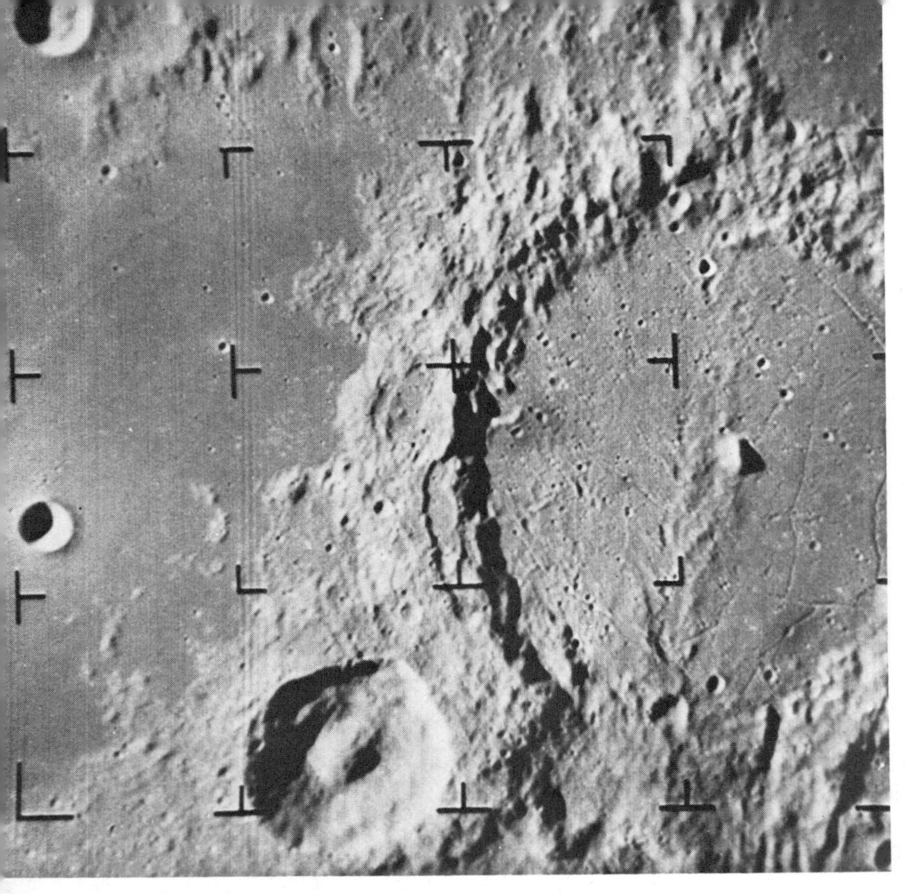

(b) View from 415 kilometers above the lunar surface taken by the spacecraft Ranger on March 24, 1965, just before it landed. The large crater Alphonsus, seen in (a), dominates the view. Compare details of the two photos. (Photo by NASA.)

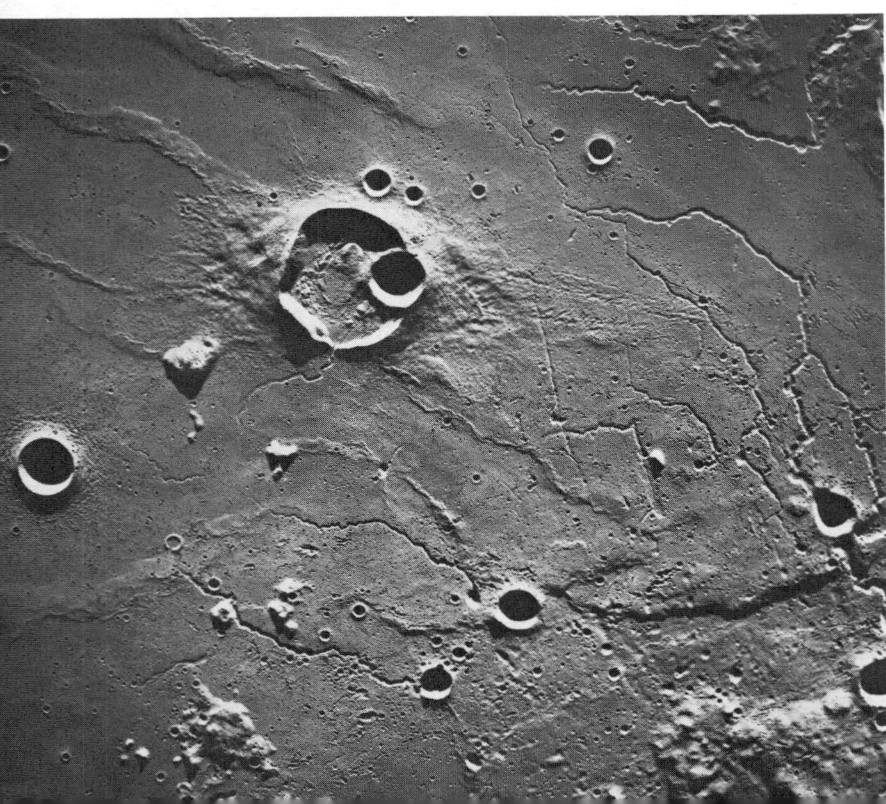

(c) View of the lava-floored Oceanus Procellarum and superimposed features. Chief interest in this area centers in the numerous narrow sinuous depressions or rilles, many of which emanate from craters. The origin of these is problematical. Also seen are straight grooves and wrinkles in the lava surface, likewise not fully explained. (Photo by NASA.)

(d) (Right) A classic photo is this first ground view of the surface and rocks of another planet. Photo is of the lunar surface near the Surveyor 1 lander on June 1, 1966. The large boulder is about 50 cm long with rounded edges, pits, and distinct fracture lines. (Photo by NASA.)

(e) (Below) The "bull's-eye" on the far side of the Moon. Officially called the Mare Orientale, this ringed feature is considered to be one of the largest impact craters on the Moon. The diameter of the outer ring of mountains averages 900 km. (Photo courtesy of NASA.)

COPERNICUS
(Ech)
(Ipm)
Ccd
Ccd?
Gay-Lussac J
Ccrr
Copernicus DA
Copernicus K
Copernicus KA
Ccrh
Cs
Ccw
Ccfs
Ccp
Ccfh
Copernicus BB
Copernicus N
Copernicus E
Copernicus G
Copernicus H
FAUTH
Fauth D
Fauth B
Fauth C
Fauth E
Reinhold E
Reinhold G
Csc
Ipm

(f) (Left) A geologist's view of the Moon. This is a geologic map of the great Copernicus crater seen from one of the landers in Figure 1-2. This map shows the different materials or formations designated by symbols and outlined in the same way that geologic features are mapped on earth. Here, however, the mapping was done from photographs. Original map is in full color. (Photo courtesy of U.S. Geological Survey.)

The very intense bombardment of the Moon by large asteroids or comets also took place rather early, and by 4.0 billion years ago most of the large maria were in existence. Partial melting at intermediate levels produced basaltic lava between about 4 billion and 3 billion years ago (Figure 1-4). This molten material rose and more or less filled the preexisting mare basins. Little has happened over the past 3 billion years except cratering on a diminishing scale. The Moon is now geologically relatively inert.

FIGURE 1-4 *Basaltic moon rock brought back by the Apollo 12 mission. Basalt is a relatively heavy igneous rock that has solidified from liquid lava. It is common on earth as well as on the moon. Specimen is about 5 cm long. (Photo by NASA.)*

Mercury

In a very successful mission, the spacecraft Mariner 10 passed near the planet Mercury on March 1974, September 1974, and March 1975. Hundreds of excellent photographs and other information were returned to Earth to clear up old misconceptions and raise new problems about the innermost planet. The surface of Mercury is very moonlike, being marked by numerous craters of varied sizes, but is unmoonlike in having no jumbled mountainous regions (Figure 1-5). The most significantly different features are long, low winding cliffs or scarps that resemble certain fault lines on earth. These are considered to have resulted from shrinkage of the core and from adjustments of the crust to fit correspondingly less area.

The density of Mercury is 5.45 (Earth, 5.5). The only way a planet so small can be so heavy is to have a large metallic core. It is calculated that the core extends to within 480 to 600 kilometers of the surface. There is obviously a crust of relatively loose material, easily stirred by meteoric impact. It has been said that Mercury possesses a moonlike crust and an earthlike core. A few volcanic features including lava flows were photographed, but these are much less common than on Mars.

Mercury was found to possess a very thin atmosphere, chiefly of helium and argon. There is a weak magnetic field, which is difficult to explain in view of the very slow rotation and probably solid core. The planet was obviously bombarded by large bodies early in its history [see Figure 1-6(a) and (b)], probably during the same episode that affected the Moon, Mars, Venus, and Earth. There is nothing to indicate that Mercury did not originate along with the rest of the planets about 4.5 billion years ago, but it has its own peculiar characteristics that will need to be integrated into a total picture of planetary geology.

FIGURE 1-5 *(Left) Photo mosaic of Mercury constructed of 18 photos taken from a distance of 130,000 miles (210,000 km) by Mariner 10 on March 29, 1974. Resemblance to the Moon is carried out not only in the varied craters but in the raylike configurations around some of them. Missing, however, are jumbled mountain masses and wide lava plains. (Photo by NASA.)*

FIGURE 1-6 *(Right) Densely cratered region of Mercury photographed by Mariner 10 spacecraft, September 21, 1974. The resemblance to the lunar surface indicates similar history and processes. (Photo by NASA.) (Below) A lightly cratered region of Mercury crossed by sinuous faultlike traces and a system of faintly polygonal cracks that may be due to adjustments of a rigid crust to a shrinking core. (Photo by NASA.)*

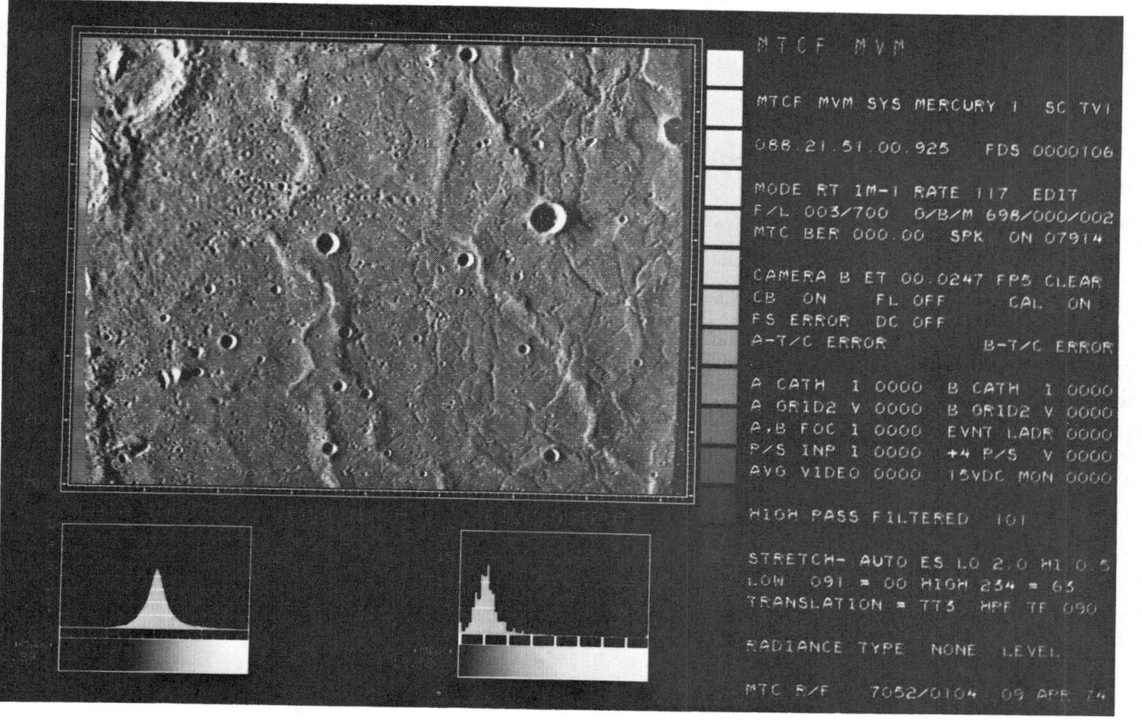

FIGURE 1-7 This view of Venus was taken from 450,000 miles (720,000 km) by Mariner 10's television cameras, February 6, 1974. The swirling pattern of high-level clouds hides a hot rocky surface below. (Photo by NASA.)

Venus

Venus has been called a sister planet to the Earth; the masses, diameters and densities of these two planets are surprisingly similar (see Table 1-1). But Venus rotates around its axis in 343 days and revolves around the Sun in 440 days, giving it the unique characteristic of retrograde motion. Other planets progress with a wheellike motion around the sun. Venus moves forward but with a rotation opposite to that of the other planets. There are other puzzling characteristics. Venus is veiled in a misty atmosphere 100 times more massive than that of Earth. This atmosphere is at least 80 percent CO_2 with a small amount of H_2O and an appreciable amount of H_2SO_4.

The temperature of the surface has been measured at 900°C, but drops in the higher levels of atmosphere to about −40°C. No one knows the meaning of these great variations, but they may be due to a "greenhouse" effect that traps solar heat within the lower atmosphere. The surface of Venus may be extremely dry and dusty with no surface bodies of liquid of any kind.

Direct photography of the surface of Venus from space appears to be out of the question; only swirling cloud patterns are seen (Figure 1-7). However radar has penetrated the clouds to give limited images of some areas. Tantalizing glimpses of craters, chaotic highlands, rilles, and canyonlike depressions have been received. Venus has been a special target of space exploration by the Soviet Union. As of November 1975, the Russians had launched seven vehicles designed to land on the surface. In October 1975, the last two missions succeeded in sending back views from the surface. Contrasting pictures of angular rocky terrain, and smooth rounded formations were recovered.

Mars

Mars has long been of surpassing interest to man. Photographs obtained by the orbiting spacecraft Mariner 9 in 1971 and 1972 greatly heightened this interest. Two Viking missions of 1976 successfully placed instruments on the surface and orbiters around the planet (Figure 1-8). Mars has proven to be literally a new world geologically speaking. It may be thought of as intermediate between Earth and Moon—less dynamic than the former, more dynamic than the latter. Mars resembles Earth in having evidence of a heavy metallic core, but this may not be liquid. The magnetic field is relatively weak. The lithosphere is apparently relatively thicker than the Earth's and there is clearly no evidence of great lateral movements such as result on Earth from the processes of sea-floor spreading and continental drift.

FIGURE 1-8 *The surface of Mars as seen at the sites of the Viking I landing (above) and Viking 2 landing (below). Scattered rocks at and below the surface seem to be fragments broken mechanically and not shaped by water. The porous pieces have the appearance of basaltic lava. The color as shown by the original photos is distinctly reddish and may be due to hydrated iron oxide. (Photo by NASA.)*

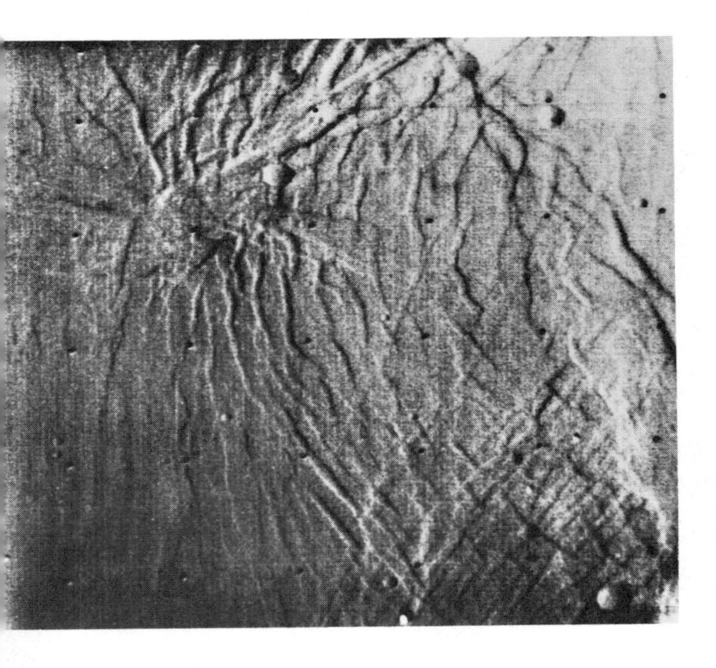

FIGURE 1-9 (a) (Above) An area broken by faults into mosaiclike sections. The depressed trenches are about $1\frac{1}{2}$ miles (2.4 km) across. (Courtesy of NASA.) (b) (Below) Rilles or cracks that are part of a system more than 1,100 miles (1,800 km) long. View taken from an elevation of 1,072 miles (1,730 km). (Photo by NASA.)

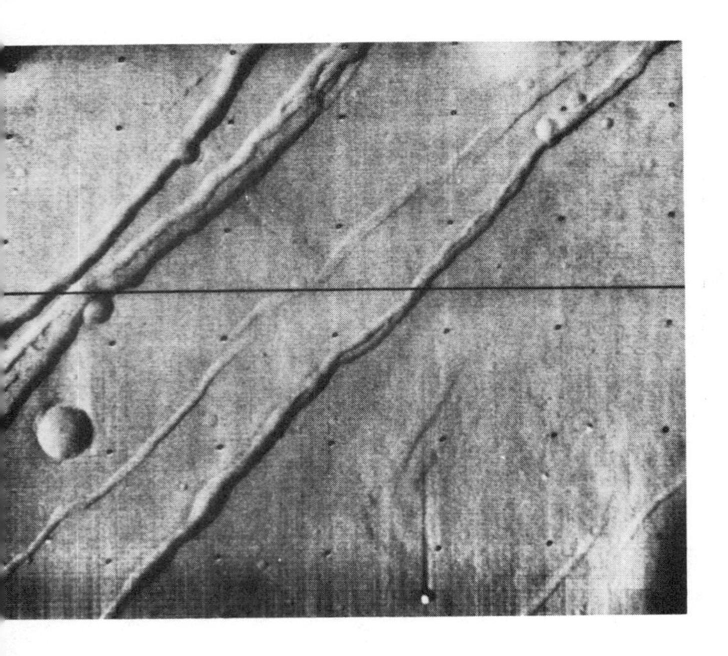

So diverse and alien is the topography of Mars that entirely new classifications of landscapes have been proposed to describe it. Included are: (a) pitted and etched forms mainly due to scouring by wind; (b) fretted forms resulting from melting of the frozen subsurface; (c) troughs caused by systematic collapse or subsidence brought on by withdrawal of melted rock, splitting under tension, or by melting of ground ice; (d) hollowed forms of uncertain origin, and (e) chaotic forms caused by slumping, collapse, and subsidence with no apparent pattern or obvious cause. Of course, dense or sparse craters of many sizes are superimposed on all types of terrane and are due to external haphazard impacts. Figure 1-9 illustrates some of the diverse topographic features.

Although there is evidence of great, even catastrophic, flooding and erosion, there are no surface bodies of water on Mars. Great canyons with branching tributaries, wide braided channels cutting in and around streamlined "islands," and areas that resemble eroded badlands on earth indicate former activity of running water. Evidence of igneous action abounds. The massive mountain Nix Olympica, 25 km high, is the greatest volcano known in the solar system. There are many smaller cones and great lava fields with superimposed flows and sheets of debris. And yet no active volcanism.

Much of the landscape has been shaped by wind, and this agent has been observed in action. There are both constructional and destructional features. Very earthlike are the small dunes and streamlined trails of sand around rock fragments seen in the vicinity of the Viking 1 lander. Possibly the least understood and most unearthly of the Martian landscapes are those caused by frozen water on the surface and in the subsurface. The polar icecaps are now known to be water ice, and it is thought that this may occasionally melt to provide liquid water. Also the thawing of frozen ground may account for the collapse and slumping of vast tracts at all latitudes.

Other findings include the presence of nitrogen, argon, and krypton in the atmosphere. Most of the experiments sent with the Viking missions were biological in nature—intended to discover life or the possibilities of life. Although conditions are in many ways favorable—the presence of water, for example, being confirmed—no positive evidence of past or present life has been discovered to date (1977).

(c) (Right) An area having a pattern much like water-eroded badlands on earth. Note the branching of channels upward toward the divides. (Photo by NASA.) (d) (Below) Gigantic volcanic mountain on Mars, called Nix Olympica, is 310 miles (500 km) broad at the base, and 18 miles (25 km) high. The crater area at the summit is 40 miles (65 km) across. The steep cliffs that rim the base of the mountain are unexplained. (Courtesy of NASA.)

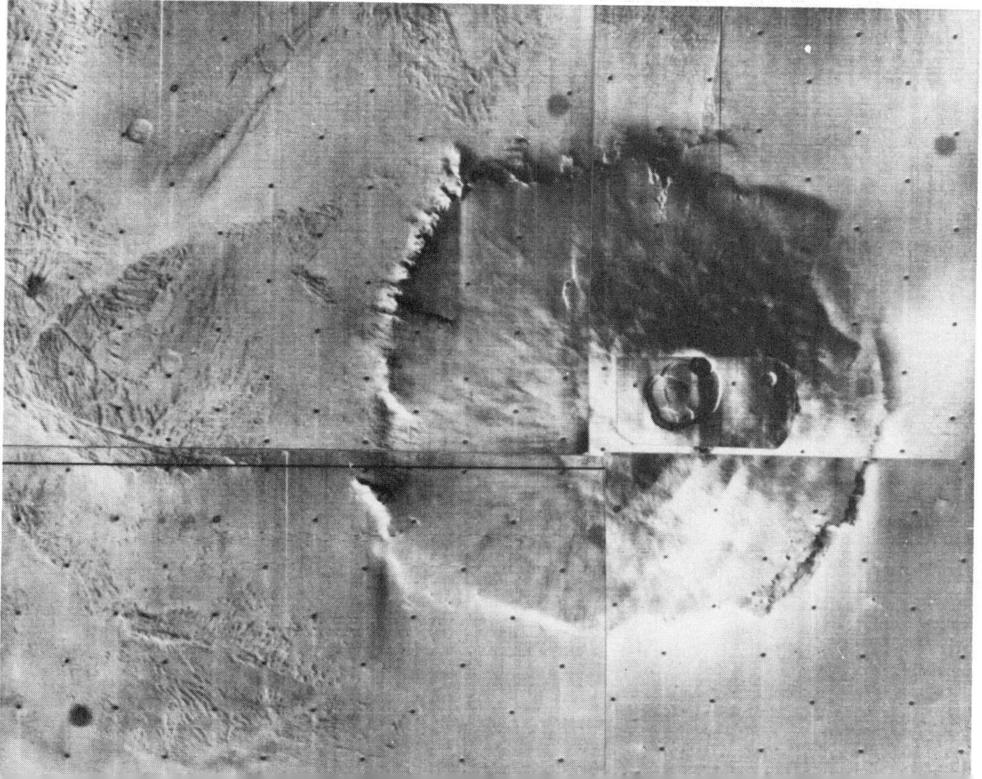

(e) Winding valley system of Mars that is thought by most students of the planet to have resulted from erosion by running water during a more humid episode in the distant past. (Photo by NASA.)

Asteroids, meteorites, comets, and tektites

Numerous relatively small bodies that move in a wide belt between the orbits of Jupiter and Mars are called *asteroids.* Largest is Ceris, 770 km in diameter. Over 1,500 other sizeable asteroids have been cataloged, and several hundred thousand smaller ones are estimated to exist. A representation of the larger ones is given in Figure 1-10. The widely held view that these are remnants of a shattered planet of considerable size is now rejected in favor of the theory that they are fragments of a large number of smaller bodies comparable to Ceris. Careful study of reflected light and mutual relations favors the thought that the parent bodies were quite planetlike in possessing metallic cores encased in lighter shells of siliceous or carbonaceous material. The asteroids probably accreted in cooler regions of the original nebula. The present shattered and fragmented condition of the asteroids is attributed to continuing collisions

FIGURE 1-10 *Diagrammatic comparison of some of the larger named asteroids and the United States plus a comparison of Eros with the island of Manhattan. [Redrawn from A. Lee McAlester, The Earth (Englewood Cliffs, N.J.: Prentice-Hall, Inc., 1973), p. 456, and used with permission.]*

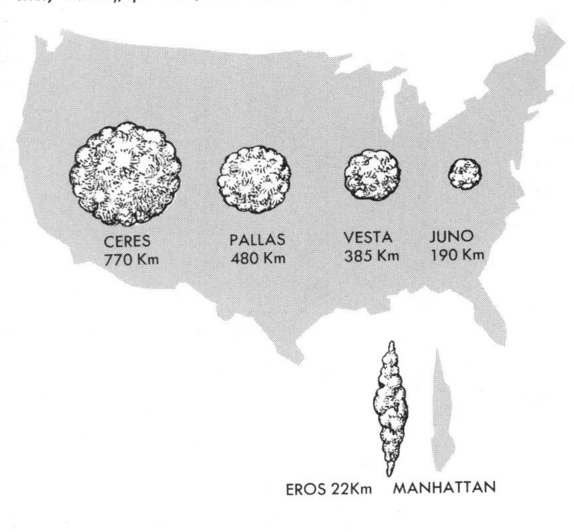

among the group and to a period of invasion by the same large bodies that left indelible marks on the other rocky planets early in the history of the solar system.

Inasmuch as the asteroid belt is the most probable source of meteorites, it is well to give brief attention to these bodies. A *meteorite* by definition is an object larger than a molecule and smaller than a planet that passes through the atmosphere and reaches the surface of the earth without being vaporized. There are two basic types—the irons and the stones. The irons are almost pure metal, iron being a chief constituent (Figure 1-11). The crystalline pattern shown by many irons indicates an origin under conditions of heat and pressure such as might exist in a large asteroid. Many of the stony meteorites are much like earth rocks. The most common type of stony meteorite comprises the *chondrites,* so named because they contain small, rounded, droplike inclusions called *chondrules* (Figure 1-12). Chondrules have a rather uniform chemical composition similar to the heavier basic rocks of the earth and may be samples of the original swarm of particles that aggregated to form the earthlike planets. In support of this theory is the discovery that they show a surprisingly uniform age of about 4.6 billion years, a figure close to that calculated for the earth.

Comets are in many ways the most spectacular bodies to circle the sun. Most of them move in exceedingly elongated orbits and are near the sun for only a short time in each revolution. It is during these close approaches that they blaze into prominence. The heat of the sun vaporizes a part of the nucleus or head and converts it into the lengthy, ever-changing tail. This behavior is illustrated by Figure 1-13.

Although *comets* show great variation and individuality, it is assumed that all are fundamentally aggregations of ice, dust, and rock. Many of the meteorites that crash to earth (including chunks of ice) are undoubtedly parts of disintegrated comets. The breakup of known comets into incoherent swarms of meteoritic fragments has been well authenticated. In the earlier stages of the solar system, comets may have been much more abundant; many have fallen into the planets, adding considerable substance to them.

FIGURE 1-11 *Iron meteorite from Henbury, Australia. (Center for Meteorite Studies, Arizona State University.)*

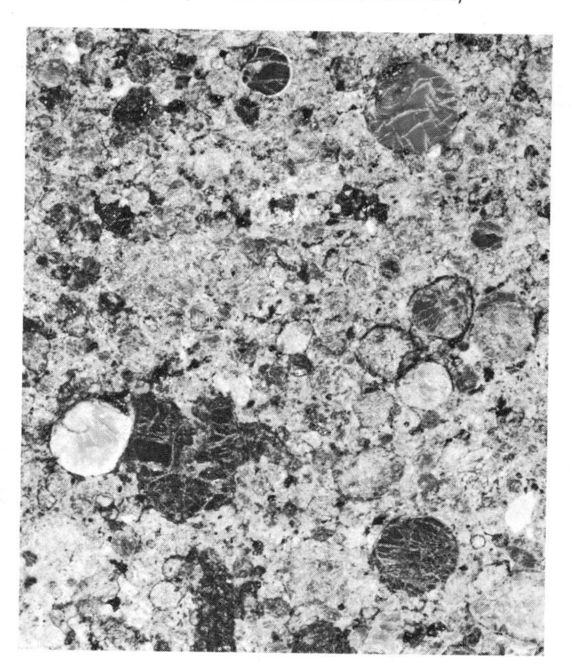

FIGURE 1-12 *Enlarged view of the interior of a gray chrondritic meteorite that fell in Indiana. This shows the spherical, variously colored chondrules. They average about 0.25 cm in diameter. (Courtesy of the Smithsonian Institution.)*

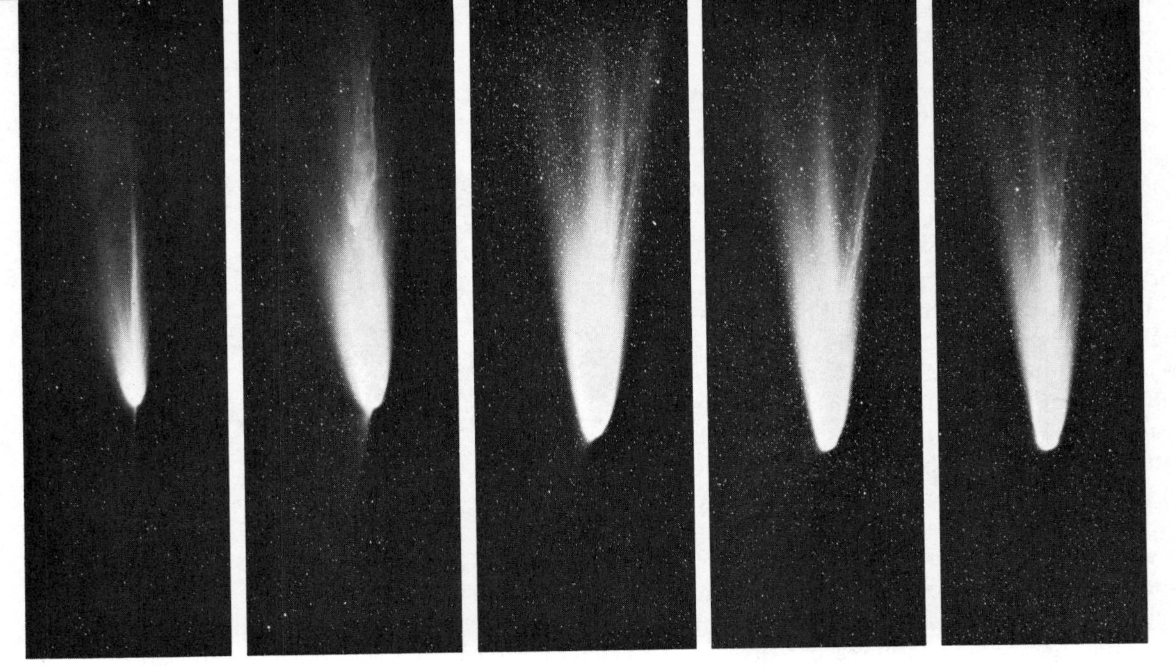

FIGURE 1-13 *Arend-Roland comet, showing changes from April 26 to May 1, 1957. (Courtesy Mt. Wilson and Palomar Observatories.)*

Tektites are small pieces of glassy material resembling the natural rock obsidian. Although they are generally regarded as a type of meteorite, none have actually been observed to fall. In shape they resemble teardrops, spheres, buttons, dumbbells, and other rounded objects. Their surfaces show clear evidence of melting, and there are internal indications of very rapid cooling. In chemical composition they resemble soil or sedimentary rocks and are very unlike ordinary meteorites (Figure 1-14).

FIGURE 1-14 *Four tektites showing variations in shape and surface markings: (a) and (b) are from the Philippine Islands, and the other two are from southern Australia. Specimens range from about 2 to 6 cm in length. (Courtesy of Smithsonian Institution.)*

(a)

(b)

(c)

(d)

Thousands of tektites have been collected, but they are not distributed uniformly over the earth's surface. Eight distinct areas of the earth yield tektites that are designated accordingly as Australites, Indochinites, Philippinites, and so forth. The origin of these bodies is a mystery, though several theories have been proposed. Some investigators claim they are of earthly origin, some believe they are of cometary origin, and others believe them to be material splashed out of the moon by large meteorites. A cometary origin now seems the most likely.

Jupiter and the other outer planets

On December 5, 1973, the spacecraft Pioneer 10 passed within 130,000 km of Jupiter, and in December 1974 Pioneer 11 also came near the giant planet. The planet seems to be composed essentially of liquid and metallic hydrogen for over 60,000 km of its 66,000-km radius. There may be a small rocky core at the center, but most of the planet is thought to consist of concentric shells of liquid and gaseous material.

It has been observed that among these numerous shells is one having temperature, pressure, and chemical components favorable to the synthesis of living things. Here evolution of the molecules basic to life might be going on, but anything higher than this would seem to be impossible.

Jupiter has 13 moons; the whole family is a miniature of the solar system. Two of the satellites, Io and Europa, are similar to the Moon in size, mass, and density. Two others, Ganymede and Callisto, are larger than the Moon but only about twice as dense as water. These four large satellites are good reflectors of light and are apparently covered with layers of ice or other frozen gases. Little is known about the fainter moons. A space picture of Jupiter is shown by Figure 1-15.

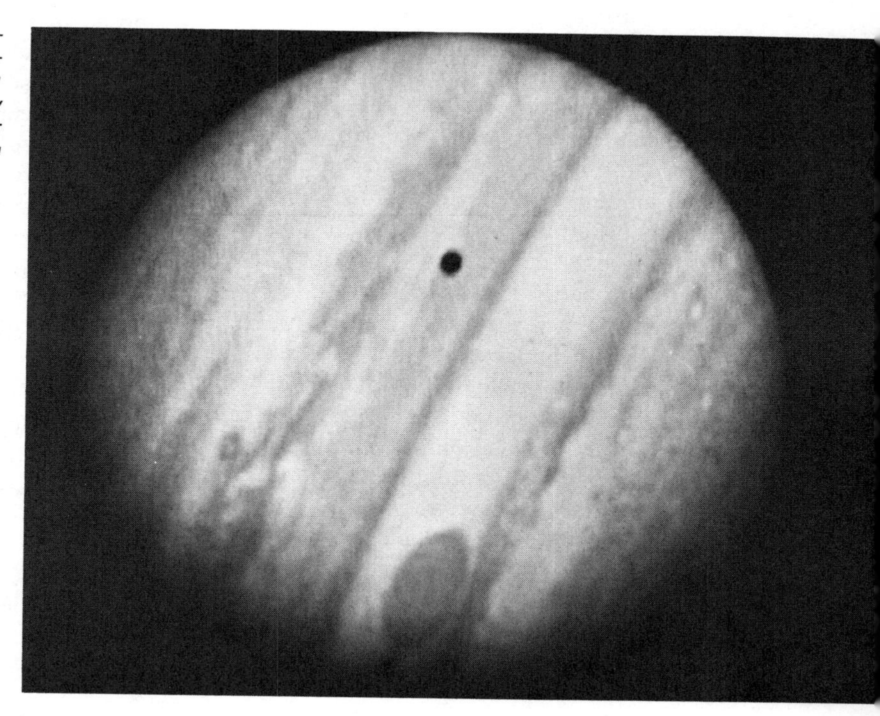

FIGURE 1-15 *The planet Jupiter as photographed from Pioneer 10 spacecraft in November 1974. The dark, large oval area known as the great red spot is a surprisingly long-lasting feature of the gaseous outer layers; the small dark spot is the shadow of a satellite passing between Jupiter and the sun. (Photo by NASA.)*

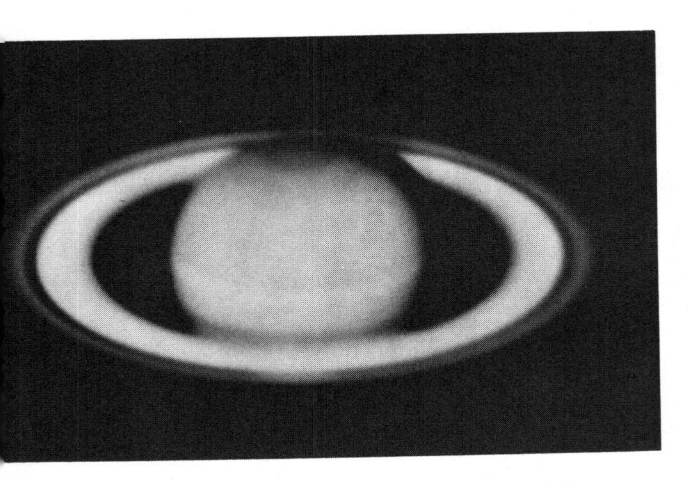

FIGURE 1-16 The planet Saturn. (Photo by NASA.)

Saturn, the most spectacular of the planets, has an extremely low mean density, being only .8 as dense as water. Even so, it probably has a good-sized core of earthy material buried under thick layers of gas, vapor, and ice. There is more methane and less ammonia than on Jupiter. Because of the low temperature, the ammonia might exist in frozen form at deeper levels.

Saturn has 11 satellites. Titan, the seventh outward, is unusual in possessing an obvious atmosphere. Since all of the measurable satellites have a density of less than 2.4, they must consist of water ice, frozen ammonia, and other light compounds mixed with lesser amounts of stony or earthy material. There is little hope of finding life in this system. Figure 1-16 is a view of Saturn taken through an earth based telescope.

The next two planets, Uranus and Neptune, are so similar that they may well be called "planetary twins." The density of Uranus is 1.53, of Neptune 2.41. Both rotate rapidly: Uranus in 12.3 hours, Neptune in 15.8 hours. Methane is abundant and ammonia is absent in the atmosphere of both planets. Pluto, the small outermost planet, is not well known. It has a high density more like that of the inner planets and may be a large satellite that escaped from Neptune.

Origin of the Earth and our solar system

The ultimate origin of the Earth is clearly an astronomical problem, and it is impossible to devote much space or time to the preplanetary aspects of the subject. Astronomers have discovered an amazing array of objects in the heavens and have had to envision an equally amazing variety of physical processes and events to explain them. Among the great number of questions that might be asked about the origin of the Earth, one stands out as a suitable starting point: How did an object as heavy and rich in iron as the earth emerge from a system that is predominately hydrogen and helium? Hydrogen and helium, the two simplest elements, make up an estimated 99 percent of all matter, and the same figure applies to the sun. The probable structure of the sun is shown by Figure 1-17.

Hydrogen is also dispersed widely in interstellar space. The well-known "big-bang" theory of the origin of the universe is supposed to have produced little more than hydrogen and helium. How then did the heavier elements, so characteristic of Earth, come into being? The answer is that evolutionary physical processes have been at work over billions of years to create and segregate the heavier elements. As a beginning, hydrogen is transformed to helium by the process of fusion in self-luminous stars. Helium has been referred to as the "ashes" of hydrogen burning. The sun is currently depleting its store of matter at the rate of four million metric tons each second in the conversion of hydrogen to helium. This is the amount of matter converted to energy according to Einstein's famous formula, $E = MC^2$.*

* E = energy in ergs
M = mass in grams
C = velocity of light in centimeters per second.

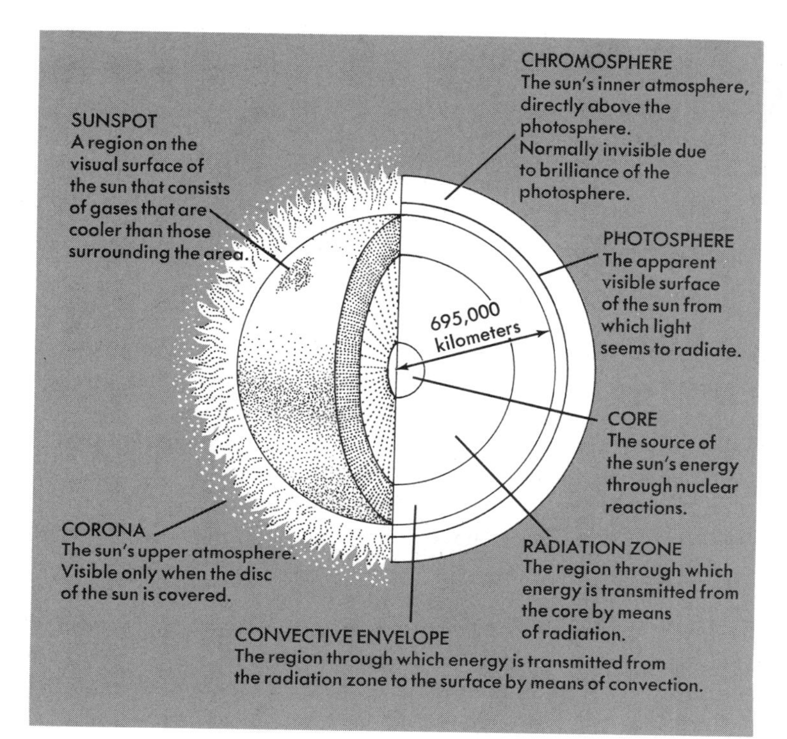

FIGURE 1-17 *The structure of the sun.*

At the exhaustion of its hydrogen fuel, a typical star becomes a red giant. Although the outer envelope of such a star expands, the center contracts and becomes hotter. Helium is now burned into carbon, and elements as heavy as iron can be formed in the outer envelope. Stars that have reached this stage, and are more than eight times as massive as the sun, may now become *supernovas*—that is, they may explode violently with the release of tremendous energy. A supernova event creates temperatures and pressures not attained in previous processes. The Crab Nebula, Figure 1-18, is the best known example. Products of such explosions include not only iron and related elements but also, on a lesser scale, all of the remaining heavier elements. Supernovas not only create heavy elements but also disperse them into surrounding space where they become available for incorporation into subsequent stellar bodies. It is entirely possible that the matter of stars such as the sun has

FIGURE 1-18 *The "Crab" nebula in Taurus. This nebula is composed of the remains of a supernova that appeared in* A.D. *1054. (Photo by Hale Observatories.)*

been constituted and reconstituted several times. This seems to be the only way in which relatively metal-rich stars can come into existence. Certain elements produced in supernova explosions are radioactive and can be used to check the age of the universe or at least give a minimum figure for it. The best estimate by this method is 20 billion years.

The sun and its attendant planets, satellites, asteroids, comets, and meteorites evidently constituted an aggregation of matter that condensed and contracted by gravitational attraction in a way no different from that of innumerable other systems. The central mass, destined to become a self-luminous body, contains all but a small fraction of the total matter of the system. In its early stages the protosun was surrounded by a far-reaching cloud of dust and gas from which the planets and smaller entities were destined to form. It can be assumed that the original cloud was essentially homogeneous and that the composition of the sun today gives a good clue as to the original mixture of elements. The sun is known to consist mainly of 15 elements. The least abundant of these 15 major elements (nickel) is more abundant than all the minor ones combined.

We now have a partial answer to the original question as to how an iron-rich planet might emerge from a predominantly hydrogen medium. Iron was brought into existence before the solar nebula and existed in a dispersed state as a small fraction of the parent mass from which both sun and planets emerged. Remaining to be explained is a further drastic winnowing out of this small fraction of iron and other heavy elements from a much greater mass of lighter elements. There are ample clues if we can but interpret them. The arrangement of planets in two distinct groups is clearly significant. Two distinct families of planets exist: an inner group consisting of Mercury, Venus, Earth, and Mars, called the *terrestrial planets;* and an outer group, Jupiter, Saturn, Uranus, and Neptune, called the *Jovian planets.*

Thermal conditions in the primitive cloud of dust and gas seem to have governed this arrangement. It is logically assumed that the central regions were hotter and the marginal areas cooler. Furthermore, there must have been a gradual cooling of the entire system with the passage of time. The rate of cooling was most important in governing the final outcome. If cooling was relatively rapid, the less volatile elements would have had one brief opportunity to solidify, after which they would have been sealed off from further reactions. Bodies with distinct onionlike shells might be expected. If cooling was slower, the period of possible reactions would have been longer, so that correspondingly more varied compounds could form. A distinction between the behavior of dust and gas must be borne in mind. Volatilized material behaves as a gas, while the same or similar material in small solid particles is dust. Gas particles cannot clump together in the same way that dust particles do. Think of ice crystals and water vapor in this connection. The passage from dust to gas or gas to dust is governed by temperature, which in the original nebula depended on the distance from the center of the system. Figure 1-19 illustrates in a general way the supposed stages in the formation of a typical galaxy.

The fate of water under conditions of slow cooling is significantly different from what happens to it under rapid cooling. Under the quick-cool theory, there would be a short period of time during which an estimated 60 percent of all material condensed would be water. Water passes from vapor to ice at 175°K. Once formed, the ice layer might be sealed off and remain henceforth ice, unless melting took place as a totally independent action.

FIGURE 1-19 Formation of galaxies. (A) Initial condensation of hydrogen and helium in the thin, expanding gas cloud; (B) Further condensation forms protogalaxies of gas and dust; rotation begins; (C) An individual galaxy with stars and nova explosions.

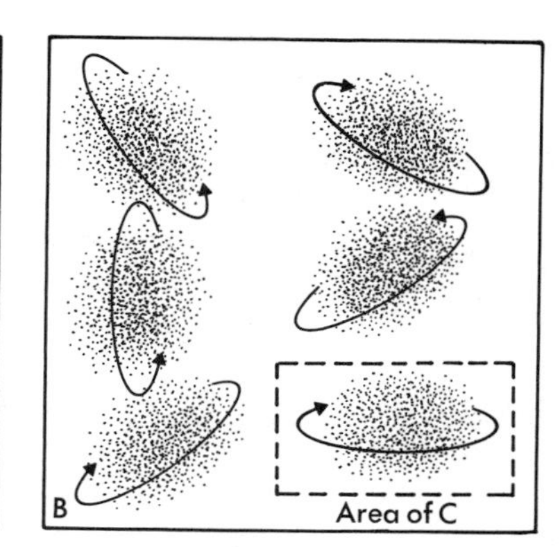

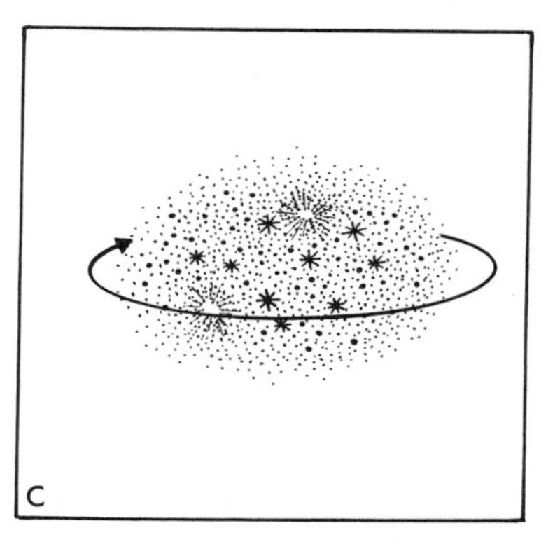

Under the slow-cool theory, water and ice are able to react for a longer time with previously formed solids and with still-free gases. At 550°K, water vapor reacts with calcium-bearing minerals to form tremolite, $Ca_2(Mg,Fe)_5Si_8O_{22}(OH)_2$. At 425°K, water combines with the mineral olivine to make serpentine, $Mg_6Si_4O_{10}(OH)_8$, which is probably the most important water-bearing mineral and rock on earth. At 150°K, after water has become ice, it can still react with ammonia (NH_3) to form the solid hydrate NH_3H_2O; and at 120°K, it reacts with methane gas (CH_4) to make another solid hydrate, $CH_4 \cdot 7H_2O$. Thus by a long-continued series of reactions, water can be bound up in a great number of compounds, and the process is much more diversified in a slowly cooling environment than in a quickly cooling one.

The process of cooling—no matter whether it was slow or rapid—would have produced heavier planets near the sun and lighter ones outward. Heavier minerals, those containing refractory oxides such as $CaO \cdot Al_2O$ together with metallic iron-nickel alloys, would form first near the sun, but lighter compounds could come only much later. If the surrounding medium had been depleted of more volatile elements by the time these could be accreted, there would be fewer lighter compounds in the inner planets. It is supposed that one method by which the lighter compounds and elements were dispersed was by radiation pressure or so-called solar wind. This is well illustrated by the behavior of comets. As they approach the sun, they are heated so that vapor is produced. This vapor is driven or pushed outward by solar radiation in such a way that a comet's tail always streams away from the sun. In the rich, dense atmosphere of the primitive sun, this mechanism would have depleted the inner regions of lighter elements and would have enriched the inner regions correspondingly in heavier ones.

Consider now the outer planets. The accretion of volatile materials could begin earlier and could continue longer in the cooler environments of the nebular margins. Large gaseous and icy planets would naturally result here. The place of water in this scheme of planet formation can be easily reconstructed.

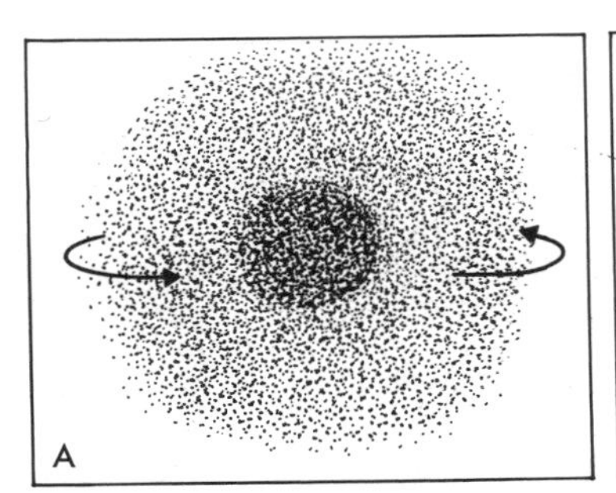

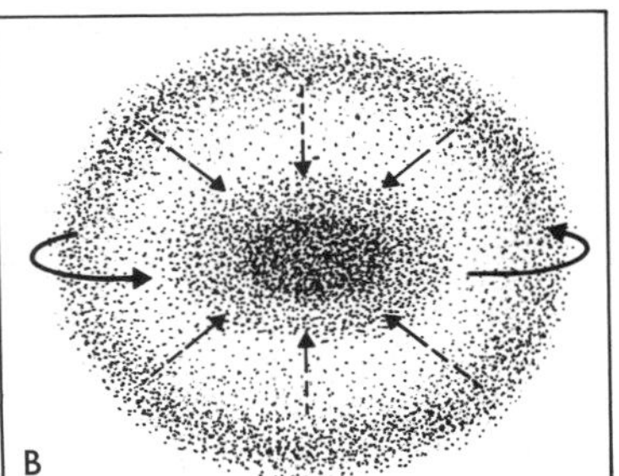

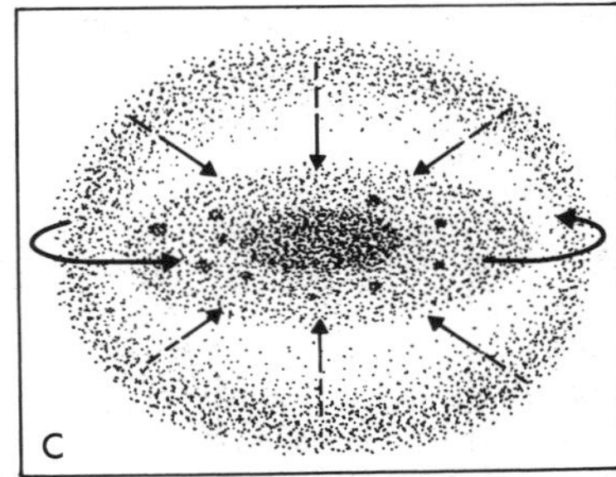

Water exists in three states and on earth the liquid state predominates. It may not be entirely coincidental that on Venus, next nearest the sun, most water is vaporized; while on Mars, farther out, the common state is ice. Earth seems to have occupied a position in space most favorable to the accumulation of liquid water and thus became the "watery planet." Figure 1-20 shows the hypothetical major steps in the formation of our solar system.

The problem of how water came to be unusually abundant on Earth is central to another theory of planetary origin. This theory commences with a homogeneous planet composed of a compacting aggregation of materials in which little or no layering existed. The density stratification—heaviest in the central core, progressive lighter spheres outward—is assumed to be a direct result of melting of the entire mass. The downsinking of iron to the center with a simultaneous rise of lighter material upward to form a multilayered mantle and crust is an expectable process that neatly explains the arrival of liquid water and the gaseous atmosphere enveloping the solid earth below. The technical name for the expulsion of water from a molten mixture is *degassing,* a term borrowed from metallurgy. That water is still being produced is demonstrated by the eruption of volcanoes such as shown in Figure 1-21.

According to this theory, the Earth is different from other terrestrial planets not because it had more water in the first place but because it has passed through a completely molten stage. The other planets, it is said, have undergone only incomplete melting, and little or no water has been expelled. Mars, according to some investigators, might now be entering the stage of water formation and in the

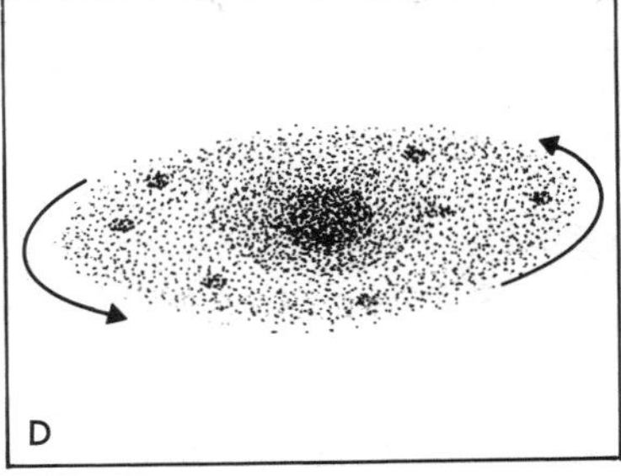

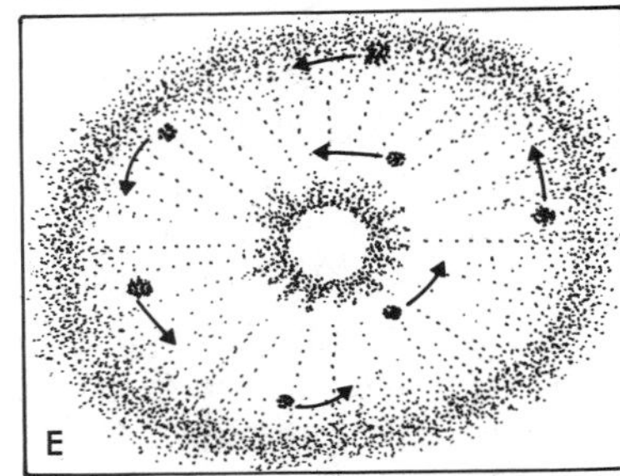

FIGURE 1-20 *Possible stages in the formation of the solar system: (A) Globular aggregation of gas and dust with a central core at higher density; (B) Central region collapses and becomes more dense with accretion of additional material from the surrounding cloud—a shell of material remains in space well beyond the collapsed central core; (C) Accretion continues with flattening and clumping of small solid asteroidlike bodies; (D) Protoplanets reach considerable size and central solar mass approaches critical density; (E) Thermonuclear reactions begin in the sun; dust and gas driven outward; planets emerge.*

future could support oceans and streams. Small bodies such as the Moon and Mercury could not retain water for long even if they did produce it. It was the discovery that Mercury seems to have acquired a heavy core and superficial crust without ever having melted that casts doubt on the hypothesis that Earth had to melt to become stratified.

The truth may belong with both theories. After all, the Earth now has a molten core that probably could not have been molten from the beginning. Also, a partly stratified Earth could have melted and become even more nicely layered by a process of melting. Even though the original Earth had a superabundance of water-rich compounds, these have by one or many means been broken down and the water liberated. A melting process seems to be the most effective mechanism of release. What we now have is not one but two effective explanations as to why the Earth is watery and other planets are not. This is not the place to discuss the numerous ways in which water has influenced the development of the Earth and its life; this should become apparent as later events are considered.

FIGURE 1-21 *Eruption of Irazu volcano, Costa Rica. In addition to the fine ash that has covered the slopes of the volcano, a great deal of water vapor and other gases are being emitted. Much of the water of the earth is thought to have originated from volcanic eruptions like this at Irazu. (Courtesy of U.S. Geological Survey.)*

SUMMARY

Earth, Mars, Venus, and Mercury constitute the inner or terrestrial planets. From a geologic viewpoint, the Moon also may be included with this group. These bodies are referred to as "stony" or "rocky" because of their high content of solid compounds of silicon, iron, and other heavy elements. All have a massive stratified or layered structure such as would result from the action of gravity settling. Each has an inner, dense, metal-rich core; an intermediate massive mantle; and an outer, relatively thin crust. The crust of the Earth is 7 to 49 km thick, the mantle 2,870 km thick, and the core has a radius of 3,480 km.

The Earth is significantly different from other terrestrial bodies in a number of ways: it is the most dense (5.5); it possesses a large quantity of surface water (12,500,000,000 cubic kilometers); the outer brittle shell (lithosphere) is in a state of dynamic motion; there are few signs of ancient craters; and the electrical and magnetic fields are exceptionally strong. These characteristics may be one result of a more thorough melting, which also produced the molten metallic core and expelled large quantities of water.

Other members of the solar family include the four large Jovian planets: Jupiter, Saturn, Uranus, and Neptune. (Pluto is poorly understood.) There are also thousands of asteroids in a belt between Jupiter and Mars, as well as numerous comets and meteorites.

The entire solar system is best explained as having condensed from an extensive nebula of gas and dust that collapsed by gravity and from which the various bodies were then segretated by complex physical and chemical processes. Thermonuclear reactions in the central sun began rather early and had a strong influence on the composition of the various planets.

SUGGESTED READINGS

HARTMANN, WILLIAM K., *Moons and Planets.* Tarrytown, N.Y.: Bogden and Quigley, Inc., 1972.

KING, IVAN R., *The Universe Unfolding.* San Francisco: W. H. Freeman and Co., 1976.

Scientific American, Vol. 233, no. 3, September 1975. The entire issue is devoted to the solar system. It contains also an excellent bibliography, pp. 204—6.

SHORT, NICHOLAL M., *Planetary Geology.* Englewood Cliffs, N.J.: Prentice-Hall, Inc., 1975.

TOULMIN, STEPHEN, and JUNE GOODFIELD, *The Fabric of the Heavens.* New York: Harper & Row, Publishers, 1965.

VAN ALTENA, W. F., *The Solar System.* San Francisco: W. H. Freeman and Co., 1975.

Plate Tectonics–New Look at an Old Earth

As the shapes and sizes of the world's continents and oceans became known in the eighteenth century anyone could look at a globe or map and ponder the meaning of what had been discovered. Order and disorder were both evident. The parallelism of the opposing shores of the Atlantic Ocean was certainly one of the most obvious signs of order that called for explanation. Many thinkers speculated, and a few were bold enough to put their thoughts in print. In 1801, Alexander Von Humbolt, pioneer naturalist and so called "father of geography," theorized that the Atlantic was a huge valley invaded by sea water. In 1858, Antonio Snider, an American living in Paris, put the whole problem into a theological context by contending that an originally continuous land mass had been split to form the Atlantic at the time of the Deluge; his reconstruction is shown in Figure 2-1.

The planet Earth in space. (Photo by NASA.)

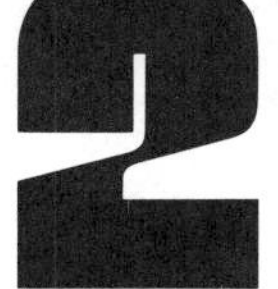

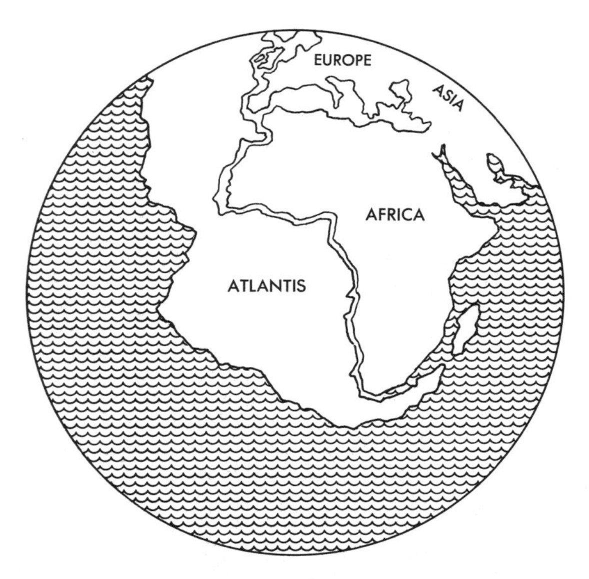

FIGURE 2-1 *Antonio Snider's 1858 reconstruction of continental placement. Redrawn from the original.*

These early thinkers took into account such information as was available. The similarity of opposing coastlines, curve for curve, weighed heavily in all theories. Snider is said to have been influenced in his thinking by the occurrence of coal fields on opposite sides of the Atlantic. Later, when more was known about the coastal areas of Africa and South America, many other geologic similarities were noted.

Scientists were busy everywhere, and other curious discoveries accumulated. Fossils of a small freshwater reptile of Permian age called *Mesosaurus* (Figure 2-2) were found in South Africa; later the same identical species were described from comparable rocks of Brazil. It seemed inconceivable that *Mesosaurus* could have traversed 3,200 km of open sea from one continent to the other. The plant world contributed clues. A peculiar group of fossil plants, the Glossopteris flora, was first discovered in India, and identical species soon turned up in South Africa, South America, and Australia. Specimens of Glossopteris were collected by the Antarctic explorer R. T. Scott within 480 km of the Pole and were found with his dead body in 1912.

Many of these earlier observations and collections had climatological implications. If Snider thought of matching coal beds of America and Europe, he was in effect also matching similar climates. Scott's fossils from Antarctica seemed to prove that this icebound continent once supported rich plant life. Also to be considered was the widespread evidence of great glaciers that were being found in South Africa, India, Australia, and South America, with some telltale deposits almost under the equator.

FIGURE 2-2 *Restoration of the small, fish-eating, freshwater reptile Mesosaurus from Triassic rocks of Brazil. Closely related species are found in South Africa suggesting former proximity of the continents.*

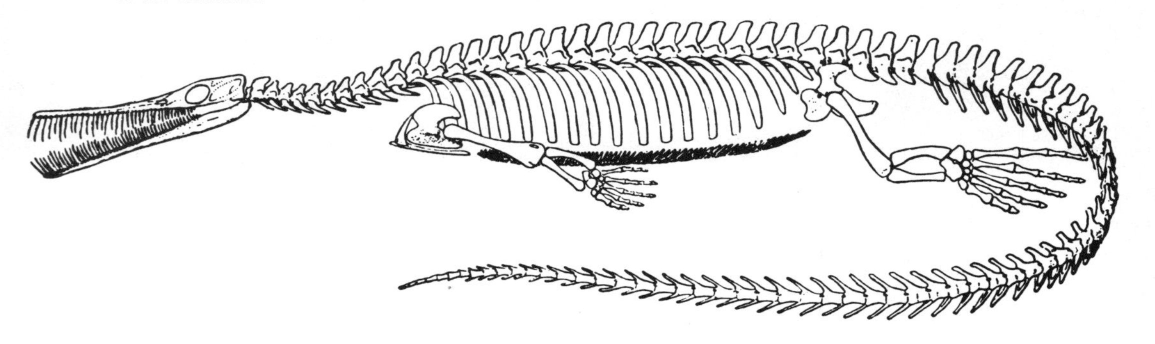

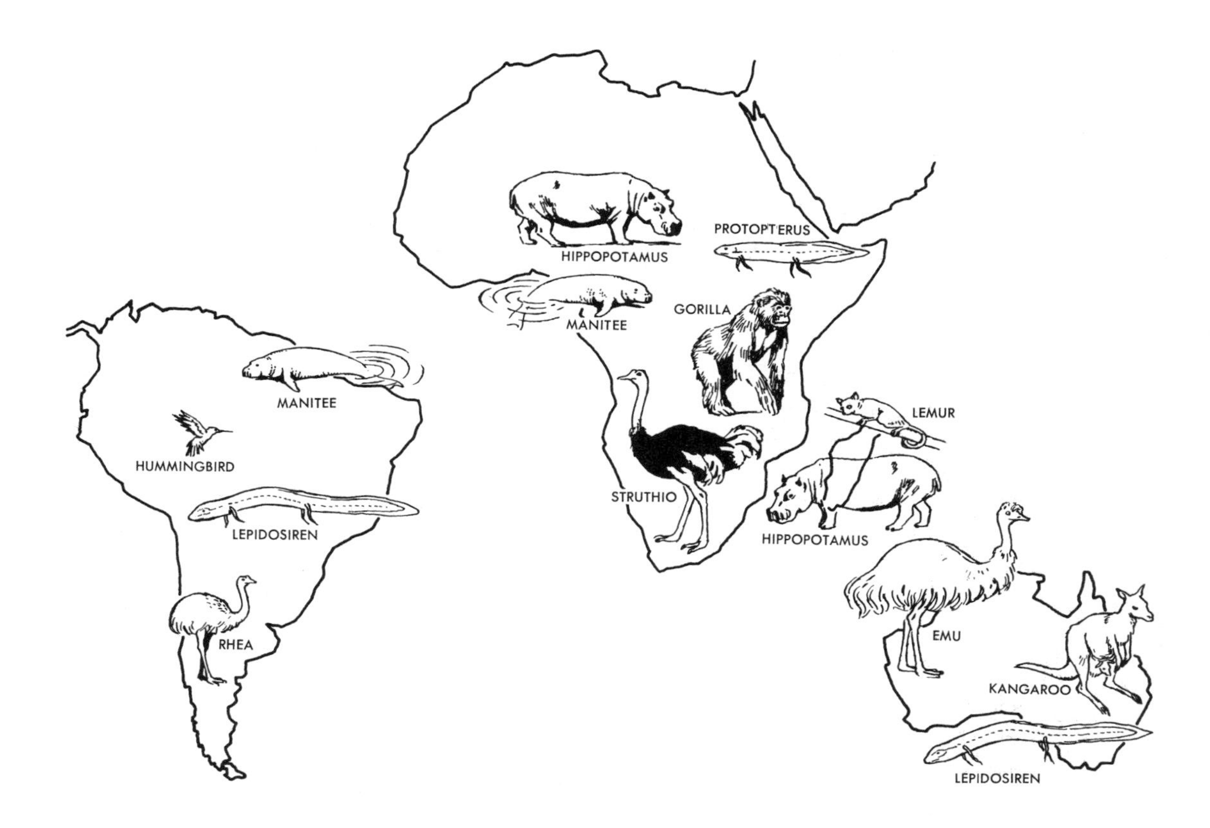

FIGURE 2-3 *The present distribution of organisms such as those illustrated here is explained on the basis of former proximity and later separation of the southern land masses. Lungfish and flightless birds are ancient types that trace back to the time when the continents were near each other. Types such as hummingbirds, gorillas, and kangaroo have evolved in isolation after the separation. The hippopotamus somehow managed to cross from Africa to Madagascar, and the manatee probably crossed the Atlantic in relatively late geologic times. Such crossings of wide seas appear to be rare.*

Students of living plants and animals had their problems too. They found surprising contrasts and comparisons among the life forms of different continents, particularly between the western and eastern hemispheres: no humming birds, cacti, or sloths in the Old World; no elephants, hippotami, or apes in the New World. Yet there were also curious similarities: large flightless birds in Africa, South America, and Australia; lungfish with the same distribution; and sea cows in both America and Africa. Consider Madagascar, only 450 km from Africa, but with no monkeys, apes, lions, leopards, hyenas, zebras, elephants, giraffes, or antelopes! Some of these distribution problems are illustrated by Figure 2-3. Thousands of other curious almost inexplicable cases of distribution were being accumulated and discussed in the late nineteenth and early twentieth centuries. At this particular time the fashionable thing to do was to call upon "land bridges," imaginary or otherwise, to get organisms from place to place.

All these diverse clues were ready and waiting to be put together. The first attempt was in 1910 by the American geologist Frank B. Taylor. Taylor's belief was that two great land masses, located originally over the polar regions, had undergone a slow creeping journey toward the equator. He made no attempt to reassemble the major pieces and left a great many features unexplained.

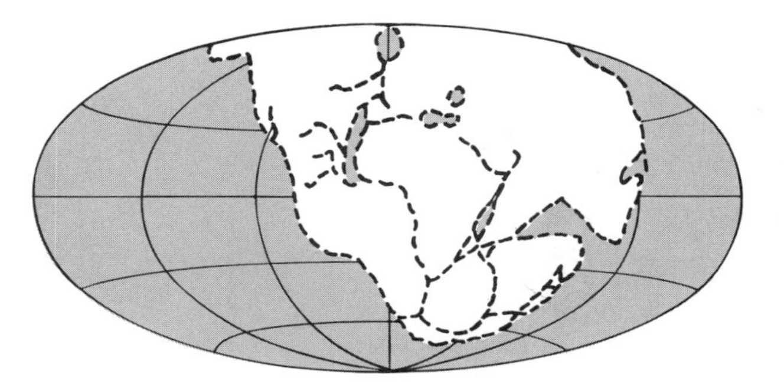

Late Carboniferous

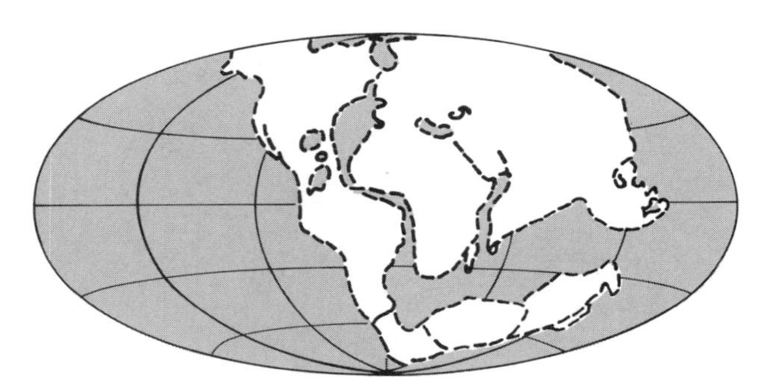

Eocene

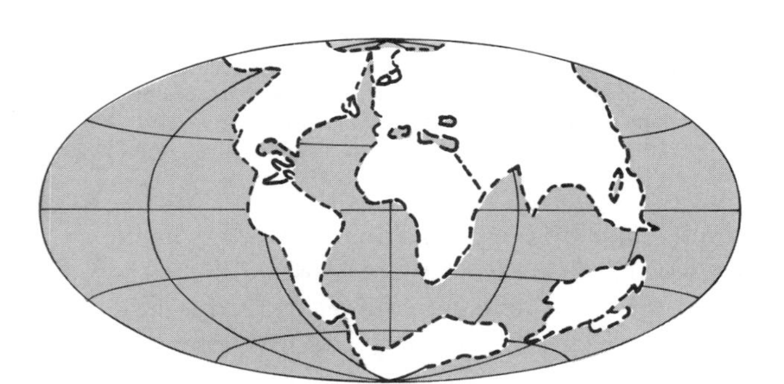

Early Pleistocene

FIGURE 2-4 *The concept of the original great landmass, Pangaea, and its breakup into present continents as proposed by Alfred Wegener.*

Taylor's work was soon overshadowed by that of Alfred Wegener, a German meteorologist. Wegener's book, *The Origin of Continents and Oceans,* was first published in 1915, with later editions until 1936. This book presented more than simply theories. Wegener gave good reasons for believing that all land had once been in one world-continent, which he referred to as Pangaa (now Pangaea). He presented maps showing how this supercontinent had broken up and drifted apart (Figure 2-4). He believed that the general movement was equatorward and westward, and that the rifted fragments moved *across* the ocean floors. He also gave detailed descriptions of many features that would match if the continents were reassembled. All in all, his explanations seemed almost perfect, but there was one weakness and this he admitted. This was the absence of a force or mechanism to bring about the splitting and drifting that he believed had surely taken place.

Wegener is the undisputed father of the concept of *continental drift.* His thinking was bold, dramatic, and wide-ranging. He may well prove to be the most outstanding earth-scientist of the early twentieth century. By all tests, his theory was an excellent one, for it brought forth literally thousands of papers aimed at either proving or disproving the idea that continents actually move. The first reactions were adverse and perhaps overly critical. Wegener died in 1930, several decades before his works bore full fruit in the greatest revolution of thinking to affect the study of the earth.

The reaction—a stalemate

Geologists were exasperated by Wegener's theory because they could neither disprove or prove it. Important personalities opposed it, calling it "a beautiful dream"; others became enthusiastic converts, working to discover new evidence. For a time in the mid-century the geological profession was split almost equally into believers and nonbelievers.

It was the geophysicists who were mostly against the theory. They could not imagine, let alone discover, any force or mechanism whereby the solid continents could move or be moved across the equally solid sea bottoms. To them the lands were solidly fixed and immobile. The permanency of continents and ocean bottoms was an article of faith to most earth-scientists trained in the earlier decades of the twentieth century. The greatest compilation of geologic and geographic facts up to that time was a monumental four-volume work, *The Face of the Earth,* written by Eduard Suess of Vienna and published in four volumes between 1884 and 1909. Suess could see many signs of dynamic movement in the earth's major features, but he related these to contraction of the globe. The crust was deformed from having to adjust to a shrinking core. Suess had great influence, and it was his ideas that had to be displaced by Wegener's. Incidentally, Suess is responsible for naming the two great ancestral land masses, Gondwanaland and Laurentia (now Laurasia).

New evidence turns the tide

Many who probably wanted to believe in continental drift simply could not do so because it seemed to be literally impossible to move solid land across solid ocean floor. But the oceans held clues that would solve this as well as many other mysteries. The incentive, equipment, and means of exploring the deep oceans came mostly as a byproduct of World War II. Americans, in particular, were forced to learn more about the oceans, and sophisticated devices for surveying the topography and shallow deposits of the ocean floor became standard on many vessels. The fact that emerged from ocean exploration is simple enough—it is not the continents alone that are mobile; it is the entire outer shell of the earth, including the floors of the ocean, that is in a state of complex relative motion. Continents must be regarded as passive riders on great plates that carry not only the lands but ocean basins as well.

The first clue to this amazing conclusion was the discovery of a great world-circling mountain chain 64,000 km long and several hundred kilometers wide, which generally bisects the oceans as a submarine ridge. This *midoceanic ridge,* as it came to be called, is not like any of the familiar dry-land mountain chains. It is composed almost exclusively of basalt flows cut by faults and capped by occasionally active volcanoes, some of which rise above the surface of the ocean. Sections of the ridge, particularly the portion in the Atlantic Ocean, have a deep narrow trench along the summit. That the midoceanic ridge is not a dormant or static feature is shown by a heavy concentration of earthquakes along its course. What goes on here is critical to the evolution and present arrangement of the earth's great first-order features—the continents and ocean basins. The basic fact is simply this: new basaltic oceanic crust is being created along the center line of the midoceanic ridges, to be gradually pushed aside or carried away to create new areas of ocean bottom on a vast scale. The startling concept that the ocean basins can be renewed fired the imagination of geologists everywhere and set the stage for confirmation of Wegener's theory.

New concepts demand new terminology, and the old term *continental drift* is being replaced by the new term *sea-floor spreading.* Even more inclusive are the terms *plate tectonics* and *global tectonics,* which are used by many. The first of these, plate tectonics, calls attention to the fact that the active agents that make up the earth's outer shell are movable platelike seg-

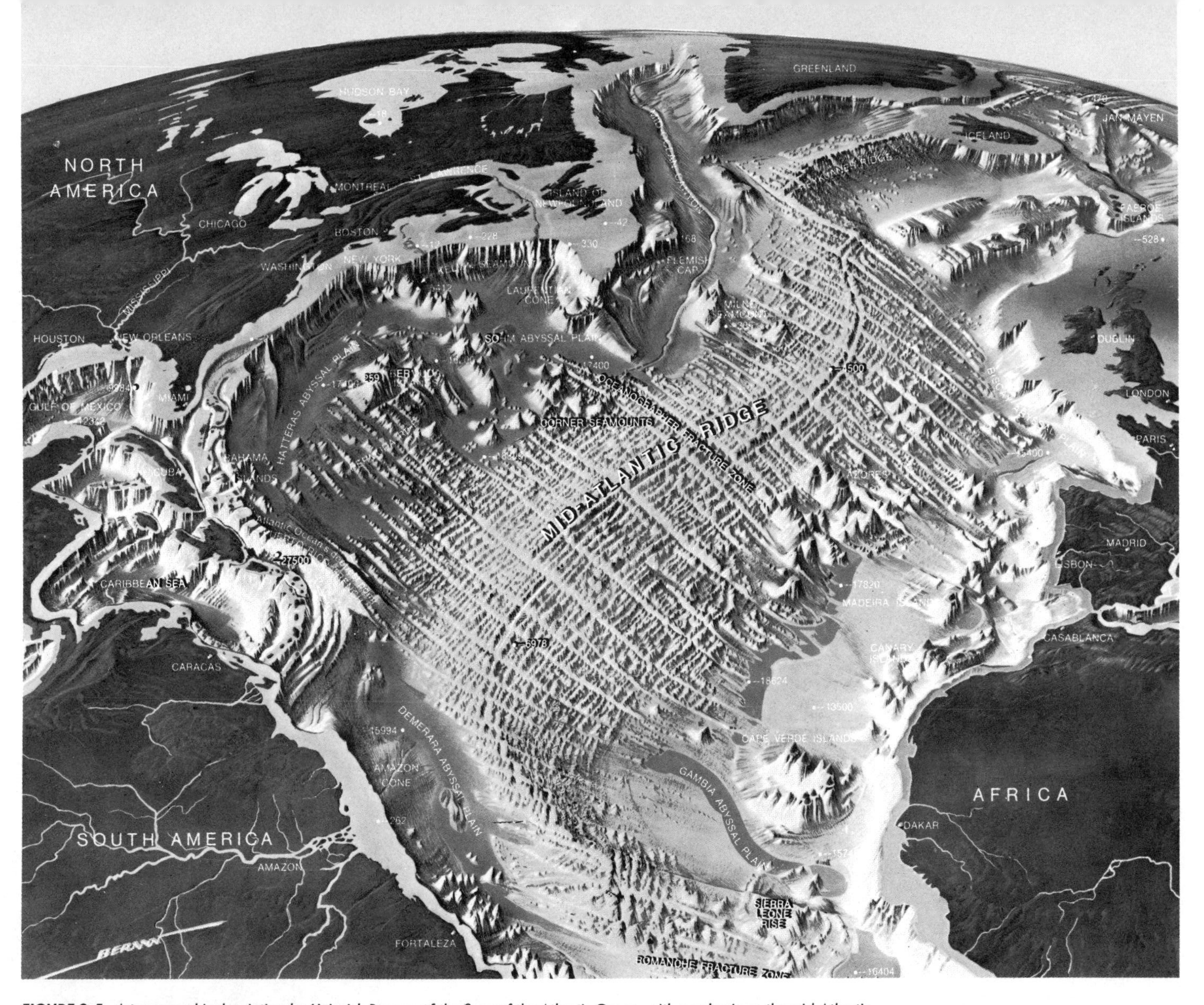

FIGURE 2-5 *A topographical painting by Heinrich Berann of the floor of the Atlantic Ocean, with emphasis on the mid-Atlantic ridge and associated structures. Numbers indicate feet above or below sea level. (Photograph courtesy of Alcoa.)*

ments. The second term, global tectonics, emphasizes the fact that all aspects of the earth's history and structure may apparently be unified by the concept of these movable surficial plates.

Structure of the earth

The clear indication that the surface of the earth is in motion leads immediately to an inquiry as to what is going on at deeper levels. Global effects seem to require global causes, and a brief digression into the overall structure of the earth now seems in order.

The internal structure of the earth and of the earthlike planets was described in Chapter 1. Recall that Earth, Moon, Mars, Mercury, and Venus have stratified or layered interiors. Earth as the best known example has three basic divisions—core, mantle, and crust. The core begins about 2,900 km below the surface and continues for 3,470 km to the center. There is an inner core and an outer core, the division being at about 1,220 km. The inner core appears to be solid, the outer molten. The density of the core is about 10.5 (water = 1). Iron and nickle apparently are essential components. The mantle, which surrounds the core but may not be solidly attached to it, is about 2,870 km thick and makes up 84 percent of the earth by volume and 67 percent by weight. It seems to be composed of silica- and metal-rich minerals such as pyroxene, $Ca(Mg,Fe,Al,)(Al,SiO_2O_6)$. The outer shell of the earth is the crust, which is 6 to 50 km thick. It is separated from the mantle by the Mohorovičić discontinuity (Moho for short).

FIGURE 2-6 The structure of the earth. Bar graph is out of scale for clarity.

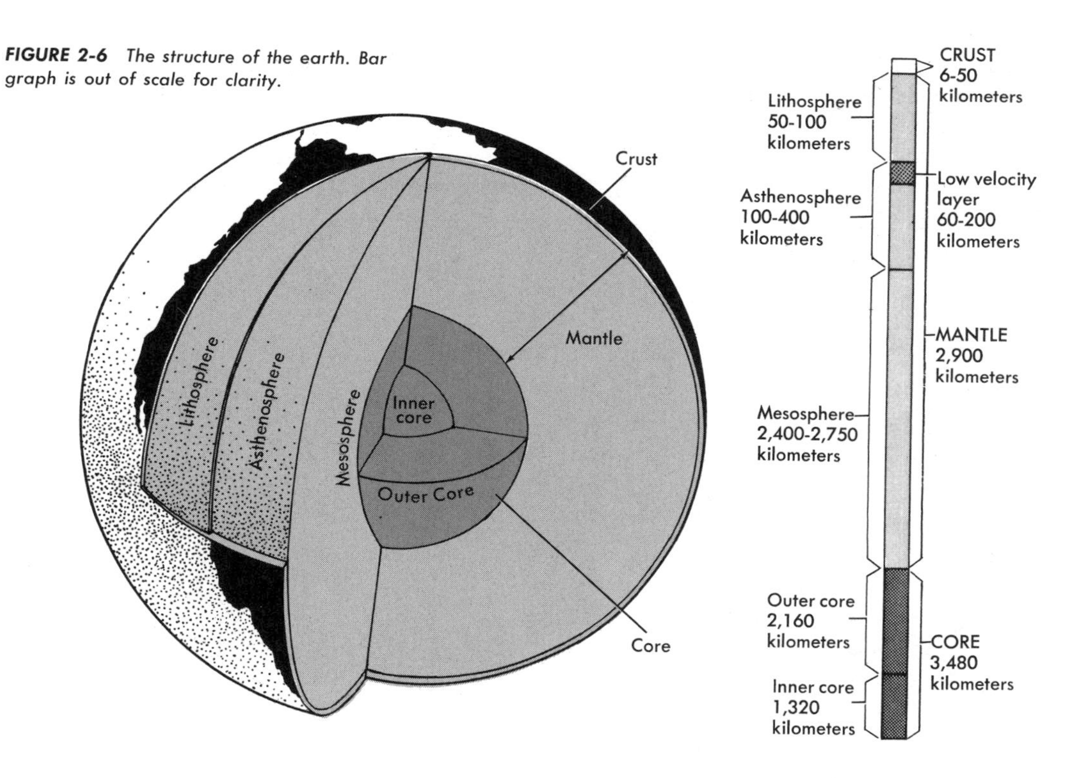

Rocks below the Moho are more dense, contain less gas and liquid, and are richer in iron than those above. The structure described above is illustrated by Figure 2-6.

The distinction between crust and mantle is much less important than that between what are called the *lithosphere* and *asthenosphere.* The term lithosphere is applied to the outer solid, brittle, and relatively cool shell of the earth, which is 50 to 100 km thick, and thus includes the entire crust, the Mohorovîcić discontinuity, and the upper levels of the mantle. Below the lithosphere is the asthenosphere, composed of soft, hot, easily-flowing material. The asthenosphere ranges in thickness from 130 to 160 km; it apparently extends unbroken around the earth. Below the asthenosphere is the main mass of the mantle or *mesosphere,* about 2,200 km thick, which is solid and dense. One of the great unsolved problems of geology is how a hot, plastic, semiliquid layer such as the asthenosphere can exist sandwiched between hard brittle layers so near the surface of the earth. High heat and low pressure do not fully explain the situation, since the outer core, much deeper, is also melted. The essential agent is probably water, which does much to lower the melting points of silicate minerals such as those of the upper mantle.

The combination of solid mesosphere, plastic asthenosphere, and solid lithosphere is somewhat analogous to a giant jelly sandwich and constitutes an unstable system. The lithosphere as a whole is movable upon the asthenosphere and is itself divided into a number of great movable plates separated by zones of complex movement. The names, relative sizes, and mutual relations of the lithospheric plates are best shown on maps of the entire earth. Figure 2-7a shows the earth's major features. The relationship of these features to the plates and their boundaries is given in Figure 2-7b. Finally, another class of features, namely, distribution of earthquake epicenters, the concentration of which is an infallible guide to heightened geologic activity, is depicted in Figure 2-7c. The relation of geologic activity to

FIGURE 2-7 *(a) Simplified sketch of the world's major geographic features.*

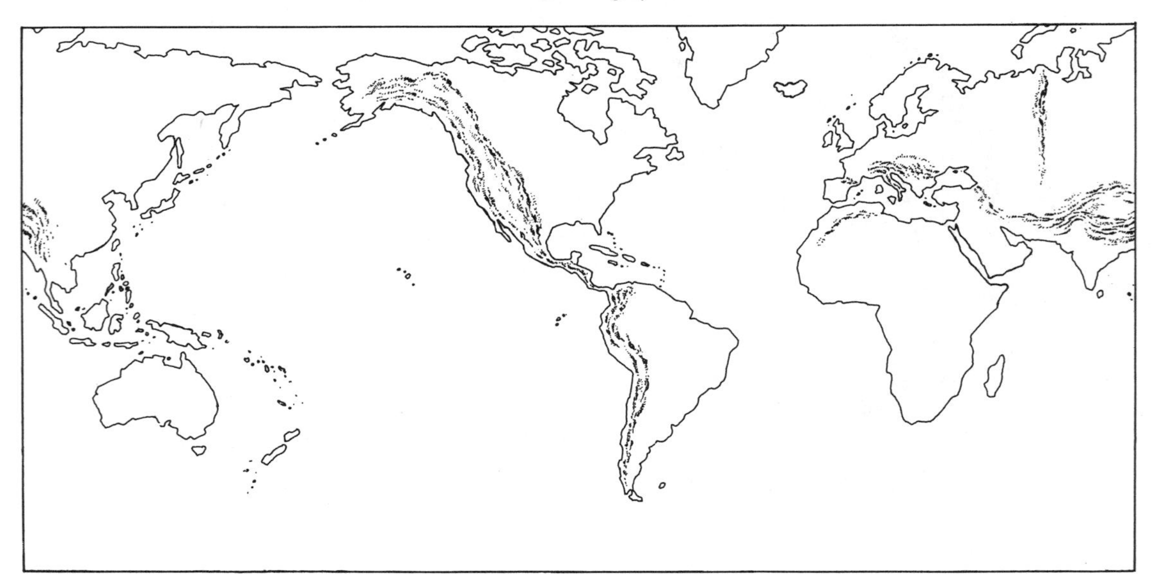

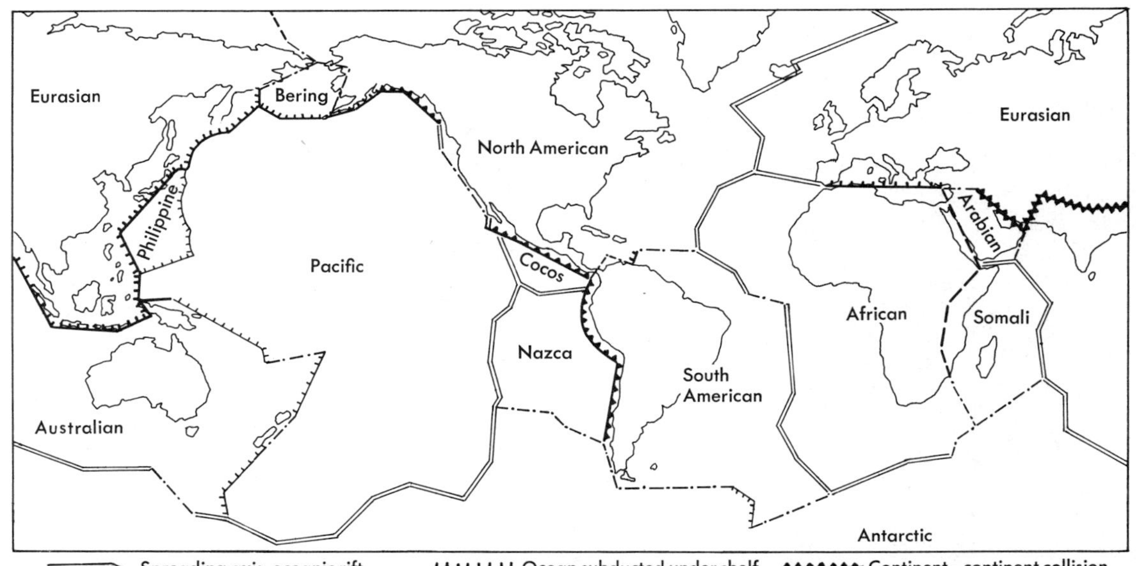

(b)

(*b*) *Major lithospheric plates of the earth. Shown on same map projection as 2-7a. Types of plate contacts are indicated by coded symbols. (Refer also to Figure 2-9.)*

(*c*) *Earthquake belts of the earth as outlined by epicenters of quakes occurring from 1961 to 1967. Active volcanoes added from other sources. [Redrawn from M. Barazangi and J. Dorman, "World Seismicity Maps Compiled from ESSA, Coast and Geodetic Survey, Epicenter Data 1961–1967,* Seismological Society of America Bulletin, *59 (1969), pp. 369–80.]*

(c)

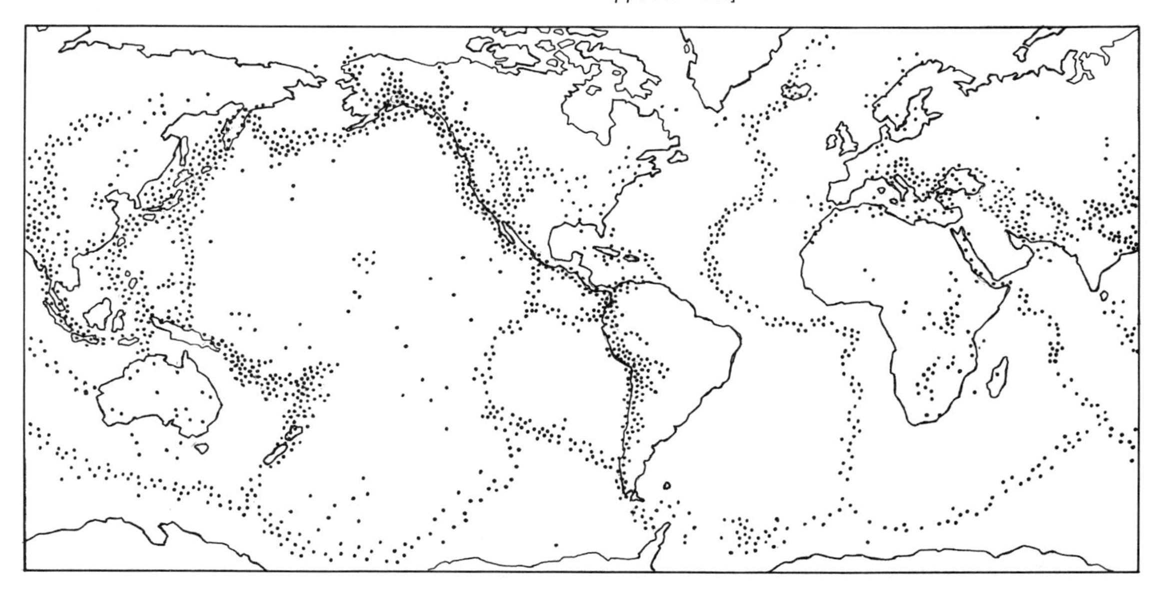

FIGURE 2-8 *Continental separation. This space photo shows the Arabian Peninsula above, the Horn of Africa to the right, and the continent of Africa lower left. The Gulf of Aden extends to the upper right, the Red Sea to the left, and a zone of complex rifting that extends across East Africa begins near the center. The parallelism of opposing coastlines is evident. The water bodies are floored with basalt flows and are cut by numerous faults such as would be expected under forceable separation of the crust. This area clearly illustrates the initial separation of land masses by the mechanism of sea-floor spreading. Many unusual geologic features are found near the junction of the Red Sea and Gulf of Aden where a section that would normally be sea floor is exposed. (Photo by NASA.)*

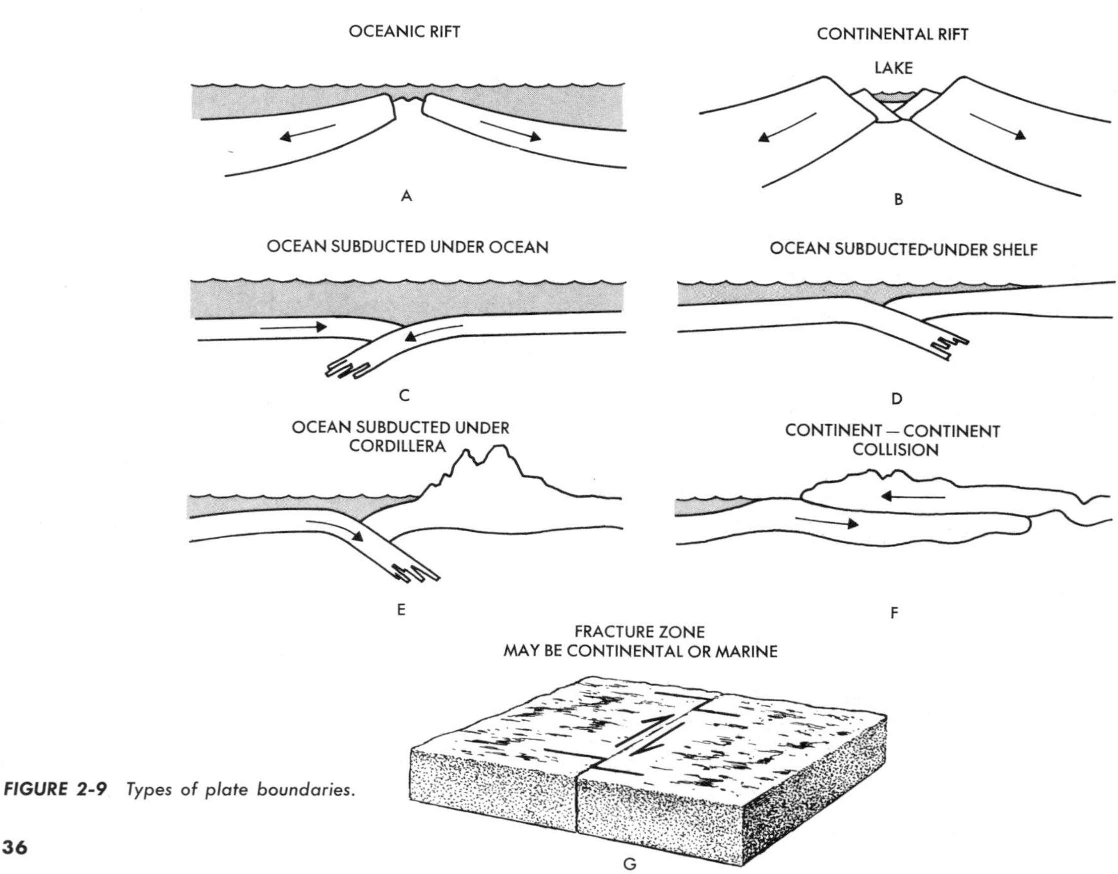

FIGURE 2-9 *Types of plate boundaries.*

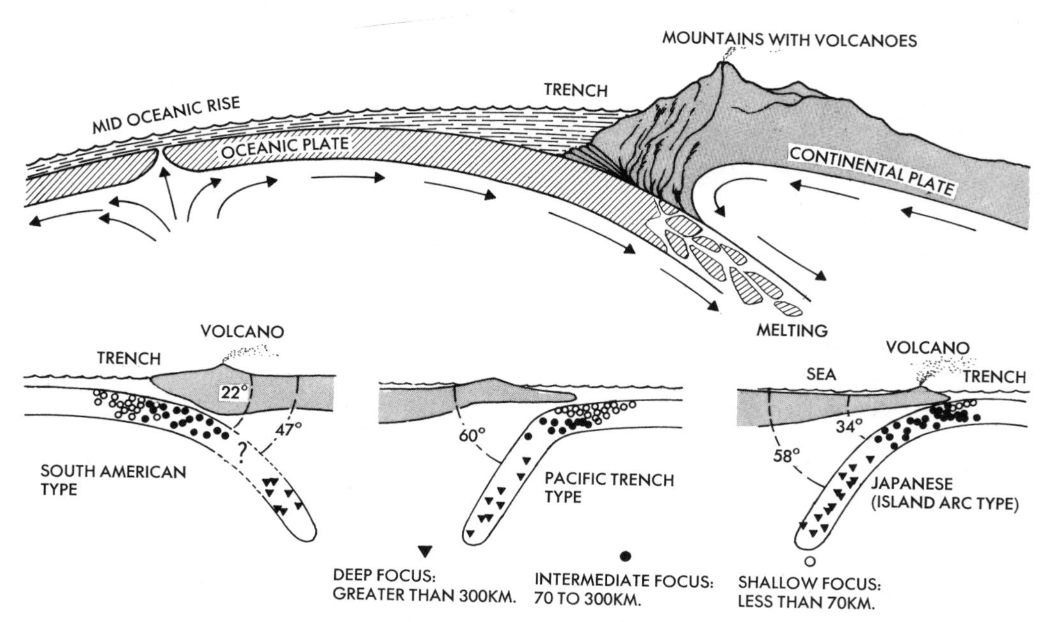

FIGURE 2-10 *Geologic activity associated with subduction zones. Upper cross section is a generalized one showing known and inferred relations where one plate descends under another; in this case an oceanic plate is being destroyed under a continent-bearing plate. Diagrammatic figures below illustrate variations and are intended to show chiefly the relation to earthquake epicenters. Left, South American or cordilleran type; center, Pacific type; right, Japanese or island-arc type. Earthquakes are classed as deep-focus if the epicenter is over 300 km deep, intermediate if 70 to 300 km deep, and shallow if less than 70 km.*

plate boundaries is obvious. The area of the Red Sea (Figure 2-8) shows clearly how certain features relate to the concept. Figure 2-9 illustrates the various types of boundaries known to exist in the plate system.

Geologists, with few exceptions, have come to believe in sea floor spreading and plate tectonics as unifying concepts of great value in understanding what they observe about the past history and present appearance of the earth. Credit for the first statement binding the necessary elements together goes to Harry H. Hess, Princeton geologist, who in 1962 published a paper "History of the Ocean Basins." Geology suddenly passed from what is best described as its descriptive phase to a more mature historical phase. Nothing illustrates this better than the changes and additions that have been made in geological and geographic terminology of the earth's major features.

The subduction zones

A little thought will convince anyone that new sea floor cannot be made and pushed aside indefinitely without creating more material than can be accommodated on the earth's surface. The problem is solved by the destruction of lithospheric material at the same rate that it is created. Ocean bottom is consumed by being returned to the asthenosphere along certain well-defined elongated belts of downbending and remelting called *subduction zones.* Figure 2-10 illustrates by the cross-section method what goes on at these important junctions.

Subduction zones are of several types. One is typified by the Peru-Chile trench, which lies near and parallel to the coastline of South America. This trench is a zone of downbending and destruction of the eastward-moving Nazca plate as it is overridden by the westward-moving American plate. This collision of major plates has produced more than the oceanic trench—the Andes range is a spectacular result. As would be expected, there have been extensive mechanical disturbances of marine sediments along the zone of compression so that deposits formed near sea level are now found higher in the mountains. More important have been the igneous effects that are also neatly explained by plate tectonics. As the oceanic lithosphere is forced downward, it enters zones of progressively higher temperatures until it undergoes complete melting to produce a complex magma made chiefly of remelted basalt and water-saturated ooze of the former sea floor. The lighter components of this mobile magma rise toward the surface where they may come to rest as great bodies of granite, or they may erupt as flows of andesite and light-colored volcanic ash. (Andesite get its name from the Andes.) Meanwhile, the heavier components continue to sink and return to the asthenosphere beneath. Another geologic byproduct at this plate boundary is a great belt of earthquakes, many of which are of the deep-seated variety. These are thought to be a direct result of adjustments that must accompany the entry of the cooler, water-charged oceanic plate into the hot and mobile asthenosphere.

Not all plate interactions result in the downward movement and destruction of the heavier member. Plates of more or less equal density may double up to create an excessive thickness of the lighter material. This is the favored theory of the origin of the Himalayan Mountains and Tibetan Plateau. Peninsular India and an unknown amount of land north of it now appear to have moved northward from a position adjacent to east Africa to collide with southern Asia. There may have been a subduction effect at the northern part of the Indian plate, but the colliding masses were both of rather low density, and the result was a doubling up of continental material to create the greatest elevated tract on earth.

An even more important type of subduction zone is that associated with what have long been known as *island arcs.* These are curving, festoonlike chains of islands, usually highly volcanic, which are typified by the Aleutians, Japanese, Marianas, Lesser Antilles, and Indonesian island groups. In terms of plate tectonics, these are the result of oceanic plates colliding with and being consumed at the edges of less active continental plates.

Island arcs show considerable variety. In general they may be described in terms of narrow parallel belts of similar structure and activity. Starting on the oceanward side is a deep trench (also called a *trough* or *foredeep*) that is the site of downbending and disappearance of the oceanic plate that is being destroyed. Next is a zone of earthquakes, which are of shallow origin on the outer edge of the zone and become progressively deeper toward the continent. Next is an elevated tract with active or extinct volcanoes and earthquakes of still deeper origin. An innermost expression of arc structure is a belt of very deep earthquakes originating 300 to 700 km below the surface. Many arcs are separated from the adjacent continent by a shallow sea called the *interarc basin;* the China Sea and Bering Sea are examples. Arcs of the southwest Pacific lie a great distance from Australia; the intervening area is one of complex geologic structures not yet fully understood.

The volcanic activity is possibly the most spectacular feature of the island arcs. Volcanoes there are mostly explosive and build up steep-sided cones such as Mount Fujiyama near Tokyo and Mount Mayon in the Philippines. Their products are rich in silica, potassium, and sodium; the typical extrusive rock is the light-colored rock called *andesite.* Andesite contrasts strongly with basalt, which is the prevalent product of almost all other volcanoes as well as of the oceanic spreading centers.

The structures and products of island arcs have been explained without much difficulty in terms of plate tectonics. As a slab or plate of oceanic lithosphere plunges downward, the effect is to introduce relatively cold, water-saturated material into a much hotter environment. The energetic reactions of this massive intrusion are shown by earthquakes that become deeper and deeper as the slab descends. It is supposed that the deepest earthquakes may be near the bottom of the asthenosphere. Important also is the fact that a good deal of sediment, mainly but not exclusively deep-sea ooze, goes down with the slab in addition to the basaltic crustal rocks. This water-rich siliceous material becomes part of the melted magma and, because of its lightness and vapor content, tends to rise through the overlying material to emerge in volcanic eruptions. The material emitted by island-arc volcanoes over periods of millions of years constitutes a steady addition to the mass of islands and continents. This is surely one of the major ways in which continents grow and become thicker.

Intraplate reactions

Any view of the earth demonstrates that junctions between plates show intense geologic activity, while areas within plates show relatively less. The lithospheric plates have been compared to great tabular icebergs or rafts moving upon the ocean and colliding

FIGURE 2-11 *A radar map of part of the San Francisco peninsula, with an index map identifying important features, including the San Andreas fault zone. Along this zone, the Pacific crustal plate (lower portion of map) is moving actively northward in relation to the American plate. (Photo courtesy of U.S. Geological Survey.)*

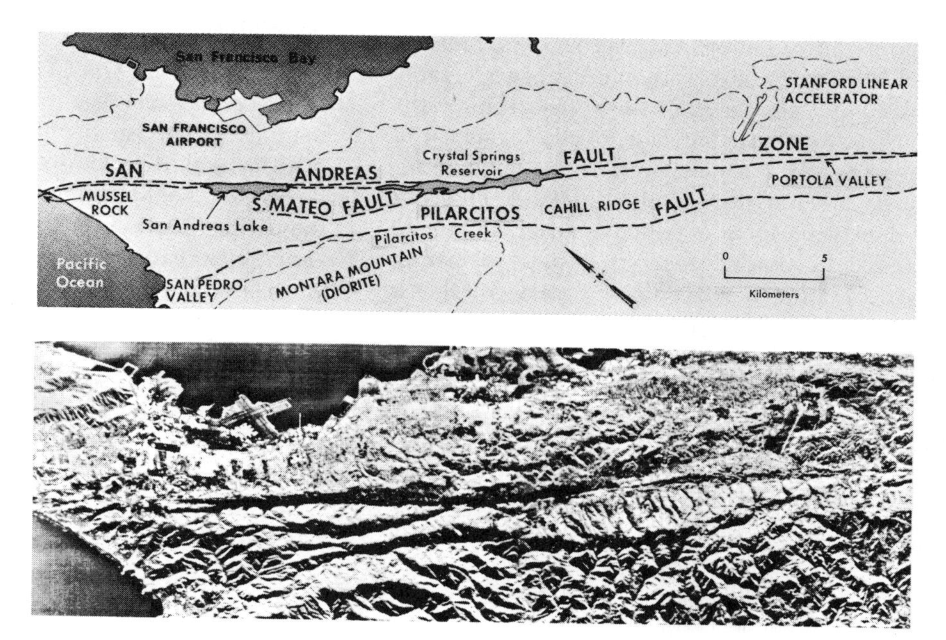

with each other in various, more or less random, violent ways. One example of this violence is the effects of movement on the San Andreas fault (Figure 2-11). The interiors of the plates are surprisingly little affected by what goes on at their margins. The inhabitants live their lives and evolve to suit prevailing conditions as though they were on solidly fixed land masses rather than on moving ones. Not that land dwellers are unaffected, but the whole process is so slow that individuals are never greatly threatened. They are like passengers on some giant unsinkable raft that appears to them either to be permanently in dock or motionless on the ocean. Only after extensive studies are geologists able to say where the rafts have been or where they are going.

Incidentally, it is the nature of rock material as such that insures the tranquillity of the interior of the continental blocks. Rocks are inherently so weak that they cannot transmit either compression or tension for great distances. If they are pushed, they will crumple, break, and slide; if they are pulled, they shatter into splinters and narrow blocks. Only if they are carried along and supported by an underlying medium can large blocks be moved over continental or oceanic distances. Translated into what may be observed geologically, we see that mountains are mostly formed at or very near the zones where blocks collide; effects are not translated far inland. Likewise when a continent is pulled apart, it is likely to split into narrow blocks such as exist in East Africa, *unless* there is something below it that is affecting entire plates.

A new line of evidence—paleomagnetism

References were made to the magnetic compass by Chinese writers of the eleventh

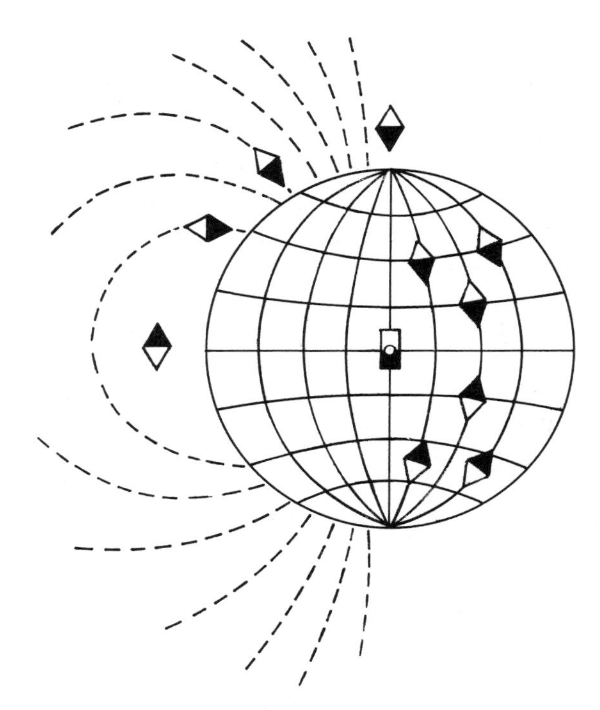

FIGURE 2-12 How magnetism relates to latitude and longitude. Small compass "needles" on right hemisphere show the pole-seeking component. This information of itself tells nothing about distance from the pole but is useful for comparing present and past orientations of rock bodies or land masses. On the left is shown how the inclination of the compass needle reveals the latitude of magnetization. This is of importance in determining the relation of a specimen to the prevailing pole and equator. Any magnetized rock sample from which the declination and inclination can be obtained is useful in locating the position of the pole in relation to the sample at the time of its formation.

century, and the Chinese were probably the first to realize that the earth itself had a magnetic field. The earth's field is very similar to one that would be produced by a giant deeply buried bar magnet aligned 11° from the axis of the earth's rotation (Figure 2-12). Many interesting observations might be made about the earth's magnetic field, but we must be brief and restrict ourselves to the topic at hand.

Magnetism was probably discovered when it was first noted that certain natural minerals attracted or repelled each other. This is most strongly shown in connection with the minerals magnetite (Fe_3O_4), hematite (Fe_2O_3), and ilmenite ($FeTiO_3$). Even rocks with a small content of these minerals are magnetic, and their magnetism can be measured. Many rocks, including the lava known as *basalt,* and sediments with iron minerals are known to acquire magnetization when they are formed. During the process of cooling of lava or magma, the susceptible iron minerals become magnetized in the earth's prevailing field. This is called *thermoremnant magnetism* (TRM). When iron minerals are being deposited as particles, they fall into place like free-swinging compass needles aligned in the direction of the earth's field. This is called *depositional remnant magnetism* (DRM).

Thus, for practical purposes, many rocks are formed with a frozen-in record of the prevailing magnetic field. This record, if properly interpreted, tells the direction and roughly the distance to the nearest magnetic pole. This idea has been proven by the demonstration that recent lava flows consistently show the pole as it is now located. This has opened up a field of study called *paleomagnetics,* the study of ancient magnetism.

As investigators began to study older iron-rich lavas and red (iron-rich) sediments, they were not surprised to find that the position of the magnetic poles has not been constant—it has shifted considerably even in historic times. What did surprise them was the fact that the earth's magnetic field has been reversed a great many times. The north magnetic pole has frequently become the south magnetic pole. The earth did not tip over—only the invisible positive and negative poles took on reverse signs. These changes are, of course, worldwide and affect every magnetic mineral formed at any given instant. That this is true was proven by obtaining the actual ages of statistically significant number of specimens. There could be no doubt of the theory; all specimens of the same age are either negatively or positively magnetized. There was no mixing of polarity signs. (See Chapter 13 for information on dating methods.) Periods of magnetization like the present are termed *normal;* those oppositely magnetized are termed *reversed.*

The study of ancient magnetic reversals has progressed at a rapid rate. It was discovered rather early that the periods of reversal were not equal in length. For example, the change to the present normal polarity took place about 690,000 years ago. Before that was a reversed period 200,000 years long, and before that a normal period of 60,000 years. About 80 reversals are known in the last 110 million years. So far, no satisfactory explanation for reversals has been proposed. They may be caused by terrestrial or extraterrestrial causes; for example, the sun is known to reverse polarity on a regular basis every 11 years.

Polarity reversals appear to be just what geologists have been looking for—a basis for ascertaining the age of large volumes of rock with a minimum of trouble and expense. Reversals are not curiosities; they are practical signposts. They have been compared with tree rings in that they succeed each other in a regular way, can be counted, and are of two kinds only, normal or reverse. In practice, a short interval of time dominated by one polarity with a duration of 10,000 to 100,000 years is called a *polarity event.* A *polarity epoch* is a longer time span, 100,000 to 1,000,000 years, that is predominantly, but not exclusively, of one polarity. A *polarity period* covers 1,000,000 to 10,000,000 years, and a *polarity era* is 10,000,000 to 100,000,000 years. Polarity periods and eras, of course, embrace many reversals. Polarity epochs are to be designated by numbers and polarity events by letters, except for a few of the most recent that have been given names taken from places where they were first discovered. For examples of the applications of paleomagnetic data and their relation to other dating methods study Figure 2-14.

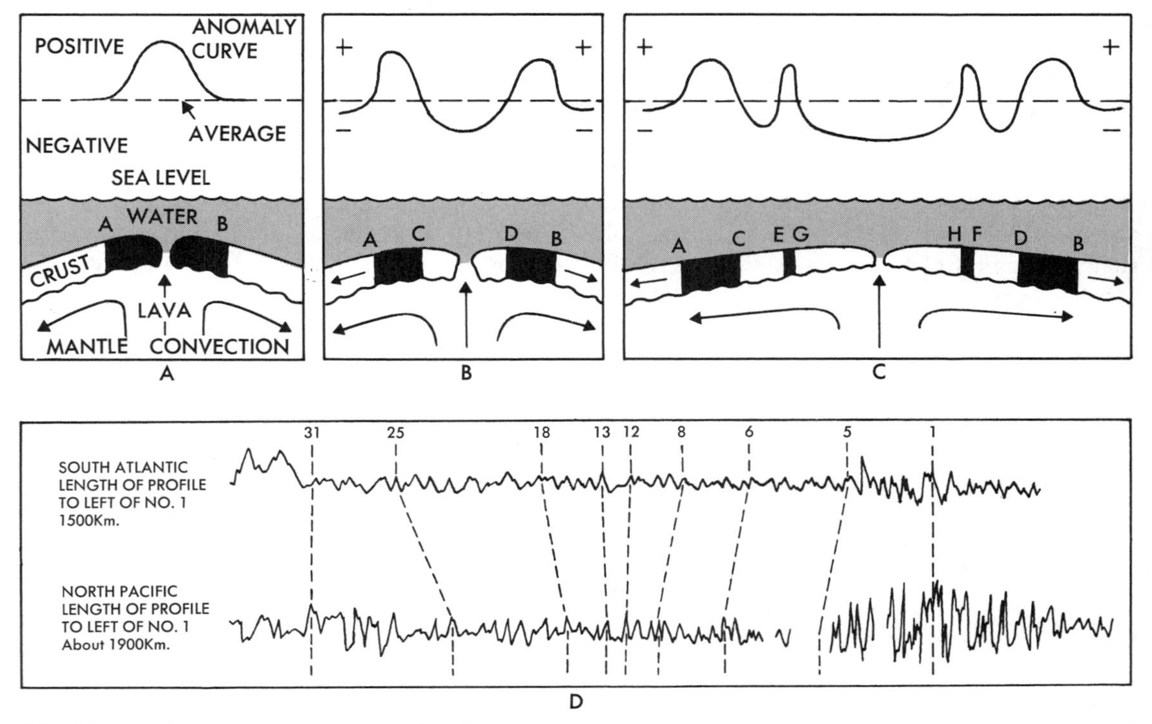

FIGURE 2-13 (Above) Magnetic anomalies and profiles of the ocean basins. Parts (A), (B), and (C) of the above illustration show how alternate strips of basalt with normal (dark pattern) and reversed (white) magnetic fields are produced. The dark strips are magnetized with the poles as they are today and give rise to positive anomalies with intensities greater than average. The alternating strips, created in periods of reversed polarity, give rise to the negative portions of the curve with less than average intensities. Variations in intensity of the prevailing field together with its positive or negative orientation have the same effect simultaneously everywhere, thus giving the possibility of worldwide correlation. The lower figure (D) shows two curves from widely separated traverses, one in the South Atlantic and the other in the North Pacific, in parallel arrangement to demonstrate correlation of the anomaly patterns. The figures above the upper curve are numbers given to certain distinctive anomalies that are diagnostic enough for matching and correlating strips of sea bed throughout the earth. Curves of (D) from Dickson, G. O., Pitman, W. C., III and Heirtzler, J. R. 1968, Magnetic anomalies in the South Atlantic and ocean floor spreading: Journal of Geophysical Research, Vol. 73, p. 2087–2100 and Pitman, W. C., III, Herron, E. M., and Heirtzler, J. R., Magnetic anomalies in the Pacific and sea floor spreading: p. 2069–2085.

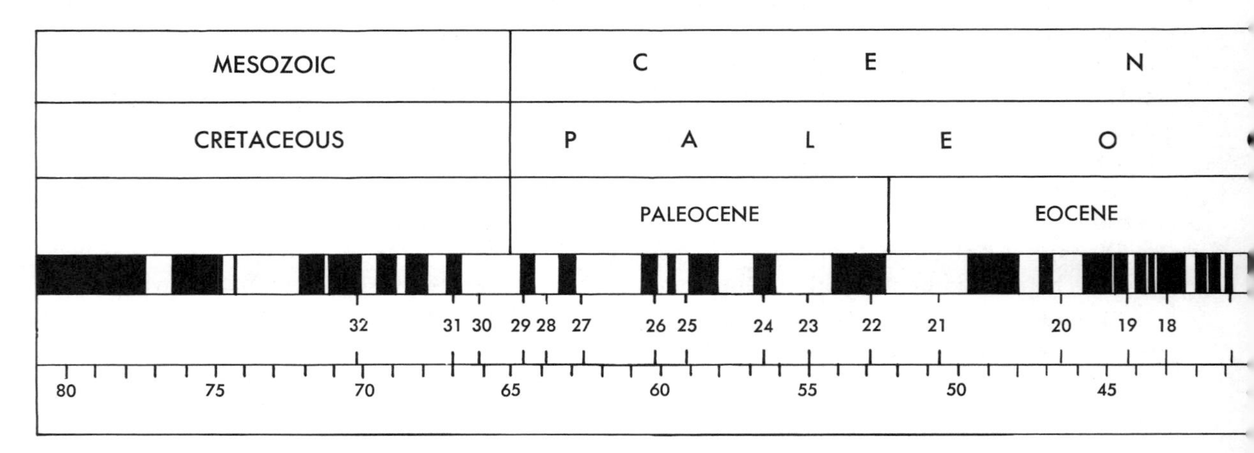

This short introduction to paleomagnetics sets the stage for a description of one of the most significant syntheses of earth science. In 1963, F. Vine at Princeton and D. H. Matthews of Cambridge University published a brief note destined to open the way to understanding the relationship of the magnetic strips of the ocean floor to the creation of new material at the midoceanic ridges. Like many great ideas, theirs was remarkably simple. They reasoned that the new lava emerging at the mid-oceanic ridges was ideally suited for acquiring and preserving a magnetic record. This record should in effect be the same on both sides of the ridge since the splitting always takes place at the exact midline. Furthermore, the magnetic record should extend backward through all the changes of polarity that had taken place while sea-floor spreading was actively operating. The magnetic strips are nothing more or less than an expression of magnetic reversals. The stronger anomalies (departures from the average) at any particular place record normal polarities, while the weaker anomalies represent reversed periods. You should understand that normal polarities are reinforced by the present field, while reversed polarities are weakened by it. Also it should be understood that the magnetism of the basalt strips is much weaker than the pervasive magnetic field of the earth but can be separated from it by fairly simple techniques. Continuous records of the magnetic strips of the oceans can be obtained by instrumental means from any properly equipped vessel, and many thousands of kilometers of records are available.

Suddenly geologists were provided with a means of dating practically all of the ocean floor. Any particular reversal is the same age everywhere, in every ocean and on land also if it can be detected. It is the detection of short-term events that is difficult; in other words, because very narrow strips are produced by events of only a few thousand years, these strips tend to get lost or be unidentifiable. The practical procedure is to divide the record into wider units (encompassing longer periods) that are identifiable. Divisions are made at characteristic anomalies of the magnetic curves. The anomalies are portions of the record with shapes that are notably different or prominent and that all workers can recognize.

FIGURE 2-14 *(Below) Relation of various paleomagnetic measurements to the standard time scale and geologic column. Note especially the polarity reversal sequence. (Modified by permission from the original in Geotimes, June 1975.)*

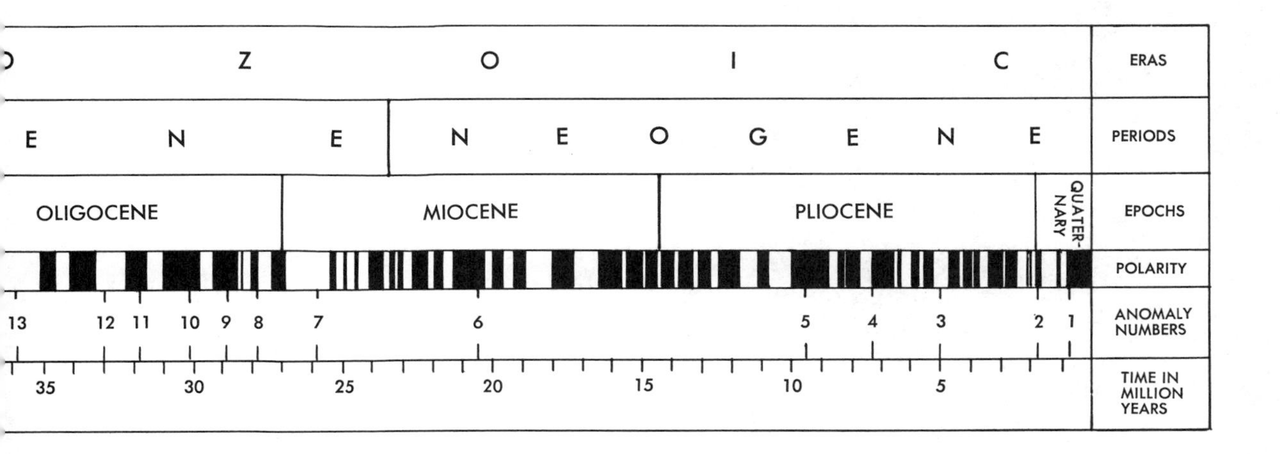

Between these anomalies there may be a number of reversals or none at all—the hope is to set off fairly equal time and rock units regardless of the reversals. The anomalies are given numbers going backward from 1 formed about 700,000 years ago. The most recent chart (July 1975) extends to anomaly 80. Not many more will be found simply because the sea floor record extends but little beyond this point.

Recall that each anomaly can be recognized in any ocean basin where it is preserved. This beautifully simple concept gives a method of relative dating and becomes near-perfect when absolute dates are added. It is possible to determine the actual age of basalt rather easily by the potassium-argon method (see Chapter 13). As an example, anomaly x is dated at y million years by a basalt sample taken at any favorable place. If everything works as it should, this date is correct for anomaly x everywhere. One date serves to give the age of thousands of square kilometers of ocean bottom. Dating of similar areas of continental rock by the use of fossils could take decades of work.

FIGURE 2-15 *A basaltic lava flow photographed 2,700 m below the surface of the ocean, Rift Mountains, Indian Ocean. Underwater eruptions usually take pillowlike forms such as this. (Official photograph courtesy of U.S. Navy.)*

Now that many age determinations have been made, it is possible to know the rate at which sea-floor spreading takes place. It ranges from about 2 cm to as high as 18 cm per year; the average is near the lower figure. Slow spreading produces less basalt, narrower magnetic bands, and more closely spaced anomalies. Correctly identified and dated anomalies give proof not only of the times when particular rocks were formed but also when seas began to open and when new ridges and trenches became active or ceased to exist.

Applications multiply

Not only does the newly evolved concept of sea-floor dating permit accurate placement of rocks and events, it unifies the fossil record also. We may have given the impression that the floor of the ocean is bare basalt. This is not so except for recently produced areas at ridge crests. Elsewhere the basalt is covered by ooze that becomes progressively thicker away from the ridge (Figure 2-15). As soon as a particular strip of sea floor is produced, it begins to receive a deposit of fine material with a content of shells and skeletons of organisms supplied from above. The age of these first deposited organisms is for all practical purposes the same age as the rock on which they fall. Thus the lowest fossils found at any spot should agree in age with the magnetic anomaly of that spot. Many observations prove that the lowest fossils do actually become progressively and systematically younger toward the ridge. Age dating of the anomalies thus provides an excellent independent scale for determining the times of arrival, exterminations, and rates of evolution for such organisms as reach the sea bed and become fossils.

Offset ridges

Lithospheric plates move apart at the midoceanic ridges and come together at the trenches. They may also slide past each other along great vertical faults so that neither side gains or loses material. The special name, *transcurrent fault,* is given to displacements of this sort. Another type of displacement, the *transform* fault is common in connection with the spreading ridges in all oceans, giving the zigzag, interlocking, or stair-step appearance that ridges show when mapped in detail. Transform faults are not recently formed offsets of continuous ridges but are formed gradually as a ridge develops.

The original ridge was probably a continuous curve, but this soon changed to a series of straight sections joined by transform faults. This mechanical arrangement allows the ridges to be at right angles to the direction of relative motion and is certainly an expression of conservation of energy. The name transform fault implies that the force ceases to be expressed by faulting and is transformed into compression or tension where a fault terminates. As shown by Figure 2-16b, there is relatively active displacement along transform faults only in the stretch between the offset portions of the ridge axis. Beyond this central area the movement of both sides is in the same direction. In confirmation of this concept it is observed that earthquakes happen only in the area between the offset segments of the ridge; the two outer ends of the fault are geologically inert. Transcurrent faults, such as the notorious San Andreas, may be active along their entire length.

Observation of a map of lithospheric plates will show many places where three plate boundaries intersect. These are known as *triple junctions.* Many complicated motions occur here, but geometric solutions seem to work well and need not concern us at this point. Those who are interested in the details of plate motion should consult the references at the end of this chapter.

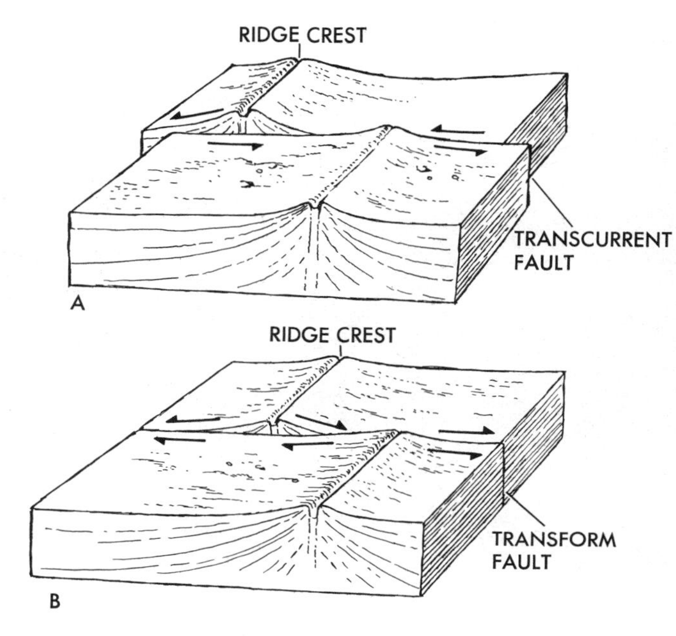

FIGURE 2-16 *A transcurrent fault (A), contrasted with a transform fault (B).*

Old clues reexamined

Ancient climates

When a sufficiently powerful effective mechanism for movement of continents was discovered in the form of sea-floor spreading, all the older evidences had to be reexamined. Consider the climatological argument based on the early discovery of fossil plants near the South Pole and of glacial deposits in tropical and semitropical lands. These phenomena are neatly taken care of by having Antarctica pass into the polar region about 50 million years ago and by tracing the previous path of this same pole across Africa and other southern lands in a systematic north to south direction between 350 to 250 million years ago.

FIGURE 2-17 Coal beds on Lignite Creek, Alaska. The vegetation making up the coal is the same kind of vegetation now found living much farther south. The seams here are of exceptional thickness. (Photo courtesy of U.S. Geological Survey.)

Glacial deposits and beds of coal (Figure 2-17) speak clearly of climatic influences. With increased perception, geologists now read other climatic messages from the rocks and are able to detect frigid, temperate, and tropical influences. Temperate regions are marked by seasonal variations, and these variations in turn may be reflected in sedimentary structures. The succession of high and low precipitation, or high and low temperature, may create banded or cyclic deposits in both fresh and salt water. Yearly deposits of this kind are called *varves,* and many varved deposits have been discovered, especially in drilling for oil. Large areas in the temperate zones are deserts and these occupy certain relative positions with respect to lands and oceans. Here wind action may be a very effective agent in moving and depositing sediment. The internal structure of wind-deposited sandstone reveals wind directions and hence climatic zonation.

The equatorial regions also have their characteristic sedimentary deposits. Here, under the vertical rays of the sun, evaporation is high, and a class of sediments called *evaporites* is commonly formed. Evaporites include salt, gypsum, and dolomite; these sediments usually occur interstratified in the same formations and, when found, are considered to be strong evidence of tropic or equatorial climates. Lack of moisture and relatively high temperatures are thought to contribute to the formation of another class of sediments known as *red beds.* In these the free access of oxygen into the sediments aided by a lack of water has led to oxidation and the production of colored iron pigments. Red beds are very frequently associated with evaporites and windblown sands. The colorful Colorado Plateau in Utah and Arizona shows rocks of this type and all evidence, biologic and lithologic, is that this area was at or near the equator for a long period of time (Figure 2-18).

Study of oceanic sediments reveals that they too are very sensitive to climatic influences. The equatorial belt is a place of heightened organic productivity because the vertical sunlight penetrates deeply into the water and a supply of nutrients is generally available. The near-surface life produces an abundance of calcareous shells and other organic material that sinks to build up an unusually thick deposit directly under the zone where sinking and rising waters mingle. Although productiv-

FIGURE 2-18 *Land-deposited formations of predominantly red color, San Rafael Swell, Utah. The Colorado Plateau is frequently referred to as "red rock country." (Photo by Parker Hamilton for Utah Tourist and Publicity Council.)*

ity falls off rapidly away from the equatorial zone, the nature of the floating life and of its dead, perhaps fossilized, remnants on the bottom may reveal much about the latitude of formation. Many forms of minute life are tied closely to temperature and salinity conditions (Figure 2-19). Of course it is realized that warm currents may penetrate poleward or cold currents may reach the equator, and these effects are taken into account as far as information permits.

FIGURE 2-19 *Minute fossils of the algae* Emiliana huxleyi *from the Black Sea. This is a marine species that invaded the Black Sea only after the last great glaciation. Species such as this are sensitive indicators of temperature and salinity variations and are distinctive guides to age as well. (Photo courtesy of David Bukry, U.S. Geological Survey.)*

Climate governs to a large extent the distribution of living things on the dry land and indirectly in the oceans as well. In fact, animals and plants are commonly classified according to the climates in which they live. This association provides invaluable clues to past climates. Coral reefs are an outstanding example; they are currently confined to warm seas and apparently have always been so. On land it is noted that trees showing annual growth rings (Figure 2-20) are typical of temperate regions; the same appears true of ancient woods. Reptiles and amphibians prefer warm climates and are more common toward the equator. Coal, being the remains of abundant vegetation, is more indicative of moisture conditions than it is of temperature. Taking into account the types of vegetation making up coal beds as well as other evidence, it is thought that most ancient coals originated in temperate zones rather than in tropical ones.

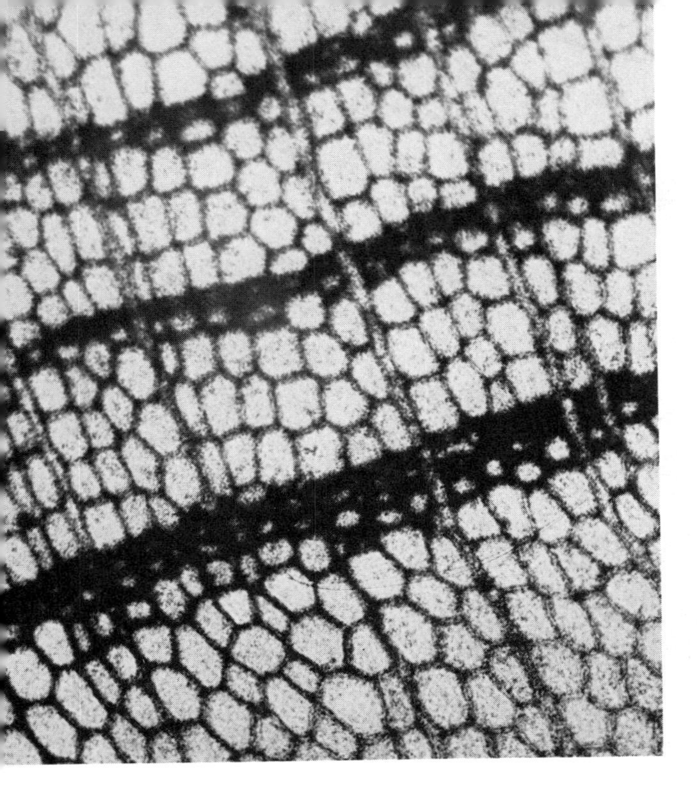

FIGURE 2-20 Enlarged view showing cell structure and annual rings of fossil Metasequoia from Yellowstone National Park. Annual rings such as these prove that growth took place in a seasonal climate almost surely in the temperate zone.

Most of the arguments for ancient climates based on what may be observed today are arguments by analogy. This manner of argument is said to be a weak one, but in its better forms it is practically irrefutable.

Plants and animals

Recall the ancient Glossopteris flora, found on all southern continents as well as India, and the little reptile, *Mesosaurus*, from both South Africa and South America. These two fossils were among the first to be brought forth as evidence of continental drift. The list of witnesses has lengthened greatly—in fact, the history of every plant and animal group can well be examined in light of plate tectonics. The reason is simply stated: land masses once in contact or in near proximity should show essentially the same plants and animals. If an originally continuous continent is broken and its pieces scattered so that communications among its organisms are no longer possible, the life forms will gradually become more and more unlike as evolutionary changes take effect. The degrees of similarity and dissimilarity between two areas provide good clues as to how close these areas once were and how long they have been separated. Thus, the life of Africa and South America had much in common 200 million years ago, but much less today.

FIGURE 2-21 Sequential relations of Africa, South America, and North America from early Mesozoic to present: (A) Triassic—there was easy access of organisms between Africa and South America; distant connections only with North America. (B) Early Tertiary—a minimum of communications among the three land masses; African record is poor; South American faunas developing many peculiar endemic forms; North America in intermittent connection with Eurasia. (C) Late Tertiary—South America makes contact with North America and a broad exchange of faunas takes place; African influences on South America almost lost.

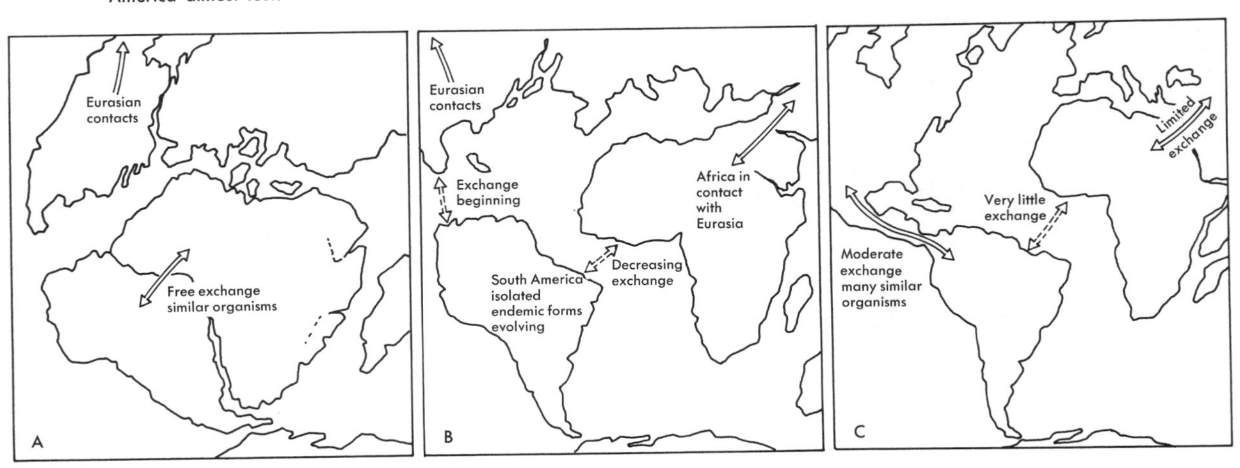

On the other hand, if two land masses that have never been in contact or have been out of contact for a very long period are brought together by continental drift, there will be an exchange of life forms when the organisms from the two land masses invade each other. This is illustrated by the approach and eventual joining of North and South America late in the Tertiary period about 3 million years ago (Figure 2-21). (Refer to time scale on the inside front cover for placement of the geologic periods.)

Several specific examples are outstanding. An aggregation of early Triassic reptiles and amphibians known by the name of its most distinctive member, *Lystrosaurus*, occurs in South Africa, India, Antarctica, and China. *Lystrosaurus* is a medium-size reptile with unusual skeletal characteristics including large down-pointing tusks. Neither it nor any of its associates could possibly cross the oceans that now separate the southern continents. Its discovery in Antarctica in 1969 was hailed as final proof for the reality of Gondwanaland. The Chinese occurrence is taken by many as evidence that the land area in which it is found may also have been a part of Gondwanaland. Indeed there is a place adjacent to Antarctica and Australia where this fragment could have been located. Just how eastern Asia may have been assembled is still unknown.

Fossil evidence for former land connections in the northern hemisphere has been more difficult to acquire, and the clues are in the form of less spectacular creatures than *Lystrosaurus*. The North Atlantic ocean is considered to have been created by splitting apart of opposing shores of North America and Europe-Africa. Before this there existed a similar but narrower body of water called the proto-Atlantic, which was closed and obliterated by continental collisions from 450 to 350 million years ago.

Marine fossils of the early proto-Atlantic include many trilobites and graptolites, with different species on the European shores than on the American shores. This is shown by comparison of fossils that now exist in Scotland and New England. As the proto-Atlantic was closed, these fossil-bearing beds were brought into proximity and were literally mashed together in a mountain-building event that created much of New England. Later the continents began to split apart and the modern Atlantic was born. But the new opening did not coincide with the old line of closure, and a small piece of North America was carried away with the European block to become part of Scotland, and a piece of Europe remains to make up part of New England. This is shown by the fossils of these misplaced areas—older European fossils in New England, older American fossils in Scotland. Other features including great faults and igneous bodies show that northeast America and northwest Europe were once very close together. Continental drift gives the best explanation. Figure 2-22 shows the sequence of events in semi-diagrammatic manner.

The most spectacular group of extinct animals, the dinosaurs, furnishes many clues favorable to ocean spreading and continental drift. Dinosaurs appeared when the continents were mostly united, and the ancestral forms reached all favorable environments by walking over dry land. Passable land connections in favorable climates existed well into the Jurassic when dinosaurs reached a high point in diversity and numbers. At this time, almost identical species lived in England, Wyoming, and East Africa. One great Late Jurassic assemblage well represented in the Morrison Formation of the Rocky Mountains is matched best in the Tendaguru Formation thousands of kilometers away in Tanzania, Africa. Indeed the greatest dinosaur of all time, *Brachiosaurus*, is found only in Colorado-Utah, southwestern Europe, and East Africa. Evidently it could have traveled from North America to Africa by way of Europe.

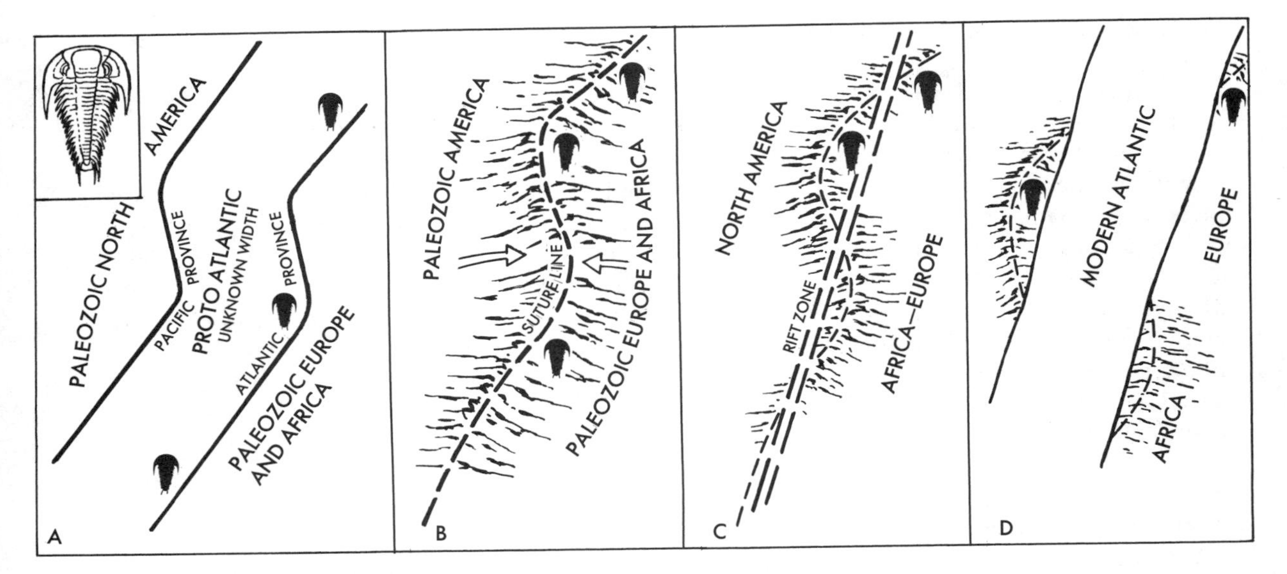

FIGURE 2-22 *A tale of two oceans: (A) The proto-Atlantic of early Paleozoic time as it existed 500 million years ago, width unknown. The trilobite Paradoxidies lived along the eastern but not the western shores. (B) The proto-Atlantic is squeezed out of existence, and the sediments are forced into a complex mountain chain known as the Caledonian Mountains in Europe and the Taconic in North America; the zone of collision is called a continental suture. (C) After a long period of contact, a new zone of separation begins to open in the Triassic Period; this zone is parallel to but does not follow the previous suture line.*

Later, in the Cretaceous, as connections were broken in wholesale fashion, dry-land bridges became fewer and fewer. Yet dinosaurs continued to travel those available. The sauropod genus *Titanosaurus* is found in Australia, Madagascar, and South America but not in North America. Finally in the Late Cretaceous the same general types of dinosaurs are found in western North America as in central Asia; in fact, the horned dinosaurs are confined to these two regions. Bering Straits was then a passable land bridge, and the climate there was favorable to dinosaur migrants. Figure 2-23 summarizes in graphic form the facts just outlined.

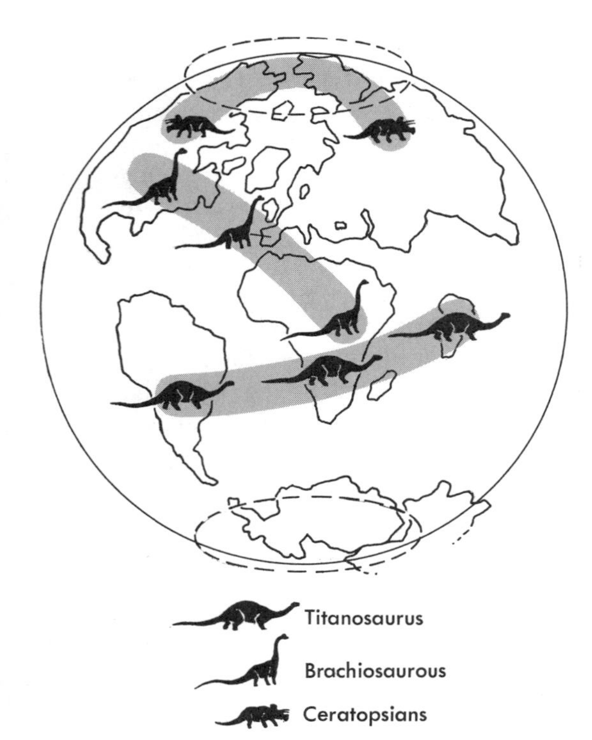

FIGURE 2-23 *Dinosaurs and continental drift. Titanosaurus, large land-living sauropod, is found in South America, Africa, and India. It seems unlikely that it could cross large water bodies to reach these localities, and its original territory may have been continuous at one time. A problem exists in that India is thought to have been far removed from Africa at the time Titanosaurus lived. When and how did it "get aboard"? Brachiosaurous, largest of the long-necked dinosaurs is found in the western United States, Portugal, and East Africa in Late Jurassic deposits. How it reached these widely separated spots is another problem because the Atlantic Ocean, although much narrower, seems to have been in existence when this species lived. Perhaps scattered islands, now destroyed, helped. Finally the distinctive horned dinosaurs (Ceratopsians) of the Late Cretaceous are found only in western North America and east-central Asia. They probably spread by a northern route.*

Polar wandering

It would be a great oversimplification to say that sea-floor spreading and plate tectonics can explain everything about the present and past distribution of land masses. Another process called *polar wandering* or, preferably, *apparent polar wandering* (Figure 2-24), is recognized as being able to bring about global effects independent of plate tectonics.

The term "apparent polar wandering" calls attention to the poles. Consider some facts and questions about the geographic and magnetic poles. Even though the present geographic pole, about which the earth rotates, is quite near the present magnetic pole, has this always been the case? Are there deep-seated reasons why the two poles must be near each other? Is it possible, for example, for the magnetic pole to lie in the tropics? Might it wander unsystematically over the earth? Without going into some very technical reasoning, we can say that most investigators believe that the magnetic pole must remain at or near the geographic pole. This relationship seems to be essential to the generation of the earth's magnetic field.

There are ways of testing the theory of "apparent polar wandering." Consider that there are climatic indicators of latitudes that are independent of magnetic ones. On any globe that rotates as does the earth there must be two opposing climatic poles where solar radiation reaches a minimum, and an equator between on which maximum radiation falls. The result of this arrangement, on earth at least, is frozen ice at the poles and luxurious organic production coupled with high evaporation under the equator. These climatic extremes leave geologic records that are locked up at the same time as are the magnetic evidences—perhaps even in the same rock. This permits a critical test. If the magnetic record places the magnetic pole at a particular place and that locality shows glacial action appropriate to the geographic pole, it is evident that for this particular time the two poles coincide. If, in contrast, the magnetic record shows a pole in the midst of geologic evidences of hot deserts, the two poles are incompatible.

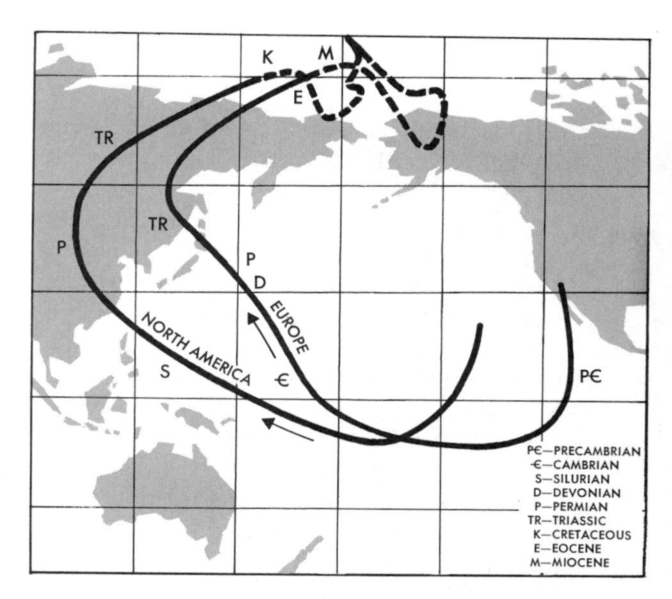

FIGURE 2-24 *Postulated "polar wandering" curves based on data from North America and Europe. The two curves coincide at the present pole, then diverge, and again converge backward in time. The paths become nearly identical if the two continents are adjusted to positions they had before, during, and after sea-floor spreading. The continents, particularly Asia, did not necessarily have the configurations shown on the map when the pole was in its successive positions. Evidence from many sources is that the North Pole was over the Pacific for most of geologic time. It became "land locked" only in the Tertiary. (Adapted from several sources.)*

To be brief, the evidence is practically all in favor of the coincidence of magnetic and climatologic poles. This coincidence reconciles most of what is known about former climates in relation to modern ones. The axis of the earth's rotation probably has remained steadfastly pointing toward the same region of space, the earth has not tipped or toppled, the area of the magnetic pole has always been near the coldest area of the globe, and the temperature requirements of plants and animals have always been roughly uniform. Therefore, any indications of past climatic zones that are out of harmony with their present global settings are the result of displacement of the outer shell of the earth into, across, and out of specific climatic zones.

Many observations show that the north magnetic pole has "wandered" over a 33,500 km-long path, taking it from a position near western North America in the late Precambrian, across the Pacific Ocean in the Paleozoic, into the regions of northeast Asia in the Mesozoic, and finally into the Arctic in the mid-Tertiary. Does this mean there has been no independent movement of lithospheric plates? Consider this line of evidence: If the lithospheric plates of the earth were solidly locked together, then all magnetic evidence would point to the same polar position for any given instant of time. As a matter of fact, indicated pole positions are different for different land masses during certain periods of the past. Thus, the pole positions are the same for Europe and America at the present time and also were the same before about 200 million years ago. Between these two times the pole positions were different. This indicates that, like great rafts, the continents had separate motions during a long intervening period. It is surely no coincidence that the period of divergence is the same period that other evidence says was one of active sea-floor splitting and drifting. The conclusion is that there has been both polar wandering and continental drifting. This is not an incompatible concept. The whole system of plates is moving in concert but with internal jostling and collision and even occasional rapid readjustments. Comparisons have been made with the pack ice of the Arctic Ocean: it is made of separate pieces, but the whole system is carried along by a strong current.

Is anything fixed in place?

At first thought one might conclude that plate tectonics and polar wandering leave nothing fixed in place on the face of the earth. There are, however, a few possible exceptions to the general motion. These exceptions are

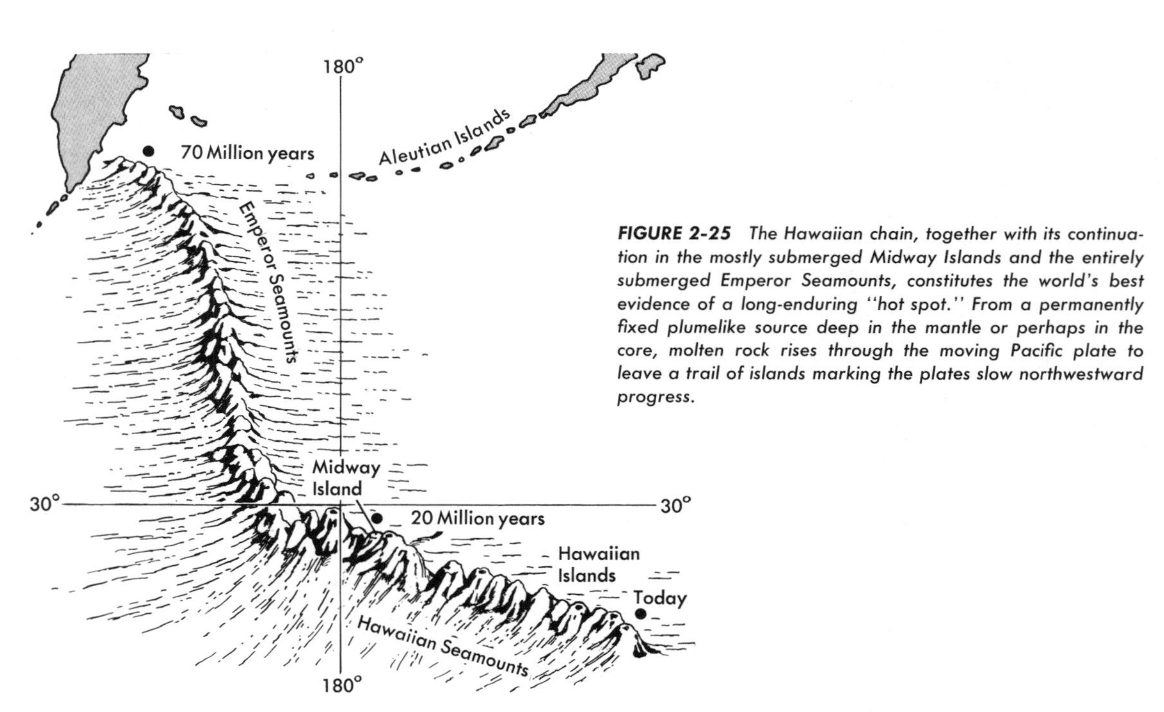

FIGURE 2-25 The Hawaiian chain, together with its continuation in the mostly submerged Midway Islands and the entirely submerged Emperor Seamounts, constitutes the world's best evidence of a long-enduring "hot spot." From a permanently fixed plumelike source deep in the mantle or perhaps in the core, molten rock rises through the moving Pacific plate to leave a trail of islands marking the plates slow northwestward progress.

***FIGURE 2-26** Steaming geyser basin suggests that Yellowstone Park is a "hot spot." (Photo courtesy of U.S. National Park Service.)*

the "hot spots" or "melting spots," the reference being to the sites of certain major volcanoes—volcanoes that are the latest in their respective chains of formerly active cones. The Hawaiian chain, which terminates in the very active volcanoes of Hawaii, is the world's best example. The Hawaiian chain extends 3,500 km northwesterly across the Pacific sea floor. The northwest end of the Hawaiian chain is near the southeast end of the entirely submerged Emperor chain of extinct volcanoes that continues northward to the western Aleutian chain. The origin of the Hawaiian-Emperor chain is thought to be a "melting spot" about 300 km in diameter and at least 65 km below the ocean floor. Other "hot spots" include Iceland, the Azores, Cape Verde Island, St. Helena Island, Prince Edward Island, Reunion Island, Easter Island, and the Galapagos Islands.

The conclusion with regard to hot spots is that all volcanoes occurring in clear-cut linear groups have their roots or sources below the lithosphere, probably below the asthenosphere, and perhaps in the stationary mantle or even in the molten core. As a lithospheric plate moves over an ascending plume of heated material, new volcanoes are created one after another and then become dormant as they are cut off from their source.

If hot spots do arise from the mantle below the asthenosphere, then they should be fixed points of reference that remain the same distances and directions from each other. The evidence seems good that lithospheric plates are overriding the hot spots. However, if volcanoes arise in the asthenosphere, they might also be moving with respect to the mantle and each other. Most of the known hot spots are in the oceans, but the Yellowstone region of Wyoming (Figure 2-26) could well be a hot spot overridden by North America. The Snake River Plain, which is largely underlain by lava flows, could mark the path of the hot spot.

What makes it go?

All the problems raised by Wegener's theory have not been solved by the concept of plate tectonics. Geologists have discovered a mechanism by which great displacements take place, but the driving force is still unexplained. The plates do move, but what moves them? Some say they are pushed from the midoceanic ridges by the expansive force of intruding magma. Some say that a plate is pulled by the momentum of the forward edge as it plunges into the subduction zone. Others say that plates are carried along by currents and eddies in the plastic asthenosphere below. One thinker believes the plates are driven ever westward by the same forces that drive the tides. Another believes that they are governed by the hot spots that elevate certain areas and from which the plates glide downhill by gravity.

The thought that plates are carried in a conveyor belt fashion by circulating currents in the asthenosphere has much to commend it. After all, this is the way the atmosphere and oceans operate to maintain equilibrium. But convection must be maintained by energy input and may not be necessary in plate tectonics. One thing is certain, heat and gravity are both involved, and when their interaction is understood, plate tectonics will probably be fully explained.

SUMMARY

The long-standing dispute as to whether or not the continents move or drift with respect to one another has at last been answered in the affirmative. Discoveries of the past decade prove not only that continents do drift but also that the entire outer brittle shell (lithosphere) of the earth is in a state of slow but complex motion. A basic discovery was that of a world-circling submarine mountain system that more or less bisects the ocean basins. This midoceanic ridge system is the site of numerous earthquakes and of the production and solidification of floods of basaltic lava and deep-seated magma. As new oceanic crust is produced, it is pushed aside, creating bands of sea floor and crust that are parallel to the ridge and move progressively away from it at an average rate of about 4 cm annually. This action is termed sea-floor spreading, and it provides the long sought mechanism of continental drift.

Oceanic crust cannot be produced indefinitely, and it is eventually destroyed along belts called subduction zones. Here the basaltic crust together with its burden of water-rich ooze plunges downward to return to the hot plastic layer (asthenosphere) below. Subduction zones may involve two actively moving plates as seen along the west coast of South America, with the continent overriding the oceanic segment. Where the continental block is stable, volcanic island-arc systems such as the Japanese and Aleutian chains are produced. When two continent-bearing plates collide (e.g., Asia and India), spectacular and complex mountain chains are produced. Crustal segments may also move past each other for long periods without loss or gain on either side; the San Andreas fault zone of California illustrates this.

Zones of production, destruction, and lateral movement outline large blocks of the earth's outer shell called lithospheric plates. Six major plates and a dozen or so small ones encase the earth like the cracked shell of an egg. Most plates bear oceans and continents; a few are essentially oceanic. The present plate configuration is ancient but apparently not permanent. The total dynamic effects of plate movement are called *plate tectonics.*

Successively formed strips of oceanic basalt can be dated by radioactive elements, by associated fossils, and by the newly discovered science of paleomagnetism. This last method depends on the fact that the earth's magnetic field varies in intensity with time and even reverses in polarity over uneven periods of the past. Newly formed iron-bearing rocks in effect "lock in" a record of the field that prevailed when they were forming. All contemporaneous magnetized rocks, continental or oceanic, have the same magnetic characteristics and can thus be correlated. By a combination of evidences it is known that few parts of the ocean basins are older than about 120 million years; older configurations have been destroyed by subduction.

Sea-floor spreading, continental drift, and plate tectonics combine to solve many problems. All great land masses were apparently once in one or two great supercontinents, which explains similarities of opposing shores. These supercontinents commenced to split apart about 200 million years ago. The Atlantic area has opened not once but at least twice. Africa is colliding with Europe, while India presses into Asia. Climatic effects become understandable. Antarctica has moved into a polar position carrying its coal beds, and most of the southern continents bear evidence of having passed into and out of frigid zones. Similarities and differences of life forms, both fossil and living, are also explainable by plate tectonics—when land or sea realms or provinces are together, their life is similar; when they are separated, their life is different.

SUGGESTED READINGS

BIRD, JOHN, and BRYAN ISACS, eds., *Plate Tectonics.* Washington D.C.: American Geophysical Union, 1972.

COX, ALLAN, ed., *Plate Tectonics and Geomagnetic Reversals.* San Francisco: W. H. Freeman and Co., 1973.

HEEZEN, BRUCE C., and CHARLES D. HOLLISTER, *The Face of the Deep.* London and New York: Oxford University Press, 1971.

MARVIN, URSULA B., *Continental Drift: The Evolution of a Concept.* Washington, D.C.: The Smithsonian Institution Press, 1973.

MATTHEWS, SAMUEL W., "The Changing Earth," *National Geographic, 143* (January 1973), 1–37.

McELHINNY, W. M., *Paleomagnetism and Plate Tectonics.* Cambridge, England: Cambridge University Press, 1973.

SULLIVAN, WALTER, *Continents in Motion: The New Earth Debate.* New York: McGraw-Hill Book Company, 1974.

TARLING, D. H., and S. K. RUNCORN, *Implications of Continental Drift to the Earth Sciences,* 2 vols. London and New York: Academic Press, 1973.

WILSON, J. TUZO, ed., *Continents Adrift—Readings from Scientific American.* San Francisco: W. H. Freeman and Co., 1972.

Matter and Minerals

The fundamental building blocks of the rocks of the earth's crust are chemical elements and chemical compounds called *minerals.* Because geologic history is recorded in the rocks, we need to know something about the minerals that compose rocks. But if we look further, we find that the minerals themselves contain smaller units—*atoms*—and these in turn are made up of still smaller units, of which protons, neutrons, and electrons are the most important to us here. So before we begin our study of minerals, we must review briefly the nature of matter, specifically as it applies to minerals. Later in this chapter we will find out what constitutes a mineral and will examine a handful of the most important minerals.

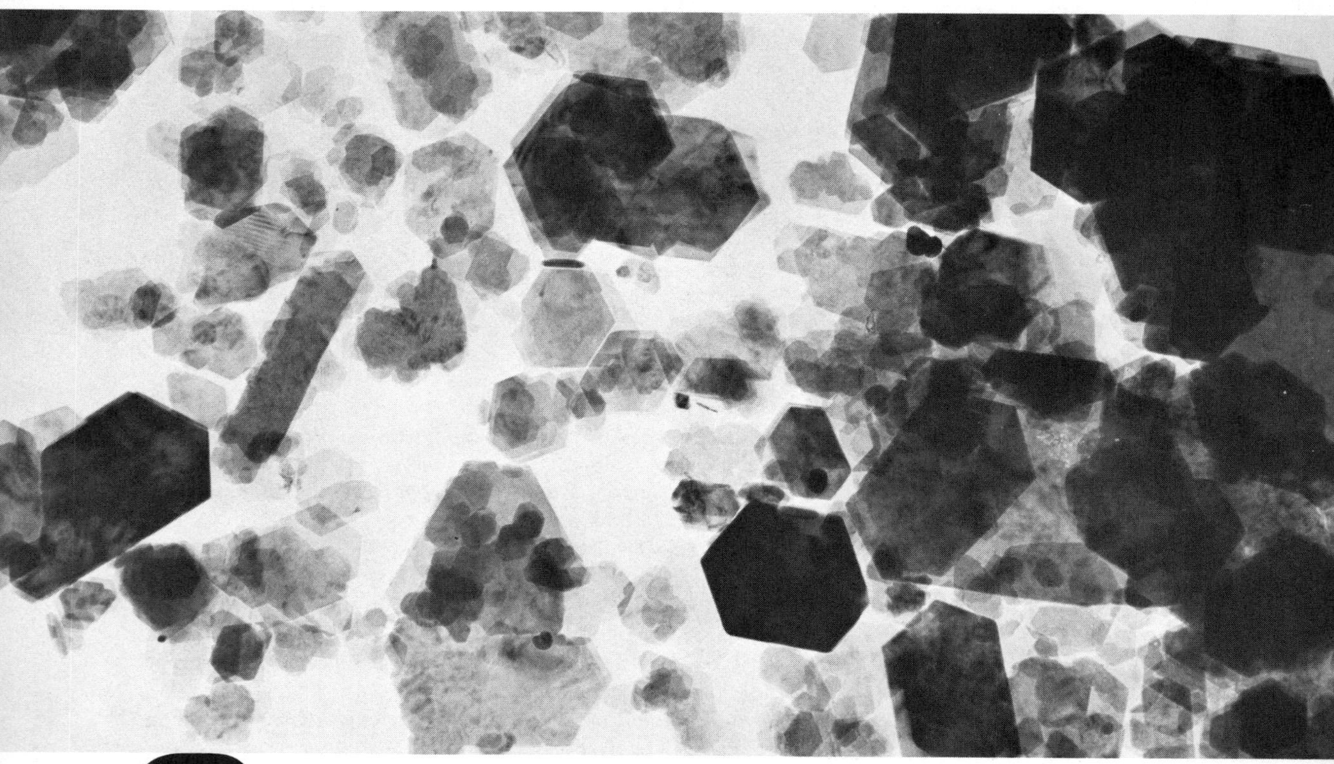

3

Crystals of the clay mineral kaolinite from Georgia. Greatly magnified (× 41,600).

Matter

All matter seems to be essentially electrical in nature, either negative or positive. This electrical nature was early illustrated by a simple experiment—a piece of fur and a piece of amber were rubbed together, and afterwards they could pick up light objects such as wool or feathers. But the material picked up by the fur was repelled by the amber. In the sixteenth century, William Gilbert, personal physician to Queen Elizabeth I, proposed that the power responsible for this phenomenon be called electricity, from the Greek word *eleck-tron,* meaning amber.

We say that *like electric charges repel each other, and unlike charges attract each other* (Figure 3-1). You have seen this principle in action in the poles of two adjacent magnets. The so-called north poles repel each other while they are attracted by the south poles, and vice versa.

Electron, proton, and neutron

When fur and amber are rubbed together, particles are said to pass from the fur to the amber. The fur, then, has a deficiency of particles and is positively charged. Therefore, the particles that moved to the amber to cause the condition must be negatively charged. These particles are called *electrons.*

Atoms are built up in part of electrons. But because electrons are negatively charged, there must be some positively charged particles to attract them if they are to be built into an atom. Scientists believe that there is such a particle, which they call the *proton.* Many additional particles of subatomic size have since been identified, but the only other one of importance to us here is the *neutron,* a particle with no electric charge.

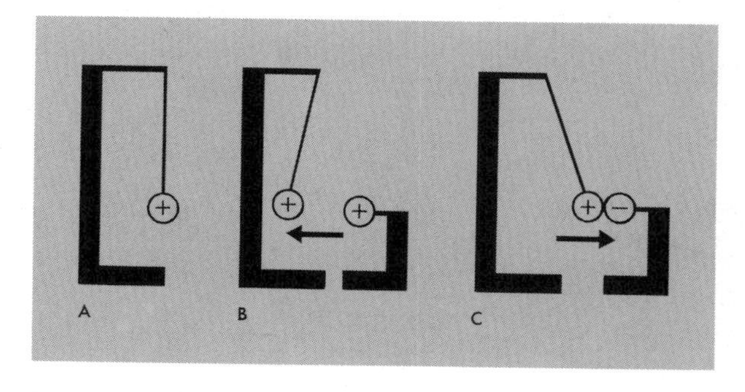

FIGURE 3-1 *Like charges of electricity repel each other; unlike charges attract each other. (A) A small, positively charged sphere. (B) This sphere is repelled by the approach of another positive charge. (C) The positive charge is attracted by a negative charge.*

The atomic model

The electron, the proton, and the neutron gather together into the atom. Our concept of the atom derives from a series of indirect observations that come from the physicist's laboratory. As a result of these observations, we believe that an atom is composed of a cloud of electrons that revolve about a central core of protons or of protons plus neutrons. Repeated experiments show that every atom has the same number of electrons as it has protons. The positively charged protons form the nucleus of the atom, and around it orbit an equal number of electrons whose negative charges balance the positive charges of the protons which are in the core of the atom.

The neutrons are also found in the nucleus of the atom, but because they are electrically neutral, they are not matched by the negatively charged electrons outside the nucleus. These electrons spin around the nucleus at a speed of hundreds of kilometers per second. Like the planets revolving about the sun, electrons revolve around the nucleus at different distances.

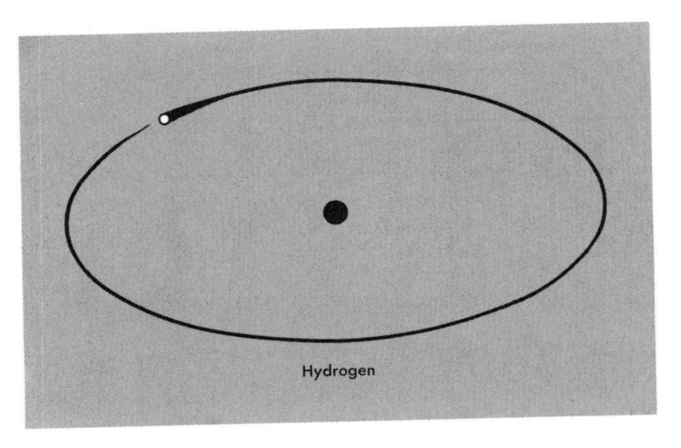

FIGURE 3-2 Schematic sketch of hydrogen atom, which consists of 1 proton and 1 electron. This is the Bohr model, introduced in 1913 by Niels Bohr.

Atomic size and mass

The electrons form a protective shield around the nucleus and give *size* to the atom. In describing atomic dimensions, we use a unit of length called the *angstrom* (abbreviated Å), which is one hundred-millionth of a centimeter (0.00000001 cm), usually written 1×10^{-8} cm.

Diameters of atomic nuclei range from a ten-thousandth to a hundred-thousandth of an angstrom—that is, from 10^{-4} to 10^{-5} Å, which is from 10^{-12} to 10^{-13} cm. Atoms of the most common elements have diameters of 2 Å, which is roughly 20,000 to 200,000 times the diameter of their nuclei. On another scale, if the sun were the nucleus of our atomic model, then the diameter of the atom would be greater than the diameter of the entire solar system.

The atom has not only size but mass. Within the atom the nucleus, which occupies only a thousandth of a billionth of the volume of the atom, contains 99.95 percent of the atomic mass (see Table 3-1).

Table 3-1 Fundamental Particles

	ELECTRIC CHARGE	REST MASS (amu)
Electron	−1	0.00055
Proton	+1	1.00760
Neutron	0	1.00894

Elements

An atom is the smallest particle of an element that can enter into a chemical combination. Ninety-two elements are known to occur in nature, and a number of others have been made by man in the laboratory. Every element is a special combination of protons, neutrons, and electrons. Each element is identified by the number of protons in its nucleus and is designated by a name and a symbol.

Element Number 1 is a combination of 1 proton and 1 electron (see Figure 3-2). Long before its atomic structure was known, this element was named *hydrogen,* or "water-former" (from the Greek roots *hydro* and *gen* meaning "water" and "to be born"), because water is formed when hydrogen burns in air. Its symbol is H. Hydrogen has first place in the list of elements because it has 1 proton in its nucleus. The number of protons, Z, determines the *atomic number* of the atom, and each element is distinguished by a different value of Z.

Element Number 2 consists of 2 protons (plus 2 neutrons in the most common form) and 2 electrons (see Figure 3-3). It was named *helium,* with the symbol He, from the Greek *helios,* "the sun," because it was first identified in the solar spectrum before it was isolated on the earth. Helium has second place in the list of elements because it has 2 protons in its nucleus.

Each addition of a proton, with a matching electron to maintain electrical balance, produces another element. Neutrons seem to be included more or less indiscriminately, although there are about as many neutrons as protons in the common form of many of the elements. The list of elements appears in Appendix A.

FIGURE 3-3 *Diagrammatic representation of an atom of helium. The nucleus consists of 2 protons and 2 neutrons, and accordingly has a mass number of 4. There are 2 electrons (negative charges) to balance the positive charges on the 2 protons. Since there are 2 protons in the nucleus, this atom is Number 2 in the table of elements. The symbol* $_2He^4$ *indicates Number 2 in the table of elements, He for the name helium, and a mass number of 4. The nucleus of helium (2 protons + 2 neutrons) without any accompanying electrons is sometimes called an alpha particle.*

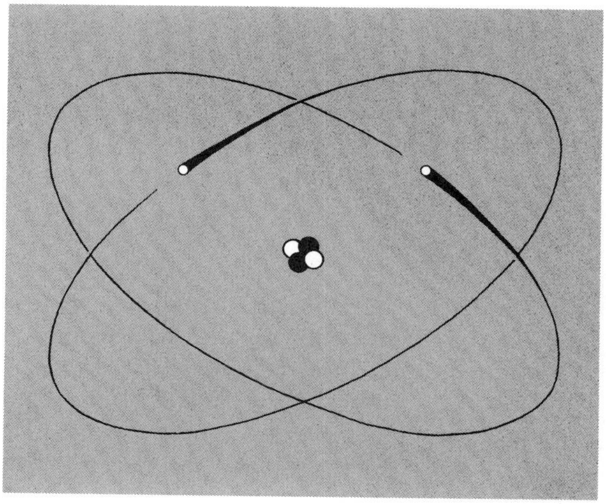

FIGURE 3-4 *Electron shells around a nucleus. In true scale, the diameter of the shells is 20,000 to 200,000 times the diameter of the nucleus. If the sun were the nucleus, the electron shells would embrace more space than the entire solar system. Yet, the nucleus contains 99.95 percent of the mass of the entire atom.*

On a statistical basis, electrons are distributed about the nucleus in specific energy levels or orbital shells. These orbitals are arranged systematically at different distances from the nucleus (see Figure 3-4), and a specific amount of energy is required to maintain an electron at a given distance from its nucleus. The discrete amounts of energy differences between energy levels are called *quanta.* For convenient reference, these orbitals are sometimes referred to as *energy-level shells.*

If you examine the list of elements in Appendix A, you will see that as electrons are added to match the increasing numbers of protons, they follow a simple pattern for the first 18 elements. After that, the system changes somewhat, but in the entire list there is *no atom with more than 8 electrons in its outermost shell.* The elements that have that maximum number are the inert gases—neon, argon, krypton, xenon, and radon—which rarely combine with other elements. The number of electrons in the outer shell of an atom determines the manner and ease with which it can join with other atoms to form more complex structures.

Isotopes

Every element has alternate forms, which, though essentially identical chemically and physically, have different masses (see Appendix A). Such forms are called *isotopes* (pronounced eye'-so-tope), from the Greek *iso,* "equal" or "the same," and *topos,* "place," since they occupy the same place in the table of elements, based on the number of protons in the nuclei. Isotopes show differences in mass as a result of differences in the number of neutrons in their nuclei. For example, hydrogen with 1 proton and no neutrons in its nucleus has a mass number of 1. When a neutron is present, however, the atom is an isotope of hydrogen with a mass number of 2, called *deuterium* (see Figure 3-5).

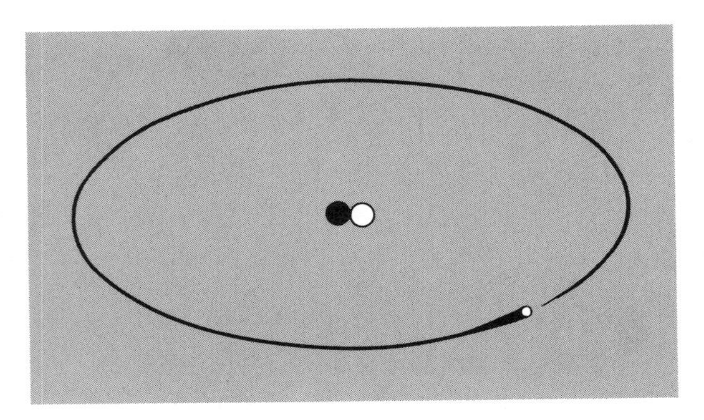

FIGURE 3-5 *Schematic sketch of deuterium, an isotope of hydrogen formed by the addition of a neutron to the nucleus. It has a mass number of 2.*

Ions

An ion is an electrically unbalanced form of an atom or group of atoms. An atom is electrically neutral. But if it loses an electron from its outermost shell, the portion that remains behind has an extra unmatched positive charge. This unit is known as a positively charged ion (a *cation*). Cations are so named because they are attracted toward a cathode, the negatively charged terminal of an electric cell.

If the outermost shell gains an electron, the ion has an extra negative charge (an *anion*). Anions are attracted by a positively charged plate, the *anode.* However, more than 1 electron may be lost or gained, as we shall see later, leading to the formation of ions with 2 or more units of electrical charge.

Compounds

Compounds are combinations of elements, effected mostly through the joining of ions. Those formed by life processes are called *organic compounds.* Others are known as *inorganic compounds.*

As we have just seen, elements with 8 electrons in their outermost shell do not combine readily with other elements. In contrast, elements that have fewer than 8 electrons in their outermost shell readily shed or pick up electrons in an effort to achieve that stable configuration.

For example, an atom of sodium has only 1 electron in its outer shell but 8 in the next shell (see Appendix A). By losing that outside electron, the atom of sodium becomes a positively charged ion (represented by the symbol Na^{+}) with 8 electrons in its outer shell (Figure 3-6). In contrast, chlorine has 7 electrons in its outer shell. By picking up 1 more, it becomes a negatively charged ion (represented by the symbol Cl^{-}) with 8 electrons in its outer shell (Figure 3-7).

If a positive sodium ion and a negative chlorine ion approach each other, the electrical attraction of their opposite charges brings them firmly together with what is called an *ionic bond,* to form a new product with properties unlike those of either sodium or chlorine.

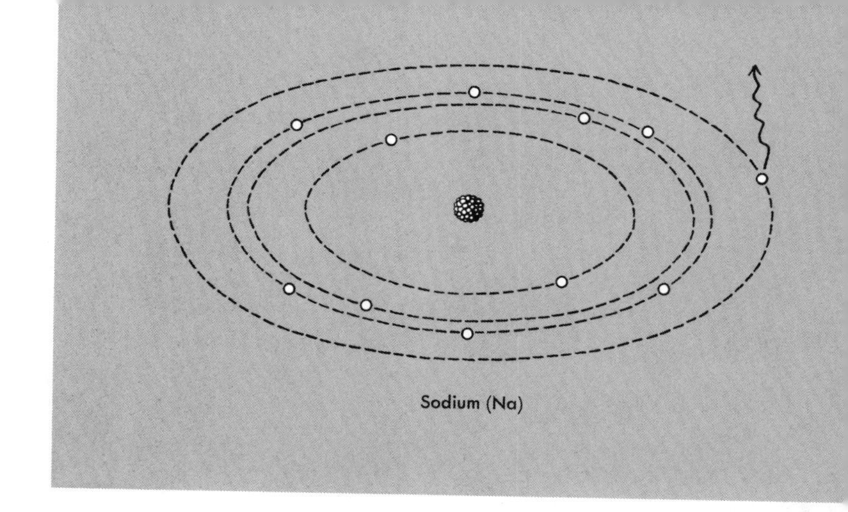

FIGURE 3-6 *Formation of the sodium ion Na^+ results when the sodium atom loses the only electron in its outermost shell.*

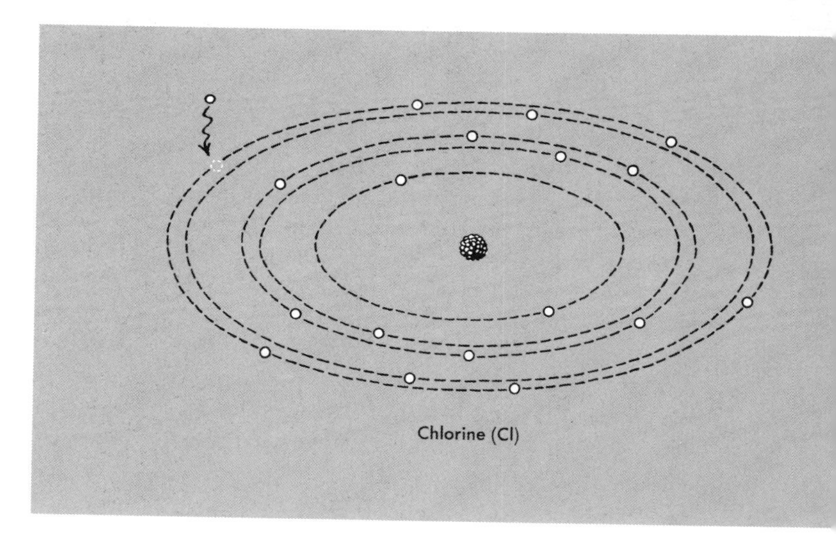

FIGURE 3-7 *Formation of the chlorine ion Cl^- results when a chlorine atom gains an electron in its outermost shell, to make the total 8.*

This product is the mineral halite (hay'-light) or common table salt. The chemical designation for this compound is simply a combination of the symbols for the elements that compose it—NaCl.

The atoms of elements can combine in other ways to form compounds, as, for example, in the formation of a water molecule. Oxygen has 6 electrons in its outer shell (see Appendix A) and therefore needs 2 more to achieve 8. If 2 hydrogen atoms, each of which has only a single electron, approach an oxygen atom, the hydrogen electrons in effect slip into the vacant slots in the outer shell of the oxygen atom but do not separate from their own protons. So the hydrogen nuclei are really *sharing* their electrons with the oxygen nucleus (Figure 3-8). When this happens, the atoms are said to be *covalently bonded.* Again, the result is a compound that is different in every way from the elements themselves. This compound is water, whose symbol, H_2O, represents the elements that make it up and the proportions in which they are present. The combination of 2 atoms of hydrogen and 1 atom of oxygen forms the smallest unit that possesses the properties of water. This unit is called a *molecule* of water. *A molecule is the smallest unit of a compound that displays the characteristic properties of that compound.*

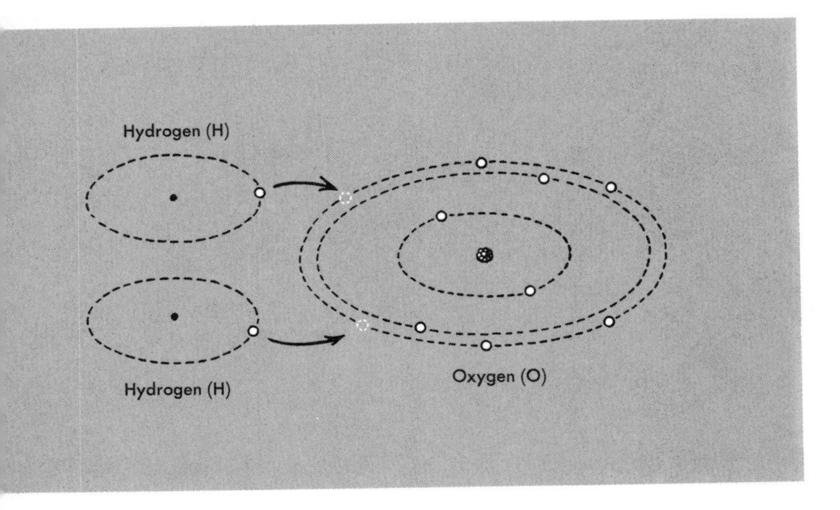

FIGURE 3-8 *Two hydrogen atoms and 1 oxygen atom join to form water, H_2O, by a covalent bond. In this bond, the hydrogen electrons do double duty in a sense, filling the two empty places in the outer shell of oxygen, yet remaining at their normal distance from their hydrogen nuclei. The result is the formation of a molecule of water, the smallest unit that displays the properties of that compound.*

Water is dipolar ("two-pole") because of the repulsion of unbonded electrons in the oxygen atom away from the 2 hydrogen atoms. The net result is a molecule with a more negative character at one end than the other. Hence, water is a *dipolar compound.* This fact gives water special properties that make it an extremely important agent in geological processes. The mechanism by which water dissolves salt (Figure 3-9) is an illustration of the ease with which water dissolves various substances and participates in weathering and other geological activities.

Organization of matter

We have seen that matter is composed of fundamental particles (protons, neutrons, and electrons) combined into atoms. In nature there are 92 different combinations, each of which is an element. Elements, in turn, combine to form compounds (see Table 3-2).

Table 3-2 The Organization of Matter

Protons + *Neutrons* → *Nucleus*

Nucleus + *Electrons* → *Atoms of Elements*

1 Proton 1 Electron	2 Protons (Neutrons) 2 Electrons	3 Protons (Neutrons) . . . . 3 Electrons	92 Protons (Neutrons) 92 Electrons
ELEMENT NO. 1	ELEMENT NO. 2	ELEMENT NO. 3	ELEMENT NO. 92

Number of Protons + Number of Neutrons = *Mass Number*

Gain or Loss of Neutrons ⟶ *Isotopes*

Gain or Loss of Electrons ⟶ *Ions*

Element + Element (etc.) ⟶ *Compounds* (Smallest Unit of Compound = Molecule)

NOTE: This table is continued in Table 3-5.

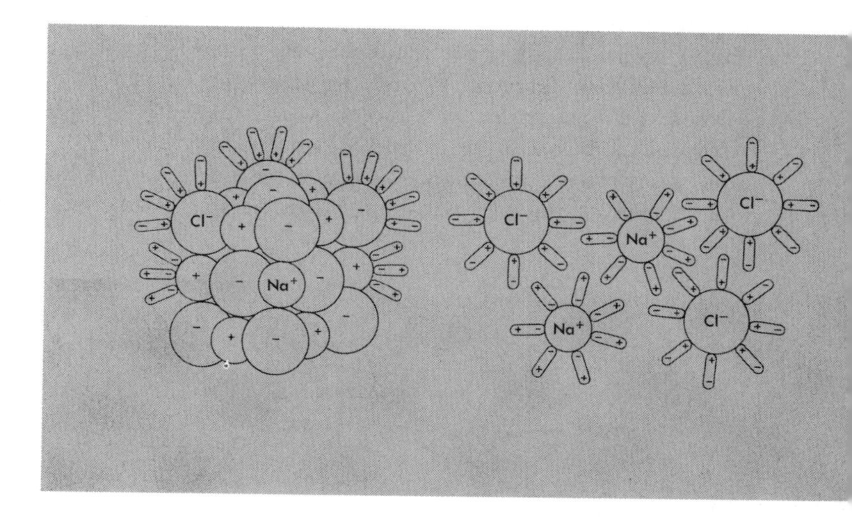

FIGURE 3-9 *Mechanism by which water dissolves salt. Water dipoles attach themselves to the ions that compose the salt and overcome the ionic attractions that hold the salt together as a solid. Each Na^+ and Cl^- ion are then convoyed by a number of water dipoles into the body of the liquid.*

Minerals

The word *mineral* has many different meanings in everyday usage. Some people use mineral to refer to anything that is nonliving, according to the old classification of all matter as animal, vegetable, or mineral. To prospectors and miners, mineral is an ore. And advertisers of pharmaceutical products associate the term with vitamins. But in our discussion, mineral refers to a *naturally occurring solid element or compound that has been formed by inorganic processes.* Later on, we will expand this definition to make it more comprehensive.

Table 3-3 Most Abundant Elements in the Earth's Crust

ELEMENT NUMBER	NAME AND SYMBOL	WEIGHT PERCENT	VOLUME PERCENT
8	Oxygen (O)	46.60	93.77
19	Potassium (K)	2.59	1.83
11	Sodium (Na)	2.83	1.32
20	Calcium (Ca)	3.63	1.03
14	Silicon (Si)	27.72	.86
13	Aluminum (Al)	8.13	.47
26	Iron (Fe)	5.00	.43
12	Magnesium (Mg)	2.09	.29

SOURCE: Based on Brian Mason, *Principles of Geochemistry,* 3rd ed. (New York: John Wiley & Sons, Inc., 1966), p. 48.
NOTE: The eight elements shown in this table make up 99 percent by weight of the Earth's crust.

Mineral composition

About 2,500 minerals are known. Some are relatively simple compounds of elements in the solid state; others are complex. The diamond (Figure 3-10) is composed of only one element—carbon. Common table salt, really the mineral halite (Figure 3-11), is composed of two elements, sodium and chlorine, in equal amounts—every sodium ion present is accompanied by one chlorine ion. The mineral pyrite (Figure 3-12), sometimes known as "fool's gold," is also composed of two elements, iron and sulfur. But in this mineral

FIGURE 3-10 *Diamond (carbon), a native element mineral. (Courtesy DeBeers Consolidated Mines.)*

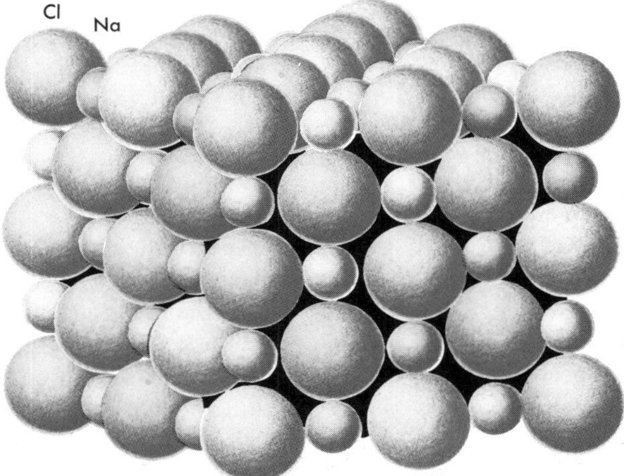

FIGURE 3-11 (Above) Cubic crystals and atomic structure of halite. (Courtesy A. Lee McAlester.)

FIGURE 3-12 (Below) Cubic crystals of pyrite. (Courtesy A. Lee McAlester.)

there are two ions of sulfur for each ion of iron, a relationship expressed by the chemical symbol for pyrite, FeS_2.

Every mineral has a characteristic chemical composition or a limited range of composition. Later on, we shall see that a mineral's composition can vary with the systematic substitution of other elements, but not enough substitution to create a new mineral.

Mineral structure

The formation of a solid can be illustrated in the laboratory by preparing a white-hot liquid composed of sodium and chlorine. So long as the temperature of the liquid is kept at a high enough level, the activity of the ions is great enough to overcome their electrical attraction for one another. Even though they come into contact from time to time, the high temperature keeps them moving about freely. Then as the temperature is reduced, they begin to lose their freedom of movement and join together to form the compound sodium-chlorine. With further cooling, larger and larger clusters develop until finally all the ions are united in fixed positions. Now the sodium and chlorine appear as solid sodium chloride, the mineral halite.

The composition of the resulting solid is the same as that of the white-hot liquid; but in the solid state, the ions of sodium and chlorine are joined together in a definite pattern. The pattern that the atoms of elements assume in a mineral is called its *crystalline structure,* the orderly arrangement of its atoms. In halite, the ions of sodium alternate with ions of chlorine.

Each mineral has a unique crystalline structure that will distinguish it from another mineral even if the two minerals are composed of the same element or elements. Consider the two minerals, diamond and graphite, for example. Each is composed of one element, carbon. In a diamond, each atom of carbon is bonded to four neighboring carbon atoms. This complete joining of all the atoms produces a very strong bond and is the reason why a diamond is so hard. In graphite, each atom of carbon is bonded in a plane to three neighboring atoms. This bonding forms sheets of layers of carbon piled one on another, but the sheets can be separated easily. Thus graphite is a soft substance.

Pyrite and marcasite are two other minerals with identical composition, FeS_2, but with different crystalline structures. In pyrite, ions of iron are equally spaced in all directions. In marcasite, they are not equally spaced. The difference in spacing accounts for the differentiation of the minerals.

Two or more minerals of the same chemical composition but with different structures are termed *polymorphs.* The most familiar examples are the polymorphs of carbon diamond and graphite.

Other minerals may have more complicated crystalline structures. They may contain more elements and have these joined together in more complex patterns. The color, shape, and size of any given mineral may vary from one sample to another, but the *internal atomic arrangement of its component elements is identical in all specimens of a particular mineral.*

After taking all these factors into account, we find it necessary to include in our definition of a mineral the fact that it has *an orderly internal atomic arrangement of its elements.*

Identification of minerals

All the properties of minerals are determined by the composition and internal atomic arrangement of their elements. So far, we have discussed chemical properties—the factors that account for the existence of specific minerals. We can identify minerals on the basis of their chemical properties, but their physical properties are the ones most often used. Physical properties include crystal form, hardness, specific gravity, cleavage, fracture, color, streak, and striations.

Crystal form

When any mineral grows without interference, it develops a characteristic crystal form that is evident as soon as the mineral is large enough. If its development is constricted or impeded, the characteristic form becomes distorted or modified.

The mineral quartz (Figure 3-13), for example, is present in many rocks as irregular grains because its growth was constricted. Yet even in these irregular grains, the ions are arranged according to their typical crystalline structure. In some parts of the world where conditions permitted the mineral to develop freely, crystals of quartz can be readily identified: these are always six-sided prisms. Whether an individual crystal of quartz is 1 or 10 cm long, the sides of the prism always join at the same angle, and the basic crystal form is always the same.

The crystal form of diamond is an eight-sided figure called an *octahedron;* graphite exhibits a flat crystal with six sides. As we know, both minerals are composed of carbon. The difference in their crystal forms comes

FIGURE 3-13 Quartz crystals.

from the arrangement of carbon atoms—one pattern in diamond, another in graphite.

The crystal form of the mineral pyrite is a cube; that of marcasite is a flattened or tubular shape. Here, again, the reason for the difference in crystal form between the two minerals lies in the internal arrangement of their atoms. *Each mineral has one or more characteristic crystal shapes*—the external form produced by its crystalline structure.

Hardness

Hardness is another property governed by the internal atomic arrangement of the elements of minerals. Again, graphite and diamond come to mind as contrasting examples. Their difference results from the ways in which the atoms of carbons are joined.

Hardness is a measure of the resistance that a smooth surface of a mineral offers to abrasion. It might be called the mineral's "scratchability." For example, if you pick up a piece of granite and try to scratch one of its light-colored grains with a steel knife blade, the granite simply refuses to be scratched. But if you drag one of its light-colored grains across a piece of glass, a scratch is easily made on the glass. Clearly, then, these particular mineral grains in granite are harder than either steel or glass. But if you have a piece of the mineral topaz handy, you can reveal the vulnerability of these light-colored mineral grains in the granite. For although they are harder than either steel or glass, they are not as hard as topaz. When you scratch a mineral, you are actually breaking the bonds between the atoms and disrupting the atomic structure.

Minerals differ widely in hardness (see Appendix C). Some are so soft that they can be scratched with a fingernail. Some are so hard that a steel knife is required to scratch them. And diamond, which is the hardest mineral known, cannot be scratched by any other mineral.

Specific gravity

Every mineral has a definite weight per unit volume. This characteristic weight is usually described by comparing it with the weight of the same volume of water at 4°C. The number that represents this comparison is called the *specific gravity* of the mineral.

The specific gravity of a mineral increases with the mass numbers of its constituent elements and with how closely these elements are packed together in their crystalline structure. Most rock-forming minerals have a specific gravity of about 2.7, although the average specific gravity of metallic minerals is about 5. The specific gravity of graphite is 2.2, that of diamond is 3.5, and gold is the highest at 19.3. The difference in specific gravity of graphite and diamond is a function of their different atomic structure—both are composed of carbon.

It is not difficult to acquire a sense of relative weight by which to compare specific gravities. We can all learn to tell the difference between two bags of equal size, one filled with feathers and one filled with lead; and the experience of hefting stones has given most geologists a sense of the "normal" weight of rocks.

Cleavage

Cleavage is the tendency of a mineral to break in certain preferred directions along smooth plane surfaces. Cleavage planes are governed by the internal arrangement of the atoms; they represent planes across which the atomic bonds are relatively weak or where there are relatively fewer bonds. Cleavage, then, is a direction of weakness, and minerals possessing cleavage tend to break along planes parallel with this direction. Cleavage planes should not be confused with crystal faces of a mineral. Quartz, for example, has a strong tendency to form crystal faces but does not have the property of cleavage.

Fracture

Fracture is the way a mineral breaks other than along cleavage planes. Several descriptive terms are used. These include *conchoidal* for fractures that display smooth, curved surfaces; *uneven* (surfaces not flat); *hackly* (rough, sharp points); *splintery* (thin pointed fragments); and *fibrous* (threadlike fragments). The fracture of a mineral is not related to the symmetry of the internal structure.

Color

Although color is not a very reliable property in identifying most minerals, it is useful in making certain general distinctions. For example, minerals containing iron are usually "dark-colored." In geologic usage, "dark" includes dark gray, dark green, and black. Minerals that contain aluminum as a predominant element are usually "light-colored," a term that includes purples, deep reds, and some browns.

Streak

The *streak* of a mineral is the color it displays in finely powdered form. The streak may be quite different from the color of a hand specimen. For example, specimens of the mineral hematite may be brown, green, or black in color, but they always have a distinctive red-brown streak. One of the simplest ways of determining the streak of a mineral is to rub a specimen across a piece of unglazed porcelain known as a *streak plate.* The color of the powder left behind on the streak plate helps to identify the mineral.

Striations

A few common minerals have parallel, threadlike lines or narrow bands called *striations* running across their surfaces. These can be seen clearly on crystals of quartz or pyrite, for example. Once again, this property is a reflection of the internal arrangement of the atoms of the crystals. The striations in plagioclase are the result of twinning. *Twinned crystals* are composite crystals of a single substance in which the individual parts are related to one another in a definite crystallographic manner.

Rock-forming minerals

Although there are as many as 2,500 minerals on the earth, only about 20 minerals are abundant. Thus, only a few minerals are *rock-forming minerals,* the minerals that constitute most of the rocks of the earth's crust. These minerals are most conveniently classified by their chemical composition.

Silicates

About 92 percent of the rock-forming minerals in the crust are silicates, compounds containing silicon and oxygen and one or more metals. Each silicate mineral has as its basic compound a complex ion called the *silicon-oxygen tetrahedron* (Figure 3-14). This is a combination of one "small" silicon ion with a radius of .42 Å surrounded as closely as geometrically possible by four "large" oxygen ions with a radius of 1.32 Å (forming a tetrahedron). The oxygen ions contribute an electric charge of −8 to the tetrahedron, and the silicon ion contributes +4. So the tetrahedron is a complex ion with a net charge of −4. Its symbol is $(SiO_4)^{4-}$.

FIGURE 3-14 The silicon-oxygen tetrahedron (SiO^{4-}). The upper view is from above and the lower from the side. Thisis the most important complex ion in geology since it is the central building unit of nearly 90 percent of the minerals of the earth's crust.

Some silicates consist of single tetrahedra alternating with positive metal ions. In others, tetrahedra are joined together into chains, sheets, or three-dimensional structures.

The most common of the silicate minerals are olivine, augite, hornblende, biotite, muscovite, feldspar, and quartz. Each one of these common rock-forming silicate minerals has a skeleton of silicon-oxygen tetrahedra.

Ferromagnesians

In the first four of these rock-forming silicates—olivine, augite, hornblende, and biotite—the silicon-oxygen tetrahedra are joined by ions of iron, magnesium, calcium, and others. Iron is interchangeable with magnesium in the crystalline structure of these silicates because the ions of both elements are approximately the same size and have the same positive electric charge. These silicate minerals are known as *ferromagnesians,* from the joining of the Latin *ferum,* "iron," with magnesium. Three of the ferromagnesians—augite, hornblende, and biotite—are very dark or black in color and have a higher specific gravity than the other rock-forming silicate minerals. Olivine is light green.

Olivine. Silicon-oxygen tetrahedra joined with positive ions of iron or magnesium, or both, form olivine; thus, its formula is best written $(Mg,Fe)_2SiO_4$. Its specific gravity ranges from 3.27 to 3.37, increasing with the amount of iron present. This mineral, named for its characteristic olive color, usually is present in grains or granular masses without well-developed cleavage.

Augite. The crystalline structure of augite is based on single chains of tetrahedra,[1] as shown in Figure 3-15, joined together by ions of iron, magnesium, and calcium. Color range is very dark green to black; the streak is colorless; the specific gravity ranges from 3.2 to 3.4; and the cleavage is along two planes almost at *right* angles to each other. This cleavage angle is important in distinguishing augite from hornblende. A good way to remember it is to recall that augite rhymes with "right."

Hornblende. The crystalline structure of the mineral hornblende is based on double chains of tetrahedra, as shown in Figure 3-16, joined together by the iron and magnesium ions common to all ferromagnesians and also joined by ions of calcium, sodium, and aluminum. Hornblende's color range is dark green to black, like that of augite; the streak is colorless; the specific gravity is 3.2. Two directions of cleavage meet at angles of approximately 56° and 124°, which helps distinguish hornblende from augite.

Biotite. Named in honor of the French physicist J. B. Biot, biotite (buy'-oh-tight) is a mica (my'-ka, from the Latin *micare,* "to shine"). Like all the other micas, it is constructed of tetrahedra in sheets, as shown in Figure 3-17. Each silicon ion shares three oxygen ions with adjacent silicon ions to form a pattern like wire netting. The fourth, unshared oxygen ion of each tetrahedron stands above the plane of all the others. The basic structural unit of mica consists of two of these sheets of tetrahedra, with their flat surfaces facing outward and their inner surfaces held together by positive ions. In biotite, those ions are iron and magnesium. These basic double sheets of mica, in turn, are loosely joined together by positive ions of potassium.

Layers of biotite, or any of the other micas, can be peeled off easily because there is perfect cleavage along the surfaces of these weak potassium bonds. In thick blocks, biotite is usually dark green, brown, or black. The specific gravity ranges from 2.8 to 3.2.

[1]The term *tetrahedra* will be used throughout this chapter to refer to silicon-oxygen tetrahedra.

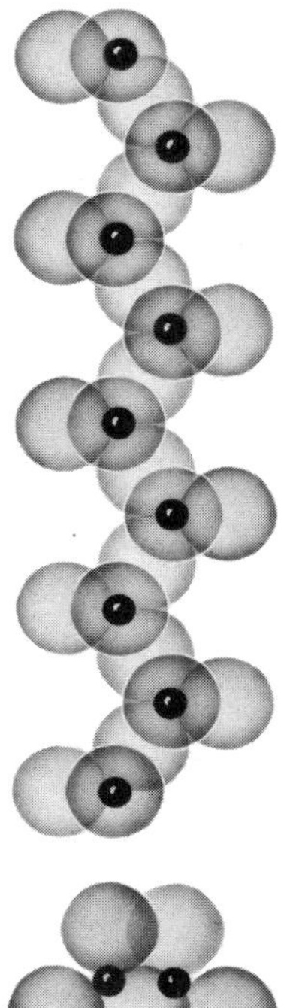

FIGURE 3-15 (Left) Single chain of tetrahedra viewed from above. Bottom: Viewed from an end. Each silicon ion (small black sphere) has 2 of the 4 oxygen ions of its tetrahedron bonded exclusively to itself, and it shares the other 2 with neighboring tetrahedra fore and aft. The resulting individual chains are in turn bonded to one another by positive metallic ions. Since these bonds are weaker than the silicon-oxygen bonds that form each chain, cleavage develops parallel to the chains.

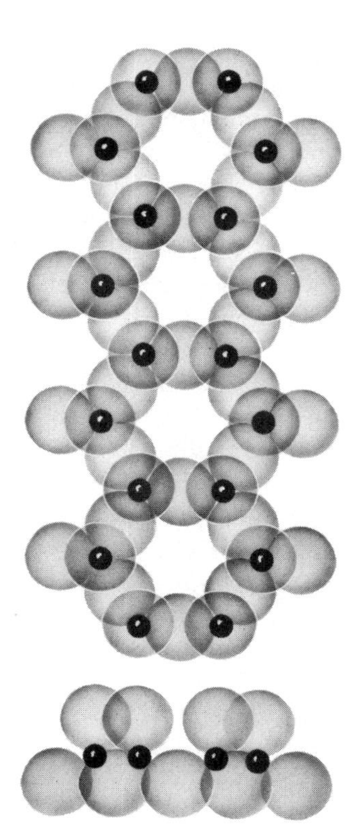

FIGURE 3-16 (Right) Double chain of tetrahedra viewed from above. Bottom: Viewed from an end. The doubling of the augite chain is accomplished by the sharing of oxygen atoms by adjacent chains.

FIGURE 3-17 (Below) Tetrahedral sheets. Each tetrahedron is surrounded by 3 others, and each silicon ion has 1 of the 4 oxygen ions to itself, while sharing the other 3 with its neighbors.

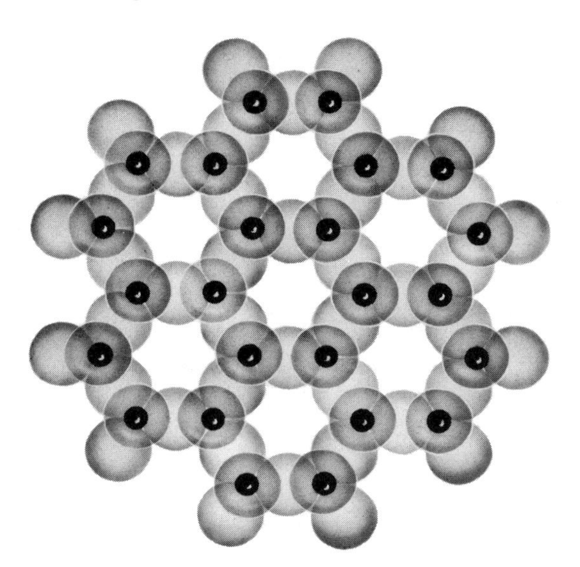

Nonferromagnesians

The other common rock-forming silicate minerals are known as the *nonferromagnesians,* simply because they do *not* contain iron or magnesium. These minerals include muscovite, the feldspars, and quartz. They are all marked by their light colors and relatively low specific gravities ranging from 2.6 to 3.0.

Muscovite. This white mica was so named because it was once used as a substitute for glass in old Russia (Muscovy). It has the same basic crystalline structure as biotite, but in muscovite each pair of tetrahedra sheets is tightly cemented together by ions of aluminum rather than iron and magnesium. As in biotite, however, the double sheets are held together loosely by potassium ions, along which cleavage readily takes place. In thick blocks, the color of muscovite is light yellow, brown, green, or red. Its specific gravity ranges from 2.8 to 3.1.

Feldspars. The most abundant rock-forming silicates are the feldspars. The name comes from the German *feld,* "field," and *spar,* a term used by miners for various nonmetallic minerals. The name reflects the abundance of these minerals. Feldspars constitute 58 percent by volume of the minerals present in all the rocks in the earth's crust.

In the feldspars, each tetrahedron shares its oxygen ions with adjoining tetrahedra ions in a three-dimensional network. However, in one-quarter to one-half of the tetrahedra, aluminum ions with a radius of .51 Å and an electric charge of +3 have replaced silicon (with its radius of .42 Å and electric charge of +4) in the centers of the tetrahedra. The negative electric charge resulting from such substitution in the tetrahedra is corrected by the entry into the crystalline structures of ions of potassium, sodium, or calcium—K^+, Na^+, or Ca^{2+}. Since the feldspars contain these elements in different proportions, they grade continuously in composition one into another.

The mineral names given the main feldspars are orthoclase ($KAlSi_3O_8$) and plagioclase. However, the plagioclase feldspars are a *solid solution* series, and the members of the series cannot be expressed in simple chemical formulas because one ion can substitute for another in the crystal lattice. The plagioclase feldspars are a continuous series from albite ($NaAlSi_3O_8$) to anorthite ($CaAl_2Si_2O_8$). Four intermediate members (oligoclase, andesine, labradorite, bytownite) between albite and anorthite are recognized. Ca^{2+} and Al^{3+} simultaneously exchange for Na^+ and Si^{4+} in the series.

The feldspars are listed in Table 3-4. In this table, the column headed Diagnostic Positive Ion indicates the ion that corrects the electrical unbalance caused by the substitution of aluminum for silicon.

Orthoclase is named from the Greek *orthos,* "straight," and *klasis,* "a breaking," because the two dominant cleavages intersect at right angles when a piece of orthoclase is broken. Aluminum replaces silicon in every fourth tetrahedron, and positive ions of potassium correct the electrical unbalance. The streak of orthoclase is white; its color is white, gray, or pinkish; and its specific gravity is 2.57.

Plagioclase ("oblique-breaking") feldspars are so named because they have cleavage planes that intersect at about 86°. One of the cleavage planes is marked by striations. In albite, aluminum replaces silicon in every fourth tetrahedron, and positive ions of sodium correct the electrical unbalance. The specific gravity of albite is 2.62. In anorthite, aluminum replaces silicon in every second tetrahedron, and positive ions of calcium correct the electrical unbalance. The specific gravity of anorthite is 2.76. Plagioclase feldspars may be colorless, white, or gray, although some samples show a striking play of colors called *opalescence.*

Table 3-4 Feldspars (Aluminosilicates)

DIAGNOSTIC POSITIVE ION	NAME	SYMBOL	DESCRIPTIVE NAME	FORMULA*
K^+	Orthoclase	Or	Potassic feldspar	$K(AlSi_3O_8)$
Na^+	Albite	Ab	Sodic feldspar } plagioclase feldspars	$Na(AlSi_3O_8)$
Ca^{+2}	Anorthite	An	Calcic feldspar } plagioclase feldspars	$Ca(Al_2Si_2O_8)$

* In these formulas, the symbols *inside* the parentheses indicate the tetrahedra. The symbols *outside* the parentheses indicate the diagnostic ions—that is, the ions that are worked in among the tetrahedra.

Quartz. The very common mineral quartz (11 percent by volume in the earth's crust) is the only rock-forming silicate mineral that is composed exclusively of silicon-oxygen tetrahedra. Every oxygen ion is shared by adjacent silicon ions, which means that there are two ions of oxygen for every ion of silicon. This relationship is represented by the formula SiO_2. The specific gravity of quartz is 2.65.

Quartz generally is smoky to clear in color, but many less common varieties include purple or violet amethyst, massive rose-red or pink rose quartz, smoky yellow to brown smoky quartz, and milky quartz. These color differences are caused by other elements that are present as impurities. The colors are not caused by, and do not affect, the crystalline structure of the quartz.

Oxide minerals

Oxide minerals are formed by the direct union of an element with oxygen. These are relatively simple minerals compared to the complicated silicates. The oxide minerals are usually harder than any other class of minerals except the silicates, and they are heavier than others except the sulfides. Within the oxide class are the chief ores of iron, chromium, manganese, tin, and aluminum. Some common oxide minerals are ice (H_2O), corundum (Al_2O_3), hematite (Fe_2O_3), magnetite (Fe_3O_4), and cassiterite (SnO_2).

Sulfide minerals

Sulfide minerals are formed by the direct union of an element with sulfur. The elements that are present most commonly in combination with sulfur are iron, silver, copper, lead, zinc, and mercury. Some of these sulfide minerals are found as commercially valuable ores, such as chalcopyrite ($CuFeS_2$), chalcocite (Cu_2S), galena (PbS), and sphalerite (ZnS).

Carbonate and sulfate minerals

We found that silicates are built around the complex ion $(SiO_4)^{4-}$—the silicon-oxygen tetrahedron. But two other complex ions also are of great importance in geology. One of these consists of a single carbon ion with three

Table 3-5 The Organization of Common Minerals (continuation of Table 3-2)

Elements — *Compounds* (*Organic*, *Inorganic*) → *Minerals*; *Others*

Elements	Elements + Oxygen	Elements + Sulfur	Elements + CO_3 Ion	Elements + SO_4 Ion	Elements + SiO_4 Ion		
ELEMENTS	OXIDES	SULFIDES	CARBONATES	SULFATES	SILICATES		
					Nonferromagnesian	*Ferromagnesian*	*Clay minerals*
Copper	Cassiterite	Chalcopyrite	Calcite	Anhydrite	Quartz	Biotite	Illite
Diamond	Corundum	Chalcocite	Dolomite	Gypsum	Feldspars	Hornblende	Chlorite
Gold	Hematite	Galena	Magnesite		Orthoclase	Augite	Kaolinite
Graphite	Magnetite	Pyrite			Plagioclase	Olivine	
Iron	Ilmenite	Sphalerite			Albite		
Platinum	Rutile				Anorthite		
Silver	Uraninite				Muscovite		
Sulfur	Ice						

oxygen ions packed around it—$(CO_3)^{2-}$. Compounds in which this complex ion appears are called *carbonates.* For example, the combination of a calcium ion with a carbon-oxygen ion produces calcium carbonate, $CaCO_3$, known in its mineral form as calcite. This mineral is the principal component of the common sedimentary rock limestone. The other complex ion is $(SO_4)^{2-}$, a combination of one sulfur ion and four oxygen ions. This complex ion combines with other ions to form sulfates; for example, it joins with a calcium ion to form calcium sulfate, $CaSO_4$, the mineral anhydrite.

Definition of minerals

We know that minerals are special combinations of elements or compounds in the solid state, and now we can complete our definition of a mineral: (1) it is a naturally occurring element or inorganic compound in the solid state; (2) it has a chemical composition that is fixed or that varies systematically; (3) it has a crystalline structure; and (4) it exhibits certain physical properties as a result of its composition and crystalline structure. The organization of the common minerals is shown in Table 3-5.

SUMMARY

Rocks are composed of minerals, minerals are composed of atoms, and atoms are composed of protons, neutrons, and electrons. Ions are electrically unbalanced forms of atoms or groups of atoms produced by the gain or loss of electrons. Atoms of the 92 elements found in nature combine in various ways to form compounds known as *minerals,* which make up the rocks of the earth.

Minerals are naturally occurring elements or compounds in the solid state and are inorganic in character. Each mineral has a chemical composition fixed within definite limits and displays a crystalline structure. The composition and structure give rise to characteristic physical properties for each mineral type.

SUGGESTED READINGS

ERNST, W. G., *Earth's Materials.* Englewood Cliffs, N.J.: Prentice-Hall, Inc., 1969.

FREY, PAUL R., *College Chemistry,* 3rd ed. Englewood Cliffs, N.J.: Prentice-Hall, Inc., 1965.

FYFE, W. S., *Geochemistry of Solids: An Introduction.* New York: McGraw-Hill Book Company, 1964.

HURLBUT, CORNELIUS, Jr., *Dana's Manual of Mineralogy,* 18th ed. New York: John Wiley & Sons, Inc., 1971.

———, *Minerals and Man.* New York: Random House, Inc., 1969.

MASON, BRIAN, *Principles of Geochemistry,* 3rd ed. New York: John Wiley & Sons, Inc., 1966.

———, and L. G. Berry, *Elements of Mineralogy.* San Francisco: W. H. Freeman and Co., 1968.

Igneous Rocks and Volcanoes

As we have seen, minerals are the units from which the rocks of the earth are built. These rocks are divided into three main groups—igneous, sedimentary, and metamorphic. In this chapter we are concerned with the igneous rocks, but before we proceed with a detailed examination of this group, it is useful to look at the general relationships of the three major rock types.

The three rock families

Igneous rocks, the ancestors of all other rocks, take their name from the Latin *ignis,* "fire." These "fire-formed" rocks were once a hot, molten, liquidlike mass known as *magma,* which subsequently cooled into firm, hard rock. Thus, the lava flowing across the earth's surface from an erupting volcano soon cools

4

Eruption of Kilauea in 1967 after one month. (Courtesy U.S. Geological Survey.)

FIGURE 4-1 (Above) Igneous rocks from when molten rock material (magma) solidifies. Here a 270-m fountain of lava spouts into the crater of Kilauea Iki, Hawaii. Hawaiian eruptions are characterized by great fluidity and small gas content. (Courtesy U.S. Geological Survey.)

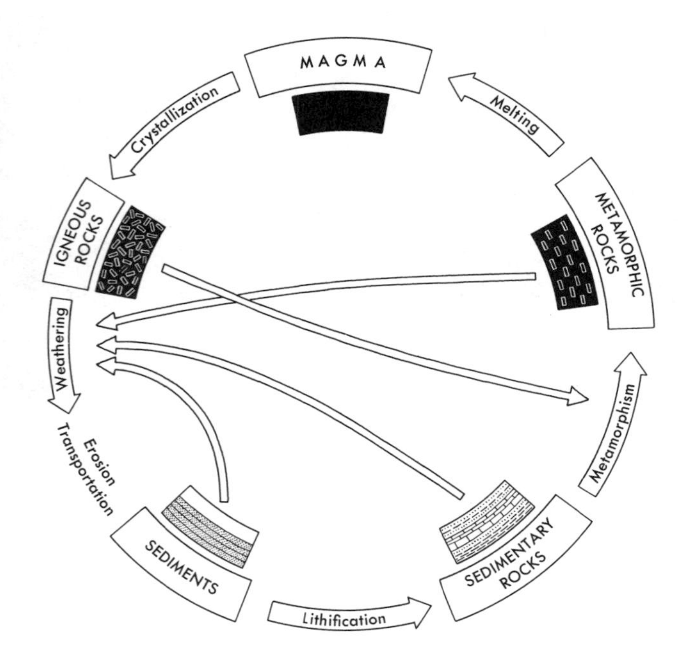

and hardens into an igneous rock (Figure 4-1). But there are other igneous rocks now exposed at the surface that actually cooled some distance beneath the surface. We see such rocks today only because erosion has stripped away the rocks that covered them during their formation.

Most sedimentary rocks (from the Latin *sedimentum,* "settling") are composed of particles derived from the breakdown of preexisting rocks. Usually these particles are transported by water, wind, or ice to new locations where they are deposited in new arrangements. For example, waves beating against a rocky shore may provide the sand grains and pebbles for a nearby beach. If these beach deposits are hardened, we have sedimentary rock. One of the most characteristic features of sedimentary rocks is the layering of the deposits that compose them.

Metamorphic rocks are the third large family of rocks. Metamorphic, meaning "changed form," refers to the fact that the original rock has been changed from its primary form to a new one. Earth pressures, heat, and chemically active fluids beneath the surface may all be involved in changing the primary rock into a new metamorphic rock.

The rock cycle

We have suggested that there are definite relationships among sedimentary, igneous, and metamorphic rocks. With time and changing conditions, any one of these rock types can be changed into some other form. These relationships form a cycle, as shown in

FIGURE 4-2 (Left) The rock cycle, shown diagrammatically. If uninterrupted, the cycle will continue completely around the outer margin of the diagram from magma through igneous rocks, sediments, sedimentary rocks, metamorphic rocks, and back again to magma. The cycle may be interrupted, however, at various points along its course and may follow the path of one of the arrows crossing through the interior of the diagram.

Figure 4-2. This is simply a way of tracing the various paths that earth materials follow. The outer circle represents the complete cycle; the arrows within the circle represent shortcuts in the system that can be, and often are, taken. Notice that the igneous rocks are shown as having been formed from a magma, and as providing one link in a continuous chain. From these parent rocks, through a variety of processes, all other rocks can be derived.

First, weathering attacks the solid rock, which either was formed by the cooling of a lava flow at the earth's surface, or is an igneous rock that was formed deep beneath the earth's surface and then was exposed by erosion. The products of weathering are the materials that will eventually go into the creation of new rocks—sedimentary, metamorphic, and even igneous. Landslides, wind, running water, and glacier ice all help to move the materials from one place to another. In the ideal cycle, this material is transported to the oceans, where layers of soft mud, sand, and gravel are consolidated into sedimentary rocks. If the cycle continues without interruption, these new rocks may, in turn, be deeply buried and subjected to heat, to pressures caused by overlying rocks, and to forces developed by earth movements. These sedimentary rocks may then change in response to the new conditions and become metamorphic rocks. If these metamorphic rocks undergo continued and increased heat and pressure, they may eventually lose their identity and melt into a magma. When this magma cools, we have an igneous rock again, and we have come full cycle.

But notice, too, that the complete rock cycle can be interrupted. An igneous rock, for example, may never be exposed at the surface and hence may be converted directly into a metamorphic rock without passing through the intermediate sedimentary stage. Other interruptions in the cycle may take place if sediments, or sedimentary rocks, or metamorphic rocks are attacked by weathering before they continue to the next stage in the outer, complete cycle.

Description of igneous rocks

We have seen that three groups of rocks—igneous, sedimentary, and metamorphic—record the history of the earth. In the course of this book, we will look at all of them in detail. Because igneous rocks seem to be the source of the other two types, we shall begin our study with them, starting with a description and then developing a classification that stems from the description.

Let us assume that we are faced with a whole binful of specimens of igneous rocks, containing representatives of the most important types. If we could determine the characteristics of these rocks, we would find a way of bringing order to our jumbled pile. What characteristics might we pick out? Here are some immediately obvious ones. First, color; second, specific gravity; third, texture; fourth, mineral composition.

Color

Color is the characteristic that we probably notice first. We immediately detect that there is a group of very light-colored, almost white rocks and a group of dark green, almost black types. In between we see rocks of varying tones of gray, as well as some of pink and even red cast. If we select the lightest- and darkest-colored rocks and let them stand at either end of our color scale, we can arrange the rest of the rocks in all tones between the two end members. For example, because of their tone, we will want to arrange the pinkish and reddish rocks on the light grey side of our scale. This is a very simple—indeed, too simple—way of classifying rocks. Yet it can be important, as we shall see.

Specific gravity

Now let's mix our rocks up and start our sorting over again. This time we will arrange our specimens according to their specific gravity. At first we heft individual rock specimens by hand. We find that we can tell the difference between some of the rocks on the basis of their apparent specific gravity. Volume for volume some of them feel heavier than do others. Now if we use a simple balance, we discover that the specific gravities range from approximately 2.3 to approximately 3.3. Furthermore, within this range, we find that the greatest number of rocks fall somewhere around 2.6 or 2.7. Another group has a density of about 3.0. We may even find a third peak halfway between 2.6 and 3.0. Looking at our pile of rocks, now separated by specific gravity, we find that again we have arranged them more or less by color. The heaviest rocks are the darkest rocks. The light-colored rocks also have the lowest specific gravity. The rocks intermediate in color are intermediate in specific gravity. Still, there are one or two exceptions. For instance, there is one glassy rock (we know it as volcanic glass or obsidian) that is dark-colored but has a specific gravity of only about 2.4. (Later, we can return to the problem of why it doesn't seem to follow the rule of dark color/high specific gravity.) We note, too, that the pink and reddish-colored rocks generally group with the light-colored ones in having specific gravities that hover around 2.6.

Having sorted our rocks by color and by specific gravity, we find that we get approximately the same order of rocks whichever method we use. We are tempted to feel, then, that there is some relationship between color and specific gravity and that these characteristics represent something important in terms of the origin and classification of igneous rocks. Let us go on and look at some of the other characteristics.

Texture

In handling and sorting our igneous rocks, we may have noted another difference: some of the rocks seem to be composed of smaller units; others seem to be homogeneous throughout. Let's try sorting on this basis now. If we look at the rocks more carefully, we see that we can further subdivide these two groups. For instance, take the homogeneous group. Some of the rocks are very glassy, reflecting light and breaking as if they were the thick bottoms of milk bottles. This type includes obsidian (Figure 4-3). Others of this homogeneous group tend to have duller surfaces, or a matte finish. So let's divide the homogeneous rocks on the basis of surface, *glassy* or *matte.*

Turning to the other pile of rocks, those with specimens made up of smaller units, we

FIGURE 4-3 *Obsidian, a rock with a glassy texture. Maximum length is 25 cm.*

find that here, too, we can subdivide into at least two groups. In one, the individual particles that we can see with the naked eye are more or less the same size. In the second, the large crystals visible to the naked eye are set in a matrix, or *groundmass,* in which finer crystals may or may not be discerned; the effect is rather like plums in a plum pudding.

Now if we look at our four new piles of rocks (the glassy, the matte, the coarse-grained, and the plum-pudding groups), we find that each pile includes rocks of different colors and specific gravities. If the characteristics described in these four piles are significant, they don't relate to color or to specific gravity.

In this third sorting of our rock pile, we have been grouping on the basis of *texture,* a term derived from the Latin *texere,* "to weave." In rocks, texture means the shape, size, arrangement, and distribution of the minerals that compose the rocks. Now, let us give formal definitions to the four textures we used to sort our rocks.

Coarse-grained texture

When the particles in a rock can be seen with the naked eye, we call the texture *coarse-grained* (Figure 4-4). If we look carefully at this rock's minerals, with a hand lens or a microscope, we see that the minerals interlock, one with the next. In speaking of coarse-grained textures, geologists usually refer only to those rocks with more or less equigranular texture, that is, one in which most of the particles are all of the same general size. When some of the particles are larger than the groundmass in which they rest, we use another term *porphyritic,* which is explained on page 78.

FIGURE 4-4 *Granite, a coarse-grained rock. Width of view is 6.5 cm.*

Fine-grained texture

When we are unable to see the individual particles with the naked eye but must instead use a microscope to identify them, we call the texture *fine-grained* (Figure 4-5). In our piles of rocks, the fine-grained igneous rocks are those with a matte surface. What we actually see are the little pinpoints of minerals that give the specimen a rather rough surface. Although we cannot see the individual particles, we could, with a microscope, make out the interlocking pattern of the individual mineral grains.

FIGURE 4-5 *Basalt, a dark fine-grained rock.*

FIGURE 4-6 *Rhyolite porphyry containing light-colored feldspar phenocrysts in a pale red aphanitic matrix. Width of view is 8 cm.*

Glassy texture

Among our homogeneous rocks, we found several that showed a shiny, glassy aspect, ranging from red to black in color. These rocks are made of *volcanic glass.* The texture is glassy. Under a microscope these rocks show few individual mineral particles. The great bulk of the rock is composed of unorganized, or noncrystalline, material.

Porphyritic texture

When larger crystals are set in a groundmass of finer crystals, we refer to the texture as *porphyritic* (Figure 4-6). These larger crystals are called *phenocrysts,* from the Greek *phainein,* "to show." The term *porphyry,* which comes from the Greek word for "purple," was originally applied to rocks containing phenocrysts set in a dark red or purple groundmass.

Preliminary classification

Based on the observations we have made so far, concerning color, specific gravity, and texture, we can construct a preliminary classification system, as suggested in Figure 4-7. Notice that in this figure the textures grade upward from glassy through fine-grained and porphyritic to coarse. The other axis is devoted to an increase in specific gravity from low to high, as well as a change in color from light to dark. Into this classification system we can fit most of our rocks. For instance, a light gray, coarse-grained rock with a specific gravity of about 2.6 generally refers to granite (see Figure 4-4). However, in actual practice, we use a somewhat more sophisticated classification than was shown in Figure 4-7. And although texture is used as indicated, for specific gravity and color we substitute mineral composition.

Composition

The material from which things are made is often fundamental to our classification of objects. We can begin determining the material of our igneous rocks by identifying the

FIGURE 4-7 *A preliminary classification system for igneous rocks based on texture (vertical axis) and color and density (horizontal axis).*

TEXTURE	LIGHT COLOR	INTERMEDIATE COLOR	DARK COLOR
Coarse-grained			
Porphyritic			
Fine-grained			
Glassy			
	Low Density	Intermediate Density	High Density

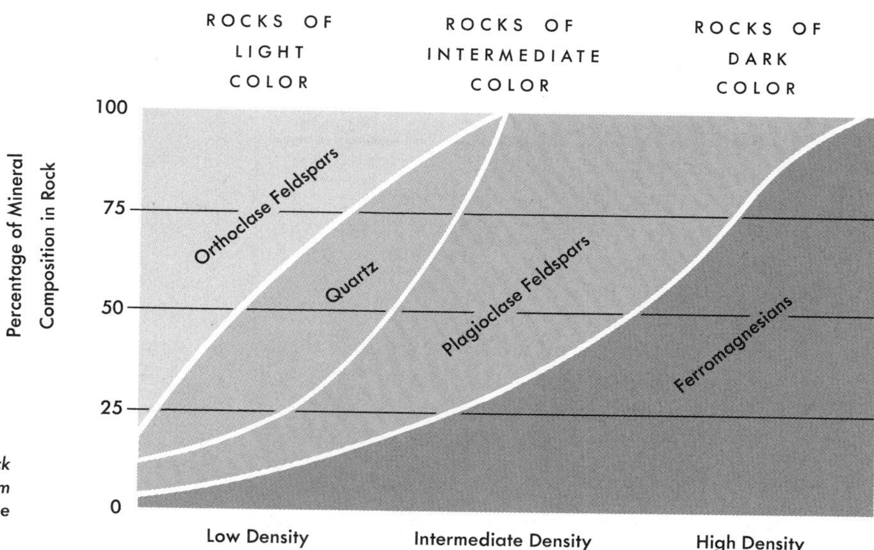

FIGURE 4-8 *As a coarse-grained rock varies from a light to dark color and from a lower to higher density, so also does the mineral composition change.*

minerals present in the coarse-grained rocks. We see that the light-colored rocks are generally very low in ferromagnesian minerals, those that are present being usually the micas and amphiboles. Furthermore, there is a recognizable amount of quartz. And there is a great deal of feldspar, both orthoclase and plagioclase, the latter tending more toward the albite side than the anorthite side. In very dark-colored, coarse-grained rocks, we find that there is no quartz and no orthoclase. But there are considerable amounts of dark-colored anorthite plagioclase and of ferromagnesian minerals, primarily augite and olivine. Coarse-grained rocks of intermediate color will have little or no orthoclase, little or no quartz, much plagioclase (intermediate in composition between albite and anorthite), and a large amount of ferromagnesian minerals, mostly hornblende with some augite. If we were to identify the major constituents for all the coarse-grained rocks, we would find that there is a complete gradation in mineral composition from light to dark, as shown in Figure 4-8.

We cannot identify with the naked eye, or even with the hand lens, the mineral components of the fine-grained igneous rocks. Use of the microscope, however, has demonstrated that the same gradation holds for the fine-grained as for the coarse-grained rocks. Turning to the porphyritic rocks, we find that if the mineral constituents are all coarse enough to be identified by the naked eye, then these rocks, too, follow the same compositional order as the fine-grained and the coarse-grained rocks. For those porphyritic rocks whose groundmass is too fine to be identified with the naked eye, the microscope again demonstrates the same pattern as found in the other types.

Classification of igneous rocks

A classification is an attempt to bring order out of a series of observations. Since our observations have been directed toward color, specific gravity, texture, and mineral composition, our classification is based on these characteristics. The classification is presented in Figure 4-9, which is a combination of Figures 4-7 and 4-8. The names of rocks are arbitrarily assigned on the basis of average mineral compositions. Actually, there are many more names in use than are shown in Figure 4-9.

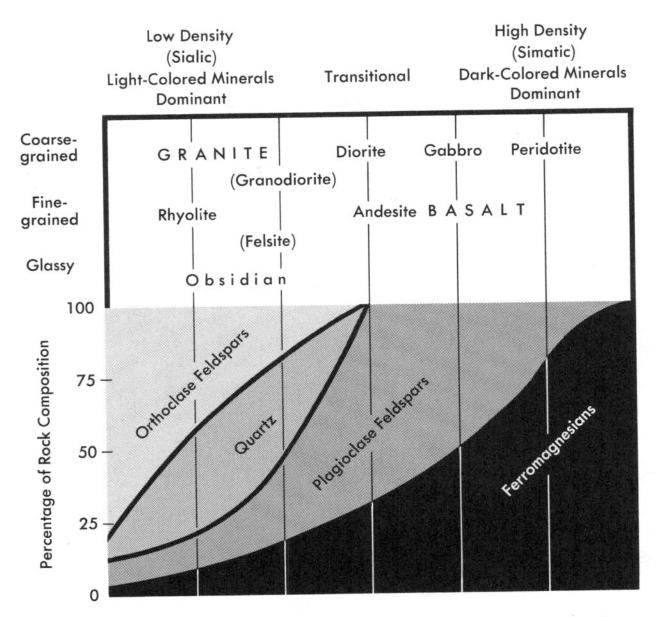

FIGURE 4-9 *General composition is indicated by a line from the name to the composition chart: Granite and rhyolite consist of about 50 percent orthoclase, 25 percent quartz, and 25 percent divided among plagioclase feldspars and ferromagnesian minerals. Relative importance is stressed by the size of the lettering for rock names: Granite is the most important coarse-grained rock, basalt the most important fine-grained rock. [Composition chart modified after L. V. Pirsson and A. Knopf, Rocks and Rock Minerals (New York: John Wiley & Sons, Inc., 1926), p. 144.]*

FIGURE 4-10 *Brownish red rhyolite showing fine granular texture. Width of view is 10 cm.*

Sometimes intermediate types are indicated by such names as *granodiorite,* a rock whose composition is between that of granite and diorite.

Light-colored igneous rocks

The igneous rocks on the light side of the classification chart are light in both color and specific gravity. They are sometimes termed *sialic* rocks. The term *"sial,"* coined from the chemical symbols for silicon and aluminum, is generally used in speaking of the composition typical of the continental areas of the earth. This composition is dominated by granite and granodiorite.

It has been estimated that granite and granodiorite together constitute 95 percent of all rocks that have solidified from magma trapped in the outer few kilometers of the earth's continental surface. The origin and history of some of these are still under debate, but we shall use the terms—granite and granodiorite—here only to indicate composition and texture, not origin.

Granite is a coarse-grained rock. Its mineral composition is approximately as follows: orthoclase feldspar 45 percent, quartz 25 percent, and small amounts of ferromagnesian minerals 5 percent. Rocks with the same mineral composition as granite, but with a fine-grained texture, are called *rhyolite* (Figure 4-10).

The glassy equivalent of granite is *obsidian.* Although this rock is listed near the light side of the composition chart, it is usually pitch black in appearance. Actually, pieces of obsidian thin enough to be translucent turn out to be smoky white against a light background. If we were to grind up obsidian and granite separately, each would produce a powder that would be very light in color and would be essentially indistinguishable as between the two rocks.

Dark-colored igneous rocks

The darker, heavier rocks are sometimes designated collectively as *sima.* This name, which was coined from *Si* for silicon and *Ma* for magnesium, is generally used in speaking of the shell of dark, heavy rock that underlies the ocean basins and the sialic rocks of the continents.

Of the total volume of rock formed from molten material that has poured out onto the earth's surface, by far the greatest amount is *basalt* (Figure 4-11). A popular synonym for basalt is "trap rock," from a Swedish word meaning "step." This name refers to the tendency of certain basalts to weather or break down into masses that look like giant stairways. Basalt is a fine-grained rock. Its mineral composition is as follows: plagioclase feldspars 50 percent; and ferromagnesian minerals, including pyroxene and olivine, 50 percent. The coarse-grained equivalent of basalt is *gabbro. Peridotite* is a coarse-grained rock that is composed largely of the mineral olivine.

FIGURE 4-11 *Dark ropy (pahoehoe) basalt flow in Hawaii. (Courtesy U.S. Geological Survey.)*

Transitional igneous rocks

Composition

Rock compositions blend continuously from one to another as we go from the light to the dark side of the classification chart. *Andesite* (Figure 4-12) is the name given to the fine-grained rocks that are roughly intermediate in composition between rhyolite and basalt. Andesites are almost always found in areas around the Pacific Ocean where active mountain-making has taken place. The coarse-grained equivalent of andesite is *diorite* (Figure 4-13). The mineral composition of these rocks is typically as follows: orthoclase and quartz, very small or missing; plagioclase feldspars 75 percent, mostly midway between albite and anorthite; and ferromagnesian min-

FIGURE 4-12 *Light gray andesite showing well-developed crystals of hornblende set in a uniform fine-grained groundmass of feldspars. Width of view is 14 cm.*

FIGURE 4-13 Close-up of diorite. Plagioclase is dominant over orthoclase. Specimen contains small amounts of quartz. Hornblende is more abundant than biotite. Width of view is 9 cm.

erals, largely hornblende, with some pyroxene and some micas, approximately 25 percent.

Texture

As we go from the top to the bottom of the chart in Figure 4-9, we find that the rock textures grade continuously from coarse-grained to fine-grained, whereas the composition remains the same. For example, as we read down along the first vertical rule, we find that granite, rhyolite, and obsidian are progressively finer-grained types, although all three have essentially the same composition. The same is true of gabbro and basalt.

Any of the rocks may have porphyritic texture (Figure 4-6). This texture means that a given rock has grains of two distinctly different sizes, that is, conspicuously large phenocrysts embedded in a finer-grained groundmass. When the phenocrysts constitute less than one-quarter of the total, the adjective *porphyritic* is used to modify the rock name, as in porphyritic granite or porphyritic andesite. When the phenocrysts constitute more than one-quarter, the rock is called a *porphyry*. The composition of a porphyry and the texture of its groundmass are indicated by using rock names as modifiers, for example, granite porphyry or andesite porphyry. These relationships are summarized for the most common rocks in Table 4-1.

Table 4-1 Porphyritic Rock and Rock Porphyry

LESS THAN 25 PERCENT PHENOCRYSTS	MORE THAN 25 PERCENT PHENOCRYSTS
Porphyritic granite	Granite porphyry
Porphyritic rhyolite	Rhyolite porphyry
Porphyritic diorite	Diorite porphyry
Porphyritic andesite	Andesite porphyry

Volcanoes and extrusive igneous rocks

During the last five hundred years, more than 200,000 people have been killed by volcanoes. The most famous of all volcanoes in the western world is Vesuvius (Figures 4-14 and 4-15), which is mostly remembered for its eruption of August 24 in A.D. 79 that destroyed Pompeii, Herculaneum, and Stabiae. Of those who perished, most were slaves, soldiers, and those too greedy to leave their possessions. Death was caused by suffocation by falling ash, by volcanic mud, and by volcanic gases. One of the best descriptions of the eruption is that in two letters from Pliny the Younger to his friend Tacitus, the Roman historian. The Elder Pliny, the world's first volcanologist, went to his death in an attempt to help the inhabitants as his nephew, the Younger Pliny, remained behind to work on his studies.

Vesuvius has continued its activity. In A.D. 472 ash drifted as far east as Constantinople. A violent eruption in 1631 killed 18,000 people. Other major eruptions took place in 1794, 1872, 1906, and 1944. Yet, life goes on at the foot of Vesuvius.

On May 8, 1902 at 8:02 A.M., the Caribbean coastal town of St. Pierre, below Mount Pelée on Martinique, was destroyed in seconds

FIGURE 4-14 (Right) Mount Vesuvius throwing out still-molten blobs of lava and clouds of ash-laden steam.

FIGURE 4-15 (Below) In A.D. 79 the Roman towns of Pompeii and Herculaneum were buried in volcanic ash from Mount Vesuvius. Here archaeological excavations have exposed portions of ancient Herculaneum. In the middle distance, modern Herculaneum stands on volcanic deposits that still cover portions of the old town. On the skyline a white plume of steam issues from Mount Vesuvius. (Photo by Ewing Galloway.)

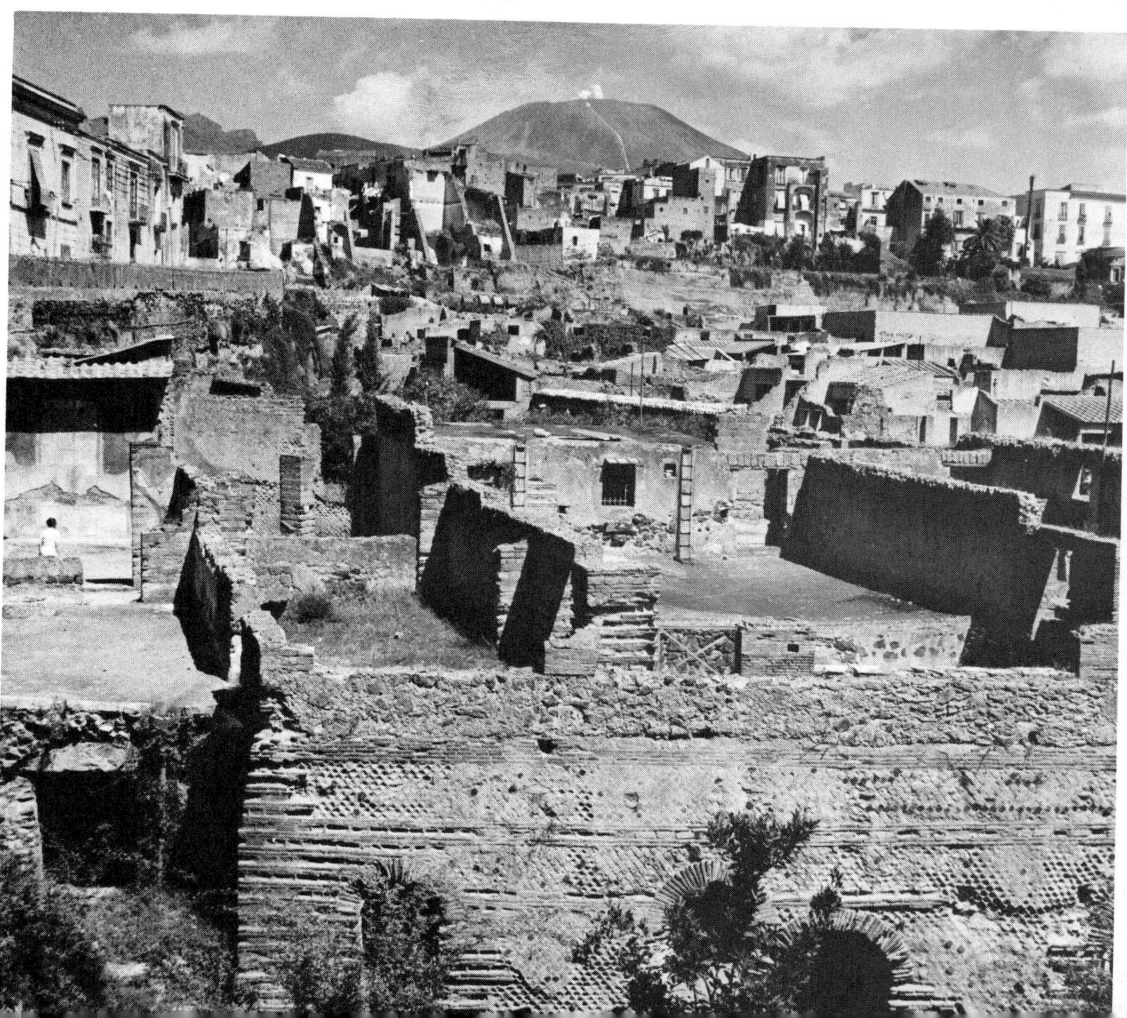

by a glowing (internal temperature of 800°C) cloud of hot volcanic ash, dust, and gas that rushed down from Mount Pelée at almost 3 km per minute. Within 60 seconds, more than 30,000 people died. For half a century the volcano had been dormant.

Violent eruptions and explosions of hot gasses and viscous magmas are now termed *peléan,* after the Mount Pelée explosion. The term *nuée ardente* is used to describe glowing avalanche eruptions of the type that occurred at Mount Pelée.

A quite different but equally famous earlier eruption was that of Krakatoa in the Sunda Straits between Java and Sumatra on August 27, 1883. The death toll on Krakatoa was zero, but the explosion generated a sea wave (*tsunami*) that reached a height of almost 40 m and drowned 30,000 to 40,000 people in nearby coastal towns. The sound of the explosion was heard in Australia, nearly 2,000 km away. Approximately 20 km^3 of debris was discharged in the air. Krakatoa was a *phreatic* eruption, the result of expansion of steam formed when magma contacted confined ground water. It has been estimated that the explosion had a TNT equivalent of about 100 megatons.

FIGURE 4-16 *Close-up of light olive-gray pumice showing fine frothy nature. Tiny pores are formed by escaping gases. Glassy nature of rock is visible. Width of view is 7 cm.*

More recently, a great explosion burst from the southern base of Hibokhibok volcano in the Philippines on December 4, 1951, and a tremendous avalanche swept silently and lethally over the villages north of Mambajo. Five hundred persons lost their lives.

Mount Arenal in Costa Rica erupted in 1964 for the first time in 450 years, killing 78 people. During May 1964, Mount Agung in Indonesia erupted, killing 1,610 people and leaving 78,000 homeless.

These and other disasters caused by volcanoes have emphasized the desirability of being able to predict eruptions. There are many early signs: earthquakes, seismic tremors, and ground tilts. Considerable further study is needed, but it is likely that the imminence of much volcanic activity can be predicted, with consequent savings in life and property. Although there are only 38 active volcanoes in the United States (33 in Hawaii), about 460 active volcanoes are present throughout the world.

Extrusive igneous rocks

Rock types

The igneous rocks formed by volcanoes originate in the cooling magma that has been extruded through the upper layers of the earth's crust onto the surface. Volcanic rocks are therefore termed *extrusive rocks.*

Actual observation has shown that the rocks produced by modern volcanoes include rhyolite, andesite, basalt, felsite, and obsidian, all rocks with fine-grained or glassy textures already described. Other products that also result from volcanic eruptions include volcanic ash, cinders, and dust, as well as three special rock types, pumice, tuff, and breccia.

Pumice. Pumice, a very light-weight rock, has a lower specific gravity than water (Figure 4-16). It is a froth of volcanic material

FIGURE 4-17 Close-up of yellowish gray volcanic tuff showing fine-grained gritty texture. Width of view is 8 cm.

that has chilled with the bubbles still intact. In composition, pumice is on the granitic or rhyolitic side of our classification. In texture, although it does not resemble obsidian, it is in fact a volcanic glass. Close observation with a hand lens reveals that the dividing walls between adjacent cells of the air are actually glassy.

Tuff. Volcanic tuff is a fine-grained rock composed of volcanic ash with fragments less than 4 mm in diameter (Figure 4-17). Most of the fragments are volcanic glass, but other kinds of volcanic fragments are also present in tuffs.

Breccia. Volcanic breccia is composed dominantly of fragments of lava more than 4 mm in diameter. Although some breccias are formed like tuff, many breccias are formed by volcanic mudflows.

FIGURE 4-18 (Right) Location of major volcanic eruptions of the world (top) and frequency of volcanic eruptions (bottom). (Courtesy Center for Short-Lived Phenomena, 1968–1974.)

Distribution

During the period from 1800 to 1914 an estimated total of 393 km^3 of volcanic material was ejected at the earth's surface. This did not originate from volcanoes scattered at random across the face of the earth. On the contrary, vents are concentrated in certain narrow bands. An examination of the map presented in Figure 4-18 indicates immediately that the greatest concentration of volcanoes still active today, or only recently extinct, runs around the rim of the Pacific Ocean. This circum-Pacific belt extends from New Zealand northward through Melanesia into eastern Indonesia, the Philippines, Japan, and Kamchatka, eastward through the Aleutian Islands and

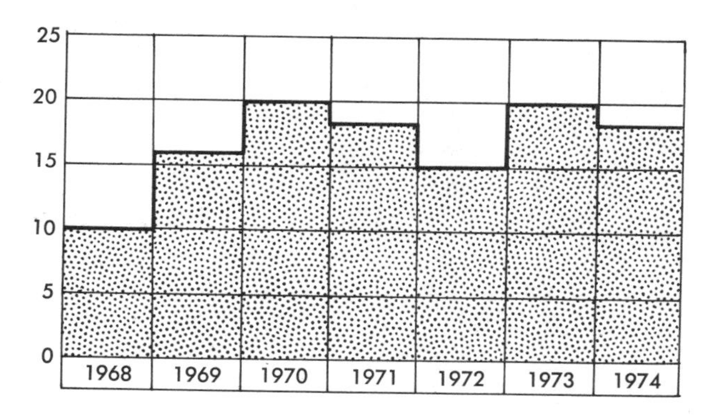

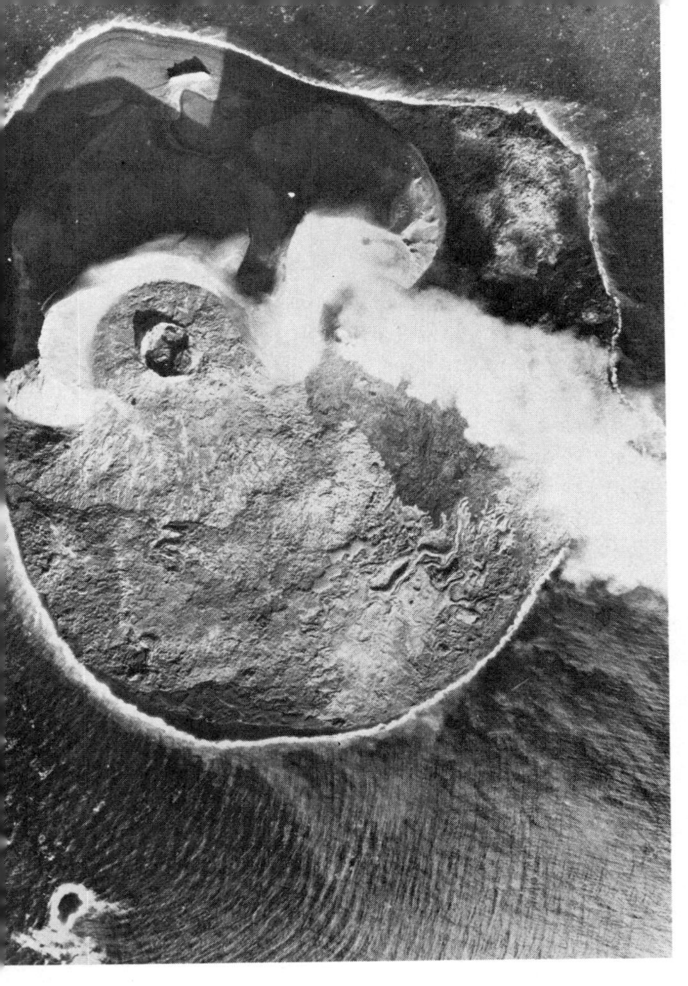

FIGURE 4-19 Vertical shot of Surtsey Volcano, Iceland. Pyroclastic material in upper part of photograph (north) and lava veneer in lower part (southern side). (Courtesy U.S. Geological Survey.)

southeastern Alaska, and southward along the western coast of the Americas to southern Chile. Outliers in Antarctica complete the belt.

Other important groups of active volcanoes are present in the Mediterranean region, northern Asia Minor, Red Sea area, and the eastern rift zone of Africa. The famous volcanoes of the Mediterranean are mostly in Italy.

Seventeen percent of the world's known active volcanoes are within ocean basins, and 83 percent are on continents. However, these statistics are misleading. It is probable that submarine volcanism has been much greater than continental volcanism. This is because much submarine volcanism has not been recorded or even witnessed. But groups of drowned volcanoes, or *seamounts,* are characteristic of the Pacific Ocean and also of other major oceans. Gordon A. Macdonald (see Suggested Readings) recently estimated that during the last 100 million years the total number of volcanoes in the ocean basins has been several hundred thousand, while on all the continents together there was only something over 100,000 volcanoes.

Types of volcanic eruption

Igneous activity as we see it at the surface often appears to be intermittent, and in some cases it seems to repeat itself at regular intervals. Actually, however, this activity is the result of a continuous series of events that are connected with the cooling of magma.

There are two principal types of volcanic activity: explosive and quiet or effusive. At any given time in its history a volcano shows a persistent tendency toward one of these types. However, it may occasionally exhibit the other type of behavior. A long-lived volcano may shift from one type to the other for extended periods of time.

The explosiveness of a volcano depends on the amount of gas trapped in its lava and on the viscosity of the lava. If the gases, mostly steam (Table 4-2), are held firmly within the lava and are prohibited from escaping to the

Table 4-2 Composition of Volcanic Gas Extracted from Lava, Mt. Pelée, West Indies

CONSTITUENT	VOLUME PERCENT
H_2O	82.5
CO_2	10.1
F_2	3.3
CO	2.0
N_2	0.9
S_2	0.5
Cl_2	0.4
H_2	0.2

surface, large pressures build up. These pressures are relieved suddenly in a great explosion. The ease with which gases are retained within a lava depends on its viscosity; and its viscosity, in turn, usually depends on the amount of silica in the melt. Turning back to our classification of igneous rocks, we find that the magmas and lavas with the highest amount of silica produce rocks high in quartz content; these are the silicic or sialic rock types. The explosive volcanoes, then, are on the sialic side of our composition chart (Figure 4-9); therefore, the rock types most commonly found in conjunction with explosive volcanoes are rhyolite, obsidian, volcanic tuff, and pumice, in addition to ashes and cinders. Less explosive volcanoes produce andesite, and the least explosive volcanoes produce basalt.

The volcanoes rimming the Pacific are by far the most violent. Quiet eruptions are found in the Atlantic and Indian Oceans as well as in the mid-Pacific. Southern Italy, Iceland, and the Africa Rift system have eruptions intermediate between the two extremes.

FIGURE 4-20 *Taal Volcano eruption in the Philippines, September 30, 1965. Volcanic ash is hurled 450 m into the sky above the erupting volcano. The effects of horizontal blasts from the volcano extended for about 2.2 km. The 1965 eruption killed 190 persons. A more violent eruption of Taal in 1911 killed more than 1,300 persons. (Courtesy U.S. Geological Survey.)*

Volcanic form

No single volcano has the precise shape of any other volcano. Yet, despite the individual differences among volcanoes, they can be grouped into the following major subdivisions: basaltic flows, shield volcanoes, and composite cones. These forms reflect the nature of the magma and type of the volcanic activity.

Basaltic lavas, which in general give rise to quieter eruptions than any other lava type, are highly fluid (Figure 4-21). They originally contain less gas than the silicate types, and what gas they do contain escapes more easily to the air than does the gas in other lavas. Basalts may flood out through long narrow fissures or groups of fissures and then spread as great sheets across the countryside. Basaltic lavas that compose the Columbia Plateau of the northeastern United States buried 130,000 km^2 of preexisting topography that had a maximum relief of more than 1.5 km. Some individual flows are more than 100 m thick, and some spread more than 60 km from their source.

When basaltic lavas erupt large amounts from a single central vent, they build a mound similar to the broad low cones of Iceland or to the much larger cone of the Island of Hawaii. Slopes are gentle, 2 to 10°, and the height is low compared with the diameter. Such volcanoes, which are made up largely of lava flows with only a few beds of cinders and ash, are termed *shield* volcanoes because they take on

very nearly the aspects of a large circular shield with a central boss.

Most of the great volcanic mountains on the continents are composed of interbedded lava, ash, and cinders. The relative proportions of *tephra* (fragmental products of explosive volcanic eruption) and lava vary greatly in composite cones. In size, such cones range in height from about 50 to 800 m and in basal diameter up to about 30 km. Mt. Rainier in Washington rises 2,400 m above the nearby peaks of the Cascade Range. Mount Hood in Oregon is a beautiful example of a composite cone; a large population lives close by, which is common throughout the world because of the fertility of volcanic lands.

FIGURE 4-21 *(Left) Lava cascade of basalt over sedimentary rocks into Grand Canyon. (Photo by Arthur Trevena.)*

FIGURE 4-22 *(Below) Shishaldin volcano, Unimak Island, Alaska, is a near-perfect volcanic cone. (Courtesy U.S. Navy.)*

Most composite volcanoes are built by eruptions that come mainly from a single central vent, and their ground plan is roughly circular. On some composite cones the principal part of the slope is nearly straight, and the slope angle ranges from 10 to 35°. On others, the slope steepens progressively upward to a narrow crater rim. To attain a large size and retain symmetry, a volcanic cone requires the support of ribs of lava interbedded with the tephra.

A single volcano may develop as a shield volcano during part of its history and as a composite one later. Mount Etna, on the island of Sicily, is an example of such a volcano.

Volcanism and plate movement

As discussed previously, convergent plate boundaries develop when two plates move toward each other and collide. In contrast, divergent plate boundaries move away from each other.

Magmas emerge along both convergent and divergent plate boundaries. As plates diverge, magma rises from the mantle and breaks through the ocean floor, creating underwater lava flows. As divergence continues, the lava (basalt) accumulates above and below the sea floor, building ridges along the plate boundary.

Where plates converge, forming a subduction zone, increases in temperature and pressure modify some of the rocks and sediments within the descending plate. As the plate moves into deeper and hotter parts of the mantle, some of the rocks and sediments partially melt and then rise as intrusive and extrusive igneous rocks. Volcanic rocks along continental margins that border convergent zones (subduction zones) are more rhyolitic than the midoceanic ridges because the rocks and sediments brought down by the descending plates contain more silica, potassium, and aluminum.

Most volcanoes are present along plate boundaries, especially along the contacts of the American and Pacific plates, the African plate with the American and Eurasian plates, and the Pacific plate with the Eurasian and Indo-Australian plates. However, to a less extent, volcanoes are present within the African plate. The mechanisms of plate movements explain why volcanoes are common along midoceanic ridges (diverging plates) and why they are parallel with continental margins located inland of submarine trenches (converging plates).

Intrusive igneous rocks

As soon as a volcano becomes extinct, erosion begins to wear it away and eventually exposes the internal structure to view. Often we can identify the plug that was formed when the magma solidified in the central vent, the reservoirs from which the magma came, and the channels through which it flowed. All the rock masses that are produced when the magma solidifies below the surface are called *plutons.* Furthermore, because these rocks are intruded into older preexisting rocks, we also refer to them as *intrusive igneous* rocks. If the rocks invaded by magma are layered, a pluton that is parallel with the layering is said to be *concordant.* A pluton that cuts across the layering is *discordant.* Plutons are classified according to their size and their shape in relationship to surrounding rocks. The two major subdivisions are tabular and massive plutons.

Tabular plutons

A pluton whose thickness is small relative to its other dimensions is a *tabular pluton.* There are two types: sills and dikes (Figure 4-23).

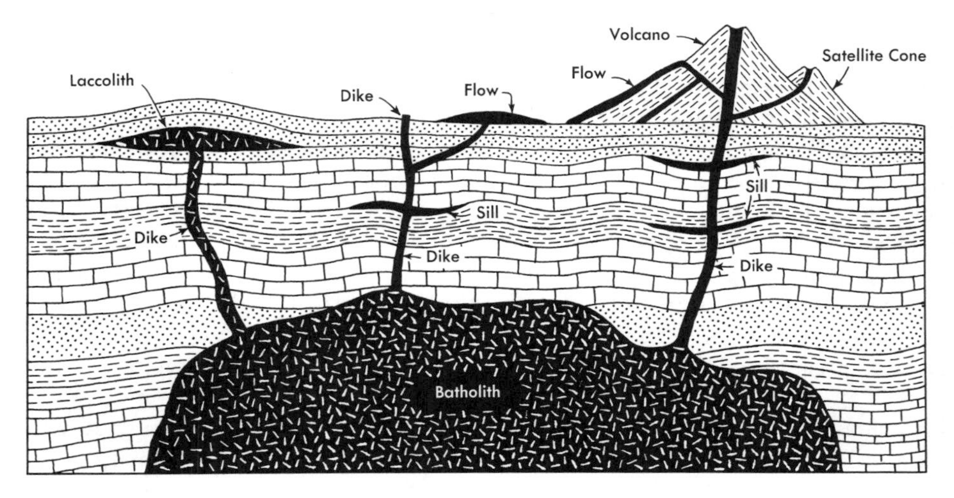

FIGURE 4-23 *Intrusive and extrusive occurrences of igneous rocks.*

Sills. A tabular concordant pluton is called a *sill.* It may be horizontal, inclined, or vertical depending on the rock structure with which it is concordant.

Sills range in size from sheets less than a centimeter in thickness to tabular masses hundreds of meters thick. A sill must not be confused with an ordinary lava flow that has been buried by other rocks later on. Since a sill is an intrusive form—that is, it has forced its way into already existing rocks—it is younger than the rocks that surround it. There are fairly reliable ways of distinguishing between the two types. Most important, a buried lava flow usually has a rolling or wavy-shaped top, pocked by the scars of vanished gas bubbles, whereas a sill has a more even surface. Also, a sill may contain fragments of rock that were broken off when the magma forced its way into the surrounding structures.

The composition of most sills is basaltic. The Palisades along the west bank of the Hudson River opposite New York City, for example, are the remants of a basalt sill that was several hundred meters thick. Here the magma was originally intruded into sedimentary rocks that are now inclined at a low angle toward the west.

Dikes. A tabular pluton that cuts across the structure of the surrounding rock formation is known as a *dike.* Some dikes can be traced for tens of kilometers. Most dikes are composed of basalt. Apparently many dikes were formed when magma forced its way out of a reservoir through fractures in the adjoining rock. Often these dikes radiate outward from a central area where the magma was presumably subjected to concentrated pressures (Figure 4-24). Others have formed in fractures that outline an inverted cone, with the apex pointing down toward the magma reservoir. Dikes of this sort are called *cone sheets.* In Scotland, for example, the angles of certain cone sheets suggest that their apex is approximately 5 km below the present surface of the earth.

When a body of magma forces its way upward into the crust, it sometimes exerts enough pressure to lift up a plug of solid rock. Later on, when some of the magma escapes by way of dikes, sills, or flows, the plug may settle back again into the remaining magma. As a result, we find plugs that are surrounded by roughly vertical, cylindrical bands of rock that mark the areas where the magma has subsequently solidified. These plutons are called *ring dikes.* Patterns of ring dikes have been mapped with diameters ranging up to 25 km. The width of individual ring dikes ranges from a few centimeters to about a meter.

Massive plutons

Any pluton that is not tabular in shape is classified as a *massive pluton.* The two types are laccoliths and batholiths.

Laccoliths. A massive concordant pluton that has pushed up the overlying rock structures into a dome is called a *laccolith* (from the Greek *lakkos,* "a cistern," and *lithos,* "stone"). A laccolith may be a blister on the surface of a sill, or it may be a local body of magma that is fed from a channel cutting up from the underlying rocks (Figure 4-25). If the

FIGURE 4-24 *Dike radiating outward from volcanic neck, Shiprock, New Mexico.*

FIGURE 4-25 *Bear Butte, near Sturgis, South Dakota, is a laccolith. Around the base are the remnants of the sedimentary rocks that once extended up over the core of the intrusion that forms the main mass of the hill. (Photo by N. H. Darton, U.S. Geological Survey.)*

ratio of width to thickness is less than 10 to 1, the pluton is arbitrarily termed a laccolith. If this ratio is more than 10 to 1, the pluton is termed a sill.

In composition, laccoliths are commonly granitic. A classic development of laccoliths is present in the Henry, LaSal, and Abajo Mountains of southeastern Utah, where these features are exposed on the Colorado Plateau, a famous geological showplace.

Batholiths. A large discordant pluton that increases in size as it extends downward and has no determinable floor is called a *batholith* (from the Greek *bathos,* "deep," and *lithos,* "stone"). "Large" in this connection is generally taken to mean a surface exposure of at least 100 km^2. A pluton that has a smaller surface exposure but that exhibits the other features of a batholith is called a *stock.* Actually, most stocks are probably offshoots from underlying batholiths.

Batholiths generally are coarse-grained granite or granodiorite because the deeply buried pluton has cooled slowly. Some batholiths originated at depths of almost 30 km. Today many of these great masses are visible because they have been uplifted and exposed by erosion that has removed the rocks that covered them (see Figure 4-23). The Coast Range batholith of Alaska and British Columbia is 30 to 250 km wide and about 1,800 km long.

Batholiths are usually located in mountainous belts where they are parallel with the axes of the ranges and form the cores of many continental mountain ranges. Although batholiths are found in mountain ranges characterized by severe crumpling and folding, it is not believed that the batholiths caused the deformation. Rather, study of batholiths indicates that the intrusion began after folding of the mountains had commenced. The close relationship between batholiths and mountain ranges does indicate, however, that batholiths are related to mountain-building processes.

A quite recent suggestion is that the formation of some batholiths was related to the subduction process at the margins of colliding crustal plates. Reheating and melting of the descending plate might provide rising hot fluids as raw materials for the formation of batholiths along the margins of overriding plates.

Some batholiths are in sharp contact with rocks that border them. These batholiths also show flow structures in which elongate crystals are arranged parallel with each other. Such features indicate that the batholiths were formed by injection of an igneous magma.

However, other batholiths give the impression of having replaced the formations into which they have intruded, instead of having pushed them aside or upward. But if that is what really took place, what happened to the great volumes of rock that the batholiths appear to have replaced? Here we encounter the problem of the origins of batholiths; in fact we meet the whole mystery of igneous activity. Some observers have even been led to question whether granitic batholiths originated from true magmas at all. The suggestion has been made that batholiths may have been formed through a process, called *granitization,* in which the solutions from magmas move into solid rocks, exchange ions with them, and convert them into the rocks that have the characteristics of granite but have never actually existed as magma.

Causes of variation in igneous rocks

We have found that there is a wide variation in igneous rocks, both in composition and texture as well as in occurrence. In the following paragraphs we examine some of the reasons for these variations.

Variations in texture

Earlier we discovered that there is a wide variation in the texture of igneous rocks, from glassy obsidian on down to very coarse-grained porphyry and pegmatite (Figure 4-26). Later we found that the volcanic rocks are in large part either very fine-grained or glassy and that the deep-seated, or plutonic, rocks are coarse-grained or porphyritic. This leads us to the conclusion that there is a relationship between the rate at which an igneous melt cools and the texture that results.

Indeed, the rate at which a magma crystallizes influences the size to which crystals grow. This is so because the cooling rate controls the time that ions have to move about and enter into combinations. If a magma cools rapidly, there is little or no time for the ions to move into the orderly arrangement of crystals. But if a magma cools slowly, there is every possibility that individual ions will seek out the pattern of crystal structures that best suits them.

The rate of crystallization varies with depth. For example, a molten mixture consisting largely of nonferromagnesians at 1100°C (2000°F), exposed to the air on top, and ranging in thickness from 0.9 to 9,000 m, would cool down to 750°C (1400°F) as shown in Table 4-3.

We can demonstrate both by experiment and by direct observation in the field that the

FIGURE 4-26 *Pegmatite body intruded into vertically foliated schist of one older Precambrian Vishnu Group, Upper Granite Gorge, Grand Canyon National Park. Pegmatites, which are characterized by very coarse-grained rocks that may contain crystals several meters across, are formed in the last stages of intrusion when magmas are less viscous and contain more dissolved gases. (Photo by Arthur Trevena.)*

Table 4-3

THICKNESS (in Meters)	TIME REQUIRED
0.9	12 days
9	3 years
90	300 years
900	30,000 years
9,000	3,000,000 years

Data adapted from R. A. Daly, *Igneous Rocks and the Depths of the Earth* (New York: McGraw-Hill Book Company, 1933), p. 63.

very glassy and fine-grained rocks are those that cool the quickest. In addition, experimentally we can show that slow cooling produces large crystals. But now we must explain about the texture that is halfway between coarse and fine, a porphyritic texture, which has large crystals set in a ground mass of smaller crystals. What does this texture mean?

The explanation lies in two different generations of cooling. The large crystals cooled first and at a slower rate than did the smaller crystals. The smaller crystals cooled later and more rapidly, and engulfed the already-formed original crystals. The way in which this double cooling takes place may vary. One method is as follows: A melt cooling at some depth beneath the surface begins to precipitate crystals of a particular type. Then this same melt becomes part of a volcanic eruption and is forced upward rapidly toward the surface and is spewed out across the ground surface as a lava flow. As it moves upward, it carries with it in suspension the already-formed crystals, and these crystals then become congealed within the groundmass of the cooling lava; also included in this groundmass will be the small crystals produced by fast cooling at the surface. The resulting texture is porphyritic.

Variation in composition

The original magma (*melt*) is actually a liquid solution of ions at high temperature. Once the heat that keeps the magma molten begins to decrease, the melt starts to crystallize. Bit by bit, solids with crystalline structure grow, and gases are released. Now we no longer have a pure liquid, but rather a liquid with solid and gaseous materials mixed in with it. As the temperature continues to drop, the mixture solidifies completely, to form an igneous rock. Exactly what sort of igneous rock is formed depends on the composition of the original magma and the conditions under which it has crystallized.

Order of crystallization

Even when a magma contains all the ions necessary to yield the major silicate minerals, these minerals do not all form at once. There is a definite temperature or temperature range at which each mineral crystallizes.

The complex ions of the silicon-oxygen tetrahedra develop first. As the mixture continues to cool, the other ions join with these tetrahedra in various ways to form the silicate minerals. The ferromagnesians form, beginning with olivine and progressing in order through augite, hornblende, and biotite. At the same time the plagioclase feldspars form, starting with anorthite and progressing through albite. After all of these minerals have crystallized, orthoclase, muscovite, and then quartz are formed.

N. L. Bowen suggested that the silicates could be arranged in two series of crystallization. Each mineral in each series is derived from the preceding mineral as the result of a reaction (that is, a chemical change) with the remaining liquid of the magma. For this reason, these series are called the *Bowen reaction series.*

In the ferromagnesian series, olivine is the first to form because it has the simplest structure: its tetrahedra are merely piled together and are not joined to one another. Augite is built around single chains of tetrahedra; hornblende is built around double chains; and biotite, the most complex of this series, consists of sheets of tetrahedra. Since each new ferromagnesian to form has a different crystalline structure and a different composition from the one that preceded it, Bowen called this a *discontinuous reaction series.*

The first feldspar to form is anorthite. This mineral crystallizes at about the same temperature as olivine. As the anorthite grows, the remaining liquid becomes impoverished in calcium, and sodium begins to substitute for it in the growing feldspar crystal. The outer rind of the crystal may thus develop a progressively more albitic composition Because of this continuous replacement of calcium ions by sodium ions in the same silicate structure, Bowen referred to the crystallization of the plagioclase feldspars as a *continuous reaction series.*

This gradual progressive change does not take place, however, between albite and orthoclase. There is no ion-by-ion replacement of sodium by potassium because, as indicated in Table 4-4, the radius of the potassium ion is so much greater than that of the sodium ion.

Despite the orderly manner in which the minerals crystallize, a single magma can produce a great variety of rocks as a result of interruptions in the reaction series. For instance, an interruption occurs when crystals settle out of the magma, for they are then separated from the liquid and cannot enter into further reactions with it. This process is *fractionation.*

If all the crystals of olivine remain in the magma after their formation, they will eventually be transformed into crystals of augite. But if some of the olivine crystals settle out of the magma, they are effectively isolated from further chemical changes. Now only the olivine left in the liquid will be changed into augite. Similarly, only the augite remaining in the magma will be transformed into hornblende, and so on through the entire reaction series. A single parent magma may produce any of the rocks shown in Figure 4-9 depending on the degree of fractionation.

Table 4-4 Ionic Radii and Electric Charges for Positive Ions of the Feldspars

		ELECTRIC CHARGE	IONIC RADIUS (Angstroms)
Calcium	Ca^{2+}	+2	0.99
Sodium	Na^{+}	+1	0.97
Potassium	K^{+}	+1	1.33

SOURCE: After Brian Mason, *Principles of Geochemistry,* 2nd ed. (New York: John Wiley and Sons, Inc., 1960), pp. 287–88.

SUMMARY

Igneous, sedimentary, and metamorphic rocks are the three types of rocks recognized in the earth's crust. Of these, we consider the igneous rocks to be the parents from which the others are eventually derived. We can relate the three types to each other by means of the rock cycle.

The chief characteristics of igneous rocks are color, density, mineral composition, and texture. Of these, color, density, and mineral composition tend to go together, whereas texture is generally independent. We classify igneous rocks on the basis of textures (coarse-grained, fine-grained, glassy, and porphyritic) and mineral compositions. Among the light-colored types are granite, and rhyolite. Dark-colored rocks include gabbro, basalt, and peridotite, and intermediate between these two groups are diorite and andesite. Igneous rocks either solidify at the surface (extrusive igneous rocks) or they crystallize beneath the surface (intrusive igneous rocks).

Volcanic eruptions vary with the composition of the magma as do the volcanic forms. Basaltic eruptions are relatively quiet, and the volcanic forms have gentle slopes. Rhyolitic eruptions are more explosive, and the forms have steeper slopes.

Intrusive igneous bodies are concordant (injected in conformity with the structure of the rock that encloses them) or discordant (injected across the structure of the enclosing rock). Concordant rocks include sills and laccoliths; discordant bodies include batholiths and dikes.

Rapid cooling produces glassy or fine-grained textures. Slow cooling means coarse-grained textures. Porphyritic textures suggest a period of slow cooling followed by a period of more rapid cooling. Bowen's reaction series explains how a magma of a given composition can give rise to igneous rocks of different compositions.

SUGGESTED READINGS

BARTH, TOM F. W., *Theoretical Petrology,* 2d ed. New York: John Wiley and Sons, Inc., 1962.

BULLARD, FRED M., *Volcanoes.* Austin, Texas: University of Texas Press, 1962.

JACKSON, KERN C., *Textbook of Lithology.* New York: McGraw-Hill Book Company, 1970.

MACDONALD, GORDON A., *Volcanoes.* Englewood Cliffs, N.J.: Prentice-Hall, Inc., 1972.

CARMICHAEL, IAN S. E., FRANCIS J. TURNER, and JOHN VERHOOGEN, *Igneous Petrology.* New York: McGraw-Hill Book Company, 1974.

Weathering and Soils

Broken rock on a mountain slope, the crumbling foundations of an old building, the blurred inscription on a gravestone—all remind us that rocks are subject to constant change and destruction (Figure 5-1). Changes of temperature, moisture soaking into the ground, the ceaseless activity of living things—all work to alter rock material. This process of alteration is known as *weathering,* which is defined as the changes that take place in minerals and rocks at or near the surface of the earth in response to the atmosphere, to water, and to plant and animal life. We will later extend this definition slightly, but it serves our needs for the time being.

Weathering leaves its mark everywhere. The process is so common, in fact, that we tend to overlook the way in which it functions and the significance of its results. Weathering plays a vital role in the rock cycle (see

5

Weathering and erosion of badlands in the Badlands National Monument, South Dakota. (Courtesy National Park Service.)

FIGURE 5-1 *Differential weathering of horizontally bedded and vertically jointed sedimentary rock has helped to produce towering turrets in Bryce National Park, Utah. Both bedding and jointing planes provide zones where weathering can proceed more rapidly. (Photo by National Park Service.)*

pp. 74–75, and Figure 4–2), for by attacking the exposed material of the earth's crust—both solid rock and unconsolidated deposits—it produces material for new rocks.

The products of weathering are usually moved by water and the influence of gravity, less commonly by wind and glacier ice. When dropped by the transporting agents, these products settle and accumulate in new places. The mud in a flooding river, for example, is weathered material that is being moved from the land to some settling basin, generally the ocean or a lake. Sometimes, however, the products of weathering remain where they are formed and so are incorporated into the rock record. Certain ores, for example, such as those of aluminum, are actually old zones of weathering.

Soil

Before examining weathering processes, let us briefly look at soil formation and erosion. Soil is a mixture of inorganic weathered mineral grains and rocks, water, air, and at least a small amount of organic material. The inorganic grains are the products of mechanical and chemical weathering of rock and sediment exposed at the earth's surface. Water and air move into the pore spaces between grains and supply support for the plants and microorganisms whose remains account for the organic content. Most soils display distinct layering or what is called a *soil profile.*

Soil formation

Soils are created over long periods of time—as much as millions of years—by complex factors involving climate, living organisms, topography (particularly slope), chemical and mechanical activity, and time itself. In the northern United States, nearly all soils are formed from the weathering of sediment transported by water, wind, or ice.

The dominant factor in soil formation is climate, and soils in areas with similar climates tend to be similar in type even though the parent material of the soil was different. Thus, similar sediment subjected to different climates leads to different soils. Optimum rainfall for soil formation is about 75 cm a year. Temperature is less important than the amount and time of rainfall, but chemical and biologic processes are increased in rate of reaction by higher temperatures.

Different types of parent material produce weathered particles of different size and different mineral composition. Sandy soils, for example, are mostly formed from quartz-rich sandstone or its weathered derivates. Soils characterized by either the abundance or the rarity of particular trace elements reflect the amounts of those elements in the original parent material.

The life in a soil is extremely important. Organisms aid in opening the soil to air so that oxygen and nitrogen can move freely through the pore spaces.

Most soils are apparently older than was long believed. Some tropical soils are millions of years old. However, soils about a meter thick have been formed within the last 10,000 to 15,000 years on glacial deposits.

The layers of a soil that are formed as water moves downward are termed the *soil profile* (Figure 5-2). The uppermost layer at the surface, the *O-horizon* (often missing), is dark colored and rich in organic material and humus. In the underlying *A-horizon,* surface water has dissolved or leached many of the less resistant minerals and has carried the material downward. The *B-horizon,* below the A-horizon, characteristically contains deposits of materials leached from above. In most places, the B-horizon marks the deepest penetration of plant roots. The *C-horizon* is slightly altered bedrock, which is broken, decayed, and mixed with clay. At its base, the C-horizon merges into solid bedrock, sometimes termed the *D-horizon.*

FIGURE 5-2 *Soil profiles in humid and semiarid climates.*

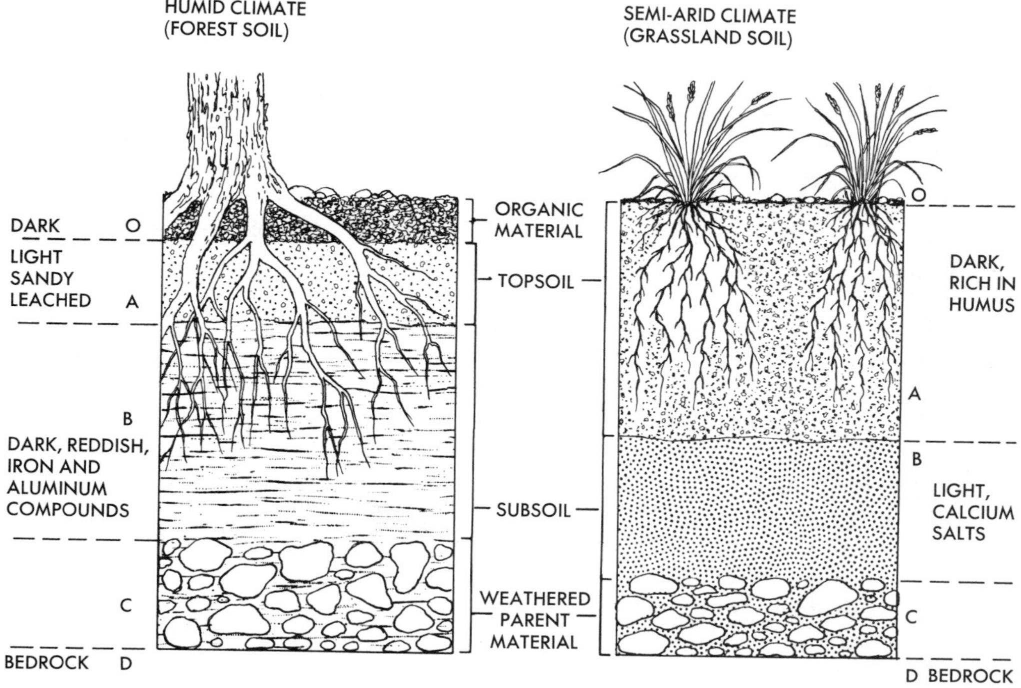

Soil groups

On the basis of the minerals present in the A- and B-horizons, three major soil groups are distinguished. In high rainfall areas, both A- and B-horizons are leached, soluble material has been carried away by water, and the soil is enriched in quartz, clay minerals, and iron oxides. Such soils are termed *pedalfers,* and most of Canada and the eastern United States are covered with these fertile soils.

In dry warm climates, soils contain soluble minerals, especially calcium carbonate as pellets and nodules in the B-horizon. The word *caliche* is applied if the horizon is cemented into a nodular, carbonate rock. Because of the low rainfall, such soils contain less clay minerals and more unaltered silicate minerals than the pedalfers. These soils, called *pedocals,* are less fertile than the pedalfers and do not contain a rich population of organisms. Pedocal soils are characteristic of the western United States.

Laterite, the deep red-brown soil of the tropics, is extremely leached because considerable water has passed through these soils. The soluble elements calcium, sodium, and potassium are highly leached. Even silica (SiO_2) has been largely removed. Silicate minerals are greatly altered, and aluminum and iron oxides predominate. A typical laterite contains much iron oxide (Fe_2O_3). However, if little iron is present in the parent material, bauxite ($Al_2O_3 \cdot nH_2O$) is formed. Lateritic soils are not very fertile, and after a few years many are depleted of humus and must be abandoned for crops. Many of the underdeveloped tropical countries of the world contain very large lateritic areas.

In some parts of the tropics, the iron content of the laterites may be great enough to constitute ore. Examples are the iron mines of Cuba and of the Surigao Peninsula in Mindanao (Philippines). The bauxite ores of Little Rock, Arkansas, are in old soils deposited in former tropical savanna climates.

Soil erosion

We have seen how immensely long is the time required to produce a soil profile. Therefore, from man's standpoint, soil is not a renewable resource. When the development of urban areas and agricultural uses destroy soil, the loss is irretrievable as far as man is concerned.

Normal erosional processes of water, wind, and ice constantly attack the earth's surface. Man's activities have greatly accelerated the normal rate of erosion. The current erosion rate for the United States is estimated at about 6 cm per thousand years, which is believed to be about twice the average rate before human activities.

The most destructive of human activities is the removal of the protective plant cover, which immediately leads to increased erosion of the soil. Thinner soil holds back less water in the pore spaces, and the destruction of the plant cover initiates a cycle of more and more runoff water with accompanying erosion and less and less soil.

Urbanization also contributes greatly to erosion of the soil. Each year, in the United States, 400,000 hectares (1 million acres) are urbanized, and fully 20 percent of this land was originally suitable for agricultural use. In addition to the direct loss of land through urban development, soil erosion is increased in nearby areas. When the construction is finished, erosion rates in the surrounding area are raised greatly because the streams now carry more water, which increases the sediment load and discharge rate of the streams. As each new urban area is developed, the erosional cycle for many kilometers around the area becomes more complex and is greatly accelerated.

Types of weathering

There are two general types of weathering: mechanical and chemical. It is hard to separate them in nature for they often go hand in hand, although in some environments one or the other predominates. Still, for our purposes, it is convenient to discuss them separately.

Mechanical weathering

Mechanical weathering, which is also called *disintegration,* is the process by which rock is broken down into smaller and smaller fragments as the result of the energy developed by physical forces (Figure 5-3). For example, when water freezes in a fractured rock, enough energy may develop to pry off individual pieces. Or a boulder moved by gravity down a rocky slope may be shattered into smaller fragments. Note that in mechanical weathering the size of the material changes from large to small, but the composition is unchanged.

Expansion and contraction resulting from heat

Changes in temperature, if they are rapid enough and great enough, may bring about slight mechanical weathering of rocks. For instance, in areas where bare rock is exposed at the surface, unprotected by a cloak of soil, forest or brush fires can generate enough heat to break up the rock. The rapid and violent heating of the exterior zone of the rock causes it to expand, and if the expansion is great enough, flakes and larger fragments are split off. Lightning often starts such forest fires, and in rare instances, may even shatter exposed rock by means of a direct strike.

Variations in temperature from day to night and from winter to summer can cause expansion and contraction of rock material.

FIGURE 5-3 Disintegration of granite in the desert region of Joshua Tree National Monument, southeastern California. Mechanical weathering is dominant and chemical weathering is slight. Note effects of weathering along joints. (Photo by Mary Hill.)

Occasionally these changes may cause mechanical weathering. But it is unlikely that temperature changes are great enough to cause extensive mechanical weathering. Polished inscriptions on granite in Egypt that have faced the tropical sun for 3,000 years show almost no discernible change and still retain their polish. However, similar inscriptions in the shade where the surface is relatively damp show much more weathering of the inscriptions, illustrating the importance of even a small amount of water in weathering.

FIGURE 5-4 Strong frost action high in the Medicine Bow Range, Wyoming, has produced this field of angular boulders. Alternate freezing and thawing of water in the crevices of the bedrock have dislodged these large fragments from the solid rock beneath. (Photo by E. N. Cameron.)

Frost action

Frost is much more effective than heat in producing mechanical weathering. When the temperature of water drops from 4°C to 0°C, its freezing point, the volume increases by 9.2 percent. Those who have encountered a cracked radiator on a crisp morning are familar with high pressures exerted by freezing water. Similarly, the water that moves into cracks, joints, and pore spaces of a rock exerts tremendous pressure when it freezes. If freezing water is completely confined, pressure up to ten times greater than the tensional strength of most rocks is generated.

Frost action is responsible for much mechanical weathering (Figure 5-4). Water that soaks into the crevices and pores of a rock usually starts to freeze at its upper surface where it is in contact with the cooling air. This means that in time the water below is confined by an ice plug. Then, as the freezing continues, the trapped water expands, and pressure is exerted outward. Rock may be subjected to this action many times a year. In high mountains, for example, the temperature may move back and forth across the freezing line almost daily. Weather records indicate that freeze-thaw cycles are most frequent in high-altitude–low-altitude regions such as the mountains of the tropics and the subtropics.

Dislodged fragments of mechanically weathered rock are angular in shape, and their size depends largely on the nature of the bedrock from which they have been displaced. Usually the fragments are only a few centimeters in maximum dimensions, but in some places—along the cliffs bordering Devil's Lake, Wisconsin, for instance—they reach sizes of up to 3 m. A deposit of large angular fragments in an apron at the base of a cliff or steep slope is called *talus*.

Another process of mechanical weathering produced by freezing water is *frost-heaving*. This action usually takes place in fine-grained, unconsolidated deposits rather than in solid rock. Much of the water that falls as rain or snow soaks into the ground, where it freezes during the winter months. If conditions are right, more and more ice accumulates in the zone of freezing as water is added from the atmosphere above and is also drawn upward from the unfrozen ground below much as a blotter soaks up moisture. In time, lense-shaped masses of ice are built up, and the soil above them is heaved upward. Frost-heaving of this sort is common on poorly constructed roads. It may also cause severe damage to the concrete. And lawns and gardens are often soft and spongy in the springtime as a result of the heaving up of the soil during the winter.

Certain conditions must exist before frost action can take place: (1) There must be an adequate supply of moisture; (2) the moisture must be able to enter the rock or soil; and (3) temperatures must move back and forth across the freezing line. The phenomenum is also much more effective in fine soils such as clay and silt than it is in coarse ones such as sand and gravel.

Exfoliation

Exfoliation is a mechanical weathering process in which curved plates of rock are stripped from large rock masses by the action of physical forces. This process produces two features that are fairly common in the landscape: large domelike hills called *exfoliation domes,* and rounded boulders usually referred to as *spheroidally weathered boulders.* It seems likely that the forces that produce these two forms originate in different ways.

Let us look first at the manner in which exfoliation domes develop. Fractures or parting planes called *joints* are present in many massive rocks. These joints are broadly curved and run more or less parallel with the rock surface. The distance between these joints is only a few centimeters near the surface but increases to several meters deeper into the rock. Under certain conditions, one after another of the curved slabs between the joints is spalled or sloughed off the rock mass. Finally, a broadly curved hill of bedrock develops as shown in Figure 5-5.

Just how these slabs of rock come into being is still a matter of dispute. Most observers believe that as erosion strips away the surface cover, the downward pressure on the underlying rock is reduced. Then as the rock mass begins to expand upward, lines of fracture develop, marking off slabs that later fall away. Precise measurements made on granite blocks in New England quarries provide some support for this theory. Selected blocks were accurately measured and then removed from the quarry face, away from the confining pressure of the enclosing rock mass. When the free-standing blocks were measured again, it

FIGURE 5-5 *Independence Rock, an exfoliation dome in central Wyoming, is one of the great natural monuments on the Oregon Trail. Inscriptions on the rock date back to the 1830s, and the number of pioneer inscriptions has been estimated to be between forty and fifty thousand. It is known as the "register of the desert." (Courtesy Wyoming Travel Commission.)*

FIGURE 5-6 Spheroidal weathering in an exposure of granite, Great Smoky Mountains, North Carolina. The granite has been fractured, and the rounding by spheroidal weathering has worked inward from the fracture planes. (Photo by L. B. Gillett.)

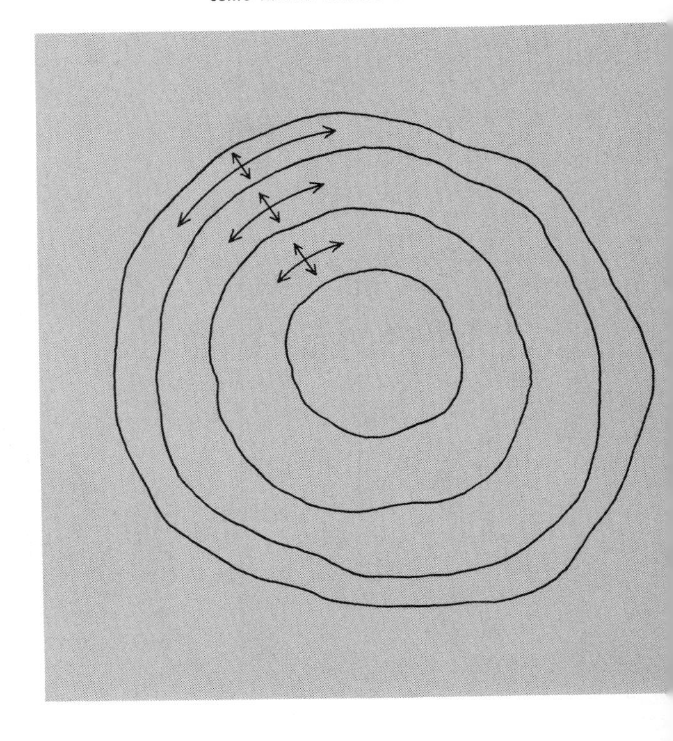

FIGURE 5-7 This cross section through a spheroidally weathered boulder suggests the stresses set up within the rock. The stress is thought to develop as a result of the change in volume as feldspar is converted to clay. Note that the shells become thinner toward the surface.

was found that they had increased in size a small, but measurable, amount. Massive rock does expand, then, as confining pressures are reduced, and this slight degree of expansion may be enough to initiate the process of exfoliation.

Among the better-known examples of exfoliation domes are Stone Mountain, Georgia; the domes of Yosemite Park, California; and Sugar Loaf in the harbor of Rio de Janeiro, Brazil.

Now let us look at a smaller instance of exfoliation—spheroidally weathered boulders. These boulders are rounded by the spalling off of a series of concentric shells of rock (see Figures 5-6 and 5-7). But here the shells develop as a result of pressures set up within the rock by chemical weathering, rather than being caused by the lessening of pressure from above by erosion.

When certain minerals are chemically weathered, the resulting products occupy a

FIGURE 5-8 Roots of tree have accelerated mechanical weathering of massive outcrop in Japan. (Photo by A. J. Eardley; courtesy of Mrs. A. J. Eardley.)

FIGURE 5-9 Lichen, shown here growing on sandstone beds, contribute greatly to mechanical weathering. (Courtesy National Park Service.)

greater volume than the original material. And it is this increase in volume that creates the pressures responsible for spheroidal weathering. Since most chemical weathering takes place in the exposed portions of the rock, it is there that we find the most expansion and hence the greatest number of shells.

Spheroidally weathered boulders are sometimes produced by the crumbling off of concentric shells. If the cohesive strength of the rock is low, individual grains are partially weathered and dissociated, and the rock simply crumbles away. The underlying process is the same in both these cases, however.

Certain rocks are more vulnerable to spheroidal weathering than others. Granite, diorite, and gabbro are particularly susceptible, for they contain large amounts of the mineral feldspar, which, when weathered chemically, produces new minerals of greater volume.

Other types of mechanical weathering

Plants also play a role in mechanical weathering. The roots of trees and shrubs growing in rock crevices sometimes exert enough pressure to dislodge previously loosened fragments of rock, much as tree roots heave and crack sidewalk pavements (Figures 5-8 and 5-9).

More important, though, is the mechanical mixing of the soil by ants, worms, and rodents. Constant activity of this sort makes soil particles more susceptible to chemical weathering and may even assist in the mechanical breakdown of the particles themselves.

Finally, such agents as running water, glacier ice, wind, and ocean waves all help to reduce rock material to smaller and smaller fragments. The role of these agents in mechanical weathering is discussed in later chapters.

Chemical weathering

Chemical weathering, sometimes called *decomposition*, is a more complex process than mechanical weathering. Mechanical weathering merely breaks rock material down into smaller and smaller particles without changing the composition. Chemical weathering, however, actually transforms the original material into something different. The chemical weathering of the mineral feldspar, for example, produces clay minerals, which have a different composition and different physical characteristics from those of the original feldspar. In some cases the products of chemical weathering have no mineral form at all, as when the mineral halite is transformed into a salty solution.

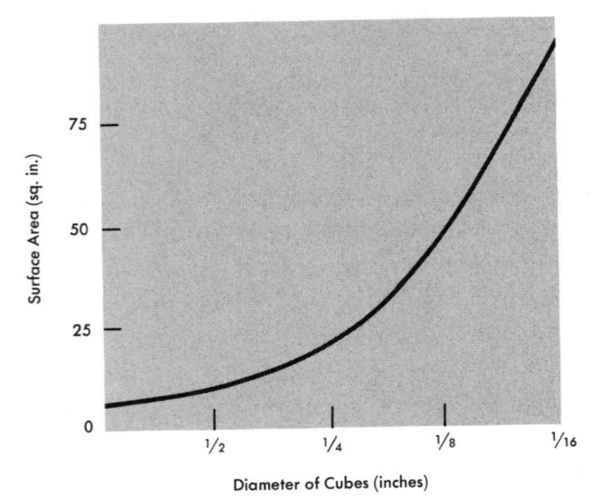

Particle size and chemical weathering

The size of individual particles of rock is an extremely important factor in chemical weathering because substances can react chemically only when they come in contact with one another. The greater the surface area of a particle, the more susceptible it is to chemical attack (Figure 5-10). If we take a pebble, for example, and grind it into a fine powder, the total surface area exposed is greatly increased. And as a result, the materials that made up the original pebble will now undergo more rapid chemical weathering.

Other factors in chemical weathering

The rate of chemical weathering is affected by other factors as well—the composition of the original material, for example. A mineral like quartz (SiO_2) responds much more slowly to chemical weathering than does a mineral like olivine ($FeMg_2SiO_4$). And to take a more familiar example, copper water pipes last longer than iron water pipes because copper weathers chemically more slowly than iron.

Climate also plays a key role in chemical weathering. Moisture, particularly when it is accompanied by warmth, speeds up chemical weathering; conversely, dryness and cold slow it down. Finally, plants and animals contribute

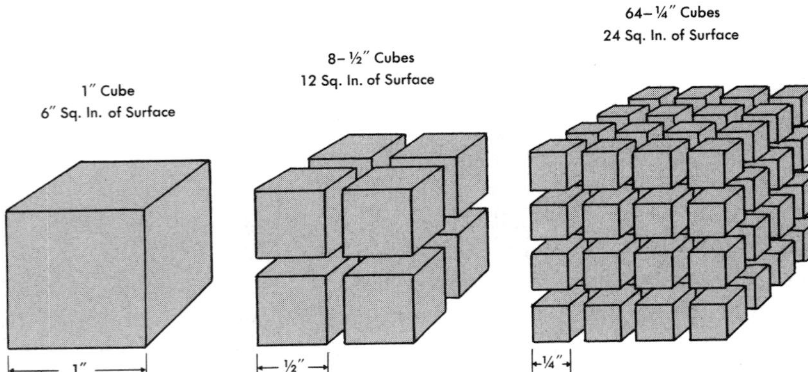

FIGURE 5-10 *Relation of volume, particle size, and surface area. In this illustration, a cube 1 in. square is divided into smaller and smaller units. The volume remains unchanged, but as the particle size decreases, the surface area increases. Because chemical weathering is confined to surfaces, the more finely a given volume of material is divided, the greater is the surface area exposed to chemical activity and the more rapid is the process of chemical weathering.*

FIGURE 5-11 Chemical weathering of limestone, Columbia State Park. Exposed by hydraulic mining. (Photo by Mary Hill.)

directly or indirectly to chemical weathering since their life processes produce oxygen, carbon dioxide, and certain acids that enter into chemical reactions with earth materials.

Carbonation and oxidation

Two examples of chemical weathering, carbonation and oxidation, are especially familiar to many of us.

Carbonation

The carbonate rocks limestone and dolomite are among the most easily weathered rocks on the earth's surface if water is present (Figures 5-11 and 5-12). Carbon dioxide dissolves readily in water and combines to form a weak acid as shown by the following reaction.

$$\underset{\text{gas}}{CO_2} + \underset{\text{water}}{H_2O} \rightleftharpoons \underset{\text{carbonic acid}}{H_2CO_3}$$

FIGURE 5-12 Solution collapse breccia in carbonate beds of the Edwards Formation of Cretaceous age, Texas. Scale in centimeters and inches. (Photo by Earle F. McBride.)

Carbonation is the reaction between carbonic acid and minerals. In the weathering of carbonate rocks, the following reaction occurs.

limestone		carbonic acid		calcium and bicarbonate ions in solution
$CaCO_3$	+	H_2CO_3	$\longrightarrow$	$Ca^{2+} + 2HCO_3^-$

Calcium bicarbonate is about 30 times more soluble than calcium carbonate in water; the carbonation reaction therefore causes rapid dissolution of limestone. The profusion of caverns, underground channels, and disappearing rivers in limestone and dolomite regions attests to the efficiency of chemical weathering by carbonation. Limestone caves may extend for hundreds of meters below the surface. Outstanding examples are Mammoth Cave, Kentucky, Carlsbad Caverns, New Mexico, and the famous Karst Region of Yugoslavia. A *karst landscape* is the term used to describe regions displaying abundant solution features on limestone.

Oxidation

Most small boys and many big men have at one time or another accidentally left their pocketknife in the yard so that rust forms on the blades. Rust quickly appears on unattended iron objects. In tropical climates, even automobiles, ships, and steel bridges are difficult to protect from the ravages of rusting.

Many rocks contain iron-bearing minerals that are oxidized when exposed to the atmosphere. A general equation expressing such a change is as follows.

ferrous iron oxide (gray green)		oxygen		ferric iron oxide (rust colored)
$4FeO$	+	O_2	$\longrightarrow$	$2Fe_2O_3$

Chemical weathering of igneous rocks

In Chapter 4, we found that the most common minerals in igneous rocks are silicates, and that the most important silicates are quartz, the feldspars, and certain ferromagnesian minerals. Let us see how chemical weathering acts on each of these three types of silicate.

Weathering of quartz

Chemical weathering affects quartz very slowly, and for this reason quartz is regarded as a stable mineral. When a rock such as granite, which contains a high percentage of quartz, decomposes, a great deal of unaltered quartz is left behind.

When quartz grains are first set free from the granite, they are sharp and angular. But since even quartz eventually responds to chemical weathering, the grains become more rounded as time passes. After many years of weathering, these grains look as though they had been abraded and worn by the water action along a stream bed or a beach. And yet the change may have come about solely through chemical action.

Weathering of feldspars

In the reaction series on p. 94, we saw that when a magma cools to form an igneous rock like granite, the feldspars crystallized before the quartz. When granite is exposed to weathering at the earth's surface, the feldspars are the first to be broken down. Mineralogists and soil scientists still do not understand the precise process by which feldspars weather, and some of the end products of this action—the clay minerals—offer many puzzles. But the general direction and results of the process (as described in the following text) seem fairly clear.

Aluminum silicate, derived from the chemical breakdown of the original feldspar, combines with water to form hydrous aluminum silicate, which is the basis for another group of silicate minerals, the clays.

The decomposition of the feldspar orthoclase is a good example of the chemical weathering of the feldspar group of silicates. Two substances are essential to the weathering of orthoclase: carbon dioxide and water. The atmosphere contains small amounts of carbon dioxide and the soil contains much greater amounts. Since carbon dioxide is extremely soluble in water, it unites with rainwater and water in the soil to form carbonic acid. Now, when orthoclase comes into contact with water containing carbonic acid, several products are formed as shown in the following equation.

2 parts orthoclase	plus	1 part carbonic acid	plus	1 part water	
$2K(AlSi_3O_8)$	+	H_2CO_3	+	H_2O	$\xrightarrow{\text{yield}}$

clay	plus	1 part potassium carbonate	plus	4 parts silica
$Al_2Si_2O_5(OH)_4$	+	K_2CO_3	+	$4SiO_2$

In this reaction, the hydrogen ion formed from the water forces the potassium out of the orthoclase, disrupting the crystal structure. Then the hydrogen ion combines with the aluminum silicate to form the new clay mineral. (The process by which water combines chemically with other molecules is called *hydrolysis.*) A second product of the original orthoclase crystal is a soluble salt, potassium carbonate, which is formed when the potassium ejected from the orthoclase combines with the carbonate ion of the carbonic acid. The third product, silica, is formed by the silicon and oxygen that remain after the hydrogen has been combined with the aluminum silicate to form the clay mineral.

Now let us examine closely each of the three products of the decomposition of orthoclase. First the clay. At the start, the clay is very finely divided. In fact, it is sometimes of colloidal size, variously estimated as between .2 micron and 1 micron.

Immediately after it is formed, the aluminum silicate may be *amorphous*—that is, the atoms of which it is composed may not be arranged in an orderly pattern. It seems more likely, however, that even at this stage the atoms are arranged according to the definite patterns of a true crystal. In any event, as time passes, the small individual particles join together to form larger particles, which, when analyzed by such means as X rays, exhibit the crystalline pattern of true minerals.

In Chapter 3, we found that the mica minerals are composed of sheets of silicon-oxygen tetrahedra. The clay minerals, too, are built of silicon-oxygen tetrahedra linked together in sheets. These sheets combine in different ways with sheets composed of aluminum atoms and hydroxyl molecules. For this reason, clay minerals are termed *hydrous aluminum silicates.*

Let us look back for a minute to the equation for the decomposition of orthoclase. Notice that the second product is potassium carbonate, which is soluble in water. We might expect that a soluble salt like this would be dissolved and would be carried off by water percolating through the ground, so that all of it would eventually find its way to rivers and finally to the sea. Yet, analyses show that not nearly as much potassium is present in river and ocean water as we would expect. What happens to the rest of the potassium? Some of it is used by growing plants before it can be carried away in solution, and some of it is absorbed by clay minerals or is even built into their crystal structures.

The third product resulting from the decomposition of orthoclase is silica, which appears either in solution (for even silica is slightly soluble in water) or as very finely divided quartz in the size range of the colloids. In the colloidal state, silica may exhibit some of the properties of silica in solution.

So far, we have considered only the weathering of orthoclase feldspar. But the products of the chemical weathering of plagioclase feldspars are much the same. Instead of potassium carbonate, however, either sodium or calcium carbonate is produced, depending on whether the feldspar is the sodic albite or the calcic anorthite (see Table 5-1). As we found in Chapter 3, plagioclase feldspar almost always contains both sodium and calcium. The carbonates of sodium and calcium are soluble in water and may eventually reach the sea. We should note here, however, that calcium carbonate also forms the mineral calcite (see p. 72), which in turn forms the greater part of the sedimentary rock limestone and the metamorphic rock marble, both of which are discussed in more detail in subsequent chapters.

Weathering of ferromagnesians

Now let us turn to the chemical weathering of the third group of the common minerals in igneous rocks—the ferromagnesian silicates. This process yields the same products as the weathering of the feldspars: clay, soluble salts, and finely divided silica. But the presence of iron and magnesium in the ferromagnesian minerals makes possible certain other products as well.

The iron may be incorporated into one of the clay minerals or into an iron carbonate mineral. Usually, however, it unites with oxygen to form hematite, Fe_2O_3, one of the most common iron oxides. Hematite commonly has a deep red color, and in powdered form it is always red; this characteristic gives it its name from the Greek *haimatités,* "bloodlike."

Sometimes the iron unites with oxygen

Table 5-1 Chemical Weathering Products of Common Rock-Forming Silicate Minerals*

	MINERAL	COMPOSITION	IMPORTANT DECOMPOSITION PRODUCTS	
			Minerals	*Others*
QUARTZ		SiO_2	Quartz grains	Some silica in solution
FELDSPARS	Orthoclase	$K(AlSi_3O_8)$	Clay Quartz (finely divided)	Some silica in solution Potassium carbonate (soluble)
FELDSPARS	Albite (sodic plagioclase) Anorthite (calcic plagioclase)	$Na(AlSi_3O_8)$ $Ca(Al_2Si_2O_8)$	Clay Quartz (finely divided) Calcite (from Ca)	Some silica in solution Sodium and calcium carbonates (soluble)
FERROMAGNESIANS	Biotite Augite Hornblende	Fe, Mg, Ca silicates of Al	Clay Calcite Limonite Hematite Quartz (finely divided)	Some silica in solution Carbonates of calcium and magnesium (soluble)
FERROMAGNESIANS	Olivine	$(Fe,Mg)_2SiO_4$	Limonite Hematite Quartz (finely divided)	Some silica in solution Carbonates of iron and magnesium (soluble)

* Based on Brian Mason, *Principles of Geochemistry,* 3rd ed. (New York: John Wiley & Sons, Inc. 1966, p. 48.)

to form another iron oxide, goethite, FeO(OH), which is generally brownish in color. (Goethite was named after the German poet Goethe, in honor of his lively scientific interests.)

Chemical weathering of the ferromagnesian minerals often produces a substance called limonite, which has a yellowish to brownish color and is commonly referred to as *rust*. Limonite is not a true mineral, however, because its composition is not fixed within narrow limits.

Limonite is actually a mixture of two minerals, goethite and lepidocrocite, each having the same chemical composition but different crystal structures. Of the two, goethite is much more abundant.

What happens to the magnesium produced by the weathering of the ferromagnesian minerals? Some of it may be removed in solution as a carbonate, but most of it tends to stay behind as newly formed minerals, particularly in the clays.

Summary of weathering products

If we know the mineral composition of an igneous rock, we can determine in a general way the products that the chemical weathering will probably yield. The chemical weathering products of the common rock-forming minerals are listed in Table 5-1. These include the minerals that constitute most of our sedimentary rocks, and we will discuss them in a subsequent chapter.

Rates of weathering

Some rocks weather very rapidly, others only slowly. This difference in rate of weathering is caused by a variety of factors, including the type of rock and mineral involved, the temperature and amount of moisture present, the topography of the land, and the degree of plant and animal activity.

Rate of mineral weathering

On the basis of field observations and laboratory experiments, the minerals commonly found in igneous rocks can be ranged according to the order in which they are chemically decomposed at the surface. We are not sure of all the details, and different investigators report different conclusions. But we can make the following general observations:

1. Quartz is highly resistant to chemical weathering.
2. The plagioclase feldspars apparently weather more rapidly than orthoclase feldspar.
3. Calcic plagioclase (anorthite) tends to weather more rapidly than sodic plagioclase (albite).
4. Olivine is less resistant than augite, and in many instances augite seems to weather more rapidly than hornblende.
5. Biotite mica weathers more slowly than the other dark minerals, and muscovite mica is more resistant than biotite.

Notice that these points suggest a pattern (Figure 5-13) similar to that of the reaction series for crystallization from magma, which was discussed in Chapter 4. But there is one important difference between these two patterns: in weathering, the successive minerals formed do not react with one another as they do in a continuous reaction series.

The relative resistance of these minerals to decomposition reflects the difference between the surface conditions under which they undergo weathering and the conditions that existed when they were originally formed. Olivine, for example, forms at high temperatures and high pressures early in the crystallization of a melt. Consequently, it is extremely unstable under the low temperatures and low pressures that prevail at the earth's surface, and it weathers quite rapidly. In contrast, quartz forms late in the reaction series, under

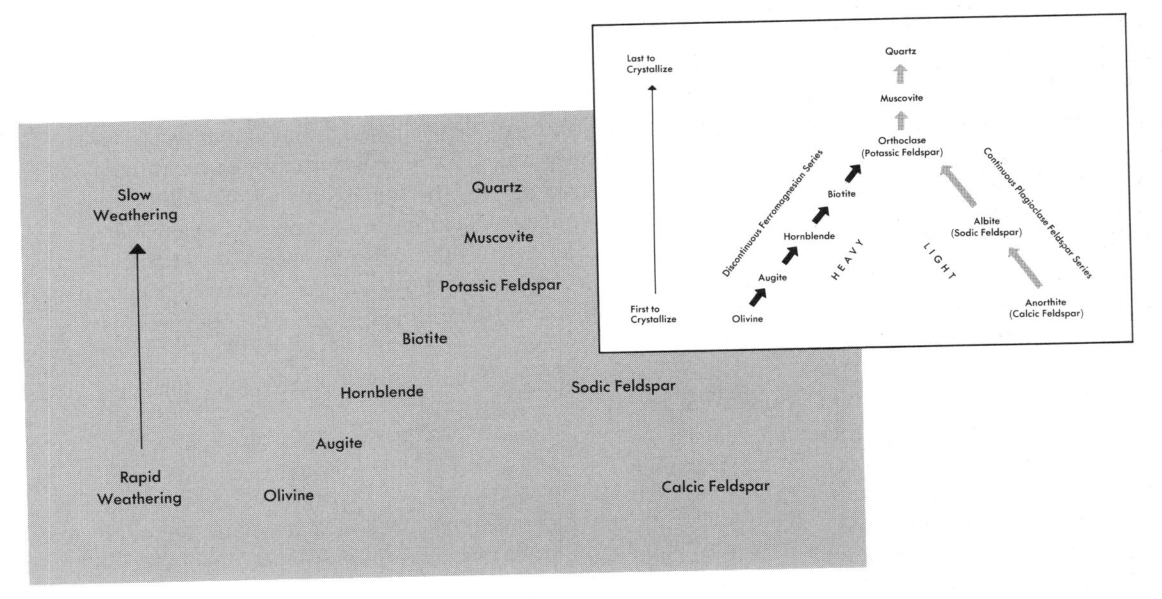

FIGURE 5-13 *Relative rapidity of chemical weathering of the common igneous rock-forming minerals. The rate of weathering is most rapid at the bottom and decreases toward the top. Note that this table is in the same order as Bowen's Reaction Series (inset). The discrepancy in the rate of chemical weathering between, for instance, olivine and quartz, is explained by the fact that in the zone of weathering, olivine is farther from its environment of formation than is quartz. It therefore reacts more rapidly than quartz to its new environment and thus weathers more rapidly.*

considerably lower temperatures and lower pressures. Since these conditions are more similar to those at the surface, quartz is relatively stable and is very resistant to weathering.

Now we can qualify slightly the definition of weathering that we gave at the beginning of this chapter. We have found that weathering disrupts the equilibrium that existed while the minerals were still buried in the earth's crust and that this disruption converts them into new minerals. We may revise our definition as follows:

> *Weathering is the response of materials that were once in equilibrium within the earth's crust to new conditions at or near contact with water, air, and living matter.*

Depth and rapidity of weathering

Most weathering takes place in the upper few meters or tens of meters of the earth's crust where rock is in closest contact with air, water, and organic matter. But some factors operate well below the surface and permit weathering to penetrate to great depths. For instance, when erosion strips away large quantities of material from the surface, the underlying rocks are free to expand. As a result, parting planes or fractures called *joints* develop hundreds of meters below the surface.

Then, too, great quantities of water move through the soil and down into the underground, transforming some of the materials there long before they are ever exposed at the surface. This is what happens to much of the rock salt that is distributed as a sedimentary rock in various places on the earth. If there is enough underground water present, the salt is dissolved and is carried off long before it can be exposed by erosion.

Weathering is sometimes so rapid that it can actually be recorded. The eruption of the volcano Krakatoa in August 1883 threw great quantities of volcanic ash into the air. A deposit of this ash over 30 m deep fell on the nearby island of Long-Eiland. By 1928, 45 years later, a soil nearly 35 cm deep had developed on top of this deposit, and laboratory analyses showed that a significant change had taken place in the original materials. Chemical weathering had removed a measurable amount of the original potassium and sodium. Furthermore, either mechanical weathering or chemical weathering, or both, had broken down the original particles so that they were generally smaller than the particles in the unweathered ash beneath.

Graveyards provide many fine examples of weathering that has taken place within recent historic time. Figure 5-14 shows a pair of marble headstones in a small cemetery near the Welsh-founded community of Remsen in central New York. The calcite that composes the two stones has weathered so rapidly that the inscriptions to the memories of Richard and Robert James are only partially legible after only a century's time has elapsed.

Undoubtedly the rate of weathering of these monuments has increased with time, for two reasons. First, continued weathering roughens the marble surface, exposing more and more of it to chemical attack and increasing the rate of decomposition. Second, the amount of carbon dioxide in the atmosphere has increased with increasing industrialization since 1900. Consequently, rainwater in the twentieth century carries more carbonic acid than it did in the nineteenth century, and thus it attacks calcite more rapidly.

In contrast to the badly weathered marble discussed above, is a headstone of red slate erected in the same cemetery and at about the same time. More than a century later the inscription is still sharp and plainly visible. Slate is usually a metamorphosed shale, which, in turn, is composed largely of clay minerals formed by the weathering of feldspars. Because these clay minerals were originally formed in the zone of weathering, they are relatively stable when exposed to additional chemical weathering. Therefore the slate headstone shows little sign of decomposition because most of the minerals of which it is composed are still in an environment similar to that in which they were originally formed.

These examples show, then, that weathering takes place at a measurable rate, which is often rapid enough to be observed within a single lifetime. Yet, although precise measurements can be made for isolated and specific situations, the factors of weathering are so numerous and so variable that exact rates of weathering are usually difficult, if not impossible, to determine.

FIGURE 5-14 *Weathering of marble headstones in a cemetery near Remsen, New York, has nearly obliterated the original inscriptions. Inscription on left-hand stone: "Richard James/Died/Feb. 6, 1862/Aged/17 yrs. and 2 mos." Inscription on the right-hand stone: "Robert B. James/born/Sept. 4, 1845/died/Jan. 29, 1870." The stones illustrate the instability and rapid weathering of calcite (predominant mineral in marble) in a humid climate. (Photo by Sheldon Judson, 1965.)*

FIGURE 5-15 *Barren, rocky outcrops of the tilted fault-block Tule Mountains, southwestern Arizona. Erosion products are being deposited in the near vicinity of their source. Area is one of low rainfall, high temperatures, and flash floods. (Courtesy National Park Service.)*

FIGURE 5-16 *Remarkable rock spheres produced by weathering. Hammer for scale. (Courtesy U.S. Geological Survey.)*

Differential weathering

Differential weathering is the process by which different sections of rock mass weather at different rates (Figure 5-15). Almost all rock masses of any size weather in this manner. The results vary from the boldly sculptured forms of Bryce Canyon, Utah (see Figure 5-1), to the isolated rounded boulders shown in Figure 5-16. Unequal rates of weathering are caused chiefly by variations in the composition of the rock itself. The more resistant zones stand out as ridges or ribs above the more rapidly weathered rock on either side.

On the other hand, a second cause of differential weathering is simply that the intensity of weathering varies from one section to another in the same rock.

SUMMARY

Weathering is the response of surface or near-surface material to contact with water, air, and living matter. Mechanical weathering (disintegration) is marked by a reduction in the size of fragments but no change in composition. Chemical weathering (decomposition) involves a change in composition of the material weathered.

Chemical weathering is hastened by a reduction in particle size and by an increase in temperature and moisture. Certain minerals weather chemically more rapidly than others: for example, olivine weathers more rapidly than orthoclase, and orthoclase more rapidly than quartz.

Weathering provides the sediment that forms sedimentary rocks. The chemical weathering of an igneous rock containing quartz, feldspar, and ferromagnesian minerals will produce the common minerals of sedimentary rocks—quartz, clay, and the iron oxides—in addition to soluble salts, some of which may be precipitated into minerals such as calcite and halite.

SUGGESTED READINGS

BIRKELAND, PETER W., *Pedology, Weathering and Geomorphology.* New York: Oxford University Press, 1974.

BLOOM, A. L., *The Surface of the Earth.* Englewood Cliffs, N.J.: Prentice-Hall, Inc., 1969.

HUNT, C., *The Geology of Soils.* San Francisco: W. H. Freeman and Co., 1972.

KELLER, WALTER D., *The Principles of Chemical Weathering,* rev. ed. Columbia, Mo.: Lucas Brothers, 1959.

KRAUSKOPF, K. B., *Introduction to Geochemistry.* New York: McGraw-Hill Book Company, 1967.

Sedimentary Rocks

River gravel, lake clay, beach sand—all are common materials. They do not remind us immediately, however, of hard, firm rock. Yet deposits of these substances, or of materials very similar to them, constitute the material from which the great majority of the rocks exposed at the earth's surface were formed.

The chronicle of these rocks, called *sedimentary rocks,* begins with the weathering processes discussed in Chapter 5, because the products of chemical and mechanical weathering provide the basic raw materials. Streams, glaciers, wind, and ocean currents then move the weathered materials to new localities and deposit them as gravel, sand, silt, or clay. The transformation of these sediments into rock is the final step.

Some sediments, particularly sand and gravel, are consolidated into rock by a process in which the individual grains are actually

6

View of well exposed sedimentary beds from Dead Horse Point, overlooking the Colorado River near Moab, southeastern Utah. (Utah Travel Council, Evans Sight and Sounds Productions.)

FIGURE 6-1 *Geological mapping of sedimentary rocks in Utah by helicopter. (Courtesy U.S. Geological Survey.)*

cemented together. Water moving through the open pore spaces leaves behind a mineral precipitate that binds the grains firmly together, giving the deposit the strength we associate with rock. Other sediments, such as clay, are transformed into rock by the weight of overlying deposits, which press or compact them into smaller and smaller spaces.

The sedimentary rock that results from either of these processes may eventually be exposed at the earth's surface (Figure 6-1). If the rock was formed at the bottom of the ocean, it may be exposed either by the slow withdrawal of the seas or by the upward motion of the sea floor forming new areas of dry land.

It is extremely difficult to give a concise, comprehensive definition of sedimentary rocks. The adjective *sedimentary,* from the Latin *sedimentum,* means "settling." So we might expect sedimentary rocks to be formed when individual particles settle from the atmosphere or from a fluid like the water of a lake or an ocean. And many sedimentary rocks are formed in this way. Fragments of minerals derived from the breakdown of rocks are swept into bodies of water where they settle out as unconsolidated sediments, to be hardened later into rocks. But others, such as rock salt, are made up of minerals that have been deposited during the evaporation of large bodies of water. Still other sedimentary rocks are composed largely of the shells and hard parts of animals, particularly of invertebrate marine organisms.

Sedimentary rocks are generally layered, or stratified (bedded). Unlike massive igneous rocks such as granite, most sedimentary rocks

FIGURE 6-2 Rhythmic stratification of micrite and clayey micrite in Cretaceous beds, Mexico. (Photo by Earle F. McBride.)

are deposited in a series of individual beds, one on top of another. The surface of each bed is essentially parallel with the horizon at the time of deposition, and in cross section the beds expose a series of layers. True, some igneous rocks, such as those formed from lava flows, are also layered. By and large, however, the stratification is the most characteristic single feature of sedimentary rocks (Figure 6-2).

About 75 percent of the rocks exposed at the earth's surface are sedimentary rocks, or metamorphic rocks derived from them (see Chapter 7).[1] Yet sedimentary rocks constitute only about 5 percent by volume of the outer 16 km of the globe. The other 95 percent (99 percent by some estimates) of the rocks in this 16-km zone are, or once were, igneous rocks (Figure 6-3). The sedimentary cover is only as thick as a feather edge where it laps around the igneous rocks of the Adirondacks, southern Canada, and the Rockies. In other places, it is thousands of meters thick. In the delta region of the Mississippi River, oil-drilling operations have cut into the crust more than 7,500 m and have encountered only sedimentary rocks. In the Ganges basin of India, the thickness of the sedimentary deposits has been estimated at between 13,500 and 18,000 m.

[1]Estimates of exposed rock types vary. Blatt and Jones recently (1975) estimated the percentages of rock types exposed on land surfaces as follows: sedimentary 66%, metamorphic and Precambrian 17%, intrusive igneous 9%, and extrusive igneous 8%. [From Blatt, Harvey, and Jones, R. L., 1975. Proportions of exposed igneous, metamorphic, and sedimentary rocks: Bull. Geol. Soc. America, v. 86, pp. 1085–1088.

Formation of sedimentary rocks

We found in Chapter 4 that igneous rock hardens from molten material (magma) that originates some place beneath the surface, under the high temperatures and pressures that prevail there. In contrast, sedimentary rocks form at the much lower temperatures and pressures that prevail at or near the earth's surface.

Origin of the material

The material from which sedimentary rocks are fashioned originates in two ways. First, the deposits may be accumulations of minerals and rocks derived either from the erosion of existing rock or from the weathered products of these rocks. Deposits of this type

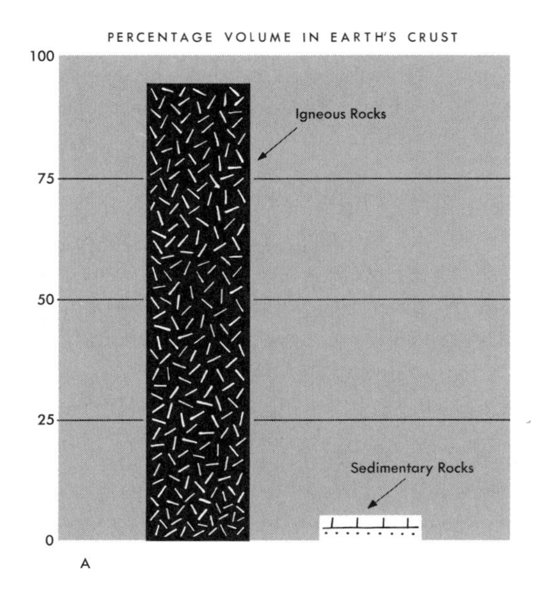

FIGURE 6-3 *Graphs showing relative abundance of sedimentary rocks and igneous rocks. Graph A shows that the great bulk (95 percent) of the outer 16 km of the earth is made up of igneous rocks and that only a small percentage (5 percent) is sedimentary. In contrast, Graph B shows that the areal extent of sedimentary rocks at the earth's surface is three times that of igneous rocks. Metamorphic rocks are considered with either igneous or sedimentary rocks, depending on their origin.*

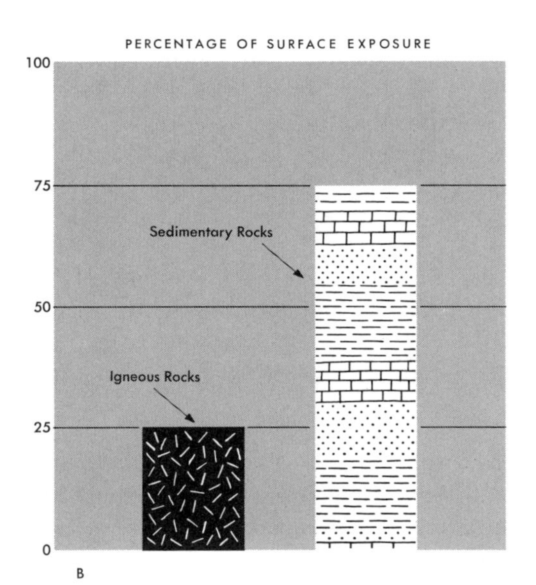

are called *detrital deposits* (from the Latin for "worn down"), and sedimentary rocks formed from them are called *detrital sedimentary rocks.* Second, the deposits can be produced by chemical processes. We refer to these as *chemical deposits* and to the rocks formed from them as *chemical sedimentary rocks.*

Gravel, sand, silt, and clay derived from the weathering and erosion of a land area are examples of detrital sediment. Let us take a specific example. The quartz grains freed by the weathering of a granite can be winnowed out by the running water of a stream and can be swept into the ocean. There the grains settle out as beds of sand, a detrital deposit. Later, when this deposit is cemented to form a hard rock, we have sandstone, a detrital rock.

Chemically formed deposits are usually deposited by the precipitation of material dissolved in water. This process may take place either directly, through inorganic processes, or indirectly, through the intervention of plants or animals. The salt left behind after a salty body of water has evaporated is an example of a deposit laid down by inorganic chemical processes. In contrast, certain organisms such as the corals extract calcium carbonate from sea water and use it to build up skeletons of calcite. When the animals die, their skeletons collect as a biochemical (from the Greek *bios* meaning "life," plus "chemical") deposit, and the rock that subsequently forms is one example of a biochemical rock—in this case, limestone.

Although detrital rocks and chemical rocks are the main divisions of sedimentary rocks based on the origin of material—indeed, as we shall see later, they form the two major divisions in the classification of sedimentary rocks—we commonly find that a chemically formed rock contains a certain amount of detrital material. In similar fashion, predominately detrital rocks include some material that has been chemically deposited.

Sedimentation

The general process by which rock-forming material is laid down is called *sedimentation,* or *deposition.* The factors controlling sedimentation are easy to visualize. To have any deposition at all, there must be something to deposit; that is, a source of sediments must exist. We also need some process to transport and distribute this sediment. And, finally, there must be some place and some process for the deposition of the sedimentary material.

Source of the material

In discussing the rock cycle (Chapter 4), we mentioned that igneous rocks are the ultimate source of the sediments in sedimentary rocks, but that metamorphic rocks or other sedimentary rocks may serve as an immediate source.

In either case, after the rock material has been weathered, it can be transported to some place of accumulation. Its movement is generally from a higher to a lower level. The energy for this movement is provided by gravity, which makes possible not only the process of mass movement itself but also the activity of such agents of transportation as running water and glacier ice. If gravity were free to go about its work without opposition, it would long ago have reduced the continents to smooth, low-lying land masses. But working against the leveling action of gravity are energies within the earth that elevate the continents and portions of the sea floor (see Chapter 9). By constantly exposing new areas of the earth's surface to weathering, movements of this sort insure a continuing supply of material for the formation of sedimentary rocks.

Methods of transportation

Water—in streams and glaciers, underground, and in ocean currents—is the principal means of transporting material from one place to another. Landslides and other movements induced by gravity also play a role, as does the wind, but we shall look more closely at these processes in Chapters 10, 11, and 12.

Processes of sedimentation

Detrital material consisting of minerals and rock fragments is deposited when the agent of transportation no longer has enough energy to move it farther. For example, a stream flowing at a certain velocity possesses enough energy to move particles up to a certain maximum size. If the stream's velocity decreases, it also loses energy, and it is unable to transport all the material that it has been carrying at the higher velocity. The solid particles, beginning with the heaviest, start to settle to the bottom. The effect is much the same when a wind that has been driving sand across a desert suddenly dies. A loss of energy accompanies the loss in velocity.

Material that has been carried in solution is deposited differently, that is, by *precipitation,* a chemical process through which dissolved material is converted into a solid and is separated from the liquid solvent. As we have noted, the precipitant may be either biochemical or inorganic.

Although at first glance the whole process of sedimentation seems simple, actually it is quite complex. Many factors are involved, and they can interact in many ways. Consequently, the manner in which sedimentation takes place and the sediments that result from it differ greatly from one situation to another.

Mineral composition of sedimentary rocks

Sedimentary rocks, like igneous and metamorphic rocks, are accumulations of minerals. In sedimentary rocks, the most common minerals are quartz, feldspars, calcite, and clay minerals.

Rarely is a sedimentary rock composed

of only a single mineral, although one mineral may predominate. The grains of many sandstones are predominantly quartz, but the cementing material that holds these grains together may be quartz, calcite, or dolomite. In general, most sedimentary rocks are mixtures of two or more minerals.

Clay

Chapter 5 described how clay minerals develop from the weathering of silicate minerals, particularly the feldspars. These clays subsequently can be incorporated into sedimentary rocks; they may, for example, form an important constituent of claystone and siltstone.

Silica

Another important component of sedimentary rock is silica (SiO_2), including the common mineral quartz as well as a number of other forms such as chert, flint, opal, jasper, chalcedony, and agate.

The mechanical and chemical weathering of an igneous rock such as granite sets free individual grains of quartz that eventually may be incorporated into sediments. These quartz grains produce the detrital forms of silica and account for most of the volume of the sedimentary rock sandstone. Silica in solution or in particles of colloidal size is also produced by the weathering of igneous rock. This silica acts as a cementing agent in certain coarse-grained sedimentary rocks. Silica may also be precipitated in other forms, such as *opal,* generally regarded as a hydrous silica ($SiO_2 \cdot nH_2O$). Opal is slightly softer than quartz and has no true crystal structure.

Silica also is present in sedimentary rocks in a form called *cryptocrystalline.* This term (from the Greek *kryptos,* "hidden," plus crystalline) indicates that the crystalline structure of this type of silica is so fine that it cannot be seen under most ordinary microscopes. The microscope does reveal, however, that some cryptocrystalline silica has a granular pattern and that some has a fibrous pattern. To the naked eye, the surface of the granular form is somewhat duller than that of the fibrous form. Among the dull-surfaced or granular varieties is *flint,* which is usually dark in color. *Chert* is similar to flint but tends to be lighter in color, and *jasper* is a red variety of the granular cryptocrystalline form of silica.

The general term *chalcedony* is often applied to the fibrous types of cryptocrystalline silica, which have a higher, more waxy luster than the granular varieties. Sometimes the term is also used to describe a specific variety of brown translucent cryptocrystalline silica.

Agate, the most popular cryptocrystalline variety of quartz, is varicolored (white, brown, black, red, blue, green), consisting of chalcedony in which the color is irregularly distributed. The colors frequently are arranged in parallel bands that may be wavy. An attractive agate, cut and polished and set in a silver ring, is inexpensive but striking in appearance.

Calcite

The chief constituent of the sedimentary rock limestone, *calcite* ($CaCO_3$), is also the most common cementing material in the detrital sedimentary rocks. The calcium is derived from igneous rocks that contain calcium-bearing minerals, such as calcic plagioclase and some of the ferromagnesian minerals. Calcium is carried from the zone of weathering as calcium bicarbonate, $Ca(HCO_3)_2$, and is eventually precipitated as calcite, $CaCO_3$, through the intervention of plants, animals, or inorganic processes. The carbonate is derived from water and carbon dioxide.

Other materials in sedimentary rocks

Accumulations of clay, quartz, and calcite, either alone or in combination, account for all but a small percentage of the sedimentary rocks. But certain other materials are present in quantities large enough to form distinct beds. The mineral *dolomite,* $CaMg(CO_3)_2$, for example, usually is intimately associated with calcite, but is far less abundant. When the mineral constitutes 50 percent or more of a rock, the rock itself is also known as dolomite.

The mineral dolomite is easily confused with calcite, and since these two often are present together, distinguishing them is important. We can do so through this chemical test: calcite effervesces freely in dilute hydrochloric acid; dolomite effervesces very slowly or not at all, unless it is finely ground or powdered. In the latter instance, the more rapid chemical activity results from the increase in surface area, an example of the general principle discussed under Chemical Weathering in Chapter 5.

Feldspars and micas are abundant in some sedimentary rocks. In Chapter 5, we found that chemical weathering converts these minerals into new minerals relatively rapidly. Therefore, when we find mica and feldspar in a sedimentary rock, chances are that it was mechanical rather than chemical weathering that originally made them available for incorporation in the rock.

Iron produced by chemical weathering of the ferromagnesian minerals in igneous rocks can be caught up again in new minerals and can be incorporated into sedimentary deposits. The iron oxide minerals that are present most frequently in sedimentary rocks are hematite, magnetite, and goethite. In some deposits, these minerals predominate, but more commonly they act simply as coloring matter or as a cementing material.

Halite (NaCl) and gypsum ($CaSO_4 \cdot 2H_2O$) are minerals precipitated from solution by evaporation of the water in which they were dissolved; they are referred to as *evaporites.* The salinity of the water—that is, the proportion of the dissolved material to the water—determines the type of mineral that will precipitate. The gypsum begins to separate from sea water when the salinity (at 30°C) reaches a little over three times its normal value. Then, when the salinity of the sea water has increased to about ten times its normal value, halite begins to precipitate.

Volcanic ash and cinders, mentioned in Chapter 4, are sedimentary rocks composed mostly of fragments blown from volcanoes. The fragments can be large pieces that have fallen close to the volcano, or extremely fine ash that has been carried by the wind and deposited hundreds of kilometers away.

Finally, organic matter may constitute or may be present in sedimentary rocks. In the sedimentary rock known as *coal,* plant materials are almost the only components. More commonly, however, organic matter is sparsely disseminated throughout sedimentary deposits and the resulting rocks.

Texture

Texture is the general physical appearance of a rock—deriving from the size, shape, and arrangement of the particles that compose it. There are two major textures in sedimentary rocks: clastic and nonclastic.

Clastic texture

Clastic is derived from the Greek for "broken" or "fragmental," and rocks that are formed from deposits of mineral and rock fragments have clastic textures. The size and shape of the original particles have a direct influence on the nature of the resulting texture. A rock formed from a bed of gravel and sand has a coarse, rubblelike texture that is very different from the sugary texture of a rock developed from a deposit of rounded, uniform sand grains. Furthermore, the process by which a sediment is deposited also affects the texture of the sedimentary rock that develops from it. Thus, the debris dumped by a glacier is composed of a jumbled assortment of rock material ranging from particles of colloidal size to large boulders. A rock that develops from such a deposit has a very different texture from one that develops from, for instance, a deposit of windblown sand in which all the particles are approximately 0.15 to 0.30 mm in diameter.

Chemical sedimentary rocks may sometimes show a clastic texture. For example, a rock made up predominantly of shell fragments from a biochemical deposit has a clastic texture that is as recognizable as that of a rock formed from sand deposits.

One of the most useful factors in classifying sedimentary rocks is the size of the individual particles. In practice, we express the size of a particle in terms of its diameter rather than in terms of volume, weight, or surface area. When we speak of "diameter," we imply that the particle is a sphere, but few if any fragments in sedimentary rocks are true spheres. In geological measurements, use of the term "diameter" means the diameter that an irregularly shaped particle would have if it were a sphere of equivalent volume. It would be a time-consuming and impractical task to determine the volume of each sand grain in a rock, and then to convert these measurements into appropriate diameters. So the diameters we use for particles are only approximations of their actual sizes. They are accurate enough, however, for our needs.

Several scales have been proposed to describe particles that range in size from boulders to minerals of microscopic dimensions. The Udden-Wentworth scale, presented in Table 6-1, is used almost universally by American and Canadian geologists. Notice that although the term *clay* is used in the table to designate all particles below $\frac{1}{256}$ mm in diameter, the same term is also used to describe certain minerals. To avoid confusion, geologists refer specifically to either "clay size" or "clay mineral," unless the meaning is clear in context.

Since determining the size distribution of particles calls for the use of special equipment, the procedure is normally carried out only in the laboratory. In examining specimens in the field, geologists make an educated guess of the main size, based on prepared standards and on their own experience.

Nonclastic texture

Some, but not all, sedimentary rocks formed by chemical processes have a nonclastic texture in which the grains are interlocked. These rocks have somewhat the same appearance as igneous rocks with crystalline texture. Most of the sedimentary rocks with nonclastic texture do have a crystalline structure, although a few of them, such as opal, do not.

Table 6-1 Udden-Wentworth Scale of Particle Sizes for Clastic Sediments

UDDEN-WENTWORTH GRADE SCALE		APPROXIMATE EQUIVALENT
Size	*Fragment*	
	Boulder	
256 mm		10 in.
	Cobble	
64 mm		$2\frac{1}{2}$ in.
	Pebble	
4 mm		$\frac{5}{32}$ in.
	Granule	
2 mm		$\frac{5}{64}$ in.
	Sand	
$\frac{1}{16}$ mm		.0025 in.
	Silt (DUST)	
$\frac{1}{256}$ mm		.00015 in.
	Clay (DUST)	

SOURCE: Modified after C. K. Wentworth, "A Scale of Grade and Class Terms for Clastic Sediments," *Journal of Geology, 30* (1922), 381.

The mineral crystals that precipitate from an aqueous solution are usually very small in size. Because the fluid in which they form has a very low density, they usually settle out rapidly and accumulate on the bottom as mud. Eventually, under the weight of additional sediments, the mud is compacted more and more. Now the size of the individual crystals may begin to increase. Their growth may be induced by added pressure causing the favorably oriented grains to grow at the expense of less favorably oriented neighboring grains. Or crystals may grow as more and more mineral matter is added to them from the saturated solutions trapped in the original mud. In any event, the resulting rock is composed of interlocking crystals. Depending on the size of the crystals, we refer to these nonclastic textures as fine-grained, medium-grained, or coarse-grained. A coarse-grained texture has grains larger than 5 mm in diameter, a fine-grained one, less than 1 mm in diameter.

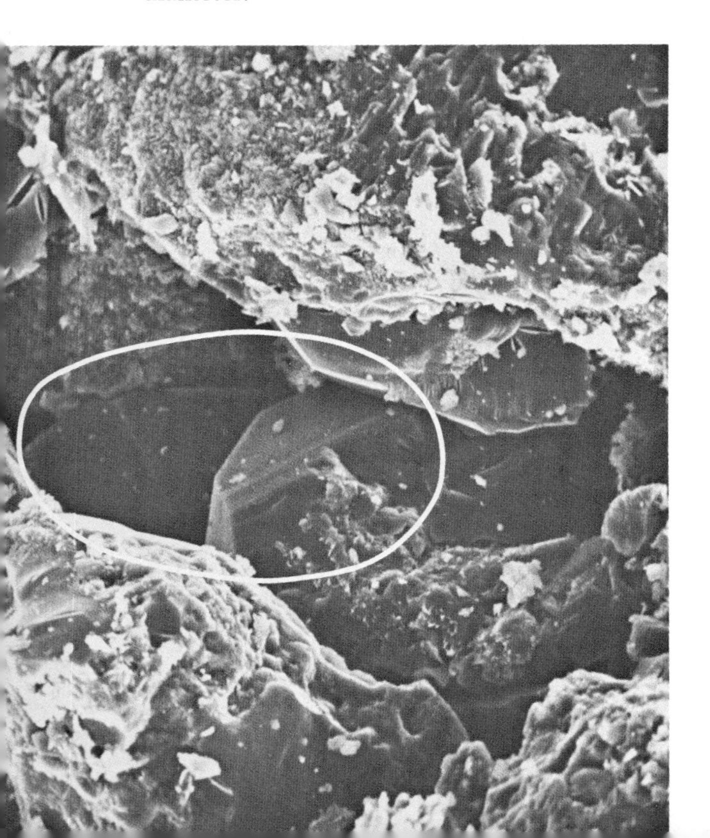

Lithification

The process of lithification converts unconsolidated rock-forming material into consolidated rock. The term is derived from the Greek for "rock" and the Latin for "to make." In the following section, the various ways in which sediment is lithified are discussed.

Cementation

In cementation, the spaces between the individual particles of an unconsolidated deposit are filled by some binding agent. Of the many minerals that serve as cementing agents, the most common are calcite, dolomite, silica, and hematite.

The cementing material is carried in solution by surface water and by deep formation water that percolates down through the open spaces between the particles of the deposit. Precipitation from the water of the primary cement results in the joining together of grains and particles by patches of cement (Figure 6-4).

In the finest-grained sedimentary rocks, such as claystone and limey mudstone, the original tiny clay or calcium carbonate particles recrystallize into a dense, interlocking network of small crystals. Siltstones, which are intermediate in grain size between claystone and sandstone, generally are cemented by the chemical cements (calcite, dolomite, silica) much in the manner of sandstone cementation.

FIGURE 6-4 *Scanning electron-microscope photograph of Miocene sandstone in Louisiana showing pore space (about 15 percent) with quartz crystal cement (white outline) in pores reducing porosity. (Courtesy Core Laboratories, Inc.)*

Table 6-2 Classification of Sedimentary Rocks

ORIGIN	TEXTURE	PARTICLE SIZE OR COMPOSITION	ROCK NAME
Detrital	Clastic	Granule or larger	Conglomerate, Breccia, Tillite
		Sand	Sandstone
		Silt and clay	Siltstone and Claystone (Shale)
CHEMICAL: Inorganic	Clastic and Nonclastic	Calcite, $CaCO_3$	Limestone
		Dolomite, $CaMg(CO_3)_2$	Dolomite
		Halite, NaCl	Salt
		Gypsum, $CaSO_4 \cdot 2H_2O$	Gypsum
CHEMICAL: Biochemical		Calcite, $CaCO_3$	Limestone
		Plant remains	Coal

Compaction

In compaction, the pore space between individual grains is gradually reduced by the pressure of overlying sediments or by the pressures resulting from earth movements. Coarse deposits of sand and gravel undergo some compaction, but fine-grained deposits of silt, clay, and carbonate mud ($CaCO_3$) respond much more readily. As individual particles are pressed closer and closer together, the thickness of the deposit is reduced, porosity is decreased, water is expelled, and coherence is increased. It has been estimated that deposits of clay-sized particles, buried to depths of 900 m, have been compacted to about 60 percent of their original volume.

Chemical changes

As sediment is buried, it is subjected to increasingly high temperatures (1°C for each 30 m) and high pressures (1 atmosphere for each 4.4 m of depth). The increased temperature and pressure lead to more rapid reactions between contained fluids (mostly water) in the pore spaces and minerals in the sediment. New minerals are precipitated, or there may be additions to existing minerals. Other changes that take place include recrystallization of unstable minerals and dissolution of the more insoluble minerals. All chemical processes that take place in sedimentary rocks after deposition are termed *diagenesis.*

Classification of sedimentary rocks

Having examined some of the factors involved in the formation of sedimentary rocks, we can consider a classification for this rock family. The classification presented in Table 6-2 represents only one of several possible schemes. Notice that there are two main groups—detrital and chemical—based on the origin of the rocks, and that the chemical category is further divided into inorganic and biochemical. All the detrital rocks have clastic texture, whereas the chemical rocks have either clastic or nonclastic texture. We use particle size to subdivide the detrital rocks, and composition to subdivide the chemical rocks.

Detrital sedimentary rocks

Conglomerate

A *conglomerate* is a detrital rock composed of more or less rounded fragments, 35 percent or more of which are of granule size (2 to 4 mm in diameter) or larger (Figures 6-5, 6-6,

FIGURE 6-5 Imbricated pebbles in recent conglomerate and conglomeratic sandstone deposit. Flow of depositing stream was from left to right. (Photo by Picard and High.)

and 6-7). If the fragments are more angular than rounded, the rock is called *breccia.* Another type of conglomerate is *tillite,* a rock formed from deposits laid down directly by glacier ice. The large particles in a conglomerate are usually rock fragments, and the finer particles are usually minerals derived from preexisting rocks.

Sandstone

Sandstone is composed of 50 percent or more of mineral grains the size of sand (between $\frac{1}{16}$ mm and 2 mm in diameter). Sandstone is thus intermediate between coarse-grained conglomerate and fine-grained siltstone and claystone. Since the size of the grains varies from one sandstone to another, we speak of coarse-grained, medium-grained, and fine-grained sandstone according to the

FIGURE 6-6 Alternating beds of sandstone and conglomerate dip inland from the sea cliff at Lobos State Park, near Carmel, California. (Photo by Sheldon Judson.)

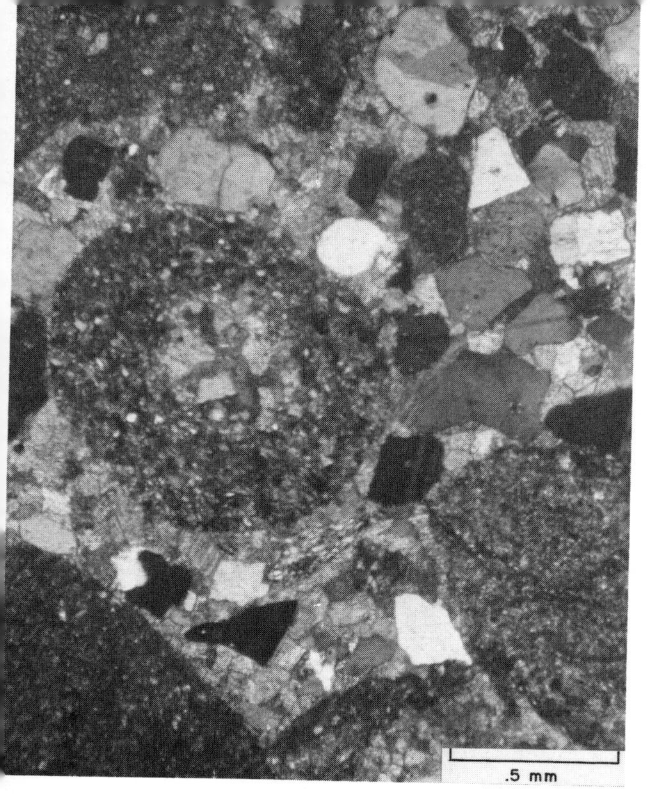

FIGURE 6-7 *Photomicrograph of limestone-pebble conglomerate. Large particles (>0.5 mm) are limestone; smaller particles (up to about 0.3 mm) are quartz and feldspar. Cement is calcite. [From McCormick, C. D., and M. D. Picard, 1969, Petrology of Garta Formation (Triassic), Uinto Mountain area, Utah and Colorado: Journal of Sedimentary Petrology, v. 39, p. 1484–1508.]*

dominant grain size in the Udden-Wentworth scale.

A sandstone is characterized by four components: grains, matrix, cement, and pore space (porosity). The grains are sand-sized minerals or rocks that constitute the framework of the sandstone. *Matrix* is defined as detrital material less than $\frac{1}{16}$ mm in diameter. Mineral composition of the matrix is similar to that of the grains, but fine-grained constituents (clay minerals and micas) are more abundant in the matrix than in the grain framework. The common cements were discussed previously. The pore space (percent of voids) is important for the fluids it contains, generally water; but in favorable locations, crude oil, natural gas, and ore minerals are also contained. In common sandstone, the matrix, cement, and pore space together equal between 25 and 35 percent of the total volume. The grains therefore constitute between 65 to 75 percent of the sandstone.

The words *quartzarenite* or *orthoquartzite* are used for sandstone that contains at least 90 percent quartz and chert grains (Figure 6-8). If the mineral grains in a sandstone are dominantly quartz and feldspar, and feldspar attains 25 percent, the French word *arkose* is applied to the sandstone. The term *lithic arenite* is applied to sandstone rich in rock fragments (more rock fragments than feldspar). Another variety of sandstone, called *graywacke,* is gray-green, poorly sorted "dirty" sandstone that was deposited in thick, cyclic sequences.

FIGURE 6-8 *Scanning electron-microscope photograph of productive quartzarenite in Tensleep Formation of Pennsylvanian age, Wyoming. Cemented by quartz, dolomite, and kaolinite. About 15 percent porosity. (Courtesy Core Laboratories, Inc.)*

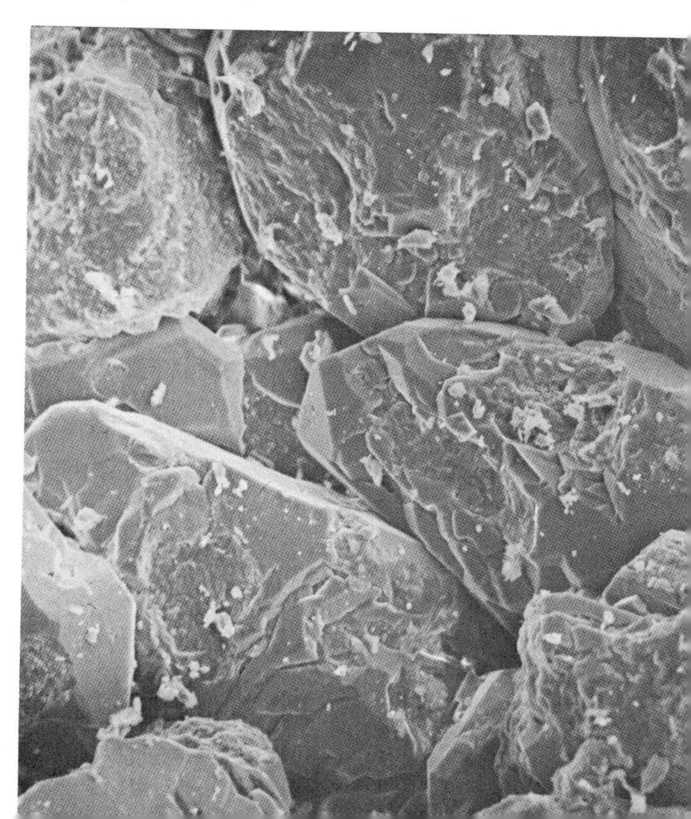

Siltstone and claystone

Fine-grained detrital sedimentary rocks are the most abundant rock on the earth's surface. Despite their abundance, they are the least known of the common sedimentary rocks because of their fine-grained size, their poor exposure in outcrops, and their slight economic value compared with sandstone and limestone.

Texture is the most significant property used in naming these rocks. The finest-grained rock of this group is termed *claystone,* which indicates the rock contains more than 50 percent material (grains) finer than $\frac{1}{256}$ mm in diameter. *Siltstone* consists of more than 50 percent of material between $\frac{1}{256}$ mm and $\frac{1}{16}$ mm. The general name *mudstone* is given to rocks composed of a mixture of clay-, silt- and sand-sized grains, none of which equals 50 percent of the rock.

Although the particles in the fine-grained detrital rocks are small, they can be identified and studied. Quartz and feldspars are the dominant minerals in most siltstone, and siltstone can be studied with the petrographic microscope and by X-ray diffraction techniques. The mineral composition of claystone can also be determined but with more difficulty than for siltstone. Clay minerals constitute about two-thirds and quartz plus feldspars about one-third of the minerals in many claystones.

The word *shale* is applied to all of the fine-grained detrital rocks by many geologists. In its broadest sense, shale thus includes claystone, siltstone, and mudstone. Shale is further restricted by some to laminated clayey rocks whose precise distribution of clay, silt, and sand is unknown.

Chemical rocks

The most abundant chemical rocks are carbonates and evaporites. Carbonate rocks are very important economically because nearly 50 percent of the petroleum and natural gas found in the world is present in pore space of carbonate rocks. Evaporites are closely associated in many basins of deposition with carbonates, and are especially common in areas where hydrocarbons are found. In addition, carbonate rocks commonly are reservoirs for groundwater, and also are hosts for many ore deposits.

Limestone

Limestone is a carbonate rock that is composed chiefly of the mineral calcite $CaCO_3$, which has been deposited by either inorganic or organic chemical processes. Most limestone is a mixture of sand-sized grains ($\frac{1}{16}$ to 2 mm) of calcium carbonate and carbonate mud (crystal size less than 10 microns). About 10 percent of the exposed sedimentary record is limestone (Figure 6-9).

Clay-size carbonate mud (*micrite*) is the most common constituent in carbonate rocks (Figure 6-2). The individual crystals in ancient

FIGURE 6-9 *Thinly bedded limestone in Hastings County, Ontario, Canada. The beds have been tilted to a vertical position from the original horizontal position. (Photo by Canadian Geological Survey.)*

rocks generally are less than 5 microns in diameter and are commonly calcite. The origin of so much fine-grained carbonate is heatedly debated, and it is not known if micrite is mainly produced by abrasion, by inorganic precipitation, or by algae.

Biochemically formed limestone is created by the action of plants and animals that extract calcium carbonate from the water in which they live. The calcium may be either incorporated into the skeleton of the organism or precipitated directly. In any event, when the organism dies, it leaves behind a quantity of calcium carbonate, and over a long period of time thick deposits are formed. Reefs, ancient and modern, are well-known examples of such accumulations. The most important builders of modern reefs are algae, molluscs, corals, and one-celled animals. The ancestors of the same animals built up the reefs of ancient seas—reefs, now old and deeply buried, that are often valuable reservoirs of petroleum.

The importance of organisms in producing carbonate sediment and rocks is very great. As much as 90 percent of the world's carbonate rocks may have been formed by living organisms. Studies of modern carbonate environments indicate that the prolific production of the necessary organisms takes place in extremely shallow water, generally less than 15 m of sea water (Figure 6-10).

Inorganically formed limestone is composed of calcite that has been precipitated from solution by inorganic processes. Some calcite is precipitated from the fresh water of streams, springs, and caves, although the total amount of rock formed in this way is negligible. When calcium-bearing rocks undergo chemical weathering, calcium bicarbonate, $Ca(HCO_3)_2$, is produced in solution. If enough of the water evaporates, or if the temperature rises, or if the pressure falls, calcite is precipitated from this solution. For example, most *dripstone*, or *travertine*, is formed in caves by the evaporation of water that has carried calcium carbonate in solution.

FIGURE 6-10 *Placer deposit of Donax variabilis on Padre Island beach, Texas. (Photo by Earle F. McBride.)*

And *tufa* (from the Italian for "soft rock") is a spongy, porous limestone formed by the precipitation of calcite from the water of streams and springs.

Dolomite

The word *dolomite* refers both to the mineral dolomite, $CaMg(CO_3)_2$, and to the carbonate rock that contains 50 percent or more of the mineral. Most dolomite rocks are probably formed from limestone by a partial replacement of Ca in a carbonate by Mg ions.

Evaporites

An *evaporite* is a sedimentary rock composed of minerals that were precipitated from a concentrated solution during evaporation.

FIGURE 6-11 Large gypsum crystals (cm divisions on hammer) in "Fer-de-Lance" beds, Sicily. (Photo by Earle F. McBride.)

Rock salt (composed of the mineral halite, NaCl), and *gypsum* (composed of the mineral of the same name, $CaSO_4 \cdot 2H_2O$), are the most abundant evaporites (Figure 6-11). *Anhydrite* (from the Greek, *anydros,* "waterless") is an evaporite composed of the mineral of the same name, which is gypsum without the water, $CaSO_4$.

Most evaporite deposits are precipitated from sea or lake water according to a definite sequence. The less soluble minerals are the first to precipitate. Thus, gypsum and anhydrite, both less soluble than halite, are deposited first. Then, as evaporation progresses, the more soluble halite is precipitated.

In the United States, extensive evaporite deposits are found in Utah, New Mexico, and Michigan. Anhydrite, halite, and bedded bittern salts (KCl) reach a thickness of at least 1,300 m in the Paradox Basin of Utah. In southeastern New Mexico, a similar sequence of evaporites has a maximum thickness of about 1,500 m; and in central Michigan there are layers of gypsum, anhydrite, and halite that reach 1,000 m in thickness.

Some evaporite deposits are mined for their mineral content, and in certain areas, particularly in the Gulf Coast states, deposits of rock salt have pushed upward toward the surface to form salt domes containing commercially important reservoirs of petroleum (Chapter 22).

Coal

Coal is a sedimentary rock rich in carbon derived from ancient, swamp-dwelling plants. Coal deposits were formed in areas of low relief where there was poor surface drainage, warm temperatures, and abundant rainfall. The pressure of successive layers of sediment compacts the plant layers and removes much of their water and volatile substances. As these are removed, the plant material is transformed into peat, lignite, bituminous coal, and anthracite coal, a progressive increase in relative carbon content. In general, coal's heat value increases as the carbon content increases. The Florida Everglades display conditions much like those present during the ancient coal-forming environments.

Relative abundance of sedimentary rocks

Siltstone and claystone (shale), sandstone, and limestone constitute about 95 percent of the total sedimentary part of the earth's crust. Of these, the fine-grained clastic rocks are by far the most abundant. Siltstone and claystone account for about 70 percent, sandstone represents 20 percent, and limestone (plus dolomite) about 10 percent.

Features of sedimentary rocks

We have mentioned that the stratification, or bedding, of sedimentary rocks is their most characteristic feature. Now we shall look more closely at this quality, along with certain other characteristics of sedimentary rocks, including sedimentary structures, fossils, and color.

Bedding

The beds or layers of sedimentary rocks are separated by parallel bedding planes, along which the rocks tend to separate or break. Bedding results from the fact that sediment accumulates in thin sheets of grains. Each sheet seldom exceeds several centimeters in thickness, but is characterized by small differences in grain size and arrangement that distinguish it from adjacent sheets.

The varying thickness of the beds in a sedimentary sequence reflects the changing environmental conditions that prevailed dur-

FIGURE 6-12 *Horizontal bedding in longitudinal stream bar. Length of scale is 15 cm. (Photo by Picard and High.)*

ing deposition of each bed. A single stream in flood or a lake or ocean current generally transports and deposits only a single bed or a thin sequence of related beds. Later deposition adds to the preceding beds, and a thickness of several hundred kilometers or more may accumulate at a single depositional site.

The most common type of bedding in sedimentary rocks is termed *horizontal bedding* (also, planar or parallel bedding) because each individual layer is parallel with the others and each group of layers is approximately horizontal (Figure 6-12). Horizontal bedding is formed under a variety of physical conditions.

The second most abundant bedding type, termed *crossbedding* (Figures 6-13 and 6-14), reveals important information about conditions of deposition and is useful in the reconstruction of ancient conditions. Crossbedding originates when the fluid motions of sediment deposition cause wavelike undulations on the surface of the sediment. The wave forms are termed *ripples* or *dunes,* depending on size, and the crossbedding is produced by sediment avalanching down the lee slope of the ripples or dunes. Avalanching takes place in response to oversteepening produced by deposition of the sediment load at the brink of the ripple or dune. The resulting structure is that of beds inclined at an angle to the boundaries of the whole group.

Crossbedding is also formed by the building of small deltas or bars into local depressions on stream or ocean floors. Several other mechanisms for the formation of crossbedding are also known.

Careful study in the field and laboratory of crossbedding has led to its use in a variety of ways. Certain types of crossbedding, for example, are now recognized as the products of wind (*eolian*) deposition. Other, quite different crossbedding is produced by stream currents.

FIGURE 6-13 *Some sedimentary beds are deposited at a noticeable angle from the horizontal, as were the sandstone layers exposed in this cliff face. They record ancient sand dunes in what is now Zion National Park, Utah. (Photo by Hillers, U.S. Geological Survey.)*

FIGURE 6-14 *Crossbedding in recent deposit. Stream current was from left to right. Fifteen-cm ruler for scale. (Photo by Picard and High.)*

Crossbedding is also much used to determine the direction of flow of the current that formed the bedding. The direction of maximum inclination of the crossbedding or the direction of inclination of the trough that contains the beds is the direction of current movement. Such studies, termed *paleocurrent analysis,* may help the geologist to determine where the sediment source areas (highlands) were located, as well as determining the orientation of ancient shorelines of seas or lakes. This information, in turn, can be used to locate oil and gas fields and ore deposits that frequently are preferentially oriented with respect to directions of stream flow or orientations of ancient shorelines.

A third type of bedding, *graded bedding* (Figure 6-15), also gives important information about geological conditions. Graded beds show a progressive decrease in particles size upward through each bed. The formation of such beds takes place when dense, sediment-ladden currents enter relatively still waters in lakes or oceans. The current velocity is decreased when this takes place, and the current deposits first larger and denser grains and then finer and less dense grains in a gradational sequence. Such debris-laden currents are called *turbidity currents.* Graded beds are widespread, occurring in all depositional settings from deep marine through eolian.

FIGURE 6-15 *Graded bedding from the Martinsburg Formation of Ordovician age, Pennsylvania. (Photo by Earle F. McBride.)*

FIGURE 6-16 Recent near-shore ripple marks from the Great Salt Lake. Lake current from left to right. (Photo by M. Dane Picard.)

Structures on surfaces of beds

Ripple marks. Ripple marks (Figure 6-16) result from the action of moving water or air on loose sediment. Under low velocities, the grains develop small wavelike features that generally are asymmetrical in form. If the small ripples in the sediment are preserved after lithification, they are termed *ancient* ripple marks. Asymmetrical ripple marks are characterized by stoss (gentle) and lee (steep) sides and by internal inclined layers (*laminae*) that are inclined in the direction of current flow. Ripple marks formed by standing or oscillatory waves have layers (laminae) that are "draped" over the ripple, parallel with the external form.

For paleocurrent studies utilizing ripple marks, the internal structure of the ripple mark should be examined. Ripples formed by currents or waves of translation are characterized by inclined layers (*foreset laminations*) in

FIGURE 6-17 Three- and four-sided mud cracks along ephemeral stream. (Photo by Picard and High.)

the down-current direction (direction of ripple migration). The external form of the ripple mark is not always indicative of the direction of flow.

Mud cracks. Mud cracks (Figure 6-17) result from a reduction in volume as fine-grained sediment dries if exposed to the air, or compacts if under standing water. It has been suggested that ideally in homogeneous material, cracks radiate from evenly spaced points at 60° angles, and the intersections of the cracks form six-sided polygons. However, in drying mud layers the ideal is rarely achieved, and most angles approximate 90°, resulting in four-sided polygons.

Raindrop impressions. Raindrop impressions are small, hemispherical impact pits left by rain falling on sediment, especially on mud or sand (Figures 6-18 and 6-19). Many of the pits have rims that project above the surrounding surface. The diameters of the pits vary. Raindrop impressions are developed in sets and individual depressions may overlap. This sedimentary structure forms soon after exposure of the sediment before it has dried and developed a hard surface rind. Preservation of raindrop impressions in rocks is unlikely, but some have been found as seen in Figure 6-18.

Nodules, concretions, and geodes

Many sedimentary rocks contain structures that were formed *after* the original sediment was deposited. Among these are nodules, concretions, and geodes.

A *nodule* is an irregular, knobby-surfaced body of mineral matter that differs in composition from the sedimentary rock in which it has formed. It usually lies parallel with the bedding planes of the enclosing rock, and sometimes adjoining nodules coalesce to form a continuous bed or network. Nodules average about $\frac{1}{3}$ m in maximum dimension. Silica, in the form of chert or flint, is the major component. Most nodules apparently formed when silica replaced some of the materials of the original deposit; others, however, may consist of silica that was deposited at the same time as the containing beds were laid down.

FIGURE 6-18 *Raindrop impressions on channel floor of ephemeral stream. Length of scale is 15 cm. Photograph on the right is of raindrop impressions in a Pennsylvanian sandstone. Specimen is 4 cm wide. (Photos by Picard and High.)*

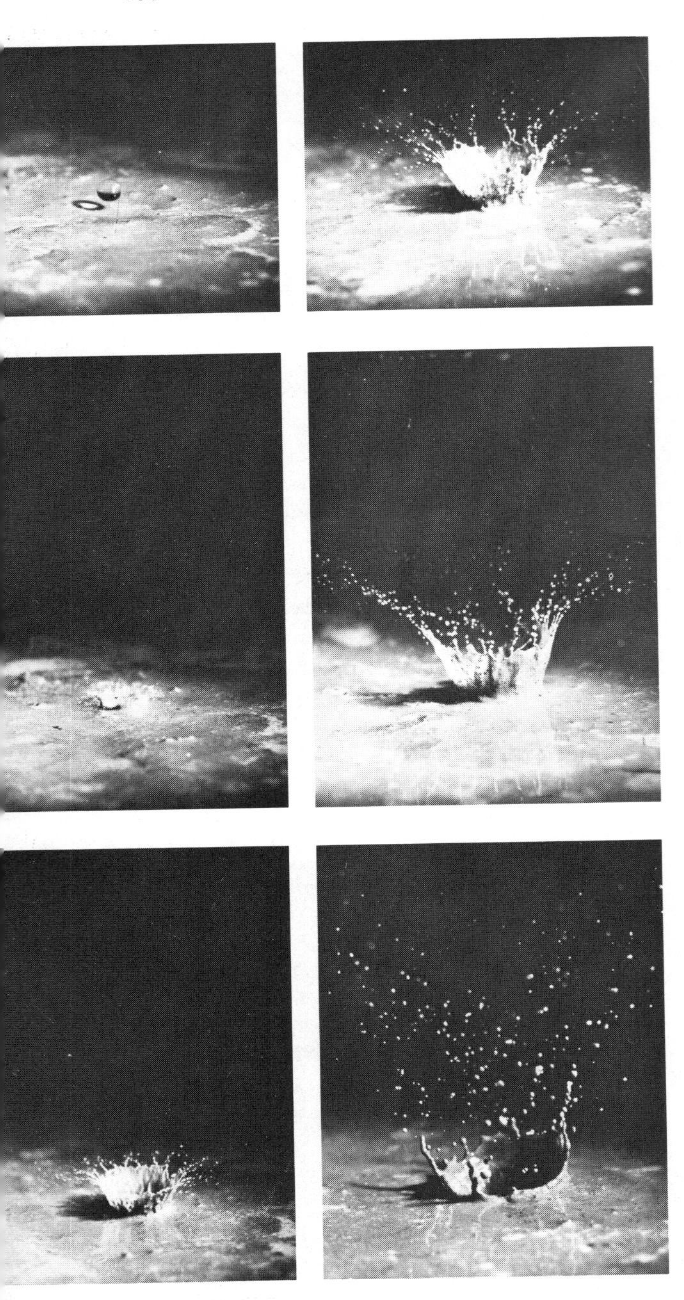

FIGURE 6-19 *Sequence of photos showing raindrop impact and formation of raindrop impression. (Courtesy U.S. Department of Agriculture.)*

A *concretion* is a local concentration of the cementing material that has lithified a deposit into a sedimentary rock. Concretions range in size from a fraction of a centimeter to a meter or more in maximum dimensions (Figure 6-20). Most are shaped like simple spheres or disks, although some have fantastic and complex forms. For some reason, when the cementing material entered the unconsolidated sediment, it tended to concentrate by spreading outward from a common center point or along a common center line. The particles of the resulting concretion are cemented together more firmly than the particles of the host rock that surrounds it. The cementing material usually consists of calcite, dolomite, iron oxide, or silica—in other words, the same cementing materials that we find in the sedimentary rocks themselves.

Geodes, more eye-catching than either concretions or nodules, are roughly spherical, hollow structures up to 30 cm or more in diameter (Figure 6-21). An outer layer of chalcedony is lined with crystals that project inward toward the hollow center. The crystals, often perfectly formed, are usually quartz, although crystals of calcite and dolomite have also been found and, more rarely, crystals of other minerals. Geodes are most commonly found in limestone, but they also are present in claystone.

Fossils

The word *fossil* (derived from the Latin *fodere,* "to dig up") originally referred to anything that was dug from the ground, particularly a mineral or some inexplicable form. "Fossil" is still used in that sense occasionally, as in the term "fossil fuel" (see Chapter 22). But today the term *fossil* generally means any direct evidence of past life—for example, the bones of a dinosaur, the shell of an ancient clam, the footprints of a long-extinct animal,

FIGURE 6-20 *Sandstone concretions in Cretaceous beds, Big Bend Park, west Texas. (Courtesy Earle F. McBride.)*

FIGURE 6-21 *External form (left) and internal structure (right) of geode. Dark outer layer of chalcedony is lined with clear quartz crystals that project inward toward a hollow center. Maximum diameter of specimen is 9 cm. (Photos by M. Dane Picard.)*

FIGURE 6-22 *Structures in the San Vicente Formation of California formed by burrowing organisms. (Courtesy Earle F. McBride.)*

or the delicate impression of a leaf (Figure 6-22).

Fossils are usually found in sedimentary rocks; they rarely turn up in igneous and metamorphic rocks. They are most abundant in claystone, siltstone, and limestone, but also are found in sandstone, dolomite, and conglomerate. Fossils account for almost the entire volume of certain limestone beds.

The remains of plants and animals become completely decomposed if they are left exposed on the earth's surface; but if they are somehow protected from destructive forces, they may become incorporated in a sedimentary deposit where they will be preserved for millions of years. In the ocean, for example, the remains of starfish, snails, and fish may be buried by sediments that settle slowly to the bottom. If the sediment is subsequently lithified, the remains of the animals are preserved as fossils that tell us about the sort of life that existed when the sediment was first deposited.

Fossils are also preserved in deposits that have settled out of fresh water. Countless remains of land animals, large and small, have been dug from the beds of extinct lakes, floodplains, and swamps.

In Chapter 13 we shall find that fossils are extremely useful in subdividing geologic time and in constructing the geologic calendar. The detailed story of the development of life through geologic time has been recorded by the fossils found for the most part in sedimentary rocks. We will consider this story in Chapters 15 through 21.

Sedimentary facies

If we examine the environments of deposition that exist at any one time over a wide

FIGURE 6-23 *Diagram illustrating a change in sedimentary facies. Here, the fine-grained muds (silt and clay) are deposited in a lagoon close to shore. A sand bar separates them from sand deposits farther away from shore. The sand in this instance has been derived from a sea cliff and transported by waves and currents.*

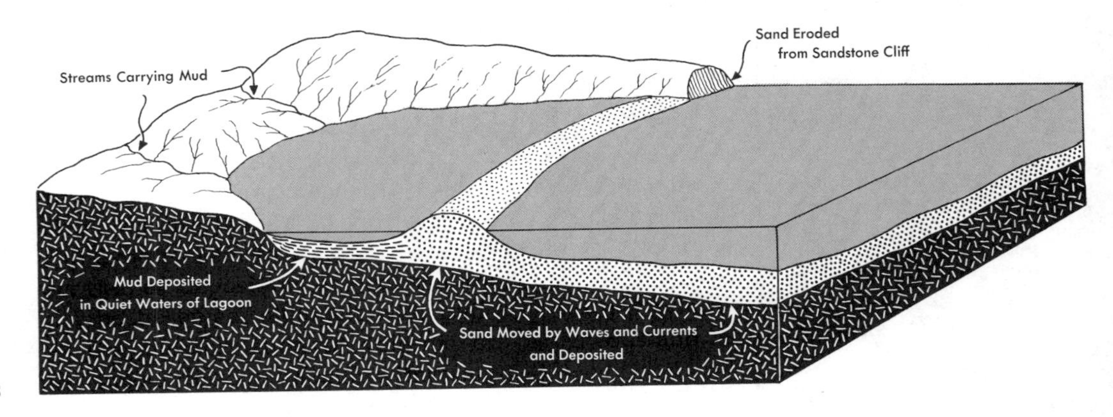

area, we find that they differ from place to place. Thus, the freshwater environment of a river changes to a brackish water environment as the river nears the ocean. In the ocean itself, marine conditions prevail. But even here, the marine environment changes—from shallow water to deep water, for example. And as the environment changes, the nature of the sediments that are deposited also changes. The deposits in one environment show characteristics that are different from the characteristics of deposits laid down at the same time in another environment. This change in the "look" of the sediments is called a change in *sedimentary facies;* the word *facies* derives from the Latin for "aspect" or "form."

Sedimentary facies is an accumulation of deposits that exhibits specific characteristics and grades laterally into other sedimentary accumulations formed at the same time but exhibiting different characteristics. The concept of facies is widely used in studying sedimentary and metamorphic rocks. The concept is not generally used in referring to igneous rocks, although there is no valid reason why it should not be used.

Let us consider a specific example of facies. Figure 6-23 shows a coastline where rivers from the land empty into a lagoon. The lagoon is separated from the open ocean by a sand bar. The fine silt and clay dumped into the quiet waters of the lagoon settle to the bottom as a layer of mud. At the same time, waves are eroding coarse sand from a nearby headland outside the lagoon. This sand is transported by currents and waves and is deposited as a sandy layer seaward of the sand bar. Different environments exist inside and outside the lagoon; therefore different deposits are being deposited simultaneously. Notice in the diagram (Figure 6-23) that the mud and the sand grade into each other along the sand bar. Now, imagine that these deposits are consolidated into rocks and are then exposed to view at the earth's surface. We would find a claystone or siltstone layer grading into sandstone—that is, one sedimentary facies grading into another.

Color of sedimentary rocks

Throughout the western and southwestern United States, bare cliffs and steep-walled canyons provide brilliant displays of a great variety of colors exhibited by sedimentary rocks. The Grand Canyon of the Colorado River in Arizona cuts through rocks that range in color from gray, through purple and red, to brown, bluff, and green. Bryce Canyon in southern Utah is fashioned of rocks tinted a delicate pink, and the Painted Desert, farther south in Arizona, exhibits a wide range of colors, including red, gray, purple and pink.

The most important sources of color in sedimentary rocks are iron-bearing compounds (red, brown, yellow, green) or free carbon (gray, black). Finely divided hematite (Fe_2O_3), for example, gives rocks a red or reddish brown color. Green rocks frequently owe their color to the silicate minerals illite and chlorite and to the absence of hematite and organic material.

Only a small amount of iron oxide is necessary to color a rock. As little as 0.5 percent iron oxide finely disseminated in a sandstone will impart a light orange color. Of the common sedimentary rocks, claystone contains the most iron oxide, about 6 to 7 percent.

Most gray-to-black rocks owe their color to the presence of organic matter. However, the darkness of the rock is not closely related to the percent of organic carbon present. Although peat and oil shale are both very rich in organic matter, their color is generally brown.

The size of the individual particles in a rock also influences the color, or at least the intensity of the color. For example, fine-grained clastic rocks are usually darker than coarse-grained clastic rocks of similar mineral composition.

SUMMARY

Sedimentary rocks cover 75 percent of the world's land surface but constitute only 5 percent of the volume of rock in the outer 16 km of the earth's crust.

Material for sedimentary rocks includes the detrital material (fragments of preexisting rocks) eroded from the land masses of the world and also material precipitated (chemically or biochemically) from solution. Three minerals—clay, quartz, and calcite—are the most important rock-forming constituents of sedimentary rocks.

The texture of sedimentary rocks depends on the size, shape, and arrangement of particles. Textures are clastic (made up of particles cemented together) or nonclastic (usually crystalline and formed by the precipitation of mineral matter from solution).

Soft sediments are turned to rock (lithified) by the processes of compaction, cementation, and diagenesis.

Sedimentary rocks can be classified as detrital or chemical. Detrital rocks have clastic textures and include conglomerate, sandstone, siltstone, and claystone. Chemical rocks may have either clastic or nonclastic textures and include limestone, dolomite, rock salt, gypsum, anhydrite, chert, flint, and coal. Siltstone, claystone, sandstone, and limestone together constitute 99 percent of the sedimentary rocks. Of these, siltstone and claystone are the most abundant and limestone the least abundant.

Bedding or stratification is the most characteristic feature of sedimentary rock. Other features include ripple marks, mud cracks, raindrop impressions, fossils, nodules, concretions, geodes, and color. Iron oxide is the most important coloring agent.

A sedimentary facies is an accumulation of deposits with specific characteristics grading laterally into other sedimentary deposits with different characteristics.

SUGGESTED READINGS

BLATT, HARVEY, GERARD MIDDLETON, and RAYMOND MURRAY, *Origin of Sedimentary Rocks.* Englewood Cliffs, N.J.: Prentice-Hall, Inc., 1972.

DEGENS, EGON T., *Geochemistry of Sediments—A Brief Survey.* Englewood Cliffs, N.J.: Prentice-Hall, Inc., 1965.

DICKINSON, WILLIAM R., ed., *Tectonics and Sedimentation.* Tulsa, Okla.: Society of Economic Paleontologists and Mineralogists, 1974.

GARRELS, R. M., and F. T. MACKENZIE, *Evolution of Sedimentary Rocks.* New York: W. W. Norton and Company, Inc., 1971.

PICARD, M. DANE, and LEE R. HIGH, JR., *Sedimentary Structures of Ephemeral Streams.* Amsterdam: Elsevier Scientific Publishing Co., 1973.

RIGBY, J. KEITH, and W. KENNETH HAMBLIN, eds., *Recognition of Ancient Sedimentary Environments.* Tulsa, Okla.: Society of Economic Paleontologists and Mineralogists, 1972.

SELLEY, RICHARD C., *Ancient Sedimentary Environments.* Ithaca, N.Y.: Cornell University Press, 1970.

Metamorphic Rocks

Having examined in some detail the nature and origin of igneous rocks (Chapter 4) and sedimentary rocks (Chapter 6), let us turn to an examination of the classification, composition, and formation of the third major rock family, the metamorphic rocks.

In our study of metamorphic rocks, two difficulties confront us immediately. The first is their great variety. In Figure 4-2 on page 74 we saw that any igneous or sedimentary rock can be changed into a metamorphic rock; thus there can be at least as many metamorphic rocks as there are igneous and sedimentary rocks. We will find that there can be even more, because a single sedimentary or igneous rock can give rise to several metamorphic rocks depending on the conditions to which the rock is subjected. Furthermore, an existing metamorphic rock may itself be changed into a new one. The number of individual rock types

7

Closeup of sculpture of President Theodore Roosevelt at Mount Rushmore National Memorial in Precambrian metamorphic rocks. (Courtesy National Park Service.)

in the metamorphic family is therefore potentially greater than in the igneous and sedimentary rock groups combined.

The second problem in our study of metamorphic rocks is that we cannot observe their formation directly. Virtually all metamorphic rocks now exposed at the surface of the earth were formed at some depth beneath the surface. We are therefore dependent on erosion to expose them to view. In contrast, the processes of deposition of sedimentary rocks can be observed, although with difficulty, as can the formation of volcanic or extrusive igneous rocks, although witnessing the formation of deep-seated igneous rocks is denied us. Thus, our information concerning the origin of metamorphic rocks comes either by inferences drawn from their composition, structure, and occurrence or by analogy with artificial conditions established in the laboratory.

A first look at metamorphic rocks

In Chapter 4, we began our study of igneous rocks by arranging a representative assemblage in several ways. We tried color, density, texture, and composition. Eventually we arrived at a classification based on texture and mineral composition, to which color and specific gravity were related. Now, let us similarly assume that we have an undifferentiated collection of representative metamorphic rocks.

Separating first by color, we can arrange our samples on a scale that varies from white to black with various tones of gray, green, and red in between. Each of our final groupings, however, contains rocks that differ in composition, specific gravity, and texture. We find, therefore, that our classification has little relation to any observable characteristics other than color. We meet much the same problem when we classify according to density.

Turning to composition, we discover almost as many categories as individual specimens. We are forced, therefore, to abandon composition as a primary basis of classification. We will see later, however, how composition can be used as a secondary characteristic in classification.

We are left, then, at least for our initial separation of the metamorphic rocks, with texture, which was so important in both the igneous and the sedimentary families.

Texture

A number of textural characteristics become evident as we examine our pile of metamorphic rocks. Many of our samples, for instance, are coarse; with the naked eye we can see individual mineral crystals, which may range in the largest dimension from about 1 to 10 mm, and sometimes more. These crystals are interlocking, even as they are in coarse-grained igneous rocks and in some of the sedimentary rocks. We find, also, that many of the metamorphic rocks are very fine-grained; in these we cannot identify by eye the constituent minerals, only flashing points of light from individual crystals or from cleavage planes. And again, as we found in some igneous rocks, there can be a mixture of both small and large crystals in the same specimen. We called this characteristic *porphyritic* in the igneous rocks, but the term is not generally used in describing metamorphic rocks.

How does the texture of these metamorphic rocks differ from that of the sedimentary and igneous rocks we have studied? It turns out that we can divide metamorphic rocks into two groups according to texture. But the decisive quality is not the size or uniformity of the grains, rather it is the arrangement of the grains.

In the larger of the two groups the texture is characterized by a preferred orientation of individual grains, or by parallel layers of minerals of different composition. In size the minerals can be large or small or "mixed," but their arrangement gives the rocks a distinctive planar aspect that is called *foliation* (from the Latin *foliatus,* "leaved or leafy," hence, consisting of leaves or thin sheets).

Foliation imparts a characteristic called *rock cleavage.* Previously, the term cleavage was used to describe the relative ease with which a mineral breaks along parallel planes. Here we use the modifier "rock" to distinguish rock cleavage from mineral cleavage. All foliated rocks tend to break or cleave most easily along planes parallel with their foliation.

In the second, smaller group, of metamorphic rocks we find that there seems to be no preferred orientation of breakage. We generally refer to these rocks as *unfoliated.*

FIGURE 7-1 Closeup of phyllite showing good but irregular cleavage. Width of view is 8.5 cm.

FIGURE 7-2 Garnet crystal in mica schist. The mold once occupied by another crystal can also be seen. Ruled 2-in. (5.1 cm) squares give scale for size. (Photo by Walter R. Fleischer.)

Types of foliated rocks

We can further subdivide the foliated rocks into groups based on the coarseness or fineness of the foliation. These divisions are as follows:

1. *Slaty* (from the French *éclat,* "fragment or splinter"). In this type the cleavage is present along planes separated by distances of microscopic dimensions. The planes of cleavage are smooth and regular.
2. *Phyllitic* (from the Greek *phyllon,* "leaf"). Here the cleavage produces cleavage fragments barely visible to the unaided eye. The fragments are thicker than those of rocks with slaty cleavage, and the surface of cleavage planes is somewhat more irregular (Figure 7-1).
3. *Schistose* (from the Greek *schistos,* "divided, divisible"). This type of cleavage produces flakes that are clearly visible. Cleavage surfaces are rougher than in the slaty or phyllitic types (Figure 7-2).

FIGURE 7-3 Metamorphic rock (a gneiss) showing alignment of previously unoriented minerals. The light-colored bands are mainly orthoclase and quartz; the dark streaks are biotite and other ferromagnesian minerals. The bulk composition is that of granite; but in contrast to the random mixing in granite, the minerals are here distributed in relatively systematic patterns. Width of view is 9.5 cm.

4. *Gneissic* (from the German *gneiss,* a term originally used to refer to the country rock containing the mineral deposits of the Erzgebirge region of southern Germany). The term *gneiss* is now given to a metamorphic rock and sometimes to an igneous rock characterized by coarse foliation or banding, in which the bands of differing mineral composition are usually a few millimeters or centimeters thick (Figure 7-3). Cleavage planes are irregular and rough.

Table 7-1 Simplified Classification of Metamorphic Rocks Based on Texture and Composition

TEXTURE	MONOMINERALIC	MULTIMINERALIC
Unfoliated	Quartzite, Marble	Hornfels
Foliated		Slate, Phyllite, Schist, Gneiss

Composition

The composition of metamorphic rocks is extremely diverse, but we can make two very simple observations and refine them later in this chapter. First, some metamorphic rocks are composed exclusively or predominately of a single mineral type. They are, therefore, *monomineralic.* Examples are marble (made up of calcite) and quartzite (composed of quartz). Monomineralic rocks are either unfoliated or weakly foliated.

Second, most metamorphic rocks are composed of two or more minerals, and so are called *multimineralic.* Usually (but not always), these multimineralic rocks are foliated.[1] An example is *gneiss.* Gneisses have differing compositions, but if one has the mineral assemblage characteristic of a granite (quartz, orthoclase, plagioclase, and a few ferromagnesian minerals), we call it a *granite gneiss.* Like the granite, it is coarse-grained and has interlocking crystals. It differs from granite in possessing a foliated texture defined by a coarse banding of the various minerals.

Table 7-1 gives a simplified classification of metamorphic rocks based on texture (foliated and unfoliated) and composition (monomineralic and multimineralic).

Many of the minerals of sedimentary and igneous rocks are found in the metamorphic group as well. These include the familiar minerals of quartz, calcite, orthoclase and plagioclase feldspars, muscovite and biotite, amphibole, and augite. But the processes of metamorphism also create minerals uncommon in the sedimentary and igneous rocks. These include diopside, tremolite, sillimanite, kyanite, andalusite, staurolite, garnet, epidote, and chlorite, which will be discussed beginning on page 147, later in this chapter.

[1]Hornfels (Table 7-1), usually unfoliated, are commonly multimineralic.

Formation of metamorphic rocks

Thus far we have confined our study to a general examination and ordering of the metamorphic rocks. Now let us consider the ways in which metamorphic rocks might form. Inference and analogy have led to the universal opinion that metamorphic rocks come into being as existing rocks in the solid state in response to alterations in temperature, pressure, and chemical environment. These changes are thought to be the same as those that fold, fault, inject magma into, and elevate or depress large masses of rock. What is distinctive is that the process of metamorphism takes place within the earth's crust, below the zone of weathering and cementation and above the zone of remelting. In this environment, rocks undergo chemical and structural changes in response to conditions that differ from those under which they originally formed.

Agents of metamorphism

We limit the term "metamorphism" to changes that take place in the texture and composition of solid rocks. Metamorphism can take place only while the rock is solid, because when the rock reaches its melting point, a magma forms and igneous activity begins. We should recognize, however, that a rock can exist in the *plastic state,* a condition transitional between the brittle character of rocks at or near the earth's surface and the molten state of subterranean magma. In this zone of change, the agents producing metamorphic rocks are *heat, deforming processes* (*or pressure*), and *chemically active fluids* in pore spaces.

Heat

Various lines of evidence suggest that metamorphism takes place within a temperature range of about 100° to 800°C. For most ordinary rocks, the temperature of incipient melting is in the interval 650° to 800°C. The lower temperature for metamorphism is more difficult to determine. Sedimentary rocks in dry drill holes, where bottom temperature exceeds 150°C, show no indication of metamorphic change. However, in hot spring areas, water near 100°C can apparently cause extensive alteration. Alterations in minerals that take place below about 100°C might be termed *diagenesis, weathering,* or possibly *hydrothermal activity.*

Many believe that temperature is the single most important agent of metamorphism. Nevertheless, it is most certainly always accompanied and abetted by changes in pressure and the presence of chemically active fluids.

Pressure

Pressure during metamorphism is due mainly to the weight of overlying material. Most metamorphic rocks have not been more than 20 km below the surface, a depth at which the lithostatic pressure is about 6,000 atmospheres. Ordinarily metamorphism takes place at pressures of a few hundred atmospheres, but some consider the lower limit to be the ordinary pressure of the atmosphere.

High confining pressures combine with differential pressures to produce two general results. First, pressure reduces the space occupied by the rock mass, thereby leading to recrystallization and the formation of new minerals with closer atomic packing. Second, differential pressures combine with confining pressures to produce a flow of rock material. This flow results in intergranular motion, formation of minute shear planes within the rock, changes in texture, reorientation of grains, and growth of crystals. It is from these processes that most of the foliation in metamorphic rock derives.

Chemically active fluids

The magma from which igneous rocks crystallize contains water, most of which does not become a part of the minerals forming from the magma but is driven off. In a volcanic eruption, for instance, much of the "smoke" from the vent is water in the form of steam. If the water is released below the surface, it moves through the surrounding rock as a *hydrothermal solution.* This solution transports ions from the igneous body into the rock, where it may deposit ions, pick up others, and generally alter the original rock. When a chemical reaction within the rock or the introduction of ions from an external source cause one mineral to grow or change into another of different composition, the process is called *metasomatism* (Figure 7-4). The term describes all ionic transfers, not just those that involve gases or solutions from magma.

Some of the chemically active fluid in the process of metamorphism is already present as liquid in the pores of a rock subjected to agents of metamorphism. It is believed that such pore liquid may often act as a catalyst; that is, it expedites changes without itself undergoing change. Most metamorphic reactions are increased in rate by fluids, and for some reactions the nature and amount of fluid determine the metamorphic minerals that will form.

FIGURE 7-4 *Chemically active fluids probably helped to form these bands of mica, amphibole, and some quartz, which define the foliation in a gneiss on Pemaquid Point, Maine. The light band that cuts across the foliation at a low angle and parallel with the pen is a thin dike of granitic material. (Photo by Sheldon Judson.)*

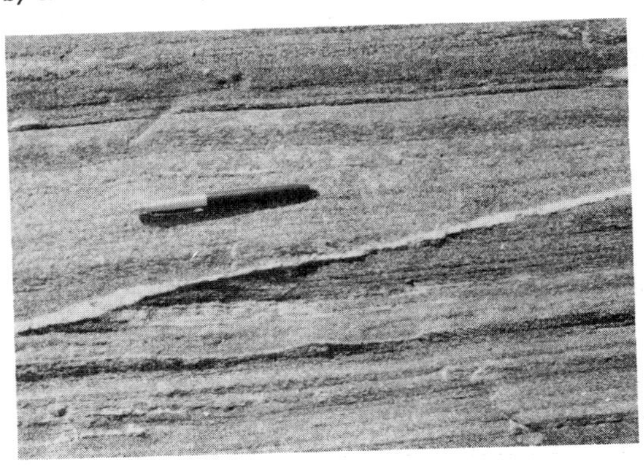

Types of metamorphism

Several types of metamorphism are known, but we concern ourselves here with the three basic ones: cataclastic metamorphism, contact metamorphism, and regional metamorphism.

Cataclastic metamorphism

Cataclasis is the metamorphic process of mechanical deformation. The shearing and grinding associated with intense folding and breaking of rocks produce a texture in which minerals are pulverized and are strung out in streaks or bands. Recrystallization and chemical changes are not common in cataclastic rocks. Instead, the original grains of the rock show the effects of plastic strain and granulation.

Several types of cataclastic rocks are recognized. *Friction breccias* contain abundant, angular, shattered rock fragments ranging from a millimeter to a meter or more in length. *Mylonites* are more completely sheared and granulated than friction breccias, and grain sizes are in the 0.01 to 0.1 mm range. *Pseudotachylites* exhibit individual grains about 0.001 mm in length. A distinctive texture, *augen,* is formed in pseudotachylites by the shearing and abrasion of large, preexisting, euhedral crystals (porphyroblasts).

Contact metamorphism

The alteration of rocks by ionic transfer brought about by high temperatures and by the introduction of magmatic solutions at or near their contact with a body of magma is

FIGURE 7-5 In many instances metamorphism involves the intimate association of igneous and metamorphic rocks. Here, near Worcester, Massachusetts, gneiss is intruded by sheets of granite injected more or less parallel to the foliation of the gneiss. The largest of the granitic intrusions is marked by the geologic pick. (Photo by Keith, U.S. Geological Survey.)

called *contact metamorphism.* At the actual surface of contact all the elements of a rock can be replaced by other elements introduced by hot gases and by solutions escaping from the magma. Farther away, the replacement may be only partial.

Contact metamorphism is present in restricted zones called *aureoles* ("halos"), which seldom measure more than a few hundred meters in width and may be only a fraction of a centimeter wide. Aureoles are found bordering laccoliths, stocks, and batholiths. During contact metamorphism, temperatures range from about 200°C to 800°C and load pressures from 100 to 3,000 atmospheres.

Contact metamorphism develops late in the mountain-building sequence (discussed in Chapter 9) and at relatively shallow depths. It is only late in the cooling of a magma that large quantities of hydrothermal solutions are released, and then only as the body of magma approaches the surface.

Considerable recrystallization is caused by the heat of contact metamorphism. Valuable deposits of copper, zinc, iron, tin, and tungsten have been formed as a result of contact metamorphism in limestone and limey claystone.

Regional metamorphism

Regional metamorphism develops over extensive areas, often involving thousands of square kilometers of rock thousands of meters thick. It is commonly believed that the huge reservoirs of melted rock associated with the formation of some mountain ranges is an expression of regional metamorphism, but the assumption has not yet been proved. The effects of regional metamorphism are best seen, however, in the root regions of old fold mountains and in the Precambrian continental shields (see Chapters 9 and 14). Thousands of meters of rock have had to be eroded in order to expose these metamorphic rocks to view.

During regional metamorphism, many new minerals are developed as rocks respond to increases in temperature and pressure. These include some new silicate minerals not found in igneous and sedimentary rocks, such as sillimanite, kyanite, andalusite, staurolite, various garnets, brown biotite, epidote, and chlorite.

The first three of these are silicates with the formula Al_2SiO_5. Their independent SiO_4 tetrahedra are bound together by positive ions of aluminum. *Sillimanite* develops in long, slender crystals that are brown, green, or white in color. *Kyanite* forms bladelike blue crystals. *Andalusite* forms coarse, nearly square prisms.

Staurolite (from the Greek *stauros,* "cross") is a silicate composed of independent tetrahedra bound together by positive ions of iron and aluminum. Staurolite has a unique crystal habit that is striking and easy to recognize: It develops six-sided prisms that intersect either at 90°, forming a cross, or at 60°, forming an X.

Garnets form a group of metamorphic silicate minerals. All have the same atomic structure of independent SiO_4 tetrehedra, but a wide variety of chemical compositions is produced by the many positive ions that bind the tetrahedra together. These ions can be iron, magnesium, aluminum, calcium, manganese, or chromium. But whatever the chemical composition, garnets appear as distinctive 12-sided or 24-sided fully developed crystals (Figure 7-2). It is actually difficult to distinguish one kind of garnet from another without resorting to chemical analysis. A common deep red garnet of iron and aluminum is called *almandite.*

Epidote is a silicate of calcium, aluminum, and iron in which the tetrahedra are in pairs that are independent of each other. On a freshly broken surface, this mineral is pistachio green in color; otherwise it is black or blackish green.

Chlorite is a sheet-structure silicate of calcium, magnesium, aluminum, and iron. The characteristic green color of chlorite was the basis for its name, from the Greek *chloros,* "green" (as in chlorophyll). Chlorite exhibits a cleavage similar to that of mica, but the small scales produced by the cleavage do not, when bent, snap back to their original position as mica flakes do. Chlorite is found either as aggregates of minute scales or as individual scales scattered throughout a rock.

Regional metamorphic zones. Regional metamorphism can be divided into zones: high-grade, middle-grade, and low-grade. Each grade is related to the temperature and pressure reached during metamorphism. Thus, low-grade metamorphism is found farthest away from the reservoir and blends into unchanged sedimentary rock.

Metamorphic zones are identified by using certain *index minerals* as diagnostic keys. Zones of regional metamorphism reflect the varied mineralogical response of chemically similar rocks to different physical conditions, especially temperatures. And each index mineral gives an indication of the conditions at the time of its formation.

The first appearance of chlorite, for example, tells us that we are at the beginning of a low-grade metamorphic zone; the first appearance of the garnet, almandite, indicates the beginning of a middle-grade metamorphic zone; and the first appearance of sillimanite marks a high-grade zone. Other minerals sometimes are present in association with each of these index minerals, but they are usually of little help in determining the degree of metamorphism of a given zone.

By noting the appearance of the minerals that are characteristic of each metamorphic zone, it is possible to draw a map of the regional metamorphism of an entire area. Of course the rocks must have the proper chemical composition to allow these minerals to form.

Regional metamorphism and plate tectonics. In volume, most metamorphic rocks of regional extent apparently are the result of horizontal stress associated with the formation of mountains. Foliation commonly is nearly vertical, indicating that horizontal stresses were responsible. Such metamorphic belts frequently are linear and are related closely to igneous intrusions of granite.

As interpreted by plate tectonic theory, large-scale metamorphism takes place in the deep roots of linear, folded mountain belts in response to the collision of crustal plates. Each mountain-building episode yields a new belt of metamorphic rocks along the continental margin, thus enlarging the area of continental crust by accretion of new material.

Types of metamorphic rocks

Metamorphic rocks are usually named on the basis of texture. A few can also be further classified by including the name of a mineral present in them, such as chlorite schist, mica schist, and hornblende schist.

Slate

Slate is a metamorphic rock produced by the low-grade metamorphism of claystone or tuff. It is fine-grained with a characteristic cleavage caused by the alignment of flat, flaky minerals under the pressures of metamorphism (Figure 7-6). Some of the clay minerals in the original rock are transformed by heat into chlorite and mica. In fact, slate is composed dominantly of small, colorless mica flakes and smaller quantities of chlorite. Dark-colored slate owes its color to the presence of carbonaceous material or iron sulfides.

Phyllite

Phyllite is a metamorphic rock with much the same composition as slate, but brighter and coarser grained. Phyllite is slate that has undergone further metamorphism. When slate is subjected to heat greater than 250° to 300°C, the chlorite and mica minerals of which it is composed develop large flakes, giving the resulting rock its characteristic phyllitic cleavage and a silky sheen on freshly broken surfaces (Figure 7-1). The predominant minerals in phyllite are chlorite and muscovite.

FIGURE 7-6 *Slate in Cornwall, England, shows both cleavage and bedding. The bands of dark and light rock running diagonally across the exposure from upper left to lower right mark bedding that was horizontal when the sediments were deposited. Earth pressures have tilted the rock and imposed a rock cleavage on it that dips in the same direction as the original bedding but at a lower angle. The resulting rock is a slate that breaks most easily parallel to the cleavage rather than to the original bedding. (Photo by British Geological Survey.)*

Schist

Of the metamorphic rocks formed by regional metamorphism, *schist* is the most abundant. There are many varieties of schist, since it can be derived from many igneous, sedimentary, or lower-grade metamorphic rocks. Table 7-2 lists some of the more common varieties of schist, together with the

Table 7-2 Common Schists

VARIETY	ROCK FROM WHICH DERIVED
Chlorite schist Mica schist	Claystone or siltstone
Hornblende schist Biotite schist	Basalt or gabbro
Quartz schist	Impure sandstone
Calc-schist	Impure limestone

FIGURE 7-7 Vertical quartz-mica schist, Mossameder Desert, southwest Angola. (Photo by Robert T. Novotny.)

names of the rocks from which they were derived. All schists are dominated by clearly visible flakes of some mineral such as mica, talc, chlorite, or hematite. Fibrous minerals are commonly present as well. Schist tends to break between the platy or fibrous minerals, giving the rock its characteristic schistose cleavage (Figure 7-7).

Schists often contain large quantities of quartz and feldspar as well as lesser amounts of other minerals, including augite, hornblende, garnet, epidote, and magnetite. A green schistose rock produced by low-grade metamorphism, sometimes called a *greenschist,* owes its color to the presence of the minerals chlorite and epidote.

Amphibolite

Amphibolite is composed chiefly of hornblende and plagioclase. There is some foliation because of the alignment of hornblende grains, but it is less conspicuous than in schists. Amphibolites, which can be green, gray, or black, sometimes contain epidote, green augite, biotite, and almandite. They are products of medium-grade to high-grade regional metamorphism of ferromagnesian igneous rocks and of some impure dolomite.

Gneiss

The coarse-grained metamorphic rock *gneiss* is most commonly formed during high-grade regional metamorphism. A banded appearance makes it easy to recognize. Although gneiss exhibits rock cleavage, the cleavage is far less pronounced than in the schists.

FIGURE 7-8 Folded gneiss, Pemaquid Point, Maine. Figures in upper right give scale. (Photo by Sheldon Judson.)

In gneiss derived from igneous rocks such as granite, gabbro, or diorite, the component minerals are arranged in parallel layers: the quartz and the feldspars alternate with the ferromagnesians. In gneiss formed from the metamorphism of clayey sedimentary rocks, such as graywacke, bands of quartz or feldspar usually alternate with layers of platy or fibrous minerals such as chlorite, mica, graphite, hornblende, kyanite, staurolite, sillimanite, and wollastonite.

Marble

Marble is composed essentially of calcite or dolomite, and originated during the contact or regional metamorphism of limestone or dolomite. It does not exhibit rock cleavage. Marble differs from the original rock in having larger mineral grains. In most marble, the crystallographic direction of its calcite is nearly parallel in response to the metamorphic pressures to which it was subjected. The rock exhibits no foliation, however, because the grains have the same color and the mineral orientation does not show up.

Although the purest variety of marble is snow-white, many marbles contain small percentages of other minerals that were formed during metamorphism from impurities in the original sedimentary rock. These impurities account for the wide variety of color in marble. *Black marble* is colored by bituminous matter; *green marble* by diopside, hornblende, serpentine, or talc; *red marble* by an iron oxide, hematite; and *brown marble* by another iron oxide, goethite. Garnets have often been found in marble and, on rare occasion, rubies. The beautiful patterns of some marbles are produced by the presence of fossilized corals in the original limestone.

Marble is found most commonly in areas that have undergone regional metamorphism, where it is often present in layers between mica schist or phyllite.

Quartzite

The metamorphism of quartz-rich sandstone forms the rock *quartzite.* The grains of quartz in the original sandstone are firmly bonded together by the entry of silica into the pore spaces. Quartzite is unfoliated and is distinguishable from sandstone in two ways: there are no pore spaces in the quartzite, and the rock breaks through the sand grains that compose it rather than around them.

The structure of quartzite cannot be recognized without a microscope. But when quartzite is cut into thin sections, we can identify both the original rounded sand grains and the silica that fills the old pore spaces.

Pure quartzite is white, but iron or other impurities sometimes give the rock a reddish or dark color. Among the minerals that often are present in small quantities in quartzite are feldspar, muscovite, chlorite, zircon, tourmaline, garnet, biotite, epidote, hornblende, and sillimanite.

Hornfels

A *hornfels* is a rock produced by the contact metamorphism of a rock such as claystone, siltstone, limestone, sandstone, tuff, or basalt. Generally fine-grained, hard and unfoliated, hornfels may contain larger crystals scattered through it. Since the process involves the high temperature attendant on igneous intrusions, the zone of hornfels defines a halo surrounding an intrusion. The temperature of metamorphism is typically higher than that of regional metamorphism and, therefore, in excess of 700°C.

Anthracite

The formation of coal, a sedimentary rock, is a process that starts with the accumulation of plant and tree remains in a swamp or a bog. There the organic matter is attacked by

FIGURE 7-9 Migmatite at Maløy, Norway, showing complex interbanding of light (felsic) and dark (mafic) layers. Hammer for scale. (Photo by R. B. Parker.)

bacteria and gradually decays beneath the water, out of reach of the air. As the decayed material builds up greater and greater thicknesses, the growing pressure of the overlying deposits compresses it and drives out the water. The cellulose and lignin of the original plant tissues are slowly altered, producing certain volatile compounds. Among these are the common methane ("marsh gas") and oxides of carbon.

If this compacted material is subjected to increased heat and pressure, many of the volatile substances are driven off and the sedimentary rock is transformed into a metamorphic rock. The volatiles carry away with them greater amounts of oxygen and hydrogen than of carbon, leaving the mass relatively richer in the carbon. The names by which the coal is known at successive stages are listed in Table 7-3.

Notice that the percentage of carbon increases as the percentage of oxygen decreases. The amount of hydrogen and nitrogen also decreases, but the effect of these elements on the formation of coal is of no importance. Sulfur, silica (SiO_2), and aluminum oxide (Al_2O_3) can also be present as impurities. These are the substances that produce most of the ashes that remain after coal has burned.

Origins of granite

The eighteenth-century geologist James Hutton once stated that granite was produced by the crystallization of minerals from a molten mass. Ever since, most geologists have accepted the magmatic origin of granite. For several decades, however, several geologists have questioned this conclusion, suggesting instead that granite is a metamorphic rock produced from preexisting rocks by a process called *granitization.*

In discussing batholiths, we mentioned that one of the reasons for questioning the magmatic origin of granite was the mystery of what happened to the great mass of rock that must have been displaced by the intrusion of the granite batholiths. This so-called "space

Table 7-3 Classification of Coals and Antecedent Material

NAME	PERCENTAGE OF CARBON	PERCENTAGE OF OXYGEN
Dry plant materials	50	40
Peat	60	30
Lignite	70	20
Bituminous coal	80	10
Anthracite	95	2

problem" has led some geologists to conclude that batholiths actually represent preexisting rocks transformed into granite by metasomatic processes.

Certain rock formations support this theory. According to this proposition, certain sedimentary rocks were originally formed in a continuous layer, but now grade into schists and then into *migmatites* ("mixed rocks"), which apparently formed when magma squeezed in between the layers of schist. The migmatites in turn grade into rocks containing the large, abundant feldspars characteristic of granite, but also seem to show shadowy remnants of schistose structure. Finally, these rocks grade into pure granite. The proponents of the granitization theory say that the granite is the result of extreme metasomatism, and that the schists, migmatites, and granitelike rocks with schistose structure are intermediate steps in the transformation of sedimentary rocks into granite.

What process is responsible for granitization? Perhaps ions migrated through the original solid rock, building up the elements characteristic of granite, such as sodium and potassium, and removing superfluous elements, such as calcium, iron, and magnesium. The limit to which the migrating ions are supposed to have carried the calcium, iron, and

FIGURE 7-10 *Ancient (Precambrian) gneisses and schists exposed by uplift and erosion in the Black Canyon of the Gunnison River, western Colorado. (Courtesy National Park Service.)*

magnesium is called the *simatic front.* The limit to which the migrating ions are supposed to have deposited the sodium and potassium is called the *granitic front.*

Geologists are still enthusiastically debating the origin of granite. But they have reached agreement on one fundamental point—namely, that various rocks with the composition and structure of granite may have different origins—some may be igneous and others metasomatic. The debate between "magmatists" and "granitizationists" has thus been reduced to a question of percentages. Those that favor a magmatic origin admit that perhaps 15 percent of the granite exposed at the earth's surface is metasomatic. But the granitizationists reverse the percentages and insist that about 85 percent is metasomatic and only 15 percent is magmatic.

While the battle rages, the magmatists are seeking a completely satisfactory explanation for the origin of magma, particularly magma of sialic composition. And the granitizationists are trying to unravel an equally knotty problem—the process by which preexisting rocks have been converted to granite.

SUMMARY

We divide metamorphic rocks on the basis of texture into two main groups: foliated and unfoliated. The foliated group—which is the largest—is characterized by rock cleavage that parallels a preferred orientation of mineral grains or layers of minerals of differing composition. The unfoliated rocks lack rock cleavage. Textures of rocks are: slaty, phyllitic, schistose, and gneissic.

A few metamorphic rocks such as marble and quartzite are dominated by a single mineral and are called *monomineralic.* Most metamorphic rocks, however, are multimineralic.

Agents of metamorphism include heat, pressure, and chemically active fluids. Metamorphism may be contact metamorphism near the edge of an igneous mass, regional metamorphism, or cataclastic metamorphism.

Metamorphic rocks include slate, phyllite, schist, amphibolite, gneiss, marble, quartzite, hornfels, and anthracite.

Some geologists believe that granite is a metamorphic rock produced by extreme metamorphism of preexisting rocks, a process called *granitization.*

SUGGESTED READINGS

BARTH, TOM F. W., *Theoretical Petrology* 2nd ed. New York: John Wiley & Sons, Inc., 1962.

ERNST, W. G., *Earth Materials.* Englewood Cliffs, N.J.: Prentice-Hall, Inc., 1969.

KRAUSKOPF, KONRAD B., *Introduction to Geochemistry.* New York: McGraw-Hill Book Company, 1967.

TURNER, FRANCIS J., *Metamorphic Petrology.* New York: McGraw-Hill Book Company, 1968.

WINKLER, H. G. F., *Petrogenesis of Metamorphic Rocks* 2nd ed. New York: Springer-Verlag, 1967.

Earthquakes and the Earth's Interior

Shortly before dawn on All Saint's Day, 1755, the hills of Lisbon were shaken violently. Within six minutes, the great stone arches and roofs of the churches collapsed, trapping and killing thousands of worshipers in their fall. The sea withdrew and then swept over the docks, plunging hundreds to their death. Landslides rushed from the mountains onto the city. Fires broke out as the sun rose, as if to further light the horror. And when night fell, more than a fourth of Lisbon's 235,000 people were dead. No one knows the number that were injured or maddened.

The effects of the Lisbon earthquake extended far beyond Lisbon (see Kenneth Clark's *The Romantic Rebellion*). Madame de Pompadour gave up rouge for a week ("She has offered it up," said Horace Walpole, "to the Demon of Earthquakes"). "Never before," said Goethe, "has the Demon of Fear so

8

Effects of the Alaska earthquake of March 27, 1964. (Courtesy U.S. Geological Survey.)

quickly and so powerfully spread horror throughout the land."[1]

The Romantic Movement perhaps first showed itself as an expression of fear, according to Clark, and 1755 is as defensible as any date can be for the beginning of a revolution. Waves of terror from the Lisbon earthquake and later earthquakes still move around the world. Let us, then, turn to these most fearsome of all natural phenomena, and later we will discuss the interior of the earth itself.

Earthquakes

During the next 24 hours after you have finished reading this chapter, more than 2,800 earthquakes will shake the earth strongly enough to be felt. Most of us once or twice in our lifetime will feel the earth tremble, but earthquake-shaking hazards vary (Figure 8-1). Earthquakes are the most destructive to man of all natural processes, and their prediction and control are much to be desired. Our knowledge of earthquakes is increasing rapidly; but that it will be sufficient in the near future to help man in earthquake-prone densely populated areas is far from certain.

[1]Kenneth Clark, *The Romantic Rebellion* (New York: Harper & Row, Publishers, 1974).

Cause of earthquakes

The immediate cause of an earthquake is the sudden movement of rock masses that have been distorted beyond the limit of their strength in a process called *faulting*. Our as-

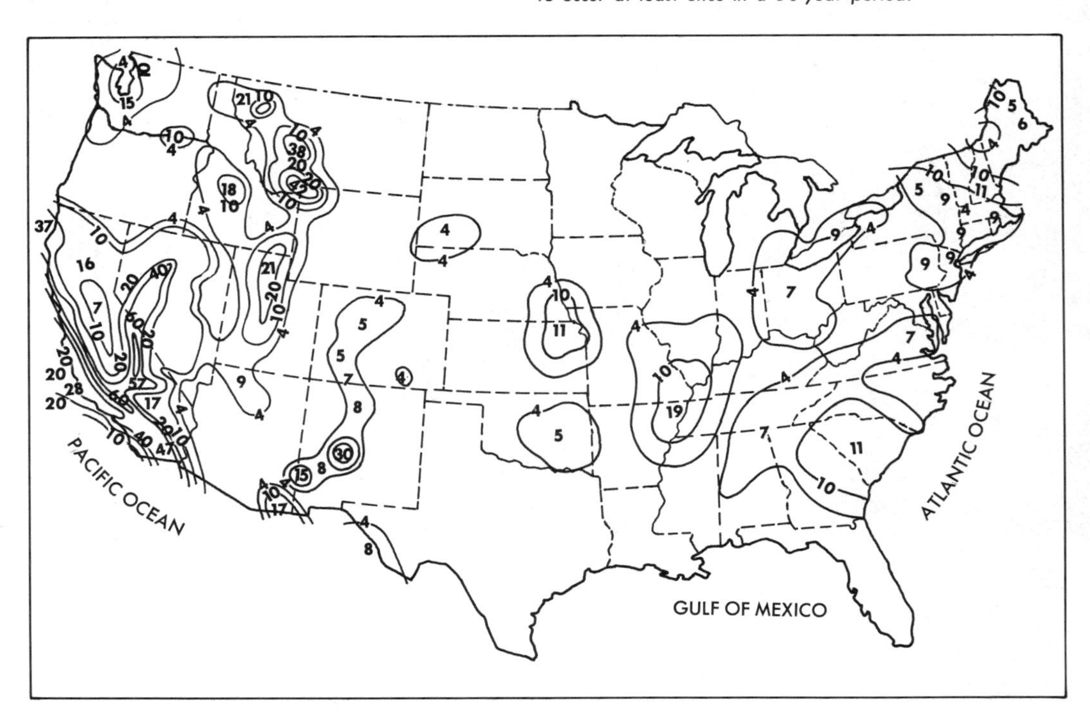

FIGURE 8-1 *Map shows expectable levels of earthquake-shaking hazards. Levels of ground shaking for different regions are shown by contour lines that express, in percentages of the force of gravity, the maximum amount of shaking likely to occur at least once in a 50-year period.*

surance that earthquakes are caused by faulting is based on actual observations of effects that can be seen directly on the earth's surface. It is also based on our knowledge of how rocks behave when they are subjected to deforming forces. Under certain pressures these rocks gradually change shape, but they resist more and more as the pressure builds up, until finally they reach the breaking point. Then they tear apart and snap back into unstrained positions. This snapping back is called *elastic rebound.* Just before the rocks reach the rupture point, small shocks sometimes announce that the stress has become critical. In actual earthquakes, these shocks are often felt as *foreshocks.* Minor adjustments take place after the break, too; these are called *aftershocks.*

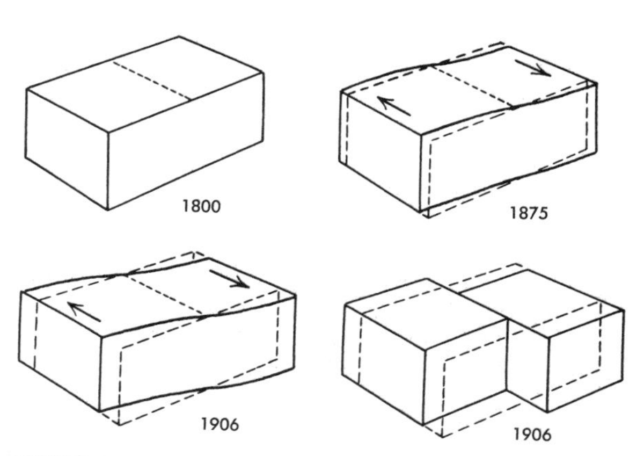

FIGURE 8-2 *Sequence of deformation leading to fracture along the San Andreas fault, which caused the California earthquake of 1906.*

San Andreas fault

Events along the San Andreas Fault in California have provided an unusual opportunity to study elastic rebound. Surveys of part of the fault had been made by the U.S. Coast and Geodetic Survey at various intervals during the years preceding the great earthquake of 1906. H. F. Reid, the American geophysicist, later analyzed the events in three groups, 1851–1865, 1874–1892, and 1906–1907.[2] From the displacements revealed by those surveys and from the displacements that occurred at the time of the actual quake, Reid reconstructed a history of the movement (Figure 8-2). Although there was no direct evidence, he assumed that the elastic energy had been stored at a uniform rate over the entire interval, and that the region had started from an unstrained condition approximately a century before the earthquake. As the years passed, a hypothetical line—which in 1800 would have cut straight across the fault at right angles—became progressively more and more warped.

[2]H. F. Reid, The California Earthquake of April 18, 1906, *Mechanisms of the Earthquake,* v. 11, Carnegie Institution of Washington, 1908.

When the relative movement of the blocks on either side of the fault became as great as 6 m, in any one place, the strength of the rock was exceeded and fracture occurred. The blocks then snapped back toward an unstrained position, driven by the stored elastic energy.

The fault zone runs roughly from northwest to southeast; its total length on land is 1,000 km. Land on the western, or Pacific Ocean, side of the fault moved northwest relative to land on the eastern side. In 1906 approximately 400 km of the fault zone were in motion, from Point Arena to San Juan Batista. The maximum offset was 6.4 m about 50 km north of San Francisco; in the neighborhood of San Francisco the offset was consistently about 4.5 m. Slight vertical displacements of less than a meter took place locally. Roads, fences, lines of trees, and buildings intersected by the fault were neatly sliced and, subsequently, could be used to measure the amount of displacement along the fault zone.

Surveys conducted at intervals since the 1906 earthquake indicate that the region around the San Andreas Fault is still warping and storing energy that may be released in future earthquakes.

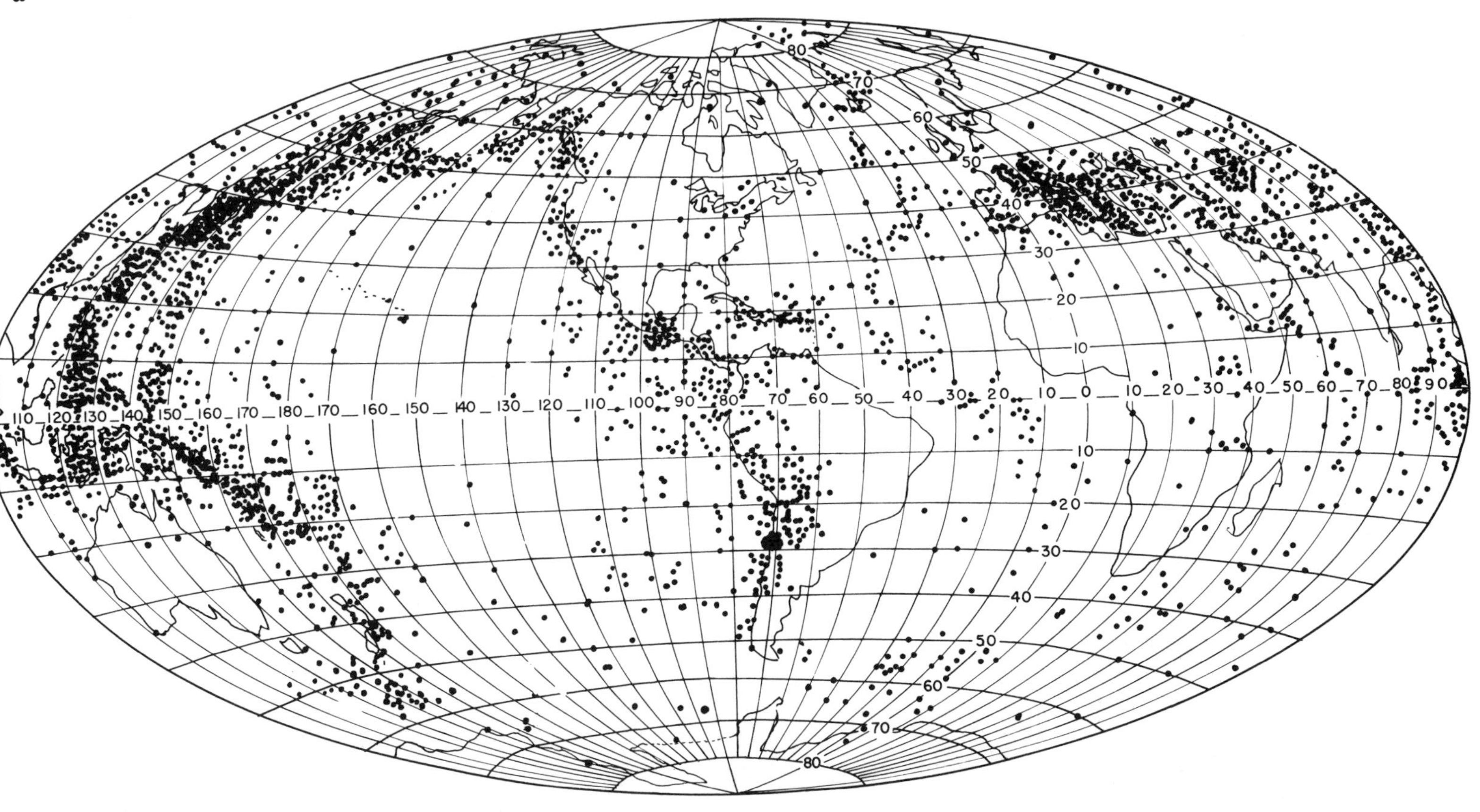

FIGURE 8-3 *Locations of 3,737 earthquakes that occurred in 30 years (1899–1910; 1913–1930, inclusive) on Aitoff's equal area projection. This map stresses the importance of regions of relatively little earthquake activity, since it does not show the number of repetitions at active centers such as Japan. (Prepared by L. Don Leet.)*

Distribution of earthquakes

Source or focus

In *seismology*, the scientific study of earthquakes, the term *focus* is used to designate the source of a given set of earthquake waves. The waves that constitute an earthquake are generated by the fracture of rock masses. When these waves are recorded by an instrument at some distant point, their pattern indicates that they originated within a certain limited region. This region is the source or focus. Most sources have dimensions closer to 50 km in length and breadth rather than 5 or 500 km in length and breadth. But trying to fix these dimensions accurately offers a real problem, still unsolved.

In any event, the focus of an earthquake is usually at some depth below the surface of the earth. The area on the surface directly above the focus is called the epicentral area, or *epicenter* (Greek *epi*, "above").

Foci have been located at all depths down to 700 km, a little more than $\frac{1}{10}$ of the earth's radius. On some continental plate margins, the foci have clustered along a plane (subduction zone) dipping toward the interior of the continent. The deepest have been limited to the Tonga-Fiji area and the Andes.

Earthquake belts

Earthquakes tend to occur in narrow belts or zones (Figures 8-3 and 8-4). On the basis of the distribution of almost 30,000 earthquakes that occurred between 1961 and 1967, the following generalizations can now be made. First of all, narrow belts of epicenters seem to coincide with the crest of the mid-Atlantic, of the east Pacific, and of other oceanic ridges—all zones where crustal plates separate. Most frequently these are shallow-focus quakes whose *first motions* (the direction of ground motion at the beginning of arrival of primary waves) indicate vertical movement (normal displacement). Shallow-focus quakes are those that originate at depths up to about 100 km.

In contrast, shallow-focus quakes with horizontal movement (reverse or thrust first motions) are common all around the Pacific and extend through Indonesia, Burma, the Himalayas, Iran, and Turkey to the Mediterranean. Scattered shocks of this type also occur in China and Mongolia.

Deep-focus earthquakes, originating at depths greater than 100 km, are largely restricted to the landward sides of deep marine

FIGURE 8-4 *Major earthquakes, magnitude 7.0 Richter and above, that have occurred throughout the world. (Reported by the Center for Short-Lived Phenomena, 1968–1974.)*

FREQUENCY OF MAJOR EARTHQUAKES OF THE WORLD

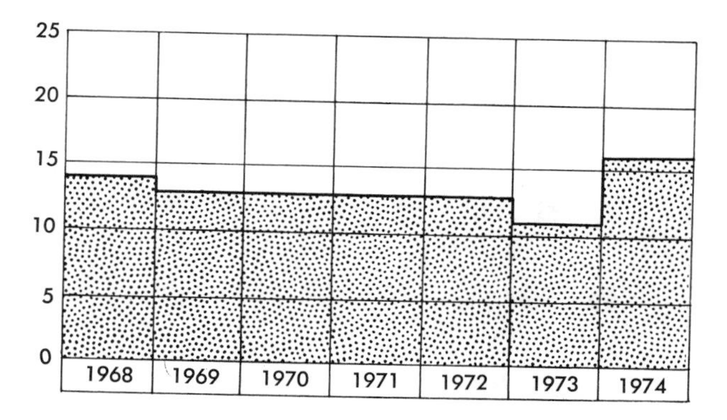

trench systems. According to crustal plate theory, deep earthquakes define the positions of subducted plates that are plunging back into the mantle beneath an overriding plate.

Although most earthquakes are located at crustal plate boundaries, and in fact are the means of defining these boundaries, a small percentage originate within plates. For example, the large Charleston (1886) and Mississippi Valley (1811–1812) earthquakes were apparently unrelated to ocean ridges, trenches, or boundaries of crustal plates.

Wave types

When rocks fracture and cause an earthquake, some of the energy released travels away by means of earth waves. The manner in which waves transmit energy can be illustrated by the behavior of waves on the surface of water.

A pebble dropped into a quiet pool creates ripples that travel outward over the water's surface in concentric circles. These ripples carry away part of the energy that the pebble possessed as it struck the water. A listening device at some distant point beneath the surface can detect the noise of impact. The noise is transmitted through the body of the water by sound waves, far different from surface waves and not visible by ordinary means.

Just as with the waterborne waves, there are two general classes of earthborn waves: (1) *body waves,* which travel through the interior of the mass in which they are generated and (2) *surface waves,* which travel only along the surface.

Body waves. Body waves are of two general types: compressional and transverse. We may also refer to them as *push-pull* and *shake* or *shear waves,* respectively. Each is defined by the manner in which it moves particles as it travels along.

Compressional or *push-pull waves,* more commonly known as *sound waves,* can travel through any material—solid, liquid, or gas. These waves vibrate rock particles back and forth; consequently the materials in the path of these waves are alternately compressed and rarefied (Figure 8-5). Because these waves reach the seismograph first, they are also called *primary,* or *P waves.* Velocities range from 6.0 to 6.7 km/sec in the crust and from 8.0 to 8.5 km/sec in the upper mantle.

FIGURE 8-5 *Push-pull wave, sometimes called compressional wave. In the top row, the balls connected by springs are all at rest. The second row shows conditions after the ball on the left-hand end has been pushed against its neighbor and the compression has started down the line. As each ball responds to this push, it compresses the next spring and pushes against the next ball. A wave of compression moves down the line, followed by a wave of pulling, or rarefaction. In this type of wave, each particle in the path of the wave moves back and forth about its starting position, along the line of the wave's advance.*

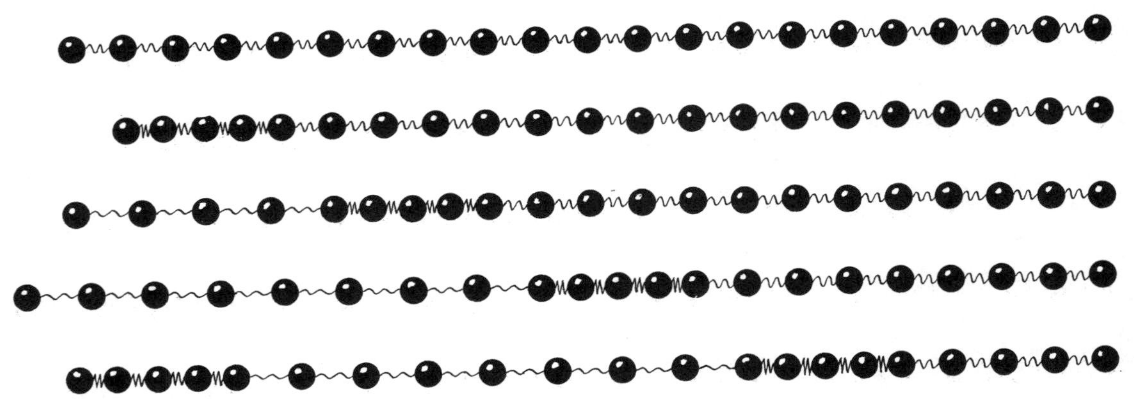

Transverse waves, or *shake* or *shear waves,* can travel only through materials that resist a change in shape. These waves shake the particles in their path at right angles to the direction of their advance. Imagine that you are holding one end of a rope fastened to a wall (Figure 8-6). If you move your hand up and down regularly, a series of waves will travel along the rope to the wall. As each wave moves along, the particles in the rope move up and down just as the particles in your hand did. In other words, the particles move at right angles to the direction of the wave's advance. The same is true when you move your hand from side to side instead of up and down.

Transverse waves are also called *secondary,* or *S waves,* because they travel slower than *P* waves, about two-thirds as fast.

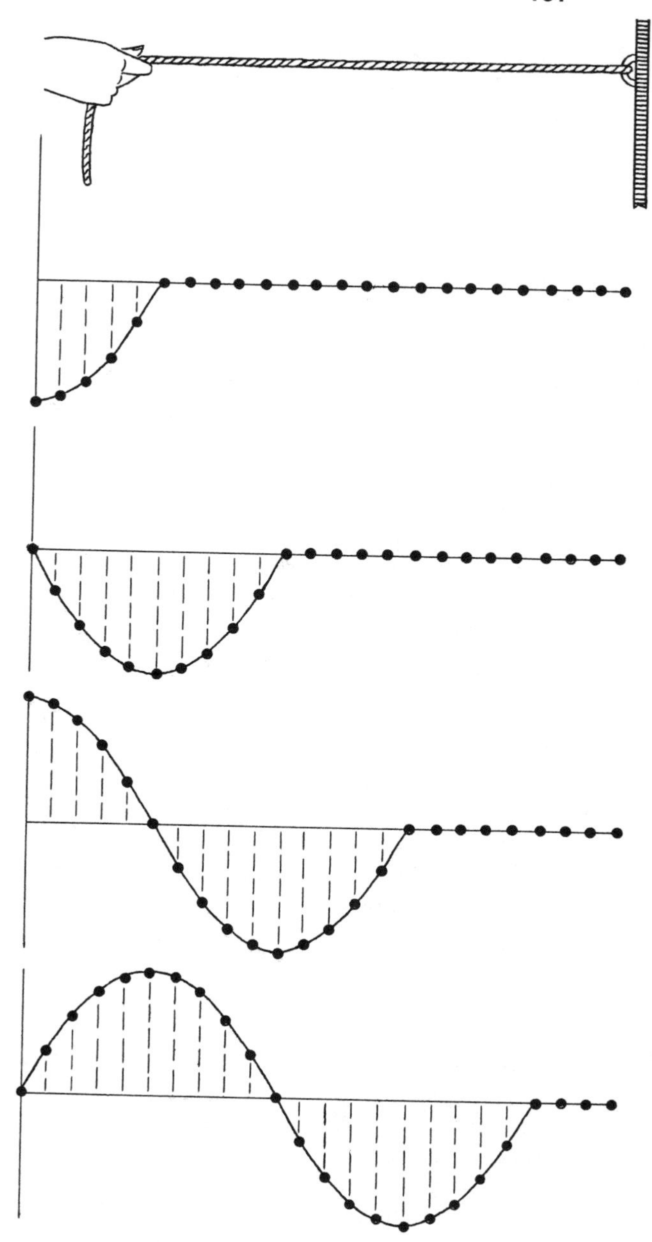

FIGURE 8-6 *Shake wave, sometimes called shear wave. If the hand is moved up and down rhythmically, adjacent particles are displaced, as shown in the sequence of diagrams, and a wave form moves along the rope. As the wave form moves forward, particles in its path move up and down. A similar result could be obtained by moving the hand from side to side or in any intermediate direction.*

Surface waves. Surface waves (or *long* waves) can travel along the surface of any material. Let us look again at the manner in which waves transmit energy along the surface of water. If you stand on the shore and throw a pebble into a quiet pool, setting up surface waves, some of the water seems to be moving toward you. Actually, though, what is coming toward you is *energy* in the form of waves. The particles of water move in a definite pattern as each wave advances—up, forward, down, and back in a small circle. We can observe this pattern by dropping a small cork into the path of waves. When surface waves are generated in a rock, one common type of particle motion is just the reverse of what we find in water—forward, up, back, and down. Surface waves are represented by a variety of waves with various complicated motions.

Surface waves are concentrated in the crust and uppermost mantle and are the ones we feel during an earthquake. They may circle the earth several times before friction damps them out.

Surface waves move more slowly than *P* or *S* waves; they transmit the initial shock along the earth's surface and are the type of wave that most persons associate with earthquakes and the destructive effects of earthquakes.

Recording earthquake waves

The seismograph and the seismogram

The waves that constitute an earthquake travel into and through the earth, as well as around its surface. The instrument that detects an earthquake is a *seismometer;* the record is a *seismogram;* the combined detector and record-writer is a *seismograph;* and the scientist who studies the records is called a *seismologist.*

A seismograph is designed to measure and record displacements of the ground. It does so by means of a mass that is either suspended on a spring to record vertical movements or attached to the end of a rod to record horizontal ones. A stylus or other device attached to the mass records the motion of the earth relative to the virtually unmoving mass. A special device marks off minutes and hours on the recording sheet. Most earthquake-recording stations have seismographs designed to respond to both horizontal (east-west and north-south) motion and vertical motion.

Seismographs are so designed that if the earth moves quickly, their mass remains at rest. For instance, a mass suspended on a spring so that it moves freely up and down, but not sideways, requires a certain length of time to complete one up-and-down cycle. This time is called a *period.* If the ground under the mass moves up and down with a shorter period (taking less time for each oscillation than the weight does if bouncing freely), the weight hangs still in space, or nearly so (Figure 8-7). This weight then serves as a point of reference from which to measure the earth's motion—a motion so small that it must be magnified in order to be recorded.

One way to record the earth's motion is to bounce a beam of light off the steady mass to a recorder that moves with the earth and that records the relative motion between the mass and the light. The farther the light is bounced, the greater the magnification.

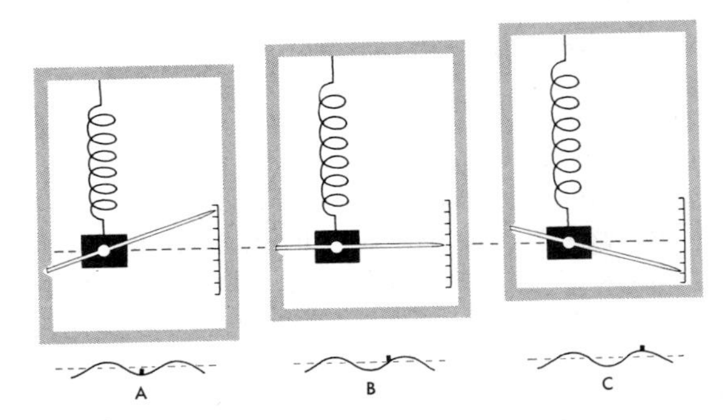

FIGURE 8-7 *The principle of a seismograph. The weight on its spring can bounce up and down, requiring a certain length of time (called a period) to complete each oscillation. But as long as the ground under the instrument moves up and down with a shorter period (taking less time for each oscillation than the weight does if bouncing freely), the weight hangs still in space, or nearly so. It can then become the fulcrum of a simple lever. Sketch B shows the seismograph in its neutral position. Sketch A shows that as the seismograph's case dips into the trough of an earth wave and the weight remains at the original level, the short arm of the lever goes down while the long arm goes up a greater amount. A record is obtained at the long arm of the lever (it is usually magnified optically or electronically, to make it large enough to be seen). Sketch C shows the opposite happening, when the case rides up onto the crest of an earth wave. The seismograph in the example records vertical movements, but the same general principle can be applied to the design of a seismograph that records horizontal movements.*

To obtain recordings of most earth waves generated by earthquakes, seismographs capable of responding to short- and long-period waves are needed. Short periods are from 1 to 5 seconds; long periods are from 5 to 60 seconds or more.

There are many kinds of seismographs. Some weigh hundreds of kilograms, or even several metric tons, and are set up in underground vaults. They can record anything from the vibrations of railroad trains kilometers away to the tremors of an earthquake any place in the world (if the earthquake is large enough). Other seismographs are small enough to be slipped into a vest pocket, or carried several in a hand, and are used to record waves generated by small dynamite charges in prospecting for minerals.

The marvels of advanced electronic technology have even led to the development of seismographs that can detect ground displacements as small as 10^{-8} cm. Such displacements are of atomic size, and in most places on the earth these sensitive seismographs would be driven off scale by "noises" (ocean waves, winds, machines). However, lunar seismographs left by the astronauts can detect seismic waves generated by a 1-kg meteorite striking anywhere on the moon's surface.

Travel times

A seismogram demonstrates that a series of three different kinds of waves are recorded after each earthquake. The first waves to arrive at the recording station are the primary waves, the second are the secondary waves, and the last are the long waves. As noted earlier, the abbreviations *P*, *S*, and *L* are commonly used for these three types.

The *P* waves are push-pull waves, or compressional waves, that travel in an elastic solid with a speed determined directly by the resistance of the material to change in volume and shape, and inversely by the density. *S* waves are shake waves, or transverse waves, whose velocity depends directly upon resistance to change in shape, and inversely in density. Both the *P* and *S* waves travel from the focus of an earthquake through the interior of the earth to the recording station. *P* waves are transmitted through liquid and solid material; *S* waves pass through solids only. The *L* waves are the surface waves that travel from the area directly above the focus, along the crust's surface to the recording station, thus taking the longest to arrive.

The *P* waves arrive at a station before the *S* waves because, although they follow the same general paths of travel, they go at different speeds. The push-pull mechanism by which the *P* waves travel produces more rapid speed than does the shake mechanism of the *S* waves. The *S* waves travel at about two-thirds the speed of *P* waves in any given earth material. The *L* waves are the last to arrive because they travel at slower speeds and over longer routes.

Table 8-1 Sample Timetable for P and S Waves

KILOMETERS FROM SOURCE	TRAVEL TIME				INTERVAL BETWEEN	
	for P		*for S*		*P and S*	*(S and P)*
	(Min)	*(Sec)*	*(Min)*	*(Sec)*	*(Min)*	*(Sec)*
1,600	3	22	6	03	2	41
3,100	5	56	10	48	4	52
4,900	8	01	14	28	6	27
6,500	9	50	17	50	8	00
8,000	11	26	20	51	9	25
9,500	12	43	23	27	10	44
11,000	13	50	25	39	11	49

Less than 11,000 kilometers

From thousands of measurements the world over, it has been learned that *P*, *S*, and *L* waves have regular travel schedules for distances up to 11,000 km. From an earthquake in San Francisco, for example, we can predict that *P* will reach El Paso, 1,600 km away, in 3 minutes 22 seconds and *S* in 6 minutes 3 seconds; *P* will reach Indianapolis, 3,220 km away, in 5 minutes 56 seconds and *S* in 10 minutes 58 seconds; *P* will reach Costa Rica, 4,800 km away, in 8 minutes 1 second and *S* in 14 minutes 28 seconds. The travel schedules move along systematically out to a distance of 11,000 km, as shown in Table 8-1.

Beyond 11,000 kilometers

Beyond 11,000 km, however, something happens to the schedule and the *P* waves are delayed. By 16,000 km, they are 3 minutes late. When we consider that up to 11,000 km, we could predict their arrival time within seconds, a 3-minute delay becomes significant.

The fate of the *S* waves is even more spectacular: They disappear altogether, never to be heard from again (Figure 8-8)!

When the strange case of the late *P* and the missing *S* was first recognized, seismologists realized they were not just recording earthquakes but were developing a picture of the interior of the globe. This subject is considered in greater detail later in the chapter.

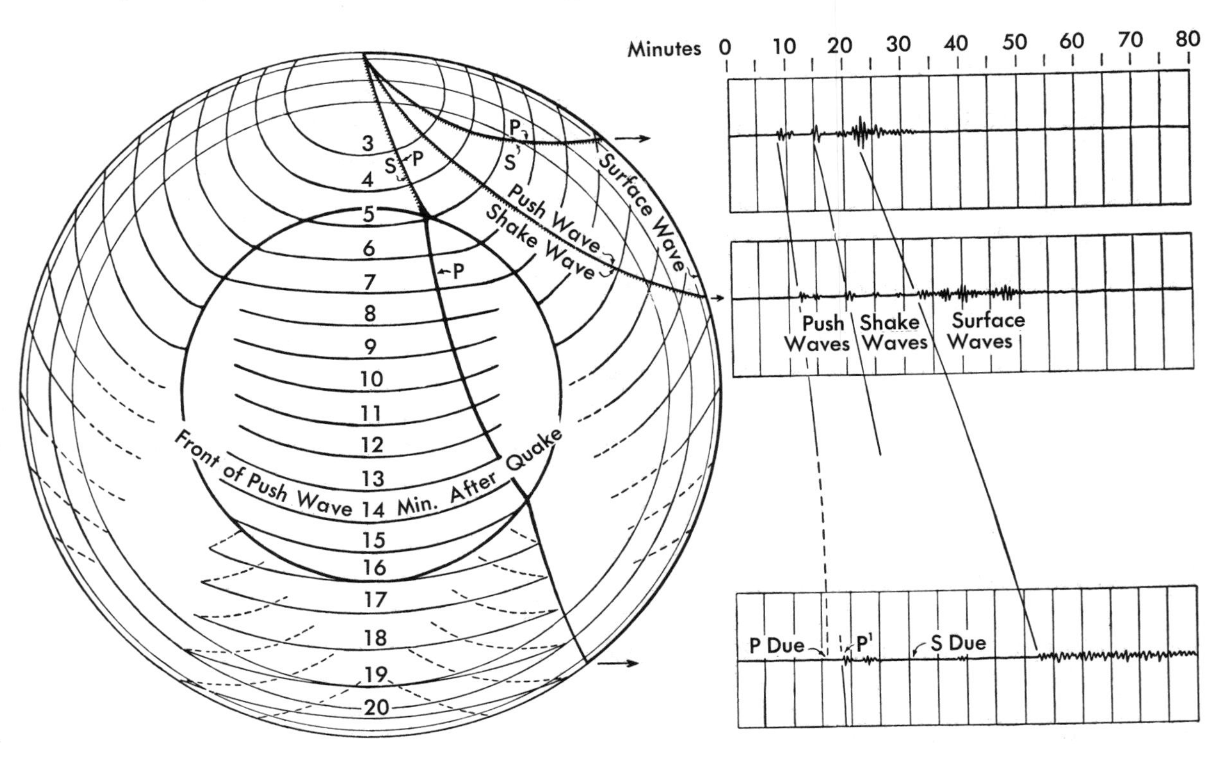

FIGURE 8-8 Successive positions of the advancing front of push-pull waves in the earth's interior. Also shown are the paths of P, S, and surface waves to three different distances, with the seismograms for those distances. The effect of the core in delaying P and eliminating S is illustrated at the greatest distance.

Earthquake magnitude and energy

How to specify the size of an earthquake has always posed a problem. Before the modern development of instrumental seismology, some of the early investigations of earthquakes led to various attempts to describe the intensity of the shaking. A missionary in some remote region would keep a diary in which he listed earthquakes as weak, strong, or very strong. This was, at best, a personal scale.

Many earthquake intensity scales have been proposed; the most useful one is the Modified Mercalli Earthquake Intensity Scale (Table 8-2). After any significant earthquake in the United States, questionnaires are circulated to people in the affected area, and their responses to the shock are tabulated. A map is then drawn showing lines of equal intensity, *isoseismal lines,* according to the Modified Mercalli scale. The isoseismals usually form crude ovals about a center that coincides with the epicenter or is close to it. Such a system of study is useful in situations where insufficient seismographs are situated, which would allow for a more sophisticated approach to intensity determinations.

It is quite unsatisfactory, however, to have to rely on the personal impressions of various individuals to compile accurate infor-

Table 8-2 Modified Mercalli Scale if Earthquake Intensities

I.	Not felt.
II.	Felt by persons resting, particularly on upper floors of buildings.
III.	Felt indoors but may not be recognized as an earthquake; vibrations like those of a passing light truck.
IV.	Felt indoors like the vibrations from a passing heavy truck; hanging objects swing; standing objects rock; windows, dishes, doors rattle; glasses clink.
V.	Felt outdoors; sleepers awaken; liquids move and some spill; small unstable objects move or fall; doors swing; shutters and pictures move.
VI.	Felt by all; many frightened and run outdoors; people walk unsteadily; windows, dishes, glassware break; books, dishes fall from shelves; pictures from walls; furniture moved or overturned; weak plaster and poor masonry crack; small bells ring; trees, bushes shake visibly and rustle.
VII.	Difficult to stand; noticed by drivers of cars; furniture breaks; poor masonry cracks; plaster, loose bricks, stones, tiles, cornices fall; small slides and caving along sand and gravel banks; water becomes turbid; large bells ring; concrete irrigation ditches are damaged.
VIII.	Difficult to steer cars; damage to good unbraced masonry or partial collapse; some damage to good, somewhat reinforced masonry but more to masonry reinforced against horizontal stresses; stucco and some masonry walls fall; chimneys, factory stacks, monuments, towers, and elevated tanks twist and fall; frame houses move on foundations if not bolted in place; loose panel walls thrown out; weak pilings broken; tree branches break; flow and temperature of springs and wells may change; steep slopes and wet ground crack.
IX.	Causes general panic; poor masonry destroyed; good masonry damaged seriously; frame structures not bolted shifted off their foundations; frame cracks; serious damage to reservoirs; underground pipes break; in areas of loose sediment, sand, silt, clay and water ejected.
X.	Most masonry, frame structures, and foundations destroyed; some well-built structures and bridges destroyed; serious damage to dams, dikes, embankments; large landslides; water thrown on banks of canals, rivers, and lakes; sand and mud shift horizontally on flat areas; railroad tracks bend slightly.
XI.	Puts underground pipe lines completely out of service; railroad tracks bent severely; many bridges destroyed.
XII.	Damage to man-made structures nearly total; large rock masses displaced; lines of sight distorted; objects thrown into air.

NOTE: Slightly changed from Don Tocher, 1964

mation. A scale based on the amplitude of seismic waves recorded by seismographs was developed by C. F. Richter of the California Institute of Technology in 1935. This system ascribes to each earthquake a number, called the earthquake's *magnitude,* which is based on the logarithm of the maximum amplitude plus an empirical factor that takes into account the weakening of seismic waves as they spread away from the focus. Empirical tables allow the magnitude of any earthquake to be calculated from any seismogram that records it. Seismologists all over the world can thus quickly study their records, compute nearly the same value for the magnitude of an earthquake, and prepare to meet the press and television to report their findings.

The largest earthquakes recorded have had magnitudes of about 8.6. The maximum range of magnitudes so far observed is from −3 to +8.6. Because magnitudes are based on a logarithmic scale, a change in magnitude of 1 unit corresponds to a change in amplitude of seismic waves by a factor of 10. A magnitude −8 earthquake, for example, is 10 times larger than a magnitude −7 earthquake. However, a 1 unit increase in magnitude is equivalent to an increase in energy released of about 30 times (Table 8-3). Although energy released is the best way of measuring the size of an earthquake, its calculation is a complicated process requiring the dimensions of the fault, the amount of slip of the fault, and other measurements.

Table 8-3 Comparison of Richter Magnitude and Energy Released

MAGNITUDE	APPROXIMATE ENERGY RELEASED
1.0	170 grams TNT
2.0	5.8 kg TNT
3.0	178.6 kg TNT
4.0	5.4 (metric) tons TNT
5.0	179.1 tons TNT
6.0	5643 tons TNT
7.0	179,100 tons TNT
8.0	5,643,000 tons TNT
9.0	179,100,000 tons TNT

An earthquake of magnitude 2 corresponds to a shallow shock barely noticeable near the epicenter. A magnitude of 7 is about the lower limit of a major earthquake. However, the San Fernando, California earthquake of 1971 had a magnitude of 6.4, resulting in property damage estimated at more than half a billion dollars.

The magnitude of an earthquake and its characteristic effects can be related roughly (Table 8-4). Although it is estimated that each year more than a million earthquakes shake the earth strongly enough to be felt, most

Table 8-4 Earthquake Magnitudes, Effects, and Frequency

CHARACTERISTIC EFFECTS OF SHALLOW SHOCKS IN POPULATED AREAS	APPROXIMATE MAGNITUDE	NUMBER OF EARTHQUAKES PER YEAR
Damage nearly total	$\geq$8.0	0.1–0.2
Great damage	$\geq$7.4	4
Serious damage	7.0–7.3	15
Considerable damage to buildings	6.2–6.9	100
Slight damage to buildings	5.5–6.1	500
Felt by all	4.9–5.4	1,400
Felt by many	4.3–4.8	4,800
Felt by some	3.5–4.2	30,000
Not felt but recorded	2.0–3.4	800,000

Table 8-5 Large Earthquakes (1906–1964)

DATE	PLACE	LOCATION	MAGNITUDE
Jan. 31, 1906	Colombia	1°N, 82°W	8.6
Aug. 17, 1906	Chile	33°S, 72°W	8.4
Jan. 3, 1911	U.S.S.R.	44°N, 78°E	8.4
Dec. 16, 1920	China	32°N, 105°E	8.5
Mar. 2, 1933	Japan	39°N, 145°E	8.5
Aug. 15, 1950	Chinese-Indian border	29°N, 97°E	8.6
Mar. 28, 1964	Alaska	61°N, 147°W	8.5

earthquakes are small. Great earthquakes (magnitudes >8) happen only once every 5 to 10 years. The few large earthquakes that happen each year release more seismic energy than the hundreds of thousands of small shocks combined. For example, the energy released by an earthquake of 8.6 (largest known magnitude) is 3 million times as great as that of an earthquake of magnitude 5. Further, seven great earthquakes (Table 8-5) accounted for nearly 25 percent of the total energy released by all earthquakes from 1906 to 1964.

Locating earthquakes

We now have timetables for earth waves for all possible distances from an earthquake. These are represented in the graph of Figure 8-9.

When the records of a station give clear evidence of the *P*, *S*, and *L* waves from an earthquake, the observer first determines the intervals between them. By using an interval table, such as that of Table 8-1, he can immediately translate the intervals into actual distances. For example, if he observes that *S* arrived 8 minutes after *P*, he knows that the earthquake must have been 6,440 km away. He then notes that *P* arrived at 4 hours, 12 minutes, 22 seconds that morning. From the travel timetable, he can then find that *P* requires 9 minutes, 50 seconds to travel

FIGURE 8-9 *Time-distance graph for earthquake waves. P^1 is split beyond 10,000 miles (16,100 km) by the effect of the solid inner core.*

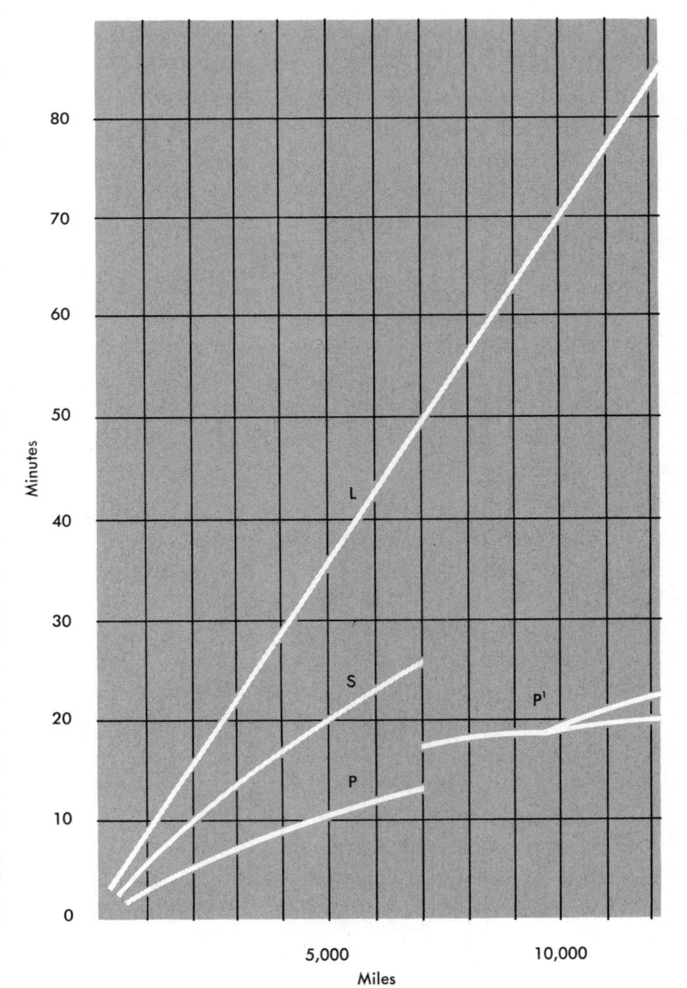

6,440 km. The earthquake, therefore, happened at 4 hours, 02 minutes, 32 seconds that morning.

Since this process is carried out at all the seismograph stations that have recorded the quake, an arc can be drawn on a globe to represent the computed distances from each station. The point at which all the arcs intersect indicates the center of the disturbance.

Effects of earthquakes

Of all the earthquakes that occur every year, only one or two are likely to produce such spectacular effects as landslides or the elevation or depression of large land masses. A hundred or so may be strong enough, near their sources, to destroy human life and property. But the rest are too small to have any serious effects.

Fire

When an earthquake happens near a modern city, fire is a greater hazard than the actual shaking of the ground (Figure 8-10). In fact, fire has caused an estimated 95 percent of the total loss caused by some earthquakes.

On September 1, 1923, an earthquake occurred underneath Sagami Bay, 80 km from Yokohama, and 110 km from Tokyo. The rupture produced vibrations that spread outward with such energy that they caused serious damage along the Japanese coast over an area 150 km long and 100 km wide.

Within 30 minutes after the beginning of the earthquake, fire had broken out in 136 places in Tokyo. In all, 252 fires were started and only 40 were extinguished. Authorities estimated that at least 44 were started by chemicals. A 20 km/hr wind from the south spread the flames rapidly. The wind shifted to the west in the evening and increased to 40 km/hr, and then shifted to the north. These changes in wind direction added greatly to the extent of the area burned. Within 18 hours, 64 percent of the houses in Tokyo had burned.

FIGURE 8-10 *City of Managua, Nicaragua, on fire after earthquake, December 27, 1972 (Table 8-6). (Photo by Lloyd S. Cluff.)*

Table 8-6 Some of the World's Worst Earthquakes in Terms of Lives Lost

YEAR	PLACE	ESTIMATED DEATHS
1290	Chihli China	100,000
1456	Naples, Italy	60,000
1531	Lisbon, Portugal	30,000
1556	Shen-shu, China	830,000
1667	Shemaka, Caucasia	80,000
1693	Naples, Italy	93,000
1693	Catania, Italy	60,000
1731	Peking, China	100,000
1737	Calcutta, India	300,000
1755	Lisbon, Portugal	60,000
1783	Calabria, Italy	50,000
1797	Quito, Ecuador	41,000
1847	Zenkoji, Japan	34,000
1908	Messina, Italy	160,000
1920	Kansu, China	180,000
1923	Tokyo and Yokohama, Japan	143,000
1932	Kansu, China	70,000
1935	Quetta, Baluchistan	60,000
1939	Erzincan, Turkey	40,000
1962	Northwestern Iran	12,000
1968	Dasht-e Bayaz, Iran	11,600
1970	Peru	50,000
1972	Managua, Nicaragua	10,000
1976	Guatemala	22,000

SOURCE: Press, Frank, and Seiver, Raymond, *Earth*. New York: W. H. Freeman and Company, 1974, 945 pp.

The fires died away finally after 56 hours with 71 percent of the houses consumed, a total of 366,262 houses lost. The spread of fire in Yokohama, a city of half a million population, was even more rapid. Within 12 hours, 65 percent of the structures in the city had burned. Eventually the city was completely destroyed. Fire was also a great killer in this earthquake. Final government statistics for this particular catastrophe were 143,000 persons killed and 43,476 missing (Table 8-6). More than half a million homes were completely destroyed.

Tidal waves and seiche

Tidal waves are more correctly referred to by the term *seismic sea waves* or by the Japanese word *tsunami*, which has the same form in both the singular and the plural.

Any earthquake that causes an abrupt change in the level of the sea bottom or produces submarine landslides may generate a tsunami. Tsunami travel across the open ocean at speeds ranging between 450 and 800 km/hr depending on the depth of the water. For example, on April 1, 1946, a severe earthquake occurred at 53.5° north, 163° west, 130 km southeast of Unimak Island, Alaska, where the ocean is 4,000 m deep on the steepest slope of the Aleutian trench. Four and one-half hours later a tsunami reached Oahu, Hawaii, after traveling 3,600 km at 800 km/hr.

The effects at Oahu and other Pacific shores were dramatic. The energy that moved thousands of meters of water in the open ocean became concentrated on moving a few meters of water at a shallow shore. There the

Table 8-7 Energy Equivalents, Deaths, and Property Damage of Several Severe Earthquakes in the United States

YEAR	LOCATION	MAGNITUDE	TNT EQUIVALENT (metric tons)	DEATHS	PROPERTY DAMAGE
1906	San Francisco, Calif.	8.25	11,295,000	700	$524,000,000
1933	Long Beach, Calif.	6.3	14,220	115	40,000,000
1940	El Centro, Calif.	7.1	225,000	9	6,000,000
1946	Hawaii	(destruction by tsunamis)		173	25,000,000
1952	Kern County, Calif.	7.7	1,791,000	14	60,000,000
1959	Hebgen Lake, Montana	7.1		28	11,000,000
1964	Anchorage, Alaska	8.5	28,395,000	131	500,000,000
1971	San Fernando, Calif.	6.6		65	504,950,000

water curled into giant crests that increased in height until they washed up over shores meters above the high tide line: 12 m on Oahu and 18 m on Hawaii. The tidal wave demolished 488 homes and damaged 936 others, with a property loss estimated at $25 million. One hundred and seventy-three people were killed.

The motion of an earthquake sets up in a lake or reservoir a feature called a *seiche,* a wave that oscillates back and forth in a basin like sloshing waves in a bathtub. The Hebgen Lake earthquake in 1959 (Table 8-7) was a striking example of this phenomenon (Figure 8-11). Hebgen Lake was a reservoir held in place by a dam. Suddenly the earthquake lowered the bottom of the reservoir. "An eyewitness standing in the moonlight on Hebgen Dam and looking down its sloping face could not see the surface of the water, so far had it receded. Then with a roar it returned, climbing up the face of the dam until it overflowed the top, and poured over it for a matter of minutes. Then the water receded again, to become invisible in the moonlit night. The fluctuation was repeated over and over, with a period of about seventeen minutes; only the first four oscillations poured water over the top of the dam, but appreciable motion was still noted after eleven hours."[3]

Damage to structures

Modern, well-designed buildings of steel-frame construction can withstand the shaking of even the most severe earthquakes. In the Tokyo earthquake of 1923, the Mitsubishi Bank Building was surrounded by many badly damaged structures of an older type of construction, but it escaped comparatively unharmed. In the July 28, 1957 earthquake, the 43-story Latino-Americano Tower in Mexico City rode the shock waves undamaged, while surrounding buildings suffered greatly.

With a death toll of 65, the San Fernando earthquake (1971) in the heavily populated Los Angeles area caused the fourth most serious loss of life in the United States in the last 200 years (Table 8-7). Densely populated urban areas where large earthquakes are possible sit on potential catastrophes.

[3]Hodgson, John H., *Earthquakes and Earth Structure.* Englewood Cliffs, N.J.: Prentice-Hall, Inc., 1964.

***FIGURE 8-11** A few minutes after midnight on August 17, 1959, an earthquake of shallow focus struck southwestern Montana. The quake triggered this landslide, which carried an estimated 2 million tons of rock into the valley of the Madison River. Scores of individuals, most of them occupants of a now-buried campsite, lost their lives, and the debris backed up the Madison River into a 30-meter-deep lake. (Photo by Lloyd Skinner.)*

The major loss of life in the San Fernando earthquake was caused by the collapse of old structures built before the passage of the Field Act, a law passed by the state legislature after the Long Beach earthquake in 1933 that established standards for school construction to prevent earthquake damage. Eleven schools in the area were so badly damaged that they had to be demolished after the earthquake; but all of these had been built before the act was passed.

The most severely damaged buildings in the San Fernando earthquake were two hospitals not constructed under the same standards as schools. The main buildings of the Olive View Hospital, only a year old, were damaged beyond repair (Figure 8-12). Vertical acceleration with the ground swell lifted the Veterans Administration Hospital, killing 48 patients (Figure 8-13).

Freeways within a 16-km radius of the rupture zone were devastated (Figure 8-14). Out of 70 spans of above the ground, prestressed concrete interchanges, 40 suffered considerable damage, and many collapsed completely. The freeways were empty at 6:01 A.M. when the quake occurred, and in this respect the people of the San Fernando area were lucky.

The Van Norman Reservoir was badly damaged, but the water level was 4.5 m below the maximum when the quake occurred. Most of the water was withdrawn from the reservoir, and 80,000 people below the dam were evacuated.

This earthquake, although it was a modest one in terms of energy released, had the highest acceleration rate ever measured, as much as 1.25 G. A wedge of rock several kilometers long was thrust southward for about 0.8 m. The surface of the wedge was raised nearly 2 m in some places (Figure 8-15).

Future destructive earthquakes can be expected in the heavily populated Los Angeles and San Francisco areas. As North America overrides the Pacific Basin in its westward drift, parts of the Pacific Ocean floor are subducting under its western border. The continent is simultaneously moving northward, relative to the Pacific Basin. The San Andreas fault zone is near the boundary between the northward-moving continent and the southward-moving Pacific Ocean Basin. There can be little comfort or tranquillity for Californians in their structural setting.

Changes in land level

Along a coast not far from Tokyo, a marine animal known as *Lithophea nasuta* (the rock-eater first described by a man named Nasuta) lives in a cigar-shaped pair of shells about 5 cm long, drilling its home in the rocky shores at mean sea level, and subsisting on organisms brought to its abode by ocean currents. At one place on the coast, abandoned sets of bore holes have been found at four different levels above the sea. Historical records suggest that each of these rises of the land coincides with a major earthquake, in the years 33, 818, 1703, and 1923, respectively. The total rise over the 1,890-year period was 14 m. This may not seem like much of a change, but if it were to continue at the same rate for the next 200,000 years, the land would rise about 1.5 km. And in geological terms that is remarkably rapid movement.

The Prince William Sound earthquake of March 27, 1964, was accompanied in Alaska by considerable vertical and horizontal displacements. These took place over more than 120,000 km^2. The U.S. Coast and Geodetic Survey found that there was a general south to north rotation of a portion of the earth's crust centered at the northern coast of Prince William Sound. A maximum uplift of 13 m and a maximum subsidence of 2 m occurred. Maximum horizontal movements were 4.5 to 6 m. At Valdez the earthquake triggered a landslide that deepened the harbor as much as 100 m in one place.

FIGURE 8-12 (Above) Collapse of the Olive View Hospital in the San Fernando earthquake, 1971. (Courtesy Lloyd S. Cluff.)

FIGURE 8-13 (Left) The destroyed Veterans Administration Hospital after the San Fernando earthquake. (Photo by Lloyd S. Cluff.)

FIGURE 8-14 (Below) Collapsed freeway after the San Fernando earthquake. (Courtesy Lloyd S. Cluff.)

FIGURE 8-15 *(Left) An example at the surface of part of the San Fernando thrust fault. (Photos by Lloyd S. Cluff.)*

(a)

(b)

FIGURE 8-16 *(a) The central plaza of Yungay, Peru, before the 1970 earthquake (Table 8-6). (b) Photograph of central plaza taken after the earthquake. (Photo by Lloyd S. Cluff.)*

Landslides

In regions where there are many hills and steep slopes, large earthquakes are often accompanied by landslides (Figure 8-16). These slides occur within a zone seldom exceeding 40 to 50 km in radius, although some very large earthquakes have affected areas as far away as 150 km. In the Alaska earthquake of 1964, for example, landslides occurred everywhere and probably caused the greatest damage to cities and towns. The earthquake reactivated geologically old slides, perhaps produced by older earthquakes.

Severe submarine slides in Alaska also destroyed dock and port facilities at Homer, Whittier, and Anchorage. At Seward, 86 houses were destroyed and 269 houses were damaged by landslides and by ground fracturing. Slide-generated ocean waves washed onto the shore and damaged additional structures.

Cracks in the ground

If the movie star Raquel Welch can fall into a rapidly opening fracture as the earth shudders, why not you? One of man's most persistent fears about earthquakes is that the earth is likely to open and swallow everyone and everything in the vicinity. Such fears have been nourished by a great many tall tales. One account of the Lisbon earthquake claimed that about 40 km from Lisbon, the earth opened up and swallowed a village's 10,000 inhabitants with all their cattle and belongings, and then closed again. The story probably got its start when a landslide buried some village in the area.

It is true that landslides and slumps do bury people and buildings, and under special conditions quakes may even open up small shallow cracks. In California in 1906 a cow did fall into such a crack and was buried with only her tail protruding. But there is no authenticated case in which solid rock has yawned open and swallowed anything.

Sound

When an earthquake happens, the vibrations in the ground often disturb the air and produce sound waves that are within the range of the human ear. These are known as *earthquake sounds.* They have been variously described, but usually as low booming noises. Very near the source of an earthquake, sharp snaps that are sometimes audible suggest the tearing apart of great blocks of rock. Farther away, the sounds have been likened to heavy vehicles passing rapidly over hard ground; the dragging of heavy boxes of furniture over the floor; a loud, distinct clap of thunder; an explosion; the boom of a distant cannon; or the fall of heavy bodies or great loads of stone. The true earthquake sound, of course, is quite distinct from the rumble and roar of shaking buildings. In some cases, however, the sounds are probably confused.

Earthquake forecasting

Man has long wished to predict earthquakes. In fact, every spring a fanatic religious zealot residing in Los Angeles or San Francisco will make such a prediction, giving the day, hour, and minute of the expected cataclysm. Although the disappointment is not large when the event fails to occur, man continues to desire the ability to foretell earthquakes, even if the prediction can be made only shortly before the actual fault movement happens.

Recent findings indicate that reliable earthquake prediction is almost a reality. Before earthquakes occur, there frequently are changes in rock behavior that affect the velocities of other earthquake waves passing through the rocks. Soviet seismologists have used earthquake waves from other unrelated earthquakes to measure the alterations in wave speed through rocks around a fault zone. For months to years before a particular earthquake, the Soviet scientists observed that the strained rock in the fault zone was deformed in a way that slowed other earthquake waves that passed through the zone. Similar behavior has preceded earthquakes in California and New York.

Changes in electrical resistance, water pressure, rock motion, and leakage of gas also can accompany the lowering of wave velocity. Fractures in the fault zone apparently open, which leads to a lowering of water pressure. When the fractures are filled by underground water, the continuing stress on the rocks is also exerted on the water in pores, which contributes to the pressure within the rocks and ultimately triggers further fault movement and earthquakes.

These preliminary events have been observed and studied for many earthquakes. The larger the earthquake, apparently the longer the time during which the preliminary events take place. Careful observation and measurement of the early events will precede reliable forecasts. The preliminary events are most evident along normal and reverse fault systems, and strike-slip faults may not produce the same effects. Although earthquake forecasting is in its infancy, both American and Soviet scientists have been able to predict the occurrence of a few earthquakes.

Earthquake control

Understanding of the causes of earthquakes has opened several possibilities for their control. Underground nuclear explosions in Nevada have released strain energy stored in certain rocks. In some instances, the shock wave from the explosion has raised the strain on nearby fractures and faults enough to initiate fault movement. All of the resulting earthquakes have been small, but a large earthquake could conceivably be initiated.

In the future, a situation may arise where it is desirable to deliberately initiate an earthquake near a heavily populated area because too large an amount of strain has accumulated on an active fault zone in the vicinity. If hazardous areas were evacuated and if emergency services were standing at the ready, such action might be deemed necessary to prevent a later much more damaging earthquake. However, the legal, environmental, and human problems would be large indeed, perhaps too great for such action to be taken.

Another possibility for earthquake control is much more exciting. Increasing water pressures can initiate faulting, as was unintentionally demonstrated by a deep well at the U.S. Army's Rocky Mountain Arsenal near Denver, Colorado, in the early 1960s. Disposal of nerve gas wastes in the well triggered movement along deeply buried inactive faults in the region. The liquid waste reduced frictional resistance along fault planes in the rocks surrounding the well, leading to movement along the faults. Some of the resulting earthquakes reached magnitudes of 3 and 4 on the Richter scale. Earthquake activity in the area correlated closely with the times of pumping of wastes into the disposal well, as was demonstrated convincingly by a Denver geologist, David Evans. Strain energy stored along the fault planes was apparently released by the fluid injection.

Experiments by the U.S. Geological Survey in the Rangely oil field of northwestern Colorado have added to the experience gained from the study of the Denver earthquakes. The Survey geologists injected water in some of the Rangely wells, causing very small earthquakes. By withdrawing the water, the earthquakes were stopped.

Although it is premature, many geologists believe we could eventually restrain earthquakes by injecting fluid into fault zones to permit slippage to take place gradually or in a series of small earthquakes. However, means must be found to control the areas affected and to be certain a major destructive quake is not triggered.

The earth's interior

So far in this chapter we have been concerned with earthquakes—phenomena that take place in the outer one-tenth of the globe. Now let us turn to the interior of the earth, down to its very center, and see what we can reasonably deduce, not only from our knowledge of the behavior of the waves generated by earthquakes but also from our knowledge of the mass, shape, and behavior of the earth and from the nature of meteorites.

The mass of the earth

The earth has a mass of 5.97×10^{27} grams, and a volume of 1.08×10^{12} km^3. The density, then, is approximately 5.5 grams per cubic centimeter (5.5 g/cm^3). In sharp contrast to this figure for the the earth as a whole are the values for the rocks that we can sample at or near the earth's surface. For instance, granite has a density of approximately 2.6, basalt an average of little less than 3. Sedimentary rocks such as sandstone, limestone, and claystone are perhaps 2.4 or 2.5. The average density of the surface rocks of the continents is 2.8. Very few minerals are as dense as 5.5.

We must therefore ask where the extra mass is in the earth. Why is the earth so much heavier than the rocks that we can actually sample at the surface? Rock density is obviously much greater at increasing depths.

As a first approximation, we might say that the lightweight skin with a density of about 2.8 completely encircles an inner zone having a density of approximately 6 or a little more. If the volume of these two zones were averaged out, we might get a density of 5.5 for the entire earth.

But if we turn to the rotation of the earth, we find that the axis of rotation changes slowly under the combined influence of sun and moon. This wobble, like that of a giant top, is known precisely and is best explained by assuming that a very large part of the earth's mass is concentrated near its center. This suggestion is in part substantiated by the shape of the earth. The extent of the equatorial bulge of the earth indicates that the mass of the earth is greatest toward its center, and that a moderately dense material extends to a considerable depth beneath the surface. Were the mass of the earth more evenly distributed through its body, the equatorial bulge would be considerably greater.

From these facts, we can conclude that there is some sort of stratification within the earth, going from lightweight material at and near the surface toward heavier material at the core. More specific deductions can be made from the use of earthquake data and from the nature of meteorites.

The evidence of earthquakes

Having studied the causes, effects, and recordings of earthquakes, we wish to use our knowledge to extend our senses and probe beyond the deepest mine and oil well, down toward the interior of the earth.

Table 8-8 Velocities of *P* Waves in Various Materials

MATERIAL	WAVE VELOCITY (in km/sec)
Sea water	1.5
Sand, silt, clay (unconsolidated sediment)	0.5–2.5
Sandstone	1.5–5.4
Limestone, dolomite	3.0–6.0
Metamorphic rocks	4.5–6.0
Granite (silicic)	3.9–6.3
Basalt	4.8–6.0
Gabbro	6.9–7.2
Peridotite, dunite (ultramafic)	7.2–8.2

Studies of the travel habits of earthquake waves through the earth and of surface waves around the earth have yielded considerable information about the structure of the globe from its surface to its center. These studies have been made possible by our knowledge of the speed at which these waves travel and of their behavior in different materials (Table 8-8). For example, waves travel at greater speeds through simatic (silicon and magnesium rich) materials than through sialic (silicon and aluminum rich) materials. When they move from one kind of material to another across a discontinuity, they are deflected, as light waves are deflected by a lens. Part of the energy of the waves is bounced back to the surface where it can be recorded. The rest of the energy travels on into the new material.

On the basis of the information assembled from studies of the travel habits of waves, the earth has been divided into three zones: the *crust*, the *mantle*, and the *core*. Recent advances have led to the recognition of finer such divisions (see Table 8-9).

The crust

Information on the earth's crust comes primarily from seismological observations. These include *P* and *S* waves from local earthquakes (within 1,127 km) and from dynamite and nuclear blasts. One of the first things revealed is that the earth's crust is solid rock. Early in the history of crustal studies a seismologist in Yugoslavia, A. Mohorovičić (Moho-ro-vee-cheech), made a special study of records of the earth waves from an earthquake on October 8, 1909, in the Kulpa Valley, Croatia. He concluded that velocities of *P* and *S* increased abruptly below a depth of about 50 km. This abrupt change in the speed of *P* and *S* indicated a change in material and became known as the *Mohorovičić (M) discontinuity*. For convenience, it is now referred to as the *Moho*. The Moho marks the bottom of the earth's crust and separates it from the mantle. It separates rocks in which *P* waves have velocities of about 6 to 7 km/sec from the underlying mantle in which *P* waves have a velocity of about 8 km/sec.

Table 8-9 General Structure of the Earth

LAYER	THICKNESS (in km)	COMPOSITION
Crust	10–12 (oceans) 30–70 (continents)	See Table 8-10. Sedimentary and metamorphic rocks, granite, basalt.
Mantle		
Lithosphere	70	Dunite, garnet peridotite, garnet pyroxenite, or eclogite.
Asthenosphere	130–160	Weak, low-velocity zone; partly molten; source of basaltic magma; peridotitic.
Transition zone	450	Solid; peridotite (olivine-rich).
Lower mantle	2198	Mg, Si, O in high-pressure, mixed mineral phases.
Outer core	2247	Liquid; iron-nickel, sulfur (?), silicon (?).
Inner core	1223 (radius)	Solid; iron-nickel, sulfur (?), silicon (?).

The continental crust

The depth to the Moho varies widely in different parts of the continents. In the United States, information collected up to the present time shows that crustal thickness ranges from about 20 km to more than 50 km. There is no relation between thickness of the crust and altitude. The Rocky Mountains in Montana, for example, have a thinner crust than the plains of Montana and North Dakota.

Composition of the continental crust is variable. Although sedimentary rocks locally attain thicknesses of 15 km, the average thickness for the United States is about 4.5 km. Large volumes of sedimentary beds throughout the world have been metamorphosed to schist, gneiss, quartzite, and marble. There also are tremendous thicknesses of volcanic rocks interbedded with the sedimentary and metamorphic sequences.

Average velocities of seismic waves in the upper continental crust are similar to those measured in granite. Granite and gneiss of approximately the same chemical composition are the principal rocks of the *shields*, which are broad continental regions of low local relief. Shields consist mainly of minerals rich in silica and aluminum (sial).

Earthquake speeds gradually increase with depth in the crust, indicating a change in the average composition. In the lower continental crust, wave speeds are variable but generally appropriate to gabbro (Table 8-8) or rocks of similar composition. However, rocks of the lower crust probably are the same average mineral composition as basalt, but are not basalts.

The oceanic crust

Seismic and gravity studies indicate that the oceanic crust is very different from the continental crust. Recent coring of samples from sea floors has provided much additional information.

Beneath thin layers of unconsolidated marine sediment and an underlying thin layer of sedimentary rocks, most ocean floors are composed of rocks whose elastic properties are like those of continental sima (Table 8-10). The sima layer of basalt or gabbro generally is less than 5 km thick above the M discontinuity. Total thickness of oceanic crust plus sea water is only 10 to 12 km, in marked contrast to the great thickness of continental crust.

Three suboceanic layers are widely recognized: (1) unconsolidated sediment, (2) sedimentary beds, and (3) basalt or gabbro. The different compositions of the layers are reflected in the different velocities of the *P* waves (Table 8-10).

The mantle

In 1926, the brilliant geophysicist Beno Gutenberg first suggested that within the upper mantle there is a low-velocity zone, now called the *asthenosphere.* From 100 to 200 km below the top of the mantle, *P* and *S* waves travel slower than they do in the overlying

Table 8-10 Structure of Oceanic Crust

LAYER	THICKNESS (in km)	*P* WAVE VELOCITY (in km/sec)
Sea water	4	1.5
Unconsolidated sediment	1 (Atlantic)	
	0.5 (Pacific)	2.0
Sedimentary rocks	1.7	5.0
Basalt or gabbro	<5	6.7
Mantle (upper part)		7.8–8.3

lithosphere. For many years Gutenberg's idea was viewed with skepticism but has now been fully accepted.

The asthenosphere is attributed to the presence of hotter than normal rocks or to pockets of magma, perhaps totalling 1 to 10 percent melt. The zone is a concentric shell, a zone of weakness present beneath both oceans and continents. It may be the place of formation of magma.

Rocks above the asthenosphere are called the *lithosphere,* the shell in which the continents are embedded. Probable composition is given in Table 8-9. The lithosphere is much stronger and more rigid than the asthenosphere, perhaps explaining the mobility of the lithospheric plates in the underlying asthenosphere. Lithospheric plates include both crust and mantle, and a single plate may include both continental and oceanic crust.

The approximately 450 km-thick transition zone is moderately complex, probably containing two intervals characterized by phase changes (that is, compositional rearrangements on the atomic level). Both intervals show rapid increases in velocity of *S* waves. In contrast, the lower mantle, from 700 to 2,898 km, changes little in composition and phase down to the outer core.

The core

For any earthquake there are some locations on the earth's surface where seismic waves are not detectable. This phenomenon can be understood by splitting the circumference of the earth into 360 one-degree arcs, each arc representing a distance of 111.1 km. If an earthquake occurs at a particular site (designated 0 degrees), a *shadow zone* exists between 103° and 143° on either side. First arriving *P* waves are difficult to detect between 103° and 143° away from the earthquake's epicenter. The waves are obvious, however, to seismic instruments on either side beyond the shadow zone. Deflection of the waves away from their normal path is caused by the earth's core, first encountered at a depth of 2,898 km.

A marked change in seismic properties takes place at 2,898 km because the outer core is a liquid. *S* waves disappear. Below the Gutenberg-Wiechert discontinuity at 2,898 km, wave velocities are far slower than in the enclosing mantle.

P waves that penetrate the core to a depth of 5,145 km have markedly higher velocities than those that merely penetrate the outer core, indicating that there is a sharp discontinuity at 5,145 km. This is the boundary between the liquid outer core and the solid inner core. A satisfactory explanation for the earth's magnetic field apparently requires fluid motion of a molten, deeply buried conductor, the liquid outer core of the earth.

Only iron and nickel are heavy enough to explain the density of the core. At the bottom of the lower mantle the density is 5.5 grams/cm^3; in the outer core the density is 9.5 grams/cm^3. Iron and nickel are concentrated in the core because they sank to the center during an early molten stage of the earth's formation, while lighter materials moved to the outer part of the planet. Some scientists believe the iron in the core is combined with sulfur or silicon or both, decreasing slightly the density of the core that would be too great if the core were pure iron.

The evidence of meteorites

Further evidence of the composition of the core of the earth comes to us from outer space in the form of meteorites. Meteorites are of two major types, stony and iron (Figure 8-17). By far the most common are the *stony meteorites.* They constitute 92 percent of all falls. (A "fall" is a meteorite that has been picked up after it has been actually seen to descend to the earth.) Stony meteorites, which have a density of between 3 and 4, are com-

FIGURE 8-17 *Small metallic meteorite from Meteor Crater, Arizona. (Photo by Willard Starks.)*

FIGURE 8-18 *(Below) This diagram summarizes the changes of properties with depth in the earth. Rather than plotting properties against the depth, each earth layer is shown in proportion to its volume. Notice that on this scale the core makes up only a modest portion of the volume of the earth even though it occupies about half of the earth's radius. (Courtesy Judson, Deffeyes, and Hargraves, 1976.)*

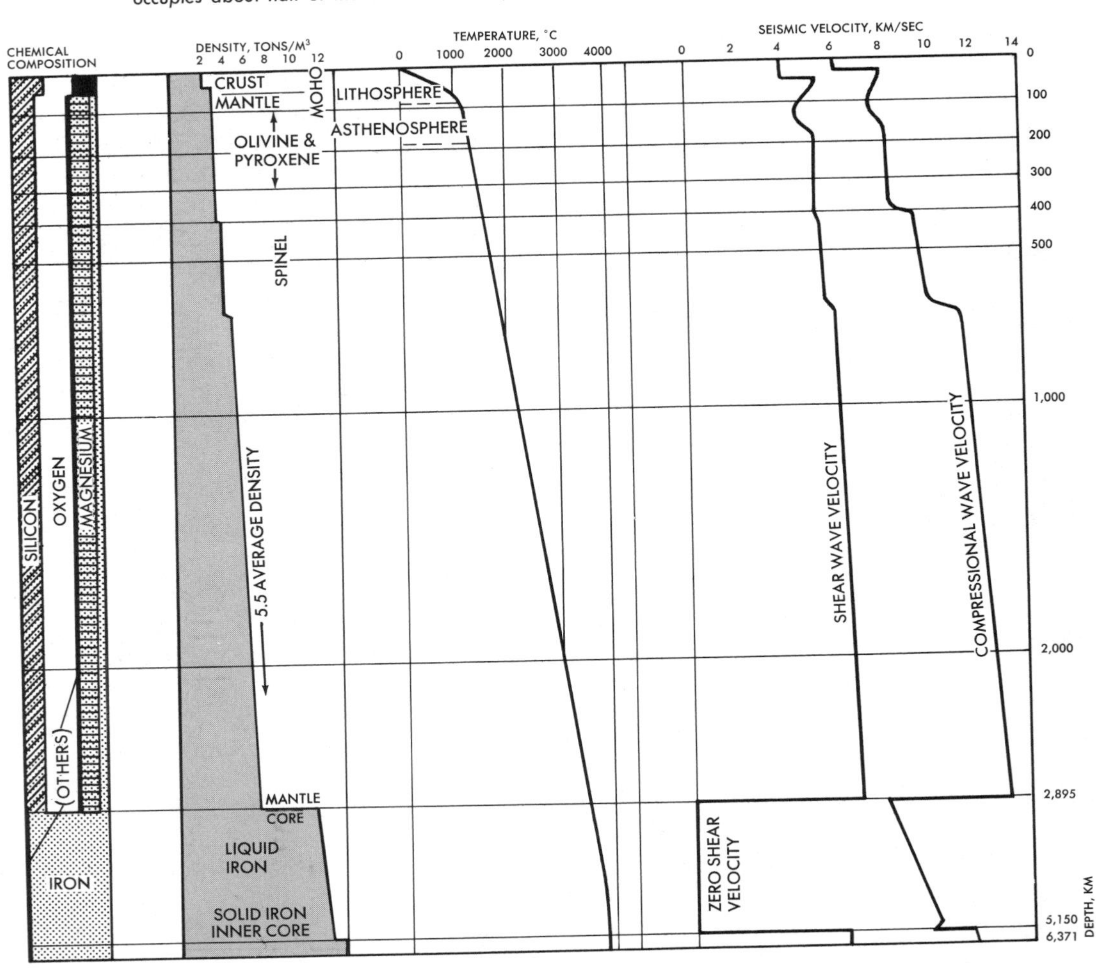

posed largely of olivine and orthopyroxene with odd mixtures of some accessory minerals, some of which are present only in meteorites.

The *iron meteorites,* in contrast, are composed dominantly of iron-nickel alloy, and have a density of about 8. Inclusions present in minor amounts are iron sulfide, iron carbide, ferrous chloride, carbon, and several silicate minerals. The meteorites are fragments of extraterrestrial origin that may be indicative of the composition of the mantle and core. If so, the composition of the mantle is largely iron-magnesian silicate, which is the general composition of the olivines and the pyroxenes. The composition of the core in this interpretation is believed to be similar to that of the iron meteorites.

An earth model

All of the available information derived from earthquake waves, rock samples, and meteorites pertaining to the earth's interior has been used to construct Figure 8-18. Such an interpretation, which summarizes the changes in chemical composition, density, temperature, and seismic velocity with depth in the earth, is popularly termed an *earth model.* The real earth is more complex than this interpretation, and changes in such interpretations are constantly being made as new information is obtained. However, it is quite useful to formulate a representation of the whole earth and to have it at hand as new problems are encountered.

SUMMARY Earthquakes are our most valuable means of probing the deep interior of the earth. They occur in narrow zones, or belts, and to depths of 700 km below the surface. Quakes generate various types of waves, of which two are called the *P* and *S* waves. These penetrate into the body of the earth and serve to map the globe's interior. The faster *P* wave is a compressional wave and is transmitted through both liquid and solid material. The slower *S* waves are transverse waves and pass through solids only. It is the behavior of these waves that leads to many of our deductions about the earth's interior. Their interpretation allows us to divide the earth into core, mantle, and crust, and to further subdivide these three major units of the earth. Thus, the difference in composition and thickness between the crust beneath the oceans and the continents is inferred from these earthquake waves. Thickness of oceanic crust is 10 to 12 km, compared with a thickness of about 35 km for the continental crust.

The earth apparently has a solid, possibly iron-nickel inner core surrounded by a liquid outer core. Radius of the core is about 3,470 km. The mantle outside the core is composed of iron-magnesium silicate material; there is a sharp transition between the mantle and the crust known as the Mohorovičić (M) discontinuity.

Earthquake forecasting and control are attracting considerable attention and study, although both are in a very early state of understanding. Forecasting is dependent on detecting changes in rock behavior from the strained to the unstrained states. Control is based on the possibility of releasing rock strain before the strain builds to destructive amounts. This is possible by triggering a small earthquake by an explosion. Another possibility for control is by increasing the water pressure in the fault zone to permit small-scale slippage.

SUGGESTED READINGS

BULLEN, K. E., *An Introduction to the Theory of Seismology*. Cambridge, England: Cambridge University Press, 1963.

HODGSON, JOHN H., *Earthquakes and Earth Structure*. Englewood Cliffs, N.J.: Prentice-Hall, Inc., 1964 (paperback).

MASON, BRIAN, *Meteorites*. New York: John Wiley & Sons, Inc., 1962.

LEET, L. DON, *Earthquakes—Discoveries in Seismology*. New York: Dell Publishing Co., 1964 (paperback).

RICHTER, C. F., *Elementary Seismology*. San Francisco: W. H. Freeman and Co., 1958.

U.S. Geological Survey, *The Alaska Earthquake* (Professional Papers 541, 542, and 543-1). Washington, D.C.: Government Printing Office, 1965–1967, 1969.

U.S. Geological Survey, *The San Fernando, California, Earthquake of February 9, 1971* (Professional Paper 733). Washington, D.C.: Government Printing Office, 1971.

WITKIND, I. S., *The Night the Earth Shook: A Guide to the Madison River Canyon Earthquake Area*. Department of Agriculture, U.S. Forest Service, Misc. Publication No. 907, 1962.

WOOD, J. H., and P. C. JENNINGS, "Damage to Freeway Structures in the San Fernando Earthquake," *New Zealand Society for Earthquake Engineering Bulletin, 4* (1971), 347–75.

Crustal Deformation and Mountain Building

The earth's crust, which seems so solid under foot, is, in reality, restless and ever-changing. The solid crust bends and breaks, yielding to forces that build mountains and raise land masses above the sea. In the previous chapter we examined one manifestation of these forces—earthquakes. Here we will look not only at the nature of rock deformation, but also at the way in which mountains are built.

Evidence of earth movement

Wherever we find the rocks of the earth's crust exposed, we see evidence of powerful stress and strain. Sometimes movement happens in a few brief moments and can be observed by man. Other movements are recorded only on the longer scale of a man's life span or of human history. As a matter of

9

Cretaceous rocks that were folded in the Alpine orogeny. View taken near St. Jean de Luz at the west end of the Pyrenees. (Courtesy, A. J. Eardley.)

fact, most of our evidence for rock deformation comes from the geologic record that stretches backward toward the dim origins of the earth.

Rapid movements

During earthquakes, rapid motion of the land has been observed on many occasions and in many different places. The motion may be vertical or horizontal or some combination thereof. During the earthquake of August 17, 1959, west of Yellowstone National Park near the Hebgen Reservoir in Montana, vertical movements up to 6 m were recorded (see Figure 9-1). During the 1923 earthquake in Tokyo, Japan, a cliff along Sagami bay moved upward by some 5 m. And as a result of the Good Friday earthquakes of 1964 in Alaska, changes in land level of several meters or more took place over a wide area. Horizontal movements were recorded during the San Francisco earthquake of 1906. In this instance, roads and fences were offset across a fracture in the earth's crust by distances of up to 7 m.

FIGURE 9-1 *A cliff about 1.5 m high formed as a result of the Hebgen, Montana, earthquake on August 17, 1959. (Photo by Lloyd Skinner.)*

Slow movements

Some cases of crustal deformation involve almost instantaneous displacements of a meter or so, but most movements of the crust apparently are going on slowly and more or less continuously. The coast of California, it is calculated, is moving northwestward an average of about 5 cm a year. In one place this slow creep has been noticeable even without engineering surveys: in Tres Pinos, a winery building was constructed across two portions of the crust that are moving relative to each other; now the building is gradually and visibly being twisted apart. In Japan, surveys have shown that the crust there is composed of a mosaic of blocks, tens of kilometers across, which are milling about and tilting one way or another like ice cakes on a stormy sea. The amount of movement in a year is minuscule, but it is measurable, and it involves the entire thickness of the crust. Undeniably powerful forces are at work, and the resulting changes in the earth's crust are by no means superficial.

Geologic evidence

In Chapter 6, we found that sedimentary beds are separated by surfaces called *bedding planes.* These bedding planes are generally horizontal or nearly so when the sedimentary rocks are deposited. Today we find that many ancient sedimentary rocks are tipped at various angles to the horizontal. Some are even vertical, and others are turned upside down. We must conclude that since the rocks were originally formed, some force has tilted them out of their normal horizontal position. As we shall see later in this chapter, this force has deformed once flat-lying rocks into folds measuring from a few centimeters to scores of kilometers in breadth. At the same time it has broken once-continuous layers and has moved the sections thus formed into new positions.

Many of the mountain ranges of the world contain sedimentary rocks that enclose the fossils of marine organisms. The presence of these marine fossils high above the ocean bottoms testifies to the extensive differential movement of the mountain masses and ocean basins since the rocks and fossils were first deposited on an ancient sea floor. The amount of this motion is measured in kilometers in some places.

The rate at which differential movement progresses is difficult to measure, but we do know that it varies considerably. Evidence indicates that a mountainous mass such as the Himalayas may move upward at a rate of 5 to 6 m per thousand years. Other rates appear to have been much less, perhaps 0.1 m per thousand years.

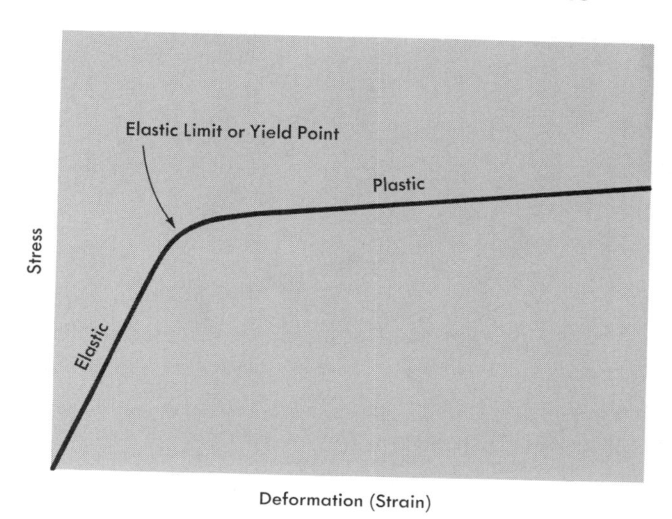

FIGURE 9-2 Graph showing that as stress increases, for a while strain is proportional to stress. This is the range of elastic deformation. Then, when the stress reaches the yield point, or elastic limit, deformation may become plastic. If the stress is removed after the material has begun to behave plastically, the material will not return to its original shape.

Behavior of rock material in response to stress

The force or pressure that produces rock movement is called *stress,* and it is usually measured in pounds per square inch or kilograms per square centimeter. As rocks react to stress, they undergo changes in shape or volume or both. Such deformation in response to stress is called *strain.*

Rocks respond to stress in three ways: (1) by *elastic deformation,* (2) by *plastic deformation,* and (3) by *rupturing.*

An elastic solid is a substance that, after undergoing a change in shape or volume when stress is applied, returns to its original condition when the stress is removed. The deformation is usually proportional to the stress.

In plastic deformation a substance undergoes a continuous change of shape. In this case the rock does not recover its original volume or shape when the stress is removed. In most instances, the deformation is elastic up to a certain critical point (called the *yield point* of a rock) after which it becomes plastic, continuing as a "flow" so long as stress is applied (Figure 9-2).

A rock may also respond by "rupturing." In this type of deformation, stresses are relieved by the actual breaking of the rock at the rupture point. Rocks that are being elastically deformed may eventually reach a point of stress in which the accumulated pressures are relieved by breaking, or rupturing. Again, if rocks are being plastically deformed at a relatively rapid rate, stress may also be relieved by rupturing (Figure 9-3). It is this rupture, or breakage, in a rock that probably causes the earthquakes discussed in Chapter 8.

The study of rocks in the laboratory under varying conditions of pressure and temperature and the observation of the deformation of rocks in nature lead to the conclusion that at the surface, rocks respond to stress first by undergoing elastic deformation, and then

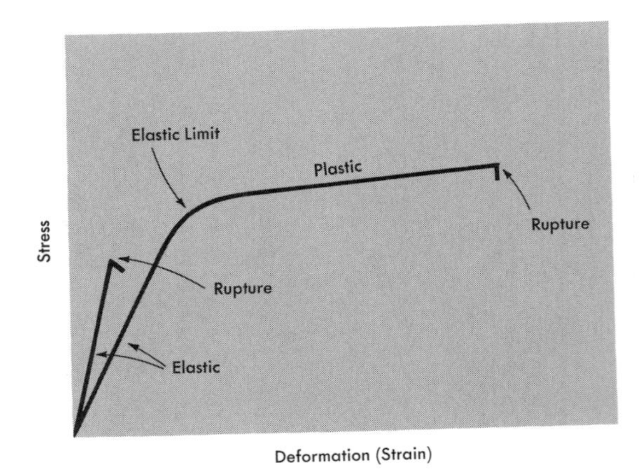

FIGURE 9-3 This graph shows the behavior of two different substances under stress. One responds elastically only, and then, when stress reaches a critical point, it deforms by rupturing. The second substance behaves elastically for a while, passes its elastic limit, and then deforms plastically. Eventually, however, this material, too, may deform by rupture.

by rupturing. At great depths, however, they undergo elastic deformation, and then plastic flow. Deep rocks rupture only if the movement is great enough and fast enough. We now know that an increase in the confining pressure of a rock increases its elastic limit and therefore the strength of the rock. At the same time, an increase in temperature weakens a rock. Long-continued stress, even if not very great, also weakens a rock's resistance to deformation. And within the earth, solutions contained within a rock also tend to weaken it.

Isostasy

We have seen that the crust and upper mantle are broken into rigid slabs of rock called *lithospheric plates.* Each plate is 75 to 125 km thick, and major plates are several thousand kilometers wide. The plates of the lithosphere include both continental and oceanic areas that float on the weaker underlying mantle (*asthenosphere*). Lithospheric plates amount to only 1 or 2 percent of the earth's diameter, a very thin skin on the big apple.

The lithosphere is rigid and strong; in comparison, the asthenosphere is weaker and flows plastically in both horizontal and vertical directions. The lithospheric plates drift in the plastic asthenosphere, and the lateral flow of the asthenosphere maintains *isostatic balance* between the weight of the lithospheric plates.

Similarly, mountain ranges, which contain relatively low density material, are much higher above the asthenosphere than are ocean basins. Continental crust that contains less dense minerals is up to 70 km thick. In contrast, the crust under the oceans is thin (10 to 12 km thick) and is characterized by an abundance of denser minerals. The mountains, with their less dense material and their deep roots, sit higher than do the ocean basins because of the effects of *isostasy* (from the Greek words meaning "equal standing").

FIGURE 9-4 An iceberg floating near the Grand Banks off Newfoundland. An estimated 0.8 to 1.1 million cubic meters of ice towers above the Coast Guard cutter "Evergreen." Nine times this amount of ice lies beneath the water level. As the exposed mass of ice is reduced, then the "root" of the iceberg will rise proportionately so that throughout the life of the iceberg a constant proportion of ice above and below the surface of the sea is preserved. (Photo by U.S. Coast Guard.)

The principle of isostasy states that below a certain depth in the earth—about 50 km—the hydrostatic pressure does not vary horizontally. On most continents, the tallest mountains do not exceed 6 km in height. This is apparently as high as they can sit and remain in isostatic balance.

A familiar analogy clarifying isostasy is that of a piece of ice floating in water: having a somewhat lower density than the water, the ice has approximately one-tenth of its volume above water and nine-tenths below (Figure 9-4). Isostatic balance also can be observed on a much grander scale. Large parts of the northern hemisphere covered by glacial ice sheets more than 1,000 m thick were depressed by the great weight of the ice. When the ice sheets melted about 12,000 years ago the depressed crust rebounded isostatically, moving upward several hundred meters.

The whole earth is subject to the force of gravity, and isostatic adjustments are familiar examples of the effects of gravitational stress. Another result of gravitational stress is the formation of salt domes in the Gulf Coast region and elsewhere in the world. Beds of salt, with a density of 2.16, may be covered by other sedimentary rocks of higher density (2.3 to 2.6). The less dense masses of salt may then rise upward piercing the higher density beds that directly overlie the salt, and arching younger beds into salt domes. Much oil and gas are produced from salt domes in the Gulf Coast region, both onshore and offshore.

Structural features

When rocks are deformed out of their original shape, they assume new patterns that we refer to as *structural features.* These features are joints, folds, faults, and unconformities.

To describe the position in space of the rocks making up such structural features, geologists have found it convenient to use two special measurements: *dip* and *strike.* These are most easily described with reference to sedimentary or layered rocks. The dip and strike of a rock layer provide a description of its orientation in relation to a horizontal plane. The *dip* is the acute angle that a tilted rock layer makes with an imaginary horizontal plane. The *direction of strike* is always at right angles to the direction of dip. Thus, a bed that has a dip either to the east or to the west has a strike north-south, usually designated simply as "north" (Figure 9-5).

FIGURE 9-5 *Dip and strike. Top: Photo showing outcropping edges of tilted beds in southwestern Colorado, a few kilometers east of Durango. Bottom: Sketch illustrating terms used to describe the attitude of these beds. The beds strike north and dip 30° east. (Photo by Soil Conservation Service, U.S. Department of Agriculture.)*

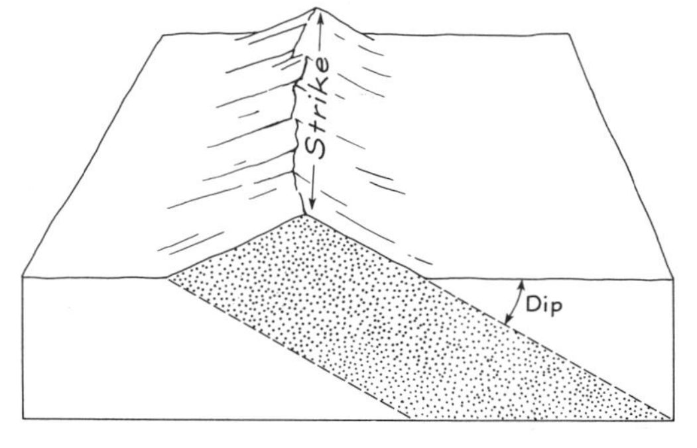

FIGURE 9-6 (Above) *Joint planes vertical to the surface and to each other cut this limestone rock on Drummond Island, Michigan. The horizontal bedding planes of the rock combine with the joint planes to produce cubic patterns on the quarry face. (Photo by Russel, U.S. Geological Survey.)*

FIGURE 9-7 (Right) *In this quarry near Washington, D.C., joint planes produce rhombic-shaped blocks in gneiss. (Photo by Diller, U.S. Geological Survey.)*

FIGURE 9-8 (Below) *Vertical view of nearly rectilinear system of joints in the Haymond Formation, Marathon, western Texas. (Photo by Earle F. McBride.)*

FIGURE 9-9 *An anticline and a syncline on St. Anne's Head, Pembroke, Wales. Width of exposure about 45 m. (Photo by Geological Survey of Great Britain.)*

Joints

The most common structural feature of rocks exposed at the surface is a joint. This is simply a fracture in the rock, without any relative movement of the rock on either side. Joints can have almost any orientation—vertical, horizontal, or at some angle—but in any given rock mass, joints tend to occur in sets, with the fractures more or less parallel to one another (Figures 9-6, 9-7, and 9-8).

Joints are formed by contraction during drying, such as in shrinkage cracks (desiccation), and by cooling of lavas as crystallization proceeds (thermal contraction). The columnar jointing of basalt flows caused by thermal contraction is spectacularly displayed at the Devil's Causeway in Ireland and at the Devil's Tower in Wyoming. Some joints are caused by tidal stress (small diurnal or semidiurnal movement) because of the influence of other bodies in space, especially the moon. Other joints and large faults that can be traced for hundreds of meters are the result of movements of the earth's crust as a result of tectonic stress.

Folds

When layered rocks are subjected to compressive stress, they may be thrown into a series of wrinkles, or *folds.* In size, folds range from small features through folds a kilometer or so across to gigantic arches and troughs 100 km or more across. The two most common types are termed *anticlines* (upfolds) and *synclines* (downfolds). In anticlines (Figure 9-9), the beds dip downward away from the central axis or ridge. In synclines (Figure 9-10), the direction is just the opposite. Small slump folds are shown in Figure 9-11.

FIGURE 9-10 *Small syncline. (Courtesy U.S. Geological Survey.)*

FIGURE 9-11 *Small slump folds in the Haymond Formation. (Photo by Earle F. McBride.)*

At the earth's surface most synclines and anticlines have been breached to a greater or lesser extent, so that rocks of differing ages are met as one crosses the fold. In an eroded anticline the oldest beds are in the center; in an eroded syncline the youngest beds are in the center. The nomenclature for a fold is given in Figure 9-12, and some of the various types of folds are illustrated in Figure 9-13.

FIGURE 9-12 *Nomenclature of a fold, shown on a plunging anticline, but applicable to a syncline, also. Axis is the line joining places of sharpest folding. Axial plane includes axis and divides fold as symmetrically as possible. If axis is not horizontal, fold is said to be plunging, and angle between axis and horizontal is angle of plunge. Sides of a fold are called the limbs.*

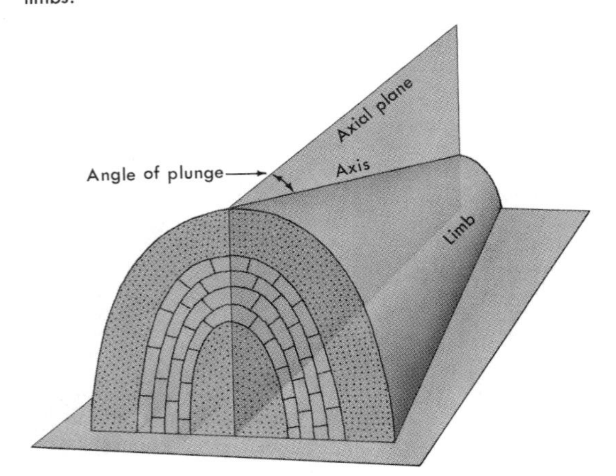

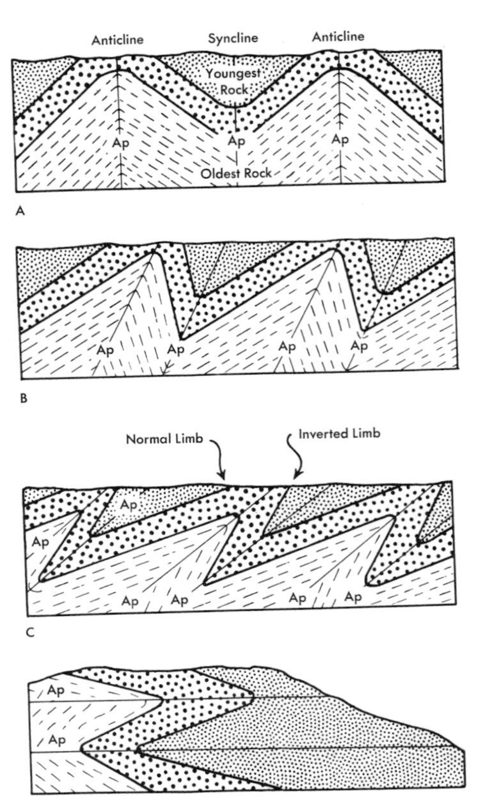

FIGURE 9-13 *Types of folds. (A) Symmetrical fold. Axial plane (Ap). Vertical, limbs dip in opposite directions at same angles. (B) Asymmetrical fold. Axial plane inclined, limbs dip in opposite directions at different angles. (C) Overturned fold. Axial plane inclined, limbs dip in same direction, usually at different angles. (D) Recumbent fold. Axial plane horizontal.* After Marland P. Billings, *Structural Geology*, 2nd ed. (Englewood Cliffs, N.J.: Prentice-Hall, Inc., 1954).

Faults

When the rupture point is exceeded, rocks break, and separate sections may be displaced. Such a structural feature is a *fault* (Figures 9-14, 9-15, and 9-16). Earthquakes, as we have noted, are caused by movements along faults. Seldom, however, do we see the results of such movement at the surface of the earth until after millions of years of erosion have exposed the old break to view.

FIGURE 9-14 (Above) A much studied section of the San Andreas fault. (Courtesy U.S. Geological Survey.)

FIGURE 9-15 (Above) Scarp formed by Garlock fault, south edge of Slate Range, San Bernardino, California. Death Valley lies beyond peak on central skyline. (Courtesy G. I. Smith, U.S. Geological Survey.)

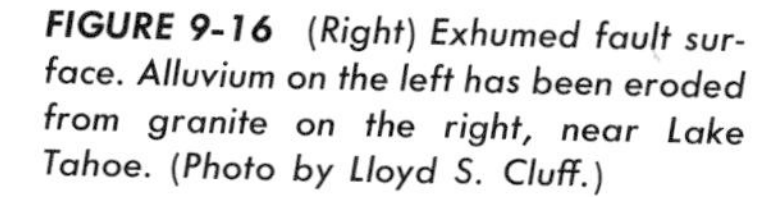

FIGURE 9-16 (Right) Exhumed fault surface. Alluvium on the left has been eroded from granite on the right, near Lake Tahoe. (Photo by Lloyd S. Cluff.)

Types of faults

If the movement along the break has been horizontal, the fault is called a *strike-slip fault,* because the slipping has been parallel with the strike of the fault. Strike-slip faults can be further designated as *right-hand* or *left-hand.* The one sketched in Figure 9-17 is right-hand: if you were walking along the road shown, and came to the offset caused by the fault, you would have to go to the right to get onto the other section of the road.

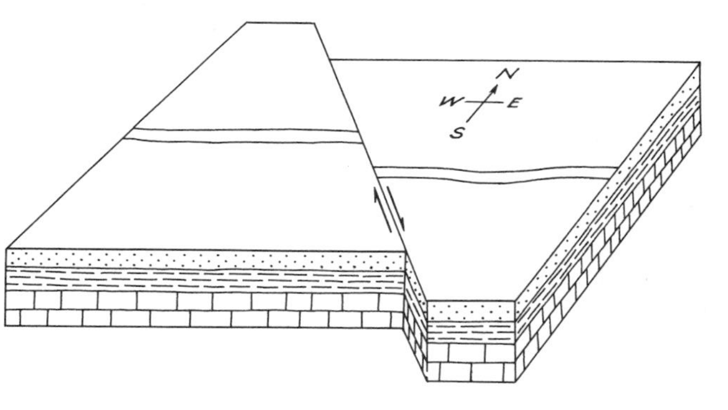

FIGURE 9-17 Sketch of right-hand strike-slip fault, depicting movement along San Andreas fault, California, in 1906.

FIGURE 9-18 *Normal fault in Cretaceous beds, Saltillo, Mexico. (Photo by Earle F. McBride.)*

Faults are classified in other ways. One of the most useful systems is based on the relative movement of the two rock masses affected by the fracture. The fracture, or *fault plane,* usually dips at some angle from the vertical. The mass of rock above the fault plane is called the *hanging wall;* the one beneath it, the *foot wall.* (These names, which come from mining, refer to the position of masses of rock relative to miners.) Faults are described on the basis of the relative movement of the foot wall to the hanging wall. A *normal fault,* for instance, is one in which the hanging wall has moved downward in relation to the foot wall (Figure 9-18). A *thrust fault,* sometimes called a *reverse fault,* is one in which the hanging wall moves upward in relation to the foot wall.

The direction of movement is relative along faults. It is not possible to designate which way or how much either wall has moved except in relation to the other.

Unconformities

In many places younger rocks are separated from older rocks by surfaces of erosion or of nondeposition. Such surfaces are called *unconformities.* They represent events in earth history, but events that are not preserved as

FIGURE 9-19 *An angular unconformity east of Cody, Wyoming. Several terraces are visible in the background. (Photo by M. Dane Picard.)*

rock material. The time span may range from a few hundred years or thousands of years up to tens and even hundreds of millions of years.

Geologists recognize three types of unconformities: (1) *angular unconformity,* (2) *disconformity,* and (3) *nonconformity.* An angular unconformity may form in the following way. Sedimentary rocks deposited on a sea floor may be tilted and raised above sea level, where they are partially removed by erosion. If this surface of erosion, underlain by partially beveled and deformed sedimentary rocks, is flooded again by the sea and if new sedimentary rocks are deposited on top of the surface, then an angular unconformity is created. If we examine a section through this sequence of rocks, we see a series of tilted rock layers overlaid at some angle by younger rocks. The surface that separates the two groups of rocks represents the period of erosion and nondeposition (Figure 9-19).

Sometimes sedimentary rocks that have been exposed to erosion undergo further deposition without being tilted or folded. An unconformity between parallel beds formed in this way is called a *disconformity* (Figure 9-20). A *nonconformity* is an unconformity that is developed when massive igneous rocks that are exposed to erosion are then covered by sedimentary rocks.

FIGURE 9-20 *A disconformity in the walls of the Grand Canyon. Line traces buried erosion surface. (Photo by U.S. Geological Survey, courtesy Robert M. Garrels.)*

Geosynclines

All major rock types—igneous, metamorphic, and sedimentary—are found in the folded mountains of the world. In many mountains, the sedimentary rocks represent thick accumulations of deposits that originally reached as much as 10,000 or 12,000 m in thickness, and were deposited close to sea level. Reconstruction of the events surrounding the accumulation of these sedimentary rocks shows that they were deposited in a slowly subsiding trough in which subsidence and deposition were concurrent. Such a subsiding trough is referred to as a *geosyncline,* and the muds, silts, and sands accumulated in it are called *geosynclinal sediments.*

The thickness of beds deposited in geosynclines contrasts sharply with rocks of the same age deposited in the interior of the continents on nearly flat areas termed *shields.* The geosynclinal sections are eight to ten times thicker than those of the shields.

Geosynclines are generally linear, sometimes thousands of kilometers in length, and from a few tens of kilometers to hundreds of kilometers in width. Both deep-water beds (turbidite sandstone) and shallow-water beds (limestone and shallow water clastic deposits) are found. However, the great thickness and wide distribution of the shallow-water deposits indicate that geosynclines were not deep trenches in the oceans. Shallow-water fossils, shrinkage cracks, ripple marks, salt casts, and coal beds are further indications that the surfaces of deposition in most geosynclinal troughs were not far below sea level. Geosynclines subside at a rate of between 150 and 300 m per million years.

Many geosynclines can be divided into an inner trough closer to the continental platform and shield, the *miogeosyncline,* and an outer trough, the *eugeosyncline.* Eugeosynclinal sequences are thick and are composed of turbidite sandstone and mudstone, deep-water limestone, basalt, and ash. In contrast, miogeosynclinal sequences are thinner, contain shallow-water marine limestone, quartz-rich sandstone, and claystone, and grade toward the continent into beds deposited in streams and on flood plains.

Folded mountain chains commonly are the result of a wave of deformation commencing in the eugeosyncline and later extending through the miogeosyncline. Lateral forces intensely compress the geosynclinal beds, and they are deformed into a series of anticlines, synclines, and thrust faults parallel with the edge of the shield. Intrusions invade the folded and compressed beds. In a later stage, batholiths are emplaced, at least in the major mountain chains along the continental margins.

Mountains and their origin

Mountains are the dominant landscape feature of the continents. In general, they are present in long chains, following curving or straight lines, and are concentrated largely toward the margins of the continents. In contrast, the interiors seem relatively lower and are devoid of extensive mountain-building activity. This apparent lack of mountains toward the interior of continents, however, is belied if we look far enough back into geologic time. For there we find that as the continents evolved, these heartlands were also marked by mountains, mountains that have ceased to exist as topographic features. Erosion has long since removed all but their roots.

Mountainous masses are not restricted to the continents. The ocean basins are crossed by a remarkable series of ridges towering above the general elevation of the ocean floors. Thus, under the Atlantic Ocean, from Iceland to the Antarctic, there is a belt of mountains called the mid-Atlantic ridge, which roughly parallels the outlines of the continents and is nearly midway between them. It stands as much as 1,800 m above the ocean bottom and is covered in places by 2,700 m of water. In a few places its peaks protrude above the water to form islands, such as the Azores, St. Helena, and Tristan da Cunha. Similar ridges characterize the other oceans.

Types of mountains

All of the geosynclinal features are well developed in the Appalachian Mountains, and in fact the great American geologist James Hall first gave the name "geosyncline" to the more than 10 km of shallow-water beds in the Appalachians. In the following paragraphs these mountains are closely examined, as well as several other types of mountains.

Appalachian mountains

The Appalachian Mountains run from Alabama and Georgia on the south, northeastward through the eastern United States, and on into the Maritime Provinces of Canada. Let us look at a section through this feature in North Carolina and Tennessee (Figure 9-21). Traveling northwestward from the seaboard we cross first the Atlantic Coastal Plain, which is underlain by a series of gently dipping, relatively unconsolidated marine beds of sandstone, siltstone, and claystone. These sedimentary rocks lap up onto older igneous and metamorphic rocks that form the Piedmont Plateau of the eastern United States. Folded and faulted, they represent the eroded roots of the ancestral Appalachian Mountains. Today they are marked in their eastern portion by low, gently rolling hills.

As we progress westward, the hills become higher, and soon we encounter the escarpment of the Blue Ridge, which separates the plateau country from the highest mountains in the eastern United States. These mountains are also underlain by igneous and metamorphic rocks of the same general age as those of the Piedmont. Evidence exists to suggest that they have been thrust upward and outward over sedimentary rocks lying farther to the west.

These sedimentary rocks of sandstone, claystone, and limestone form what is called the Valley and Ridge Province of the Appalachian Mountains. In this section of the Appalachians, the mountains are marked by a series of thrust faults indicating stresses from the east. Farther to the north, in Pennsylvania, the pressures have not been as intense, and the same rocks, instead of being thrust one over the other in great slices of material, now stand as folded mountains. But in both sections of the Valley and Ridge Province the topography is characterized by a series of nearly parallel ridges and valleys. The valleys were formed as the processes of erosion etched out areas of nonresistant rock, leaving behind as ridges the resistant rocks, which are usually composed of quartzite and sandstone. Still farther to the west the rocks that are faulted in the Valley

FIGURE 9-21 A geologic cross section through the Appalachian Mountains in North Carolina and Tennessee.

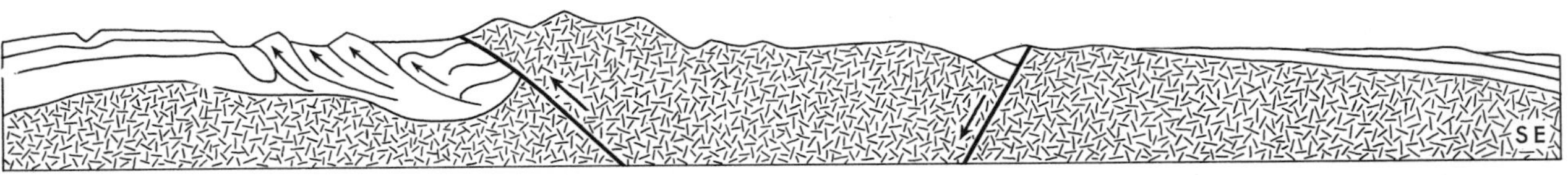

and Ridge Province stand as relatively undeformed sedimentary rocks underlying plateau country. In parts of Pennsylvania and New York the plateaus are known as the Catskill Mountains. In the area of Tennessee and Virginia they form the Cumberland Plateau. Erosion has dissected these plateaus to a greater or lesser extent, creating low mountains or high hills depending on one's frame of reference.

Looking back, then, across the Appalachian Mountains, we find that there are three distinct forms within this range (Figure 9-21). (1) On the extreme western side are the hills or mountains carved in flat-lying rocks, such as the Catskill Mountains. (2) Next eastward lie the parallel features of the Valley and Ridge Province underlain by folded and thrust-faulted beds that alternate in their resistance to erosion. (3) Still farther to the east lie the high mountains carved on the massive and resistant beds of crystalline rocks both metamorphic and igneous. Progressing toward the sea is the rolling Piedmont section, which is also underlain by igneous and metamorphic rocks. It passes under the gently dipping younger strata of the Coastal Plains.

Plate tectonic interpretation. The formation of the Appalachian Mountains has been interpreted according to plate tectonic theory by R. S. Dietz,[1] who recognizes several distinct stages. According to Dietz, in the first stage, more than 600 million years ago, a continuous North American-African supercontinent split apart, and an ocean, the ancestral Atlantic opened. As North America receded from Africa and Europe, the North American continental shelf received sediments from the northwest.

During the second stage, about 500 million years ago, the ancestral Atlantic began to close, a subduction zone developed, and the geosynclinal beds deposited on the continental margins were deformed, intruded by magma, and uplifted. In the third stage, about 375 million years ago, the ancestral Atlantic completely closed, North America collided with Africa and Europe, and the ancient Appalachian Mountains were formed. The convergence of the two plates deformed the eugeosynclinal deposits, magmatic intrusion and volcanism took place, and the ancient Appalachians were uplifted. Subsequently, but as a phase of the same sequence of events, the miogeosynclinal deposits were folded.

The final stage began about 200 million years ago when the modern Atlantic Ocean opened. Geosynclines once again developed on the continental shelf and rise; and North America, Africa, and Europe drifted to their present positions.

Other mountains

Let us turn now to the West Coast of the United States and begin a traverse from central California eastward across the coastal ranges, through the Sierra Nevada, across Nevada and Utah, and into Colorado. The coastal ranges are underlain by a complexly folded and faulted series of sedimentary rocks and are separated from the Sierra Nevada on the east by the Great Valley of California. The Sierra Nevada are asymmetric, with a gentle western slope and a steep eastern scarp. The mountains themselves are cored with granitic rocks. Most outstanding is the great fault that bounds and defines the eastern scarp of the mountain. It is along this fault that the mountain mass of the Sierra Nevada has risen upward to its present height and now towers above the lower desert valleys and basins on the east.

As we continue eastward toward the Colorado Plateau, we cross a series of ranges separated by broad basins, the Basin and

[1] After "Geosynclines, Mountains and Continent-Building" by R. S. Dietz. Copyright © 1972 by Scientific American, Inc. All rights reserved.

Range Province. This area is characterized not only by great aridity with dry lake basins and a few salt lakes, but also by a series of mountains termed fault-block mountains. Even as the Sierra Nevada were thrown up along a long fault, so also have the various ranges in the Basin and Range Province been uplifted along bordering, generally north-south fault lines. While these blocks moved upward to expose igneous, metamorphic, and sedimentary rocks, the intervening basins moved downward to receive the sediments carried by desert and mountain streams from the nearby mountain masses.

To the east the Colorado Plateau stands as a great block of sedimentary rocks overlying older igneous and crystalline materials, forming not really mountains, but an interior plateau. Nonetheless, the Colorado Plateau will undoubtedly someday be carved into a series of hills or mountains not dissimilar to the Catskill Mountains, for instance, in the eastern United States.

The Rocky Mountains in Colorado are characterized by two long parallel ridges of igneous and metamorphic rock. The western ridge is called the Park Range, and the eastern ridge, the Front Range. Lapping up onto the flanks of these mountains are Paleozoic and Mesozoic beds (see Chapters 17–19) that stretch from the Front Range eastward onto the Great Plains and from the Park Range westward toward the Colorado Plateau. Between these two ranges lie the somewhat lower areas, such as South Park and Estes Park, which are filled with sediments washed from the higher mountains. The two ranges are upbucklings of the granitic and other crystalline rocks that underlie much of the center of the continent.

Volcanic mountains are still another type. Northward from the Sierra Nevada in Oregon and Washington lie the Cascade Mountains, which support a series of high, largely aligned, towering peaks of volcanoes, extinct or dormant or only recently active.

The Black Hills and the Adirondacks are examples of what may be called *upwarped mountains.* These mountains are composed mainly of very old metamorphic and intrusive rocks; they are not elongated as are most folded geosynclines but display independent trends. Mountains of this type apparently are upwarped parts of the shield, probably on the sites of former fold mountains but created long after the original folding and independent of it.

This brief survey by no means lists all of the various mountains of the United States. But it does tell us that mountains have different forms and origins. The rocks within mountains range from sedimentary through igneous to metamorphic. These rocks can be folded or faulted or merely flat-lying. The mountains can be lifted upward in great buckles or arches, or they may be thrust laterally from one place to another, or they may be bounded by fault scarps along which they have moved upward toward their present positions.

Mountain building and plate tectonics

Although plate tectonic theory cannot be used to explain the origin of all mountains, the emergence of this revolutionary idea has led to some reasonable explanations for many mountain systems throughout the world. Plate tectonics is perhaps most useful in explaining the origin of those arcurate, long, narrow, young mountain belts on the margins of continents.

According to this theory, the sea floor is ephemeral, and nearly all oceanic lithosphere returns to the mantle by subduction at convergent plate boundaries. In contrast, continental lithosphere is a mobile but permanent feature once it is created. Continents are too buoyant and too thick to be subducted. Although continents can be broken, moved, reassembled and deformed, their bulk over a long period of time is apparently not much decreased.

Also, according to this theory, mountain building (*orogeny*) takes place at the boundaries of colliding plates. Such zones of crustal convergence are characterized by volcanism and earthquakes. At the boundaries, the foci of earthquakes may be very deep, and they define an inclined seismic zone (the *Benioff zone*) that extends from the oceanic trench down under the continents to depths as great as 700 km. These relations are typical of all trenches that contain thick marine sediments (mostly turbidite deposits) eroded from the continent or the island arc. The Benioff zones are interpreted to be subduction zones along which crust, formed at the oceanic ridges, moves down and is reincorporated into the mantle.

However, plate convergence zones that result in orogeny also present other special circumstances. As noted previously, the light continental plate is not subducted. Rather, the stacking of one continent over another leads to a thickening of continental crust and to the formation of high mountain ranges. The thrusting of Africa under Europe to form the Alps is believed to be an example of such continental collisions.

Each mountain system is complex and different and must be studied in detail to determine its own particular history. Nevertheless, there are several general stages in the development of mountain belts on the margins of continents that can be recognized in terms of plate tectonic theory.

Where a convergence zone develops, a trench may form as the plate bends down into a subduction zone, especially if the rate of plate movement is faster than about 5 cm per year. Most present trenches are close to land and are deepening as they receive thick marine sediments, mostly turbidites, eroded from the continent or from an island arc that has formed as melting takes place along the descending lithospheric plate. Volcanism is pronounced along the island arc.

Continuing plate movement compresses and deforms the accumulated sediment. Deformation commences in the trench. However, deformation then proceeds to beds (limestone, sandstone, siltstone, claystone) deposited in shallow-water environments between the continents and the island arcs. In terms of geosynclinal nomenclature, the shallow-water beds are called *miogeosynclinal,* and the deep water turbidite deposits, *eugeosynclinal.*

As the sedimentary and volcanic rocks continue to descend, the whole rock mass becomes thicker and is plastically deformed. The growing rock mass attains greater elevations and also pushes deeper into the mantle where it is partly metamorphosed. Magma intrudes the deformed and metamorphosed beds and a batholith may form. Vertical uplift of the thickened rock mass exposes the rising mountains to erosion, and the sculpturing of peaks proceeds as sediment is removed and transported to new depositional sites.

Origin of plate motions

The driving mechanism of plate motions is still unknown. Full understanding of the processes involved is probably the most interesting and difficult problem in modern geology. Thousands of geologists and geophysicists are actively pursuing this study, trying to comprehend the cause of plate motions.

Of the two major energy sources—the sun and the heat flow from the interior of the earth—only the heat flow can have a direct effect on the crustal plates. The temperature of the earth's crust increases with depth. This heat comes from the decay of radioactive isotopes in the crust and in the upper mantle, and from the residual heat of planetary formation in the lower mantle and core.

The average heat flow for the continents is about 1.5 μcal/cm^2/sec.[2] For the major oceanic provinces, measurements indicate that the average heat flow expressed in μcal/cm^2/sec is as follows: for ocean ridges, more than 2; for ocean basins, about 1.3; and for ocean trenches, less than 1.0. Rising magma at the ocean ridges contains large amounts of heat from the mantle. As the magma plume cools, crystallizes, and becomes part of a lithospheric plate, it moves away from the ridge and continues to cool as heat is lost by conduction to the sea floor. At the ocean trenches the cold plate sinks back into the mantle at a subduction zone.

Heat flow on continents is largely from the crust where granites contain three to four times the amounts of the radioactive elements uranium, potassium, and thorium as does basalt. Heat flow from the ocean floor is mostly from the mantle, and the average heat flow for sea floor and continents is much the same, as noted above in the statistics for heat flow. The reasons for the approximate equality of average heat flow remains an enigma.

Before radioactivity was discovered, shortening of the crust in folded mountain belts was believed by many geologists to be the result of deformation of the crust due to shrinking of the earth as heat was lost. This theory is no longer popular, and it is not known for certain whether the earth as a whole is actually heating or cooling.

The lithosphere can be moved by thermally driven convection in the mantle, as first suggested by Arthur Holmes. The mantle rises beneath the trailing edge of a plate and sinks beneath the leading edge. Where the currents descend, the crust is squeezed together and a mountain belt is formed. However, recent studies indicate that this pattern of convection is unlikely.

Other scientists have proposed that uplift of the oceanic ridges provides a gradient sufficient for sliding without the help of dragging by a convection current. But this mechanism has also encountered pertinent objections and is not generally accepted as a complete mechanism.

To summarize, the origin of plate motions is still an exciting question. Although plate tectonic theory can be used to explain many diverse observations of mountains, our ignorance of the mechanism is still great and much remains to be learned.

[2] μcal/cm^2/sec means microcalories per square centimeter of surface area per second.

SUMMARY

The earth's solid crust is subject to movement that varies in rate from several meters per second to a few centimeters per thousand years. Stress, strain, and strength are, respectively, the pressure on the material, the resulting deformation, and the pressure at which the material deforms permanently. Rocks in the earth's crust undergo both elastic and plastic deformation.

Dip and strike are measurements that define the position of a rock or bed in space. The strike is the intersection of a plane with an imaginary horizontal plane, and the dip of the plane is the maximum acute angle that it makes with the horizontal.

Stress on rocks produces features that include joints, anticlines, synclines, and normal and reverse faults. In many places, younger rocks are separated from older rocks by surfaces of erosion or nondeposition. These are called unconformities.

The term isostasy is applied to the state of balance in which different parts of the crust and mantle maintain their elevation according to their density. Although never fully achieved, mountain ranges that contain relatively low density material are much higher above the asthenosphere than are ocean basins. Continental crust is up to 70 km thick; in contrast, crust under the oceans is only 10 to 12 km thick and is characterized by an abundance of denser minerals.

Geosynclines are basins of deposition, sometimes thousands of kilometers in length, and from a few tens of kilometers to hundreds of kilometers in width. They have accumulated as much as 10,000 to 12,000 m of deposits. Many geosynclines become the location of future mountains.

Mountains have different forms and origins. They can be lifted upward in great folds, or they may be thrust laterally from one place to another, or they may be bounded by fault scarps along which they have moved upward toward their present positions. Volcanic mountains are still another type of mountain.

The stress that causes deformation of rocks may be tidal stress (small diurnal or semidiurnal movement), gravitational stress (slow sliding of large portions of unstable rock material or upward motion of low-density rocks through high-density rocks), or tectonic stress (the most important type).

Plate tectonic theory is useful in explaining the origin of arcuate, long, narrow, young mountain belts on the margins of continents. According to this theory, mountain building takes place at the boundaries of colliding plates, boundaries that are characterized by volcanism and earthquakes. The driving mechanism of plate motions is still unknown.

SUGGESTED READINGS

BILLINGS, MARLAND P., *Structural Geology,* 3rd ed. Englewood Cliffs, N.J.: Prentice-Hall, Inc., 1972.

COMPTON, ROBERT R., *Manual of Field Geology.* New York: John Wiley & Sons, Inc., 1962.

EARDLEY, A. J., *Structural Geology of North America.* New York: Harper & Row, Publishers, 1951.

HILLS, E. SHERBON, *Elements of Structural Geology,* 2nd ed. New York: John Wiley & Sons, Inc., 1972.

SPENCER, EDGAR W., *Introduction to the Structure of the Earth.* New York: McGraw-Hill Book Company, 1969.

Streams and Underground Water

Water is the earth's most important natural resource, and probably the resource that is deteriorating most rapidly. The volume of water composing the hydrosphere is enormous, but most of the world's water is not readily available because it is bound up in the oceans and in glaciers (Table 10-1).

Industry uses more than one-half of the total water used. Although considerable water is involved, most of the water used by industry is returned to surface waters or is recycled. However, some surface waters are being used to their maximum capacities, and the volume of water in the Great Lakes, for example, apparently is generally decreasing because of overuse, pollution, and the ageing cycle of lakes. Table 10-2 gives some typical water requirements for present industrial uses and for some possible future needs.

10

Great Falls of the Potomac during flooding. (Courtesy U.S. Geological Survey.)

Table 10-1 Distribution of the Earth's Waters

	VOLUME (10^6 km^3)	PERCENT
Oceans	317,000	97.1
Glaciers, polar ice caps	7,300	2.24
Underground water	2,000	0.61
Streams, lakes, inland seas	55.28	0.017
	326,355	99.967

SOURCE: Nace, R. L., 1960, water management, agriculture, and ground-water supplies: U.S. Serlogical Survey Circular 415, 12 p.

The United States does not have as much water per unit area as does the rest of the world, but we are using more water per person. Further, our demands for water are booming; it is estimated that water use will rise from 1,200 billion liters daily in 1960 to about 3,400 billion liters by 2000. Although this prediction may be much too high, a substantial increase in use will take place, and there is not much we can do to increase the water naturally available. For much of the United States, the demand in the year 2000 will exceed the supply. Right now it is hard to get as good a drink of water as your grandfather had, and there are many other problems as well. We clearly need much better management of our water resources.

Running water

Floods take place somewhere in the world every year (Figure 10-1). With little warning, man, his animals, and his land are threatened with sudden destruction by streams. Great floods have been recorded throughout history.

Rapid City, a community of 43,000 at the foot of the Black Hills in South Dakota, was a recent victim. By late evening on June 9, 1972, there had been as much as 38 cm of rain at some locations in less than 6 hours. Walls of water up to 6 m high were reported. Shortly before midnight the mayor was told that Rapid City had about 20 minutes to prepare for a wall of water moving down Rapid Creek to-

Table 10-2 Water Used or Needed in the Production of Various Goods

ITEM PRODUCED	WATER USED OR NEEDED (*liters*)
Sunday newspaper	760
Wheat for loaf of bread	1,140
Tankful of gasoline	1,520
One pound of beef	15,200
Automobile	190,000
One ton of alfalfa	760,000
One ton of synthetic rubber	2,280,000
Needed each day for 50,000 barrels/day shale-oil plant	26,600,000
Needed each day for 250 million cubic feet/day of synthetic natural gas from coal-gasification plant	380,000,000

SOURCE: Laporte, L. F., *Encounter with Our Earth*. San Francisco: Canfield Press, 1975.

(a) **FIGURE 10-1** *(a) Flooding on small stream in northeastern Utah. (b) Channel of small stream shown in (a). Photograph was taken on the day after the flood. Note high-water mark on sagebrush. (Photo by Picard and High.)* (b)

ward the city. Most residents were asleep. As a result, at least 238 people died.

Cars, tents, houses, and trailers bobbed as corks in the muddy torrent. One-half of the houses in Rapid City were damaged; ten percent of them were completely destroyed. The drinking water was polluted, power failed, and fires broke out. The area inundated lies completely within the known flood plain of Rapid Creek. Total damage was more than $128 million.

Although flash floods may cause extreme results, they are only one of the processes that attack highlands, wearing them down and reducing them again to near sea-level elevations. This constant *gradation,* or leveling, of the land involves weathering, which was considered in Chapter 5. Weathering helps to prepare material for erosion and for transport to depositional sites. This leveling movement, though slow, is constantly going on. Were it not for processes working to raise the land, erosion would long ago have reduced continents to featureless plains. At the present rate of erosion and assuming no uplift of the land, the North American continent could be reduced to sea level in about 12 million years, a remarkably short span indeed when considered against the 4 to 5 billion years of earth history.

There is a struggle, then, between those processes that build up the land above sea level, and those that tear it down and, particle by particle, move it back toward the oceans. This process has been going on for as far back as we can understand the earth's record.

Of all the agents at work leveling the earth's surface, running water is the most important one. Year after year, the streams of the earth move staggering amounts of debris and dissolved material through their valleys to the oceans.

Precipitation, runoff, and stream flow

Nearly all the water that runs off the slopes of the land in rills or in thin sheets, and then travels on in streams and rivers, is originally derived from the oceans by way of the *hydrologic cycle.* There is only one exception: volcanic eruptions bring water to the surface from deep beneath the earth. But once this water has reached the surface, it also follows the general pattern of water movement from sea to land and back again to the sea.

Once water has fallen on the land, it follows one of the many paths that constitute the hydrologic cycle. By far the greatest part of the precipitation is evaporated back to the air directly or is taken up by the plants and transpired (breathed back) by them to the atmosphere. A smaller amount follows the path of

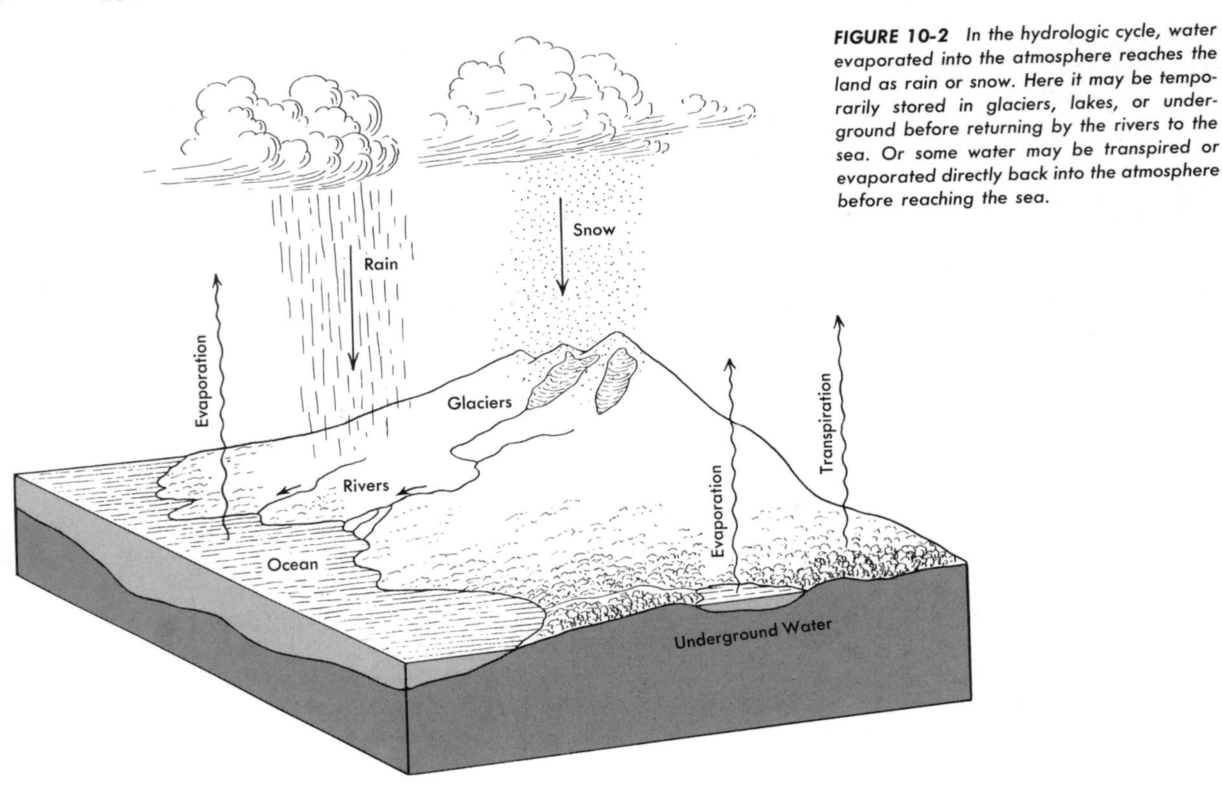

FIGURE 10-2 *In the hydrologic cycle, water evaporated into the atmosphere reaches the land as rain or snow. Here it may be temporarily stored in glaciers, lakes, or underground before returning by the rivers to the sea. Or some water may be transpired or evaporated directly back into the atmosphere before reaching the sea.*

runoff, the water that flows across the land. And an even smaller amount of precipitation soaks into the ground through *infiltration* (Figure 10-2).

Figure 10-3 shows how infiltration, runoff, and evaporation-transpiration vary in six widely separated localities in the United States. In the examples given, between 54 and 97 percent of the total precipitation travels back to the atmosphere through transpiration and evaporation. About 2 to 27 percent drains into streams and oceans as runoff, and between 1 and 20 percent finds its way into the ground through infiltration. Bearing in mind the ways in which water proceeds through the hydrologic cycle, we can express the amount of runoff by the following generalized formula:

$$\text{Runoff} = \text{Precipitation} - (\text{Infiltration} + \text{Evaporation and Transpiration})$$

Laminar and turbulent flow

When water moves slowly along a smooth channel or through a tube with smooth walls, it follows straight-line paths that are parallel with the passage. This type of movement is *laminar flow.*

If the rate of flow increases, however, or if the confining channel becomes rough and irregular, this smooth, streamline movement is disrupted. The water in contact with the channel is slowed down by friction, whereas the rest of the water tends to move along as before. As a result (Figure 10-4), the water is deflected from its straight paths into very irregular or criss-crossing paths that tend to mix the water thoroughly. This type of movement is *turbulent flow.*

Streams, winds, and ocean currents are nearly always turbulent. The average velocity

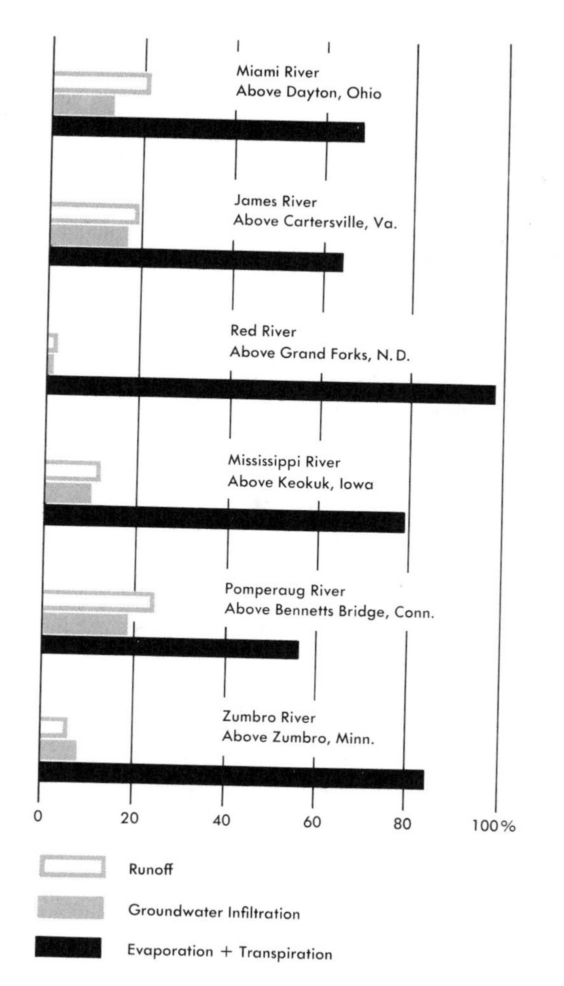

FIGURE 10-3 *Distribution of precipitation in selected drainage basins. Notice that in all cases 50 percent or more of all moisture that falls is returned to the atmosphere by evaporation and transpiration. Runoff from the surface is comparatively small, and infiltration of water in the underground is still less. [Data from W. G. Hoyt et al.,* Studies of Relation of Rainfall and Run-off in the United States, *U.S. Geological Survey, Water Supply Paper 772 (Washington, D.C.: Government Printing Office, 1936).]*

of stream flow above stream beds is 1 to 3 m/sec, but is much lower near stream beds because of frictional effects. Stream velocity is determined by the slope of the stream bed, the shape and roughness of the channel, and the discharge. These concepts are discussed in the next section.

Velocity, gradient, and discharge

The *velocity* of a stream is measured in terms of the distance its water travels in a unit of time, usually in feet or meters per second. A velocity of 0.15 m/sec is relatively low, and a velocity of 6 to 8 m/sec is relatively high.

A stream's velocity is determined by many factors, including the amount of water passing a given point, the nature of the stream banks, and the *gradient* or slope of the stream bed. In general, a stream's gradient decreases from its headwaters toward its mouth; as a result, a stream's longitudinal profile is more or less concave upward. The gradient of a stream is the vertical distance a stream descends during a fixed distance of horizontal flow. The Mississippi River from Cairo, Illinois, to the mouth of the Red River in Arkansas has a low gradient, for along this stretch the drop varies between 2 and 10 cm/km. In contrast, the Arkansas River in its upper reaches through the Rocky Mountains in central Colorado has a high gradient, with an average drop of 7.5 m/km. The gradients of other rivers are even higher. The upper 20-km stretch of the Yuba River in California, for example, has an average gradient of 42 m/km; and in the upper 6.5 km of the Uncompahgre River in Colorado, the gradient averages 66 m/km.

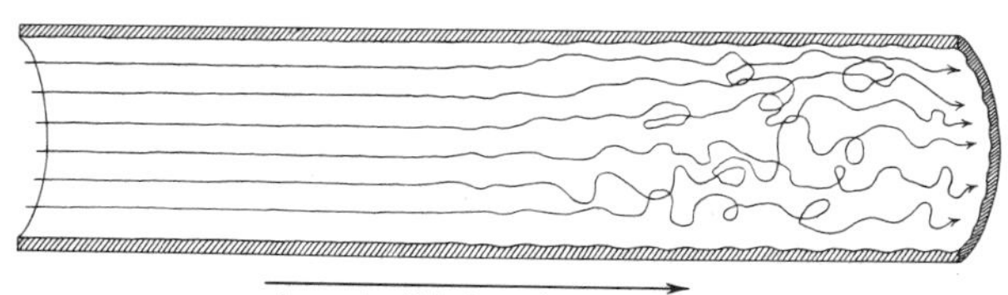

FIGURE 10-4 *Diagram showing laminar and turbulent flow of water through a section of pipe. Individual water particles follow paths depicted by the black lines. In laminar flow, the particles follow paths parallel to the containing walls. With increasing velocity or increasing roughness of the confining walls, laminar flow gives way to turbulent flow. The water particles no longer follow straight lines but are deflected into eddies and swirls. Most water flow in streams is turbulent.*

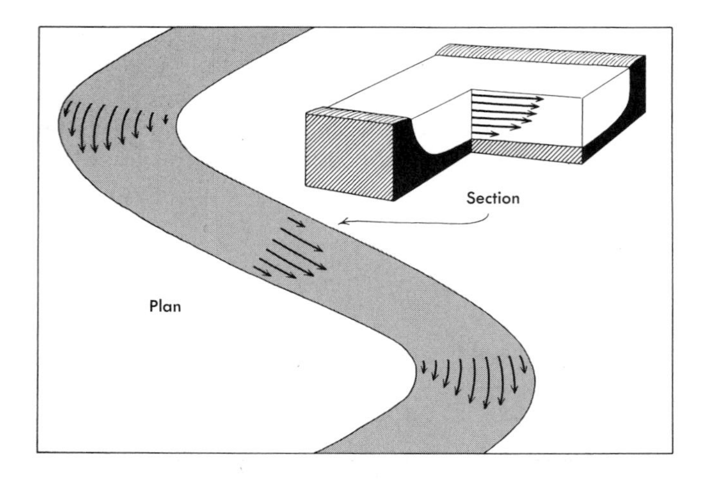

FIGURE 10-5 Velocity variations in a stream. Both in plan view and in cross section, the velocity is slowest along the stream channel, where the water is slowed by friction. On the surface, velocity is most rapid at the center in straight stretches and toward the outside of a bed where the river curves. Velocity increases upward from the river bottom.

The velocity of a stream is checked by friction along the banks and bed of its channel and, to a much smaller extent, by friction with air above. The velocity therefore varies from point to point. Along a straight stretch of a channel, the greatest velocity is achieved toward the center of the stream at, or just below, the surface, as shown in Figure 10-5.

There are, then, two opposing forces: the *forward flow* of the water under the influence of gravity, and the *friction* developed along the walls and bed of the stream. These two forces create different velocities, and zones of maximum turbulence occur where the different velocities come into the closest contact (Figure 10-6).

There is one more term that is helpful in discussing running water—that is, *discharge,* or the *quantity* of water that passes a given point in a unit of time. Discharge is usually measured in cubic meters per second. Discharge varies not only from one stream to another, but also within a single stream from time to time and from place to place.

The equilibrium of a stream

Elsewhere, we noted that earth processes tend to seek a balance, to establish an equilibrium. Weathering is a response of earth materials to the new and changing conditions they are exposed to at or near the earth's surface.

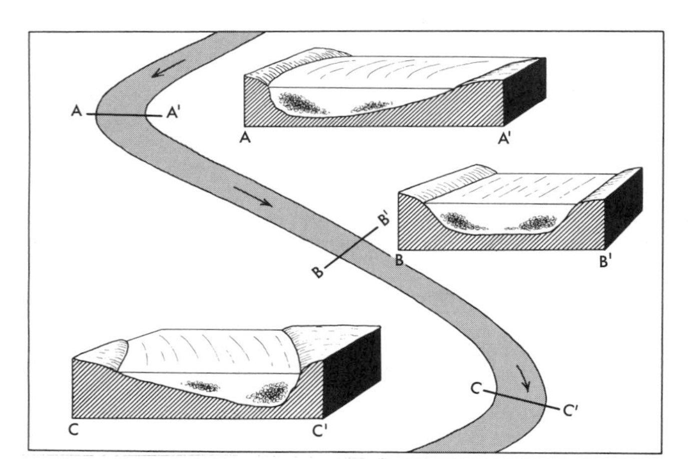

FIGURE 10-6 Zones of maximum turbulence in a stream are shown by the shaded areas in the sections through a river bed. They occur where the change between the two opposing forces—the forward flow and the friction of the stream channel—is most marked. Note that the maximum turbulence along straight stretches of the river is located where the stream banks join the stream floor. On bends, the two zones have unequal intensity; the greater turbulence is located on the outside of a curve.

Water running off the land in streams and rivers is no exception to the universal tendency of nature to seek this equilibrium.

Adjustments of discharge, velocity, and channel

The discharge of a river is related to its width, depth, and velocity as follows:

$$\underset{\text{(c.m.s.)}}{\text{Discharge}} = \underset{\text{(m)}}{\overset{\text{Channel}}{\text{width}}} \times \underset{\text{(m)}}{\overset{\text{Channel}}{\text{depth}}} \times \underset{\text{(m/sec)}}{\overset{\text{Water}}{\text{velocity}}}$$

If the discharge at a given point along a river increases, then the width, depth, or velocity, or some combination of these factors must also increase. These variations in width, depth, and velocity are neither random nor unpredictable. In most streams, if the discharge increases, then the width, depth, and velocity each increase at a definite rate. The stream maintains a balance between the amount of water it carries on the one hand, and its depth, width, and velocity on the other. Moreover, it does so in an orderly fashion.

Let us turn from the behavior of a stream at a single locality to the changes that take place along its entire length. Generally, the discharge of a stream increases downstream as more and more tributaries contribute water to its main channel. The width and depth also increase downstream. Accurate data on the width, depth, velocity, and discharge of a stream from its headwaters to its mouth for a particular stage of flow—say, flood or low-water—indicate that changes follow a definite pattern and that depth and width increase downstream as the discharge increases.

Table 10-3 Downstream Changes in Flow Properties

FLOW PROPERTIES	CHANGES IN DOWNSTREAM DIRECTION
Slope of stream bed	Decreases
Discharge	Increases
Width	Increases
Depth	Increases
Velocity	Increases or decreases depending on changes in slope and discharge.
Frictional resistance	Increases or decreases depending on cross-sectional perimeter and load.
Sediment load	Increases
Size of grains	Decreases

However, in many streams, the discharge does not increase greatly downstream because the lower inclination of the downstream slope reduces the velocity. Nonetheless, most streams are able to transport more material of finer grain size at a higher discharge rate in spite of the lower slope. Table 10-3 gives the normal changes in flow properties in the downstream direction.

Base level of a stream

An important concept in the study of stream activity is the *base level,* or the lowest

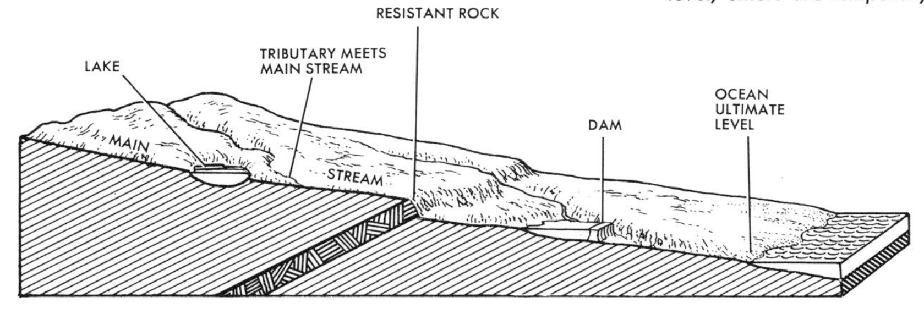

FIGURE 10-7 *Base level for a stream may be determined by natural and artificial lakes, by a resistant rock stratum, by the point at which a tributary stream enters a main stream, and by the ocean. Of these, the ocean is considered ultimate base level; others are temporary base levels.*

point to which a stream can erode its channel. Base level is reached when the slope of the stream approaches zero. At this point the velocity is checked or slowed. The stream can erode no further.

The *ultimate base level* of a stream is located at its mouth, which is usually slightly below sea level. However, a lake, beds of resistant rock, or the level of the main stream into which a tributary drains may form a *temporary base level* (Figure 10-7).

Work of running water

The water that flows along through river channels does a variety of jobs: (1) it transports debris, (2) it erodes the river channel deeper into the land, and (3) it deposits sediment at various points along the valley or delivers sediment to lakes or oceans.

Transportation

The material that a stream picks up directly from its own channel—or that is supplied to it by slope wash, tributaries, or mass movement—is carried downstream toward its eventual goal, the ocean. The amount of material that a stream carries at any one time is known as its *load,* which is usually less than its *capacity*—the total amount it is capable of carrying under any given set of conditions. The maximum size of particle that a stream can move measures the *competency* of a stream.

The competency of a stream depends primarily on its velocity. At low velocities, a stream may run clear; at high velocities, larger grains begin to move; at very high velocities during floods, some fantastically large objects can be moved (Figure 10-8). Railroad locomotives, along with houses and people, were swept along during the Johnstown Flood of 1889. When the San Francisquito dam in California failed in 1928, a wall of water 38-m high raced down the canyon at a velocity of 80 km/hr and carried 10,000-ton blocks of concrete almost a kilometer downstream. A photograph of this bed load would be enough to frighten strong men and women.

There are three ways in which a stream can transport material: (1) by solution, (2) by suspension, and (3) by bed load.

Solution. When water falls and filters into the ground, it dissolves some of the soil's compounds. Then the water may seep down through openings, pores, and crevices in the bedrock and dissolve additional matter as it moves along. Much of this water eventually finds its way to streams at lower levels. The amount of dissolved matter contained in water varies with climate, season, and geologic setting and is measured in terms of parts of dissolved matter per million parts of water. Sometimes the amount of dissolved material exceeds 1,000 parts per million (ppm), but usually it is much less. The average is about 200 ppm. By far the most common compounds found in solution in running water, particularly in arid regions, are calcium and magnesium carbonates. In addition, streams carry small amounts of chlorides, nitrates, sulfates, and silica, with perhaps a trace of potassium. The total load of dissolved material delivered to the seas every year by the streams of the United States is nearly 300 million metric tons. All the rivers of the world together deposit about 3.9 billion metric tons of dissolved material into the oceans each year.

Suspension. Particles of solid matter that are swept along in the turbulent current of a stream are said to be in *suspension.* This process of transportation is controlled by two factors: (1) the turbulence of the water, and (2) a characteristic known as the *terminal velocity* of each individual grain. The terminal velocity is the constant rate of fall that a particle eventually attains when the acceleration caused by gravity is balanced by the resistance of the

FIGURE 10-8 *Telephone pole deposited downstream during flooding of small ephemeral stream. Flow was from right to left. (Photograph by M. Dane Picard.)*

fluid through which the grain is falling. If we drop a grain of sand into a quiet pond, it will settle toward the bottom at an ever-increasing rate until the friction of the water on the grain just balances this rate of increase. Thereafter, it will settle at a constant rate, its terminal velocity. If we can set up a force that will equal or exceed the terminal velocity of the grain, we can succeed in keeping it in suspension. Turbulence supplies such a force. The eddies of turbulent water move in a series of orbits, and grains caught in these eddies will be buoyed up, or held in suspension, so long as the velocity of the turbulent water is equal to, or greater than, the terminal velocity of the grains.

Terminal velocity increases with particle size, given the same general shape and density. The bigger a particle, the more turbulent the flow needed to keep it in suspension. And since turbulence increases when the velocity of stream flow increases, it follows that the greatest amount of material is moved during flood time when velocities and turbulence are highest.

Silt and clay-sized particles are distributed fairly evenly through the depth of a stream, but coarser particles in the sand-size range are carried in greater amounts lower down in the current, in the zone of greatest turbulence.

Bed load. In a stream there generally is a complete gradation in grain size of those particles small enough to remain suspended and those fragments too large to be lifted from the bottom. The largest fragments are rolled and dragged along the bottom by *traction.* The term *bed load* is applied to grains that move along, or move close to, the stream bed, but there is no definite boundary between suspended grains and grains moving along the bed.

Bed-load grains roll, slide, and bounce, or momentarily vault into the suspended load. As the velocity of a stream flowing on a sand bottom is gradually increased, the motion of the grains progresses from short leaps of a few grains, through spasmodic movement and deposition of groups of grains, to general transport of many grains. At higher velocities, many more grains are transported, and the bed load may become indistinguishable from the suspended load.

Bed forms. Sand grains moving along a stream bed accumulate in different forms. At low flow-velocities, ripple marks develop on the bottom of the stream bed. These *asymmetrical ripples* have a gentle slope (*stoss side*) upstream and a steep slope (*lee side*) downstream, and usually are less than 30 cm long and 0.5 to 6 cm high. They migrate slowly downstream.

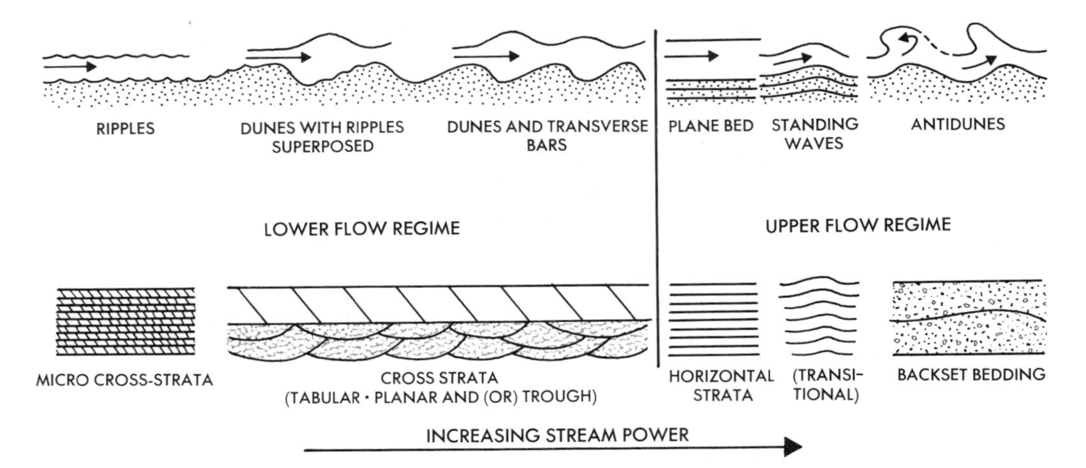

FIGURE 10-9 *Bed forms and stratification in alluvial channels.*

As flow velocity increases, *dunes* form on the bed. Dunes have the same general form and structure as ripples but are larger, ranging in size from 60 cm to several hundred meters in length and from 6 cm to many meters in height. As dunes grow larger, small ripples form and move up their stoss sides and disappear over their lee sides. Ripples and dunes formed in sandy stream bottoms are characteristic of streams in what is termed the *lower flow regime* (Figure 10-9).

If there is a further increase in velocity, there is a change from *streaming flow* (tranquil flow) to *shooting flow* (high-velocity flow). The ripples and dunes then disappear, and the bed of the stream becomes essentially a plane. At still higher velocities, the smooth bed is transformed into *antidunes* in which the wave forms of dunes and of surface water waves travel upstream. Plane beds and antidunes typically develop in streams in the *upper flow regime.* Finally, at very high velocities, a complete washout of the whole bed of sand takes place, and the grains are thrown violently into suspension.

FIGURE 10-10 *Crossbedded sandstone beds deposited by streams. Inclination of foreset beds indicates stream flow was dominantly from upper right to lower left. (Photo by M. Dane Picard.)*

Preserved ripple marks, dunes, and rare antidunes account for the crossbedded sandstone, siltstone, and conglomerate that are characteristic of stream-deposited beds. The inclination of the laminae composing the beds can be used to determine the direction of stream flow (Figure 10-10), and the position of the highlands that were eroded to furnish the detritus. The laminae (foresets) of both crossbedding and ripple marks are inclined downstream; in contrast, antidune crossbedding is inclined upstream.

Erosion

A stream does more than simply transport material that has been brought to it by other agencies or erosion, for it is an effective agent of erosion in itself. In various ways, an actively eroding stream may remove material from its channel or banks.

Direct lifting. In turbulent flow, water travels along paths that are not parallel to the stream bed. The water eddies and whirls, and if an eddy is powerful enough, it dislodges particles from the stream channel and lifts them into the stream. Particles in the fine sand-size range move with the lowest velocities. Compacted finer particles and coarser particles require increasing velocities to set them in motion. Unconsolidated silt and clay begin to move at about the same low velocities as fine sand (Figure 10-11).

Abrasion, impact, and solution. The solid particles carried by a stream may themselves act as erosive agents, for they are capable of abrading the bedrock itself or abrading larger fragments in the bed of the stream. When the bedrock is worn by abrasion, it usually develops a series of smooth, curving surfaces, either convex or concave. Similarly, when the individual cobbles or pebbles on a stream bottom are rubbed together as they are moved and rolled about by the force of the current, they become rounder and smoother.

FIGURE 10-11 *The velocities at which particles of varying size are set in motion. The colored zone represents the range of experimental data. [From Ake Sundborg, Geografis Ann, 38(1956), 197, Figure 16]*

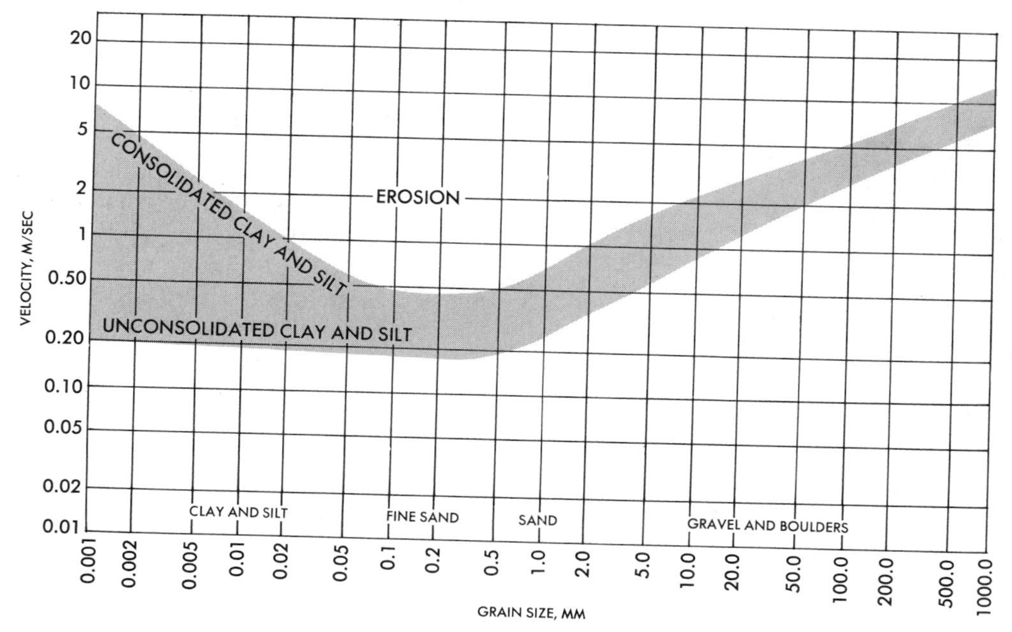

Also erosive is the impact of large particles against the bedrock or against other particles, an impact that knocks off fragments that are then added to the load of the stream.

Some erosion also results from the solution of channel debris and bedrock in the water of the stream. Most of the dissolved matter carried by a stream, however, is probably contributed by the underground water that drains into it.

Slope erosion by running water. So far, we have only considered the erosion that takes place along a stream channel—certainly the most conspicuous form of erosion by water—but the total area reached by stream channels is only a small proportion of the total land surface drained by streams, perhaps about 1 percent. Furthermore, most of the flood water carried by streams originates as runoff from neighboring slopes. The runoff—flowing as a sheet of water called *slope wash,* or in closely spaced, shallow channels called *rills*—is sometimes powerful enough to overcome the soil's natural resistance to erosion, and consequently the runoff manages to transport a great deal of material downslope toward the stream channels.

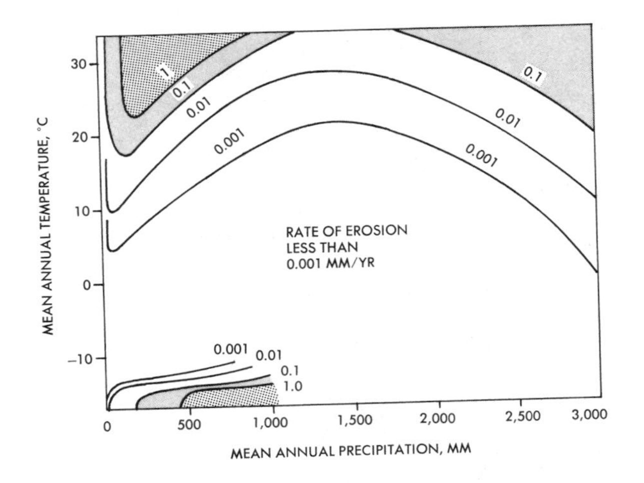

The muddy water running off a plowed field or off a newly graded slope during a heavy rain is a familiar example of the erosive power of runoff. Although the importance of slope erosion by running water is often overlooked, such action plays a significant role in the general process of erosion. A general rate of erosion with different amounts of rainfall is given in Figure 10-12.

Time of most rapid erosion. Other things being equal, the greater the stream's velocity, the greater its erosive power. The greatest erosive (and transporting) power of any stream is developed during floods. When a stream is at flood stage, the water level rises and the channel is deepened. The fast-moving water picks up the layer of sand and gravel that usually lies on the bedrock of the stream channel during nonflood stages and sweeps it downstream. If the flood is great enough, the bedrock itself is exposed and eroded. A new layer of debris collects as the flood waters subside, but by that time great masses of material have been moved downstream toward the oceans, and the bedrock channel of the stream has been permanently lowered.

Deposition

As soon as the velocity of a stream falls below the point necessary to hold material in suspension, the stream begins to deposit its suspended load. Deposition is a selective process. First, the coarsest material is dropped; then, as the velocity (and hence the energy) continues to slacken, finer and finer material settles out. We shall consider stream deposits in more detail elsewhere in this chapter.

FIGURE 10-12 *Generalized rates of erosion in millimeters of lowering per year as a function of temperature and precipitation. [Modified from M. A. Carson and M. J. Kirkby,* Hillslope and Soil Process *(London: Cambridge University Press, 1972), p. 219, Figure 8.18.]*

Features of valleys

Cross-Valley profiles

Earlier in this chapter we mentioned the longitudinal profile of a stream. Now let us turn to a discussion of the cross-valley profile—that is, the profile at right angles to the stream's line of flow. In Figure 10-13A, notice that the channel of the river runs across a broad, relatively flat *flood plain.* During floods, when the channel can no longer accommodate the increased discharge, the stream overflows its banks and inundates this area. On either side of the flood plain, valley walls rise to crests called *divides,* which are separations between the central valley and the other valleys on either side. In Figure 10-13B, no flood plain is present, for the valley walls descend directly to the banks of the river. This diagram also illustrates two different shapes of divides. One is broad and flat; the other is narrow, almost knife-edged. Both contrast with the broadly convex divides shown in A.

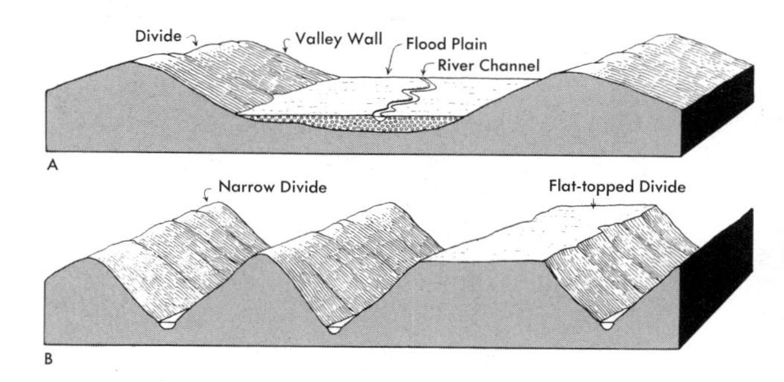

FIGURE 10-13 Cross-sectional sketches of typical stream valleys. The major features of valleys in cross section include divides, valley walls, river channel, and, in some instances, a flood plain. Divides may be either flat-topped or broadly rounded.

Drainage basins

A *drainage basin* is the entire area from which a stream and its tributaries receive their water. The Mississippi and its tributaries drain a tremendous section of the central United States reaching from the Rockies to the Appalachians. Each tributary of the Mississippi has its own drainage area, and each of these small areas forms a part of the larger basin. Every stream, even the smallest brook, has its own drainage basin, shaped differently from stream to stream, but characteristically pear-shaped, with the main stream emerging from the narrow end (Figure 10-14).

Rivers with the largest drainage basins are listed in Table 10-4. And areas of some of the world's largest lakes and their maximum depths are given in Table 10-5. Because of the effects of differences in climate, the volume,

Table 10-4 Areas of Drainage Basins of Large Rivers

	DRAINAGE BASIN	DISCHARGE RATE
	(1,000 km²)	*(m²/sec)*
Amazon, Brazil	6,150	175,000
Congo, Africa	3,822	39,000
Mississippi, USA	3,222	17,270
Plata, Argentina-Uruguay	3,100	22,900
Nile, Egypt	2,802	2,640
Yenisey, USSR	2,619	18,000
Lena, USSR	2,478	16,100
Ob, USSR	2,470	10,200
Niger, Nigeria	2,092	5,700
Yangtze, China	1,827	32,190

Table 10-5 Size of Some of the World's Great Lakes

	AREA	GREATEST DEPTH
	(km²)	*(m)*
Caspian Sea, Iran–USSR	371,000	980
Aral Sea, USSR	64,500	68
Baikal, Siberia	31,500	1,620
Victoria, Africa	62,940	81
Tanganyika, Africa	32,000	1,471
Nyasa, Africa	22,490	706
Superior, Canada–USA	83,270	397
Huron, Canada–USA	60,700	229
Michigan, USA	58,020	281
Great Bear, Canada	31,790	413
Great Slave, Canada	28,440	614
Erie, Canada–USA	25,680	64
Winnipeg, Canada	24,510	28
Maracaibo, Venezuela	14,300	60

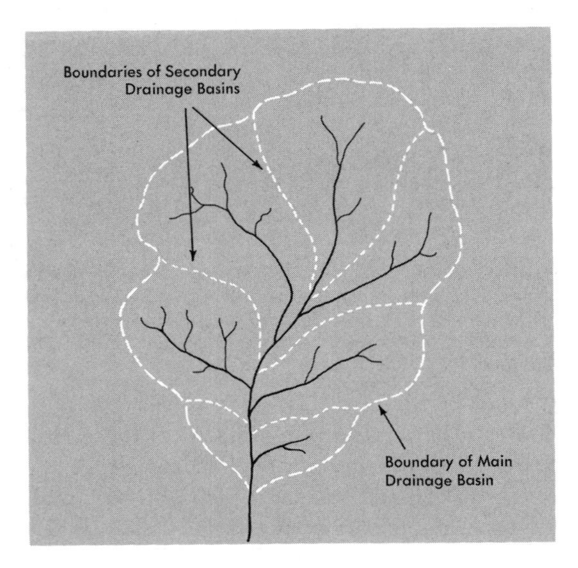

FIGURE 10-14 Each stream, no matter how small, has its own drainage basin, the area from which the stream and its tributaries receive water. This basin displays a pattern reminiscent of a tree leaf and its veins.

flow, and drainage area of a large river system may have little in common with these same features of another river system. For example, the drainage basin of the Amazon River is almost twice that of the Mississippi River, but the discharge rate of the Amazon is ten times that of the Mississippi (Table 10-4), mainly because of the much greater rainfall in the Amazon drainage basin. And in desert areas, for example, the Nile, discharge rates are smaller than might be expected because of the very low rainfall and very high evapotranspiration rates (Table 10-4).

Enlargement of valleys

It is not known with certainty how running water first fashioned the great valleys and drainage basins of the continents, for the record has been lost. However, certain processes are now at work in widening and deepening valleys, and it seems safe to assume that these processes also operated in the past.

If a stream were left to itself in its attempt to reach base level, it would erode its bed straight downward, forming a vertical-walled chasm in the process. But the stream is not the only agent at work in valley formation. As the stream cuts downward and lowers its channel into the land surface, other agents of erosion—weathering, slope wash, and mass movement—come into play, constantly wearing away the valley walls, pushing them farther back. Under the influence of gravity, material is carried down from the valley walls and dumped into the stream, to be moved onward toward the seas. The result is a valley whose walls flare outward and upward from the stream in a typical cross-valley profile. In time, the cliffs of even the steepest gorge will be angled away from the axis of its valley.

Features of narrow valleys

Waterfalls and rapids. Waterfalls are among the most fascinating spectacles of the landscape (Figure 10-15). Thunderous and powerful as they are, however, they are actually short-lived features in the history of a stream. They owe their existence to a sudden drop in the river's longitudinal profile—a drop that will be eliminated with the passing of time.

Waterfalls are caused by many different conditions. Niagara Falls, for instance, is held up by a relatively resistant bed of dolomite underlain by beds of nonresistant shale (Figure 10-16). This shale is easily undermined by the swirling waters of the Niagara River as they plunge over the lip of the falls. When the undermining has progressed far enough, the dolomite lip collapses and tumbles to the base of the falls. The same process is repeated over and over again as time passes, and the falls slowly retreat upstream. Historical records suggest that the Horseshoe or Canadian Falls (by far the larger of the two falls at Niagara)

FIGURE 10-15 Spectacular waterfall in Yellowstone National Park. (Courtesy National Park Service.)

have been retreating at a rate of 120 to 150 cm/yr, whereas the smaller American Falls have been eroded away at a rate of 5 to 6 cm/yr. The 11 km of gorge between the foot of the falls and Lake Ontario are evidence of the headward retreat of the falls through time.

Yosemite Falls in Yosemite National Park, California, plunge 770 m over the Upper Falls, down an intermediate zone of cascades, and then over the Lower Falls. The falls leap from the mouth of a small side valley high above the main valley of the Yosemite. The Upper Falls alone measure 430 m, nine times the height of Niagara. During the Ice Age, glaciers scoured the main valley much deeper than they did the small side valley. Then, when the glacier ice melted, the river in the main valley was left far below its tributary, which now joins it after a drop of nearly 0.8 km.

Rapids, like waterfalls, occur at a sudden drop in the stream channel. Although rapids do not plunge straight down as waterfalls do, the underlying cause of formation is often the same. In fact, many rapids have developed directly from preexisting waterfalls.

FIGURE 10-16 *Niagara Falls tumbles over a bed of dolomite, underlain chiefly by shale. As the less resistant shale is eroded, the undermined ledge of dolomite breaks off, and the lip of the falls retreats. [Redrawn from G. K. Gilbert, Niagara Falls and Their History (New York: American Book Company, 1896), p. 213.]*

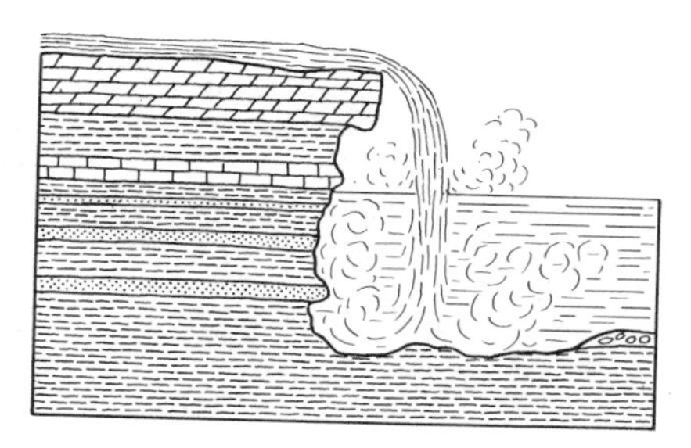

A very youthful stream whose bed is choked with boulders is shown in Figure 10-17.

Potholes. As the bedrock channel of a stream is eroded away, *potholes* sometimes develop (Figure 10-18). These are deep holes, circular to elliptical in outline, and a few centimeters to meters in depth. They are most often observed in the stream channel during low water or along the bedrock walls, where they have been left stranded after the stream has cut its channel downward. Potholes are most common in narrow valleys, but they often are present in broad valleys as well.

A pothole begins as a shallow depression in the bedrock channel of a stream. It may be caused by ordinary abrasion, or by some irregularity in the bedrock. Then, as the swirling, turbulent water drives sand, pebbles, and even cobbles around and around the depression, the continued abrasion wears the potholes ever deeper, as if the bedrock were being bored by a giant drill.

Features of broad valleys

If conditions permit, the various agents working toward valley enlargement ultimately produce a broad valley with a wide level floor. During periods of normal or low water, the river running through the valley is confined to its channel; but during high water, it overflows its banks and spreads over the flood plain.

Meanders. The channel of the Menderes River in Asia Minor curves back on itself in a series of broad hairpin bends. In fact, the very name of the river is derived from the Greek *maiandros,* "a bend." Today, all such bends are called *meanders,* and the zone along a valley floor that encloses a meandering river is called a *meander belt* (Figures 10-19, 10-20, and 10-21).

Both erosion and deposition are involved in the formation of a meander. First, some obstruction swings the current of a stream

FIGURE 10-17 *Very youthful stream. Note big boulders. (Photo by Mary Hill.)*

FIGURE 10-18 *This nearly dry stream bed is marked by a series of potholes. Coarse sand and gravel, caught up in the eddies of turbulent water, served as the cutting materials that carved the holes. (Photo by Paul MacClintock.)*

FIGURE 10-19 Flooding of meandering Red Lake River at Crookston, Minnesota. (Courtesy of U.S. Department of Agriculture.)

against one of the banks, and then the current is deflected to the opposite bank. Erosion takes place on the outside of each bend, where turbulence is greatest. The material detached from the banks is moved downstream, there to be deposited in zones of decreased turbulence—either along the center of the channel or on the inside of the next bend. As the river swings randomly from side to side, the meander continues to grow by erosion on the outside of the bends and by deposition on the inside. Growth ceases when the meander reaches a critical size, a size that increases with an increase in the size of the stream.

FIGURE 10-20 Meander belt in Alaska showing oxbows, oxbow lakes, and chutes. (Photo by A. J. Eardley.)

FIGURE 10-21 *Pronounced meander belt on flood plain, Virginia. Note deposits on insides of several meanders. (Courtesy U.S. Geological Survey.)*

Because a meander is eroded more on its downstream side than on its upstream side, the entire bend tends to move slowly downvalley. This movement is not uniform, however, and under certain conditions the downstream sweep of a series of meanders is distorted into cutoffs, meander scars, and oxbow lakes.

In its down-valley migration, a meander sometimes runs into a stretch of land that is relatively more resistant to erosion. But the next meander upstream continues to move right along, and gradually the neck between the two meanders is narrowed. Finally, the river cuts a new, shorter channel, called a *neck cutoff,* across this neck. The abandoned meander is called an *oxbow,* because of its characteristic shape. Usually both ends of the oxbow are gradually silted in, and the old meander becomes completely isolated from the new channel. If the abandoned meander fills up with water, an *oxbow lake* results. Although a cutoff will eliminate a particular meander, the stream's tendency toward meandering still exists, and the rest of the stream bed will continue in its meander pattern.

As we found, a meander grows and migrates by erosion on the outside of the bend and by deposition on the inside. This deposition on the inside leaves behind a series of low ridges and troughs. Swamps often form in the troughs, and during flood the river may develop an alternate channel through one of the troughs. Such a channel is called a *chute cutoff,* or simply a *chute.*

The meandering river demonstrates a unity in ways other than the balance of erosion and deposition. The length of a meander, for example, is proportional to the width of the river, and this is true regardless of the size of the river. It holds for channels a few meters wide as well as for those as large as the Mississippi. This principle also is true of the Gulf Stream, even though this "river" is unconfined by solid banks. A similar relationship holds between the length of the meander and the radius of curvature of the meander.

Braided streams. On some flood plains, particularly where streams are heavily loaded with coarse bed loads, a stream may build up a complex tangle of converging and diverging channels separated by sand bars or islands. A stream of this sort is called a *braided stream* (Figure 10-23). When the velocity is checked either by a decrease in the stream's gradient or by a loss of water through infiltration into porous deposits, the energy of the stream also

FIGURE 10-22 *Crossbedding in sandstone and in conglomeratic sandstone beds that were deposited by meandering and braided streams. (Photo by M. Dane Picard.)*

decreases. Consequently, a large part of the stream's suspended load is suddenly dropped. The deposited material then deflects the current into different channels. This braided pattern is commonly found on alluvial fans (see below), glacial outwash deposits, and along certain rapidly depositing rivers.

Natural levees. In many flood plains, the water surface of the stream is held above the level of the valley floor by banks of sand and silt known as *natural levees,* a name derived from the French verb *lever,* "to raise." These banks, or levees, slope gently, almost imperceptibly, away from their crest along the river and down toward the valley wall. The levees are created during floods when the water spills over the river banks onto the flood plain. Since the muddy water rising over the stream bank is no longer confined by the channel, its velocity and turbulence drop immediately, and much of the suspended load is therefore deposited close to the river. The rest of the load is carried farther along, to be distributed across the flood plain. The deposit from a single flood is a thin wedge tapering away from the river; but over many years, the cumulative effect of many floods produces a natural levee that is considerably higher alongside the river banks than away from it. On the Mississippi delta, for instance, the levees stand 5 to 6 m above the *back swamps,* which are the marshy areas of the low-lying flood plain.

Although natural levees tend to confine a stream within its channel, each time the levees are raised slightly, the bed of the river is also raised. In time, the level of the bed is raised above the level of the surrounding flood plain. If the river manages to escape from its confining walls during a flood, it will assume a new channel across the lowest parts of the flood plain toward the back swamps.

Flood plain deposits. The floors of most flood plains are covered by two and sometimes three different types of deposits. The coarsest material is deposited directly by the stream along its channel. During flood periods, finer sand, silt, and clay are spread across the flood plain, away from the river banks. In addition, relatively small amounts of debris of various types and sizes move down the valley walls under the influence of slope wash and mass movement and are spread along the sides of the valley floor.

FIGURE 10-23 *Complex braided stream and its deposits, Sicily. (Photo by Earle F. McBride.)*

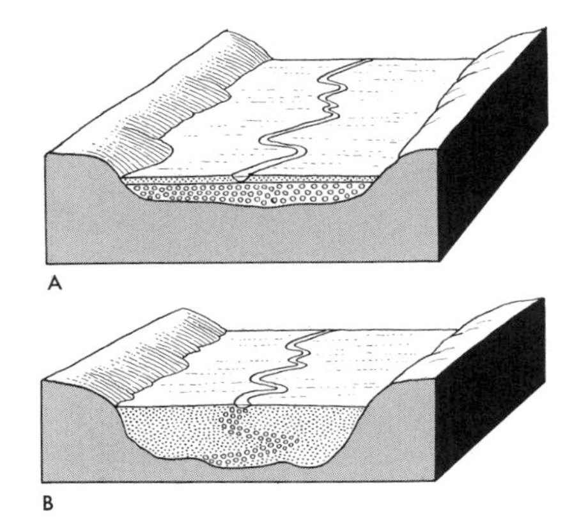

FIGURE 10-24 (Above) *Deposits underlying the flood plain of a slowly aggrading river or beneath a flood plain formed by erosion differ from those beneath a rapidly aggrading river's flood plain.* (A) *Deposits are those to be expected beneath an erosional flood plain or one being slowly aggraded. Coarse river-channel deposits underlie the entire flood plain and are veneered with fine-grained sediment.* (B) *The type of deposit beneath the flood plain of a rapidly aggrading river, the great bulk being fine-grained sediment deposited during flood periods. Coarse channel deposits may form a ribbon of gravel within the finer deposits.*

The distribution of the channel and flood deposits across a flood plain depends on the rate at which a stream builds up its valley floor. A meandering stream that builds up its flood plain at a slow rate is constantly shifting its channel, and over a period of time it may succeed in occupying every possible position across the plain. A cross section through the flood plain developed by such a stream reveals a cover of gravel capped by fine-grained sediments deposited during overflow (Figure 10-24A). This pattern of sediments is typical of erosional flood plains and of very slowly aggraded flood plains.

FIGURE 10-25 (Below) *In this picture taken from Gemini 4, the delta of the Nile River stands out because it is well-vegetated and thus contrasts with the desert country that surrounds it.* (Photo by NASA.)

But a meandering stream that builds up its flood plain at a rapid rate has less opportunity to occupy each spot across the valley floor. Consequently, its flood plain will be covered for the most part by fine sediments deposited during times of overflow. A cross section reveals an irregular band of coarse material, marking successive positions of the channel (Figure 10-24B). Rapidly aggraded flood plains show this pattern of deposition.

Deltas and alluvial fans. For centuries, the Nile River has been depositing sediments as it empties into the Mediterranean Sea, forming a great triangular plain with its apex upstream. This plain came to be called a *delta* because of the similarity of its shape to the Greek letter Δ (Figure 10-25).

When a stream flows into a body of standing water, such as a lake or an ocean, its velocity and transporting power quickly decrease. If it carries enough debris and if conditions in the body of standing water are favorable, a delta is gradually built up. An ideal delta is triangular in plan, with the apex pointed upstream, and with the sediments arranged according to a definite pattern. Coarse material is deposited at the front of the delta slope at the angle of repose in *foreset beds.* The finest material remains longer in suspension and finally settles on the sea or lake floor as *bottomset beds.* The flat-lying thin overbank sediments that top the delta are termed *topset beds* (Figure 10-26).

FIGURE 10-26 The ideal arrangement of sediment beneath a delta. Some of the material deposited in a water body is laid on the bottom of the lake or sea as bottomset beds. Other material is dumped in inclined foreset beds, built farther and farther into the body of water and partly covering the bottomset beds. Over the foreset beds the stream lays down topset beds.

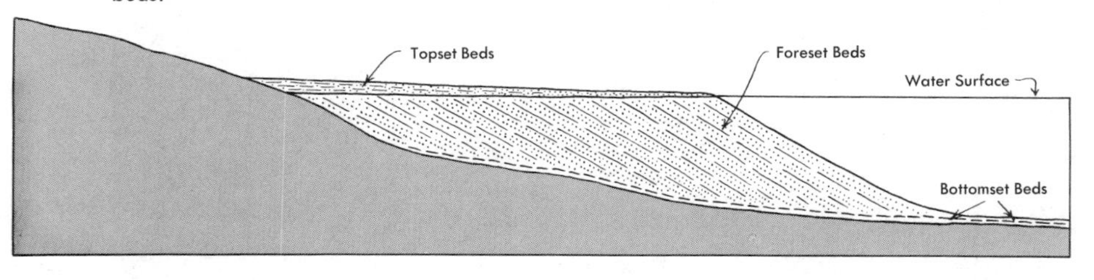

Very few deltas, however, show either the perfect delta shape or this regular sequence of sediments. Many factors, including lake and shore currents, varying rates of deposition, the settling of delta deposits as a result of their compaction, and the down-warping of the earth's crust, may modify the typical form and sequence.

Across the top delta deposits, the stream spreads seaward in a complex of channels radiating from the apex. These *distributary channels* shift their position from time to time as they seek more favorable gradients.

An *alluvial fan* is the terrestrial counterpart of a delta. Such fans are typical of arid and semiarid climates, but they can form in almost any climate if conditions are favorable. A fan marks a sudden decrease in the carrying power of a stream as it descends from a steep gradient to a flatter one—for example, when the stream flows down a steep mountain slope onto a plain. As the velocity is checked, the stream rapidly begins to deposit its load. In the process, it builds up its channel, often with small natural levees along its banks. Eventually, as the levees continue to grow, the stream may flow above the general level. Then, during a time of flood, it seeks a lower level and shifts its channel to begin deposition elsewhere. As this process of shifting continues, an alluvial fan is constructed.

Stream terraces

A *stream terrace* is a relatively flat surface running along a valley, with a steep bank separating it either from the flood plain or from a lower terrace. It is a remnant of the former channel of a stream that now has cut its way down to a lower level.

either side of the new one; these are the cut-and-fill terraces. Terraces that face each other across the stream at the same elevation are termed *paired terraces.*

Sometimes the downward erosion by streams creates *unpaired terraces* rather than paired ones. This occurs when the stream encounters resistant rock beneath the unconsolidated deposits. The exposed rock will then deflect the stream and prevent further erosion. Thus a single terrace is left behind with no corresponding terrace on the other side.

Terraces, either paired or unpaired, may be cut into bedrock as well. A thin layer of sand and gravel usually rests on the beveled bedrock of these terraces.

FIGURE 10-27 *A Roman mausoleum, partially buried in stream deposits, is exposed in this stream bank just north of Rome, Italy. In A.D. 50, when the mausoleum was built, the stream flowed at the same level as it does today, but the steep banks were not then present. Sometime after the third century A.D., this stream began to build up its flood plain until the valley floor stood at a level marked by the top of the modern bank. Thereafter the stream cut down to its present level, reexposing the partially buried structure and leaving its old valley floor and flood plain standing as a low terrace. (Photo by C. T. Stifter.)*

Cut-and-fill terraces are created when a stream first clogs a valley with sediments and then cuts its way down to a lower level (Figure 10-27). The initial aggradation may be caused by a change in climate that leads either to an increase in the stream's load or to a decrease in its discharge. Or the base level of the stream may rise, reducing the gradient and causing deposition. In any event, the stream chokes the valley with sediment and the flood plain gradually rises. Now if that regime is upset and the stream begins to erode, it will cut a channel through its deposits. The level of flow will be lower than the old flood plain, and at this lower level the stream will begin to carve out a new flood plain. As time passes, remnants of the old flood plain are left standing on

FIGURE 10-28 *Terraces along the Madison River, Montana. The various levels have been formed by the river as it simultaneously swung laterally across, and cut downward into, deposits of sand and gravel deposited in front of a now-vanished glacier farther upstream. (Photo by William C. Bradley.)*

Stream patterns and stream types

The overall pattern developed by a system of streams and tributaries depends partly on the nature of the underlying rocks and partly on the history of the streams. Almost all streams follow a branching pattern in the sense that they receive tributaries; the tributaries, in turn, are joined by still smaller tributaries. But the manner of branching varies widely (see Figure 10-29).

A stream that resembles the branching habit of a maple, oak, or similar deciduous tree is called *dendritic,* or "treelike." A dendritic pattern develops when the underlying bedrock is uniform in its resistance to erosion and exercises no control over the direction of valley growth. This situation occurs when the bedrock is composed either of flat-lying sedimentary rocks or massive igneous or metamorphic rocks. The streams can cut as easily in one place as another; thus the dendritic

FIGURE 10-29 *Patterns of stream networks. (A) Dendritic drainage is developed on flat-lying sedimentary rocks of the midcontinent region. (B) Radial drainage flows outward from Mt. Hood, a volcanic cone. (C) East of Lake Piseco, New York, the right-angle bends in the streams reflect the rectangular nature of structures in the crystalline rocks of the area. (D) Along the border between West Virginia and Virginia trellis drainage has developed in folded sedimentary beds of differing resistance to erosion. (From Judson, Sheldon, Deffeyes, K. S., and Hargraves, R. B., Physical Geology. Englewood Cliffs, N.J.: Prentice Hall, Inc., 1976.)*

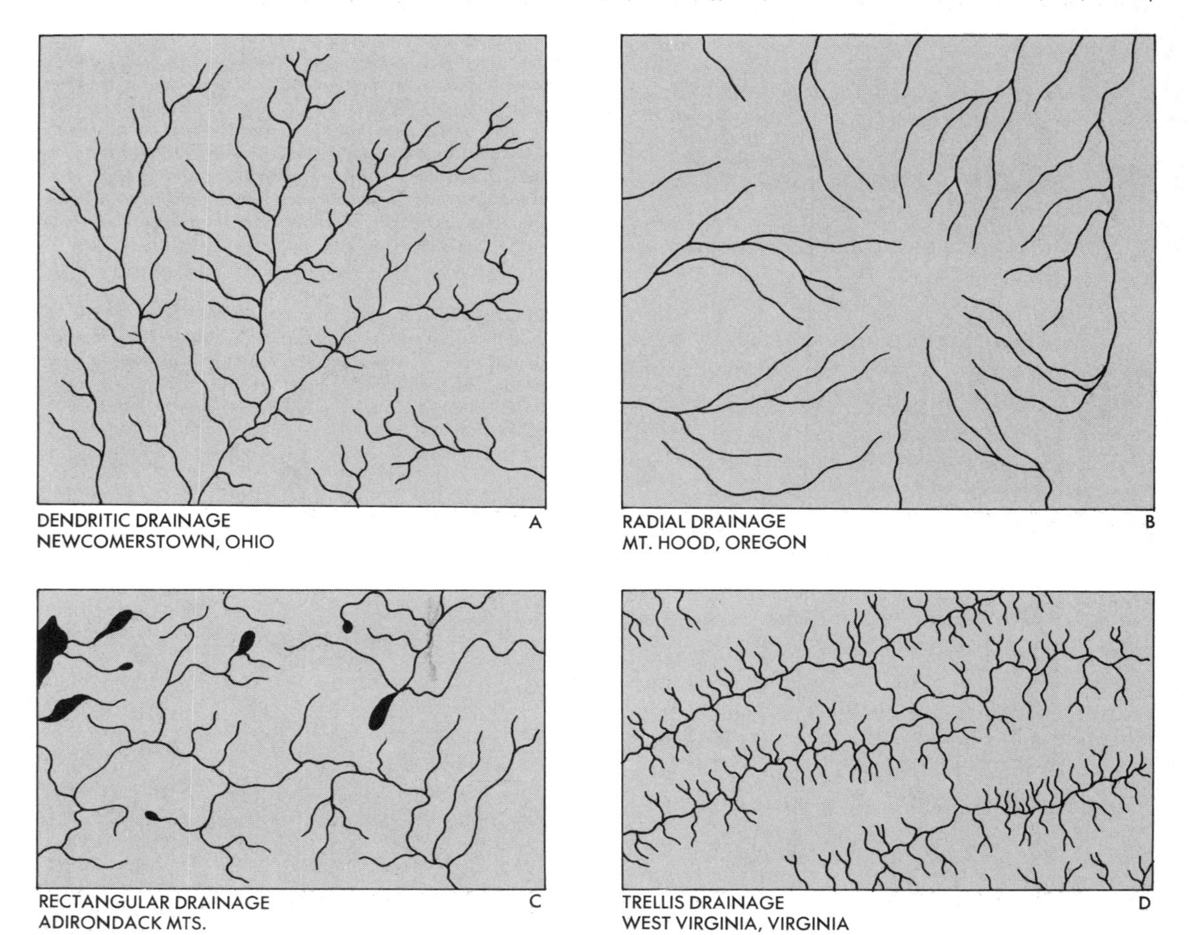

FIGURE 10-30 (Right) Recent erosion has produced this dendritic stream pattern on a Wisconsin farm. (Photo by Wisconsin Conservation Department.)

FIGURE 10-31 (Below) Stream channel on Mars with dendritic pattern. (Courtesy NASA.)

pattern is, in a sense, the result of the random orientation of the streams (Figures 10-30 and 10-31).

Another type of stream pattern is *radial* or *centrifugal,* where streams radiate outward from a high central zone. Such a pattern is likely to develop on the flanks of a newly formed volcano, where the streams and their valleys radiate outward and downward from various points around the cone.

A *rectangular* pattern is developed when the underlying bedrock is crisscrossed by fractures that form zones of weakness particularly vulnerable to erosion. The master stream and its tributaries then follow courses marked by nearly right-angle bends.

Some streams, particularly in the Appalachian Mountains, follow a *trellis* pattern. This pattern is common where the outcropping edges of weak and resistant, folded sedimentary rocks form long, nearly parallel bands.

Underground water

Earlier in this chapter we found that a portion of the water that falls on the surface of the earth seeps into the ground. In the following pages let us consider the movement and activity of this water.

Basic distribution

Underground water, subsurface water, and *subterranean water* are all general terms used to refer to water in the pore spaces, fractures, tubes, and crevices of the consolidated and unconsolidated material beneath our feet. The study of underground water is largely an investigation of these openings and of what happens to the water that moves into them.

Zones of aeration and saturation

Some of the water that moves down from the surface is caught by rock and earth materials and is checked in its downward progress. The zone in which this water is held is known as the *zone of aeration,* and the water itself is called *suspended water.* The spaces between particles in this zone are filled partly with water and partly with air. Two forces operate to prevent suspended water from moving deeper into the earth: one is the molecular attraction exerted on the water by the rock and earth materials, and the second is the attraction exerted by the water particles on one another (Figure 10-32).

The zone of aeration is subdivided into three belts: (1) *the belt of soil moisture,* (2) *the intermediate belt,* and (3) *the capillary fringe.* Some of the water that enters the belt of soil moisture from the surface is used by plants, and some is evaporated back into the atmosphere. But some water also passes down to the intermediate belt, where it may be held by molecular attraction (as suspended water). Little movement takes place in the intermediate belt, except when rain or melting snow sends a new wave of moisture down from above.

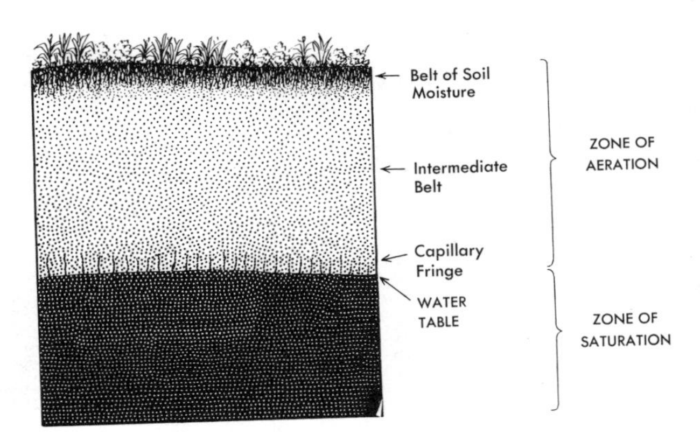

FIGURE 10-32 *Underground water's two major zones: zone of aeration and zone of saturation. The water table marks the upper surface of the zone of saturation. Within the zone of aeration is a belt of soil moisture, the source of moisture for many plants. From here, also, some moisture is evaporated back to the atmosphere. In many instances, this belt lies above an intermediate belt where water is held by molecular attraction, and little movement occurs except during periods of rain or melting snow. In the capillary fringe, just above the water table, water rises a few centimeters to a meter or so from the zone of saturation, depending on the size of the interstices.*

Beneath the zone of aeration lies the *zone of saturation.* Here the openings in the rock and earth materials are completely filled with *ground water.* The surface between the zone of saturation and the zone of aeration is the *ground-water table,* or simply the *water table.* The level of the water table fluctuates with variations in the supply of water coming down from the zone of aeration, with variations in the rate of discharge in the area, and with variations in the amount of ground water drawn off by plants and human beings.

In marshes, lakes, seas, and streams the water table is at the surface. On land, the water table is generally at a depth of several meters, but it may be at depths of more than a kilometer in deserts.

It is the water below the water table, within the zone of saturation, that we shall focus on for the rest of this chapter.

The water table

The water table is an irregular surface of contact between the zone of saturation and the zone of aeration. Below the water table lies the ground water; above it lies the suspended water. The thickness of the zone of aeration differs from one place to another, and from time to time, and the level of the water table fluctuates accordingly. In general, the water table tends to follow the irregularities of the ground surface, reaching its highest elevation beneath hills and its lowest elevation beneath valleys. It should be noted, however, that although the water table reflects variations in the ground surface, the irregularities in the water table are much less pronounced.

In looking at the topography of the water table, let us consider an ideal situation. Figure 10-33 shows a hill underlain by completely homogeneous material. Assume that, initially, this material contains no water at all. Then a heavy rainfall occurs, and the water soaks slowly downward, filling the interstices at depth. In other words, a zone of saturation begins to develop. As more and more water seeps down, the upper limit of this zone continues to rise. The water table remains horizontal until it just reaches the level of the two valley bottoms on either side of the hill. Then as additional water seeps down to the water table, some of it seeks an outlet into the valleys. But this added water is "supported" by the material through which it flows, and the water table is prevented from maintaining its flat surface. The water is slowed by the friction of its movement through the interstices and even, to some degree, by its own internal friction. Consequently, more and more water is piled up beneath the hill, and the water table begins to reflect the shape of the hill. The water flows away most rapidly along the steeper slope of the water table near the valleys, and most slowly on its gentler slope beneath the hill crest.

We can modify the shape of the ground-water table by providing an artificial outlet for the water. For example, we can drill a well on the hill crest and extend it down into the saturated zone. Then, if we pumped out the ground water that flowed into the well, we would create a dimple in the water table. The more we pumped, the more pronounced the depression—called a *cone of depression*—would become.

FIGURE 10-33 *Ideally, the water table is a subdued reflection of the surface of the ground. In (A) and (B), the water table rises as a horizontal plane until it reaches the level of the valley bottoms on either side of the hill. Thereafter, as more moisture soaks into the ground, the water seeks an outlet toward the valleys. If the movement of the water was not slowed down by the material making up the hill, it would remain essentially horizontal. The friction caused by the water's passing through the material (and even to some extent the internal friction of the water itself) results in a piling up of water beneath the hill; the bulge is highest beneath the crest and lowest toward the valleys (C). The shape of the water table may be altered by pumping water from a well (D). The water flows to this new outlet and forms a cone of depression.*

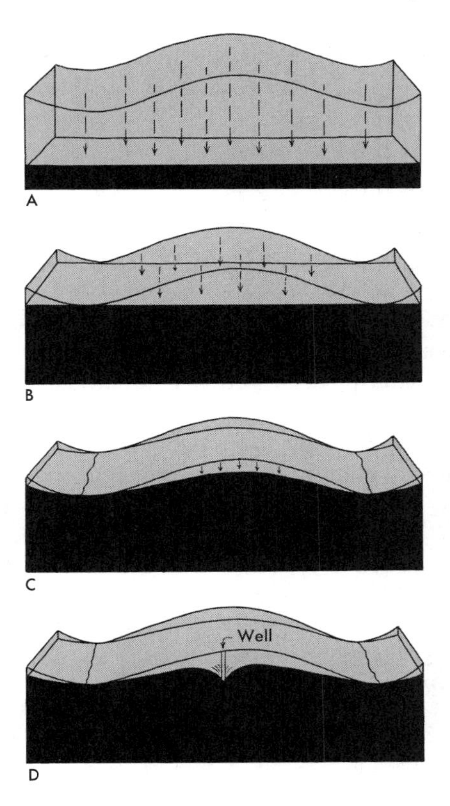

Returning to our ideal situation, we find that if the supply of water from the surface were to be completely stopped, the water table under the hill would slowly flatten out as water discharged into the valleys. Eventually it would almost reach the level of the water table under the valley bottoms; then the flow would stop. This condition is common in desert areas where rainfall is sparse.

Movement of underground water

We previously considered the flow of water in stream channels at the earth's surface. In dealing with the underground we find that water seldom has such large channels in which to flow. Underground rivers do exist, but they are rare and generally are present only in areas of soluble rock such as limestone where water circulating through the rock can dissolve large caverns and tunnels.

Underground water flows at an average rate of a few centimeters per day in most aquifers. Rates of about 15 cm/day may be reached in exceptionally permeable gravel beds near the surface.

Porosity

Porosity is the ratio of pore volume to total volume, expressed as a percentage. The more porous a rock is, the greater the amount of open space it contains. Through these pore spaces, underground water moves.

The range of porosity in earth materials is great. Recently deposited muds (called *slurries*) may hold up to 90 percent by volume of water, whereas unweathered igneous rocks such as granite, gabbro, or obsidian may hold only a fraction of 1 percent. Unconsolidated deposits of clay, silt, sand, and gravel have porosities ranging from about 20 to as much as 50 percent. But when these deposits have been consolidated into sedimentary rocks by compaction and cementation, their porosity is sharply reduced. In general, a porosity of less than 5 percent is low; from 5 to 15 percent represents medium porosity; and over 15 percent is high (Figure 10-34).

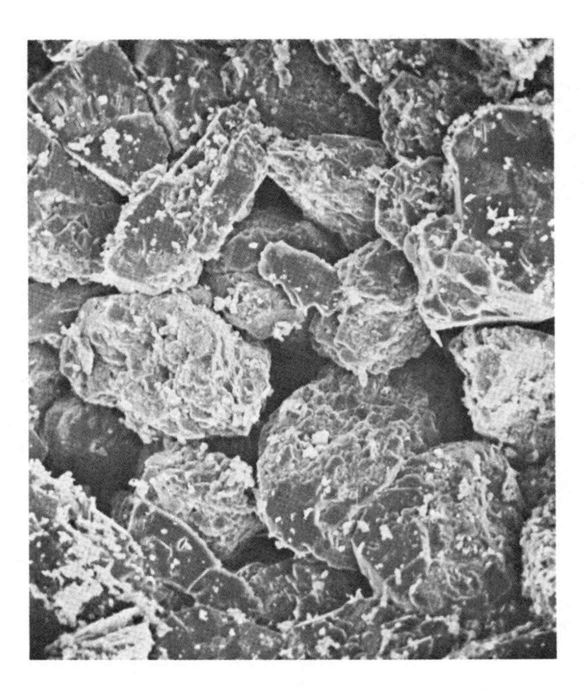

FIGURE 10-34 *Excellent porosity (23.6 percent) and permeability (768 millidarcies) in petroleum productive sandstone. Scanning electron microscope photograph of Muddy Sandstone, Montana. (Courtesy Core Laboratories, Inc.)*

Permeability

Whether we find a supply of fresh ground water in a given area depends on the ability of the earth materials to transmit water, as well as on their ability to contain it. The ability to transmit fluids is termed *permeability.*

A permeable material that carries underground water is an *aquifer,* from the Latin for "water" and "to bear." The most effective aquifers are unconsolidated sand and gravel, sandstone, and some limestone and dolomite. The permeability of carbonate rocks is usually the result of secondary solution that has enlarged fractures, bedding planes, and pore spaces. Fractured zones of some of the denser rocks such as granite, basalt, and gabbro also act as aquifers, although the permeability of such zones decreases rapidly with depth. Claystone and most metamorphic and crystalline igneous rocks are generally poor aquifers.

Since the flow of underground water is usually very slow, it is largely laminar; in contrast, the flow of surface water is largely turbulent. There is one exception, however—the turbulent flow of water in large underground passageways formed in cavernous limestone or dolomite.

The energy that causes underground water to flow is derived from gravity. Gravity draws water downward to the water table; from there it flows through the ground to a point of discharge in a stream, lake, or spring. Just as surface water needs a slope to flow on, so must there be a slope for the flow of ground water. This is the slope of the water table, the *hydraulic gradient.* It is measured by dividing the length of flow (from the point of intake to the point of discharge) into the vertical distance between these two points, a distance called *head.* Therefore, hydraulic gradient is expressed as h/l, where h is head and l is length of flow from intake to discharge. Thus, if h is 10 m and l is 100 m, the hydraulic gradient is 0.1 or 10 percent.

An equation to express the rate of water movement through a rock was proposed by the French engineer Henri Darcy in 1865. What is now known as Darcy's Law is essentially the same as his original equation. The law may be expressed as follows:

$$V = P\left(\frac{h}{l}\right)$$

where V is velocity, h is head, l is the length of flow, and P is a coefficient of permeability that depends on the nature of the rock in question. Because h/l is simply a way of expressing the hydraulic gradient, we may say that in a rock of constant permeability, the velocity of water will increase as the hydraulic gradient increases. Remembering that the hydraulic gradient and the slope of the ground water table are the same thing, we may also say that the velocity of ground water varies with the slope of the water table. Other things being equal, the steeper the slope of the water table, the more rapid the flow. In ordinary aquifers, the rate of water flow has been estimated as not faster than 1.5 m per day and not slower than 1.5 m per year. However, rates of over 120 m per day and as low as a few centimeters per year have been recorded.

The movement of underground water down the slope of the water table is only part of the picture, for the water is also in motion at depth. Water moves downward from the water table in broad looping curves toward some effective discharge agency, such as a stream, as suggested in Figure 10-35. The water feeds into the stream from all possible directions, including straight up through the bottom of the channel. We can explain this curving path

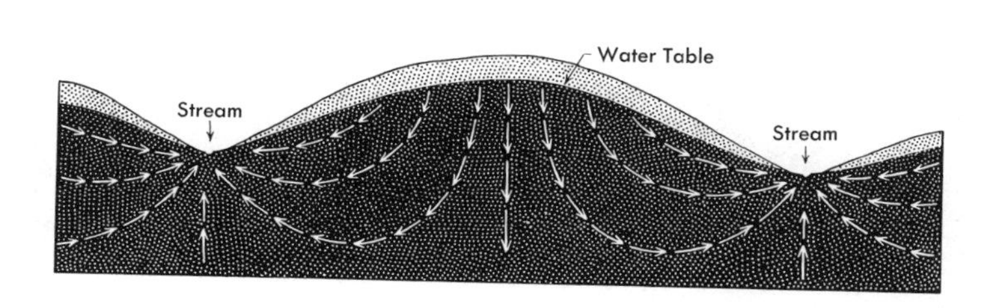

FIGURE 10-35 *The flow of ground water through uniformly permeable material is suggested here. Movement is not primarily along the ground-water table; rather, particles of water define broadly looping paths that converge toward the outlet and may approach it from below. [Redrawn from M. King Hubbert, "The Theory of Ground-water Motion," Journal of Geology, 48 (1940), 930.]*

as a compromise between the force of gravity and the tendency of water to flow laterally underground in the direction of the slope of the water table. This tendency toward lateral flow is actually the result of the movement of water toward an area of lower pressure (i.e., the stream channel in Figure 10-35). The resulting movement is neither directly downward nor directly toward the channel, but is rather along curving paths to the stream.

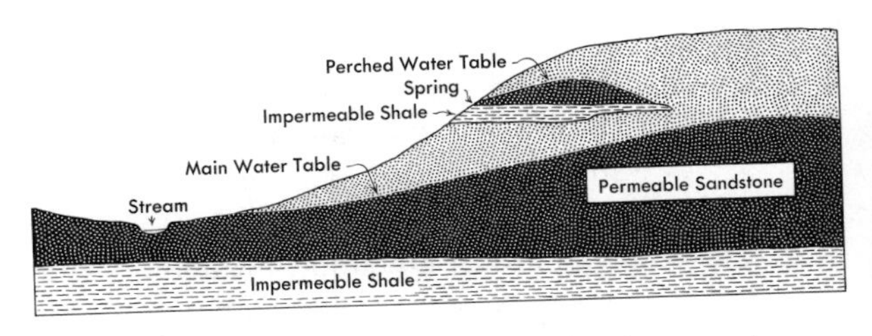

FIGURE 10-37 *A perched water table results when ground water collects over an impermeable zone and is separated from the main water table.*

Springs, wells, and geysers

We have been assuming that ground water is free to move on indefinitely through a uniformly permeable material of unlimited extent. Actually subsurface conditions fall far short of this ideal situation. Some layers of rock material are more permeable than others, and thus the water tends to move rapidly through these beds in a preferred direction more or less parallel to the bedding planes. Even in a rock that is essentially homogeneous, the ground water tends to move in some preferred direction.

Simple springs and wells

Underground water generally moves freely downward from the surface until it reaches an impermeable layer of rock or until it arrives at the water table. Then it begins to move laterally. Sooner or later it may flow out again at the surface of the ground in an opening called a *spring.*

FIGURE 10-36 *Nature seldom, if ever, provides uniformly permeable material. In this diagram, a hill is capped by permeable sandstone and overlies impermeable shale. Water soaking into the sandstone from the surface is diverted laterally by the impermeable beds. Springs result where the water table intersects the surface at the contact of the shale and sandstone.*

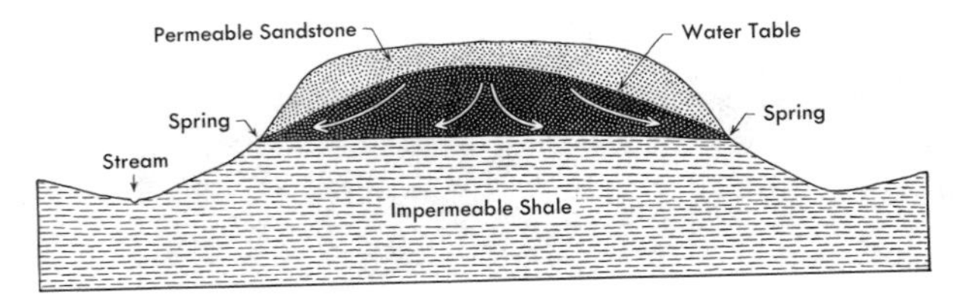

Springs range from intermittent flows that disappear when the water table recedes during a dry season, through pint-sized trickles, to an effluence of 3.8 billion liters daily—which is the amount of discharge of springs found along a 16-km stretch of the Fall River in California.

This wide variety of spring types is the result of underground conditions that vary greatly from one place to another. As a general rule, however, a spring results wherever the flow of ground water is diverted to a discharge zone at the surface (Figure 10-36). For example, a hill made up largely of permeable rock may contain a zone of impermeable material, as shown in Figure 10-37. Some of the water percolating downward will be blocked by this impermeable rock, and a small saturated zone will be built up. Since the local water level here is actually above the main water table, it is called a *perched water table.* The water that flows laterally along this impermeable rock may emerge at the surface as a spring. However, springs are not confined to points where water from a perched water table reaches the surface, and it is clear that if the main water table intersects the surface along a slope, then a spring will form.

Even in impermeable rocks, permeable zones may develop as a result of fractures or solution channels. If these openings fill with water and are intersected by the ground surface, the water will issue forth as a spring.

A spring is thus the result of a natural intersection of the ground surface and the water table. But a *well* is an artificial opening cut down from the surface into the zone of saturation. A well is productive only if it is drilled into permeable rock and penetrates below the water table. The greater the demands that are made on a well, the deeper it must be drilled below the water table. Continuous pumping creates the cone of depression previously described, which distorts the water table and may reduce the flow of ground water into the well (Figure 10-38). Wells drilled into fractured crystalline rock, such as granite, may produce a good supply of water at relatively shallow depths. But the yield of such wells cannot be increased appreciably by deepening them since the number and size of the fractures commonly decrease the farther down the well is drilled.

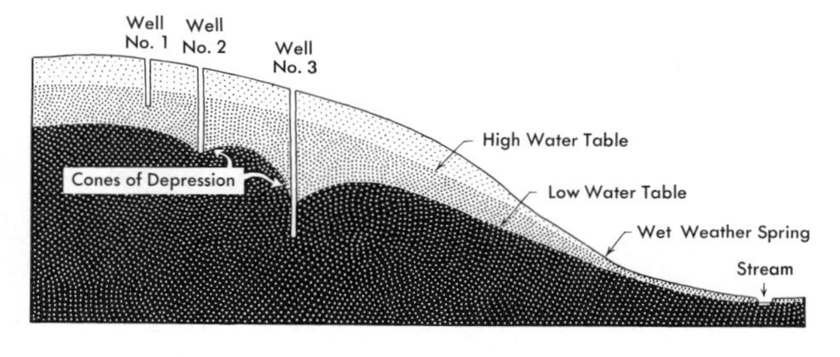

FIGURE 10-38 To provide a reliable source, a well must penetrate deep into the zone of saturation. In this diagram, Well No. 1 reaches only deep enough to tap the ground water during periods of high water table; a seasonal drop of this surface will dry up the well. Well No. 2 reaches to the low water table, but continued pumping may produce a cone of depression that will reduce effective flow. Well No. 3 is deep enough to produce reliable amounts of water even with continued pumping during low water-table stages.

Wells drilled into limestone beds that have been riddled by large solution passages may yield a heavy flow of water part of the time and no flow the rest of the time, simply because the water runs out rapidly through the large openings. Furthermore, because the water soaks down from the surface to the well in a very short time, the water may be contaminated because there has not been enough time for impurities to be filtered out as the water passes from the surface to the well. In sandstone, on the other hand, the rate of flow is slow enough to permit the elimination of impurities even within a very short distance of underground flow. Harmful bacteria are destroyed in part by entrapment, in part by lack of food and by temperature changes, and in part by hostile substances or organisms met along the way, particularly in the soil.

Artesian water

Contrary to common opinion, artesian water does not necessarily come from great depths. But other definite conditions (Figure 10-39) characterize an artesian water system, as follows: (1) the water is contained in a permeable layer, the aquifer, inclined so that one end is exposed to receive water at the surface; (2) the aquifer is capped by an impermeable layer; (3) the water in the aquifer is

FIGURE 10-39 The wells in the diagram meet the conditions that characterize an artesian system: (1) an inclined aquifer, (2) capped by an impermeable layer, (3) with water prevented from escaping either downward or laterally, and (4) sufficient head to force the water above the aquifer wherever it is tapped. In the well at the right, the head is great enough to force water out at the surface.

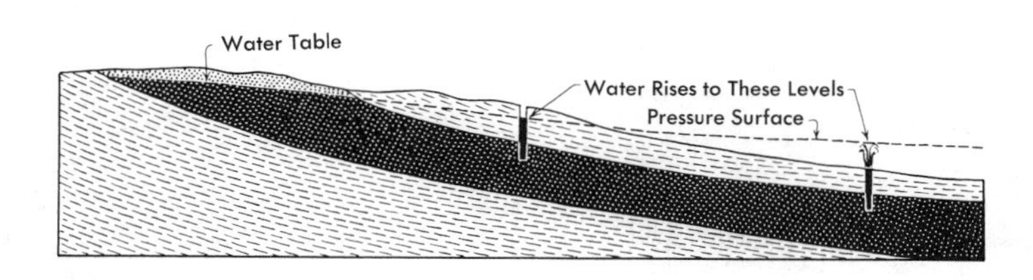

prevented from escaping either downward or off to the sides; and (4) there is enough head to force the water above the aquifer wherever it is tapped. If the head is great enough, the water will flow out to the surface either as a well or a spring. The term *artesian* is derived from the name of a French town, Artois (originally called *Artesium* by the Romans), where this type of well was first studied.

Thermal springs

Springs that bring warm or hot water to the surface are called *thermal springs, hot springs,* or *warm springs.* A spring is usually regarded as a thermal spring if the temperature of its water is 6°C higher than the mean air temperature. There are more than 1,000 thermal springs in the western mountain regions of the United States, 46 in the Appalachian Highlands of the east, 6 in the Ouachita area in Arkansas, and 3 in the Black Hills of South Dakota.

Most of the western thermal springs derive their heat from masses of magma that have pushed their way into the crust almost to the surface and are now cooling. Either the ground water comes into contact with the magma and is heated; or *juvenile* water, freed by igneous bodies at depths, cools somewhat on its way to the surface. In the eastern group of hot springs, however, the circulation of the ground water carries it to depths great enough for it to be warmed by the normal increase in earth heat (see "Thermal gradient" in Glossary).

FIGURE 10-40 Minerva Hot Spring ("Terrace Mound") at Yellowstone National Park. (Courtesy of National Park Service.)

FIGURE 10-41 *Large hot spring deposits at Thermopolis, Wyoming. Big Horn River (flow from right to left) in center of photograph. (Photo by M. Dane Picard.)*

Spectacular hot springs deposits are present at Yellowstone National Park (Figure 10-40) and at Thermopolis in northwestern Wyoming (Figure 10-41). The material deposited directly from the mineral-rich springs or geysers is composed of silica, and these deposits are termed *siliceous sinter.* White to gray in color, the siliceous sinter contains fine particles of silca very much like opal, which were precipitated from solutions that originally dissolved the silica from silicate rocks deep below the surface near hot igneous rocks.

Travertine is a calcium carbonate deposit made by calcareous algae that survive in the temperatures of hot springs and pools. Many hot waters contain large amounts of sulfur, formerly considered to be emanations from hell. In modern times, the mineral-rich, sulfurous water has been bottled and briskly sold to tourists for medicinal purposes.

Geysers

A *geyser* is a special type of thermal spring that ejects water intermittently with considerable force (Figure 10-42). The word *geyser* comes from the name of a spring of this type in Iceland, *geysir,* probably based on the verb *geysa,* "to rush furiously."

FIGURE 10-42 *Grand Geyser, Yellowstone National Park, a manifestation of the internal heat of the earth. Geysers occur mainly in areas of waning volcanic action. (Courtesy Union Pacific Railroad.)*

The details of geyser action are not completely understood. In general, a geyser's behavior is caused by the arrangement of its plumbing and the proximity of a good supply of heat. Ground water moving downward from the surface fills a natural pipe, or conduit, that opens upward to the surface. Hot igeneous rocks, or the gases given off by such rocks, gradually heat the column of water in the pipe and raise its temperature toward the boiling point. Now, we know that the higher the pressure of water, the higher its boiling point. And since water toward the bottom of the pipe is under the greatest pressure, it must be heated to a higher temperature than the water above before it will boil. Eventually the column of water becomes so hot that either a slight increase in temperature or a slight decrease in pressure will cause it to boil. At this critical point, the water near the base of the pipe is heated to the boiling point. The water then changes to steam and as it does so, it expands, pushing the water above it toward the surface. But this raising of the heated column of water reduces the pressure acting upon it, and so it, too, begins to boil. The energy thus developed throws the water and steam high into the air, producing the spectacular action characteristic of many geysers. After the eruption has spent itself, the pipe is again filled with water and the whole process begins anew.

We can compare this theoretical cycle with that of Old Faithful Geyser in Yellowstone National Park. The first indication of a coming eruption at Old Faithful is the quiet flow of water in a fountain some 1 to 2 m high. This preliminary activity lasts for a few seconds and then subsides. It represents the first upward push of the column of water described in our theoretical case. This push reduces the pressure and thereby lowers the boiling point of the water in the pipe. Consequently, the water changes to steam; and in less than a minute after the preliminary fountain, the first of the violent eruptions takes place. Steam and boiling water are thrown 45 to 50 m into the air. The entire display lasts about 4 minutes. Emptied by the eruption, the tube then gradually refills with ground water; the water is heated, and in approximately 1 hour the same cycle is repeated. The actual time between eruptions of Old Faithful averages about 65 minutes but may be from 30 to 90 minutes.

Recharge of ground water

As we have seen, the ultimate source of most underground water is precipitation that finds its way below the surface through either natural means or artificial means.

Some of the water from precipitation seeps into the ground, reaches the zone of saturation, and raises the water table. Continuous measurements taken over long periods of time at many places throughout the United States show an intimate connection between water level and rainfall. However, since water moves relatively slowly in the zone of aeration and the zone of saturation, fluctuations in the water table usually lag behind fluctuations in rainfall.

In many localities, the natural recharge of the underground supplies cannot keep pace with man's demands for ground water. Consequently, attempts are sometimes made to recharge these supplies artificially. On Long Island, New York, for example, water that has been pumped out for air-conditioning purposes is returned to the ground through special recharging wells, or, in winter, through the idle wells that are used in summer for air conditioning. And in the San Fernando Valley, California, surplus water from the Owens Valley aqueduct is fed into the underground in an attempt to keep the local water table at a high level.

FIGURE 10-43 *(Above) The floor of this section of the Luray Caverns, Virginia, is partially flooded by a lake. Both stalagmites and stalactites are present. Stalagmites, stalactites, and columns are the most common shapes developed by the precipitation of dripstone in caverns. (Photo by permission of the Luray Caverns, Virginia.)*

FIGURE 10-44 *(Below) Beautiful column in Carlsbad Caverns National Park, New Mexico. (Courtesy National Park Service.)*

Caves and related features

The great Athenian hero Theseus marked his trail with a ball of thread provided by the lovely Ariadne as he entered the Labyrinth to kill the Minotaur. After battering the monster to death with his fists, Theseus followed the thread out, with the prospective victims of the Minotaur following safely behind him. This was good safety practice in cave exploration for the Greeks, and it remains so today.

Caves are spectacular examples of the handiwork of underground water. In dissolving great quantities of solid rock in its downward course, the water fashions large rooms, galleries, and underground stream systems as the years pass. In many caves, the water deposits calcium carbonate as it drips off the ceilings and walls, building up fantastic deposits known as *dripstone* (Figure 10-43).

The beautiful and gigantic Carlsbad Caverns in New Mexico, where the main caverns are 250 m below the surface, are the most famous caves in the United States (Figure 10-44). Caves of all sizes can develop in limestone ($CaCO_3$), and small caves are also found in dolomite [$CaMg(CO_3)_2$]. Rock Salt (NaCl), gypsum ($CaSO_4 2H_2O$), and similar rocks are subject to such rapid solution that underground caverns usually collapse under the weight of overlying rocks before erosion can open them to the surface, so no notable caverns are found in these rocks.

Calcite, the main component of limestone, is highly insoluble in pure water. But when the mineral is attacked by water containing small amounts of carbonic acid, it undergoes rapid chemical weathering. Most natural water contains carbonic acid (H_2CO_3), the combination of water with carbon dioxide. This carbonic acid reacts with the calcite to form calcium bicarbonate, $Ca(HCO_3)_2$, a soluble substance that is then removed in solution.

The first signs of solution in limestone usually appear along original lines of weakness, such as bedding surfaces or fractures. As water seeps into these areas, it dissolves some of the rock and enlarges the openings. The dissolved material is moved onward by the underground water and is either redeposited or discharged into streams. As time passes, the openings grow larger and larger, until finally

FIGURE 10-45 *Tower karst developed in Paleozoic limestone exposed along the Li-Kiang, south of Kweilin, Kwangsi Province, Peoples Republic of China.* (Photo by Frank Brown.)

they form large passageways. Whether this cave-forming activity goes on above the water table, at the water table, or at some distance beneath it is a question that is still being argued. Most geologists, however, believe that caves are generally formed below the water table and are exposed later when the water table is lowered by downward-cutting surface streams.

Regardless of where caves are originally formed, the weird rock formations so characteristic of most of them must have developed above the water table, when the caves were filled with air. These bizarre shapes are composed of calcite deposited by underground water that has seeped down through the zone of aeration. They develop either as *stalactites,* looking like stony icicles hanging from the cave roof (Greek, *stalactos,* "oozing out in drops") or as *stalagmites,* heavy posts growing up from the floor (Greek, *stalagmos,* "a dropping or dripping"). When a stalactite and a stalagmite meet, a *column* is formed (Figure 10-44).

A stalactite forms as water charged with calcium bicarbonate in solution seeps through the cave roof. Drop after drop forms on the ceiling and then falls to the floor. But during the few moments that each drop clings to the ceiling, a small amount of evaporation takes place; some carbon dioxide is lost, and a small amount of calcium carbonate is deposited. Over the years, a large stalactite may gradually develop. A stalagmite forms in much the same fashion. Part of the water that falls to the cave floor runs off, and part is evaporated. This evaporation again causes the deposition of calcite, and a stalagmite begins to grow upward to meet the stalactite hanging down.

Sinkholes

On the ground surface above the soluble rock material, depressions sometimes develop that reflect areas where the underlying rock has been carried away in solution. These depressions, called *sinkholes* or *sinks,* usually form when the surface collapses into a large cavity beneath. Surface water may then drain through the sinkholes to the underground, or if their subterranean outlets become clogged, these sinkholes may fill with water to form lakes. An area with numerous sinkholes is said to display *karst topography* (Figure 10-45).

Water witches

In medieval times, "water witches," or "dowsers," firmly grasped a forked stick in both hands and walked about a field until the sweat broke out. An irresistable force apparently pulled the end of the stick down, causing the dowser to exclaim, "Drill here for water!" Ground water was thought to flow in rivers like surface water, and a correctly placed well would penetrate to one of these rivers. Although scientific knowledge of ground water occurrences is much greater in the twentieth century than formerly, water witches still walk the land, and more than one farmer still "hears" underground rivers flowing below the surface of his field.

Water witches do enjoy some success and frequently are paid for their efforts. This is because aquifers are common in basins or plains where farmlands occur, and most wells drilled *on any basis* will eventually encounter some water. In the practice of their magic, water witches attain considerable practical knowledge of the location and depths of good aquifers in particular regions. Thus they are usually more successful than the farmer would be or a local expert from the community pool hall. The water witch frequently is also the person who drills the water wells in a region. This dual role adds to his chances of success in finding water, and at most sites drilling can be continued until water is found. However, for difficult problems of water location, an experienced ground-water geologist is necessary.

Major uses of water in the United States

About 70 percent of precipitation in the United States is returned to the atmosphere by evapotranspiration. More than one-half of this takes place in agricultural and forested lands. Thirty percent of yearly precipitation reaches streams, lakes, and ground water. We currently withdraw and use about 8 percent of the total annual precipitation. Twenty-two percent is not withdrawn and returns to the oceans unused.

Three-fourths of the withdrawn water for all uses comes from surface supplies and one-fourth from ground-water supplies. The greatest individual use by far of water is for electrical power generation. Almost 90 percent of daily use is for this purpose, but only a small amount of this water is actually consumed. Most is returned to surface waters or is recycled.

The next largest user is industry, which consumes about 670 million cubic meters per day. Most of industrial use is for cooling power plants in the eastern states that burn oil, gas, and coal (Table 10-2). Other industrial uses include metal processing and refining, food processing, paper production, and petroleum refining. The quality of the water is changed in these latter uses and may contain many toxic chemicals or organic matter that consumes dissolved oxygen. Discharge of this polluted water back into surface water supplies constitutes a serious problem today, and not only in the United States.

Agriculture, although only the third largest user of water, accounts for 80 percent of all water that is consumed annually in the United States. Such loss takes place in seepage and evaporation in moving water to the fields and in spraying, flushing, and spreading the water in irrigation ditches. Demand for irrigation water is greatest during months when rainfall is least. Evaporation rates are very high in those regions where irrigation is most important.

Urban uses and rural domestic and livestock uses require only about 10 percent of the total daily use of water. Most of this water is either polluted or consumed.

Ground water and the future

The amount of ground water in the United States is staggering, many times the amount of water in all its streams and freshwater lakes. About one-half of this large resource is in unconsolidated aquifers: river gravel, sand, silt, and glacial outwash. In the Pacific Northwest and in Hawaii, joints, lava tubes, and brecciated tops and bottoms of basalt flows contain large amounts of ground water. Cavernous limestone and porous sandstone beds also contain huge volumes of ground water in various parts of the United States.

One-third of the water for cities, towns, and farms comes from ground water; two-thirds are obtained from surface waters. In the west and southwest, ground water is being used in some areas faster than it is being replenished. Recharge (the addition of water to the zone of saturation) rates are less than the amount being withdrawn for farming and ranching purposes. This is essentially mining of a nonrenewable resource, at least in terms of man's near future.

Heavy use of ground water can have damaging environmental side effects. Land may subside from excessive withdrawal, and has done so in many parts of California, Las Vegas, Nevada, and Mexico City. The settling, tilting, and cracking of houses and buildings are no small problems to their owners.

Another bad side effect is that fresh ground water may become contaminated and invaded by sea water. In coastal regions, ground water extends seaward and merges with marine, saline ground water. If excessive water is pumped from wells along a coast, sea water moves into the fresh water in the aquifers. Well waters deteriorate in quality as more and more sea water mixes with the fresh water. Serious problems have already arisen in Long Island, the New Jersey coast, Florida, and southern California.

Careful management of ground water is necessary. For the most part, present efforts are inadequate, poorly supported financially, and shortsighted. Our ground-water resource could be seriously damaged and depleted in many areas because of increased demands placed upon it. Mining of ground water or damaging its quality is an irrevocable act. There will be no new supply under existing climatic conditions. We must be wary and suspicious of the short-term gains through increased ground-water use that have been proposed by some entrepreneurs.

SUMMARY

Water near and at the earth's surface moves from oceans to atmosphere to land and back to the oceans in the hydrologic cycle.

The flow of most water is turbulent, but sometimes, particularly beneath the surface, the flow is laminar. Water in a stream flows down a gradient, has a velocity, and is measured in terms of discharge per unit time. As discharge increases, so also does width, depth, and velocity of a stream.

The base level of a stream is the point below which it cannot erode. Material carried by a stream may be in solution, or in suspension, or may be moved along the channel floor as bed load. Erosion, which is most rapid in periods of flood, is accomplished by direct lifting, abrasion, impact, and solution. Deposition takes place when a stream's velocity (and hence its energy) decreases.

In general, stream valleys are characterized by divides, valley walls, stream channels, and usually by flood plains. A valley widens as a stream erodes in its channel and as mass movement and slope erosion push back the valley walls. Narrow valleys include waterfalls and rapids. Broad valleys have flood plains, meanders, braided streams, natural levees, and stream terraces.

Most valleys are cut by the streams that flow through them. The patterns of stream channels reflect the nature of the underlying rock and the stream history.

Water beneath the ground is distributed largely in a zone of saturation separated from the surface by a zone of aeration. The top of the zone of saturation is the ground-water table and is very often a subdued reflection of the topography.

The porosity of earth materials is a function of the total void space it contains. Permeability is the ability of a material to transmit fluids.

Most ground-water flow is laminar and is driven by gravity. Velocity equals $P(h/l)$, where P is the coefficient of permeability, h is the head, and l is the length of flow.

Wells draw water from the zone of saturation. Artesian wells are those in which water rises above the top of the aquifer. Springs arise when the ground surface intersects the water table. Thermal springs are heated by still-warm igneous rocks and by the normal increase of earth's temperature with depth. Geysers are thermal springs marked by periodic, violent eruptions.

Most caves result from the solution of limestone by underground water. After the caves have been formed, they may be decorated by the deposition of new limestone in the form of stalactites and stalagmites. Karst topography is characterized by sinkholes that mark the collapse of rock into underlying caves.

The greatest individual use of water is for electrical power generation. About 90 percent of daily use is for this purpose but most of this water is returned to surface waters or is recycled. The next largest user is industry, followed by agriculture. Urban uses and rural domestic and livestock uses require only about 10 percent of the total daily use of water.

One-third of the water for cities, towns, and farms comes from ground water; two-thirds are obtained from surface waters. In some regions, ground water is being used faster than it is being replenished. Therefore, careful management of ground water is necessary.

SUGGESTED READINGS

CARSON, M. A., and M. J. KIRBY, *Hillslope Form and Process.* New York: Cambridge University Press, 1972.

DAVIS, S. N., and R. J. M. DEWIEST, *Hydrogeology.* New York: John Wiley & Sons, Inc., 1966.

HOYT, WILLIAM G., and WALTER B. LANGBEIN, *Floods.* Princeton, N.J.: Princeton University Press, 1955.

LEOPOLD, LUNA B., M. GORDON WOLMAN, and JOHN P. MILLER, *Fluvial Processes in Geomorphology.* San Francisco: W. H. Freeman and Co., 1964.

LEOPOLD, LUNA B., K. S. DAVIS, and the Editors of Life, *Water.* New York: Time, Inc., 1966.

MOORE, C., and T. POULSON, *The Life of the Cave.* New York: McGraw-Hill Book Company, 1966.

MORISAWA, MARIE, *Streams: Their Dynamics and Morphology.* New York: McGraw-Hill Book Company, 1968.

MURRAY, C., and E. REEVES, *Estimated Uses of Water in the United States, U.S. Geological Survey Circular 676. Washington, D.C.: Government Printing Office, 1972.*

SWEETING, MARJORIE M., Karst Landforms. New York: Columbia University Press, 1973.

TODD, D. K., *Ground Water Hydrology.* New York: John Wiley & Sons, Inc., 1959.

The Work of Glaciers, Mass Movements, and Winds

The seas and rivers of moving ice known as *glaciers* have attracted inquisitive men deep into the Arctic, Antarctic, and mountainous regions of the world. There it has been discovered that glaciers are active agents of erosion, transportation, and deposition, and that these impressive masses of ice were far more widespread in the past than they are now. Geologists have learned, too, that the ice of the last great glacial period has modified and molded great stretches of landscape in what are now the temperate zones.

Formation of glacier ice

A *glacier* is a mass of ice that has been formed by the recrystallization of snow, and that flows forward, or has flowed at some time in the past, under the influence of gravity. This definition eliminates the pack ice formed from

11

Coalescing glaciers, Mount McKinley National Monument (South Unit), Alaska. (Courtesy National Park Service.)

sea water in polar latitudes and—by convention—icebergs, even though they are large fragments broken from the seaward end of glaciers.

Like surface streams and underground reservoirs, glaciers depend on the oceans for their nourishment. Some of the water drawn up from the oceans by evaporation falls on the land in the form of snow. If the climate is suitable, part of the snow lasts through the summer without melting. As the years pass, the accumulation gradually grows deeper and deeper, until a glacier is born.

In areas where the winter snowfall exceeds the amount of snow that melts during the summer, stretches of perennial snow known as *snowfields* cover the landscape. The lower limit of a snowfield is the *snow line.* Above the snow line, glacier ice collects in the more sheltered areas of the snowfields. The exact position of the snow line varies from one climatic region to another. In polar regions, for example, it reaches down to sea level, but near the equator it recedes to the mountain tops. In the high mountains of East Africa, for instance, it ranges from elevations of 4,500 to 5,400 m. The highest snow lines in the world are in the dry regions known as the "horse latitudes," which lie between the parallels 20° to 30° north and south of the equator; here the snow line reaches higher than 6,000 m.

FIGURE 11-1 *Snowflakes exhibit a wide variety of patterns, all hexagonal and all reflecting the internal arrangement of hydrogen and oxygen. It is from snowflakes that glacier ice eventually forms.*

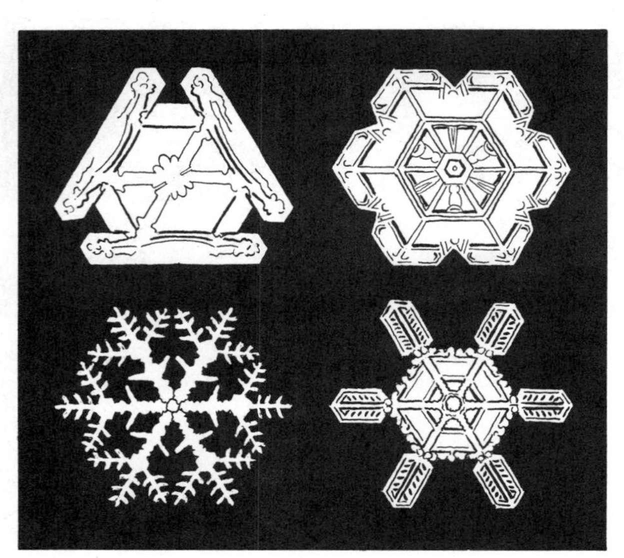

Fresh snow falls as a feathery aggregate of complex and beautiful crystals with a great variety of patterns. All the crystals are basically hexagonal, however, and all reflect their internal arrangement of hydrogen and oxygen atoms (Figure 11-1). Snow is not frozen rain; rather, it forms from the condensation of water vapor at temperatures below freezing.

After snow has lain on the ground for some time, it changes from a light, fluffy mass to a heavier, granular material called *firn,* or *névé* (pronounced *nay-vay*). *Firn* derives from a German adjective meaning "of last year," and *névé* is a French word derived from the Latin for "snow." We shall use the term *firn* for future discussion. Solid remnants of large snow banks, those tiresome vestiges of winter, are largely firn.

Several processes are at work in the transformation of snow into firn. The first is *sublimation,* a general term for the process of a solid material changing into a gaseous state without first becoming a liquid. In sublimation, molecules of water vapor escape from the edges of the snowflakes, and most evaporate. However, some of these molecules attach themselves to the center of the flakes, where they adapt themselves to the structure of the snow crystals. Then, as time passes, one snowfall follows another, and the granules that have already begun to grow as a result of sublimation are packed tighter and tighter under the pressure of the overlying snow.

Water has the unique property of increasing in volume when it freezes; conversely, it decreases in volume as the ice melts. But the cause and effect may be interchanged: if added pressure on the ice squeezes the molecules

FIGURE 11-2 Mountain glaciers in the St. Elias Range, Alaska. (Photo by Austin Post.)

closer together and reduces its volume, the ice may melt. In fact, if the individual granules are in contact at 0°C, they begin to melt with only a slight increase in pressure. The resulting meltwater trickles down and refreezes on still lower granules at points where they are not yet in contact. All through this process, however, the basic hexagonal structure of the original snow crystals is maintained.

A layer of firn granules, ranging from a fraction of a millimeter to approximately 3 or 4 mm in diameter, is gradually built up. The thickness of this layer varies, but 30 m is average on many mountain glaciers.

The firn undergoes further change as continued pressure forces out most of the air between the granules, reduces the space between them, and finally transforms the firn itself into *glacier ice,* a true solid composed of interlocking crystals. This glacier ice takes on a blue-gray color from the air and from the fine dirt that it contains.

The ice crystals that compose glacier ice are minerals; the mass of glacier ice, made up of many interlocking crystals, is a metamorphic rock, for it has been transformed from snow into firn and eventually into a true solid—glacier ice.

Classification of glaciers. The glaciers of the world are of three principal types: (1) valley glaciers, (2) piedmont glaciers, and (3) ice sheets.

Valley glaciers are streams of ice that flow down the valleys of mountainous areas (Figure 11-2). Like streams of running water, they vary in width, depth, and length. A branch of the Hubbard Glacier in Alaska is 120 km long, whereas some of the valley glaciers that dot the higher reaches of the western mountains of the United States are only a few hundred meters in length. Valley glaciers that are nourished on the flanks of high mountains and that flow down the mountain sides are sometimes called *mountain glaciers* or *Alpine glaciers.* Very small mountain glaciers are referred to as *cliff glaciers, hanging glaciers,* or *glacierets.* A particular type of valley glacier sometimes grows up in areas where large masses of ice are dammed by a mountain barrier along the coast; some of this ice escapes through valleys in the mountain barrier to form an *outlet glacier,* as it has done along the coasts of Greenland and Antarctica.

Piedmont glaciers form when glaciers emerge from their valleys and spread out to form an apron of moving ice on the plains below.

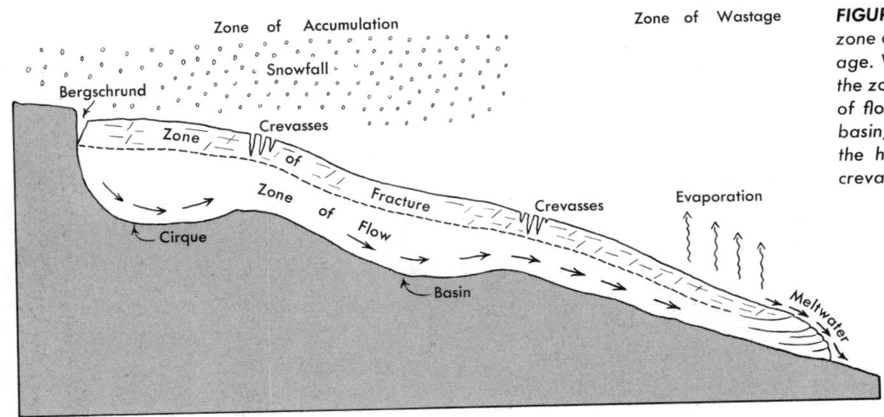

FIGURE 11-3 A glacier is marked by a zone of accumulation and a zone of wastage. Within a glacier, ice may lie either in the zone of fracture or deeper in the zone of flow. A valley glacier originates in a basin, the cirque, and is separated from the headwall of the cirque by a large crevasse, the bergschrund.

Ice sheets are broad, moundlike masses of glacier ice that tend to spread redially under their own weight. The Vatna Glacier of Iceland is a small ice sheet measuring about 120 by 160 km and 225 m in thickness. A localized sheet of this sort is sometimes called an *icecap.* The term *continental glacier* is usually reserved for great ice sheets that obscure the mountains and plains of large sections of a continent, such as those of Greenland and Antarctica. On Greenland, the ice exceeds 3,000 m in thickness near the center of the icecap.

Distribution of modern glaciers

Modern glaciers cover approximately 10 percent of the land area of the world, i.e., about 17.9 million km^2; of this, the Greenland and Antarctica ice sheets account for about 96 percent. The Antarctica ice sheet covers approximately 15.3 million km^2, and the Greenland sheet covers about 1,735,000 km^2. Small icecaps and numerous mountain glaciers scattered around the world account for the remaining 4 percent.

Nourishment and wastage of glaciers

When the weight of a mass of snow, firn, and ice above the snow line becomes great enough, movement begins and a glacier is created. The moving stream of ice flows downward across the snow line until it reaches an area where the loss through evaporation and melting is so great that the forward edge of the glacier can push no farther.

A glacier can be divided into two zones: (1) *a zone of accumulation* and (2) *a zone of wastage* (see Figure 11-3). The position of the front of a glacier depends on the relationship between these two zones—that is, between the glacier's rate of nourishment and its rate of wastage. When nourishment just balances wastage, the front is stationary and the glacier is in equilibrium. This balance seldom lasts for long, however, since a slight change in either nourishment or wastage will cause the front to advance or retreat.

At the front of the glacier, wastage takes place through a double process of evaporation and melting known as *ablation.* If a glacier terminates in a body of water, great blocks of ice break off and float away in a process called *calving.* This is the action that produces the icebergs of the polar seas.

Glacier movement

Glaciers move only a few centimeters or at most a few meters per day, except in rare cases. That they actually do move, however, can be demonstrated in several ways. The most conclusive test is to measure the movement directly, by emplacing a row of stakes across a valley glacier. As time passes, the stakes move down-valley with the advancing ice, with the center stakes moving more rapidly than those near the valley walls.

A second source of evidence is provided by the distribution of rock material on the surface of a glacier. Many of the boulders and cobbles lying along a valley glacier could not have come from the walls immediately above; the only possible source lies up-valley. The boulders must have been carried to their present position on the back of the glacier. Another indication of glacier movement is that when a glacier melts, it often exposes a rock floor that has been polished, scratched, and grooved by the glacier.

Clearly, then, a glacier does move (Figure 11-4). In fact, different parts of it move at different rates. But although we know much about how a glacier flows forward, certain phases of the movement are not yet clearly understood. In any event, two zones of movement can be distinguished: (1) an upper zone between 30 and 60 m thick, which reacts like a brittle substance—that is, it breaks sharply rather than undergoing gradual, permanent distortion; and (2) a lower zone, which, because of the pressure of the overlying ice, behaves like a plastic substance. The first is the *zone of fracture;* the second is the *zone of flow.*

As plastic deformation takes place in the zone of flow, the brittle ice above is carried along. But the zone of flow moves forward at different rates—faster in some parts, more slowly in others—and the rigid ice in the zone of fracture is unable to adjust itself to this irregular advance. Consequently, the upper part of the glacier cracks and shatters, giving rise to a series of deep, treacherous *crevasses.*

The central part of a glacier moves faster than the sides or bottom, for the sides and bottom are retarded by friction against the valley walls and beds. In this respect, the

FIGURE 11-4 *Patterns of ice flow on the Seward lobe of the Malaspina Glacier, Alaska. (Photo by Austin Post.)*

movement of an ice stream resembles that of a stream of water.

The mechanics of ice flow are still a matter of study—a study that is difficult because we cannot actually observe the zone of flow, which lies concealed within the glacier. Yet the ice from the zone of flow eventually emerges at the snout of the glacier, and there it can be studied. We find that by the time it has emerged it is brittle, but it retains the imprint of movement by flow. Also the individual ice crystals are now several centimeters in size; in contrast, crystals in ice newly formed from firn measure but a fraction of a centimeter. These ice crystals (in the zone of flow) have grown by recrystallization as they passed through the zone of flow. The ice at the snout is also marked by bands that represent shearing and differential movement within the glacier. Recrystallization has taken place along many of the old shear planes, and along others the debris carried forward by the ice has been concentrated. These observations clearly indicate that some movement in the zone of flow has taken place as a result of shearing.

Results of glaciation

Methods of erosion

Glaciers have special ways of eroding, transporting, and depositing earth materials. A valley glacier, for example, acquires debris by means of *frost action, landsliding,* and *avalanching.* Fragments pried loose by frost action clatter down from neighboring peaks and come to rest on the back of the glacier. And great snowbanks unable to maintain themselves on the steep slopes of the mountainsides, avalanche downward to the glacier, carrying along quantities of rock debris and rubble. This material is buried beneath fresh snow or avalanches, or else tumbles into gaping crevasses in the zone of fracture and is carried along by the glacier.

When a glacier flows across a fractured or jointed stretch of bedrock, it may lift up large blocks of stone and move them off. This process is known as *plucking* or *quarrying.* The force of the ice flow itself may be strong enough to pick up the blocks, and the action may be helped along by the great pressures that operate at the bottom of a glacier. Suppose the moving ice encounters a projection of rock jutting up from the valley floor. As the glacier ice forces itself over and around the projection, the pressure on the ice is increased and some of the ice around the rock may melt. This meltwater trickles toward a place of lower pressure, perhaps into a crack in the rock itself. There it refreezes, forming a strong bond between the glacier and the rock. Continued movement by the glacier may then tear the block out of the valley floor.

At the heads of valley glaciers, plucking and frost action sometimes work together to pry rock material loose. Along the back walls of the collection basins of mountain glaciers, great hollows called *cirques* (pronounced *sirks*) or *amphitheaters* develop in the mountainside. As the glacier begins its movement downslope, it pulls slightly away from the back wall, forming a crevasse known as a *bergschrund.* One wall of the bergschrund is formed by the glacier ice; the other is formed by the nearly vertical cliff of bedrock. During the day, melting water pours into the bergschrund and fills openings in the rock. At night, the water freezes, producing pressures great enough to loosen blocks of rock from the cliff. Eventually, these blocks are incorporated into the glacier and are moved away from the headwall of the cirque.

Glaciers pick up rock material by means of abrasion. As the moving ice drags rocks, boulders, pebbles, sand, and silt across the glacier floor, the bedrock is cut away as though by a great rasp or file. And the cutting tools themselves are abraded. It is this mutual abra-

FIGURE 11-5 *Rock fragments embedded in glacier ice often gouge scratches or striations in bedrock as the ice moves across it. This exposure in northeastern Wisconsin reveals that ice movement was parallel to the orientation of the striations. (Photo by Raymond C. Murray.)*

FIGURE 11-6 *A tarn occupies the basin in this cirque in the Rocky Mountains. (Photo by Austin Post.)*

sion that produces *rock flour*—very fine particles of pulverized rock—and gives a high polish to many of the rock surfaces across which a glacier has ridden. The streams that drain from the front of a melting glacier are charged with rock flour. So great is the volume of this material that it gives the water a characteristically grayish blue color similar to that of skim milk. Here, then, is further evidence of the grinding power of the glacier mill. But abrasion sometimes produces scratches, or *striations,* on both the bedrock floor and on the grinding tools carried by the ice. More extensive abrasion creates deep gouges, or *grooves,* in the bedrock. The striations and grooves along a bedrock surface show the direction of the glacier's movement (Figure 11-5).

Erosional forms

The erosional effects of glaciers are not limited to the fine polish and striations just mentioned, however. For glaciers also operate on a grander scale, producing spectacularly sculptured peaks and valleys in the mountainous areas of the world.

Cirques. As we have seen, a cirque is the basin from which a mountain glacier flows, the focal point for the glacier's nourishment. After a glacier has disappeared and all its ice has melted away, the cirque is revealed as a great amphitheater or bowl, with one side partially cut away. The back wall rises a few scores of meters to over 900 m above the floor, often as an almost vertical cliff. The floor of a cirque lies below the level of the low ridge separating it from the valley of the glacier's descent. A lake that forms in the bedrock basin of the cirque floor is called a *tarn* (Figure 11-6).

Horns, arêtes, and cols. A *horn* is a spire of rock formed by the headward erosion of a ring of cirques around a single high mountain. When the glaciers originating in these cirques finally disappear, they leave a steep, pyramidal mountain outlined by the headwalls of the cirques. The classic example of a horn is the famous Matterhorn of Switzerland.

An *arête* (pronounced *a-ret',* from the French for "fishbone," "ridge," or "sharp edge") is formed when a number of cirques gnaw into a ridge from opposite sides. The ridge becomes knife-edged, jagged, and serrated (Figure 11-7).

A *col* (from the Latin *collum,* "neck"), or pass, is fashioned when two cirques erode headward into a ridge from opposite sides. When their headwalls meet, they cut a sharp-edged gap in the ridge.

Glaciated valleys. Rather than fashion their own valley, glaciers probably follow the course of preexisting valleys, modifying them in a variety of ways; but usually the glaciated valleys have a broad U-shaped cross-profile (Figure 11-8), whereas mountain valleys created exclusively by streams have narrow, V-shaped cross-profiles. Since the tongue of an advancing glacier is relatively broad, it tends to broaden and deepen the V-shaped stream valleys, transforming them into broad, U-shaped troughs. And since the moving body of ice has difficulty manipulating the curves of a stream valley, it tends to straighten and simplify the course of the original valley. In this process of straightening, the ice snubs off any spurs of land that extend into it from either side. The cliffs thus formed are shaped like large triangles or flatirons with their apex upward, and are called *truncated spurs.*

Glaciers also give a mountain valley a characteristic longitudinal profile from the cirque downward. The course of a glaciated valley is marked by a series of *rock basins,* probably formed by plucking in areas where the bedrock was shattered or closely jointed. Between the basins are relatively flat stretches underlain by rock more resistant to plucking. As time passes, the rock basins may fill up

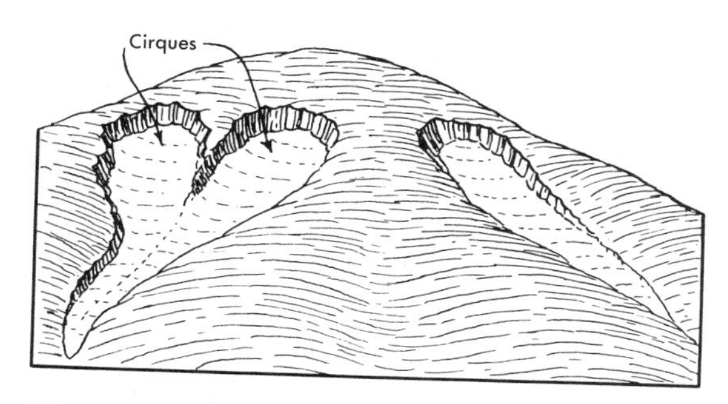

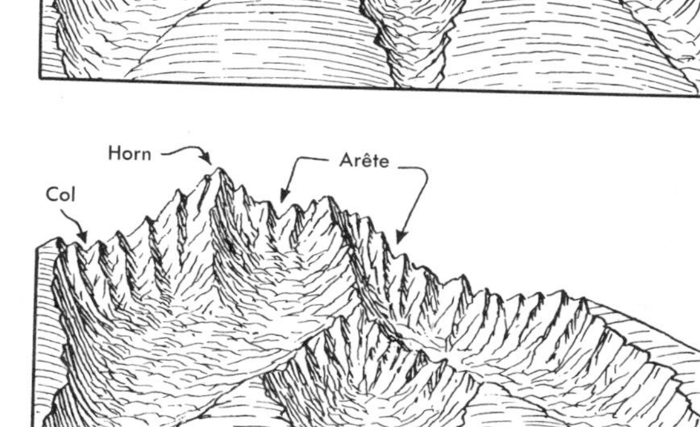

FIGURE 11-7 *The progressive development of cirques, horns, arêtes, and cols. In the first diagram, valley glaciers have produced cirques; but since erosion has been moderate, much of the original mountain surface has been unaffected by the ice. The result of more extensive glacial erosion is shown in the second diagram. In the final drawing, glacial erosion has affected the entire mass and has produced not only cirques but also a matterhorn, cols, and jagged, knife-edged arêtes. [Redrawn from William Morris Davis, "The Colorado Front Range," Annals Association of American Geologists, 1 (1911), p. 57.]*

FIGURE 11-8 *(Below) U-shaped glacial valley. (Courtesy U.S. Geological Survey.)*

with water, producing a series of lakes that are sometimes referred to as *pater noster lakes* because they resemble a string of beads (see Figure 11-8).

Hanging valleys are another characteristic of mountainous areas that have undergone glaciation. The mouth of a hanging valley is left stranded high above the main valley through which a glacier has passed. As a result, streams from hanging valleys plummet into the main valley in a series of falls and plunges. Hanging valleys may be formed by processes other than glaciation, but they are almost always present in mountainous areas that formerly supported glaciers and are thus very characteristic of past valley glaciation.

What has happened to leave these valleys stranded high above the main valley floor? During the time when glaciers still moved down the mountains, the greatest accumulation of ice would tend to travel along the central valley. Consequently, the erosive action there would be greater than in the tributary valleys with their relatively small glaciers, and the main valley floor would be cut correspondingly deeper. This action would be even more pronounced where the main valley was underlain by rock that was more susceptible to erosion than the rock under the tributary valleys. Finally, some hanging valleys were probably created by the straightening and widening action of a glacier on the main valley. In any event, the difference in level between the tributary valleys and the main valley is not apparent until the glacier has melted away.

Cutting deep into the coasts of Alaska, Norway, Greenalnd, Labrador, Chile, and New Zealand are deep, narrow arms of the sea known as *fiords.* Actually, these inlets are stream valleys that were modified by glacier erosion and then partially filled by the sea (Figure 11-9). The deepest known fiord, Vanderford in Vincennes Bay, Antarctica, has a maximum depth of 2,287 m.

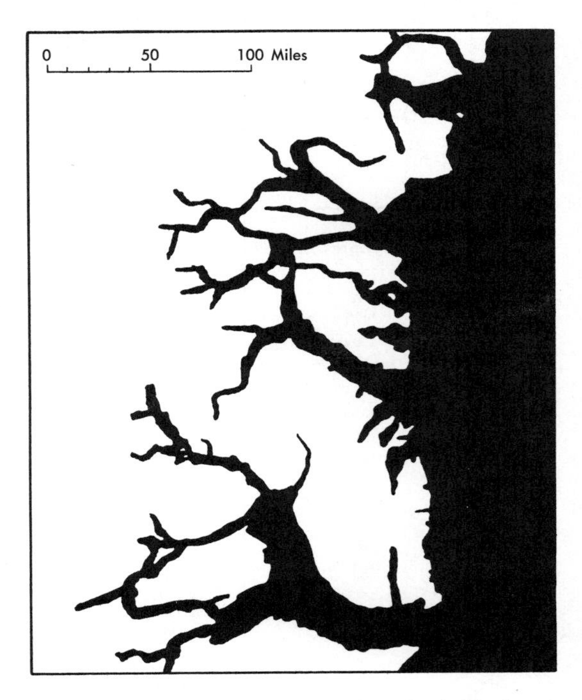

FIGURE 11-9 Glaciated valleys have been flooded by the sea (black) to produce these fiords along the coast of Greenland. [Redrawn from Louise A. Boyd et al., "The Fiord Region of East Greenland," American Geographical Society, *Special Publication No. 18 (1935), p. xii.]*

Some valleys have been modified by continental glaciers rather than by the valley glaciers that we have been discussing so far. The valleys occupied by the Finger Lakes of central New York State are good examples. These long narrow lakes lie in the basins that were carved out by the ice of a continental glacier. As the great sheet of ice moved down from the north, its progress seems to have been checked by the northern scarp of the Appalachian Plateau. But some of the ice moved on up the valleys that had previously drained the plateau. The energy concentrated in the valleys was so great that the ice was able to scoop out the basins that are now filled by the Finger Lakes.

Asymmetric rock knobs and hills. Glacier erosion of bedrock in many places produces small, rounded, asymmetric hills with gentle, striated, and polished slopes on one side and steeper slopes lacking polish and striations on the opposite side. An assemblage of these undulating knobs is referred to as *rôches moutonnées.* The now-gentle slope faced the advancing glacier and was eroded by abrasion. The opposite slope has been steepened by the plucking action of the ice as it rode over the knob. Large individual hills have the same asymmetric profiles as the smaller hills. Here, too, the gentle slope faced the moving ice.

Types of glacial deposits

The debris carried along by a glacier is eventually deposited, either because the ice that holds it melts or, less commonly, because the ice smears the debris across the land surface.

The general term *drift* is applied to all deposits that are deposited directly by glaciers or that, as a result of a glacial activity, are laid down in lakes, oceans, or streams. The term dates from the days when geologists thought that the unconsolidated cover of sand and gravel blanketing much of Europe and America had been made to drift into its present position either by the sea or by icebergs. Drift can be divided into two general categories: (1) *stratified* and (2) *unstratified.*

Deposits of unstratified drift. Unstratified drift deposited directly by glacier ice is known as *till.* It is composed of rock fragments of all sizes mixed together in random fashion, ranging all the way from boulders weighing several metric tons to tiny clay and colloid particles. Many of the large pieces are striated, polished, and faceted as a result of the wear they underwent while being transported by the glaciers. Some of the material picked up along the way was smeared across the landscape during the glacier's progress, and some of it was dumped when the rate of wastage began to exceed the rate of nourishment and the glacier gradually melted away.

Till is deposited by receding and advancing glaciers in a great variety of topographic forms, including moraines and drumlins.

Moraines. *Moraine* is a general term used to describe many of the landforms that are composed largely of till. A *terminal moraine,* or *end moraine,* is a ridge of till that marks the utmost limit of a glacier's advance (Figure 11-10). This type of ridge ranges in size from a rampart scores of meters high to a very low, interrupted heap of debris. A terminal moraine forms when a glacier reaches the critical point of equilibrium—the point at which it wastes away at exactly the same rate as it is nourished. Although the front of the glacier is stable, ice continues to rush down from above, delivering a continuous supply of rock debris. As the ice melts in the zone of wastage, the debris is dumped and the terminal moraine grows. At the same time, water from the melting ice pours down over the till and sweeps part of it out in a broad flat fan that butts against the forward edge of the moraine like a giant ramp (Figure 11-11).

The terminal moraine of a mountain glacier is crescent-shaped, with the convex side extending down-valley. The terminal moraine of a continental ice sheet is a broad loop or series of loops traceable for many kilometers across the countryside.

Behind the terminal moraine, and at varying distances from it, a series of smaller ridges known as *recessional moraines* may build up. These ridges mark the position where the glacier front was stabilized temporarily during the retreat of the glacier.

Not all the rock debris carried by a glacier finds its way to the terminal and reces-

FIGURE 11-10 *(Above) Details of terminus of Ruth Glacier, Alaska. Tokositna Mountain at left; Mts. Foraker, Hunter, and McKinley in distance. (Courtesy National Park Service.)*

sional moraines, however. As the main body of the glacier melts, a great deal of till is deposited to form gently rolling plains across the valley floor. Till in this form, called a *ground moraine,* may be a thin veneer lying on the bedrock, or it may form a deposit scores of meters thick, partially or completely clogging preglacial valleys.

Finally, valley glaciers produce two other special types of moraines. While a valley glacier is still active, large amounts of rubble keep tumbling down from the valley walls, collecting along the sides of the glacier. When the ice melts, all this debris is stranded as a ridge along each side of the valley, form-

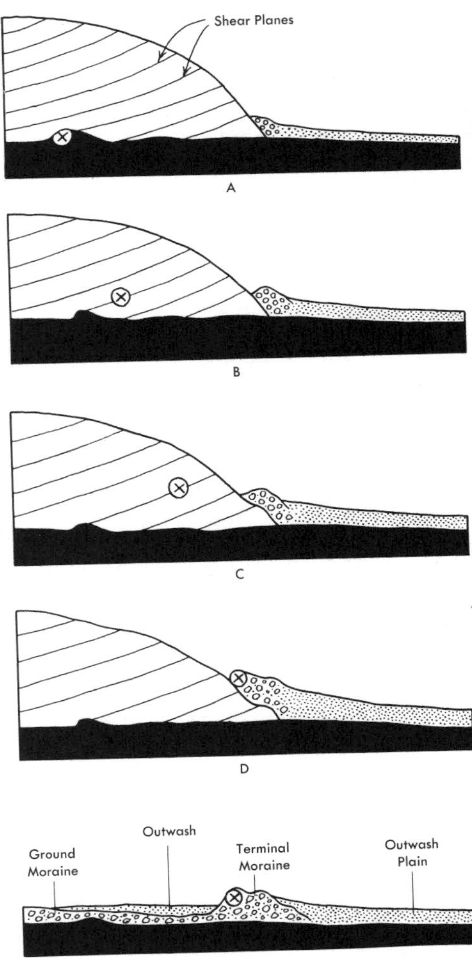

FIGURE 11-11 *(Right) A sequence of diagrams to suggest the growth of a terminal moraine at the edge of a stable ice front. The progressive movement of a single particle (X) is shown. In (A), it is moved by the ice from the bedrock floor. Forward motion of ice along a shear plane carries it ever closer to the stabilized ice margin, where finally it is deposited as a part of the moraine in diagram (D). (E) represents the relation of the terminal moraine, ground moraine, and outwash after the final melting of the glacier.*

ing *lateral moraines.* At their down-valley ends, the lateral moraines grade into the terminal moraine.

The other special type of deposit produced by valley glaciers is a *medial moraine,* created when two valley glaciers join to form a single ice stream. Material formerly carried along on the adjacent edges of the two separate glaciers is combined in a single moraine near the center of the enlarged glacier (Figure 11-12). A streak of this kind builds up whenever a tributary glacier joins a larger glacier in the main valley. Although medial moraines are characteristic of living glaciers, they are seldom preserved as topographic features after the disappearance of the ice.

Drumlins. *Drumlins* are smooth, elongated hills composed largely of till. The ideal drumlin shape has an asymmetric profile with a blunt nose pointing in the direction from which the vanished glacier advanced, and with a gentler, longer slope pointing in the opposite direction. Drumlins range from about 8 to 60 m in height, the average somewhat less than 30 m. Most drumlins are between $\frac{1}{2}$ and 1 km in length and are usually several times longer than they are wide.

In most areas, drumlins are present in clusters known as *drumlin fields.* In the United States, these are most spectacularly developed in New England, particularly around Boston; in eastern Wisconsin; in west-central New York State, particularly around Syracuse; in Michigan; and in parts of Minnesota. In Canada, extensive drumlin fields are located in western Nova Scotia and in northern Manitoba and Saskatchewan.

Just how drumlins are formed is still not clear. Since their shape is a nearly perfect

FIGURE 11-12 *Formation of medial moraines as several valley glaciers join to form a single glacier. (Courtesy U.S. Geological Survey.)*

FIGURE 11-13 This outwash plain in North Dakota was produced by a now-vanished glacier. It is underlain by stratified sand and gravel. The crop is flax. (Photo by Saul Aronow, U.S. Geological Survey.)

example of streamlining, it seems probable that they were formed deep within active glaciers in the zone of plastic flow.

Deposits of stratified drift. *Stratified drift* is ice-transported material that has been washed and sorted by glacial meltwaters according to particle size. Since water is a much more selective sorting agent than ice, deposits of stratified drift are deposited in recognizable layers, unlike the random arrangements of particles typical of till. Stratified drift is present in outwash and kettle plains, eskers, kames, and varves—all discussed in the following sections.

Outwash sand and gravel. The sand and gravel that are carried outward by meltwater from the front of a glacier are termed *outwash.* As a glacier melts, streams of water heavily loaded with reworked till, or with material *washed* directly from the ice, weave a complex, braided pattern of channels across the land in front of the glacier.

These streams, choked with clay, silt, sand, and gravel, rapidly decrease in velocity and deposit their load of debris as they flow away from the ice sheet. In time, a vast apron of bedded sand and gravel is built up that may extend for kilometers beyond the ice front. If the zone of wastage is located in a valley, the outwash deposits are confined to the lower valley and compose a *valley train.* But along the front of a continental ice sheet the outwash deposits extend for kilometers, forming an *outwash plain* (Figure 11-13).

Kettles. Sometimes a block of stagnant ice becomes isolated from the receding glacier during wastage and is partially or completely buried in till or outwash before it finally melts. When it disappears, it leaves a *kettle,* a pit or depression in the drift. These depressions range from a few meters to several kilometers in diameter, and from a few meters to over 30 m in depth. Many outwash plains are pockmarked with kettles and are termed *pitted outwash plains.* As time passes, water sometimes fills the kettles to form lakes or swamps, features found through much of Canada and northern United States.

Eskers and crevasse fillings. Winding, steep-sided ridges of stratified gravel and sand, sometimes branching and often discontinuous, are called *eskers* (Figure 11-14). They usually vary in height from about 3 to 15 m, although a few are over 30 m high. Eskers range from a fraction of a kilometer to over 160 km in length, but they are only a few meters wide. Most investigators believe that eskers were formed by the deposits of streams running through tunnels beneath stagnant ice. Then, when the body of the glacier finally disappeared, the old stream deposits were left standing as a ridge.

FIGURE 11-14 Esker near Tweed, Ontario, Canada. (Photo by Geological Survey of Canada.)

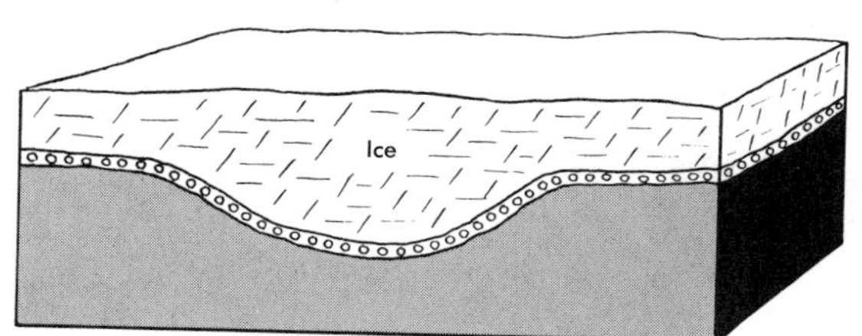

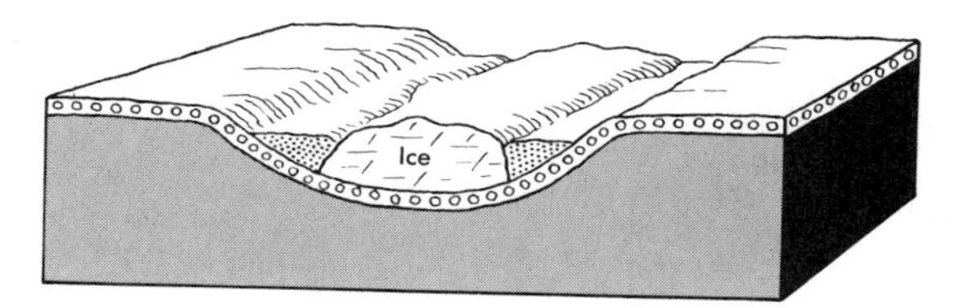

FIGURE 11-15 The sequence in the development of a kame terrace. Ice wasting from an irregular topography lingers longest in the valleys. While the ice still partially fills one of these valleys, outwash may be deposited between it and the valley walls. The final disappearance of the ice leaves the outwash in the form of terraces along the sides of the valley.

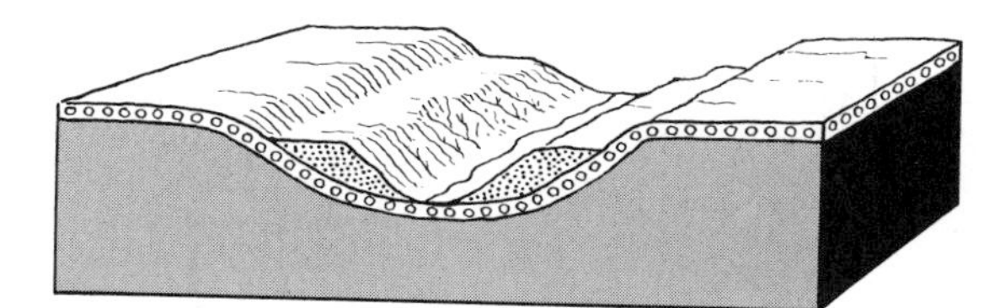

FIGURE 11-16 Varves deposited in a now-extinct glacial lake near Hanover, New Hampshire. (Photo by R. W. Sayles.)

Crevasse fillings are similar to eskers in height, width, and cross profiles, but unlike the sinuous and branching pattern of eskers, they run in straight ridges. As their name suggests, they were probably formed by the filling of a crevasse in stagnant ice.

Kames and kame terraces. In many areas, stratified drift has built up low, relatively steep-sided hills called *kames,* either as isolated mounds or in clusters. Unlike drumlins, kames are of random shape, and the deposits that compose them are stratified. They were formed by the material that collected in openings in stagnant ice. In this sense they are similar to crevasse fillings, but without the linear pattern.

A *kame terrace* is a deposit of stratified sand and gravel that has been deposited between a wasting glacier and an adjacent valley wall. When the glacier disappears, the deposit stands as a terrace along the side of the valley (Figure 11-15).

Varves. A *glacial varve* is a pair of thin sedimentary beds, one coarse, one fine (Figure 11-16). This pair of beds is usually interpreted as being the deposits of a single year and is thought to form in the following way. During the period of summer thaw, waters from a melting glacier carry large amounts of clay, silt, and fine sand into lakes along the ice margin. The coarser particles sink fairly rapidly and blanket the lake floor with a thin layer of silt and silty sand. But as long as the lake is unfrozen, the wind creates currents strong enough to keep the finer clay particles in suspension. When the lake freezes over in the winter, these wind-generated currents cease, and the fine particles sink through the quiet water to the bottom, covering the somewhat coarser summer layer. A varve is usually a few millimeters to 1 cm thick, though thicknesses of 5 to 8 cm are common. In rare instances thicknesses of 30 cm or more are known. Some varves in carbonate and evaporite beds are shown in Figure 11-17.

Table 11-1 Features of Valley and Continental Glaciations

FEATURES	VALLEY	CONTINENTAL
Striations, polish, etc.	Common	Common
Cirques	Common	Absent
Horns, arêtes, cols	Common	Present in marginal areas
U-shaped valleys, truncated spurs, hanging valleys	Common	Rare
Fiords	Common	Absent
Till and stratified drift	Common	Common
Terminal moraines	Common	Common
Recessional moraines	Common	Common
Ground moraines	Common	Common
Lateral moraines	Common	Absent
Medial moraines	Common, easily destroyed	Absent
Drumlins	Rare or absent	Locally common
Kettles	Common	Common
Eskers, crevasse fillings	Rare	Common
Kames	Common	Common
Kame terraces	Common	Present in hilly country

Comparison of valley and continental glaciation features. Some of the glacial features that we have discussed are more common in areas that have undergone valley glaciation; others usually are present only in regions that have been overridden by ice sheets; many other features, however, are found in both types of area. Table 11-1 lists and compares the features that are characteristic of the two types.

Development of the glacial theory

Geologists have made extensive studies of the behavior of modern glaciers and have carefully interpreted the traces left by glaciers that disappeared thousands of years ago. On the basis of these studies, the following *glacial theory* has been developed. This holds that *in the past, great ice sheets covered large sections of the earth where no ice now exists, and that existing glaciers once extended far beyond their present limits.*

The beginnings

The glacial theory took many years to evolve, years of trying to explain the occurrence of the vast expanses of drift strewn across northern Europe, the British Isles, Switzerland, and adjoining areas. The exact time when inquisitive minds began to seek an explanation of these deposits is unknown. But by the beginning of the eighteenth century, explanations of what we now know to be glacial deposits and features were being published. According to the most popular early hypothesis, a great inundation had swept these deposits across the face of the land with cataclysmic suddenness, or else had drifted them in by means of floating icebergs. Then, when the flood receded, the material was stranded in its present location.

By the turn of the nineteenth century, a new theory was born—the theory of ice transport. We do not know who first stated the idea or when it was first proposed, but it seems

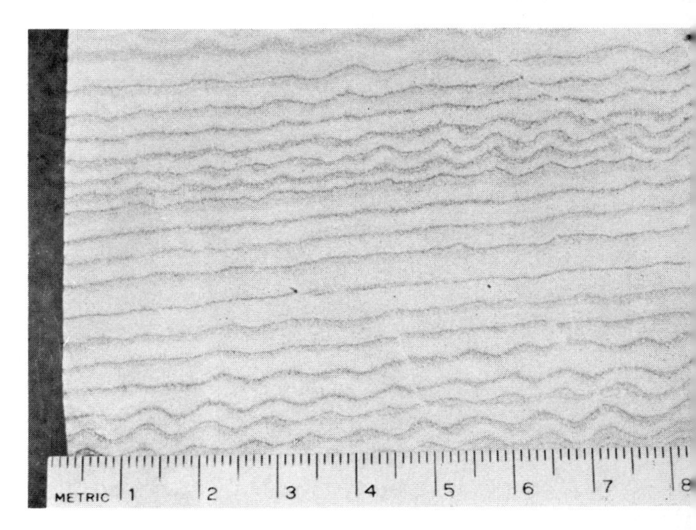

FIGURE 11-17 Varves in Castile Formation of Permian age, Texas. Gypsum (light) and organic-rich carbonate (dark). (Photograph by Earle F. McBride.)

clear that it was not hailed immediately as a great truth. As the years passed, however, more and more observers became intrigued with the idea. The greatest impetus came from Switzerland, where the activity of living glaciers could be studied.

In 1821, J. Venetz, a Swiss engineer delivering a paper before the Helvetic Society, presented the argument that Swiss glaciers had once expanded on a great scale. It has since been established that from about 1600 to the middle of the eighteenth century there actually was a time of moderate but persistent glacier expansion in many localities. Abundant evidence in the Alps, Scandinavia, and Iceland indicates that the climate was milder during the Middle Ages than it is at the present, and that communities existed and farming was carried on in places later invaded by advancing glaciers or devastated by glacier-fed streams. We know, for example, that a silver mine in the valley of Chamonix was being worked during the Middle Ages and that it was subsequently buried by an advancing glacier, where it lies to this day. And the village of St. Jean de

Perthuis has been buried under the Brenva Glacier since about 1600.

Although Venetz's idea did not find immediate acceptance, by 1834 Jean de Charpentier was arguing in its support before the same Helvetic Society. Yet the theory continued to have more opponents than defenders. It was one of the skeptics, Louis Agassiz, who did more than anyone else to develop the glacial theory and bring about its general acceptance.

Louis Agassiz

Louis Agassiz (1807–1873), a young zoologist, had listened to Charpentier's explanation; afterwards he undertook to demonstrate to his friend and colleague the error of his ways. During the summer of 1836, the two men made a trip into the upper Rhone Valley to the Getrotz Glacier. Before the summer was over, it was Agassiz who was convinced of his error. In 1837, he spoke before the Helvetic Society championing the glacial theory and suggesting that during a "great ice age" not only the Alps but much of northern Europe and the British Isles was overrun by a sea of ice.

Agassiz's statement of the glacial theory was not accepted immediately, but in 1840 he visited England and won the support of leading British geologists. In 1846 he arrived in America, where in the following year he became professor of zoology at Harvard College and later founded the Museum of Comparative Zoology. In this country, he convinced geologists of the validity of the glacial theory, and by the third quarter of the nineteenth century the theory was established. The last opposition expired with the turn of the century.

Proof of the glacial theory

What proof is there that the glacial theory is valid? The most important evidence is that certain features produced by glacier ice are produced by no other known process. Thus, Agassiz and his colleagues found isolated stones and boulders quite alien to their present surroundings. They noticed, too, that boulders were actually being transported from their original location by modern ice. Some of the boulders they observed were so large that rivers could not possibly have moved them, and others were perched on high places that a river could have reached only by flowing uphill. They also noticed that when modern ice melted, it revealed a polished and striated pavement unlike the surface fashioned by any other known process. To explain the occurrence of these features in areas where no modern glaciers exist, they postulated that the ice once extended far beyond its present limits.

The proof of glaciation lies not in the authority of the textbook or of the lecture. It lies in observing modern glacial activity directly and in comparing the results of this activity with features and deposits found beyond the present extent of the ice.

Gravity and the mass movement of surface material

Water and ice erode and move material and model the earth's surface. These two agents are driven by gravity. But gravity also operates without any obvious association of water and ice, and it is to this process that we now turn.

Gravity acts to move the products of weathering, and even unweathered bedrock, to lower and lower levels. This movement of surface material caused by gravity is termed *mass movement* or *mass wasting.* Sometimes it takes place suddenly in the form of great landslides (Figure 11-18) and rock falls from precipitous cliffs, but often it occurs almost

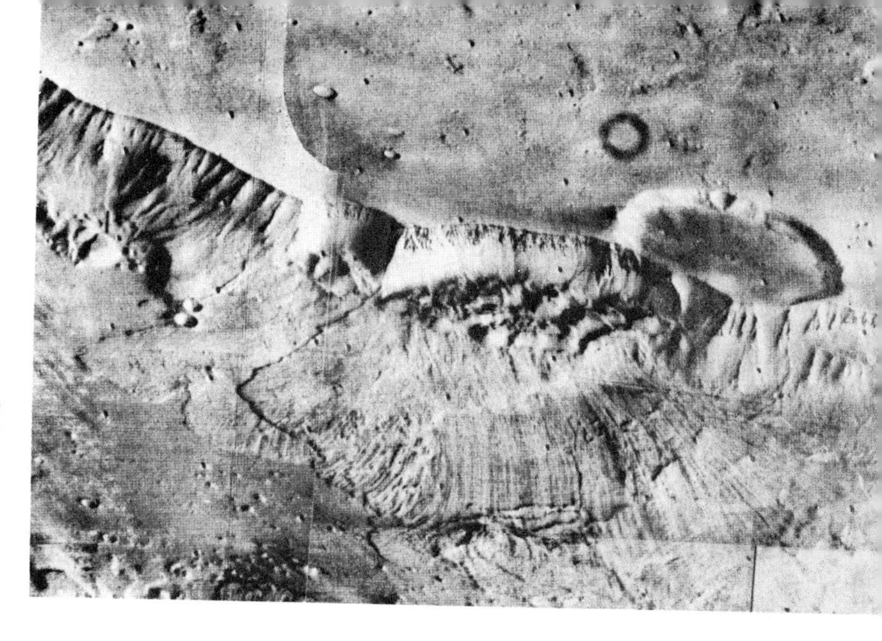

FIGURE 11-18 Massive landslide on Mars at Valles Marineris. Wall of canyon is more than $1\frac{1}{2}$ km high. The wall has collapsed at different periods. On Mars, such landslide collapse, together with wind erosion, has enlarged canyons and created complex jumbled slide masses in many channels and canyons. (Courtesy NASA.)

imperceptibly, as in the slow creep of soil across gently sloping fields. Mass movement, then, is one type of adjustment that earth materials make to their physical environment; it is one of the many ways in which erosion acts to wear down the land masses of the earth.

Factors of mass movement

Gravity provides the energy for the downslope movement of surface debris and bedrock. But several other factors, particularly water, augment gravity and ease its work.

Immediately after a heavy rainstorm, you may have witnessed a landslide on a steep hillside or on the bank of a river. Movement of this sort is often mistakenly attributed to the "lubricating action" of water. But water does not "grease the skids" in the strict sense of the phrase. With many minerals, water actually acts not as a lubricant but as an *anti*lubricant. Heavy rains do not promote movement by "lubrication."

Water does aid in downslope movements, however. In many unconsolidated deposits, the pore spaces between individual grains are filled partly with moisture and partly with air. And so long as this condition persists, the surface tension of the moisture gives a certain cohesion to the soil. But when a heavy rain comes along and forces all the air out of the pore spaces, this surface tension is completely destroyed, the cohesion of the soil is reduced, and the whole mass becomes more susceptible to downslope movement. The presence of water also adds weight to the soil on a slope, although this added weight is probably not a very important factor in promoting mass movement.

Water that soaks into the ground and completely fills the pore spaces in the slope material contributes to instability in another way. The water in the pores is under pressure, pressure that tends to push apart individual grains or even whole rock units and to decrease the internal friction or resistance of the material to movement. Here again, water assists in mass movement.

Gravity can move material only when it is able to overcome the material's internal resistance against being set into motion. Clearly, then, any factor that reduces this resistance to the point where gravity can take over contributes to mass movement. The erosive action of a stream, an ocean, or a glacier may so steepen a slope that the earth material no longer can resist the pull of gravity and is forced to give in to mass movement. In regions of cold climate, alternate freezing and thawing of earth materials may be enough to set these materials in motion. The impetus needed to initiate movement may also be furnished by earthquakes, excavations, or blasting operations, or even by the gentle activities of burrowing animals and growing plants.

Behavior of material

Earth materials under stress within the earth behave as elastic, plastic, or fluid substances. Materials in motion down a slope can also behave in the same ways.

We could study any type of mass movement and classify it on the basis of these three types of movement. But we would have to assemble an excessive amount of technical information, and we would find the picture complicated by the fact that material often behaves in different ways during any one movement. So we shall simply classify mass movement as either *rapid* or *slow*. (See Table 11-2.)

Rapid movements

Catastrophic and destructive movements of rock and soil, the most spectacular and easily recognized examples of mass movement, are popularly known as *landslides*. But geologists subdivide this general term into *slump, rock slides, debris slides, mudflows, earthflows,* and *talus falls.*

Landslides

Landslides include a wide range of movements, from the slipping of a stream bank to the sudden, devastating release of a whole mountainside (Table 11-3). Some landslides involve only the unconsolidated debris lying on bedrock; others involve movement of the bedrock itself. Movement takes place when the strength of the material is exceeded by the force of gravity.

Table 11-2 Description of Mass-Movement Processes

PROCESS	DESCRIPTION
Rapid Movements	
Rock and debris fall, talus fall	Rapid fall of debris from a cliff or by bounding down slope. Velocity up to the terminal velocity in air of particular fragment.
Rock avalanche, rock slide, debris slide	Rapid descent of rock mass down a slope. Frank, Alberta, 96 km/hr; Elm, Switzerland, 161 km/hr.
Slump	Downward and outward movement of rock or unconsolidated material as a unit or series of units.
Earthflow	Combination of slump and plastic movement of regolith. Slow but perceptible movement.
Mudflow	Flow in which material has consistency of mud. Contains fine grains and large amount of water. Mayflower Gulch mudflow near Denver, Colorado, moved at velocity of 16 m/sec.
Slow Movements	
Creep	Slow downward movement of surface material. From a few mm/yr to 70 mm/yr. Vaiont area of the Italian Alps had creep rate of 1 cm/week before avalanche of October 9, 1963.
Solifluction	Downslope movement of debris under saturated conditions where soil is strongly affected by alternate freezing and thawing. As fast as 12 cm/yr.

Table 11-3 Volumes of Some Large Landslides

LOCALITY OF LANDSLIDE	DATE	VOLUME (millions of cubic meters)
Elm, Switzerland	1881	7.6–8.4
Madison Canyon, Montana	1959	32.0
Frank, Alberta, Canada	1903	36.5
Gros Ventre, Wyoming	1925	37.0
Vaiont, Italy	1963	296.4
Tin Mountain, California	Prehistoric	1,786.0
Saidmarreh, Iran	Prehistoric	16,000–20,000

SOURCE: After Morton, D.M., and Robert Streitz, "Landslides," in *Man and His Physical Environment*, 1972, pp. 64–73.

We usually think of a landslide as breaking loose without warning, but it is more accurate to say that people in the area simply fail to detect and heed the warnings.

For example, a disastrous rock slide at Goldau, Switzerland, in 1806, destroyed a whole village, killing 457 people. The few who lived reported that they had no warning of the coming slide, but that animals and insects in the region may have been more observant or more sensitive. For several hours before the slide, horses and cattle were extremely nervous, and bees abandoned their hives. Some slight preliminary movement probably took place before the rock mass actually broke loose.

During the spring of 1935, slides occurred in clay deposits along a German superhighway that was being built between Munich and Salzburg. The slides came as a complete surprise to the engineers, but for a full week beforehand the workmen had been murmuring, "Der Abhang wird Lebendig" ("The slope becomes alive."). Landslides are often preceded by slowly widening fractures in the rock near the upward limit of the future movement.

There is some evidence that landslides may recur periodically in certain areas. For example, in southeastern England, not far from Dover, extensive landslides have been happening once every 19 to 20 years. Some observers believe there may be some correlation between such periodical mass movements and periods of excessive rainfall. On steep slopes in very moist tropical or semitropical climates, for instance, landslides apparently follow a cyclic pattern. First, a landslide strips the soil and vegetation from a hillside. In time, new soil and vegetation develop, the old scar heals, and when the cover reaches a certain stage, the landsliding begins again. Although landslides may happen in cycles, information is too scanty to support any firm conclusions at this time.

Slump. Sometimes called *slope failure*, *slump* is the downward and outward movement of rock or unconsolidated material traveling as a unit or as a series of units. Slumps usually occur where the original slope has been sharply steepened, either artificially or naturally (Figure 11-19). The material reacts to the pull of gravity as if it were an elastic solid, and large blocks of the slope move downward and outward along curved planes. The upper surface of each block is tilted backward as it moves.

Once a slump begins to move, it is often helped along by rainwater collecting in basins between the tilted blocks and the original slope. The water drains down along the surface on which the block is sliding and promotes further movement.

FIGURE 11-19 (Above) *Destruction through slumping of houses at Orinda, California, 1968. (Photograph by Lloyd S. Cluff.)*

(a)

(b)

FIGURE 11-20 (Left) (a) *Yungay, Peru. View looking east from Cemetary Hill showing village as it existed before May 31, 1970 earthquake and rock avalanche. (b) Same view as in (a) except that photograph was taken after avalanche. Rock and debris fell 3,600 m and traveled 11.3 km at an average speed of about 322 km per hour. (Courtesy Lloyd S. Cluff.)*

Rock slides. The most catastrophic of all mass movements are *rock slides*—sudden, rapid slides of bedrock along planes of weakness (Figures 11-20 and 11-21). A great rock slide took place in 1925 on the flanks of Sheep Mountain, along the Gros Ventre River in northwestern Wyoming, not far from Yellowstone Park (Figure 11-22). An estimated 37 million cubic meters of rock and debris plunged down the valley wall and swept across the valley floor. The nose of the slide rushed some 110 m up the opposite wall and then settled back, like liquid being sloshed in a great basin. The debris formed a dam between 68 and 75 m high across the valley, the dammed-up river creating a lake almost 8 km long. The spring floods of 1927 raised the water level to the lip of the dam, and in mid-May the water flooded over the top. So rapid was the downcutting of the dam that the lake level was lowered about 15 m in 5 hours. During the flood that followed, several lives were lost in the town of Kelly, in the valley below.

The Gros Ventre slide was a long time in the making, and there was probably nothing that could have been done to prevent it. Conditions immediately before the slide are shown in the top diagram in Figure 11-23. In this part of Wyoming, the Gros Ventre valley cuts through sedimentary beds inclined between 15° and 21° to the north. The slide took place on the south side of the valley. Notice that the sandstone bed is separated from the limestone

FIGURE 11-21 *The London Road landslide in Oakland, California, violently moved houses to lower elevations.*

FIGURE 11-22 *(Above) Aerial photograph of the Gros Ventre rock slide in northwestern Wyoming. The lake in the lower left of the picture has been dammed by a landslide that moved down into the valley of the Gros Ventre River. The area from which the material slid is about $2\frac{1}{2}$ km long and is well marked by the white scar down the center of the photograph. The adjoining slopes appear dark because of a vegetative cover of trees and bushes, a cover that has not yet reestablished itself in the slide area. (Photo by U.S. Army Air Force.)*

FIGURE 11-23 *(Below) Diagrams to show the nature of the Gros Ventre slide. (A) represents the conditions existing before the slide took place. (B) shows the area of the slide and the location of the debris in the valley bottom. Note that the sedimentary beds dip into the valley from the south. The large section of sandstone slid downward along the clay bed. [Redrawn from William C. Alden, "Landslide and Flood at Gros Ventre, Wyoming,"* Transactions, American Institute of Mining and Metallurgical Engineers, *76 (1938), 348.]*

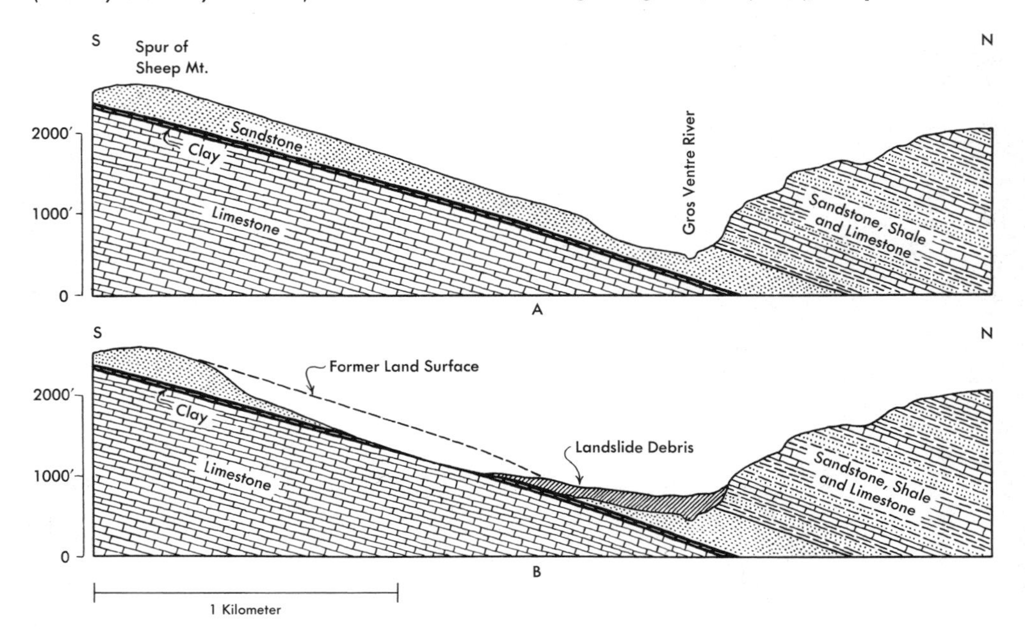

strata by a thin layer of clay. Before the rock slide occurred, the sandstone bed near the bottom of the valley had been worn thin by erosion. The melting of winter snows and the heavy rains that fell during the spring of 1925 furnished an abundant supply of water that seeped down to the thin layer of clay, soaking the clay and reducing the adhesion between it and the overlying sandstone. When the sandstone was no longer able to hold its position on the clay bed, the rock slide roared down the slope. The lower diagram of Figure 11-23 suggests the amount of material that was moved from the spur of Sheep Mountain to its present resting place on the valley floor.

A more recent rock slide happened in southwestern Montana a few minutes before midnight on August 17, 1959. An earthquake whose focus was located just north of West Yellowstone, Montana, triggered a rock slide in the mouth of the Madison Canyon, about 32 km to the west. An estimated 32 million cubic meters of rock slid from the south wall of the canyon down to the valley bottom, where it dammed up a lake 8 km long and over 30 m deep. Campers in the path of the slide felt a terrifying blast of air before the rush of debris engulfed them. Some had their clothes torn from them, and at least 28 were killed.

By far the most destructive series of slides occurred in Kansu Province, China in 1920. Gigantic masses of wind-deposited silt (loess) were shaken loose during an earthquake and moved down mountainsides into populated valleys. More than 100,000 people perished.

Debris slides. A *debris slide* is a small rapid movement of largely unconsolidated material that slides or rolls downward, producing a surface of low hummocks and small intervening depressions. Movements of this sort are common on grassy slopes, particularly after heavy rains, and in unconsolidated material along the steep slopes of stream banks and shorelines.

Mudflows. A *mudflow* is a well-mixed mass of rock, earth, and water that flows down valley slopes with the consistency of newly mixed concrete. In mountainous, desert, and semiarid areas, mudflows manage to transport great masses of material.

The typical mudflow originates in a small, steep-sided gulch or canyon where the slopes and floor are covered by unconsolidated or unstable material. A sudden flow of water, from cloudbursts in semiarid country or from spring thaws in mountainous regions, flushes the earth and rocks from the slopes and carries them to the stream channel. Here the debris blocks the channel until the growing pressure of the water becomes great enough to break through. Then the water and debris begin their course down-valley, mixing together with a rolling motion along the forward edge of the flow. The advance of the flow is intermittent, for sometimes it is slowed or halted by a narrowing of the stream channel; at other times it surges forward, pushing obstacles aside or carrying them along with it.

Eventually, the mudflow spills out of the canyon mouth and spreads across the gentle slopes below. No longer confined by the valley walls or the stream channel, it splays out in a great tongue, spreading a layer of mud and boulders that ranges from a few centimeters to a meter or so in thickness. Mudflows can move even large boulders weighing 80 metric tons or more for hundreds of meters, across slopes as gentle as 5°.

Earthflows. *Earthflows* are a combination of slump and plastic movement of unconsolidated material. They move slowly but perceptibly and may involve from a few to several million cubic meters of earth material. Some of

FIGURE 11-24 Earthflow in coast ranges, near Pleasanton, California, 1966. (Courtesy Lloyd S. Cluff.)

FIGURE 11-25 Talus has accumulated at the base of the cliff as rock and debris have moved down the steep curved slope or fallen from the cliff face. (Photo by M. Dane Picard.)

the material behaves like an elastic solid, and some like a plastic substance, depending on its position in the moving mass.

The line at which a slump pulls away from the slope is marked by an abrupt *scarp* or cliff, as shown in Figure 11-24. Notice that the slump zone is composed of a series of blocks that move downward and outward, tilting the original surface back toward the slope. Farther down, the material tends to flow like a liquid, often beneath the vegetative cover. At the downslope limit of an earthflow, the sod often bulges out and fractures. Earthflows occur in unconsolidated material lying on solid bedrock and are usually helped along by the presence of excessive moisture.

Talus falls. Strictly speaking, a *talus* (Figure 11-25) is a slope built up by an accumulation of rock fragments at the foot of a cliff or a ridge. The rock fragments are sometimes termed *rock waste* or *slide-rock*. In practice, however, talus is widely used as a synonym for rock debris itself.

In the development of a talus, rock fragments are loosened from the cliff and clatter downward in a series of free falls, bounces, and slides. As time passes, the rock waste finally builds up a heap or sheet of rock rubble. An individual talus resembles a half-cone with its apex resting against the cliff face in a small gulch. A series of these half-cones often forms a girdle around high mountains, completely obscuring their lower portions. Eventually, if the rock waste accumulates more rapidly than it can be destroyed or removed, even the upper cliffs become buried, and the growth of the talus stops. The slope angle of the talus varies with the size and shape of the rock fragments. Although angular material can maintain slopes up to 50°, rarely does a talus ever exceed angles of 40°.

A talus is subject to the normal process of chemical weathering, particularly in a moist climate. The rock waste is decomposed, especially toward its lower limit, or toe, and the material there may grade imperceptibly into the soil.

Slow movements

Slow mass movements of unconsolidated material are harder to recognize and are less fully understood than rapid movements; yet they are extremely important in the sculpturing of the land surface. Since these slow movements operate over long periods of time, they are probably responsible for the transportation of more material than are rapid and violent movements of rock and soil.

Before the end of the nineteenth century, William Morris Davis aptly described the nature of slow movements.

> The movement of land waste is generally so slow that it is not noticed. But when one has learned that many land forms result from the removal of more or less rock waste, the reality and the importance of the movement are better understood. It is then possible to picture in the imagination a slow washing and creeping of the waste down the land slopes; not bodily or hastily, but grain by grain, inch by inch; yet so patiently that in the course of ages even mountains may be laid low.[1]

Creep

In temperate and tropical climates, a slow downward movement of surface material known as *creep* operates even on gentle slopes with a protective cover of grass and trees. It is hard to realize that this movement is actually taking place. Since the observer sees no break in the vegetative mat, no large scars or hummocks, he has no reason to suspect that the soil is in motion beneath his feet.

Yet this movement can be demonstrated by exposures in rock soil profiles (Figure 11-26), and by the behavior of tree roots, of large blocks of resistant rock, and of man-built objects such as fences and railroads. Figure 11-27 shows a section through a hillside underlain by flat-lying beds of limestone, claystone, sandstone, and coal. The slope is covered with rock debris and soil. But notice that the beds near the base of the soil bend downslope and thin our rapidly. These beds are being pulled downslope by gravity and are strung out in ever-thinning bands that may extend for hundreds of meters. Eventually, they approach the surface and lose their identity in the zone of active chemical weathering.

The same diagram shows other evidence that the soil is moving. Although when viewed from the surface the tree appears to be growing in a normal way, it is actually creeping slowly down the slope. Since the surface of the soil is moving more rapidly than the soil beneath it, the roots of the tree are unable to keep up with the trunk. Consequently, they are spread out like great streamers along the slope.

[1] William Morris Davis, *Physical Geography* (Boston: Ginn & Company, 1898), p. 261.

FIGURE 11-26 *Near-vertical beds are bent downslope in their upper portion by gravity. (Photo by U.S. Geological Survey.)*

We can discover evidence of the slow movement of soil in displaced fences and tilted telephone poles and gravestones. On slopes where resistant rock layers crop up through the soil, fragments are sometimes broken off and distributed down the slope by the slowly moving soil.

Many other factors cooperate with gravity to produce creep. Probably the most important is moisture in the soil, which works to weaken the soil's resistance to movement. In fact, any process that causes a dislocation in the soil brings about an adjustment of the soil downslope under the pull of gravity. Thus, the

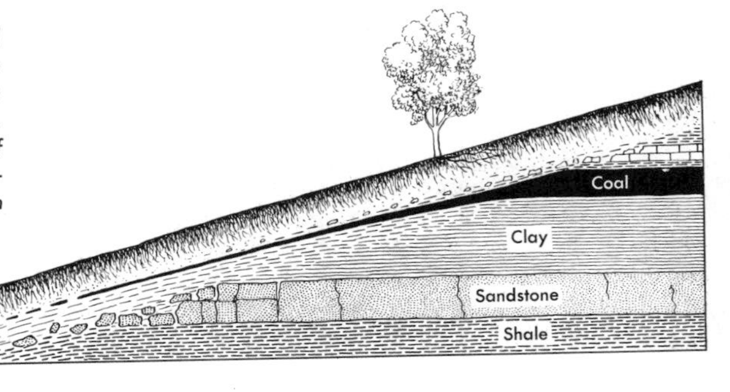

FIGURE 11-27 *The partially weathered edges of horizontal sedimentary rocks are dragged downslope by soil creep. The tree is also moving slowly downslope, as is indicated by the root system spread out behind the more rapidly moving trunk. [Redrawn from C. F. S. Sharpe and E. F. Dosch, "Relation of Soil-Creep to Earthflow in the Appalachian Plateaus," Geomorphology Journal, 5 (December, 1942), 316, by permission of Columbia University Press.)*

burrows of animals tend to increase movement downslope, and the same is true of cavities left by the decay of organic material, such as the root system of a dead tree. The prying action of swaying trees, the tread of animals, and even of men, may also aid in the motion. The end result of all these processes, aided by the influence of gravity, is to produce a slow and inevitable downslope creep of the surface cover of debris and soil.

Solifluction

The term *solifluction* (from the Latin *solum,* "soil," and *fluere,* "to flow") refers to the downslope movement of debris under saturated conditions in high latitudes where the soil is strongly affected by alternate freezing and thawing. Solifluction is most pronounced in areas where the ground freezes to great depths. But even moderately deep seasonal freezing promotes solifluction.

Solifluction takes place during periods of thaw. Since the ground thaws from the surface downward, the water that is released cannot percolate into the subsoil and adjacent bedrock, which are still frozen and therefore impermeable to water. As a result, the surface soil becomes sodden and water-laden and tends to flow down even the gentlest slopes. Solifluction is an important process in the reduction of land masses in arctic climates, where it transports great sheets of debris from higher to lower elevations.

During the glacier advances of the Pleistocene (see Chapter 21), a zone of intense frost action and solifluction bordered the southward moving ice. In some places there still is evidence of these more rigorous climates preserved in distorted layers of earth material just below the modern soil.

Frost action plays weird tricks in the soils of the higher elevations and latitudes.

FIGURE 11-28 This rock glacier is in the Wrangell Mountains, Alaska. (Photo by Austin Post.)

Strange polygonal patterns made up of rings of boulders surrounding finer material, stripes of stones strewn down the face of hillsides, great tabular masses of ice within the soil, and deep ice wedges that taper downward from the surface—all are found in areas where the ground is deeply frozen. The behavior of frozen ground is one of the greatest barriers to the settlement of Arctic regions. The importance of these regions has increased in recent years, and studies begun by Scandinavian and Russian investigators are now being intensively pursued by American scientists.

Rock glaciers

Rock glaciers are long tongues of rock waste that form in the valleys of certain mountainous regions. Although these glaciers consist almost entirely of rock, they bear a striking resemblance to ice glaciers (Figure 11-28). A typical rock glacier is marked by a series of rounded ridges, suggesting that the material has behaved as a viscous mass.

Observations on active rock glaciers in Alaska indicate that movement takes place because of interstitial ice within the mass. Favorable conditions for the development of rock glaciers include a climate cold enough to keep the ground continuously frozen, steep cliffs to supply debris, and coarse blocks that allow for large interstitial spaces.

Work of the wind

Not infrequently in the spring the storm door is ripped from your hand and is smashed against the side of the house. Insurance companies will reluctantly accept such an explanation for the glass damage, but generally will not replace the whole door.

The wind is a turbulent stream of air quite capable of eroding, transporting, and depositing large amounts of sediment. The low density of air limits its competence to move large particles, but a person caught in a dust storm is not acutely aware of this limitation. Winds are effective agents not only in semiarid and arid regions but in humid regions as well. The common term used to describe wind processes and deposits is *eolian,* from Aeolus the Greek god of the winds.

Movement of material

Winds are greatly variable in direction and power (Table 11-4). Their velocities increase rapidly with height above the ground surface. The general movement of wind is forward, across the surface of the land. But within this general movement, the air is moving upward, downward, and from side to side. In the zone about 1 m above ground surface, the average velocity of upward motion in an

Table 11-4 Terminology of Wind Speeds

WIND SPEED (km/hr)	INTERNATIONAL DESCRIPTION	U.S. WEATHER BUREAU
<1	Calm	Light wind
1–11	Light air–light breeze	Light wind
12–28	Gentle–moderate breeze	Gentle–moderate
29–38	Fresh breeze	Fresh wind
39–61	Strong breeze–moderate gale	Strong wind
62–88	Fresh–strong gale	Gale
89–117	Whole gale–storm	Whole gale
>117	hurricane	Hurricane

air eddy is approximately one-fifth the average forward velocity of the wind. This upward movement greatly affects the wind's ability to transport small particles of earth material.

Right along the surface of the ground there is a thin but definite zone where the air moves very little or not at all. Field and laboratory studies have shown that the depth of this zone depends on the size of the particles that cover the surface. On the average, the depth of this "zone of no movement" is about one-thirtieth the average diameter of the surface grains (Figure 11-29). Thus, over a surface of evenly distributed pebbles with an average diameter of 30 mm, the zone of no movement would be about 1 mm deep. This fact, too, has a bearing on the wind's ability to transport material.

FIGURE 11-29 *In a thin zone close to the ground there is little or no air movement, regardless of the wind velocity immediately above. This zone is approximately one-thirtieth the average diameter of surface particles. Two zones are shown in the graph: one for surfaces on which the particles average 1 mm in diameter, and one for surfaces with 1-cm particles. Diagonal lines represent the increase in velocity of a wind of given intensity blowing over surfaces covered with particles of 1 mm and 1 cm average diameter, respectively. [Reproduced by permission from R. A. Bagnold,* The Physics of Blown Sand and Desert Dunes *(London: Methuen & Co., Ltd., 1941), p. 54.]*

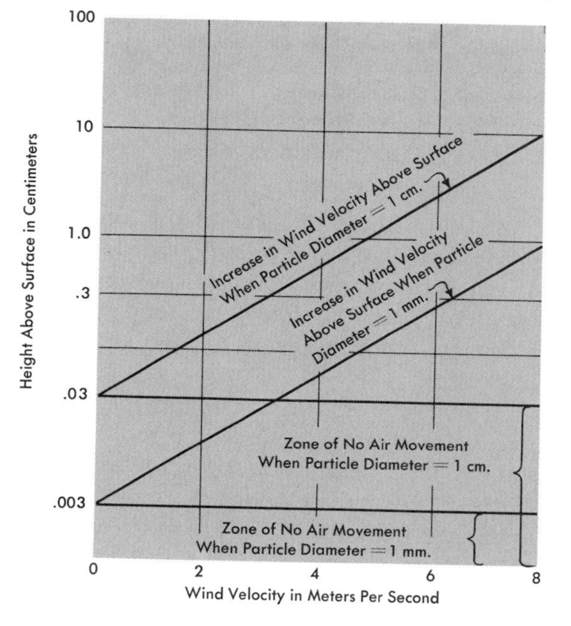

Dust storms and sandstorms

Material blown along by the wind usually falls into two size groups. The diameter of wind-driven sand grains averages between 0.15 mm and 0.30 mm, with a few grains as fine as 0.06 mm. All particles smaller than 0.06 mm are termed *dust.*

In a true *dust storm,* the wind picks up fine particles and sweeps them upward hundreds or even thousands of meters into the air, forming a great cloud that may blot out the sun and darken the sky. In contrast, a *sandstorm* is a low, moving blanket of wind-driven sand with an upper surface 1 m or less above the ground. Actually, the greatest concentration of moving sand is usually just a few centimeters above the ground surface, and individual grains seldom rise even as high as 2 m. Above the blanket of moving sand, the air is quite clear, and a man on the ground appears to be partially submerged, as though he were standing in a shallow pond. Often, of course, the dust and sand are mixed together in a wind-driven storm. But the wind soon sweeps the finer particles off, and eventually the air above the blanket of moving sand becomes clear.

Apparently, then, the wind handles particles of different sizes in different ways. A dust-sized grain is swept high into the air, and a sand-sized grain is driven along closer to the ground. The difference arises from the strength of the wind and the terminal velocity of the grain.

The terminal velocity of a grain is the constant rate of fall attained by the grain when the acceleration due to gravity is balanced by the resistance of the fluid—in this case, the air—through which the grain falls. Terminal velocity varies only with the size of a particle when shape and density are constant. As the particle size increases, both the pull of gravity and the air resistance increase too. But the pull of gravity increases at a faster rate than the air resistance: a particle with a diameter of 0.01 mm has a terminal velocity in air of about 0.01 m per second; a particle with 0.2 mm diameter has a terminal velocity of about 1 m per second; and a particle with a diameter of 1 mm has a terminal velocity of about 8 m per second.

To be carried upward by an eddy of turbulent air, a particle must have a terminal velocity that is less than the upward velocity of the eddy. Close to the ground surface, where the upward currents are particularly strong, dust particles are swept up into the air and carried in suspension. Sand grains, however, have terminal velocities greater than the velocity of the upward moving air; they are lifted for a moment and then fall back to the ground. But how does a sand grain get lifted into the air at all if the eddies of turbulent air are unable to support it?

Movement of sand grains

Careful observations, both in the laboratory and on open deserts, show sand grains moving forward in a series of jumps, in a process known as *saltation.* We used the same term to describe the motion of particles along a stream bed. But there is a difference: an eddy of water can actually lift individual particles into the main current, whereas wind by itself cannot pick up sand particles from the ground.

Sand particles are thrown into the air only under the impact of other particles. When the wind reaches a critical velocity, grains of sand begin to roll forward along the surface. Suddenly, one rolling grain collides with another; the impact may either lift the second particle into the air or cause the first to fly up.

Once in the air, the sand grain is subjected to two forces. First, gravity tends to pull it down to earth again (and eventually it will succeed). But even as the grain falls, the horizontal velocity of the wind drives it forward. The resulting course of the sand grain is parabolic from the point where it was first thrown into the air to the point where it finally hits the ground. The angle of impact varies between 10° and 16° (Figure 11-30).

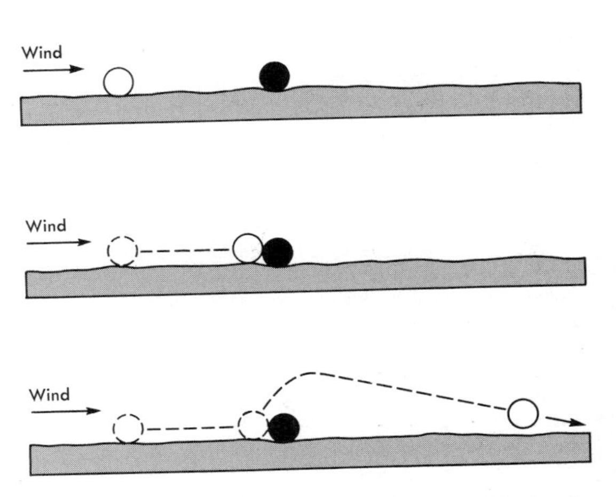

FIGURE 11-30 *A sand grain is too heavy to be picked up by the wind but may be put into the air by saltation. Here a single grain is rolled forward by the wind until it bounces off a second grain. Once in the air, it is driven forward by the wind, then pulled to the ground by gravity. It follows a parabolic path, hitting the ground at an angle between 10° and 16°.*

When the grain strikes the surface, it may either bounce off a large particle and be driven forward once again by the wind, or it may bury itself in the loose sand, perhaps throwing other grains into the air by its impact.

In any event, it is through the general process of saltation that a sand cloud is kept in motion. Countless grains are thrown into the air by impact and are driven along by the wind until they fall back to the ground. Then they either bounce back into the air again or else pop other grains upward by impact. The initial energy that lifts each grain into the air comes from the impact of another grain, and the wind contributes additional energy to keep it moving. When the wind dies, all the individual particles that compose the sand cloud settle back down to earth.

Some sand grains, particularly the large ones, never rise into the air at all, even under the impact of other grains. They simply roll forward along the ground, very much like the rolling and sliding of particles along the bed of a stream of water. It has been estimated that between one-fifth and one-quarter of the material carried along in a sandstorm travels by rolling; the rest travels by means of saltation.

Notice that once the wind has started the sand grains moving along the surface, initiating saltation, the wind no longer acts to keep them rolling. The cloud of saltating grains obstructs the wind and shields the ground surface from its force; thus, as soon as saltation begins, the velocity of near-surface wind drops rapidly. Saltation continues only because the impact of the grains continues. The stronger the wind blows during saltation, the heavier will be the blanket of sand, and the less the possibility that surface grains will be rolled by the wind.

Movement of dust particles

As we have seen, dust particles are small enough and have low enough terminal velocities to be lifted aloft by currents of turbulent air and to be carried along in suspension. But just how does the wind lift these tiny particles in the first place?

Laboratory experiments show that under ordinary conditions, particles smaller than 0.03 mm in diameter cannot be swept up by the wind once they have settled to the ground. In dry country, for example, dust may lie undisturbed on the ground even though a brisk wind is blowing. But if a flock of sheep passes by and kicks loose some of the dust, a dust plume will rise into the air and move along with the wind.

The explanation for this seeming reluctance of dust particles to be disturbed lies in the nature of air movement. The small dust grains lie within the thin zone of negligible air movement at the surface. They are so small that they do not create local eddies and disturbances in the air, and the wind passes them by. Alternatively, the dust particles may be shielded by larger particles against the action of the wind.

Some agent other than the wind must set dust particles in motion and lift them into a zone of turbulent air—perhaps the impact of larger particles, or sudden downdrafts in the air movement. Irregularities in a plowed field or in a recently exposed stream bed may help the wind to begin its work by creating local turbulence at the surface. Also, vertical downdrafts of chilled air during a thunderstorm sometimes strike the ground with velocities of 40 to 80 km/hr and churn up great swaths of dust.

Erosion

Erosion by the wind is accomplished through two processes: *abrasion* and *deflation.*

Abrasion

Like the particles carried by a stream of running water, saltating grains of sand driven by the wind are highly effective abrasive agents in eroding rock surfaces. As we have seen, wind-driven sand seldom rises more than 1 m above the surface of the earth, and measurements show that most of the grains are concentrated in the $\frac{1}{2}$ m closest to the ground. In this layer the abrasive power of the moving grains is concentrated.

Although evidence of abrasion by sand grains is rather meager, there is enough to indicate that this erosive process does take place. For example, we sometimes find fence posts and telephone poles abraded at ground level and bedrock cliffs with a small notch along their base. In desert areas the evidence is more impressive, for here the wind-driven sand has in some places cut troughs or furrows in the softer rocks. The knife-edged ridges between these troughs are called *yardangs,* a term used in the deserts of Chinese Turkestan, where they were first described; the furrows themselves are called *yardang troughs.* The cross-profile of one of these troughs is not unlike that of a glaciated mountain valley in miniature, the troughs ranging from a few centimeters to perhaps 8 m in depth. They run in the usual direction of the wind, and their deepening by sand abrasion has actually been observed to take place during sandstorms.

The most common products of abrasion are certain pebbles, cobbles, and even boulders that have been eroded in a particular way. These pieces of rock are called *ventifacts* (from the Latin for "wind" and "made"). They are found not only on deserts but also along modern beaches—in fact, wherever the wind blows sand grains against rock surfaces.

FIGURE 11-31 *A dune field on Mars with features remarkably similar to many seen in the deserts of earth. Large boulder at left is about 8 m from the Viking I Lander and measures 1 × 3 m. Wind direction across the dune is from upper left to lower right. (Courtesy NASA.)*

The surface of ventifacts is characterized by a relatively high gloss or sheen and by a variety of facets, pits, gouges, and ridges.

The face of an individual ventifact may display only 1 facet, or it may display 20 facets —or more. These facets are sometimes flat, but more commonly they are curved. Where two facets meet, they often form a well-defined ridge, and the intersection of 3 or more facets gives the ventifact the appearance of a small pyramid. Apparently, the surface becomes pitted when it lies across the direction of wind movement at an angle of 55° or more; it becomes grooved when it lies at angles of less than 55°.

Deflation

Deflation (from the Latin "to blow away") is the erosive process of the wind carrying off unconsolidated material. The process creates several recognizable features in the landscape. For example, it often scoops out basins in soft, unconsolidated deposits ranging from a few meters to several kilometers in diameter. These basins are known as *blowouts,* for obvious reasons.

In arid and semiarid country we sometimes see finely honeycombed rocks, and others called *pedestal rocks,* which have been fashioned into weird pillars resembling toadstools. Although wind has often been cited as the cause of these formations, differential weathering is primarily responsible. The wind has merely removed the loose products of weathering.

Deflation removes only the sand and dust particles from a clastic sediment deposit and leaves behind the larger particles of pebble or cobble size. As time passes, these "stones" form a surface cover, known as a *desert pavement,* that cuts off further deflation.

Deserts

A third of the world is desert or near-desert, one of the most hostile environments to humans. Vegetation is sparse in deserts, droughts are common, and precipitation is less than 25 cm per year. The cowboys we admire galloping along in scenes shot in southern California and Nevada must watch their water bags.

The desert regions of the world are mainly in the tropics and subtropics (Sahara, Kalahari of South Africa, Great Australian Desert), where rainfall may be less than 25 mm/yr, and in the middle latitudes (central Asia, Mohave, Great Basin). Deserts develop below high-pressure cells in the atmosphere where descending air is heated, along cold-water coasts where moisture-laden air bypasses the coast and is transported to warmer land, and behind high mountain ranges that force the prevailing winds to rise, cool, and precipitate moisture before crossing the range.

The majority of deserts are not sand-covered but rather are broad areas of barren rock with only poorly developed soil profiles. In many desert regions, streams begin and end in the desert, creating a pattern of *interior drainage.* Only the greatest rivers such as the Nile and Colorado can persist through wide deserts to the sea.

Most desert streams flow for only a few hours, depositing alluvial fans at their mouths. Adjacent fans coalesce along mountain fronts to form alluvial aprons, or *bajadas.* A dead-ended, intermontane basin surrounded by bajada slopes is a *bolson.* Such basins may contain temporary lakes, *playa lakes* formed by storm runoff.

Some desert lakes do persist for long periods of time. Although Great Salt Lake fluctuates considerably with wet and dry climate cycles, it and its predecessors have been large interior lakes for at least 700,000 years. Such lakes accumulate large amounts of salts, and the salinity of Great Salt Lake is 24 percent.

Desert erosion

In arid and semiarid regions the characteristic landform produced by erosion is the *pediment,* so named because it resembles the triangular unit at the front of Greek temples. Pediment surfaces slope away from desert mountains, are thinly covered with gravel, and grade downward toward desert streams, playas, or bolsons.

As weathering and erosion proceed, mountain slopes retreat parallel with their original slopes, waste material accumulates in bajadas at the mountain fronts, and playas develop in the bolson. Streams are ephemeral, flowing only for short periods, and the drainage is unintegrated. The bolson gradually is filled with sediment, and a thick alluvial plain is formed at the mountain front. Pediments become more and more extensive as the mountain slopes retreat until only small isolated hills, *inselbergs,* remain. Continued erosion would ultimately produce a wide rock plain, a *pediplain,* but such a stage has never been recognized over a large desert area.

Deposition

Whenever the wind loses its velocity, and hence its ability to transport the sand and dust particles it has picked up from the surface, it drops them back to the ground. The landscape features formed by wind-deposited materials are of various types, depending on the size of particles, the presence or absence of vegetation, the constancy of wind direction, and the amount of material available for movement by the wind. We still have a great deal to learn about this sort of deposit, but there are certain observations and generalizations that seem quite valid, which will be discussed below.

Loess

Loess is a buff-colored, unstratified deposit composed of small, angular mineral fragments. Loess deposits range in thickness from a few centimeters to 10 m or more in the central United States to over 100 m in parts of China. A large part of the surface deposits across some 0.75 million square kilometers of the Mississippi River basin is made up of loess, and this material has produced the modern fertile soils of several midwestern states, particularly Iowa, Illinois, and Missouri (Figure 11-32).

Most geologists believe loess to be dust originally deposited by the wind. They base their conclusion on several facts. First of all, the individual particles in a loess deposit are very small, strikingly like the particles of dust carried by the wind today. Moreover, loess deposits stretch over hill slopes, valleys, and plains alike, an indication that the material has settled from the air. And the shells of air-breathing snails present in loess strongly impugn the possibility that the deposits were laid down by water.

Many exposures in the north-central United States reveal that loess deposits there are intimately associated with glacial deposits built up during the Great Ice Age. Since the loess lies directly on top of the glacial deposits in many areas, it seems likely that the loess was deposited by the wind during periods when glaciation was at its height, rather than during interglacial intervals. Also, since there is no visible zone of weathering on underlying deposits, the loess probably was laid down on the newly formed glacial deposits before any soil could develop on them (Figure 11-33).

All loess, however, is not derived from glacial deposits. In one of the earliest studies of loess, it was shown that the Gobi Desert has provided the source material for the vast stretches of yellow loess that blanket much of northern China, and this gives the characteristic color to the Yellow River and the Yellow Sea. Similarly, much of the land used for cotton growing in the eastern Sudan of Africa is believed to be composed of particles blown from the Sahara Desert on the west. Finely divided mineral fragments are swept up in suspension during sandstorms and carried along by the wind far beyond the confines of the desert. Clearly, then, the large amounts of very fine material present in most deserts would make an excellent source of loess.

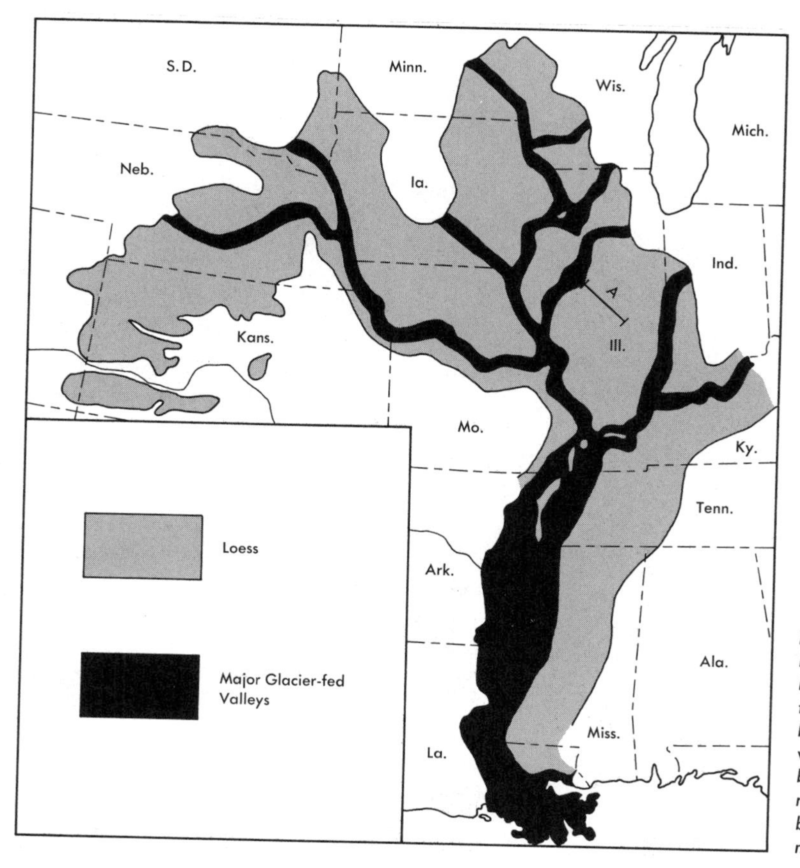

FIGURE 11-32 The great bulk of the loess in the central United States is intimately related to the major glacier-fed valleys of the area and was probably derived from the floodplains of these valleys. In Kansas and parts of Nebraska, however, the loess is probably nonglacial in origin and has presumably been derived from local sources and the more arid regions to the west.

Sand deposits

Unlike deposits of loess, which blanket whole areas, sand deposits assume certain characteristic and recognizable shapes. Wind often heaps the sand particles into mounds and ridges called *dunes,* which sometimes move slowly along in the direction of the wind. Some dunes are only a few meters in height, but others reach tremendous sizes. In southern Iran, dunes have grown to 200 m with a base 1 km wide.

As the velocity of the wind decreases, so does the energy available for the transportation of material; consequently, deposition takes place. We need to examine this relationship more closely and to determine why sand is deposited in the form of dunes rather than as a regular continuous blanket.

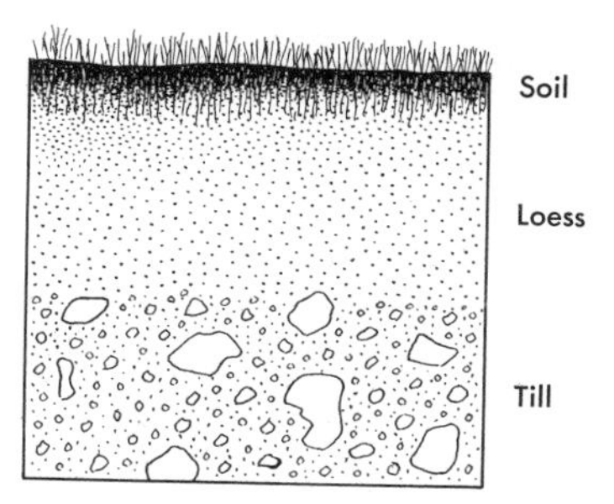

FIGURE 11-33 In many places, unweathered till is overlain by loess on which a soil zone has developed. The lack of a weathering zone on the till beneath loess often indicates rapid deposition of the loess immediately after the disappearance of the glacier ice and before weathering processes could affect the till. Not until loess deposition has slowed or halted is there time available to allow weathering and organic activity capable of producing a soil.

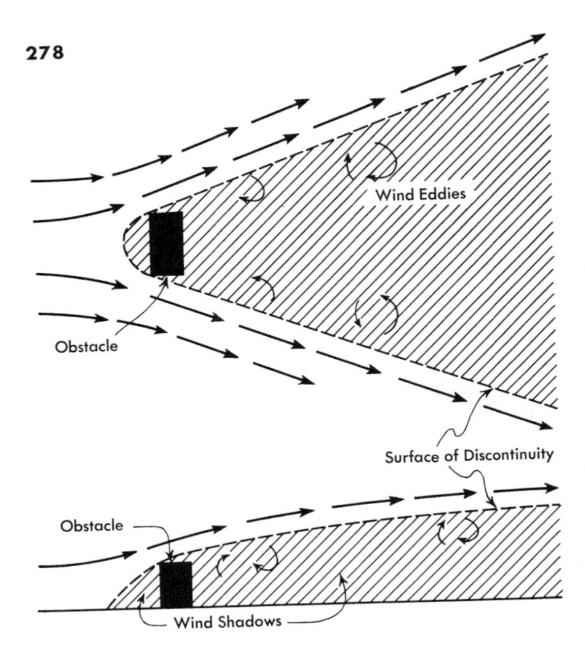

FIGURE 11-34 (Above) *The shaded area indicates the wind shadow created by an obstacle. The wind is diverted over and around the obstacle. Within the wind shadow, wind velocity is low and air movement is marked by eddies. A surface of discontinuity separates the air within the wind shadow from the air outside.* (Reproduced by permission from Bagnold, *The Physics of Blown Sand and Desert Dunes*, p. 190.)

The wind shadow. Any obstacle—large or small—across the path of the wind will divert moving air and create a "wind shadow" to the leeward, as well as a smaller shadow to the windward immediately in front of the obstacle. Within each wind shadow the air moves in eddies, with an average motion less than that of the wind sweeping by outside. The boundary between the two zones of air moving at different velocities is called the *surface of discontinuity* (Figure 11-34).

When sand particles driven along by the wind strike an obstacle, they settle in the wind shadow immediately in front of it. Because the wind velocity (hence energy) is low in this wind shadow, deposition takes place and gradually a small mound of sand accumulates. Other particles move past the obstacle and cross through the surface of discontinuity into the leeward wind shadow behind the barrier. Here again the velocities are low, deposition takes place, and a mound of sand (a dune) builds up—a process aided by eddying air that tends to sweep the sand in toward the center of the wind shadow (Figures 11-35 and 11-36).

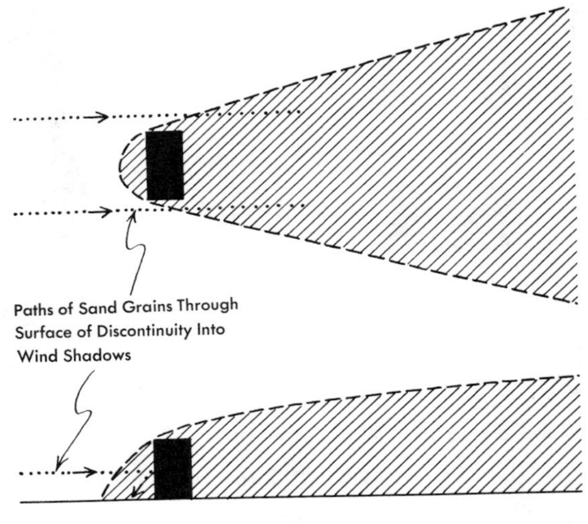

FIGURE 11-35 *Because of its momentum, the sand in the more rapidly moving air outside the wind shadow either passes through the surface of discontinuity to settle in the wind shadow behind the obstacle or strikes the obstacle and falls in the wind shadow in front of the obstacle.* (Reproduced by permission from Bagnold, *The Physics of Blown Sand and Desert Dunes*, p. 190.)

FIGURE 11-36 *Sand falling in the wind shadow tends to be gathered by wind eddies within the shadow to form a shadow dune, as shown in this sequence of diagrams.* (Reproduced by permission from Bagnold, *The Physics of Blown Sand and Desert Dunes*, p. 190.)

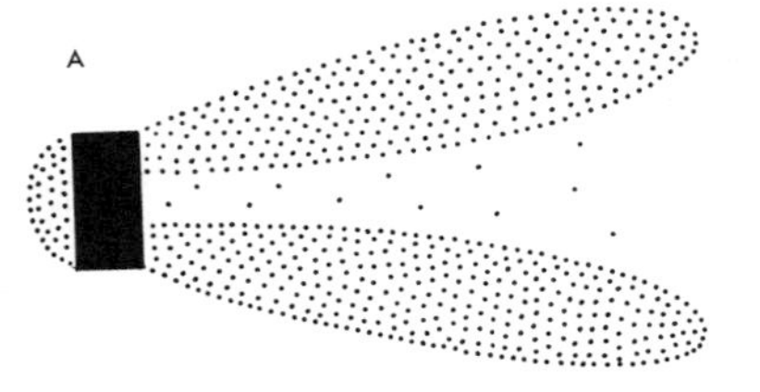

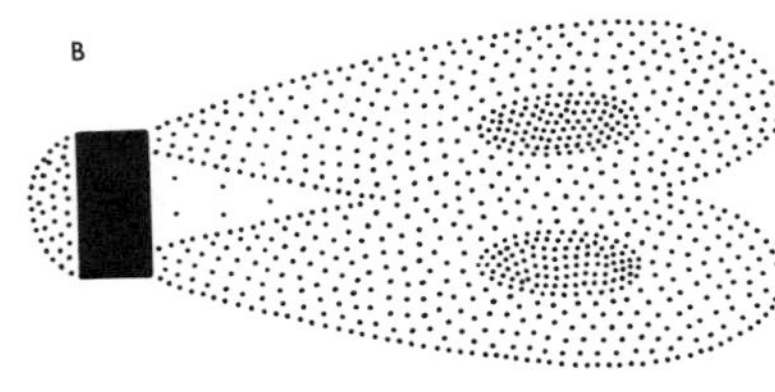

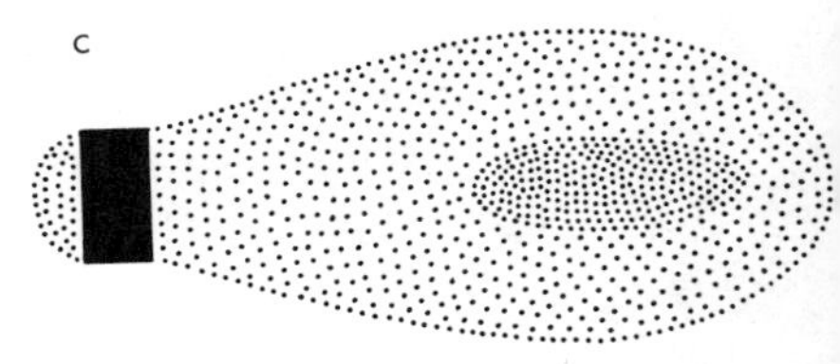

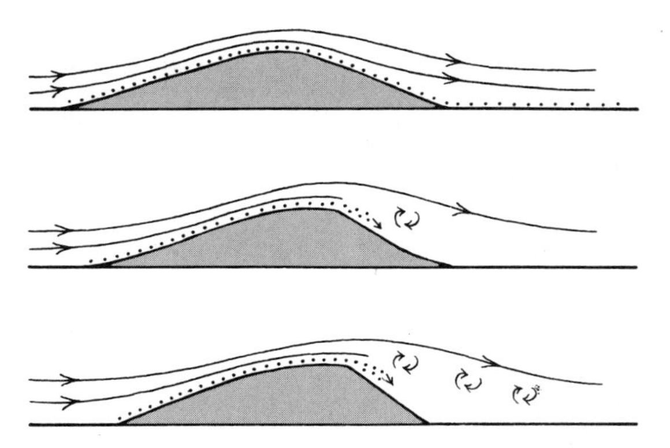

FIGURE 11-37 *The development of a slip face on a dune. Wind converges on the windward side of the dune and over its crest, and diverges to the lee of the dune. The eventual result is the creation of a wind shadow in the lee of the dune. In this wind shadow, sand falls until a critical angle of slope (about 34°) is reached. Then a small landslide occurs, and the slip face is formed. (Reproduced by permission from Bagnold, The Physics of Blown Sand and Desert Dunes, p. 202.)*

Wind shadow of a dune. Actually, a sand dune itself acts as a barrier to the wind, and by disrupting the flow of air, it may cause the continued deposition of sand. A profile through a dune in the direction toward which the wind blows shows a gentle slope facing the wind and a steep slope to the leeward. A wind shadow exists in front of the steep leeward slope, and it is here that deposition is active. The wind drives the sand grains up the gentle windward slope to the dune crest and then drops them into the wind shadow. The steep leeward slope is called the *slip face* of the dune because of the small sand slides that take place there.

The slip face is necessary for the existence of a true wind shadow. Here is how the slip face is formed. A mound of sand affects the flow of air across it, as shown in the topmost diagram of Figure 11-37. Notice that the wind flows over the mound in streamlined patterns. These lines of flow tend to converge toward the top of the mound and diverge to the leeward. In the zone of diverging air flow, velocities are less than in the zone of converging flow. Consequently, sand tends to be deposited on the leeward slope just over the top of the mound where the velocity begins to slacken. This slope steepens because of deposition, and eventually the sand slumps under the influence of gravity. The slump usually takes place at an angle of 34° from the horizontal. A slip face is thus produced, steep enough to create a wind shadow in its lee. Within this shadow, sand grains fall like snow through quiet air. Continued deposition and periodic slumping along the slip face account for the slow growth or movement of the dune in the direction toward which the wind blows.

Shoreline dunes. Not all dunes are found in deserts. Along the shores of oceans and of large lakes, ridges of windblown sand called *fore dunes* are built up even in humid climates. These dunes are well developed along the southern and eastern shores of Lake Michigan, along the Atlantic coast from Massachusetts southward, along the southern coast of California, and at various points along the coasts of Oregon and Washington.

These fore dunes are fashioned by the influence of strong onshore winds acting on the sand particles of the beach. On most coasts, the vegetation is dense enough to check the inland movement of the dunes, and so they are concentrated in a narrow belt that parallels the shoreline. These dunes usually have an irregular surface, sometimes pockmarked by blowouts (see previous section on "deflation").

Sometimes, however, in areas where vegetation is scanty, the sand dunes move inland in a series of long, wavy ridges at right angles to the wind. These *transverse dunes* exhibit the gentle windward slope and the steep leeward slope characteristic of other dunes. Transverse dunes are also common in arid and semiarid regions where sand is abundant and vegetation sparse (Figure 11-39).

FIGURE 11-38 *(Left) Leeward side of dune near d'Arcachon, France, showing advance of the dune on the forest, engulfing it at the rate of about 18 to 20 m a year. (Courtesy U.S. Department of Agriculture.)*

FIGURE 11-39 *(Below) Transverse gypsum dunes at White Sands National Monument, New Mexico. Gypsum is brought to the surface by ground water. The area on the right between dunes is as flat as the ground-water table. The wind that shaped the dunes blew from left to right. (Courtesy National Park Service.)*

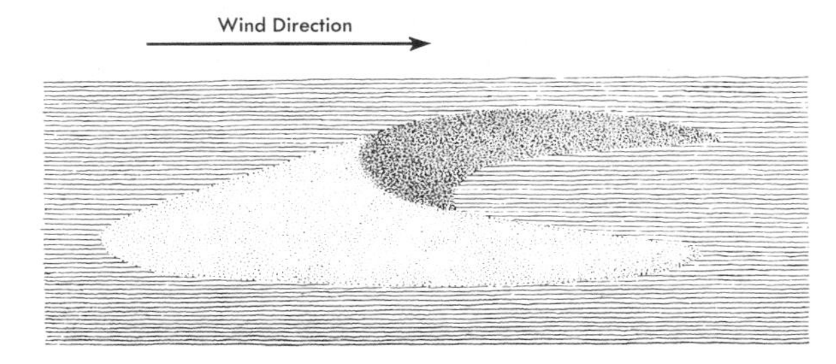

FIGURE 11-40 A barchan is a crescent-shaped dune with its horns pointed downwind and its slip face on the inside of the crescent.

Barchans. *Barchans* (pronounced *barkans*) are sand dunes shaped like a crescent, with their horns pointing downwind. They move slowly with the wind, the smaller ones at a rate of about 15 m/yr, the larger ones at about 7.5 m/yr. The maximum height obtained by a barchan is about 30 m, and maximum spread from horn to horn is about 300 m (Figure 11-40).

Just what leads to the formation of a barchan is still a matter of dispute. Certain conditions do seem essential, however: a wind that blows from a fixed direction, a relatively flat surface of hard ground, a limited supply of sand, and a lack of vegetation.

Parabolic dunes. Long, scoop-shaped, *parabolic dunes* look rather like barchans in reverse—that is, their horns point upwind rather than downwind. They are usually covered with sparse vegetation that permits limited movement of the sand. Parabolic dunes are quite common in coastal areas and in various places throughout the south-western states. Ancient parabolic dunes, no longer active, exist in the upper Mississippi Valley and in Central Europe.

FIGURE 11-41 Large-scale, crossbedded Nugget Sandstone in northeastern Utah. (Photograph by M. Dane Picard.)

Longitudinal dunes. *Longitudinal dunes* are long ridges of sand running in the general direction of wind movement. The smaller types are less than 3 m high, and about 60 m long. In the Libyan Desert, however, they commonly reach a height of 100 m and may extend for 100 km across the country. There they are known as *seif dunes* (rhymes with "strife"), from the Arabic word for "sword."

Ancient dune deposits

Careful study of modern sand dunes by geologists has made it possible to recognize ancient dune deposits. The internal stratification of recent dunes is very complex, and similar high-angle, thick, large-scale, concave-upward crossbedding is characteristic of ancient dune deposits. The carbonate-cemented Nugget Sandstone of Jurassic age (Figure 11-41) contains other diagnostic features indicating its ancient dune origin. Quartz grains are dominant, sorting and rounding of grains are good, mica minerals are rare, and ventifacts are found.

The small fauna that has been discovered in the Nugget also is indicative of dune (eolian) settings. The dinosaur remains and footprints, the rare plant remains, the tracks and trails, and the crustaceans suggest a specialized and transient fauna closely associated with small lakes between the sand dunes. No evidence of marine life has been discovered.

Some of the ripple marks in the cross-bedded sandstone of the Nugget Sandstone are types formed only by winds. Such ripple marks are very asymmetrical, and have long wave lengths and low amplitudes.

Even the surface texture of quartz grains can be used to interpret the environment of deposition. The surface texture of such grains has been modified by mechanisms and processes characteristic of a particular environment of deposition. In the Nugget Sandstone, for example, quartz grains frequently are frosted, pitted, and display parallel plates when viewed at high magnifications. The impact of quartz grains on each other during their transport and chemical etching in desert dew impart these features. Many ancient dune deposits laid down in arid desert settings have been recognized throughout the world on the basis of the criteria we have used here to interpret the formation of the Nugget Sandstone.

SUMMARY

Snow metamorphoses to ice, and if enough ice accumulates, it will flow under the influence of gravity. Glacier ice is present in valley glaciers, piedmont glaciers, and ice sheets. Glaciers and ice sheets advance when the accumulation of snow exceeds wastage, and shrink when wastage exceeds accumulation. A thin brittle zone overlies a plastic zone.

Erosion by ice takes place by plucking and abrasion. Such erosion produces striations, polish, and grooves on rock surfaces, and cuts cirques, horns, arêtes, cols, U-shaped valleys, fiords, and asymmetric rock knobs and hills. Glacial deposits include till in the form of moraines and drumlins, and outwash plains, eskers, crevasse fillings, kames, and kame terraces.

Gravity works directly on earth materials to carry them to lower and lower levels. The material may respond as an elastic solid, a plastic substance, or a fluid. Rapid movements include landslides, mudflows, earthflows, and talus falls. Slow movements are those of creep, solifluction, and rock glaciers.

Wind moves two different sizes of material—sand and dust. Sand grains are moved by saltation and rolling. Dust grains move in suspension. Erosion by wind produces ventifacts and blowouts. Deposition of dust produces loess, a deposit derived both from the deserts and from glacial outwash. Sand collects into dunes, including barchans, parabolic dunes, and longitudinal dunes.

SUGGESTED READINGS

BAGNOLD, R. A., *The Physics of Blown Sand and Desert Dunes*. New York: William Morrow, 1942.

CHARLESWORTH, J. K., *The Quaternary Era,* 2 vols. London: Edward Arnold (Publishers) Ltd., 1957.

DENNY, C. S., "Fans and Pediments," *American Journal of Science, 265* (1967), 81–105.

ECKEL, E. B., ed., *Landslides in Engineering Practice* (Highway Research Board, Special Report 29). Washington, D.C.: National Research Council, 1958.

FLINT, R. F., *Glacial and Quaternary Geology*. New York: John Wiley & Sons, Inc., 1971.

GAUTIER, E. F., *Sahara, The Great Desert,* trans. by D. F. Mayjew. New York: Columbia University Press, 1935.

GLENNIE, K. W., *Desert Sedimentary Environments*. Amsterdam: Elsevier Scientific Publishing Co., 1970.

HADLEY, R. F., "Pediments and Pediment-Forming Processes," *Journal of Geological Education, 15* (1967), 83–89.

JOPLING, A. V., and B. C. McDONALD, eds., *Glaciofluvial and Glaciolacustrine Sedimentation*. Tulsa, Okla.: Society of Economic Paleontologists and Mineralogists, 1975.

LEGGET, R. F., *Geology and Engineering*. New York: McGraw-Hill Book Company, 1962.

PATERSON, W. J. B., *The Physics of Glaciers*. London: Pergamon Press, 1969.

PICARD, M. D., "Facies, Petrography and Petroleum Potential of Nugget Sandstone (Jurassic), Southwestern Wyoming and Northeastern Utah," *Rocky Mountain Association of Geologists,* 1975 Symposium, pp. 109–27.

PRICE, R. J., *Glacial and Fluvioglacial Landforms*. New York: Hafner Publishing Co., Inc., 1973.

SHARP, R. P., "Kelso Dunes, Mojave Desert, California," *Geological Society American Bulletin, 77* (1966), 1045–73.

SHARPE, C. F. S., *Landslides and Related Phenomena*. New York: Columbia University Press, 1938.

WAHRHAFTIG, CLYDE, and ALLAN COX, "Rock Glaciers in the Alaska Range," *Geological Society America Bulletin, 70* (1959), 383–436.

YOUNG, A., *Slopes*. Edinburgh: Oliver and Boyd, 1972.

Ocean Processes

More than 70 percent of the surface of the earth lies deep in mystery beneath the oceans, largely unknown and until recently totally unexplored. Yet these regions are of the utmost importance to the geologist (Figure 12-1), for it was in the oceans and seas of the past that most sedimentary rocks formed—rocks that today cover three-quarters of the continental land masses. According to the modern view, the great ocean basins seem clearly linked to the origin of the continents.

We are only beginning to assemble the complex picture of the ocean floors—their topography, composition, and history—and to understand the nature of the chemical and physical processes that operate within them. We shall trace here, in briefest outline, some of the facts that have been gathered and some of the problems that have arisen concerning the oceans and ocean basins.

12

Otter cliffs, facing the ocean, at Acadia National Park, Maine. (Courtesy National Park Service.)

FIGURE 12-1 Scientists at work at sea. Large structure is A-frame; box corer in foreground. (Photo by Peter Roth.)

Ocean water

The distribution of sea water

The Northern Hemisphere is sometimes called the "land hemisphere" because north of the equator the oceans and seas cover only about 60 percent of the earth's surface, whereas in the Southern Hemisphere over 80 percent is flooded by marine waters (Figure 12-2). Between 45° N and 70° N, the ocean occupies only 38 percent of the surface; in contrast, 98 percent of the surface is covered by the ocean between 35° S and 65° S.

The greatest ocean depths so far recorded are from an area in the Pacific Ocean near the island of Guam. Here the depth is more than 11,000 m, which is considerably greater than the height of Mt. Everest, the world's highest mountain, which rises about

FIGURE 12-2 On the land hemisphere map, centered on western Europe, land and sea are about evenly divided. But an indisputable predominance of the seas is revealed on the water hemisphere map, centered on New Zealand.

Land Hemisphere

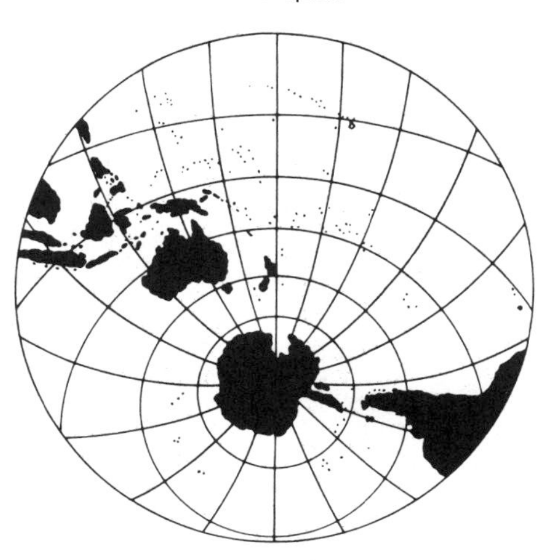

Water Hemisphere

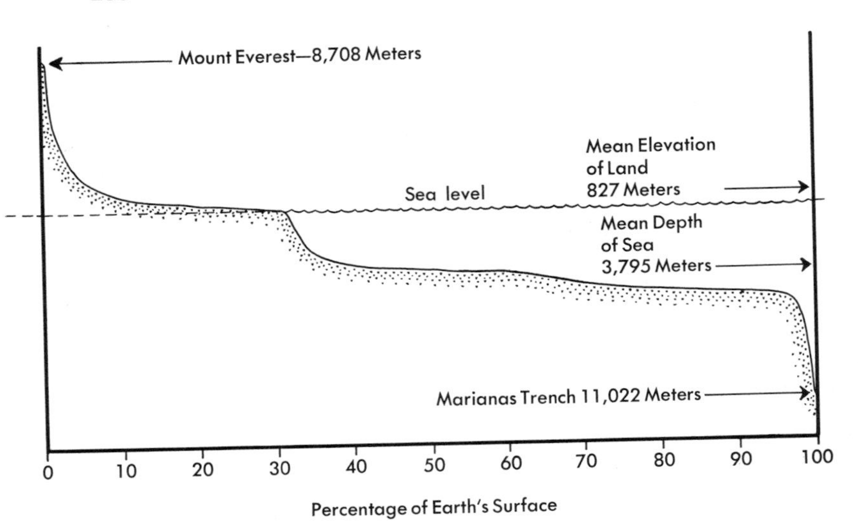

FIGURE 12-3 The relative distribution of land and sea. Note that the mean ocean depth is 3,795 m, whereas the mean elevation of the land is only 827 m.

8,840 m above sea level. The average ocean depth is about 3,800 m (Table 12-1); the mean elevation of the continents is only 827 m (Figure 12-3).

It has been estimated that the globe would be covered with a layer of water about 1½ to 3 km thick if all the irregularities of the surface were eliminated. Such a situation probably never existed in the past, nor need we worry about its happening in the future. Modern oceans are confined to great basins, and presumably, so were the oceans of the past. We will discuss the characteristics of these basins in some detail later in the chapter.

The nature of sea water

About 70 elements have been identified in sea water, and others certainly are present. Included among the materials known to be dissolved in the sea are the chlorides that give sea water its familiar saltiness, all the gases found in the atmosphere, and a large number of less abundant materials, including such rare elements as uranium, gold, and silver. Figures for the major constituents are given in Table 12-2.

Table 12-1 Information on the Oceans

Volume	1,370 $\times$ 10^6 cubic kilometers
Salinity	34.482 parts per thousand
Dissolved material	5 $\times$ 10^{22} grams
Average sea water density	1.025 grams per cubic centimeter
Area	361 ($\times$ 10^6 square kilometers)
Pacific	180
Atlantic	107
Indian	74
Average depth	3,795 (in meters)
Pacific	4,028
Atlantic	3,332
Indian	3,897

Table 12-2 The Major Constituents Dissolved in Sea Water

ION	CL = 19%	PERCENTAGE OF ALL DISSOLVED MATERIAL
Chlorine, Cl^-	18.980	55.5
Sodium, Na^+	10.556	30.61
Sulfate, SO_4^{2-}	2.649	7.68
Magnesium, Mg^{2+}	1.272	3.69
Calcium, Ca^{2+}	0.400	1.16
Potassium, K^+	0.380	1.10
Bicarbonate, HCO_{3^-}	0.140	.41
Bromine, Br^-	0.065	0.19
Total		99.89

Salts dissolved in sea water

Through millions of years the rivers of the world have slowly transported tremendous quantities of dissolved material to the oceans. Some of this material, such as iron, silicon, and calcium, is used by plants and animals in their life processes and is thus constantly being removed from the sea water. As a result, the amount of these elements present in solution is less than expected, judging from the rate at which rivers are currently supplying them to the oceans. In contrast, we find a relatively high percentage of the "salt" ions, notably Cl^-, even though rivers are presently supplying these materials at a relatively low rate. Salt ions have continued to collect in sea water because plants and animals do not concentrate them and because they are extremely soluble.

Since the proportions of the various salt ions are constant throughout the oceans, in a given place a measurement of any one of them enables us to compute the abundance of the others. The total concentration of salt ions—that is, the salinity of the sea water—varies from place to place, however. At the equator, heavy precipitation dilutes the sea water, reducing its salinity. In the Arctic and Antarctic areas, the melting of glacier ice also serves to reduce the saltiness of the seas. But in the subtropical belts on the north and south, low rainfall and high evaporation tend to increase the salinity, as is indicated in Figure 12-4. In the open ocean the salinity averages about 3.5 percent.

Gases dissolved in sea water

Although all of the gases found in the atmosphere are also present in water, probably the most important are oxygen and carbon dioxide. Near the surface of the oceans the water is saturated with both gases, but their concentration and relative proportions vary with depth. As the surface water circulates downward through the first few tens of meters, intense plant activity depletes the supply of carbon dioxide. At the same time, oxygen is given off by the plants. This near-surface zone is deficient in carbon dioxide and tends to be oversaturated with oxygen.

Below the depth to which light can penetrate effectively, however, plant activity falls off and the amount of oxygen in solution decreases. Of the oxygen present, some is used by animals and some becomes involved in the oxidation of organic matter settling toward the bottom. At the same time, the relative amount of carbon dioxide increases because there is no plant activity to deplete it. Thus, with increasing depth, oxygen becomes relatively less abundant and carbon dioxide relatively more abundant.

Were it not for the slow circulation of sea water through the ocean basins, water at the greatest depths would be devoid of oxygen. Actually, at the bottom of some ocean basins, the circulation of water is so slow that almost no oxygen is present, and the water is stagnant. Here, there are high concentrations of hydrogen sulfide. This is true, for example, in

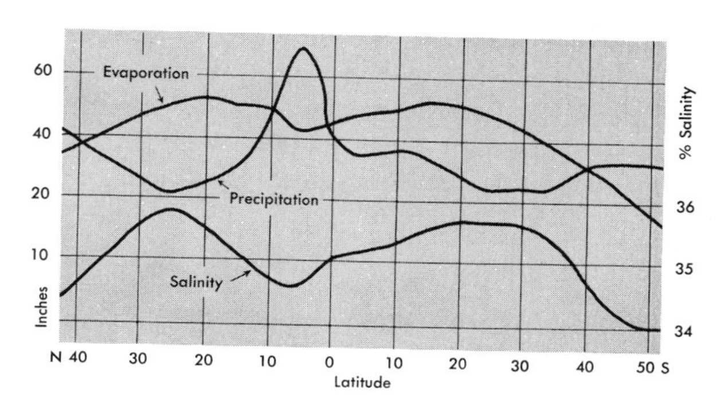

FIGURE 12-4 *Total salinity varies in the oceans from south to north across the equator. The low salinity of surface waters in the vicinity of the equator is attributed to the freshening effect of heavy tropical rains. North and south of this zone the rainfall decreases and evaporation increases; as a result, the total salinity of the surface waters increases. [Redrawn from R. H. Fleming and Roger Revelle, "Physical Processes in the Oceans," in Trask, Recent Marine Sediments—A Symposium (Tulsa, Okla.): American Association of Petroleum Geologists. (1939), p. 88.]*

the Black Sea below a depth of about 150 m, and in many of the Norwegian fiords whose glacially deepened basins lie below the general level of the adjacent floor of the North Atlantic. Because of the almost complete absence of oxygen at the bottom of these basins, sediments lying there oxidize very slowly, if at all. Consequently, a high content of organic matter and the hydrogen sulfide compounds that are produced give these deposits their characteristic black color. Petroleum probably was formed in environments such as this in ancient seas.

Origin of oceans

In the late 1940s a brilliant field and theoretical geologist, William W. Rubey, began to wonder if the saltiness of the oceans had been substantially different in the past. Such speculation, as is frequently the case in geology, led to other questions: Where did the water and salt come from and why is the ocean salty?

Rubey, in 1951, compared the average chemical composition of igneous, metamorphic and sedimentary rocks, and found, as had others, that sedimentary rocks contain much more water and carbon dioxide than the other two groups of rocks.[1] He then computed the amounts of water, carbon dioxide, chlorine, and several other substances present in sedimentary rocks, in the atmosphere, and in ocean waters. A comparison of these totals with estimates of the amounts that could have originated through the weathering of igneous rocks revealed large differences. There is far too much water, carbon dioxide, and chloride in the atmosphere, oceans, and sedimentary rocks for these substances to have accumulated solely from the weathering of igneous rocks.

The process called upon by Rubey to furnish the excesses is termed *degassing*, which refers to the release from the earth's interior of gasses by volcanoes, hot springs, and geysers. The ratios of water to carbon dioxide and other elements in gases from volcanoes are roughly in proportion to the same ratios for the crust of the earth. Therefore degassing apparently is the process mainly responsible for the formation of the oceans and the atmosphere.

Rubey also found that marine life has not been adversely affected by changes in saltiness of the oceans. After the evolution of life, the volcanic release of gases must have been in nearly uniform proportions. Rubey concluded that neither the oceans nor the atmosphere has varied much in composition during geologic time. Apparently the volcanic gases have accumulated slowly, steadily, and in about the same relative compositions as they are at present.

The movement of sea water

The movement of sea water is of two general types: (1) the movement of the average level of the surface either up or down relative to the land—called *changes of sea level*, and (2) the transfer of water from one place to another in ocean basins by movements generally termed *currents*.

Changes of sea level

The level of the sea in relation to dry land is constantly changing. The daily changes caused by tides are familiar; but much slower fluctuations are no less real or important; in fact, in the geologic past such changes have been extensive and significant.

[1] W. W. Rubey, 1951, Geologic history of sea water. An attempt to state the problem: Geol. Soc. America Bull., v. 62, p. 1111–1148.

Eustatic and tectonic changes of sea level

It is easy to measure the daily changes in sea level caused by the rise and fall of tides along the coastline (Figure 12-5). But there are other movements so slow that they are revealed only by long-continued records of mean sea level, or so local and rare that they are not generally recognized.

A change in sea level relative to land can be caused by the upward or downward movement of either the ocean or the land, or by their combined movement (Figures 12-6 and 12-7). If the movement is confined to the ocean, change of level is *eustatic,* a term that refers to the static condition or continuing stability of the landmass. When the relative sea level changes because of land movement, the change is *tectonic* (from the Greek for "builder" or "architect"), in reference to the movements that shape the earth's surface (see Chapter 9).

It is frequently difficult to distinguish between eustatic and tectonic movements of sea level by observing only a small section of the shore. Eustatic changes of level, however, are worldwide; and when recording stations

FIGURE 12-5 *(Above) Twice a day the tides rise and fall along this section of the Connecticut coast at Buffalo Bay near Madison. (Photos by Sheldon Judson.)*

FIGURE 12-6 *(Right) In the recent geologic past, ocean waves beveled this platform across tilted rocks on the California coast south of San Francisco. Today the platform stands slightly above sea level as evidence of a change in level between land and sea. In this instance, geologists believe that the motion has been tectonic and that the land has moved upward relative to the sea. (Photo by William C. Bradley.)*

FIGURE 12-7 *Torghatten Island off the coast of Norway testifies to a 120-m drop in sea level, the distance from the dotted line to the surface of the present-day ocean. Note the benches that were cut at the time of higher sea level. A tunnel was cut by wave action when the sea stood at the higher level. Here the earth's crust has moved upward after being warped down by the weight of the now vanished Scandinavian ice sheet. (Photo by C. A. Ericksen.)*

over an extensive area report a long-continued movement of the sea, the movement can safely be termed eustatic. In contrast, tectonic movements tend to be local and spasmodic, controlled as they are by the forces that deform solid earth materials.

Eustatic movements may be caused in various ways. If the amount of water locked up in glaciers and lakes increases, the sea level falls; then, if the glaciers melt or if the lakes are drained, the sea level rises. Or sea level may rise or fall as a result of changes in the size of ocean basins, either because of continuing deposition of sediments on their floors or because of actual deformation by earth forces. Still another cause of eustatic change lies in the addition or removal of water from the earth's surface. Volcanoes are constantly adding to our atmosphere new water that eventually finds its way to the sea. And water is constantly being trapped in sedimentary deposits and incorporated in such minerals as clay, causing at least a temporary loss to the oceans.

Tectonic and eustatic changes are continuously taking place around the globe; mean sea level is endlessly rising and falling. Although recent changes in sea level are slight, we have ample evidence that such changes are in fact actually taking place at present.

Recent changes in sea level

On the basis of tidal measurements made at various points around the world, observers have concluded that an eustatic rise in sea level is now taking place. Beginning around 1850, sea level began to rise at the rate of about 11.4 cm per century. The invasion of the sea is attributed to the melting of mountain glaciers and to the depletion of the Greenland and Antarctic icecaps. There is still enough water stored in modern glaciers to raise the sea at least an additional 60 m.

Modern changes in sea level caused by tectonic movements have been observed in localities where the crust of the earth is known to be undergoing deformation at the present day.

Currents

Sea water is in constant movement, in some places horizontal, in others downward, and in still others upward. The rate of movement varies from spot to spot, but it has been estimated that there is a complete mixing of all the water of the oceans about once every 1,800 years. If we assume that movements similar to those of the present have been going on throughout the long history of the earth, by studying modern seas we can gain an insight into the history recorded in sedimentary rocks that were once mud and sand on the floor of ancient seas.

Although we still cannot explain completely the movements of the modern oceans, we do know that they are caused chiefly by tides, by the changing density of sea water, by wind, and by the rotation of the earth.

Tidal currents

The attractive forces that operate between the sun, the moon, and the earth set the waters of the ocean in horizontal motion to produce tidal currents. The speed of these currents may reach several kilometers per hour if local conditions are favorable. Velocities in excess of 20 km/hr develop during the spring tides in Seymour Narrows, between Vancouver Island and British Columbia, and tidal currents of half this velocity are not uncommon. The swiftest currents usually build up where a body of sea water has access to the open ocean only through a narrow and restricted passage. Such currents are capable of moving particles up to, and including, those of sand size, and these currents may be strong enough to scour the sea floor.

Density currents

The density of sea water varies from place to place with changes in temperature, salinity, and the amount of material held in suspension. Cold, heavy water sinks below warmer and lighter water; water of high salinity is heavier than water of low salinity and sinks beneath it; and heavy muddy water sinks beneath light, clear water.

In the Straits of Gibraltar, the water passage between the Atlantic Ocean and the Mediterranean Sea, differences in density are partially responsible for a pair of currents flowing one above the other. The Mediterranean, lying in a warm, dry climatic belt, loses about 1.5 m of water every year through evaporation. Consequently the saltier, heavier water of the Mediterranean moves outward along the bottom of the Straits and sinks downward into the less salty, lighter water of the Atlantic. At the same time, the lighter surface water of the Atlantic moves into the Mediterranean basin. The water flowing from the Mediterranean settles to a depth of about 1,000 m in the Atlantic and then spreads slowly outward beyond the equator on the south, the Azores on the west, and Ireland on the north. It has been estimated that as a result of this activity, the water of the Mediterranean basin is changed once every 75 years.

The density of water is also affected by variations in temperature. As a result of such variations, water from the cold Arctic and Antarctic regions sinks and creeps slowly toward the warmer environment near the equator. The cold, relatively dense water from the Arctic sinks near Greenland in the North Atlantic and can be traced to the equator and beyond as far as 60° S. Denser and colder water moves downward to the sea floor off Antarctica and creeps northward, pushing beneath the North Atlantic water. In fact, the Antarctic water reaches well north of the equator before it loses its identity.

A third type of density current, known as a *turbidity current,* results because turbid or muddy water (Figure 12-8) heavy with sediment has a greater density than clear water and therefore settles beneath it. The turbid water of the Ganges-Brahmaputra, for example, flows under and displaces the less dense sea water of the Bay of Bengal. Turbidity currents can attain high velocities and can travel for thousands of kilometers if the density contrast and the bottom slope are great enough.

On November 18, 1929, an earthquake with an epicenter at the edge of the Grand Banks off Nova Scotia triggered an enormous slump (160 by 320 km) of the continental

FIGURE 12-8 This photograph shows a muddy Alaskan stream entering the clearer waters of the Gulf of Alaska. Although turbid, the stream water does not sink because it is still lighter than the salt water of the ocean.

FIGURE 12-9 Large flute ("basoon") casts on the base of a graded bed. Flow was from right to left. (Photo by Earle F. McBride.)

slope that snapped 16 transatlantic cables almost immediately and 5 others in succession downslope. The turbidity current that formed, as the slump collapsed and mixed with the water, raced downslope over a distance of nearly 500 km in 13½ hours, attaining a velocity of about 100 km/hr near the base of the continental slope but slowing to less than 25 km/hr when it cut the last cable. Graded beds were deposited over a large area of the North Atlantic floor.

It is estimated that normally turbidity currents may flow at the rate of 30 to 50 km/hr and may extend as thin sheets for a thousand kilometers. .Such currents are the principal mechanism for spreading sand and silt from the continents to the abyssal plains.

As early as 1936, R. A. Daly suggested that turbidity currents might be responsible for cutting submarine canyons (discussed on p. 295) and for producing the associated submarine channels. Many subsequent studies, especially since 1950, have indicated the value of the turbidity current hypothesis in explaining many features associated with graded bedding, as well as such events as the broken cables associated with the Grand Banks earthquake. However, direct observations of the formation of modern turbidity currents have not yet been made.

Major surface currents

The major movements of water near the ocean's surface take place in such currents as the Gulf Stream, the Japanese Current, and the Equatorial Currents. These great currents are caused by a variety of factors, including the prevailing winds, the rotation of the earth, variations in the density of sea water, and the shape of ocean basins. Let us examine, by way of illustration, the surface currents of the Atlantic Ocean in both the Northern and Southern Hemispheres.

The *equatorial currents* lie on each side of the equator, and they move almost due west. They derive their energy largely from the trade winds that blow constantly toward the equator, from the northeast in the Northern Hemisphere and from the southeast in the Southern Hemisphere. The westerly direction of the currents is explained by the *Coriolis effect.* This effect is produced by the rotation of the earth, which causes moving objects to veer to the right in the Northern Hemisphere and to the left in the Southern Hemisphere. As the water driven by the trade winds moves toward the equator, it is deflected west in both hemispheres.

As a result, the North and South Equatorial Currents are formed. As these currents approach South America, one is deflected north and the other mainly south. This deflection is caused largely by the shape of the ocean basins, but it is aided by the Coriolis effect and by the slightly higher level of the oceans along the equator where rainfall is heavier than elsewhere.

The North Equatorial Current moves into the Caribbean waters and then northeastward, first as the Florida Current and then as the Gulf Stream. The Gulf Stream, in turn, is deflected to the east (to the right) by the Coriolis effect. This easterly movement is strengthened by prevailing westerly winds between 35° N and 45° N, where the Gulf Stream becomes the North Atlantic Current.

FIGURE 12-10 *Turbidite and pelagite deposits from the Apennine Mountains, Italy. (Photo by Earle F. McBride.)*

As it approaches Europe, the North Atlantic Current splits. Part of it moves northward as a warm current past the British Isles and parallel to the Norwegian coast. The other part is deflected southward as the cool Canaries Current, and eventually it is caught up against the northeast trade winds, which drive it into the North Equatorial Current.

In the South Atlantic, the picture is very much the same—a kind of mirror image of the currents in the North Atlantic. After the South Equatorial Current is deflected mainly southward, it travels parallel with the eastern coast of South America as the Brazil Current. Then it is bent back to the east (toward the left) by the Coriolis effect and is driven by prevailing westerly winds toward Africa. This easterly moving current veers more and more to the left until finally, off Africa, where it is known as the Benguela Current, it is moving northward. This stream in turn is caught up by the trade winds and is turned back into the South Equatorial Current.

The cold surface water from the Antarctic regions moves along a fairly simple course, uncomplicated by large landmasses. It is driven in an easterly direction by the prevailing winds from the west. In the Northern Hemisphere, however, the picture is complicated by continental masses. Arctic water emerges from the polar seas through the straits on either side of Greenland, to form the Labrador Current on the west and the Greenland Current on the east. Both currents subsequently join the North Atlantic Current and are deflected easterly and northeasterly.

We need not examine in detail the surface currents of the Pacific and Indian oceans. We can note, however, that the surface currents of the Pacific follow the same general patterns as those of the Atlantic. Furthermore, the surface currents of the Indian Ocean differ only in detail from those of the South Atlantic.

The ocean basins

Most geologists now agree that the continents of the world are composed largely of sialic rock overlying a layer of heavier, crystalline simatic rock. The sialic layer is missing from the deep ocean basins, however, and the ocean floor is composed of simatic rock with a covering of sediments that ranges from zero to about 3 km, probably averaging about 0.5 km.

Most of the sea water surrounding the continents is held in one great basin that girdles the Southern Hemisphere and branches northward under the Atlantic, Pacific, and Indian oceans. The Atlantic and Pacific oceans in turn are connected with the Arctic Ocean through narrow straits. But the sea water still floods over the margins of the continents, for even this great fingered basin cannot contain all the water of the earth.

Topography of the sea floor: continental shelves, slopes, and ocean deeps

The margins of the continents that lie flooded beneath the seas are termed *continental shelves.* The average width of these shelves is 65 km, but there are many local variations. Shelves generally are narrower on the leading edge and wider on the trailing edge of a continental plate, depending on whether the plate is moving toward or away from another plate.

Average water depth on the continental shelves is small, about 130 m. During periods of glaciation the sea level drops as ice accumulates, and substantial portions of the shelves are exposed to erosion. During interglacial periods the sea level rises as glaciers melt, and water is returned to the seas. The range in sea level variation was about 200 m during the last major glacial-interglacial cycle, varying from 140 m below to 60 m above the present position of sea level.

The surface of almost all the shelves is irregularly marked by hills, valleys, and depressions of low to moderate relief. Furthermore, soundings along the shelf bordering the eastern coast of North America show the presence of submarine terraces that record former lower levels of the ocean, just as higher levels are recorded by terraces stranded above sea level. A thin veneer of sediment derived from the continents has accumulated on these shelves.

The continental shelves generally are inclined gently toward the ocean basins until they are abruptly terminated by the steeper *continental slopes,* which descend into the largely featureless, flat-lying surface of the *abyssal plains* under several thousands of meters of water. Continental slopes are steepest in their upper portion, and commonly extend more than 3,600 m downward. In certain places, where earth movements have created deep trenches in the ocean floor, the continental slopes reach to much greater depths. Off the island of Mindanao, in the Philippines, the slope drops down 10,000 m. Scarring the face of the continental slopes at various places around the world are deep submarine canyons, but the floor of the great world-encircling basin that contains the oceans is even more irregular than the surface of the continental shelves and slopes. The more soundings that are made, the more complex and spectacular the topography appears. The ocean floor is divided into innumerable smaller basins and is marked by plains, plateaus, valleys, towering peaks, and mountain ranges.

Submarine canyons

The surfaces of the continental shelves, slopes, and ocean deeps are furrowed by *submarine canyons* of varying width, depth, and length, rather like the valleys of the continents. The majority of these canyons begin on the continental shelf far from shore, have steep gradients, concave long profiles, and few tributaries. Most of them are cut into the continental slope.

One of the best known submarine canyons is the submerged extension of the Hudson Valley off the eastern coast of the United States. This valley extension is relatively straight, cutting down about 60 m into the continental shelf, and widening from almost 5 km to approximately 24 km at its seaward end.

Probably the most studied canyon is the LaJolla Submarine Canyon and Fan Valley (off the coast of San Diego, California), with its major tributary, Scripps Canyon. A gently sloping, sand-covered terrace extends from the beach seaward. There, about 215 m from shore at a depth of 12 m, the bottom suddenly drops away in the steep 24-m headwall of LaJolla Canyon. From the head of the canyon seaward the valley widens on each side and then narrows again into a rock-walled gorge whose sides are covered with marine plants and animals. The canyon floor is covered with sand, seawood, and other organic material.

The LaJolla Fan Valley is cut into its own fanlike deposits. Farther seaward its depth decreases until it merges with the sea floor. Such fans, which are deposits of sand, silt, and other detritus that moved down the associated canyons, are common elsewhere. If several adjacent canyons or valleys deposit material onto the sea floor, a submarine alluvial plain is formed, similar to coalesced alluvial deposits on land.

In addition to these larger stream canyons, there are smaller troughs in the continental shelves that are presumed to have been cut by tidal scouring. Several troughs of this sort are present off the northeastern coast of the United States. Other canyons were apparently cut on the continental shelves by glacier ice that may simply have deepened already existing valleys.

The origin of submarine canyons has been hotly debated for decades. Some canyons, such as those on the slopes off the Hudson and Congo rivers, apparently are extensions of valleys on the land or on the continental shelves. However, most canyons have no association with such valleys.

Most canyons probably were formed by submarine erosion by turbidity currents and by more regular currents carrying sand down the canyon. Scuba divers have observed spectacular sand streams at Scripps Canyon and at other sites. At irregular intervals Scripps Canyon clears itself of detritus by slides.

Another suggestion is that the canyons originated through subaerial erosion when sea level was lowered 100 m or so during the Pleistocene glaciation. The upper parts of some canyons possibly could have been cut by streams on land, but such an explanation is not reasonable for the Hudson and Congo canyons. Sea level could not have been lowered to the base of the continental slope without noticeable disastrous effects on the abundant marine life of the continental slope.

Seamounts

Dotting the deep sea floor, between fracture lines and along them, are drowned, isolated, steep-sloped volcanic peaks called *seamounts* (Figure 12-11). They stand at least 1,000 m above the surrounding ocean floor, their crests covered by depths of water measured in hundreds, even thousands, of meters. The base of a large seamount may cover more than 2,000 km^2, with its summit rising 3 or 4 km above the abyssal floor. Seamounts have been identified in all the oceans, but the greatest number by far are reported from the Pacific Ocean. There, by the late 1960s, some 1,700 such peaks had been mapped; it is estimated that ten times this number remain to be discovered.

Hawaii and Tahiti are exposed peaks of huge submarine volcanoes. In long, essentially uninterrupted slopes, the Hawaiian volcanoes extend from $4\frac{1}{2}$ km above the sea level to $5\frac{1}{2}$ km below sea level, a combined relief greater than that of Mt. Everest (9 km).

FIGURE 12-11 *Seismic profile of Horizon Guyot (on the left) and several seamounts from the Central Pacific. (Courtesy Arthur D. Raff, Deep Sea Drilling Project.)*

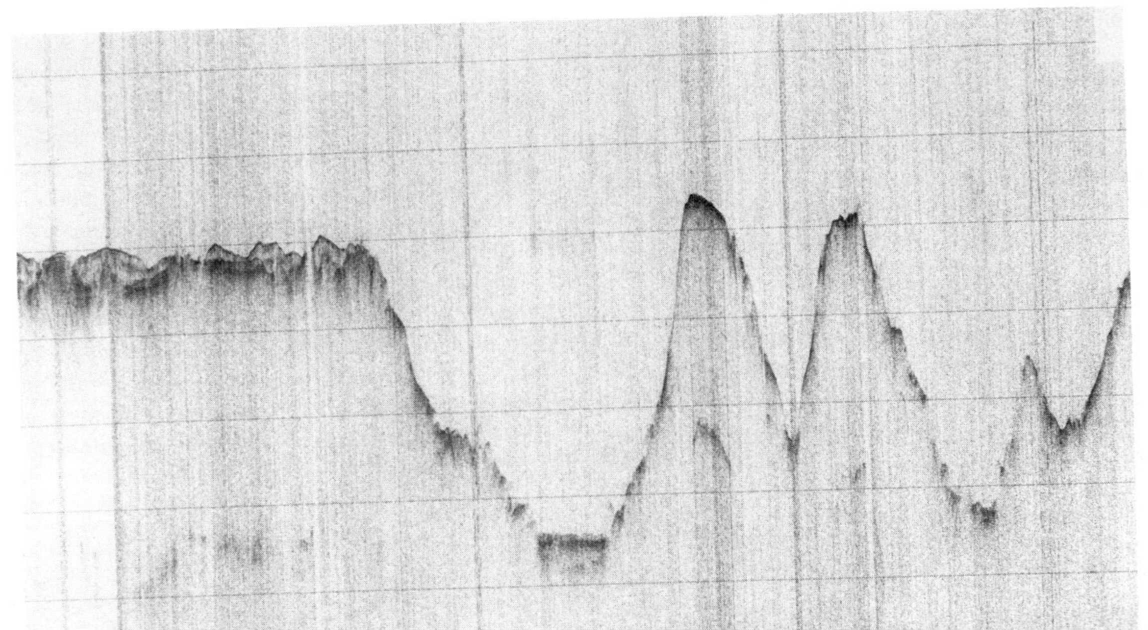

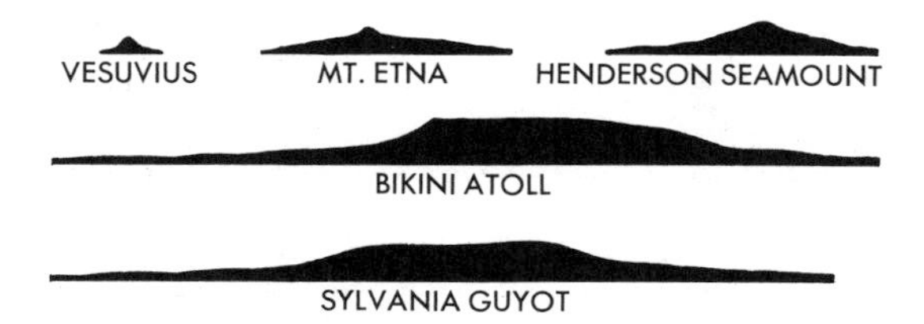

FIGURE 12-12 *Comparison of sizes of guyot, atoll, seamount, and two land volcanoes, Vesuvius and Mt. Etna. All drawn to same scale.*

Most seamounts have sharp peaks, and all are apparently of volcanic origin. Some of the peaks have flat tops and are called *table-mounts,* or *guyots* (pronounced *gee-yoz*—hard "g" as in geese). Guyots (see Figure 12-12) are volcanic cones whose tops have been cut off by the action of surface waves.

During its life history, a seamount has a short phase of constructive eruption, generally of less than a million years. It then experiences a long period of subsidence, which may in a few million years drag its peak down into deeper water.

Many seamounts and guyots originate in narrow zones at crustal plate edges. Others were formed, however, far from plate edges and were constructed on much older parts of the lithosphere. Volcanic chains are common in older parts of the western Pacific. A speculative idea of the moment is that *hot spots* deep within the mantle generate magma and spasmodically inject it into the lithosphere. As a migrating plate drifts over the hot spot, a line of volcanoes will presumably form as the plate moves along. Much clearly remains to be learned about the history and formation of seamounts.

Submarine ridges, rises, and fractures

Among the major features of the deep ocean basins are long, submerged ridges and rises. In general, ridges have steep sides and irregular topography. The rises differ in being broader and gentler in form (Figure 12-13). They rise thousands of meters above the deep ocean floor and in some places actually appear above the surface to form islands. Many continental slopes, such as that off eastern North America, grade into *continental rises,* which in turn grade into the deep ocean floors. In other places, such as off the southern side of the Aleutian Islands or off western South America, the continental slopes lead directly into the deep trenches in the ocean floor. Ridges and rises may form a more or less integrated system of high topography that segments the deep oceans into smaller basins.

In addition to ridges and rises, long towering escarpments caused by earth movements scar some sections of the ocean floor. Thus, the

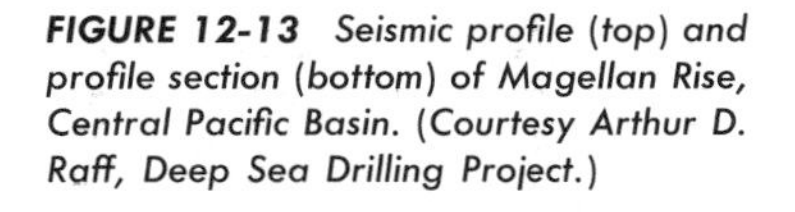

FIGURE 12-13 *Seismic profile (top) and profile section (bottom) of Magellan Rise, Central Pacific Basin. (Courtesy Arthur D. Raff, Deep Sea Drilling Project.)*

Table 12-3 Dimensions of Large Trenches of the Pacific

TRENCHES	DEPTH (meters)	LENGTH (km)	MEAN WIDTH (km)
Mariana	11,022	2,550	70
Tonga	10,800	1,400	55
Kermadec	10,047	1,500	60
Mindanao (Philippines)	10,030	1,400	60
Japan (Idzu-Bonin)	9,810	800	90
Kurile-Kamchatka	10,542	2,200	120
New Hebrides	9,165	1,200	70

SOURCE: Data from Rhodes W. Fairbridge, 1966, Trenches and related deep sea troughs: *in The Encyclopedia of Oceanography,* Van Nostrand Reinhold Co., p. 929–939.

Mendocino, Murray, Clarion, and Clipperton fracture zones stretch westward into the Pacific from the coasts of the United States and Central America (Figure 12-14). The vast Mendocino Escarpment reaches heights of 2,400 m and extends more than 3,000 km into the central Pacific.

Less well-known but similar fractures and related escarpments lie to the south of this system. And fracture systems of similar magnitude are present along many of the oceanic ridges.

H. W. Menard believes that fracture zones are linear mountain ranges that offset and are roughly perpendicular to spreading centers on the ocean floor.[2] These fracture zones include long ridges and troughs, cliffs separating regions with different depths, and lines of volcanoes. The characteristic ridges and troughs of fracture zones are 1 to 3 km high, 10 to 20 km wide, and a few hundred kilometers long. The cliffs and regional differences in depth across the fracture zones reflect the elevation of the crust at spreading centers and its subsidence away from the center. The fracture zones are thus interpreted to be transform faults that sometimes spread slowly and leak a little lava.

[2]H. W. Menard, *Geology, Resources, and Society.* San Francisco: W. H. Freeman and Co., 1974.

Deep-sea trenches

The greatest ocean depths are termed *trenches;* the Tonga and Mariana trenches reach about 11,000 m below sea level. In general pattern, trenches are great, arcuate or bow-shaped troughs on the sea floor, some of them 200 km in breadth and 24,000 km or more in length (Table 12-3). In cross section, trenches are steep-walled and V-shaped, and the slope on the ocean side is less than that on the landward side. Some trenches contain considerable sediment; others do not. Trenches are distributed along the margins of many ocean basins, particularly in the Pacific Ocean.

FIGURE 12-14 *Major fracture zones in the eastern Pacific.*

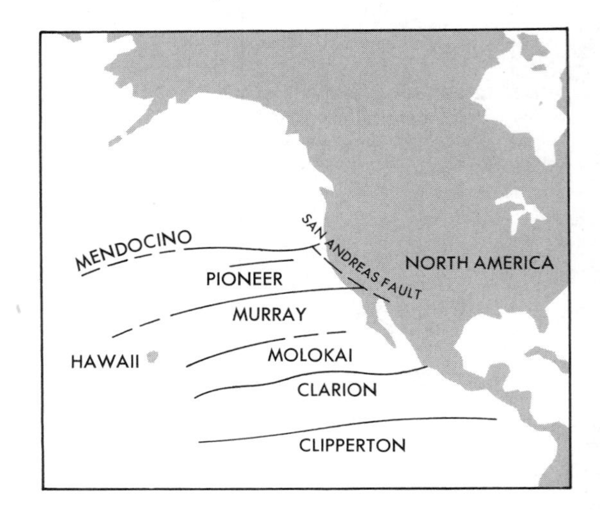

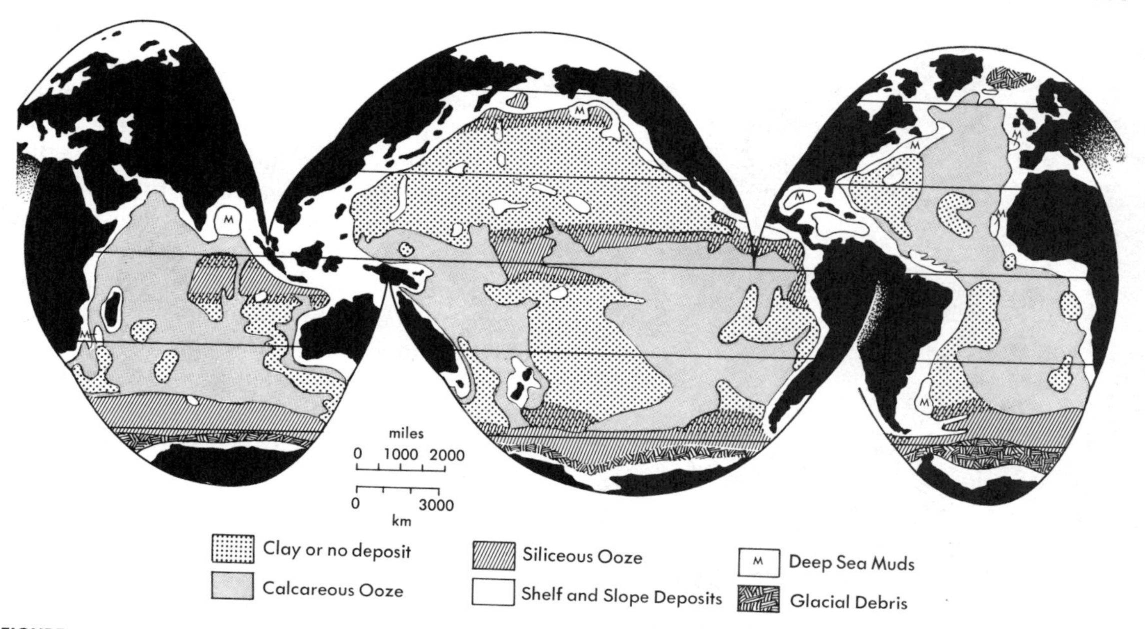

FIGURE 12-15 Distribution of recent sediments on the deep-sea floor. (From W. H. Berger.)

According to plate tectonic theory, deep-sea trenches form where crustal plates converge at a relatively rapid rate, perhaps at twice the average rate of plate movement. The topographic expression of the underthrusting of one plate by another is a trench.

Sediments of the ocean

In earlier chapters, we discussed the processes by which earth materials are weathered, eroded, transported, and finally deposited to be transformed into sedimentary rocks.

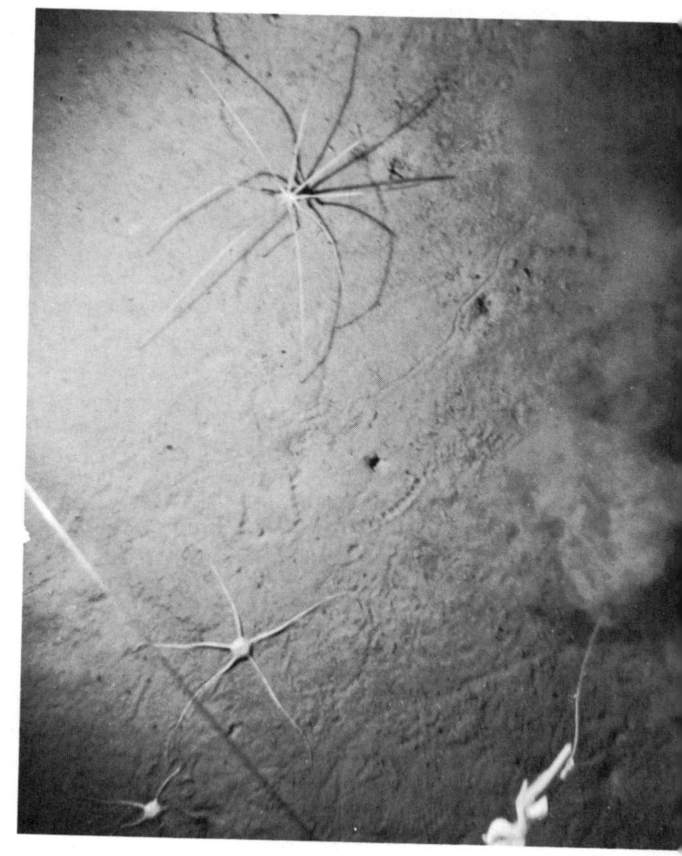

FIGURE 12-16 This muddy sea floor is south of Cape Cod, Massachusetts, at a depth of 1,800 m. The cloud of sediment was raised by a fishing line and sinker. The larger animal is a sea spider measuring 71 cm. The smaller, five-armed animals are brittle stars. (Photo by D. M. Owen, Woods Hole Oceanographic Institution.)

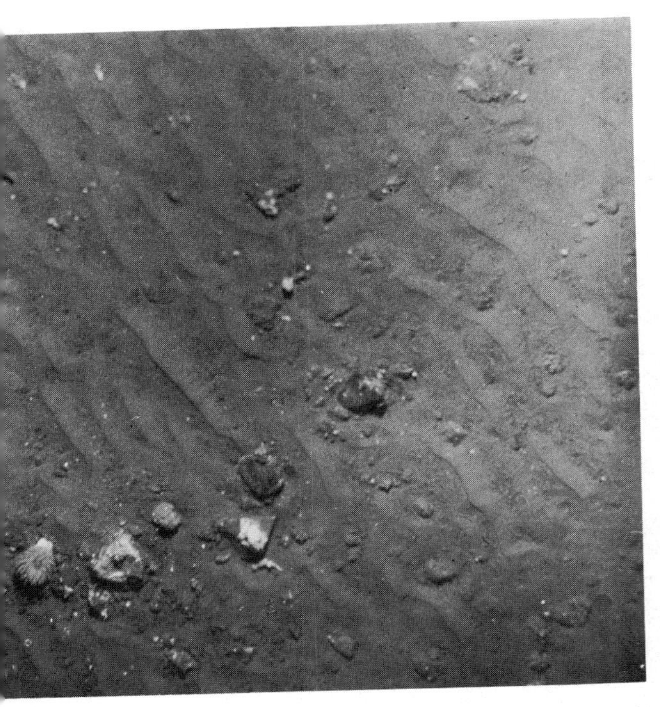

FIGURE 12-17 Ripple-marked sand on the continental shelf between George's and Brown's Banks, off the New England coast. The ripple marks indicate currents moving diagonally from the upper right toward the lower left. The "sand dollar" in the lower left is 5 to 8 cm in diameter. (Photo by D. M. Owen, Woods Hole Oceanographic Institution.)

The great ocean basins of the world constitute the ultimate collection area for the sediments and dissolved material that are carried from the land. The great bulk of the sedimentary rocks found on our modern landmasses were once deposits on sea floors of the past. In this section we will review briefly the sediments being deposited in modern oceans and seas—the sediments destined to become the sedimentary rocks of the future. Figure 12-15 shows the distribution of recent sediments on the deep-sea floor, and Figures 12-16 and 12-17 depict the details of several depositional settings in the oceans.

Deposits on the continental shelves

When particles of solid material are carried out and deposited in a body of water, the largest particles generally should settle nearest the shore and the finest particles farthest away from the shore, in a neatly graduated pattern. But there are a great many exceptions to this generalization. Many deposits on the continental shelves show little tendency to grade from coarse to fine away from the shoreline. We would expect to find sand close to shore only, but it shows up from place to place all along the typical continental shelf right up to the lip of the continental slope. It is particularly common in areas of low relief on the shelf. In fact, on glaciated continental shelves, sand mixed with gravel and cobbles constitutes a large part of the total amount of deposited material.

Also common on the continental shelves are deposits of mud, especially off the mouths of large rivers and along the course of ocean currents that sweep across the river-laid deposits. Mud also tends to collect in shallow depressions across the surface of the shelves, in lagoons, sheltered bays, and gulfs.

Where neither sand nor mud collects on the shelves, the surface is often covered with fragments of rock and gravel. This is commonly the case on open stretches of shelf where strong ocean currents can winnow out the finer material, and off rocky points and exposed stretches of rocky shoreline. In narrow straits running between islands or giving access to bays, the energy of tidal and current movements is often so effectively concentrated that the bottom is scoured clean and the underlying solid rock is exposed.

The geologic record indicates that many calcareous deposits were laid down in ancient seas, eventually to give rise to limestone. However, only a few calcareous deposits are being built upon the continental shelves of modern seas. And most of these limy mud deposits are being built up in warmer waters, particularly near coral reefs. No satisfactory explanation accounts for the apparent deficiency.

Deposits on the continental slopes

Although we have less information about the deposits being laid down on the continental slopes than we have about the shelf deposits, evidence indicates that here, too, gravel, sand, mud, and bedrock are all found on the bottom. And deposition, it seems, is taking place even more rapidly on the slopes than on the shelves.

Deposits on the deep-sea floor

The deposits that spread across the floors of the deep sea are generally much finer than those on the slopes and shelves lying off the continents, although occasional beds of sand have been found even in the deeps. Deep-sea deposits of material derived from the continents are referred to as *terrigenous* ("produced on the earth") deposits. Those formed of material derived from dissolved or suspended matter in the ocean itself are *pelagic* ("pertaining to the ocean") deposits (Table 12-4). Small amounts of *extraterrestrial* sediment (meteorite particles, cosmic dust) are universally present in all deep-sea deposits.

Although seldom red in color and containing only small amounts of clay minerals, the fine-grained sediment that dominates the floor of the North and South Pacific and most very deep oceanic areas is termed *red clay*. The principal minerals present are quartz, mica, several clay minerals, chlorite, and complex silicate minerals. The depositional rate is very slow, probably less than 1 mm per 1,000 years.

The finest particles of continental detritus, which may have been in suspension for many years, is the main component of red clay. Turbidity currents, rivers, icebergs, and wind transport the detritus. Meteorite particles and fossils also constitute a small percentage of the red clay. Beer cans thrown into the sea by careless seafarers now rest on red clay at depths greater than 5 km.

Table 12-4 Classification of Deep-Sea Sediments

PELAGIC	TERRIGENOUS
Inorganic	*Turbidite Deposits*
"Red" clay (brown clay, pelagic clay, oxypelite)	Slide deposits (gravity displacement)
Volcanic ash and glass	Mud
Meteorite particles, cosmic dust	Material from icebergs (glacial-marine sediment)
Tecktites	
Authigenic deposits (phillipsite, clinoptilolite, manganese nodules)	
Organic	
Calcareous ooze (foraminiferal ooze)	
Globigerina ooze	
Nannoplankton ooze	
Pteropod ooze	
Siliceous ooze	
Diatom ooze	
Radiolarian ooze	
Coral Reef debris	
Fish debris (phosphatic)	
Organic matter	

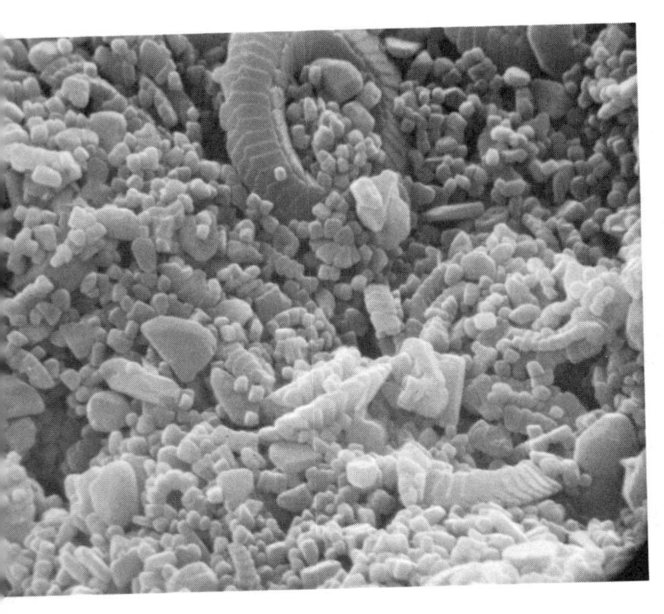

FIGURE 12-18 (Above) Deep-sea chalk composed of calcareous ooze (coccoliths and coccolith fragments) from the northwest Pacific. (Photo by Peter Roth.)

FIGURE 12-19 (Below) Centric diatoms (large discs) and radiolarians (spherical, conical and lacy structures, as well as the large spines) from the tropical east Pacific. (Photo by Peter Roth; specimen collected by W. H. Berger.)

The most common rock found in dredging the ocean bottoms is basalt, particularly on oceanic ridges and seamounts. Major eruptions may spread extensive layers of volcanic ash, frequently less than a centimeter thick, over thousands of square kilometers of the ocean floor. Submarine volcanic eruptions contribute particles of basaltic glass. Pumice, a volcanic glass, may float in surface waters for many days, eventually falling to the underlying abyssal mud.

At low pressures and temperatures, water and the dissolved materials of sea water may convert basaltic lavas to *zeolites* (complex silicate minerals) and clay mineral assemblages. *Phillipsite,* the potassium-rich zeolite, is common in sediment of the Pacific Ocean. In contrast, *clinoptilolite,* the sodium-calcium zeolite, is present in the Atlantic Ocean, the western Indian Ocean, and the Antarctic Ocean where phillipsite is essentially absent.

Biological debris (*ooze*) dominates abyssal deposition throughout large areas of the open ocean. Calcium carbonate ($CaCO_3$) oozes are the most abundant of the biologically produced sediments on the sea floor (Figure 12-18). Two calcareous oozes are common. In one—*globigerina ooze*—the deep-sea sediment is rich in foraminiferal tests and coccoliths. In the other—*pteropod ooze*—the shells of tiny marine molluscs are dominant. Pteropod remains are preserved in deep-sea sediments at water depths of less than 3,500 m.

Oozes of silica from the siliceous tests of several groups of organisms are present mainly at high latitudes and in the equatorial Pacific. The most important of the silica oozes are *diatom oozes* (shells of green unicellular algae) and *radiolarian oozes* (shells of minute single-celled animals). Figure 12-19 shows diatoms and radiolarians from the tropical East Pacific.

Chemical precipitation produces the spectacular *manganese nodules* (Figure 12-20);

Table 12-5 Abundance by Weight Percent of Significant Elements in Fifty-four Samples of Manganese Nodules from the Pacific Ocean

ELEMENT	MINIMUM	MAXIMUM	AVERAGE
Mn	8.2	41.1	24.2
Fe	2.4	26.6	14.0
Ni	0.16	2.0	0.99
Ti	0.11	1.7	0.67
Cu	0.028	1.6	0.53
Co	0.014	2.3	0.35
Pb	0.02	0.36	0.09
Zn	0.04	0.08	0.047
Ag		0.0006	0.003

SOURCE: Data from Mero, J. L., *The Mineral Resources of the Sea:* New York: American Elsevier, 1965, p. 180.

these are rounded, concentric masses ranging in size from a few centimeters to about 25 cm. In addition to containing significant amounts of manganese or iron, these nodules contain nickel, titanium, copper, zinc, gold, silver, cobalt, lead, and other elements (Table 12-5). In the Pacific Ocean, the volume of manganese nodules is estimated at more than 1.5 billion metric tons. It frequently is suggested that these nodules are potential ores of manganese that will be recovered in the near future. However, manganese from this source will not soon be competitive with rich ores found on land. Origin of the nodules generally is believed to be submarine springs associated with volcanoes, but leaching of submarine volcanic rocks may be more important. The nodules apparently accumulate in areas of very slow deposition.

Table 12-6 Depositional Rates of Recent Deep-Sea Sediment

SEDIMENT	RATE OF DEPOSITION (millimeters per thousand years)
Terrigenous mud	50–2,000
Calcareous ooze (globigerina, nannoplankton, and pteropod ooze)	
Pacific	5–60
Atlantic	10–60
Siliceous ooze (diatom and radiolarian ooze)	
Pacific	2–5
Atlantic	2–7
Indian	2–10
Red clay	
North Pacific	0–2
South Pacific	2–3

FIGURE 12-20 *The first successful recovery of large amounts of manganese nodules by Kennecott in the Pacific in 1972. (Courtesy Kennecott Copper Corporation.)*

Depositional rates of deep-sea sediment

Most sediment is deposited in the shallow water of the continental shelf or on the slope leading down to the deep sea. Little sediment gets to the deep-sea floor, where the rate of deposition varies according to sediment type (Table 12-6). In contrast, on shelves and slopes the rate is rapid and appears to be 10 to 20 times the depositional rate on the deep-sea floor.

Shorelines

Few people have occasion to make a detailed study of the ocean currents or of the topography of the ocean floor, but most of us have many opportunities to observe the activity of water along the shorelines of oceans or lakes. The nature and results of wave action along such shorelines can be a drama of power and persistence.

The processes

The energy that works upon and modifies a shoreline comes largely from the movement of water produced by tides, by wind-formed waves, and, to a lesser extent, by tsunami. Since we have discussed tidal currents (and tsunami were discussed in Chapter 8), we may now turn to the nature and behavior of wind-formed waves as they advance against a shoreline.

Wind-formed waves

Most water waves are produced by the friction of air as it moves across a water surface. The harder the wind blows, the higher the water is piled up into long *wave crests* with intervening troughs; both crests and troughs are at right angles to the wind. The distance between two successive wave crests is the *wave length,* and the vertical distance between the wave crest and the bottom of an adjacent trough is the *wave height* (Figure 12-21). When the wind is blowing, the waves it generates are called a *sea.* But wind-formed waves persist even after the wind that formed them dies. These waves, or *swells,* may travel for hundreds or even thousands of kilometers from their zone of origin.

We are concerned with both the movement of the wave form and the motion of water particles in the path of the wave. Obviously the wave form itself moves forward at a measurable rate. But in deep water, the water particles in the path of the wave describe a circular orbit: any given particle moves forward on the crest of the wave, sinks as the following trough approaches, moves backward under the trough, and rises as the next crest advances. Such a motion can be visualized by imagining a cork bobbing up and down on the water surface as successive wave crests and troughs pass by. The cork itself makes only

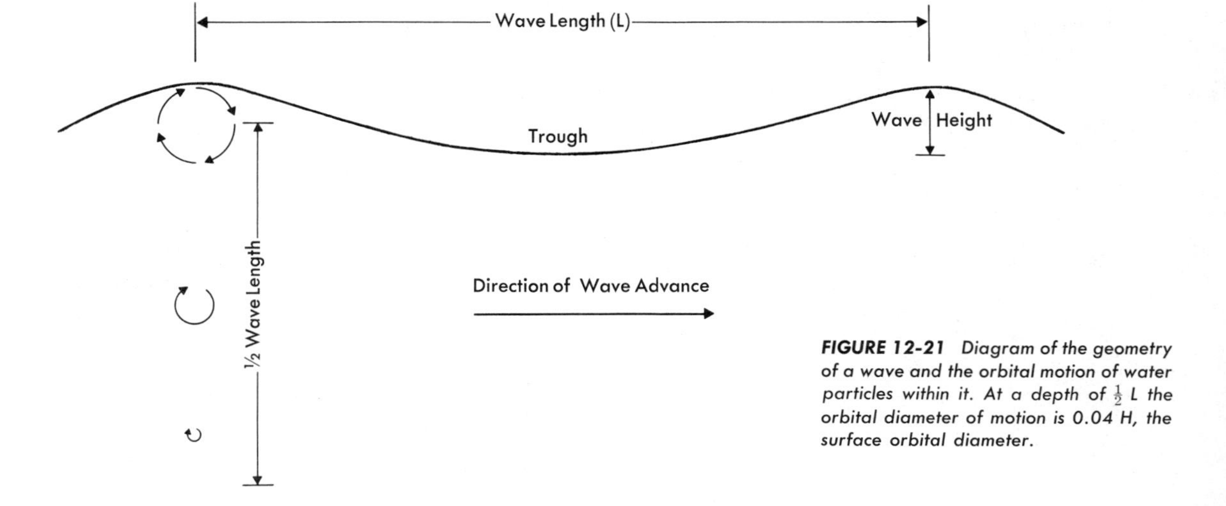

FIGURE 12-21 *Diagram of the geometry of a wave and the orbital motion of water particles within it. At a depth of $\frac{1}{2}$ L the orbital diameter of motion is 0.04 H, the surface orbital diameter.*

very slight forward progress under the influence of the wind. Wave motion extends downward, until at a depth equal to about one-half the wave length it is virtually negligible. But between this level and the surface, water particles move forward under the crest and backward under the trough of each wave, in orbits that decrease in diameter with depth (see Figure 12-21).

As the wave approaches a shoreline and the water becomes shallower, definite changes take place in the motion of the particles and in the form of the wave itself. When the depth of water is about half the wave length, the bottom begins to interfere with the motion of water particles in the path of the wave, and their orbits become increasingly elliptical. As a result, the length and velocity of the wave decrease and its front becomes steeper. When the water becomes shallow enough and the front of the wave steep enough, the wave crest falls forward as a breaker, producing *surf.* At this moment, the water particles within the wave are thrown forward against the shoreline. The energy thus developed is then available to erode the shoreline or to set up currents along the shore that can transport the sediment produced by the wave erosion.

Wave refraction and shoreline profile

Most waves advance obliquely toward the shoreline until they begin to encounter the sea bottom. The influence of the sea floor then tends to bend or refract them so that they approach the shore approximately head-on.

For example, let us assume that we have a relatively straight shoreline with waves approaching it obliquely over an even bottom that grows shallow at a constant rate. As a wave crest nears the shore, the section closest to land feels the effect of the shelving bottom first and its velocity is decreased, while the seaward part continues along at its original rate. The effect is to swing the wave around and to change the direction of its approach to the shore, so it then approaches head-on.

As the wave breaks, however, not all of its energy is expended on erosion of the shoreline. Some of the water thrown forward is deflected and moves laterally, parallel with the shore. The energy of this water movement is used up partly by friction along the bottom, and partly by the transportation of material parallel with the shoreline by longshore currents.

Refraction also explains why, on an irregular shoreline, the greatest energy is usually concentrated on the headlands and the least energy is directed along the bay. Figure 12-22 shows a bay separating two promontories, and a series of wave crests sweeping in to the shore across a bottom that is shallow off the headlands and deep off the mouth of the bay. Where the depth of the water is greater than one-half the wave length, the crest of the advancing wave is relatively straight. Closer to shore, off the headlands, however, the depth of water becomes less than half the wave length, and the velocity of the wave begins to slow down. In the deeper water of the bay the wave continues to move rapidly shoreward until there, too, the water grows shallow and the wave crest slows. This differential bending of the wave tends to make the crest of the wave parallel with the orientation of the shoreline. Thus, the wave energy is concentrated on the headlands and dispersed around the bay, as suggested in Figure 12-22.

A composite profile of a shoreline, from a point above high tide and then seaward to some point below low tide, reveals features that change constantly as they are influenced by the nature of waves and currents along the shore (Figure 12-23). All features are not present on all shorelines, but several are present in most shore profiles. First, the *offshore* section extends seaward from low tide. Next, the *shore,*

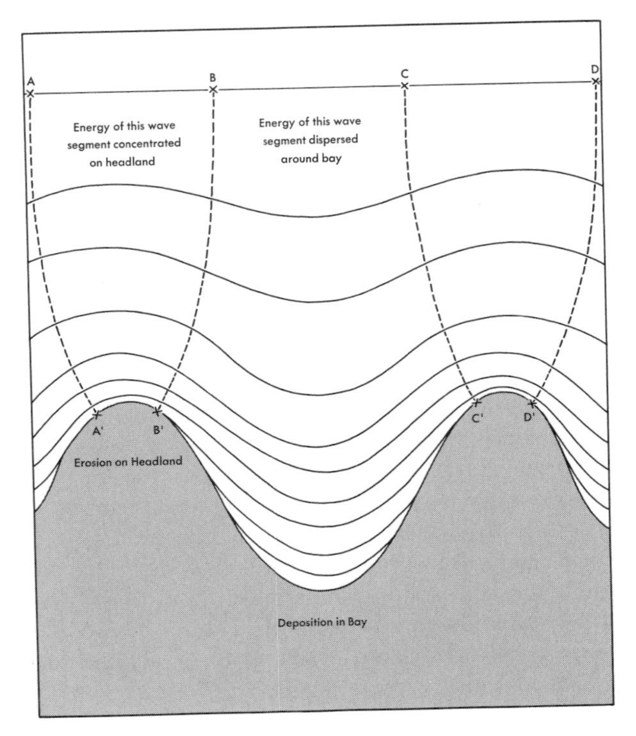

FIGURE 12-22 *Refraction of waves on an irregular shoreline. It is assumed that the water is deeper off the bay than off the headlands. Consider that the original wave is divided into three equal segments, A-B, B-C, and C-D. Each segment has the same potential energy. But observe that by the time the wave reaches the shore, the energy of A-B and C-D has been concentrated along the short shoreline of headlands A′-B′ and C′-D′, whereas the energy of B-C has been dispersed over a greater front (B′-C′) around the bay. Energy for erosion per unit of shoreline is therefore greater on the headlands than along the bay.*

or *beach,* section reaches from low tide to the foot of the *sea cliff* and is divided into two segments—the backshore and the foreshore. The *backshore* is in front of the sea cliff and is characterized by one or more *berms,* which resemble small terraces with low ridges on their seaward edges built up by storm waves. Seaward from the berms is the *foreshore.* Finally, inland from the shore lies the *coast.* Deposits of the shore may veneer a surface that is cut by the waves on bedrock and is known as a *wave-cut terrace.* In the offshore section, too, there may be an accumulation of unconsolidated deposits composing a *wave-built terrace.*

The shoreline profile is ever-changing. During great storms the surf may pound directly against the sea cliff, eroding it back and at the same time scouring down through the beach deposits to abrade the wave-cut terrace. As the storm (and hence the available energy) subsides, new beach deposits build up out in front of the sea cliff. The profile of a shoreline at any one time, then, is an expression of the available energy: it changes as the energy varies. This relation between profile and available energy is similar to the changing of a stream's gradient and channel as the discharge (and therefore the energy) of the stream varies (see Chapter 10).

FIGURE 12-23 *Some of the features along a shoreline, and the nomenclature used in referring to them. [In part, after F. P. Shepard,* Submarine Geology *(New York: Harper & Row, Publishers, 1963), p. 168.]*

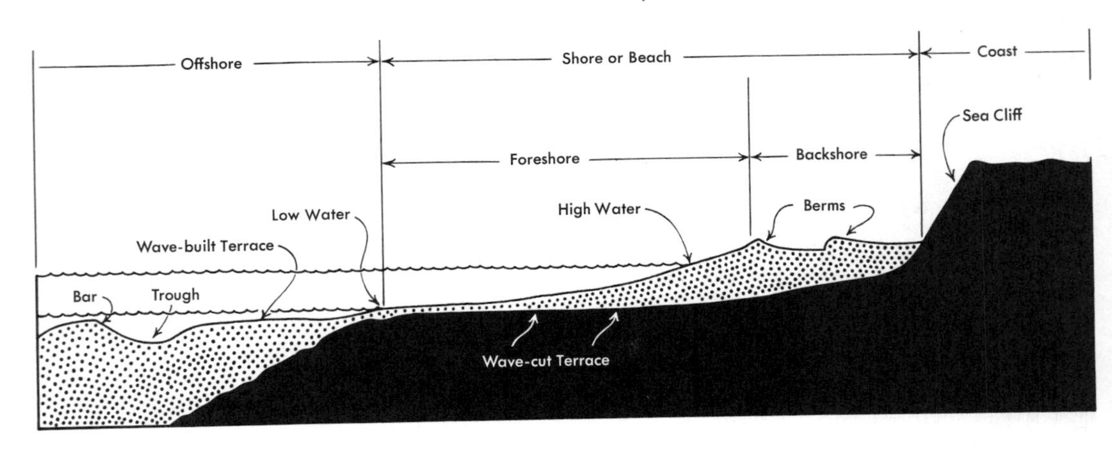

Shoreline features

Not even shorelines have escaped man's constant desire for classification. But to date no completely acceptable system of classification for shorelines has been devised. For many years it was common practice to group shorelines as *emergent* or *submergent,* depending on whether the sea had gone down or risen in relation to the landmass. Thus, large sections of the California coast, having emerged from the sea during geologically recent times, would be termed emergent. Across the continent, the shoreline of New England indicates that it has been drowned by a slowly rising sea, and thus it would be referred to as submergent. This system has been criticized because some of the features that were thought to represent emergence of land actually form where the land is being submerged. Conversely, along shorelines of submergence, features thought to characterize emergence may also develop.

Another attempt at a classification is based on the processes that form shorelines. For example, some shorelines have major features traceable to glacial erosion, others to glacial deposition. A system divided along these lines has much to recommend it, but there are problems here too. Therefore, we shall examine some of the individual shoreline features without attempting to fit them into an all-inclusive system.

Erosion and deposition work hand in hand to produce most of the features of shorelines. An exception to this generalization is an offshore island that is merely the top of a hill or a ridge that was completely surrounded by water as the sea rose in relation to the land. But even islands formed in this way are modified by a certain amount of erosion and deposition.

Features caused by erosion

As mentioned earlier, *wave-cut cliffs* are common erosional features along a shore, particularly where the shore slopes steeply down beneath the sea. Here waves can break directly on the shoreline, and thus they expend the greatest part of their energy in eroding the land. Wave erosion pushes the wave-cut cliff steadily back, producing a wave-cut terrace or platform at its foot. Since the surging water of the breaking waves must cross this terrace before reaching the cliff, the water loses a certain amount of energy through turbulence and friction. So the farther the cliff retreats, and the wider the terrace becomes, the less effective are the waves in eroding the cliff. And

FIGURE 12-24 *The sea has cut arches through this promontory. To the far right is a stack, a rock mass that erosion has cut off from the mainland. Arches State Park, California. (Photo by Sheldon Judson.)*

if sea level remains constant, the retreat of the cliffs becomes slower and slower.

Waves pounding against a wave-cut cliff produce various features as a result of the differential erosion of the weaker sections of the rock. For example, wave action may hollow out cavities, or *sea caves,* in the cliff, or if this erosion should cut through a headland, a *sea arch* is formed. The collapse of the roof of a sea arch leaves a mass of rock, a *stack,* which is isolated in front of the cliff (Figure 12-24).

Features caused by deposition

Features of deposition along a shore are built of material eroded by the waves from the headlands, and of material brought down by the rivers that carry the products of weathering and erosion from the land masses. For example, part of the material eroded from a headland may be drifted by currents into the protection of a neighboring bay, where it is deposited to form a sandy beach (Figure 12-25).

The coastline of northeastern New Jersey (Figure 12-26) illustrates some of the features caused by deposition. Notice that the Asbury Park-Long Branch section of the coastline is a

FIGURE 12-25 *Pocket beach formed between cliffed headlands, Bonassola, Italy. (Photo by Earle F. McBride.)*

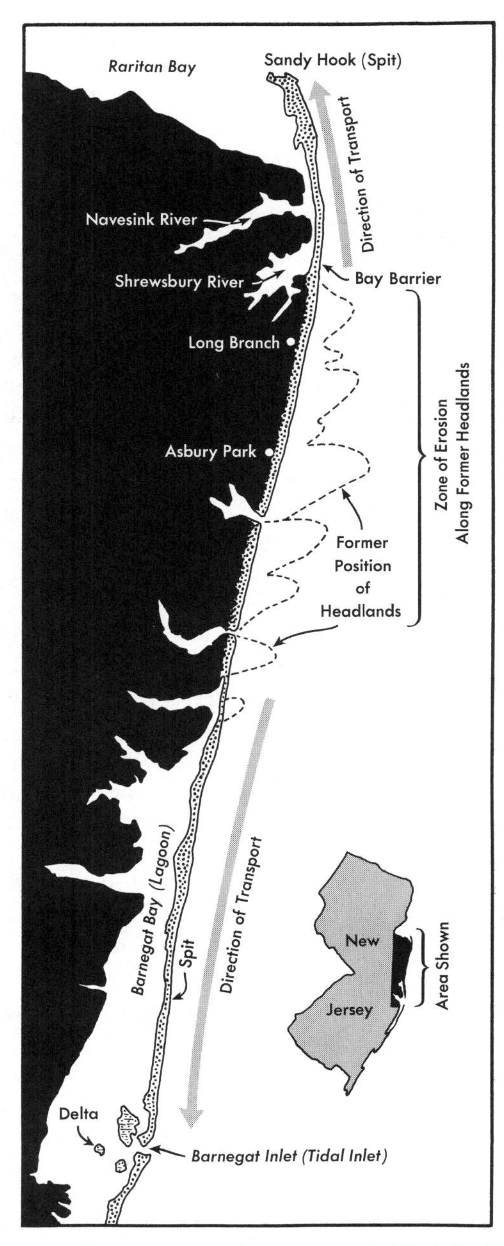

FIGURE 12-26 Erosion by the sea has pushed back the New Jersey coastline as indicated on this map. Some of the material eroded from the headlands has been moved northward along the coast to form Sandy Hook, a spit. To the south, a similar but longer feature encloses Barnegat Bay, a lagoon with access to the open ocean through a tidal inlet. (After an unpublished map by Paul MacClintock.)

zone of erosion that has been formed by the destruction of a broad headland area. Erosion still goes on along this part of the coast, where the soft sedimentary rocks are easily cut by the waves of the Atlantic. The material eroded from this section is moved both north and south along the coastline. Sand swept northward is deposited in Raritan Bay and forms a long sandy beach projecting northward, a *spit* known as Sandy Hook.

Just south of Sandy Hook, the flooded valleys of the Navesink River and of the Shrewsbury River are bays that have been almost completely cut off from the open ocean by sandy beaches built up across their mouths. These beaches are called *bay barriers.*

Sand moved southward from the zone of erosion has built up another sand spit. Behind it lies a shallow lagoon, Barnegat Bay, that receives water from the sea through a tidal inlet, Barnegat Inlet. This passage through the spit was probably first opened by a violent storm, presumably of hurricane force. Just inside the inlet a *delta* has been formed of material partly deposited by the original breakthrough of the bar and partly deposited by continued tidal currents entering the lagoon.

Long stretches of the shoreline from Long Island to Florida, and from Florida westward around the Gulf Coast, are marked by shallow, often marshy lagoons separated from the open sea by narrow sandy beaches. Many of these beaches are similar to those that enclose Barnegat Bay, apparently elongated spits attached to broad headlands. Others, such as those that enclose Pamlico Sound at Cape Hatteras, North Carolina, have no connection with the mainland. These sandy beaches are best termed *barrier islands.* It has been suggested that these islands originated from spits that were detached from the mainland as large storms breached them at various points. Some geologists think that they may represent spits isolated from the mainland by a slowly rising sea level. Still a third possibility is that over a long period of time wave action has eroded sand from the shallow sea floor and has heaped it up in ridges that lie just above sea level.

Another depositional feature, a *tombolo,* is a beach of sand or gravel that connects two islands, or connects an island with the mainland. Numerous examples exist along the New England coastline, fewer off the West Coast, although Morro Rock, a small, steep-sided island, is tied to the California mainland by a tombolo.

Lost beaches

After many years of anticipation, Picard recently visited Miami Beach and was shocked to see that the beach had very nearly disappeared. Lost beaches are sad to see. Their demise frequently is related to man's interference with a natural system in equilibrium.

For example, dams that are built upstream accumulate behind them sand that formerly reached the shoreline. Or builders of beach-front hotels use considerable sand from the beach in their various "constructive" activities, thereby not only depleting beach sand but disturbing the beach-forming process.

The building of jetties and breakwaters, supposedly to improve harbors and marinas, also interferes drastically with the movement of sand-sized material into and along the coast. Such man-made structures completely disrupt the processes that originally deposited our beaches.

Sea walls are often built to protect the beach, but these structures are seldom helpful. The sea breaks against the sea wall and undermines it. Water that comes over sea walls stays behind them, which creates still other problems.

Beach erosion problems are too numerous, important, and costly for them to be in the hands of realtors and hotel owners. Local, state, and federal agencies working in concert with geologists and oceanographers must become much more involved in these questions. The cost and complexity of such projects prohibit very much private participation.

Coral-reef shorelines

In tropical and semitropical waters lying within a belt between about 30° N and 25° S, many shorelines are characterized by coral reefs of varying sizes and types. These reefs are built up by individual corals with calcareous skeletons, as well as by other lime-secreting animals and plants. Reef-building organisms require water of a least 20°C, normal salinity, and nearly clear of mud. The reef-building corals do not grow below a depth of about 50 m or much above the low-tide level.

The coral-reef shorelines are of three types: the *fringing reef,* the *barrier reef,* and the *atoll.* A fringing reef grows out directly from a landmass, whereas a barrier reef is separated from the main body of land by a lagoon of varying width and depth opening to the sea through passes in the reef. An atoll is a ring of low coral islands arranged around a central lagoon.

The origin of atolls has been debated for well over a century, ever since Charles Darwin first advanced his explanation in 1842. Darwin postulated that an atoll begins as a fringing reef around a volcanic island. Since the island rests as a dead load on the supporting material, it begins to subside but at a rate slow enough for the coral to maintain a reef. With continued subsidence, the island becomes smaller and smaller, and the actively growing section of the reef becomes a barrier reef. Then, with the final disappearance of the island below the sea, the upward-growing reef encloses only a lagoon and becomes a true atoll (Figure 12-27). In support of this theory are many volcanic islands in the Pacific now surrounded by barrier reefs. Furthermore, in-

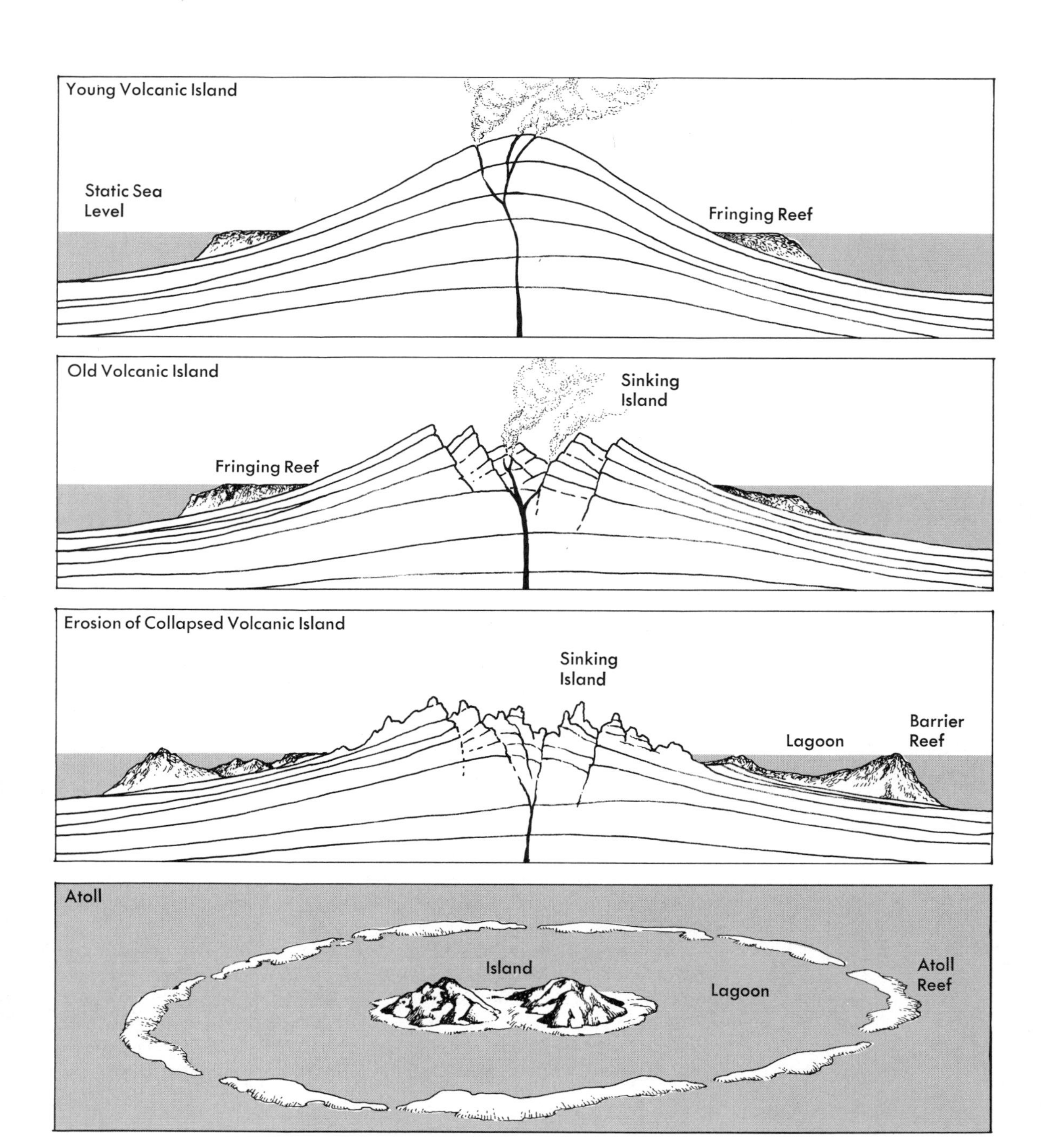

FIGURE 12-27 *Cross sections showing four stages in the formation of an atoll, according to the subsidence theory of Charles Darwin.*

vestigations on Bikini Island (an atoll) indicate that the volcanic rock core is surmounted by several hundred meters of coral rock. Finally, geophysical evidence suggests that there actually has been subsidence of some of the volcanic islands of the Pacific.

The subsidence theory originally advanced by Darwin, however, does not explain the nearly constant depth of countless modern lagoons within the atolls. In part to overcome this difficulty, the so-called *glacial-control theory* has been advanced. Proponents of this explanation of atoll formation have postulated that volcanic islands were truncated at a lower sea level during one or more of the Pleistocene ice advances. Around the edges of such wave-planed platforms the coral reefs began to grow as the continental glaciers melted and sea level rose. The coral islands around the edge of the platforms would then enclose lagoons of more or less constant depth. In summary, there is evidence in support of each theory. When the final answer is known, both theories—subsidence and glacial control—may be applicable.

SUMMARY

Oceans cover more than 70 percent of the earth's surface. The level of the sea rises and falls not only with the tides but also because of the eustatic and tectonic processes.

Degassing, the release from the earth's interior of gases by volcanoes, hot springs, and geysers, apparently is the process mainly responsible for the formation of ocean water and the atmosphere. After the evolution of life the volcanic release of gases must have been in nearly uniform proportions.

About 70 elements have been identified in sea water, and others certainly are present. Chlorides give sea water its familiar saltiness, and all the gases found in the atmosphere are present in the oceans. A large number of less abundant materials, including such rare elements as uranium, gold, and silver are also found. The major constituents dissolved in sea water are chlorine, sodium, sulfate, magnesium, calcium, potassium, bicarbonate, and bromine.

Ocean water is circulated by tidal currents, by wind-driven surface currents, and by density currents that move to the deep ocean bottoms. Density variations are caused by changes in temperature, salinity, and turbidity of the water.

Ocean basins are divided into areas of continental shelves, slopes, and deep-sea floors. Among the major topographic features are submarine canyons and valleys, deep-sea trenches, rises and ridges, and seamounts.

Shorelines are modified by energy derived from the wind-driven waves. Erosional features include cliffs, stacks, caves, and arches. Depositional features include spits, beaches, bay barriers, barrier islands, deltas, and tombolos.

Sediments in the oceans are deepest on the shelves and slopes and thinnest on the deep-sea floors. Sediments are divided into pelagic and terrestrial, the latter dominating the shelves and slopes.

SUGGESTED READINGS

CARSON, RACHEL, *The Sea Around Us* rev. ed. New York: Oxford University Press, 1961.
COTTER, CHARLES H., *The Physical Geography of the Oceans.* New York: American Elsevier Publishing Co., 1965.
GROSS, M. G., *Oceanography, A View of the Earth.* Englewood Cliffs, N.J.: Prentice-Hall, Inc., 1972.
HEEZEN, B. C., and C. D. HOLLISTER, *The Face of the Deep.* London: Oxford University Press, 1971.
KING, CUCHLAINE A. M., *An Introduction to Oceanography.* New York: McGraw-Hill Book Company, 1963.
MENARD, H. W., *Geology, Resources, and Society.* San Francisco: W. H. Freeman and Co., 1974.
PICKARD, GEORGE L., *Descriptive Physical Oceanography.* New York: The Macmillan Company, 1964.
SHEPARD, F. P., *Submarine Geology,* 2nd ed. New York: Harper & Row, Publishers, 1963.
SVERDRUP, H. U., MARTIN W. JOHNSON, and RICHARD H. FLEMING, *The Oceans.* Englewood Cliffs, N.J.: Prentice-Hall, Inc., 1942.
TUREKIAN, K. K., *Oceans.* Englewood Cliffs, N.J.: Prentice-Hall, Inc., 1968.

Time in Geology

The importance of time in geologic processes has been touched upon in connection with almost every topic discussed in previous chapters. It has been suggested and implied repeatedly that lengthy spans of time seem to have been necessary to bring the earth to its present condition. We shall now devote our attention specifically to the subject of time and its measurement as applied to geology.

We may think of geologic time in two ways: relative and absolute. *Relative time* relates only to whether one event in earth history came *before or after* another event, years not considered. *Absolute time* is stated in terms of years and is tied to the calendar system currently in use.

Naturally, we would like to be able to date geologic events with absolute precision. But so far this has been impossible, and the accuracy achieved in determining the dates of

Symbolic of man's attempt to measure and understand time is ancient and mysterious Stonehenge on Salisbury Plain, southern England. (British Information Service.)

human history, at least written human history, will probably never be reached in geology. Still, we can determine approximate dates for many geologic events. For instance, we can say that the dinosaurs became extinct about 65 million years ago, and that about 11,000 years ago the last continental glacier was receding from New England and the area bordering the Great Lakes.

Years and seasons

The rotation and revolution of the earth provide us with our most important and useful measure of time—the year and the day. Commencing with crude observations of the changing seasons and positions of the heavenly bodies, the length of the year has been quite accurately fixed as 365 days, 5 hours, 49 minutes, and 12 seconds. Figure 13-1 illustrates the astronomical basis for common time units. Although such measures as the week, hour, minute, or second (defined as 1/31,556,925.9 of the year) are accepted standards of time measurement, these are purely artificial fractional measures invented by man. Obviously, if the length of the year should change or vary, so would the second and all of its multiples.

Much greater accuracy is achieved by measuring the vibrations of the atoms of certain elements such as cesium. In 1972 a system of integrated time keeping by the atomic method was initiated.

For present purposes it is sufficient to gain an understanding of how long the familiar cycles of the year and day may have been in effect. Evidence for the regular yearly journey around the sun exists in the form of seasonal effects. Because the axis of the earth is inclined to the plane along which it travels, characteristic changes of light, temperature, and precipitation are repeated with the seasons. These effects exercise a profound influence on the food supply and growth patterns of plants and animals and, locally at least, on the erosion, transportation, and deposition of sediments. Seasonal effects are recorded in living and nonliving materials in many ways and have thereby become permanently recorded in the earth's crust.

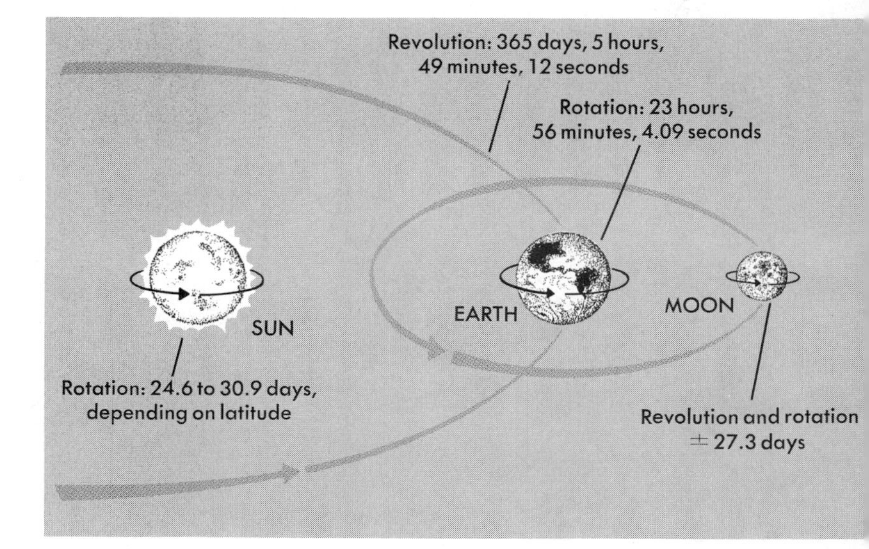

FIGURE 13-1 *Astronomical "clocks," some more reliable than others. Notice the times of revolution and rotation. This diagram is not intended to portray true relative sizes of the earth, moon, or sun, or their orbits.*

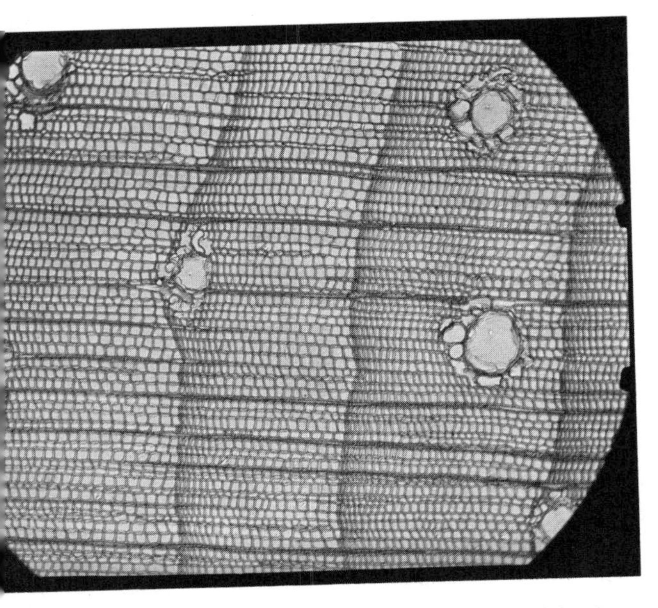

FIGURE 13-2 *Enlarged view of wood structure of the pinyon pine (Pinus edulis) showing several annual rings. (U.S. Forest Service.)*

Growth rings and varves

Growth rings in plants and animals

The best-known seasonal records that are preserved in living organisms are *tree rings.* The width and spacing of the rings depend on temperature, light, and moisture variations that are largely of a seasonal nature. Each ring consists of two parts: the so-called *summer wood,* which has small cells and thick walls, and the *spring wood,* which has larger cells and thinner walls (Figure 13-2). The study of tree rings, called *dendrochronology,* has helped archaeologists to date various archaeological sites, especially in the arid portions of the American Southwest where wood is easily preserved. A continuous chronology going back to 273 B.C. has been pieced together, and many important ruins have been dated by pieces of structural wood.

An even longer sequence of rings can be found in the trunks of the larger sequoia trees of California and in a less spectacular tree, the

FIGURE 13-3 *Bristlecone pines (Pinus aristata) growing in the Inyo National Forest, California. Portions of a tree may continue to grow after other parts have died. Specimens from this area are among the oldest known living things. (U.S. Department of Agriculture, photo by Leland J. Prater.)*

bristlecone pine, which is found in various drier parts of California, Nevada, and Utah (Figure 13-3). One tree of this species so dated is more than 4,800 years old, and by use of dead trees a chronology going back to 6225 B.C. has been established.

Animals also respond to seasonal changes, usually by variations in growth rates corresponding with variations in the favorability of growth conditions. Shells of clams and other aquatic organisms show growth "rings" much like those of trees. Fish scales reveal their age by similar marks. Less well known examples are layering in the spines and otoliths (ear bones) of fish, in the horns of mammals, and in the limb bones of certain reptiles.

An amazing discovery is that certain corals and molluscs lay down *daily* rings (Figure 13-4). Counting of the rings of ancient specimens shows that there were more days in a year in past ages, meaning that the earth's rotation must be slowing down. About 1.5 billion years ago the length of the day was calculated to be 11 hours, and in the Devonian period, 400 million years ago, it was slightly more than 21 hours.

FIGURE 13-5 *Varved sediments. The large cylindrical core is from the Paradox Formation in southern Utah. The dark-colored varves are salt, whereas the thin, light-colored ones are dolomite. These relatively thick varves measure about 2.54 cm. The small core at the right is from the Green River Formation in eastern Utah. In this specimen the individual varves are only about 0.01 cm thick. The bands visible in the photograph are aggregations of many varves. The rough specimen at the left is from the Castile Formation in Texas. These varves are chiefly pure and impure gypsum. The varve-based estimate of the total period of deposition in the Castile is 300,000 years.*

FIGURE 13-4 *Shell of the modern pelecypod* (Pecten diegensis) *showing daily growth rings. From a point near the center of the shell to its outer edge there are 50 growth lines, known to have been laid down in 51 days. (Courtesy George R. Clark, II.)*

The effects of seasonal change are not confined to living things. A great deal of the earth is exposed to recurrent variations in precipitation during wet and dry seasons. Under favorable conditions these variations are reflected in the erosion and deposition of sediment. Ideally, for the leaving of records, there should be an interval of little or no deposition followed by one of rapid sedimentation, corresponding to seasons of low and high stream flow.

Any deposit that reflects a yearly cycle is called a *varve.* Several types are illustrated and described in Figure 13-5. The most clearly marked and easily interpreted varves are asso-

FIGURE 13-6 *Varves formed of alternating layers of silt and clay. This site is near the mouth of Sherman Creek, Ferry County, Washington. (U.S. Geological Survey.)*

ciated with glacial activity. In and near most ice fields, the seasons of melting and freezing are sharply marked, and there are abundant lakes and ponds in which deposits may be preserved. During the warm season, when snow and ice are melting, a large quantity of sediment is deposited in bodies of water, and a relatively thick layer of coarse sediment is laid down. With the onset of winter, when water ceases to flow and the ponds and lakes are covered by ice, deposition slows down. During this quiet period, very fine clay particles and some dead organic matter slowly settle to the pond and lake bottoms, forming a thinner layer of finer and usually darker material. As a result of these conditions, each varve consists of two gradational parts—the thicker, coarser for the summer, and a thinner, finer layer for the winter.

Glacial varves are usually relatively thin, ranging from a few millimeters to several centimeters in thickness. Thus the record of hundreds of years may readily accumulate in one lake or pond. An example of multiple varves is shown by Figure 13-6.

The width of glacial varves, like tree rings, will vary, depending on the length and relative warmth of the seasons. As a matter of fact, the variation in the spacing of thick and thin varves permits geologists to correlate one set of varves with another, just as one tree may be correlated with another. The most complete sequences of glacial varves were created in lakes that formed as large continental glaciers melted away. The oldest varve accumulation lies near the point of maximum extent of the glacier, and the latest may still be forming at the glacier's edge. Between the two points lies a series of sediment-filled lakes with overlapping varve histories.

Through great effort, geologists have traced the record of retreat of the last ice sheets in Europe and America. Unfortunately, there are gaps in the records that can be filled

this: there is a discrepancy between dates from varves, tree rings, and artifacts on one hand and ^{14}C dates on the other. For the past 3,000 years the differences are less than 200 years. Beyond 1000 B.C., the ^{14}C dates become progressively younger than the true dates until about 5000 B.C., when the discrepancy is 600 to 800 years.

The lack of agreement of ^{14}C and other dates seems to be due to the fact that ^{14}C has been created and supplied at an uneven rate. The observed departures, however, have not discouraged anthropologists, geologists, and historians from continued use of the ^{14}C method. It is still the best dating method they have for prehistoric materials less than 50,000 years old.

Potassium-argon

One of the three isotopes of potassium, ^{40}K, is radioactive. It decays to argon-40 with a half-life of 1,310,000,000 years. Many potassium-bearing minerals can be used—orthoclase, biotite, hornblende, and glauconite are examples. Whole rocks that contain potassium (including volcanic glass) can be dated without the individual minerals being segregated. The potassium-argon method is probably the most important age-dating method; the range of usefulness is 10,000, to over 3 billion years.

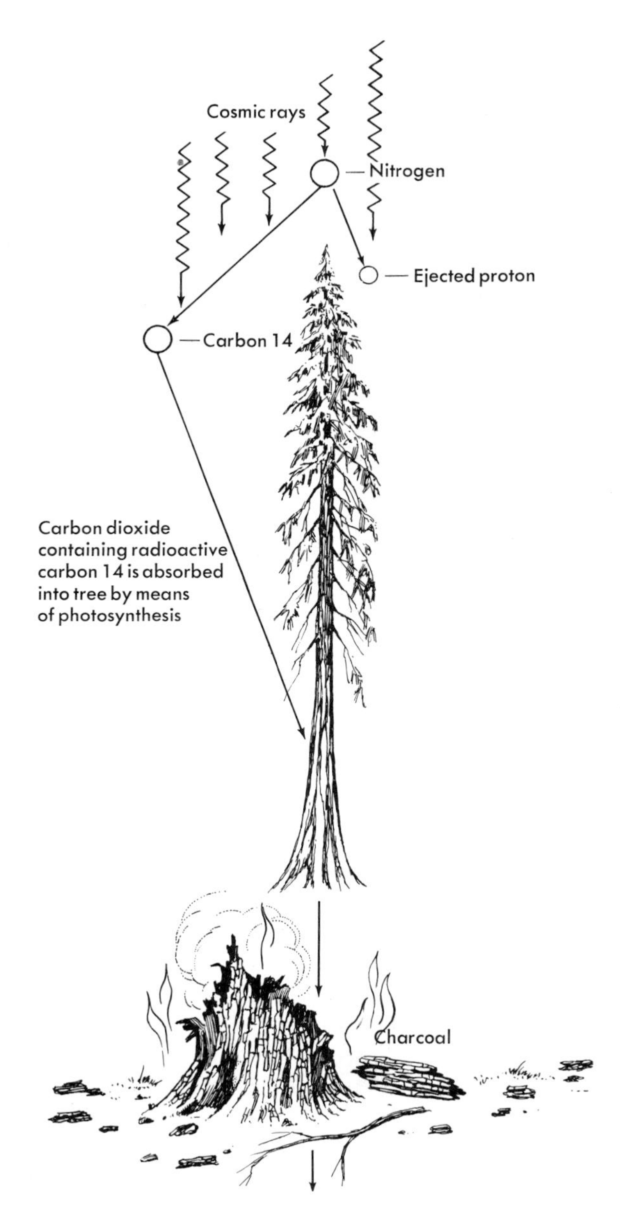

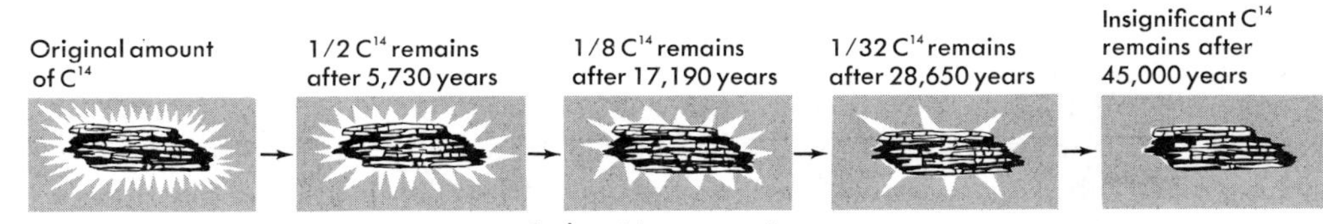

FIGURE 13-10 Formed from nitrogen in the atmosphere, ^{14}C is incorporated in all living things. The rate of addition and disintegration is assumed to be constant and in equilibrium. By reverting to nitrogen, ^{14}C disappears from dead organic material; the amount of ^{14}C remaining in a dead specimen is a measure of age.

Rubidium-strontium

Rubidium-87 is radioactive and decays to strontium-87 with a half-life of 47,000,000,000 years. Usable minerals include plagioclase, biotite, hornblende, and apatite, but rubidium is much less common than potassium. The method is successful in the age range of from 30 million years to the oldest rocks.

Uranium-lead and thorium-lead

The transmutation of uranium to lead is the classical example of radioactive change. Uranium-238 decays to lead-206 with a half-life of 4,530,000,000 years; uranium-235 decays to lead-207, half-life 713,000,000 years; and thorium-232 decays to lead-208, half-life of 13,890,000,000 years. Although many rocks contain uranium, there are few really satisfactory minerals for age dating. The method is most successful in the range of 100 million years to 5 billion years. However, use of the lead methods has greatly declined as the potassium-argon method has improved.

The production of radiogenic lead from uranium or thorium is a steplike process in which a series of radioactive products are produced, each one of which has its peculiar half-life and chemical properties. Two members of the uranium radioactive decay series, thorium-230 and protactinium-231, have low solubilities in natural waters and hence are not readily available for inclusion in shells or skeletons of aquatic animals. Uranium on the other hand has been found to enter into living coral skeletons, and the extent to which thorium-230 or protactinium-231 has been produced in a specimen can be used as a measure of age. The method is mostly restricted to dating of corals in the 10,000 to 400,000-year age range. Another application of the method is the dating of stalactites in caves.

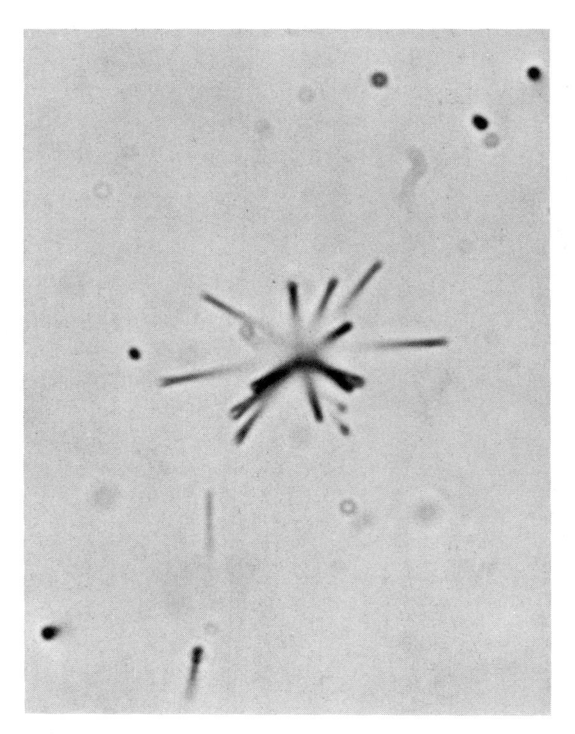

FIGURE 13-11 *Fossil fission tracks in biotite mica. The crystal was etched in hydrofluoric acid to enlarge the tracks, which radiate from a microscopic impurity containing a larger number of uranium atoms. Magnification about 3,000X. (Courtesy of P. Buford Price.)*

Fission-track dating

Natural fission of uranium-238 produces two smaller nuclei that fly apart with enough energy to penetrate surrounding material with visible destructive effects (Figure 13-11). This is not a decay process; but it is spontaneous. When isolated atoms of U-238 have been included within certain mineral crystals or other homogeneous material such as man-made glass, the effects of subsequent fission appear as minute tubes or tunnels, called tracks, that radiate from the parent atoms. These tunnels can be enlarged by acid so as to be countable. The count gives a measure of activity since the beginning of the process. The remaining material may be activated artificially in a neutron field to produce a new set of tracks that are also countable. The ratio of naturally occurring fission tracks to potential total tracks is a measure of age.

Tritium dating

Tritium is an isotope of hydrogen with two neutrons in the nucleus. Like carbon-14, it is chiefly formed in the atmosphere by cosmic radiation. The half-life is about $12\frac{1}{2}$ years, and it is useful in checking the time required for rainwater to penetrate and travel through rocks and to check on the age of wine or other stored foods.

Obsidian hydration

Obsidian is a natural amorphous glass, chemically the same as lava. When a fresh surface is created, as in the making of a spear point, a slow absorption of water begins, which in time creates a visible rind or coat. If conditions have been reasonably stable, the thickness of this rind gives a measure of the age of the artifact (not the obsidian as a whole). The method is of most use in connection with human artifacts and has been used successfully in the range of tens of thousands of years.

Amino-acid dating

All amino acids can exist in two mirror-image forms: left and right-handed or L- and D-isomers. A puzzling fact is that only L-isomers are found in living proteins. Upon death of an organism, a slow change of the L to D form begins and continues until the ratio of right-handed to left-handed molecules is one to one. The process is known as *racemization,* and a complete transition may take hundreds or thousands of years. Ratios can be measured and the age of an organism calculated. A serious drawback is that heat speeds up the process and upsets the calculation. The process is theoretically applicable well beyond the range of carbon-14.

Limitations and criticisms of age-dating methods

All methods of radiometric age dating rest on the following assumptions: (1) the transmutation of one atom to another is a spontaneous or random process, (2) the rate of transmutation is constant under terrestrial conditions and can be determined, and (3) the system has been closed—nothing added or subtracted—during the life of the specimen being examined. It is the last assumption that is the most difficult to maintain. The amounts of original (parent) material and derived (daughter) material are frequently fantastically small, and addition or subtraction of either one during the history of the systems renders any measurement useless. A cracked hourglass from which sand is leaking is not a fit timekeeper.

Some of the products of radioactive decay are gases (i.e., argon, helium, radon), which can escape through minute cracks, especially under elevated temperatures. With carbon-14 it is the addition or subtraction of carbon that is upsetting. Plant roots for example may reach a buried bone or piece of charcoal and add enough carbon to give an incorrect age. Fission tracks are extremely sensitive to heating effects and may be completely erased at even moderate temperatures. Success of the amino-acid method is also dependent on uniform temperatures.

In spite of these difficulties, age dating of ancient rocks is being vigorously carried out in laboratories throughout the world. Investigators have learned to search out and recognize those materials that are most likely to give reliable dates. At the same time, instruments and procedures have steadily improved.

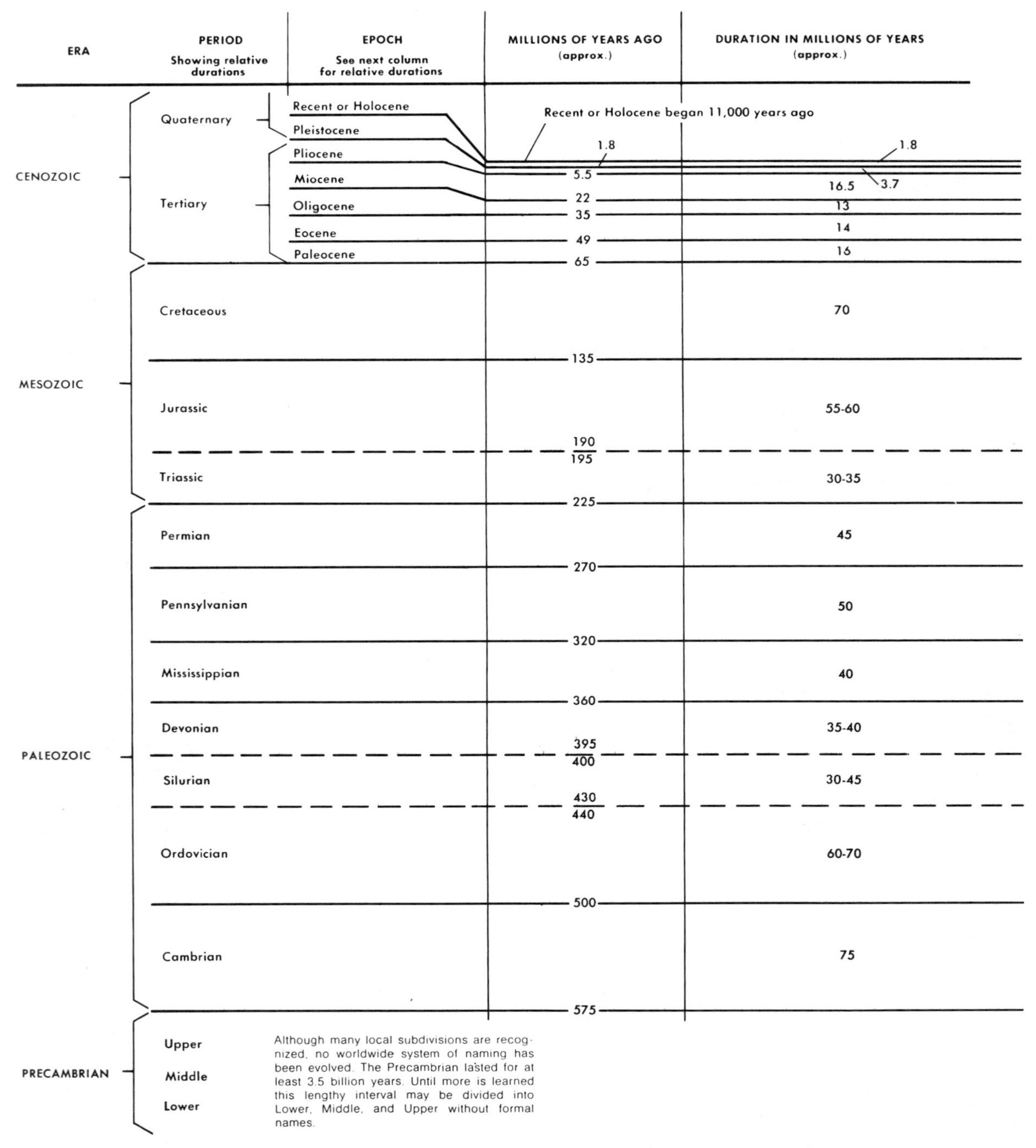

FIGURE 13-12 *The geologic column and time scale with dates and durations as currently estimated. For more information on the relation of this scale to important geologic and biologic events, see inside front cover of the book.*

The geologic column and the time scale

The essential facts of any historical subject are often more comprehensible if presented in tabular chronological form. Geologists have found it useful to relate their information about the past history of the earth to a fairly simple arbitrary outline called the *geologic column* or *geologic time scale* (Figure 13-12). The beginnings of this arrangement go back to the eighteenth century when observers first saw the need for naming the various related groups of rocks they found in nature. It soon became clear that the best method would have to be based on the relative ages of the rocks involved; in other words, the column should portray the chronological order of origin from oldest to youngest.

The most difficult problem was, and still is, the matter of subdividing the record. The units that were first recognized and named are based on what early observers considered to be natural interruptions in the rock record. These interruptions are made evident, at least locally, by actual physical discontinuities (unconformities) and by interruptions in the orderly evolution of life forms. Progress was slow, since there was no general plan, no central authority, and no clear idea about how much or what might ultimately be classified.

The divisions that gradually emerged and are currently used were proposed chiefly in Europe, where geology had its beginning. Although early workers believed their units to be firmly based on worldwide "natural" interruptions in depositions, it is now clear that these interruptions are quite arbitrary. The difficulty of adapting a local classification to a complex worldwide situation has become all too apparent. Adjustments have been made, however, and the present geologic column is, with few exceptions, a product of investigations carried on in Europe during the nineteenth century.

Derivation of names

The derivation of the *period* names will illustrate the unsystematic way in which a framework grew or simply accumulated. The *Cambrian, Ordovician,* and *Silurian* are named for ancient native tribes of Wales and England. The *Devonian* comes from Devonshire, England. The *Mississippian* and *Pennsylvanian* are American names taken from the Mississippi Valley and the State of Pennsylvania, respectively. Europeans do not use these two names, favoring instead one term, *Carboniferous,* to include both periods and so named from the coal content of the rocks. The *Permian* is derived from the Perm Province on the flanks of the Ural Range in Russia. The term *Triassic* comes from a typical threefold association of distinctive formations in Germany, but no particular place is implied. The *Jurassic* gets its name from the Jura Mountains. The *Cretaceous* is not named for a particular locality but for the characteristic high content of chalk (*creta* being the Latin for "chalk"). The *Tertiary* and *Quaternary* derive from still another naming procedure. In one of the earlier schemes, the terms *Primary, Secondary, Tertiary* and *Quaternary* were used for successively older deposits. Of these, only the last two names have survived in general usage. The subdivisions (*epochs*) of the Tertiary and Quaternary get their names from the so-called presence-absence method based on fossil content. For example, of the fossil species found in the *Eocene* (Greek *eos,* "dawn," plus *kainos,* "recent," meaning "dawn of the recent"), from 1 percent to 5 percent are still alive in the eastern Atlantic. Fossils of the *Pliocene* (Greek *pleion,* "more," plus *kainos,* "recent," meaning "more of the recent") include 50 percent to 90 percent of still-living species. Figure 13-13 shows the localities from which the period names are derived and Figure 13-14 is a map of the Paris Basin where so many of the Tertiary Epochs were first distinguished.

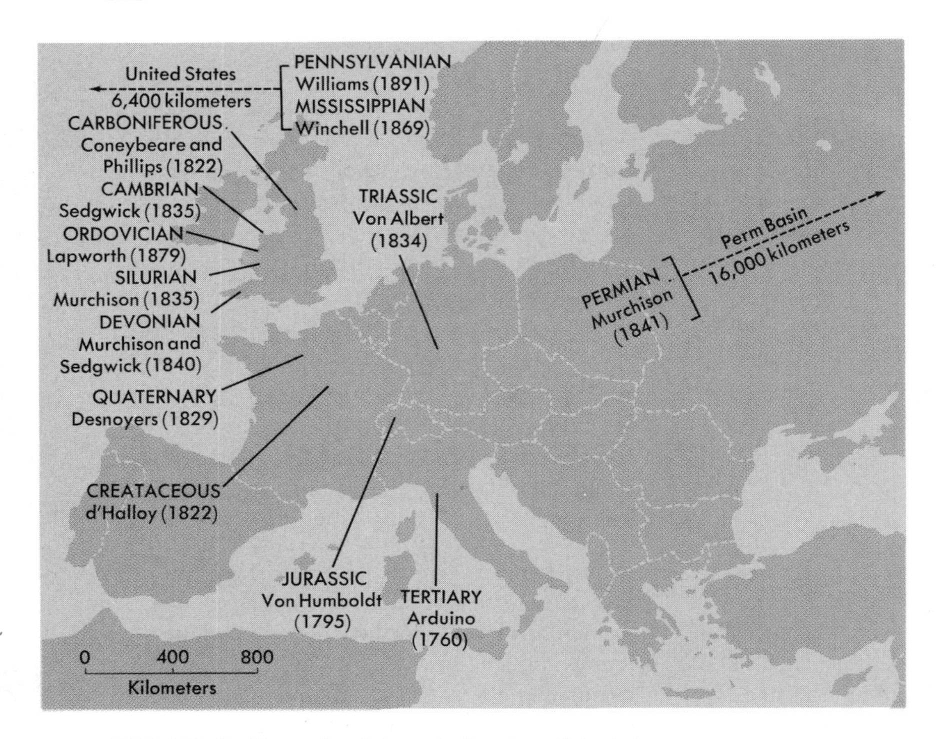

FIGURE 13-13 The naming of the geologic systems. [Adapted from D. L. Eicher, Geologic Time (Englewood Cliffs, N.J.: Prentice-Hall, Inc., 1968).]

Reference to the time scale shows that the periods have been combined in lengthy intervals called *eras.* These are based on the stage of evolution of characteristic life forms. *Paleozoic* means "ancient life," *Mesozoic,* "middle life," and *Cenozoic,* "recent life." Like the naming of the months of the year or the days of the week, the development of the geologic time scale was neither well-planned nor consistent; nevertheless, we seem to have no alternative now but to use these schemes in spite of their imperfections.

What has been said to this point applies to the divisions of actual rock masses into groups ranging from oldest to youngest as these occur around the globe. When we add actual ages in years to these units, we have an absolute geologic time scale. In effect, the terms we apply to time units are the terms that were originally used to distinguish rock units. Thus, we speak either of Cambrian *time* or of Cambrian *rocks.* When we speak of time units, we are referring to the geologic time scale. When we speak of rock units, we are referring to the geologic column.

The geologic time scale is given on the inside front cover. Notice the terms *eras, periods,* and *epochs* across the top of the table. These are general time terms. Thus, we speak of the Paleozoic Era, or the Permian Period, or the Pleistocene Epoch. The rock terms *system* and *series* correspond with the time terms *period* and *epoch,* respectively. There is no generally accepted rock term equivalent to *era,* but the term *erathem* has been suggested.

FIGURE 13-14 *Geologic map of the Paris Basin, site of many important early studies of Mesozoic and Tertiary stratigraphy and paleontology. Here Lyell, in 1833, designated the basic subdivisions of the Tertiary that are now in worldwide use. Adapted from several sources.*

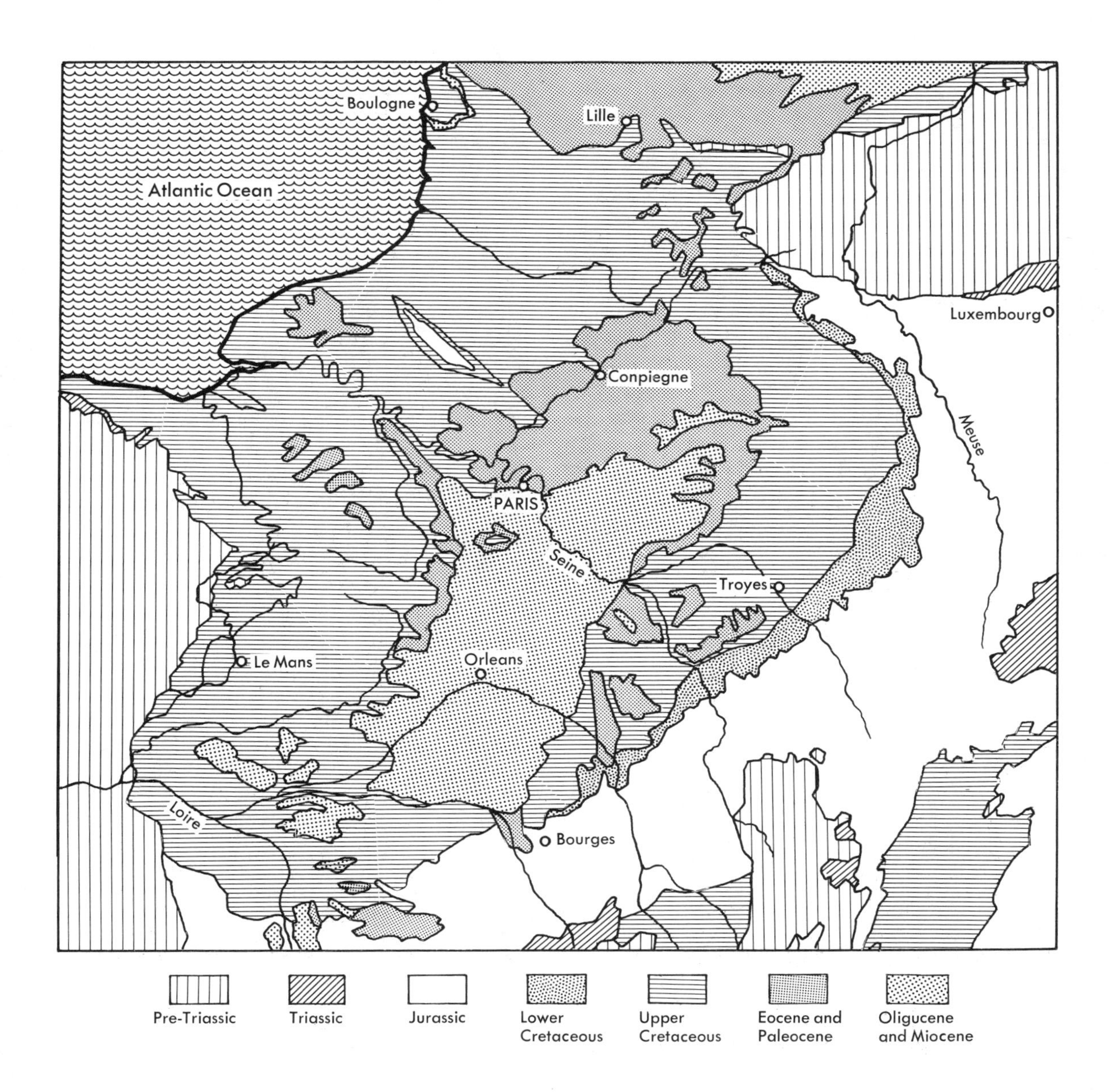

SUMMARY

A major contribution of geology is the concept that the earth is very old. From evidence of seasonal changes, such as tree rings and from annual sedimentary deposits called *varves*, it is thought that the earth has been revolving and rotating in a relatively steady manner for at least 3 billion years.

Attempts to answer the question, "How old is the earth?" on a scientific basis began in the late eighteenth century. Crude estimates based on the amount of salt in the ocean, the rate of deposition of sediments, and the rate of cooling of the earth did little more than indicate that the traditional idea of a 6,000-year-old creation was almost certainly in error.

The discovery of the spontaneous breakup of certain elements—known as *radioactivity*—opened up a new and more certain method of finding not only the age of the earth but also the ages of many rocks and minerals of its crust. Since the rate of breakup of radioactive atoms is not influenced by ordinary terrestrial conditions, a number of natural dating methods are provided by measuring the amount of material already disintegrated (as known from the daughter products) as compared with material that is yet to disintegrate. The chief radioactive transformations that are useful in age determination are uranium to lead, thorium to lead, potassium to argon, rubidium to strontium, and carbon-14 to nitrogen.

Analysis of thousands of specimens has given the following "time table of creation": (1) the earth solidified about 4.5 billion years ago; (2) first extensive permanent rock masses formed about 3.5 billion years ago; (3) fossil record begins about 3 billion years ago; and (4) abundant fossil record commences with the beginning of the Cambrian period about .6 billion years ago.

The geologic time table is an arbitrary arrangement of divisions and subdivisions of geologic time. When the names of the time divisions are applied to actual rocks, the standard geologic column results.

SUGGESTED READINGS

BERRY, WILLIAM B. N., *Growth of a Prehistoric Time Scale*. W. H. Freeman and Co., San Francisco: 1968.

EICHER, DONALD L., *Geologic Time,* 2nd ed. Englewood Cliffs, N.J.: Prentice-Hall, Inc., 1975.

HARLAND, E. B., A. GILBERT SMITH, and B. WILCOCK, eds., "The Phanerozoic Time-Scale," *Quarterly Journal Geological Society of London, Vol. 120 S* (suppl.) (1964).

LIBBY, W. F., *Radiocarbon Dating,* 2nd ed. Chicago: The University of Chicago Press, 1955.

RALPH, ELIZABETH, and HENRY N. MICHAEL, "Twenty-five Years of Radioactive Dating," *American Scientist, Vol. 62* (1974).

TOULMIN, STEPHEN, and JUNE GOODFIELD, *The Discovery of Time*. New York: Harper & Row, Publishers, 1965.

Keys to the Past

Geology is usually regarded as a derived science, meaning that it is based on the more fundamental sciences of physics and chemistry. Even so, it is marked by certain principles and techniques not widely used in other sciences. Geologists employ exact physical, chemical, and mathematical terms in describing the elements and aggregations of elements that make up the earth, but there remain the aspects of time and slow change that require somewhat different treatment. Geology is, in fact, largely historical and is deeply concerned with happenings in the distant past. But the past is not observable and is never repeatable; much of it is too complex for satisfactory mathematical treatment, and even more of it cannot be reproduced by experiment. These difficulties apply especially to historical geology, which deals almost exclusively with change through time.

14

Keys to the past as seen on the eastern shore of Hudson Bay. Light-colored granite on the left is dated at about $2\frac{1}{2}$ billion years. It has been penetrated by a dike of dark diabase of obviously later age. Both rock types have been deeply eroded and given a high polish by the most recent ice age only a few thousand years ago. Finally, a geologist leaves evidence that he has taken note of the past history by taking this photograph. (Photo courtesy of Geological Survey of Canada.)

Faced with the difficulties of reconstructing and interpreting the past, the geologist may have to be satisfied with inconclusive or uncertain explanations. The student of human history faces the same problems. But in both cases the lack of absolute certainty is no excuse for ignoring vital questions that obviously exist. About all that the historian or geologist can do is to "keep digging" and make sure that his explanations take into account what can now be observed and that these explanations do not violate established laws and principles.

The most important general concepts that have guided our thinking about the past may be reviewed at this point.

Uniformitarianism

History shows that there have been a number of ways of thinking about the origin and past history of the earth. To earlier thinkers the past was controlled by all-powerful gods who created all things for largely unknowable ends. With a decline in belief in the supernatural, the earth appears to many modern thinkers as a product of natural processes without plan or purpose. The time element has also been viewed in various ways. When little was known about the earth, its past was considered as of infinite duration, and the possibility that it had passed through an infinite variety of past conditions or cycles appears in many ancient philosophies. In contrast, modern science has discovered evidence that there was a specific beginning and hence a limited duration of the earth.

The idea of an extremely limited past derives chiefly from literal interpretations of Judeo-Christian scriptures, which compress all origins into six 24-hour days. But even the substitution of thousand-year "days" for shorter ones doesn't really help much. The choice is between an extremely long past or an extremely short one.

Belief in a "quick creation" was widespread in the prescientific era. An essential corollary of this belief was the idea that the earth was shaped by vast cataclysmic events unlike anything witnessed in modern time. Noah's flood is held to be one of the last of the universal catastrophes (Figure 14-1). The belief in sudden, world-shaking events is called *catastrophism.* This view, in its original form, is now held by very few informed people.

Catastrophism was opposed by original thinkers such as Leonardo da Vinci, who interpreted what he could see in the rocks and soils in terms of everyday processes. The idea that the best way to understand the past is to study the present has gradually gained ground. A now famous maxim, "The present is the key to the past," expresses the new concept in simplest form.

The term *uniformitarianism,* which came to be applied to the way of thinking opposed to catastrophism, comes from the word *uniform.* Uniform is defined as: "always the same, regular, even and not varying." Some of these synonyms give a more correct idea of uniformitarianism than do others. Geologists do not believe, for example, that conditions of the past have always been the same as they now are, for traces of change are plainly in evidence. It is not the earth or any part of it that is changeless; it is the laws of nature that are assumed to be unvarying in their operations. Uniformitarianism rejects supernatural (miraculous or incomprehensible) effects as long as known natural ones will suffice. It appeals to known laws or principles rather than to unproven or unprovable suppositions. It seeks explanations based on processes that can be observed in action at the present and not those based on pure imagination. It would also favor having these everyday, natural processes acting at approximately the same rates and scales as they now do, provided that these rates and scales could accomplish the observed results in the time available.

FIGURE 14-1 A scene after the Deluge, from a famous painting by Filippo Palizzi. Animals are pictured leaving the Ark to repopulate the devastated world.

Few students care to argue against uniformitarianism insofar as it pertains to the well-established "laws of nature," but there are differences of opinion about the rates or intensity of action of these laws in the past. A catastrophist might contend that the twisting and breaking of strata, the transportation of huge blocks of rock, the violent cutting of canyons, and the wholesale destruction of life are all within the power of a great universal flood—and he would be right. The uniformitarianist, however, believes that these same effects can be explained by less violent operations spread over a longer time span; he thinks he is probably more nearly correct for he has the evidence of actual observation of the present on his side.

There have obviously been more violent and extensive volcanic eruptions and floods in the past than any recorded in human history. And the evidence of a great ice age shows that modern glaciers are feeble compared with those of the past. Craters on the Moon, Mars, and Mercury signify collisions unlike any recorded by man. Yet the principles governing water action, volcanic eruption, ice movement, and falling bodies have remained the same—only the intensity of action has varied.

In the sense that uniformitarianism implies the operation of timeless, changeless laws or principles, we can say that nothing in our incomplete but extensive knowledge disagrees with this idea. It has been by application of this concept (we hesitate to call it a "law") that we have had our greatest success in unraveling the past. By uniformitarianism we achieve the most reasonable and rational explanations requiring not only the least expenditure of energy but also a minimum of hypotheses. Until we have good reason to do otherwise, we will continue to explain evidences of past events by comparing them with what can be seen at present. Our case is well stated by the pioneer geologist, Lyell (1797–1875) (see Figure 14-2),

FIGURE 14-2 Sir Charles Lyell (1797–1875). His writings established historical geology on a firm basis and greatly influenced later workers, including Charles Darwin. (Crown Copyright Geological Survey photograph. Reproduced by permission of the Controller of H. M. Stationery Office.)

who assembled the first comprehensive textbook of geology:

> In attempting to explain geological phenomena, the bias has always been on the wrong side; there has been a disposition to reason *a priori* on the extraordinary violence and suddenness of changes, both in the inorganic crust of the earth, and in the organic types, instead of attempting strenuously to frame theories in accordance with the ordinary operations of nature.

Superposition

The ways in which sedimentary rocks are deposited and preserved have been described in Chapter 6. Sedimentary formations are the chief documents of geologic history. If properly interpreted, each bed of rock or geologic formation reveals a specific event or environmental condition that existed in the past. Thus, one formation may be a product of a shallow sea, another of a dry desert, another of a flood plain. Obviously, one formation tells only a small part of the total story and is as relatively incomplete as one page of a book. To get as much of the story as possible, we must consider all the pages or formations, *and we must consider them in their proper order.* Books are bound with the pages in sequence, and these pages are provided with numbers to aid in keeping them properly arranged. However, the "book of the earth" is neither neatly arranged nor clearly numbered; it is more like an ancient scroll ravaged and fragmented by the effects of time. Patient study is needed to arrange the pieces and read the story.

From observations and reasoning about the relations among sedimentary rock layers comes the second fundamental tenet of historical geology, the *law of superposition,* which is: *In any undeformed sequence of sedimentary rocks (or other surface-deposited material such as lava), each bed is younger than the one below it and older than the one above it.* This statement expresses such a simple and self-evident fact that it seems scarcely worth emphasizing. Nevertheless, it is the most important generalization in the realm of earth history. Figure 14-4 may clarify the concept of superposition. The arrangement of the playing cards in the diagram corresponds very closely to a typical geologic map of a section of the earth with flat-lying sediments. The analogy is good because we are dealing in both cases with relatively thin, but wide, sheetlike bodies that are piled upon one another in an overlapping manner. Note the following features:

1. No one spot on the "map" has all the cards (or formations) present in the area. By ap-

FIGURE 14-3 *Formations in the walls of the Grand Canyon. Because of the scanty soil and vegetation, the formations and contacts between them are well exposed. Differences in resistance to erosion leave the harder formations standing as cliffs, the softer ones forming slopes. The Kaibab Limestone is especially resistant, and its top forms the level plain above the canyon rim. (Courtesy Union Pacific Railroad.)*

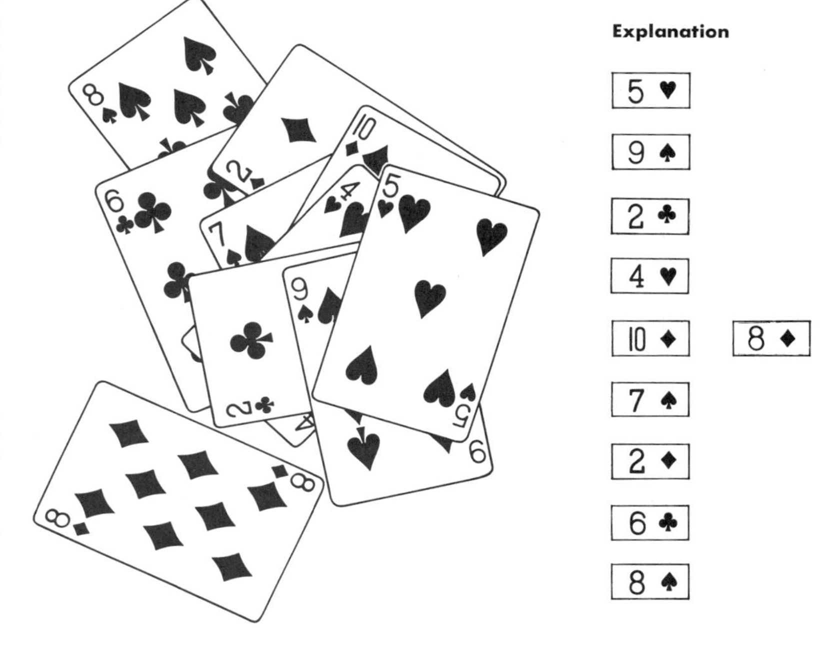

FIGURE 14-4 *A stack of playing cards illustrates the law of superposition. The arrangement from bottom to top shows the order in which the cards were laid down. The boxes on the right give the order of superposition for all the cards except the one that does not touch the others. The "age" of this card cannot be determined by superposition alone. Compare this with the simplified map in Figure 4-5, which shows how superposition is used to determine the relative ages of stratified rocks.*

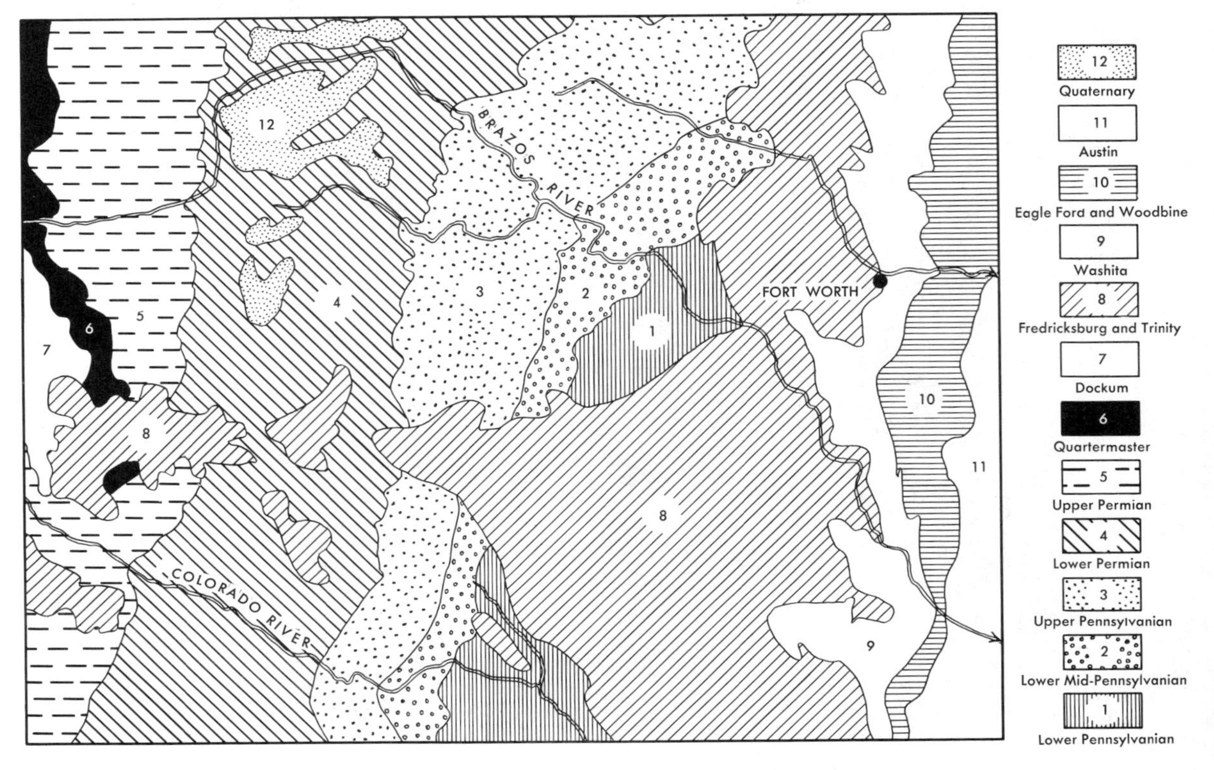

FIGURE 14-5 A geologic map of part of north-central Texas. Here the rocks are all of sedimentary origin and are almost flat-lying. The correct order of deposition can be determined by superposition of the various formations. Problem: How do we know that the rocks on the left side are older than those on the right side, and which way do the rocks of the older and younger sets of formations dip, or incline? Hint: If you were standing on the banks of one of the rivers where it passes from one formation to another, would you see younger or older beds in the hills on either side? Study the explanation for the correct order of superposition and see if you can determine how it was arrived at.

plying the principle of superposition, however, we can determine the correct sequence of all cards except the one isolated by itself. The correct order of cards is given by the key, or legend.

2. Once we understand the arrangement of the cards, we can predict with fair accuracy which ones will be found under any given spot. This procedure is comparable to the methods a geologist uses in predicting what formations will be penetrated by an oil well.
3. A card (or formation) that does not touch or interfinger with other units cannot be placed in its correct position by superposition alone. Such areas must be "dated" by other means, mainly by the use of fossils, as we shall see in the next chapter.

Variations and extensions of the law of superposition

The accumulation of sedimentary layers is a perfect geologic example of superposition and provides indisputable proof of the order in which the layers were formed. But the geologist does not depend entirely on simple superposition in working out geologic history.

Any other relationship that might indicate which of two events occurred first is also useful and significant. Thus, we not only look for masses of rock lying one upon another, but we also observe the relationships of any earth material or structures that may lie within, against, around, or upon other materials. Study Figure 14-5 as a practical example. Other illustrations follow.

Cross-cutting relationships

Although sedimentary beds do not intersect or cut across each other, faults and intrusive igneous bodies commonly show such *cross-cutting relationships.* Faults may cross faults, igneous bodies may cross igneous bodies, and of course, both may also transect sedimentary layers. Conditions can be very complex, but careful observation and mapping usually leave little doubt about the correct order of events.

FIGURE 14-6 *Diagrammatic cross sections showing the effect of two faults on a section of horizontal beds. Notice how the second fault cuts and offsets the first one so that their relative ages can be determined.*

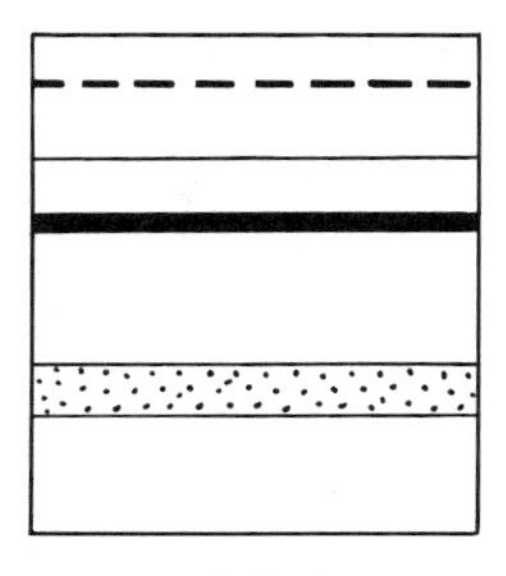

Unfaulted
Sedimentary Beds

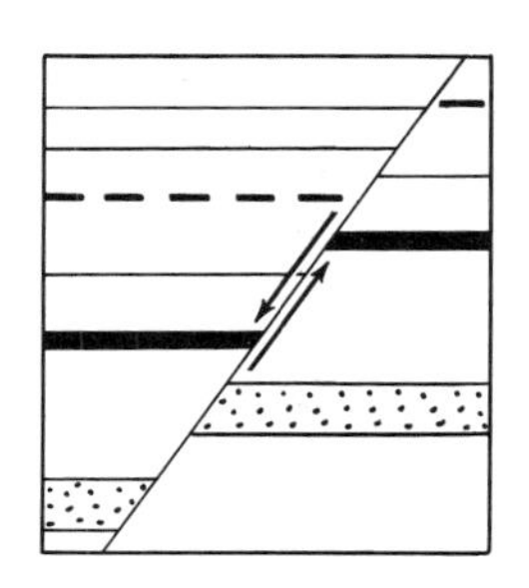

Normal Fault
Lowers Left Side

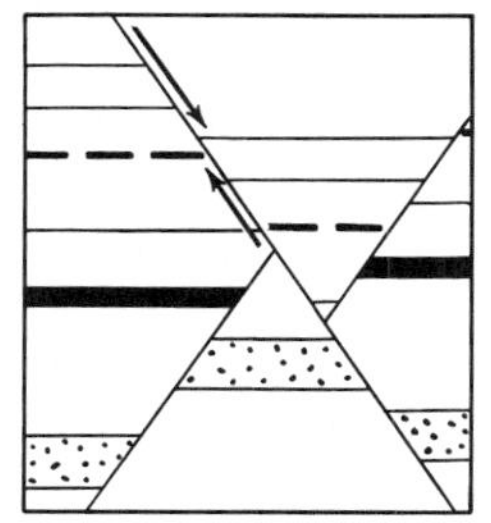

Second Normal Fault
Lowers Right Side

When two igneous bodies intersect or cross, it is obvious that the older one must have been opened up to provide space for the younger one to intrude. In other words, at the place of juncture the younger one is continuous, the older one discontinuous, regardless of the size of the bodies involved.

Faults are merely planes of movement and have no mass. When one fault crosses another, their relative ages must be determined by considering offsetting relationships (Figure 14-6). The effects of earlier faults are disrupted by later ones. Fault relationships are determined by using *key beds,* which in sedimentary rocks are usually thin, distinctive units that can be traced and recognized over wide areas. In general, the youngest fault follows a straight course; older ones are offset or displaced.

Order of growth of crystals and organisms

Geologists are interested in finding out the proper succession of events in the formation of mineral deposits, for such information may yield clues to the location of other deposits. Many minerals form in open spaces that provide ample opportunity for successive generations of crystals to grow one upon another. Many different periods of deposition may thus be recorded in veins and crevices (Figure 14-7). The study of ore deposits is complicated by the fact that certain minerals may grow within other minerals as well as upon them, and that heat and pressure may completely obliterate a great deal of the evidence. The order of events eventually may have to be decided from microscopic study of minute specimens.

Organisms, like crystals, also grow over and upon each other in various ways. Great coral reefs are impressive examples of super-

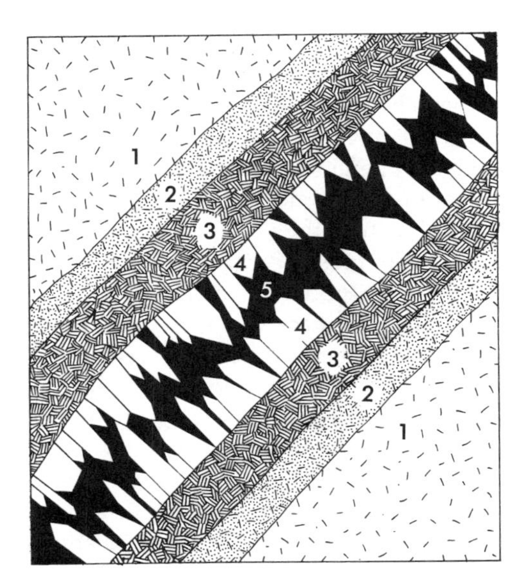

FIGURE 14-7 Succession of mineral deposits within an open fracture to produce a vein: (1) indicates original wall rock, (2) and (3) are successive coatings deposited in the fracture, (4) is a growth of crystalline material, and (5) is unfilled space.

imposed growth. The shells of dead animals become overgrown by living organisms and lime secreting plants, and layers of many generations may accumulate on even a small object.

Intrusive relationships

Igneous material arises from the depths of the earth and may solidify either below or on the surface. Lava flows and beds of volcanic ash are commonly deposited on the surface in distinct layers, and their correct order of formation can readily be determined by superposition. But superposition does not apply to subterranean igneous bodies that have been forcibly injected between or among preexisting sedimentary beds. The igneous bodies known as *sills* originate from molten material that has been forced along bedding planes so as to be parallel with the enclosing layers. After a sill has solidified, it may be folded and faulted in the same manner as the enclosing sediments. Although the rock types composing sills are easily distinguished from sediments, it is difficult to distinguish a lava flow deposited on the surface and then buried from a sill intruded between older sediments. In making this distinction, we note that the heat of a sill will affect adjacent rock both below and above, whereas the heat of a flow affects only the rock below. Then, too, the surface of a lava flow may show bubblelike cavities and cracks, indicating exposure to the air; these effects are absent in sills.

Included, or derived, fragments

Relatively large rock fragments are frequently found embedded in other rocks in a manner that indicates that they must have come from some preexisting source and are therefore older than the matrix in which they are embedded. *Conglomerate,* which is a consolidated gravel (Figure 14-9), is composed mostly of these so-called derived fragments. A typical conglomerate has pieces of all the harder rocks in its source area that were picked up and brought together by stream action. These fragments are usually large enough to contain diagnostic minerals, rocks, or even fossils that can be related specifically to older deposits.

A geologist may have to examine certain fossils very closely before he can say whether they came from an earlier deposit or whether they originated with the formation in which they now occur. Derived or reworked fossils are usually broken and waterworn. Of course, entire skeletons or large logs could not be exhumed, transported, and reburied intact.

Another type of derived material is dislodged and engulfed by igneous material as it moves upward through the earth. Obviously, such pieces must be older than the surround-

FIGURE 14-8 *Exposure of intrusions in the House Range, Utah. A tapering sill of igneous rock extends into the well-stratified sedimentary formation from the left. The sill terminates at a dike that cuts upward to join the large parent mass of igneous rock forming the top of the view. (Photo by Peter Varney.)*

FIGURE 14-9 *Gravel—the winnowed debris of vanished formations. A typical beach or stream bed contains rock fragments derived from all the older formations in the vicinity. A study of the pebbles in gravel or conglomerate frequently yields valuable clues concerning the order of geologic events. (Photo courtesy of U.S. Geological Survey.)*

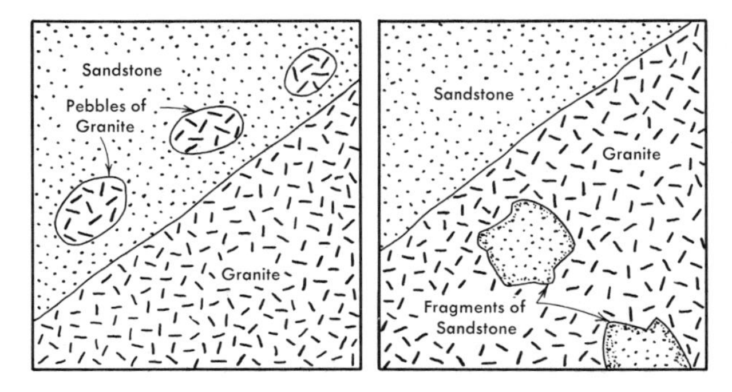

FIGURE 14-10 *Age relationships of intrusive sedimentary rocks. In the sketch on the left, pebbles of the granite rock are found in the overlying sandstone, proving that the igneous rock is older. In the situation on the right, unmelted pieces of sandstone are shown within the granite, indicating that the sandstone is older.*

ing matrix. Unmelted pieces of sedimentary rock in a large granite mass, for instance, prove the sediment to be older. If, on the other hand, we find rounded pebbles of granite in a sedimentary bed, we know the granite is older. Figure 14-10 illustrates examples mentioned above.

Succession in landscape development

What we have already said about methods of determining the order of geologic events applies chiefly to solid rocks. Our discussion would be incomplete without some mention of the methods used in unraveling the history of the surface features of the earth's crust. Even the most monotonous landscapes show the effects of successive events and changing climates, and we know that the features were not all produced at the same time. The geologist has the interesting problem of reconstructing the history of landscapes from the various forms that compose them. For example, the peaks of many high mountain ranges were sharpened and furrowed by glaciers that have melted away, and the areas once occupied by ice are now being reshaped by running water. In many areas there are steplike terraces or beaches that were cut by the waves of lakes that have since dried up. These old lake beds may be invaded by sand dunes or cut by streams, depending on subsequent climatic changes.

A study of the changes in rivers, lakes, deserts, and glaciers involves the fundamental idea of superposition, but there is the added complication of explaining the surface forms. River terraces, especially, are highly instructive but difficult to interpret, for streams are very sensitive to climatic changes, and the same river may run slowly or swiftly, swell in volume or dwindle away, and remove or deposit sediment, all within the same valley, or flood plain. Thus a terrace may have been cut on bedrock or on previous river deposits; or it may have been made by a small river meandering from side to side or by a large river occupying the entire valley. The river valley may have been filled and partly emptied of sediment several times, leaving a variety of deposits to record its changes.

Lakes in arid regions expand and shrink as climates change. The rise and fall of water levels are recorded by terraces and beaches, but it is surprisingly difficult to determine just what the true order of events has been. Likewise, glaciers in expanding and retreating leave surface evidences and thin layers of debris to record their fluctuations.

In dealing with the history of landscapes, we look chiefly for evidence indicating which of two events has effaced or disturbed the other. The sequence of cutting and filling of river valleys can be determined by noticing which deposits cut into or lie against or on other deposits; contrary to the situation in canyons where bedrock is exposed, we may find that older deposits make up higher terraces and younger ones lower terraces. For examples of cut-and-fill relationships, see Figure 14-11.

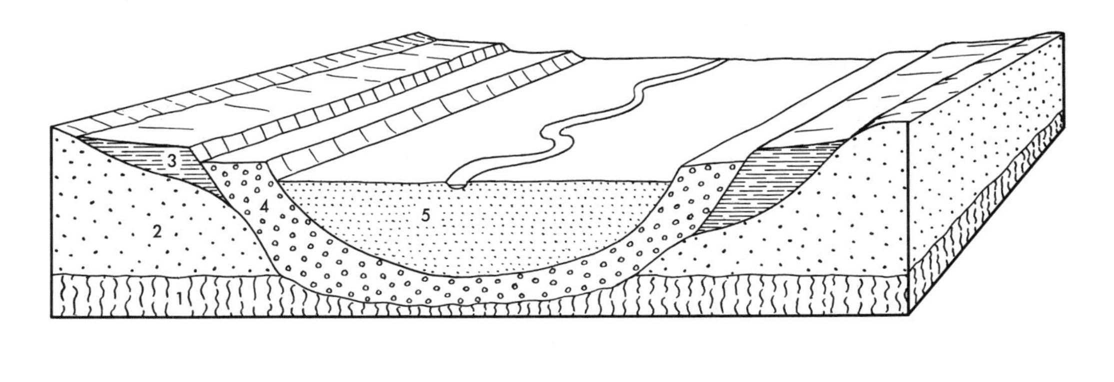

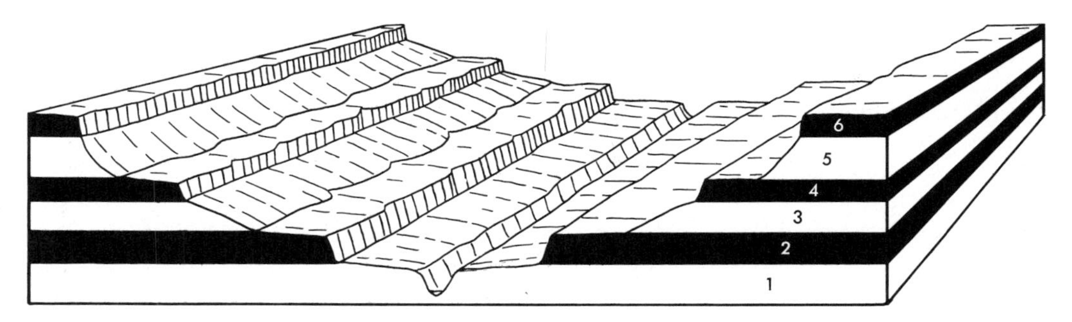

FIGURE 14-11 *Terrace formation. In the lower figure the valley profile is steplike, as in the Grand Canyon. The different ledges and slopes are arranged according to superposition: oldest at the bottom, youngest on top. In the upper diagram the effects of cut and fill show that the successive benches are older above and younger below. The true order of age is determined by superposition in all cases. Thus, in the upper figure the valley was once filled with material (3), which was cut into and mostly removed. Then, number 4 was laid down on and against number 3, and the process was repeated. Terrace formations and cut and fill can be very complex processes.*

Problems of applying the law of superposition

We have already inferred that the law of superposition does not apply to rock layers that have been greatly disturbed after their deposition. Beds that were originally perfectly level may be broken, tilted, and even turned completely over during mountain-building movements. It is common in mountainous areas to see beds that have been tipped up at steeper and steeper angles until they are standing on end. If they go beyond this angle, they are said to be *overturned,* and the bedding planes that originally faced upward now face downward. Superposition cannot apply to rocks in this disturbed position, and they must be restored, in theory at least, to their original position before they can be correctly interpreted. The problem of determining which side of a bed of rock was originally up and which down presents interesting possibilities for geologic detective work. For example, the sharp crests of ripple marks point upward and their rounded troughs curve downward if the bedding planes occur in their original position. The reverse appearance indicates overturning (Figure 14-12). Also small channels that were cut and filled before burial are convex downward if undisturbed. If turned over, they are convex upward and appear unnatural.

Folding is not the only action by which the order of rock layers is reversed. Actual breaking or fracturing of strata may also force

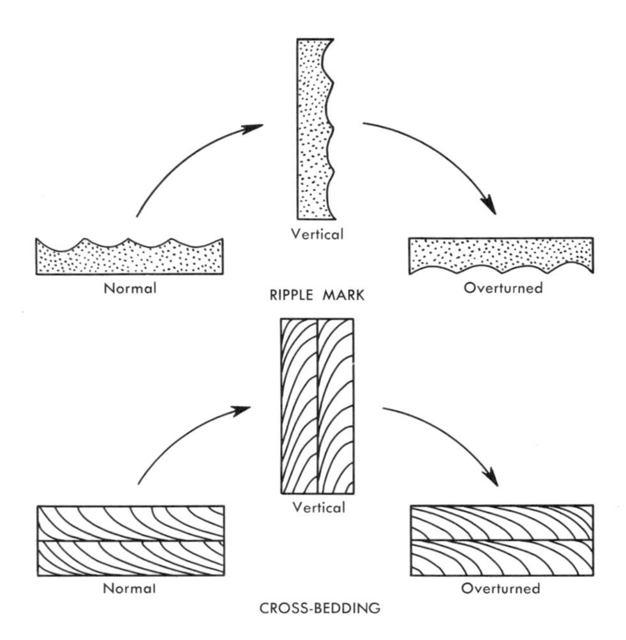

FIGURE 14-12 *(Left) The overturning of sedimentary layers illustrated by ripple marks and crossbedding. The distinctive appearance of certain original features enables us to know whether they are still in their original position or if they have been tipped up and overturned.*

FIGURE 14-13 *(Below) Chief Mountain, Glacier National Park, Montana. A great overthrust fault has carried the older mountain rocks forward over the younger rocks of the foothills. Chief Mountain is but a small erosional remnant of a much larger sheet of rocks. The movement on the fault has been estimated at 15 miles.*

FIGURE 14-14 *(Below) The man is standing on a large block of Devonian sandstone, which fell into and became part of a formation of Jurassic age, Port Gower, Southerland, England. Difference in age of the block and the enclosing sediment is about 200 million years. (NERC Copyright. Photograph reproduced by permission of the Director, Institute of Geological Sciences, London.)*

older rocks upon younger ones. If vast sections of the earth are powerfully compressed, faults will appear along which great masses are pushed upward and forward. These faults are called *thrust faults* or *reverse faults.* There are areas in which dislocated masses have moved forward many miles to occupy positions on top of rocks of entirely different age and origin. If, during these movements, rocks that were originally older come to rest on younger ones, the law of superposition obviously cannot apply. All such instances must be carefully examined; and if the rocks are well exposed, fault planes will be discovered with evidences of intense mechanical disruption, such as crushed and broken rock material and gouged or grooved surfaces that indicate movement.

Closely allied to the blocks moved by thrust faults are large landslide masses. The force of gravity acting on elevated rock masses, especially those that are saturated with water or those in submarine environments, may cause them to slip downward over lower beds. The process of sliding may be slow or rapid, and the masses involved may cover many square miles. If older rocks slide over or fall on younger rocks, we have another instance of deceptive superposition (Figure 14-14). Landslides and similar materials usually show many indications of breakage and contortion and may even occur in the form of broken fragments called *breccias.* Careful study may reveal the source of the disrupted rocks and may provide a logical reason why they have moved from their original positions.

Reconstructing past events and ancient environments

Even though a long series of events may have disturbed the sedimentary rocks of a region, it is still possible to reconstruct the major features of the past and to apply the law of superposition. The reconstruction must be done diagrammatically or with models, and it always commences in the present and proceeds backward in time. In effect, the geologist graphically removes the evidences of each significant event to reveal the condition of things before the event took place. Thus, if the last event in a particular area was a lava flow, the geologist "removes" the igneous rock to reveal a picture of conditions as they were before the eruption occurred. If faulting had preceded the lava flow, he moves the faulted strata to their original unbroken position. If there is evidence of folding, he graphically straightens out the beds to a horizontal position. Eroded material he "replaces" as far as possible to recreate mountains or larger land areas. Unconformities and losses in the record will leave many gaps, but substitute evidence may be recovered in nearby areas. Thus, through painstaking mapping and study, the broad features of the history of any area of sedimentary rocks may be reconstructed. A graphic illustration of geological reconstruction is presented by Figure 14-15.

The application of what can be learned about the present to an understanding of the past is becoming increasingly important in very practical ways. In the search for coal, oil, uranium, and other essential commodities in stratified rock geologists have developed a major discipline called *environmental reconstruction.* By this is meant that an attempt is made to reconstruct the total natural setting in which a particular deposit originated. Some coal beds, for example, are long and narrow while others are broad and sheetlike. Why the difference? It is obviously important to know the probable shape of a coal bed before an area is considered for purchase, development, or mining. Geologists know that long narrow coal deposits have accumulated in sunken troughs between parallel faults that are likely to form under certain stresses and in certain areas.

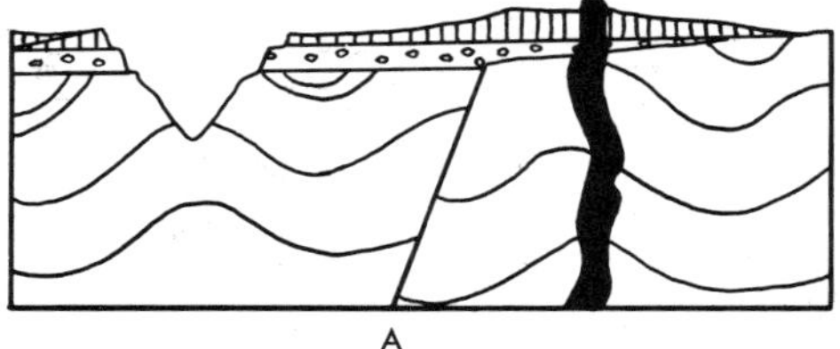
A
Present Condition

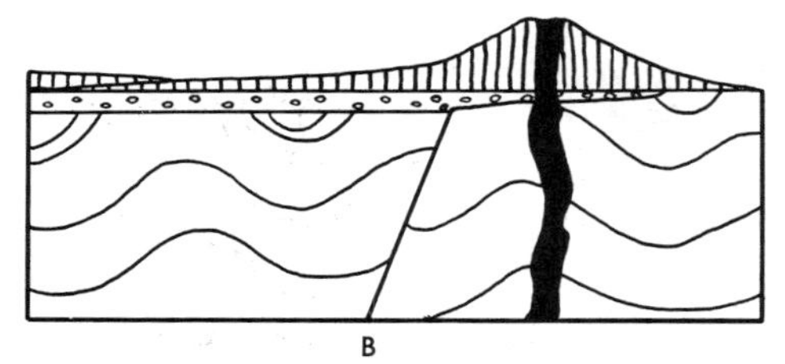
B
Effects of Erosion Removed

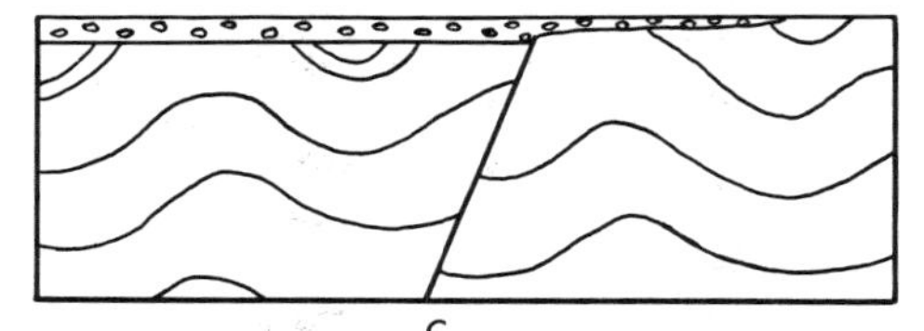
C
Effects of Volcanism Removed

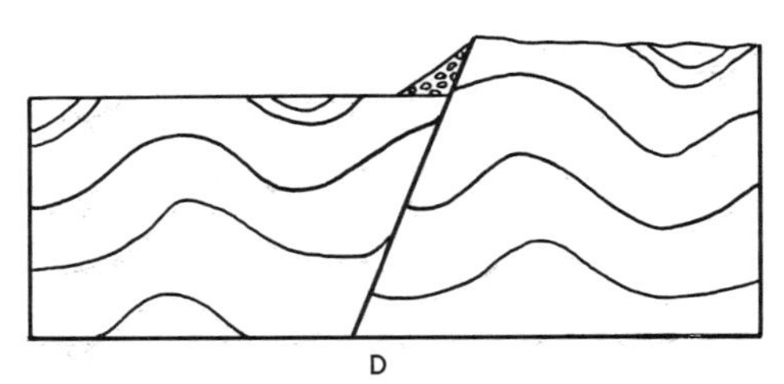
D
Eroded Material Removed and Fault Restored

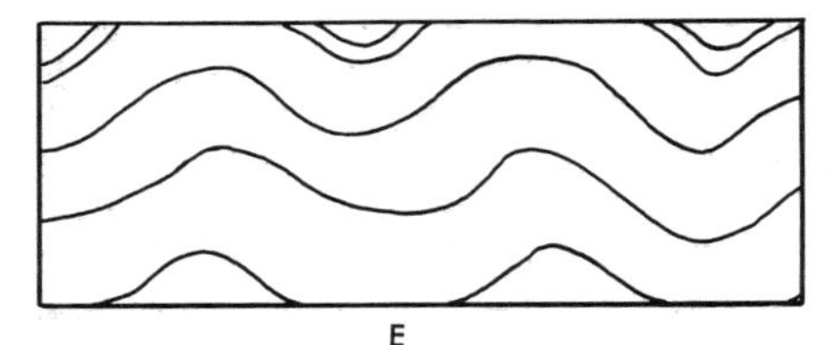
E
Effects of Faulting Removed

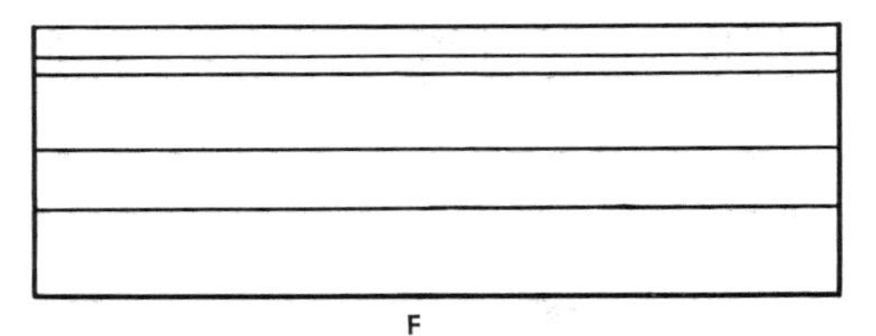
F
Effects of Folding Removed

FIGURE 14-15 *Determining the sequence of past events. (A) shows the present landscape and subsurface conditions. Removing the evidence of successively older events reveals the past history of the area. In (B), the effects of erosion (as shown by the canyon) are removed, showing that the lava flow was continuous across the area. In (C), the volcano and its subterranean conduit and surface flows are removed, exposing what was once a gravel-covered plain. In (D) the original height of the fault is removed by sliding the beds back to their original positions so that corresponding beds match. This reveals the folded condition of the beds that preceded the faulting. Finally, in section (F), the beds are unfolded, and the original condition of the area is revealed.*

Blanketlike beds are produced on shallow shelving areas fronting the ocean under relatively quiet conditions.

Many great oil fields are found in association with ancient reefs built up by algae, corals, and other organisms. Modern reefs are intensively studied by geologists in order to better understand their ancient counterparts. An outstanding and much-studied example is the Guadalupe Mountains, West Texas (Figure 14-16). From such studies it is known that reefs may take many forms: long and narrow, moundlike, blanketlike, and highly irregular depending on their relation to the sea bottom or shoreline to which they are attached. Through study of the enclosing sediments and organisms a petroleum geologist can go far in predicting favorable sites of oil accumulation in ancient buried reefs.

Still another example: the uranium deposits of the Colorado Plateau are found in discontinuous sand bodies that have been identified as the filling of ancient river channels. Furthermore the ore is associated with fossil plants buried in and along the ancient streams (Figure 14-17). By study of the habits of modern rivers and the shapes of the deposits the location of uranium ore bodies can be made less risky and expensive. Geologists who are trained and successful in analyzing ancient environments are currently much in demand and their studies have saved a great deal of time and energy in exploration programs.

FIGURE 14-16 *The Guadalupe Mountains, west Texas. Known to geologists as a Permian reef complex, the rocks and fossils have been studied intensively. It should be understood that the reef was actively growing in Permian time, was subsequently buried, and is now being revealed by erosion. Underground, such complex porous masses constitute important oil reservoirs. (Photo courtesy U.S. Geological Survey.)*

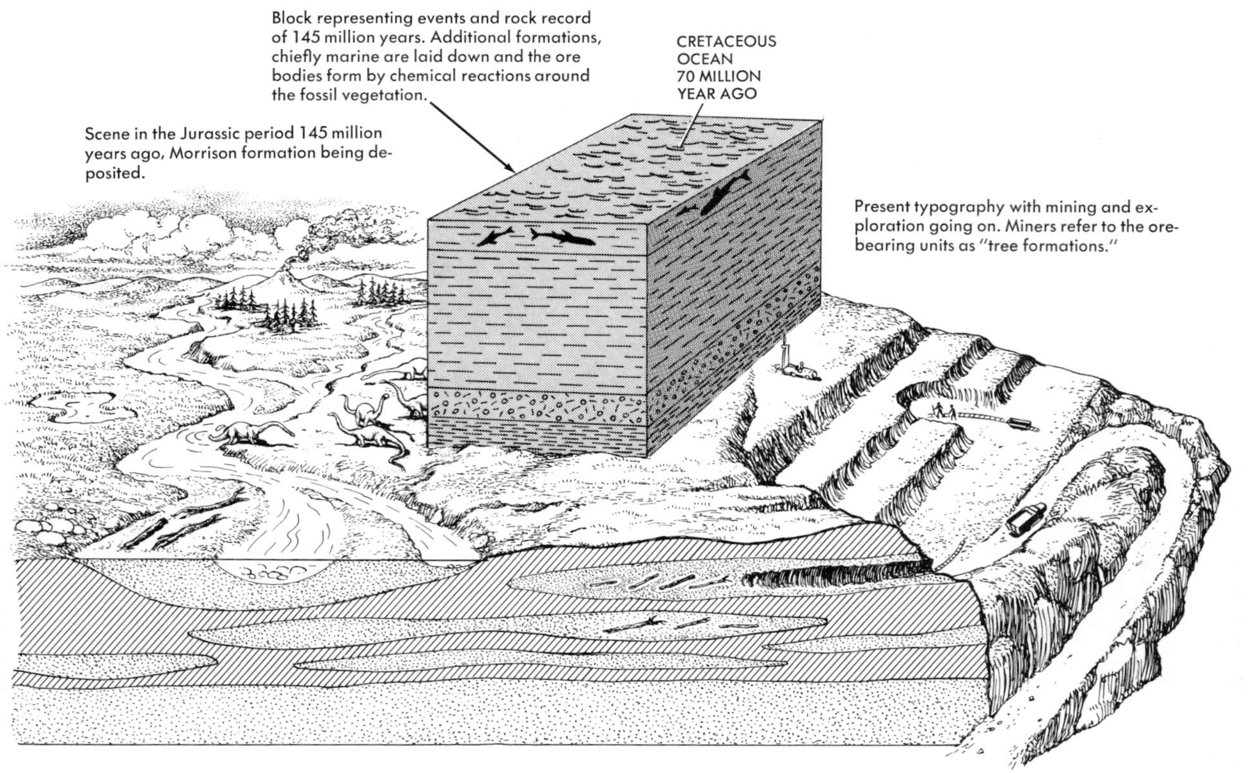

FIGURE 14-17 *The environmental reconstruction of uranium ore bodies in the Morrison Formation of the Colorado Plateau. By study of the ore deposits as they are discovered and mined as shown in the right of the diagram, the previous geologic history including the environment in which the sediments were laid down has been reconstructed. An understanding of the role of fossil logs and other organic material in localizing the ore bodies aids in exploring in the maze of sandstone channels that make up much of the formation.*

Faunal succession

A third generalization of the greatest importance in thinking about the past history of the earth is the *principle of faunal succession,* which is: *Groups of fossil plants and animals succeed one another in a definite and determinable order and each period of time can be recognized by its respective fossils.* Note that this statement does not specify the manner of origin of the different groups of fossils; it is a fact regardless of whether or not organic evolution is true.

The principle of faunal succession was not conceived as an unsupported theory; it was discovered through the accumulation and study of thousands of fossil collections from all parts of the earth. Anyone who is interested can repeat the process of collection and comparison if he wishes, as long as he is careful to apply the law of superposition to his work. When superposition is taken into account it is always found that the oldest abundant metazoan fossils are the trilobites. An Age of Trilobites can be recognized on all continents. In successively higher and younger rocks we recognize an Age of Fishes, an Age of Coal Forests, an Age of Reptiles, an Age of Mammals, and an Age of Man (Figure 14-18). These terms, of course, pertain to the groups of organisms that were especially plentiful and characteristic during certain periods. Within each of the great "ages" there are numerous minor subdivisions marked, for example, by certain species of trilobites, certain kinds of fish, and so on. That the same succession of ages is found on each major landmass, never out of order and never repeated in the same area, proves uniform worldwide similarity and correlation.

It should be understood that not all fossil-bearing periods are present everywhere. The Grand Canyon, for example, is a magnificent display of rock layers, but it shows only five of the twelve major fossil-bearing periods of earth's history. Records of two of the missing periods were never deposited in this area, but were laid down in southern Nevada a short distance away. Here they are found in proper order "sandwiched" between the formations that are present in the Grand Canyon.

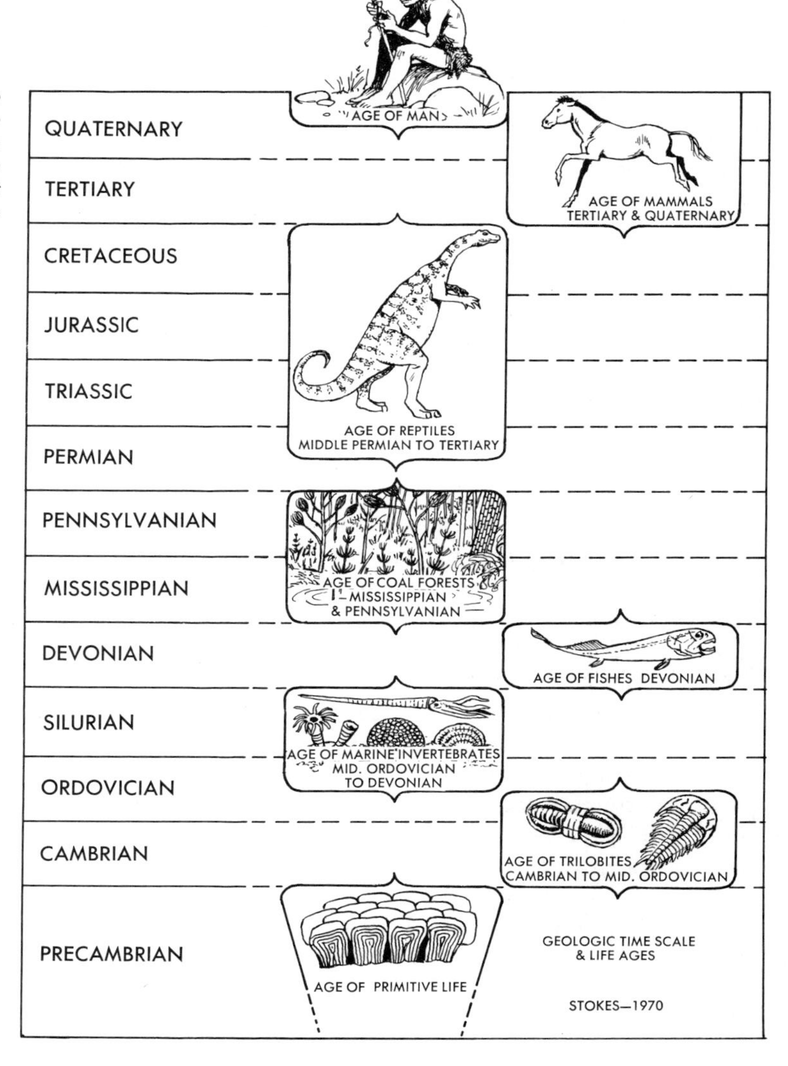

FIGURE 14-18 *Relation of the great life ages to the standard geologic time scale. The successive ages are strictly informal designations for certain intervals dominated by certain life forms from which they get their names. The ones given here are not universally used, and, of course, others may be designated for additional subdivisions.*

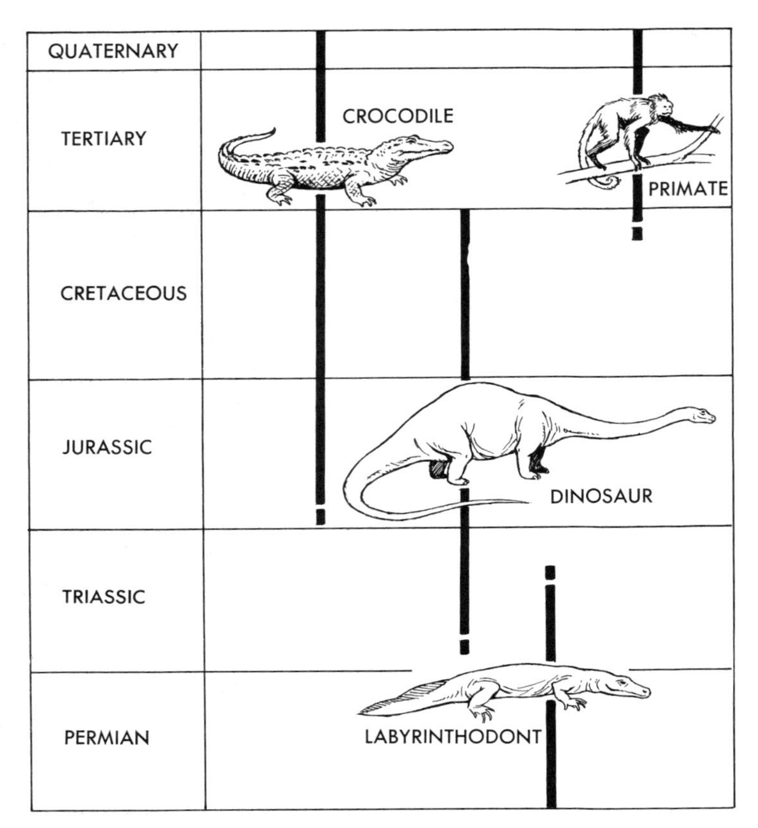

FIGURE 14-19 *Diagram illustrates the usual method of showing the "life lines" or times of existence of biologic groups. The standard geologic periods, named on the left, give proper time frame. This illustrates the discovery that certain groups are found to have lived at the same time, and that others did not. Thus, early dinosaurs lived with labyrinthodonts and late dinosaurs lived with primates, but primates did not live with labyrinthodonts. New discoveries may add to the length of a line but seldom shorten it. Broken lines indicate where possible slight lengthening may be expected.*

The other five periods are too young to be represented in the canyon; they are found in the high cliffs and plateaus of southern Utah above and some distance back from the canyon rim.

Sedimentary layers do not go entirely around the earth like the rings of an onion; they resemble more the leaves of a cabbage which overlap and interfold. Even in a cabbage the relation and relative age of any leaf can be determined by applying the idea of superposition. Superposition and faunal succession go hand in hand and the geologist in the field uses both simultaneously. Nevertheless, superposition is basic and proves faunal succession.

Biotic association

The principle of faunal succession can be proven in a second way, by *biotic association,* or the fact that specific fossils or groups of fossils do or do not occur consistently together. Proof comes again from study and comparison of thousands of collections of fossils. For example, trilobites are commonly found with cup corals but never with oysters; dinosaurs occur with cycads but never with horses. We must understand that there is not a complete replacement of one group of life forms by an entirely new group as once supposed, rather there are overlapping histories throughout the entire record. Thus, the earliest crocodiles lived with dinosaurs while later crocodiles lived with man, but man and dinosaurs did not live together. Figures 14-19 and 14-20 illustrate this fact.

The progressive evolutionary changes of life on earth can thus be demonstrated by two methods: the study of superposition and the study of associations of organisms.

Understand clearly that the principle of life succession does not mean that simple or unspecialized animals are not found in young rocks or that complex and highly specialized organisms are not present in old rocks. The principle applies only to comparisons made between the total life of successive geologic intervals and between members of the same general group. Thus, a simple sponge from a later period of time cannot be compared with a complex crustacean from an early period. Sponges must be compared with sponges and crustaceans with crustaceans. When we trace successive members of the same group through time or compare the total life of a later period with that of an earlier one, we see that significant changes have occurred. Whether or not the changes are always progressive might be debated as a philosophical point, but it seems safe to say that there is an ever-increasing tendency toward a more complete utilization of all energy sources, better adaptation to the environment, and greater complexity of organization with the passage of time.

Faunal succession as revealed by the study of fossils does not constitute positive

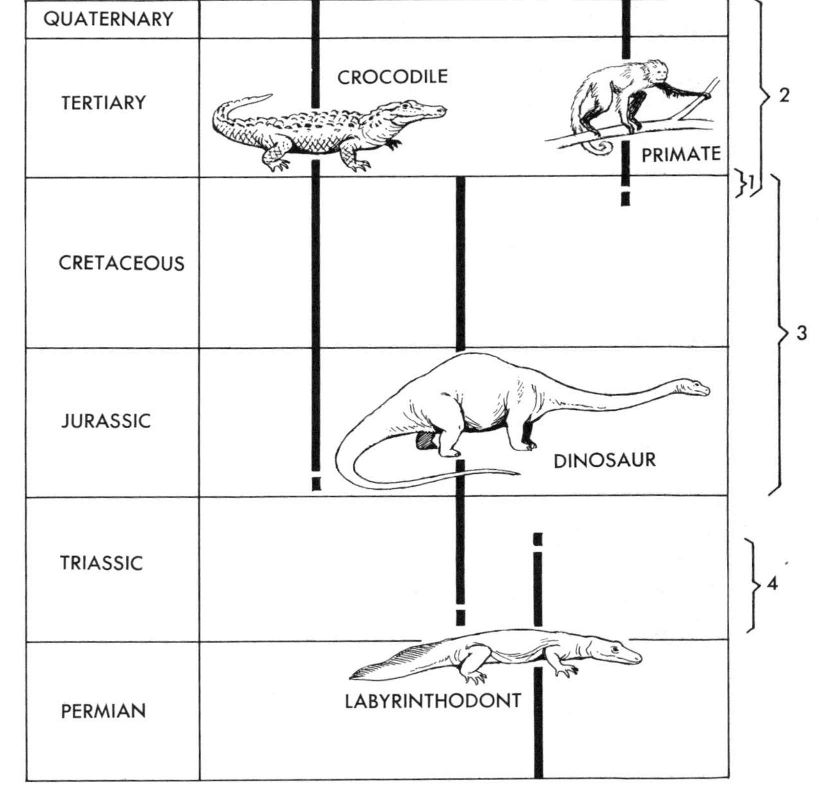

FIGURE 14-20 This is the same diagram used in Figure 14-19, repeated to illustrate the concept of fossil zones and their use in correlating time periods. The rock laid down during the time of existence of any species or other designated aggregation of living things is called a zone. Material laid down during the total existence of a specific group, as, for example, the dinosaurs, is called a range zone (# 3 of diagram). Material laid down during the time when two specified groups lived together is called a concurrent range zone, e.g., the overlap of the dinosaur-bearing and labyrinthodont-bearing deposits (# 4 of diagram). A third zone is the assemblage zone in which a number, perhaps all, of the contained animals are used to identify it. Thus, there is a zone of dinosaurs, primates, and crocodiles in the above diagram (# 1 of diagram).

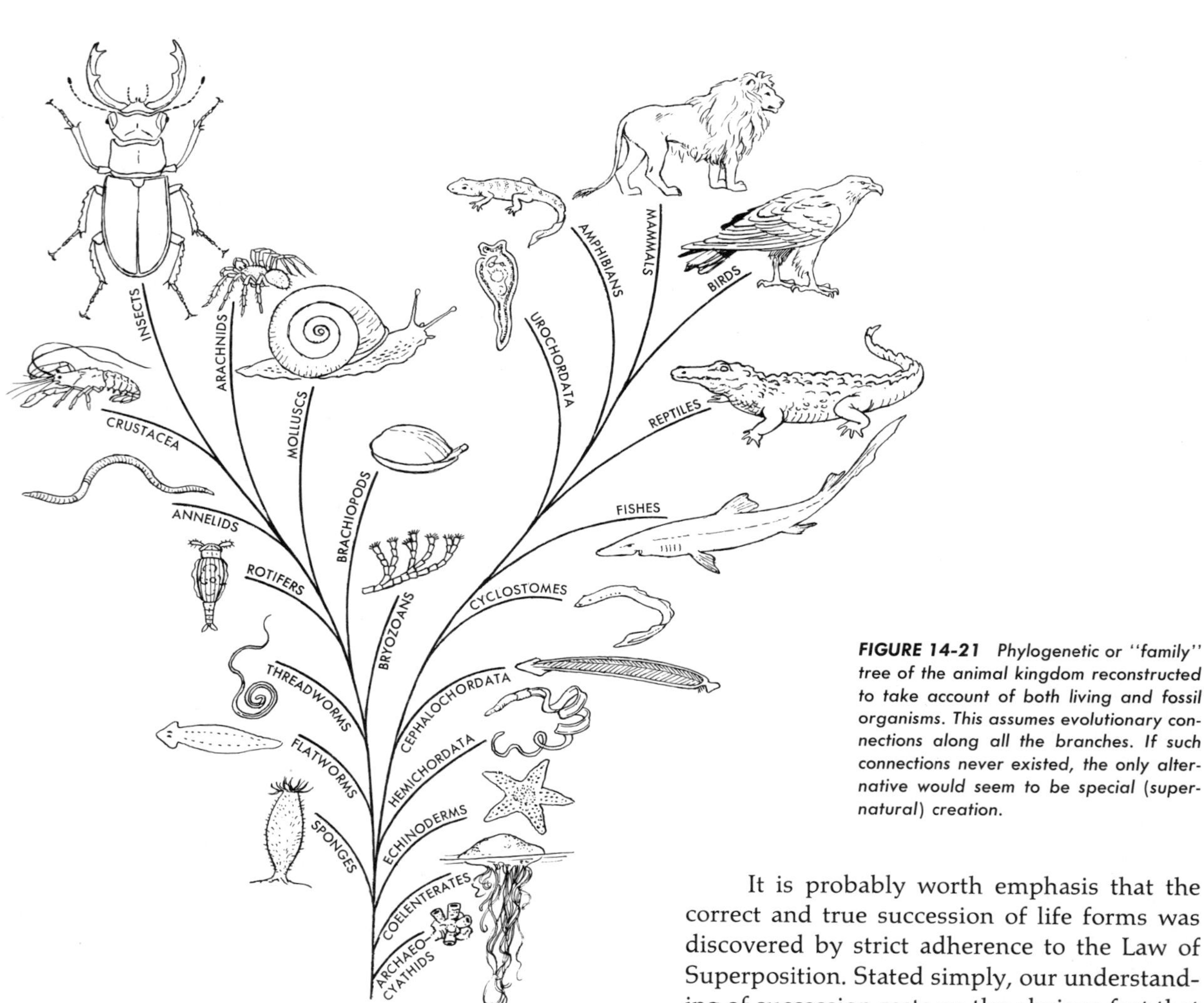

FIGURE 14-21 *Phylogenetic or "family" tree of the animal kingdom reconstructed to take account of both living and fossil organisms. This assumes evolutionary connections along all the branches. If such connections never existed, the only alternative would seem to be special (supernatural) creation.*

proof of organic evolution, in spite of the fact that most students regard it as such. Absolute proof of organic evolution would mean the positive demonstration that all life forms descended naturally from preceding ones, back to a simple beginning. This would require an endless succession of individuals, each with "birth certificates" that can never be supplied. Faunal succession is more than a theory, however, for it shows that life has indeed changed and progressed with the passage of time.

It is probably worth emphasis that the correct and true succession of life forms was discovered by strict adherence to the Law of Superposition. Stated simply, our understanding of succession rests on the obvious fact that if rock layers are progressively older downward so also are their fossil contents—the older the strata the older the fossils. Contrary to what many critics of evolution may claim, geologists do not invent faults or other devices to force rock strata into an order which agrees with their preconceived ideas of evolution. The law of superposition reigns supreme no matter where it leads. Once the fossils are arranged in their true natural order according to superposition it is obvious that they show an orderly progression from simple to complex, an order that is best explained by organic evolution. Figure 14-21 is a simplified "family tree" drawn to take account of both living and dead organisms.

SUMMARY

That the universe is governed by law is accepted by all scientists. There are, however, many differences of opinion about what the laws are. A well-founded natural law ought to be universal and timeless in its application and one that can be relied on to govern similar situations in the same way at all times and places. We have learned by experience that many so-called laws are not really as fundamental and reliable as once supposed. For example, "the law of the conservation of matter" now is stated as "the law of the conservation of matter *and energy.*" Scientists are much more cautious in proposing laws than they were a few decades ago. We may do better by referring to generalizations of an unprovable nature as principles, explanations, rules, or beliefs.

In line with this note of caution we refer to the *concept of uniformitarianism* and the *principle of faunal succession,* leaving open the possibility that these "laws" are not to be applied as absolute and invariable at all times and places. On the other hand (and the writer's colleagues may not all agree), it is safe to refer to the *law of superpositions* as invariable and changeless. Be this as it may, geologists base their conclusions chiefly on these three rules of thought. So far we have not thereby been led into impossible contradictions or unprofitable avenues of investigation.

After the concept of uniformitarianism, a second important guide is the law of superposition. Simply stated this is: In any sequence of undisturbed sediments, any bed is older than the one above it and younger than the one below it. This concept may be extended to include any situation where it can be determined from physical evidence that one event preceded or succeeded another. Igneous bodies, faults, and surface features can be placed in order of origin.

The third guide in the study of the past is faunal succession: Groups of plants and animals succeed each other in a definite and determinable order, and any period of time can be recognized by its respective fossils. This was discovered to be a fact as fossils were collected over ever-expanding parts of the earth. The same succession of life forms is found on every continent, always in the same order and never repeated in the same area. The application of the principle has made possible worldwide unified mapping and correlation of sediments. Faunal succession is proven also by biotic association, by the fact that certain organisms are found to have lived consistently together at the same time while others did not.

SUGGESTED READINGS

ADAMS, FRANK DAWSON, *The Birth and Development of the Geological Sciences.* New York: Dover Publications, Inc., 1954.

ALBRITTON, C. C., JR., ed., *The Fabric of Geology.* Reading, Mass.: Addison-Wesley Publishing Co., Inc., 1963.

CLOUD, PRESTON E., ed., *Adventures in Earth History.* San Francisco: W. H. Freeman and Co., 1970.

LAPORTE, LEO F., *Ancient Environments.* Englewood Cliffs, N.J.: Prentice-Hall, Inc., 1968.

Origin of Life and the Meaning of Fossils

Men have conceived many theories as to the origin of life. Those who believe that life is here through chance combinations of the necessary elements, without intelligent outside direction or influence, are called *mechanists.* The mechanists are opposed by the *vitalists,* who do not deny that the proper combinations are essential but believe that the combination and preservation are controlled by supernatural or divine influences. One need not belong to either of these schools; an agnostic simply hasn't decided one way or the other. Only recently has it been possible to cast aside myth and guesswork so as to study the origin of life objectively. Many sciences—including astronomy, biology, chemistry, and geology—are involved. Obviously such a vast and intricate subject cannot be fully discussed in the limited space of this text, and priority must therefore be given to certain geological aspects. We may

15

Disarticulated fossil bones tell of mass death about 15 million years ago in western Nebraska. (Photo courtesy of Chicago Museum of Natural History.)

begin with an obviously leading geological question: What was the physical condition of the earth when life first appeared on its surface?

Note that we speak of the *appearance* of life rather than the *origin* of life. The possibility that life came to earth from elsewhere is the basis for the theory of "*panspermia.*" According to this theory, life is widespread in the universe and arrived on earth from space in the form of complex key molecules or sporelike entities. The discovery in space of a great variety of molecules that are basic to the formation of living things opens up interesting speculations that are not entirely unfavorable to the theory of "panspermia." Obviously the question of life on other planets is related to the problem of how it came to be on earth.

Assuming as a point of departure that life originated at least once on earth, we are led to inquire what the prerequisites were that existed at that time. The problem is not greatly different from that confronting a chemist seeking to produce life in the laboratory. Prerequisites in either case would be: (1) essential elements, (2) proper temperature relations, and (3) energy sources. Consider these in order.

Essential elements

Disregarding any intangible constituents, all organisms seem to require at least 20 chemical elements and perhaps traces of several others. Hydrogen, carbon, nitrogen, oxygen, phosphorous, and sulfur make up at least 95 percent of protoplasm; other less prominent constituents are potassium, sodium, magnesium, calcium, and chlorine. Also required are traces of iron, copper, zinc, manganese, molybdenum, boron, fluorine, silicon, and iodine.

The mere presence of these elements is not enough to insure the appearance and subsequent well-being of life. All essential elements are present on the moon but not in the right proportions. On earth the elements that constitute water and other atmospheric constituents are notably abundant and in a constant state of motion and flux. On other planets water might be vaporized or frozen solid.

Proper temperature relations

Life as we know it cannot exist if conditions are too hot or too cold. It is no mere coincidence that life thrives only in the temperature range of liquid water—0° to 100°C, or 32° to 212°F. That this temperature now prevails on earth results from the earth's planetary relations with the sun. As a matter of fact, the possibility of life existing on any planet anywhere would be largely governed by that planet's relation to a source of radiant energy.

Temperature must not only be right for the origin of life, it must also remain favorable for a period lengthy enough for organisms to evolve to their greatest potential complexity. The earth meets this requirement. If the presence of water is a clue to the right temperature range, we find that oceans of water have been present for at least 3.5 billion years. A wide temperature range is provided by hot springs (Figure 15-1). Incidentally, the sun is a typical star of the G2 class whose potential existence is on the order of 13 billion years.

Energy sources

The earth is obviously supplied with energy sufficient to support life in all its many forms. Present energy sources of the earth are solar radiation, including ultraviolet radiation at various wavelengths, electrical discharges, cosmic rays, natural radioactivity, volcanoes and hot springs, shock waves (earthquakes, etc.), and diffuse heat flow from subterranean sources. Obviously there has always been

FIGURE 15-1 *Life may have appeared spontaneously in the water of a hot spring where the essential components existed in the right proportions, at the proper temperature, and at optimum concentrations. Exposure to radiation at certain wave lengths may have been essential. (Photo courtesy of J. R. Howell, III.)*

plenty of energy. The question is, Which form was actually involved at the origin of life and how did it operate? From the chemical point of view, a very small amount of energy applied at exactly the right place would seem to be sufficient. This should be evident from descriptions of laboratory experiments that follow.

Experimental evidence

Although the true origin of life must seemingly remain in the realm of speculation, this subject is not beyond experiment and scientific analysis. An early example of the application of scientific method was the demonstration by Redi, Spallanzani, and Pasteur that spontaneous generation as understood by their contemporaries was not possible. The forms of life as we know them must come from previous forms. Paradoxically, science after having discredited spontaneous generation, finds itself trying to prove it. But the question now is not the spontaneous appearance of highly organized beings such as was envisioned in prescientific time; it is the appearance of the first life, an event that could have been entirely unique and unrepeatable.

Early in the 1920s the Russian scientist, A. I. Oparin, published the first of a number of very influential works outlining his belief as to how life could be produced by natural processes.[1] Oparin's books stimulated a vast

[1]Oparin's most important book is *The Origin of Life* (Macmillan, New York, 1938). There have been many editions and revisions.

amount of work and speculation. Some amazing results have been achieved with laboratory experiments based on materials and conditions simulating those of early earth. In 1953, the American scientist, S. L. Miller, circulated methane (CH_4), ammonia (NH_3), hydrogen (H_2), and water (H_2O) through a heated system and subjected the vapors to electrical discharges (simulating lightning) for 125 hours (see Figure 15-2). This experiment produced 20 compounds including 4 amino acids and a surprising variety of other substances that occur in living organisms.

FIGURE 15-2 *S. L. Miller's apparatus for producing amino acids under primitive earth conditions.*

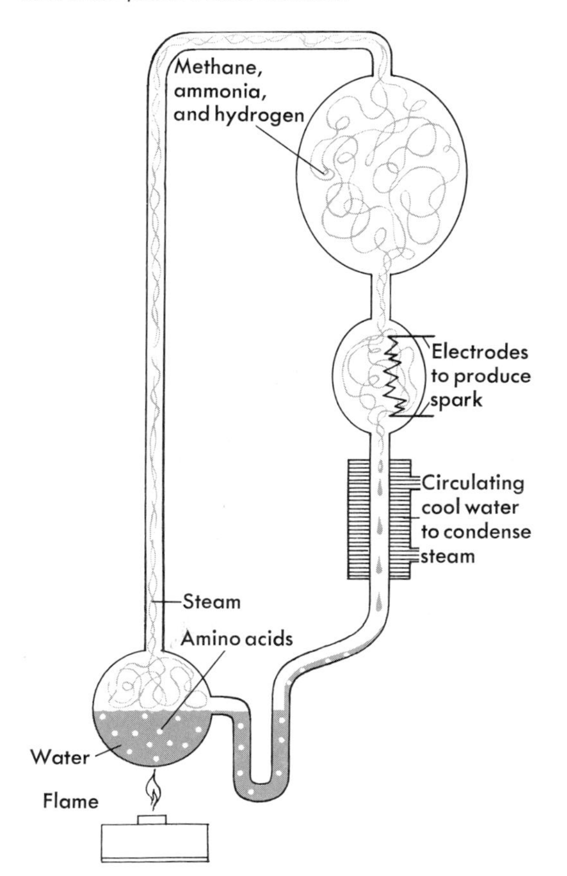

Many investigators followed Miller's lead. When other gases are added to the mixture, results are significantly different. From a mixture of methane (CH_4), nitrogen (N_2) and water (H_2O) the sparking has produced 35 important compounds including 20 amino acids. Hydrogen cyanide (HCN), one of the most important compounds leading to living things, is produced in greater abundance if carbon dioxide, carbon monoxide, and nitrogen are added to the list. If a small amount of hydrogen sulfide (H_2S) is added, the yield of certain compounds is greatly increased.

If ultraviolet light at various wavelengths is substituted for electrical discharges, the yields are similar but usually less. It must be pointed out that ultraviolet radiation from the sun is now screened out by the ozone layer and could have been much more powerful in the early history of the earth.

Assuming that a variety of amino acids and other important compounds was produced by lightning, ultraviolet radiation, and perhaps other agents in the primitive atmosphere, what would be their subsequent history? Mostly they would enter the ocean or other water bodies where they could accumulate. Amino acids are mostly relatively stable in water. It is entirely conceivable that very large amounts of organic material, amounting to nearly 1 percent of the ocean, could have accumulated.

Another important step toward living things must be the concentration of necessary compounds into an environment where they can interact effectively. In living organisms this is accomplished by confining protoplasm within a cell wall or membrane with very selective properties. Many processes for concentrating all sorts of substances are found in nature. Extremes of temperature are most effective. A natural solution such as the theoretical "primeval soup" could easily be subjected to drying and consequent concentration as in a tropical tidal pool or dwindling stream. Water solutions can also be concentrated by freezing. Consider how salt water is concentrated when ice is formed from sea water. All possible levels of salinity can be duplicated on earth in the variety of lakes, ponds, and pools that exists. Thermal regions such as Yellowstone also display temperature ranges linked to chemical variations.

Surprising results have been obtained in drying out organic solutions in the presence of mineral particles, particularly clay minerals such as montmorillonite. The organic molecules are absorbed on the mineral faces with a high degree of concentration and selectivity. Another means of concentration is furnished by liquid droplets or even by such ordinary structures as air bubbles or foam. Much research has been carried on with coacervates, which are soluble colloidal particles that can be produced in celllike form.

Next above amino acids in level of organic complexity are proteins. Unlike amino acids, they are not known to be produced by inorganic agencies. The problem of organizing proteins is therefore a complex one. A preliminary step is the linking together of amino acids into configurations called *polypeptide chains* (Figure 15-3). These linkages can be formed and broken when amino acids are brought into proximity in aqueous solutions. The process could well have operated in a random way to produce many stable or metastable configurations floating passively in the "primeval soup." A protein consists of one or a number of polypeptide chains. Only 20 amino acids are utilized in building proteins. To be operative, a protein must have amino acids in a specific sequence. To build highly structured molecules of any kind, by a random process, is a matter of trial and error with only a few chances of success. Here then is the crucial question: Could a very specific combination of molecules have occurred at least once to produce the first living thing? Some say yes, others say no. For our present purposes we can assert only that all the essential elements are present today and must have been present then, that there were a number of satisfactory energy sources, and that the environment was in a state of flux capable of bringing randomly formed molecules together in statistically significant numbers. The belief of most scientists is that, at least once, an entity was created that could be called *alive.* To say this was a chance event depends on which of the many definitions of chance one accepts.

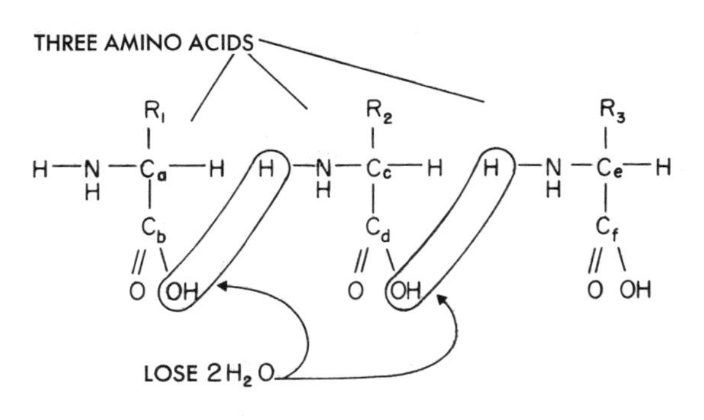

FIGURE 15-3 *The formation of peptide chains. Several amino acids, of one or several kinds, join together as illustrated. Water is lost, and the atoms of carbon and nitrogen have the configuration of —C—C—N—C—C—N— to constitute the polypeptide chain or central axis of a protein. The specific nature of the protein is determined by the hydrogen and oxygen atoms and the side chains or residues (designated R in the diagram). The carbon atoms have been designated individually from a to f to aid comparison between the amino acids and the peptide chain. [After Herbert H. Ross, Understanding Evolution (Englewood Cliffs, N. J.: Prentice-Hall, Inc., 1966) p. 30]*

Chemical fossils

No one seriously expects to find specimens of the first fragile organisms that are the supposed ancestors of all subsequent life. Speculation, guided by sound principles, is favorable to the thought that life did originate in a spontaneous way from nonliving chemicals at least once, and that most of the basic biological reactions necessary to the survival of life had evolved before substantial, preservable forms were possible. The first signs of life are not fossils in the strict sense, but they may be called *fossil residues,* or *chemical fossils.*

A great number of sedimentary rocks, including the oldest known, have been systematically analyzed for organic materials—that is, for molecules that could have been produced only by living things. One analysis requires only carbon, which is common enough in rock of all ages. It is known that plants carrying out photosynthesis use relatively more of the light stable carbon-12 than they do of the heavy isotope, carbon-13. For rocks up to 3.3 billion years old, there is evidence of enrichment in carbon-12; before that there is a greater proportion of the heavy isotope.

This is strong evidence for the existence of life as we know it, but there were almost certainly older life forms not yet capable of utilizing the photosynthetic process. These older life forms did not make their own food but depended for their energy on the limited supply of organic molecules remaining in the primitive soup. There are still a few such "chemical eaters" (Figure 15-4) that live by extracting energy from minerals such as hematite (Fe_2O_3) and gypsum ($CaSO_4 \cdot 2H_2O$). They do not need sunlight and can operate thousands of meters below the earth's surface. These organisms seem to be dead-ended so far as evolution is concerned. Obviously if photosynthesis had not been "discovered," evolution would have halted almost before it began.

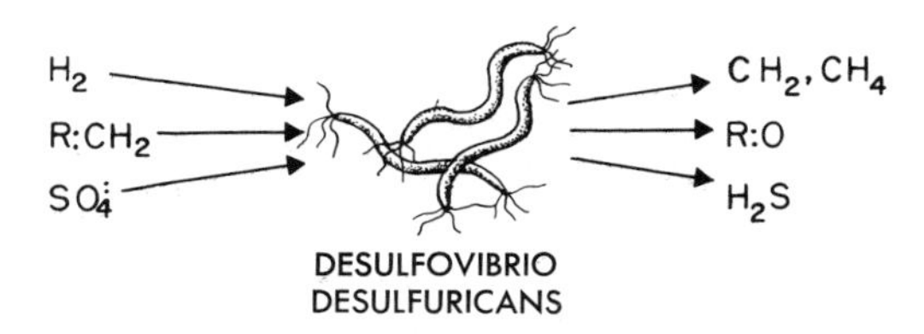

FIGURE 15-4 Diagrammatic representation of metabolic activity of the anerobic chemosynthetic bacterium Desulfovibrio desulfuricans. *The R in the chemical formulas is carbon. (Courtesy M. L. Jensen.)*

The earliest true fossils will be described in Chapter 16. We must here digress briefly to describe fossilization and the meaning of fossils generally.

Meaning of fossils

A *fossil* is the remains or evidence of an ancient organism preserved by natural means in materials of the earth's crust. Fossils exist because dead organisms are not always entirely decomposed or disintegrated. The processes of destruction that begin with death may be halted at any stage, so that the resulting remains may be entire and perfect or fragmentary and imperfect. Naturally, hard parts such as bones, teeth, shells, and wood will resist destruction longer than will soft tissues; therefore these hard parts are more common in the fossil record. The chief methods of preservation are described briefly below.

Entire organism preserved

Complete fossils of large organisms are spectacular. Most notable are the frozen carcasses of ice age mammals, particularly mammoths, which occasionally come to light in frozen ground of the Arctic regions (Figure 15-5). Hair, skin, internal organs, and stomach contents are included. Complete specimens of

FIGURE 15-5 *Part of the carcass of an extinct bison taken from frozen silt during placer gold mining operations 13 miles north of Fairbanks, Alaska. A date of about 31,400 years has been obtained from a piece of the hide by carbon 14 analysis. (Photo courtesy T. L. Pewe.)*

FIGURE 15-6 *Hard parts favor fossilization. The large gastropod shell and the molar tooth of an elephant are both old enough to be regarded as fossils. In both, the original material has been bleached and softened by removal of organic constituents. Fossilization of such objects could proceed through stages of petrifaction if they are buried in proper surroundings.*

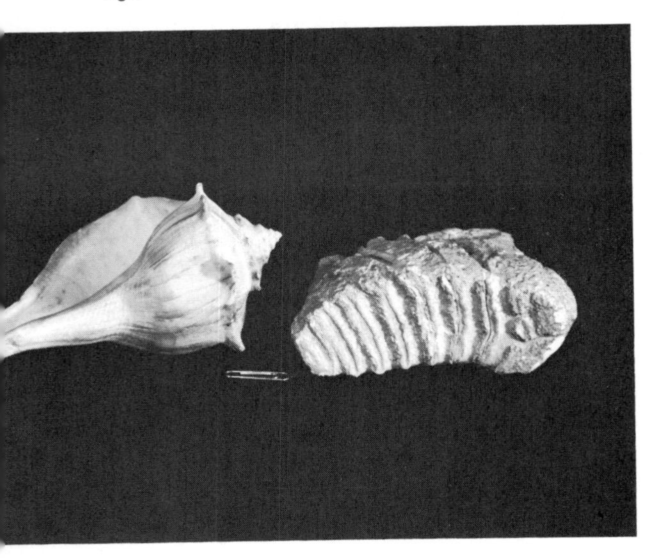

large animals are also occasionally taken from bogs; a woolly rhinoceros from the district of Sarunia, Poland, came from an oil seep. In the category of entire organisms we may also list the dessicated or mummified remains that occur in caves or dry sand.

Unaltered hard parts preserved

Bone, wood, teeth, and shells are in themselves fairly resistant and durable. They may lie unchanged in dry sediment for thousands of years, and even in moist sediment they may exist indefinitely if oxygen is excluded. It is common for the organic matter to be removed from bone, leaving only the mineral constituents. These "dry" bones may then remain unchanged or they may be petrified or replaced by mineral matter (Figure 15-6). See also section on petrifaction below.

Calcite, a common mineral, is secreted by many organisms in building shells and skeletons; once laid down it may be very resistant. A few organisms secrete silica, which is also a relatively stable mineral.

Carbon residues preserved

When an organism or part of an organism is buried in moist sediment or in quiet water, the volatile constituents—nitrogen, hydrogen, and oxygen—are lost, and a carbon-rich residue remains. This process is called *carbonization* or *distillation.* Leaves, insects, fish, and soft-bodied aquatic organisms are commonly preserved by this method (Figure 15-7).

Petrifaction or replacement

Under favorable conditions, hard, porous substances (such as bone, wood, and shell) may be replaced by mineral matter so that more or less perfect stony reproductions re-

FIGURE 15-7 A fossil leaf (Aralia) preserved as carbonaceous residue in fine-grained shale. Specimen measures about 9 cm.

FIGURE 15-8 A silicified tree trunk in position of growth in Yellowstone National Park. (Courtesy of Erling Dorf.)

sult. The original material may be partly or entirely replaced. Commonly, only the minute internal cavities and canals are filled with mineral matter; such specimens are said to be *permineralized.* If the process goes to completion until cell walls and solid matter are also replaced, the object is completely *petrified.* Even the microscopic structure may be retained.

Circulating mineral-bearing solutions are essential for petrifaction. The most common materials carried by ground water are calcite ($CaCO_3$) and silica (SiO_2); fossils replaced by these compounds are accordingly said to be *calcified* or *silicified* (Figure 15-8). A great variety of other minerals—including pyrite, marcasite, dolomite, barite, fluorite, gypsum, hematite, galena, sulfur, and talc—act as petrifying agents.

Casts and molds

Shells of invertebrate animals, which are commonly preserved as casts and molds, provide the best-known fossils. Slowly moving underground water may dissolve a shell entirely, leaving only a hollow space or mold. This may faithfully preserve the shape and surface markings but give no internal structure. If later on, such a hollow mold is filled with mineral matter, a cast is created (Figure 15-9). Again, no internal structures are preserved.

Other interesting examples of molds are natural footprints of animals, the hollow spaces left by tree trunks buried in lava, and the insects in amber that have been known for centuries from the Baltic region of Europe. Of course, if any of these molds could be filled naturally by solid material, we could get casts of footprints, trees, or insects, as the case may be. Shallow molds, such as those made by leaves or thin shells, are commonly referred to as *impressions.*

Trace fossils

Organisms need not die and leave actual remains of their bodies in order to make themselves known. A separate class of fossils has been recognized and is coming under increased study that includes objects or evidence that has never been alive. These objects are called *trace fossils* or *ichnofossils.* Included are tracks, trails, burrows (Figure 15-10), borings, and even such unusual evidence as tooth marks on bone, wounds on bone and shell, *coprolites* (fossilized excrement), and *gastroliths*

FIGURE 15-9 Casts of brachiopod shells in limestone. Traces of the original shell material adhere to these impressions, but the fossils and matrix are otherwise the same. This type of preservation is typical of most marine fossils. Scale represents $2\frac{1}{2}$ cm.

FIGURE 15-10 *Fossil burrows. The soft sediment through which the organism burrowed has eroded away leaving in relief the hardened sand that filled the burrows.*

(stomach stones). Much may be learned from trace fossils about such things as the foot structure and locomotion of large animals, and the food-gathering habits of mud-loving creatures that burrow or plow through sediments or "graze" on muddy surfaces. The shapes and distribution of animals that live more or less permanently in tubes or burrows may also be revealed. Needless to say, many of the creatures that have made trace fossils are soft-bodied and unsubstantial and have left no other evidences of their existence. However, a particular advantage of trace fossils as compared with body fossils is that they are found in the exact environment in which they were formed and cannot have been carried from one place to another either before or after death. They are thus excellent guides in environmental studies.

The evolutionary significance of fossils

Fossils have been referred to as the *corpus delecti* of evolution. They are, in other words, the actual dead remains that prove that life on earth has changed or evolved. Without fossils the past history of life might have been vaguely imagined, but the forms that actually existed would never have been known. To the average person, fossils may be mere curiosities; to the expert, they are not only clues to past conditions and events but also essential links in the great evolutionary chain of organic beings. No study of fossils is complete unless they are not only related to the environment in which they lived but are also assigned to a place in their particular family trees.

The study of organic evolution is shared by biology and geology, more specifically by molecular biology and paleontology. Paleontology is subdivided according to subject matter; thus the study of fossil invertebrates is invertebrate paleontology; of fossil vertebrates, vertebrate paleontology, and of fossil plants paleobotany. Micropaleontology is the field of study of all organisms, or parts of organisms, that must be examined with the microscope. We cannot here delve into the many important contributions of biology, but a brief mention of the areas in which geology has contributed will be worthwhile, before we commence a review of the actual fossil record.

Proof of change

Fossils for the most part belong to species no longer alive. This fact by itself proves that life on the earth has changed with time. The realization of this probably stirred theological opposition to the study of fossils because the recognition of myriads of extinct and inferior beings appears to reflect adversely on the power and intelligence of God. Even more disturbing was the theory of evolution conceived by Charles Darwin (Figure 15-11). To Darwin, the evidence seemed to require that the ancestors of still-living forms did exist in the past and that all organisms ultimately derived from one or a few simple beginnings. Evolution in this sense is now almost universally accepted by informed people. Fossils constitute the prime evidence of evolution, and evolution is one of the most far-reaching discoveries of modern time. Here we shall make only a brief mention of the many significant generalizations that can be drawn from the study of fossils. You should note the following topics and seek for examples as we continue the story of life.

FIGURE 15-11 *Photographic portrait of Charles Darwin by Julia M. Cameron. (Smithsonian Institution.)*

Energy sources and food chains

It is obvious that any creature that cannot make its own food from available inorganic materials must take it from other living things. Nothing can exist unless it is preceded by a suitable food (energy) source. All present-day food chains or pyramids depend on plants whose energy is derived by photosynthesis from solar radiation. There are other manifestations of solar energy that are of interest to geologists. For example, thermal differences give rise to wind and water currents so that these currents flow continuously over the earth as major factors of weather and climate. Water and wind erode and spread rock particles to create sedimentary rocks. Geological evidences such as ripple marks, mud cracks, and raindrop impressions reveal liquid water in action. The point of these remarks is that sedimentary rocks give evidence that at the time of their formation solar energy was falling on the earth, that liquid water existed, and that life had a suitable energy source with which to operate.

Once plants were in existence, animals could follow. Herbivores harvest plants, and one or several types of carnivores feed, in turn, on the herbivores. Man, of course, sits more or less comfortably on top of nearly all food pyramids (Figure 15-12). Paleontology verifies the existence of food chains from the very beginning of life. Before the successive types of dependent animals have appeared, there is always evidence of a preceding adequate food supply. It is no coincidence, then, that the first fossils were plants, that the earliest abundant plants were algae, that the earliest animals were passive, nonaggressive types, and that with the passage of time more efficient carnivores equipped with alert senses, claws, and teeth dominated the scene.

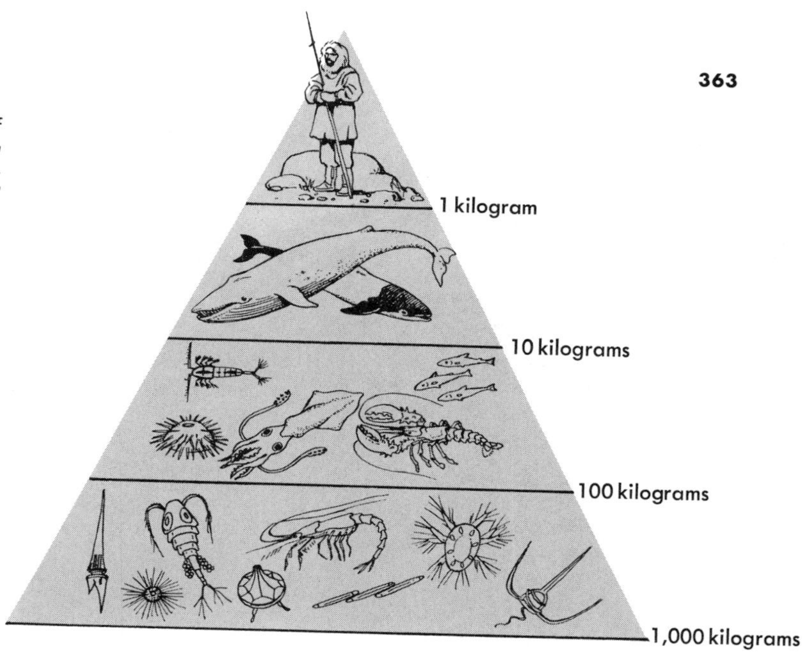

FIGURE 15-12 A simplified food pyramid of Antarctic life. The basic food supply is floating plant plankton. This is eaten by animal plankton, which is in turn consumed by whales and other marine vertebrates. Man occupies the summit of the pyramid when he captures whales or fish. The numbers indicate that for a gain of 1 kg of human flesh, from consumption of whales and other marine invertebrates, approximately 1,000 kg of plant plankton must have been produced. [From Wm L. Stokes, Essentials of Earth History, 3rd Ed., Englewood Cliffs, N. J.: Prentice-Hall, Inc., 1973, p. 428.]

Physical change and biological opportunity

What is behind the evolutionary changes that have produced abundant and diversified organic beings from a few simple ancestors? The answer, as seen by the paleontologist, is that biologic change has been forced on organisms by physical changes in the earth. Both the physical and biological worlds have changed and are still changing, but physical change is dominant and has dictated the general direction of biological change. One of the chief problems of paleontology is to detect how organisms have come to "fit" the niches created by nature. It is easier to understand this correlation if it can be demonstrated that both the organism and its environment have coexisted for periods sufficiently long for natural selection to have taken place.

Geological study shows that all major types of environments have characterized the earth in the prehistoric past. Evidence of glaciers, tropical forests, deserts, lakes, caves, and other habitats can be read from the rocks, proving that life has always had a variety of opportunities or challenges to meet. Even more important, no habitat has remained static. Life could not rest or stagnate; it has been forced to react or perish.

React or perish

When any organism, including man, is confronted by a critical change in its surroundings, it must, in theory at least, change, move, or perish. By changing, it may adapt to the new conditions; by moving, it may leave an unfavorable situation so as to follow its customary environment; by remaining without change, it may invite extinction.

If we read the record correctly, the most common reaction of life forms has been to move or emigrate as conditions change. This action is not, of course, the result of conscious analysis; adaptations exist whereby plants and animals will be dispersed so as to escape local areas into new environments whether or not such dispersal is necessary or beneficial.

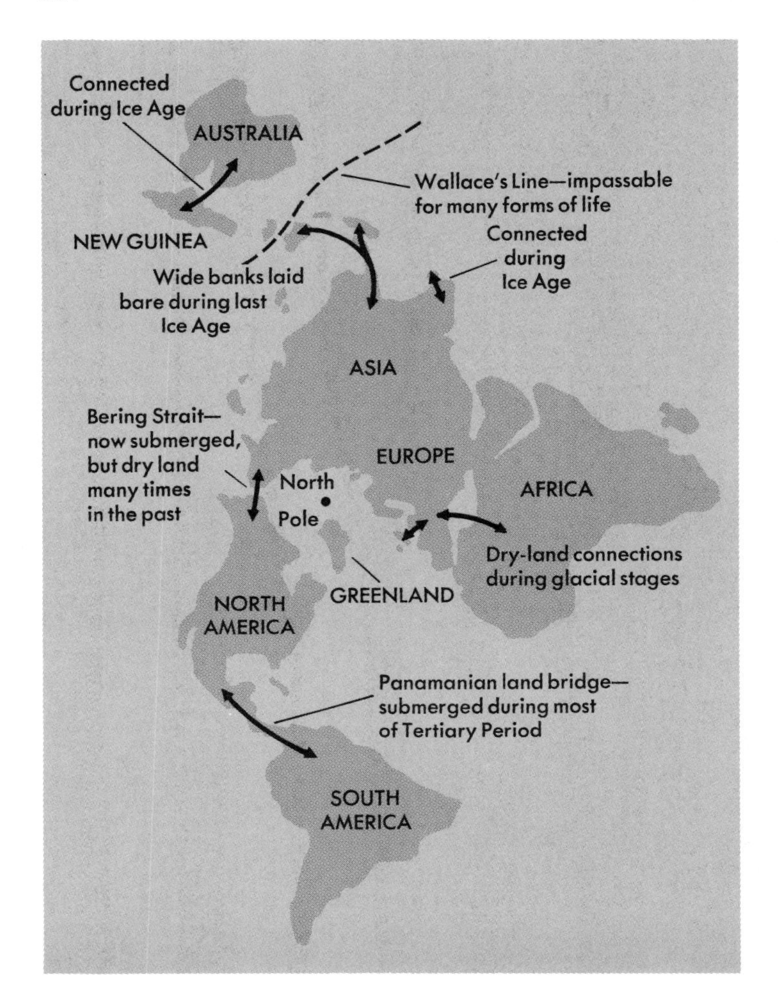

FIGURE 15-13 *Migration routes make it possible for most forms of life to disperse widely. Land bridges and shallow shelves connect all the great landmasses, and plants and animals can spread widely if climatic conditions are favorable.*

Migration and dispersal of organisms

So far as is known, each species of plant or animal came into being once and only once, and at one place only. Thus, any organism, living or extinct, that is found in more than one locality has had to migrate from its place of origin. The problem of how organisms have managed to cross wide bodies of water so as to reach islands or even distant continents is difficult to answer without the assistance of geology. The fact that North America and Asia have been joined on a number of occasions by way of the Bering Straits explains how elephants and men passed from the Old World to the New, and how camels and beavers could spread in the opposite direction (Figure 15-13). Another important land bridge is the Isthmus of Panama. Geological information shows that this tract was submerged during much of the Tertiary Period, thereby allowing the uninhibited evolution of many peculiar forms of life such as ground sloths; these creatures existed nowhere except in South America until the land bridge appeared and allowed them to escape.

The distribution of present-day vegetation would be impossible to explain except on the basis of the great ice sheets whose history is inferred from geological evidence. Examples could be multiplied, but the field of *biogeography*, or study of the distribution of organisms, is a good example of a borderline field that depends heavily on what can be learned about the past to explain the present.

Emigration does not always prove successful, however. The world has always been full of dead-end streets and blind alleys into which organisms may drift or be driven, and thus be separated for good or bad from the mainstreams of existence.

Isolation

The formation of new species is favored by the isolation of relatively small breeding populations that are cut off from other contemporary populations by one means or another. Darwin perceived this when he studied the separate races or species of plants, birds, and reptiles that were isolated on the various islands of the Galapagos Archipelago (see Figure 15-14).

Islands, mountain peaks, valleys, bays, and many other geographic features have their peculiar inhabitants, formed, it is evident, by their having been cut off or separated from former continuous populations.

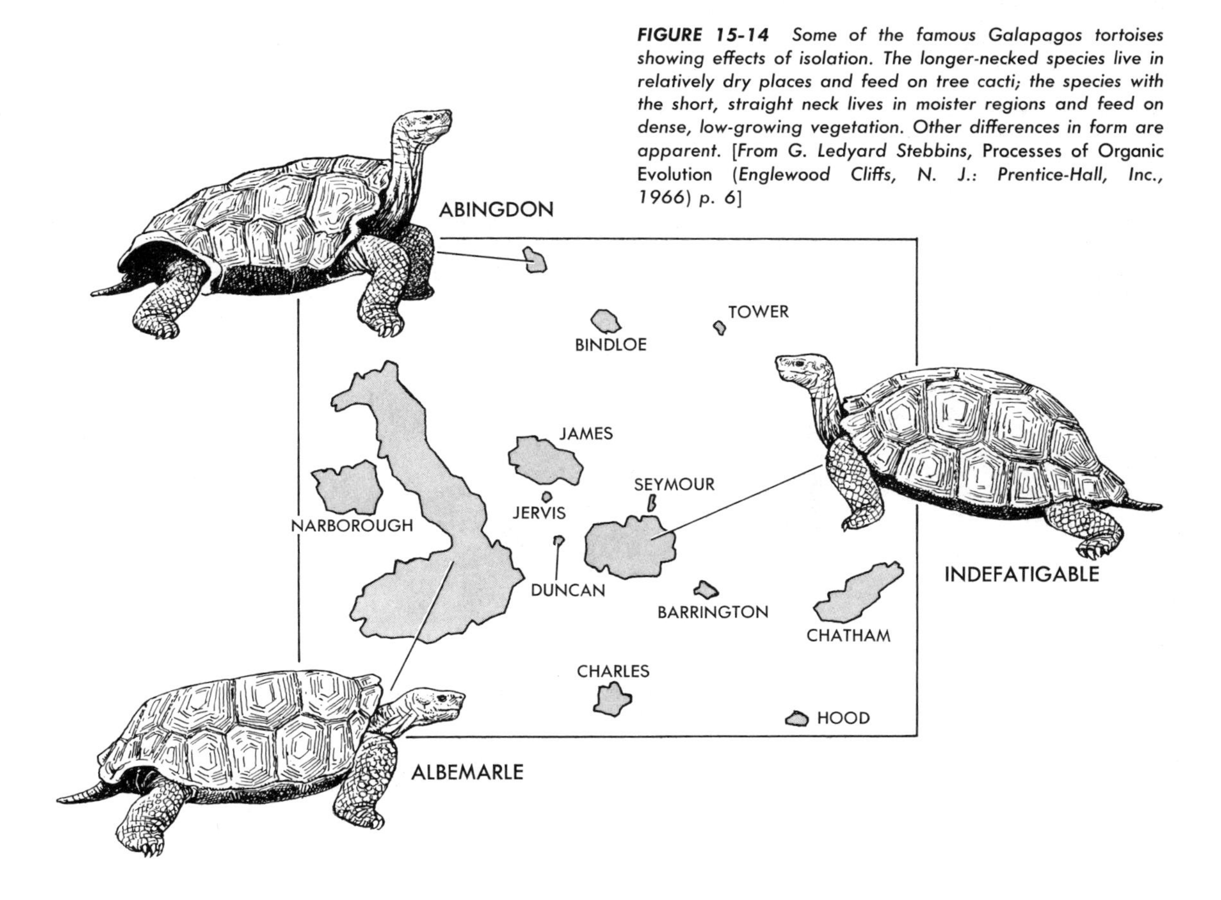

FIGURE 15-14 *Some of the famous Galapagos tortoises showing effects of isolation. The longer-necked species live in relatively dry places and feed on tree cacti; the species with the short, straight neck lives in moister regions and feed on dense, low-growing vegetation. Other differences in form are apparent. [From G. Ledyard Stebbins,* Processes of Organic Evolution *(Englewood Cliffs, N. J.: Prentice-Hall, Inc., 1966) p. 6]*

FIGURE 15-15 *Convergence among aquatic life.*

A very important contribution of geology is the demonstration of how populations may be separated by natural means. Thus, the Galapagos Islands arose by volcanic action about 13 million years ago; Ceylon was cut off from India, and Sicily from Italy, and England from the continent of Europe by rising seas of the glacial period. Mountain glaciers, on the other hand, force organisms to lower levels, perhaps into isolated valleys. Always the scene is an ever-shifting landscape in which life can exist only by constant change. The changes may be minor, or they may be so great that new species are created. No other explanation seems to account for what is seen at present or what is reconstructed from the fossil record of the past.

Pathways of survival

Among the repeating reactions that have affected many types of organisms are convergence, divergence, and radial adaptations. *Convergence* applies to organisms that were originally unlike but came to resemble each other by slow evolutionary changes (Figure 15-15). Thus, the extinct ichthyosaur, a reptile, resembles the living swordfish and porpoise. All have achieved forms that succeed in the open ocean where they live or have lived.

FIGURE 15-16 *Adaptive radiation and divergence of birds. From an ancestor even more ancient than the Archaeopteryx, a variety of forms suited to many environments has evolved.*

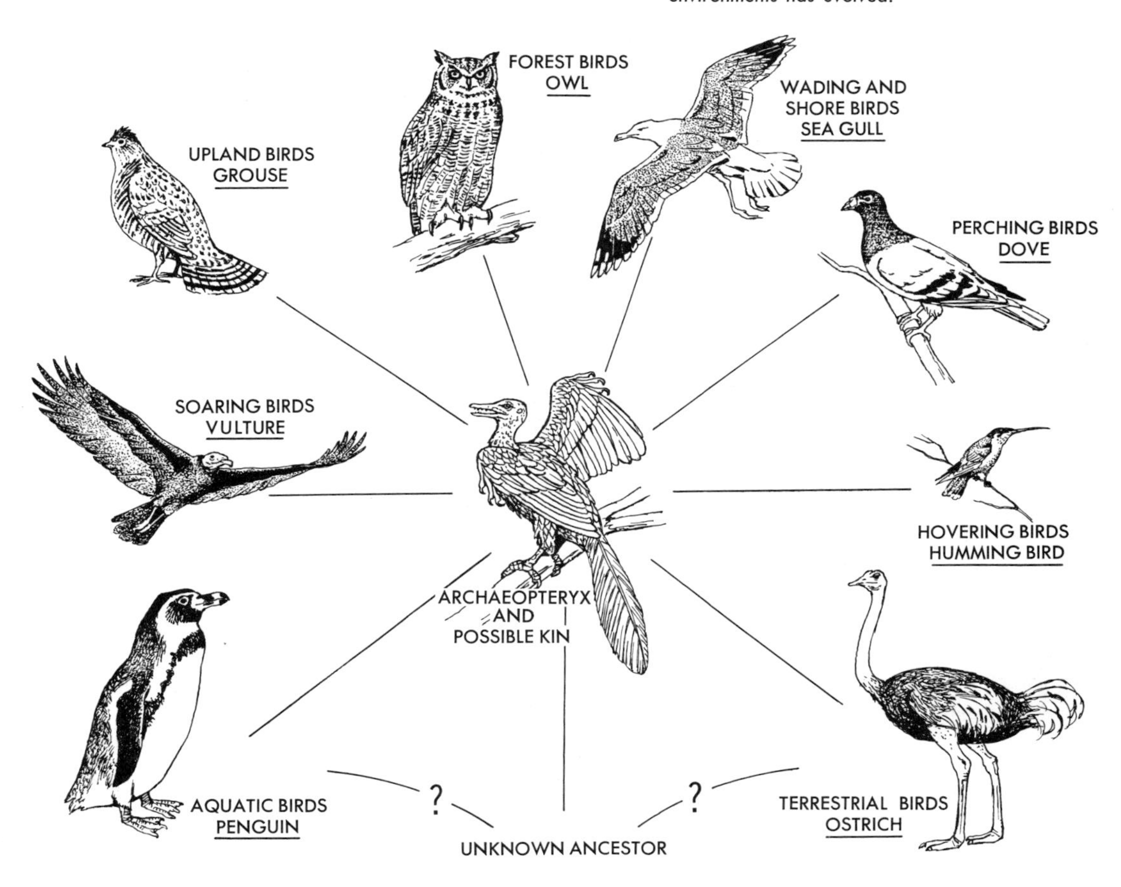

Divergence is the opposite of convergence; its effect is to create unlike organisms from the same or similar ancestors (Figure 15-16). Thus, from an ancient, primitive flesh-eating ancestor, we derive dogs, bears, and seals, each adapted to a different food and way of life.

Radial adaptation is a term that may be applied to the whole grand process of evolution or to significant segments of it. Radial adaptation may be visualized in a crude way by an imaginary view looking downward into a large spreading tree in such a way that the major branches are seen to radiate outward from the central trunk. In actuality, radial adaptation is the spreading of any group into all possible environments (Figure 15-17). Every phylum of plants or animals has followed this pattern—some with great success, some in a limited way only. You will note that the essential story of any group can be shown by a branching diagram, its "family tree," or more technically, its *phylogenetic chart.* Some groups show only a few simple branches; others are best illustrated by many spreading and subdividing lines. It is the constructing of family trees and the relating of these trees to changing environments that is the heart of paleontological research.

FIGURE 15-17 *Radial adaptation of reptiles and mammals.*

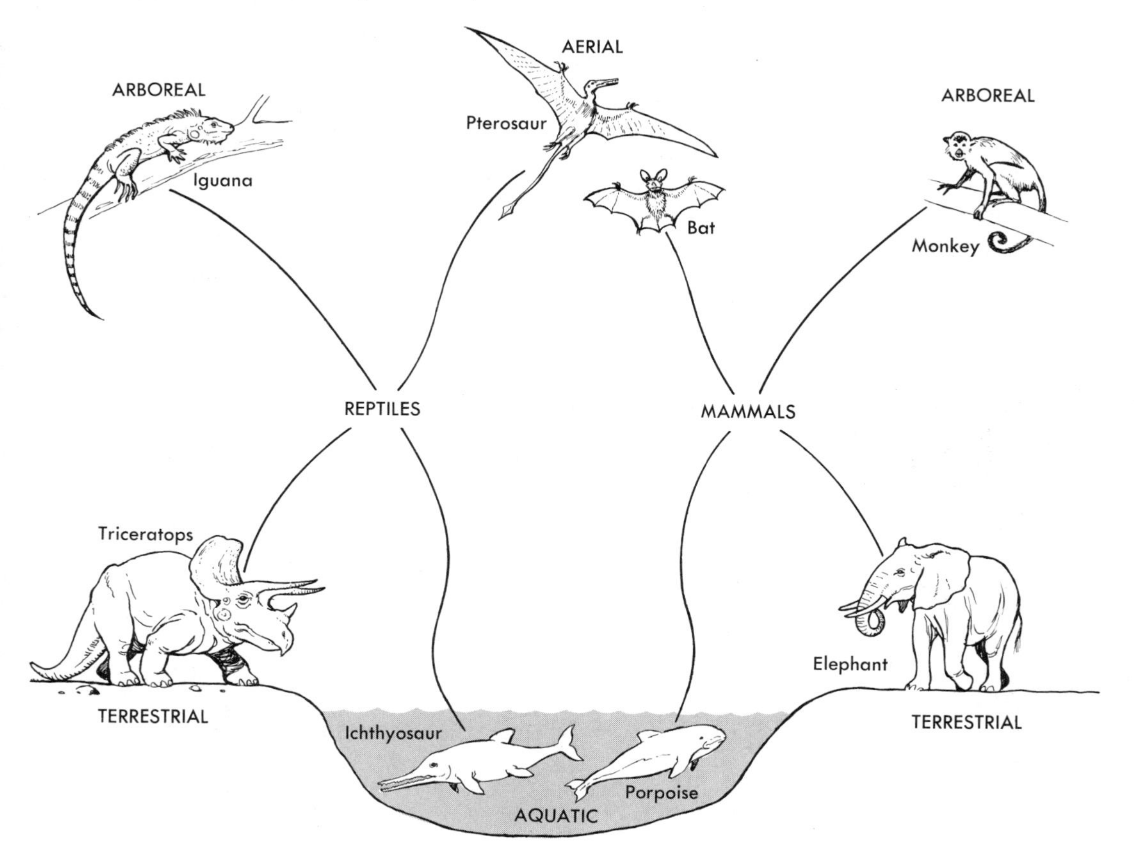

SUMMARY

An important geologic and biologic event was the appearance of life on earth. The synthesis of the first simple organisms may have been a spontaneous event made possible by the simultaneous occurrence of water solutions of essential elements with a suitable temperature range, an oxygen-poor atmosphere, abundant hydrocarbons, and effective energy sources. Remains of bacterialike organisms have been reported in rocks estimated at 3 billion years old. Limy structures and organic residues also indicate very ancient life, but multicellular animals do not appear until 1 billion to .6 billion years ago when worms, brachiopods, and coelenterates are found.

Fossils are the remains or evidences of ancient organisms. They may be preserved as impressions, casts, and molds, or as petrifactions, carbonaceous residues, or rarely, in unaltered forms. The study of fossils is called *paleontology;* chief subdivisions are invertebrate paleontology, vertebrate paleontology, paleobotany, and micropaleontology.

The study of fossils lends strong support to the idea of organic evolution; it proves that past life, like present life, was based on energy and that food pyramids always existed. Other important generalities are that alterations in the physical environment have forced organisms to adapt, to migrate, or to face extinction. The most common reaction appears to have been migration, and organisms have been dispersed throughout the globe to find new living space, or in many instances, to be isolated in small groups where accelerated evolution has taken place. The many examples of convergence, divergence, and radial adaptation that have been discovered go far toward explaining what is seen in the present world. Many thousands of species have been exterminated in the past, but the major phyla have persisted with few exceptions for hundreds of millions of years.

Knowledge of living and extinct plants and animals is extensive and has been compressed into a brief appendix to this chapter.

CLASSIFICATION AND BASIC CHARACTERISTICS OF ORGANISMS

KINGDOM MONERA

Unicellular organisms with procaryotic cell structure (cells lack nuclear membrane, plastids, and mitochondria): nutrition predominantly by absorption, but some groups are photosynthetic (energy from solar radiation) or chemosynthetic (energy from inorganic compounds); reproduction by fission or budding; locomotion by simple flagella (whiplike organelles) or by gliding.

Phylum Schizophyta: Bacteria: extremely small, with spherical, rodlike, or spiral form; fossils up to 3.2 billion years old have been discovered.

Phylum Cyanophyta: Blue-green algae: photosynthetic, numerous; stromatolites and other fossils up to 3.2 billion years old.

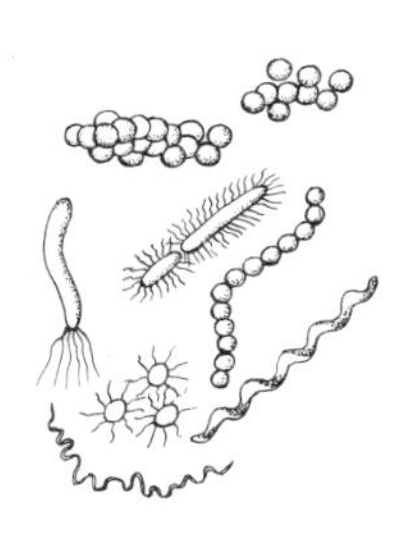

Schizophyta

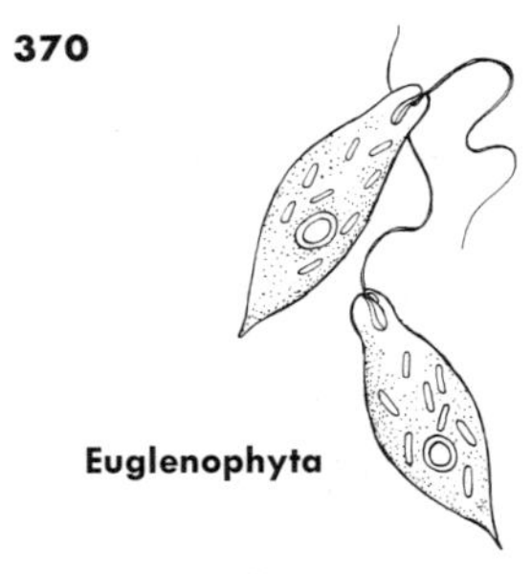

Euglenophyta

Chrysophyta (diatoms)

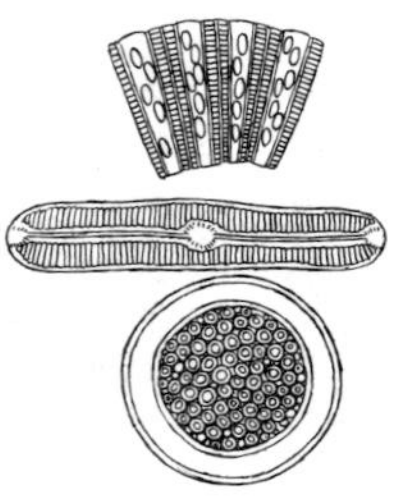

KINGDOM PROTISTA

Unicellular organisms with eucaryotic cell structure (cells have nuclear membrane, mitochondria, and, in many forms, plastids): nutrition by photosynthesis, absorption, ingestion, or a combination of these; sexual and asexual reproduction; locomotion by flagella or other means, but some are nonmotile; plantlike or animallike.

Phylum Euglenophyta: Motile, photosynthetic: no known fossils.

Phylum Chrysophyta: Diverse group that includes the following: (1) Diatoms, most abundant of the algal groups: secrete siliceous skeletons and inhabit salt and fresh water; fossils from Cretaceous onward. (2) Silico-flagellates: marine; Cretaceous onward. (3) Coccoliths: calcareous, marine; Cambrian onward. (4) Dinoflagellates: marine; Late Jurassic onward. Diatoms and coccoliths are important rock-making organisms.

Phylum Sporozoa: Chiefly parasitic; no known fossils.

Phylum Zoomastigina: Animal flagellates: similar to euglenoids but lack chlorophyll; no known fossils.

Phylum Sarcodina: Amoeboid forms with animallike metabolism: includes the shell- or skeleton-bearing foraminifera (calcareous, Cambrian(?) onward) and radiolarians (siliceous, Precambrian(?) onward); both are of great value in geologic correlation.

Phylum Ciliophora: The ciliated protozoans: one small family, the Tintinnidae, has left a fossil record.

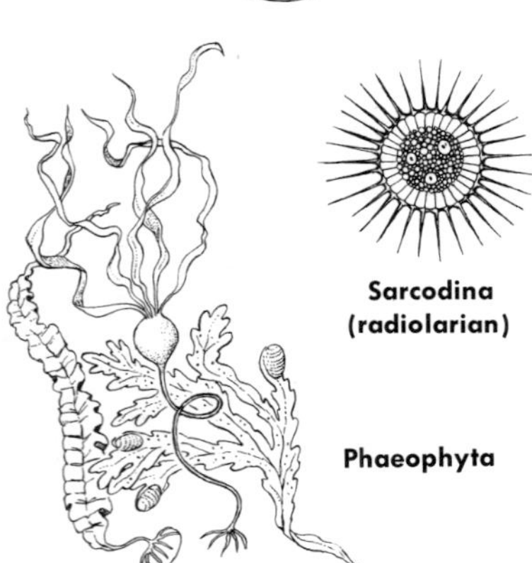

Sarcodina (radiolarian)

Phaeophyta

KINGDOM PLANTAE

Multicellular organisms with walled eucaryotic cells: almost all are photosynthetic and nonmotile; reproduction chiefly sexual; many fossils throughout the geologic record.

Phylum Rhodophyta: Red algae: may occur fossilized as indeterminate "seaweeds."

Phylum Phaeophyta: Brown algae: may occur fossilized as indeterminate "seaweeds."

Phylum Chlorophyta: Green algae: fossils about 1 billion years old are known.

Phylum Charophyta: Stoneworts: some (charophytes) secrete calcium carbonate and are represented as fossils from Devonian onward.

Phylum Bryophyta: Mosses and kin: soft-tissued; very rare as fossils.

Phylum Tracheophyta: Vascular plants: many are hard-tissued; abundant as fossils.

Subphylum Psilopsida: Psilophytes: dominant in Devonian, very rare at present.

Subphylum Lycopsida: Club mosses: dominant in the Carboniferous, rare at present.

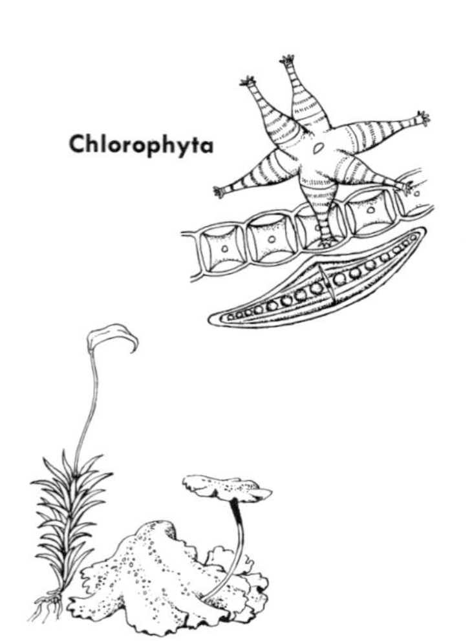

Chlorophyta

Bryophyta

Subphylum Sphenopsida: Horsetails: dominant in the Carboniferous; one genus, *Equisetum,* survives.

Subphylum Pteropsida: Ferns and seed plants.

Class Filicae: Ferns: fossils since the Carboniferous, common in the Mesozoic; subordinate to seed plants at present.

Class Gymnospermae: Conifers, cycads, ginkgoes; mostly wind-pollinated, nonflowering plants.

Class Angiospermae: Mainly insect-pollinated flowering plants; known from the Jurassic onward, become very common in the Late Cretaceous. *Dicotyledons:* over 50 orders: net-veined leaves (oak, rose, cactus); includes almost all the flower-bearing plants; abundant fossils. *Monocotyledons:* 14 orders: parallel-veined leaves (grasses, palms, orchids); abundant fossils.

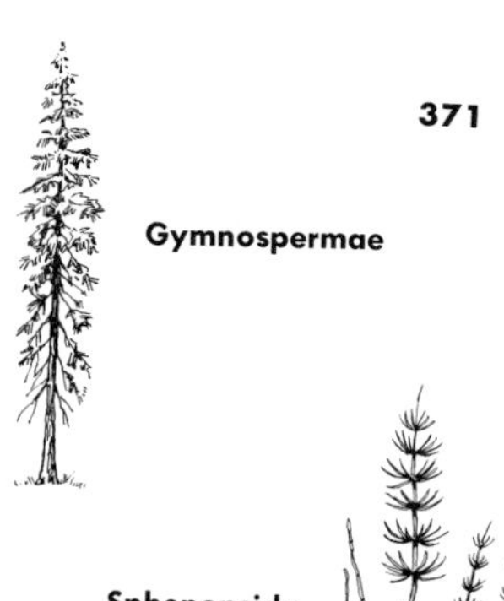
Gymnospermae

Sphenopsida

KINGDOM FUNGI

Mostly multinucleate organisms with eucaryotic nuceli; nutrition exclusively by absorption, hence essentially parasitic; mostly nonmotile, may be embedded in the food supply; reproduction both sexual and asexual; very few fossils.

Phylum Myxomycophyta: No fossils.

Phylum Schizomycophyta: No fossils.

Phylum Eumycophyta: Fossils up to 2 billion years old.

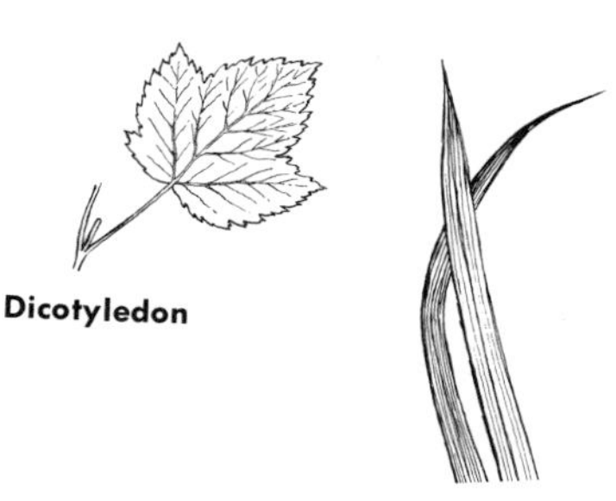
Dicotyledon

Monocotyledon

KINGDOM ANIMALIA

Multicellular organisms with wall-less eucaryotic cells lacking plastids and mitochondria: feeding is mainly by ingestion or, rarely, by absorption; level of organization complex; locomotion based on fibrils of many forms; reproduction predominantly sexual.

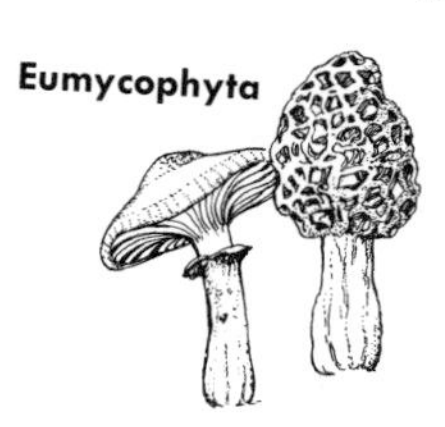
Eumycophyta

Phylum Mesozoa: Obscure wormlike, parasitic forms; no known fossils.

Phylum Porifera: Primitive multicellular animals: no true tissues or organs; skeletons of calcite, silica, or spongin; freshwater and marine habitats; fossils from Precambrian(?) onward.

Class Demospongiae: Horny sponges: skeletons of organic fibers.

Class Hexactinellida: Glass sponges: siliceous skeletons.

Class Calcaria: Chalky sponges: calcium carbonate skeletons.

Angiospermae

Phylum Archaeocyatha: Extinct; soft tissues unknown, skeleton double-walled, conical, calcareous; Cambrian only.

Phylum Coelenterata (Cnidaria): Polyps and medusae: tissues well developed; tentacles with stinging cells; large central "gut," but no true body cavity (coelom); mostly radial symmetry; many are colonial, mostly marine; fossils from Late Precambrian onward.

Porifera

Class Hydrozoa: Soft-bodied; mostly colonial and stationary; few fossils.

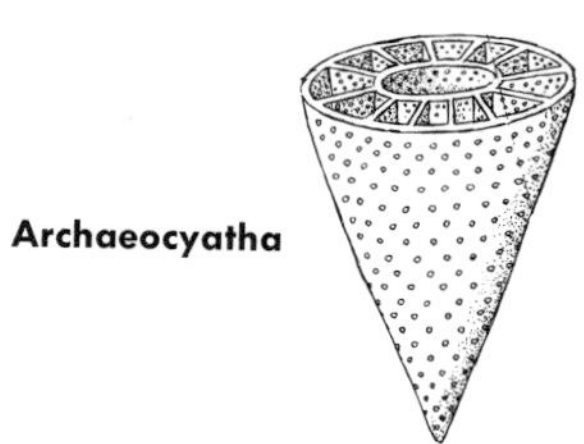
Archaeocyatha

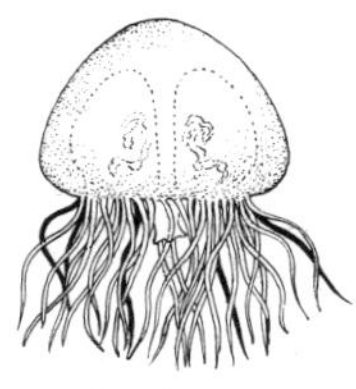

Scyphozoa

Class Scyphozoa: Jellyfish: few fossils.

Class Anthozoa: Sea anemones and corals: essentially nonmotile as adults; the corals generally have calcareous skeletons, and many are reef builders; abundant fossils; from latest Cambrian onward.

Phylum Ctenopora: Comb jellies: fragile; no fossils.

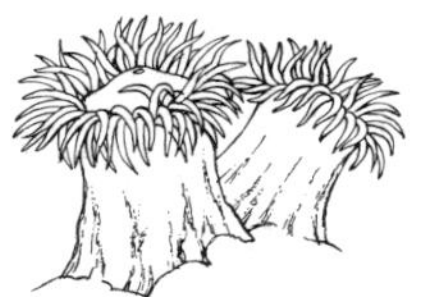

Anthozoa

NOTE: The following wormlike phyla are unrecognized or extremely rare in the fossil record: Phoronida, phoronid worms; Sipunculoidea, peanut worms; Echiuroidea, spoon worms; Acanthocephala, spiny-headed worms; Platyhelminthes, flatworms; Nemertinea, ribbon worms; Aschelminthes, roundworms, threadworms, and rotifers; Chaetognatha, arrowworms; Tardigrada, "water bears"; Pentastomida, bloodsucking parasites; Pogonophora, beard worms. Many of these live buried in sediment, and although they may have left fossil burrows, borings, or trails, these are mostly unidentifiable as to origin.

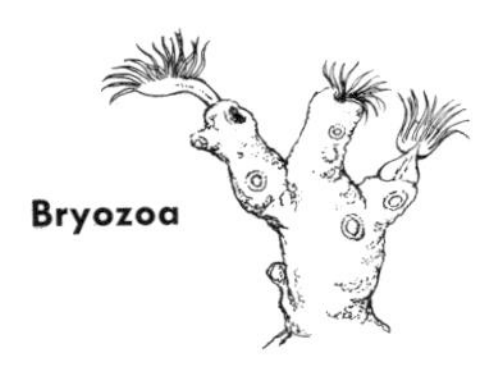

Bryozoa

Phylum Entoprocta: Very small sedentary animals, closely related to bryozoans; no known fossils.

Phylum Bryozoa: Small, stationary, colonial, coelomate animals: ciliated tentacles around mouth; skeleton calcareous, chitinous, horny, or membranous; fossils from Late Cambrian(?) onward; freshwater and marine; important as rock builders and useful in geologic correlation.

Inarticulata

Phylum Brachiopoda: Exclusively marine bivalves: one shell (dorsal) is slightly larger; stalklike pedicle in most species attaches the animal to the bottom, feeds by ciliated tentacles supported by a specialized, semirigid lophophore enclosed in the shell; fossils from Precambrian(?) onward, especially abundant in the Paleozoic.

Class Inarticulata: Chitinophosphatic shells and no tooth-and-socket hinge between the valves; fossils from earliest Cambrian onward.

Class Articulata: Calcareous shell and well-developed hinge structure; fossils much more abundant than *Inarticulata* and range from Early Cambrian onward.

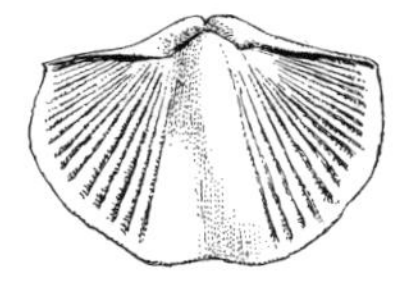

Articulata

Phylum Mollusca: Extremely diverse, mostly shell-bearing animals having an alimentary canal with two openings and a true body cavity: the shell may be in one, two, or several pieces and is secreted by an underlying soft tissue, the mantle; marine, freshwater, and terrestrial habitats; fossils from Early Cambrian onward.

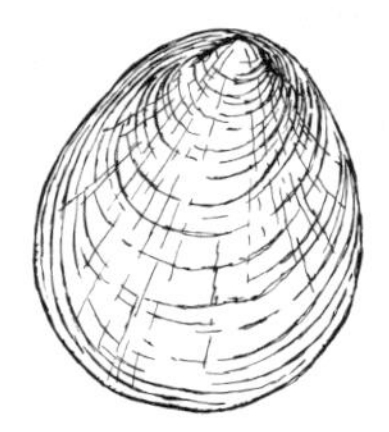

Monoplacophora

Class Monoplacophora: Primitive mollusks with cap-shaped single shells; known as rare fossils from Early Cambrian onward.

Class Aplacophora: Elongate, with eight-piece shell; marine only; rare as fossils; Late Cambrian to Recent.

Class Gastropoda: The most abundant class of mollusks (100,000 species): a few are shell-less, others have cap-shaped or coiled shells; locomotion by a broad creeping foot; distinct head, sense organs; marine, freshwater, and terrestrial habitats; abundant as fossils from Early Cambrian onward.

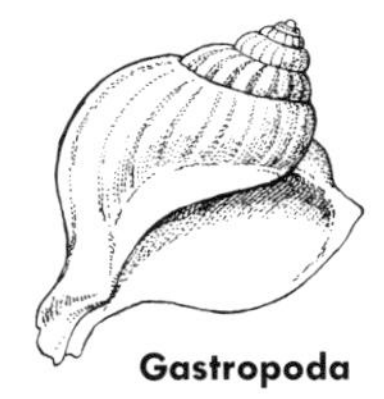

Gastropoda

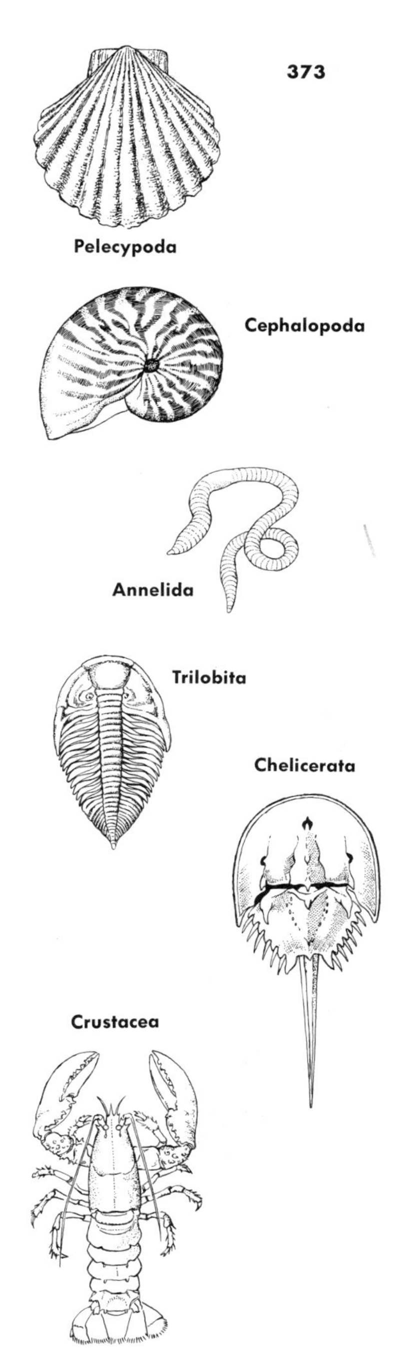

Class Pelecypoda: Bivalved mollusks (30,000 species): most have a laterally flattened body between two hinged mirror-image shells; mostly marine, a few are freshwater; locomotion primarily by a muscular, creeping, hatchet-shaped foot; some (for example, oysters) are stationary; fossils from Early Cambrian to Recent.

Class Scaphopoda: "Tusk-shelled" mollusks: shell hollow, tapering, open at both ends; marine only; uncommon as fossils except locally; Middle Ordovician to Recent.

Class Cephalopoda: A diverse group of shelled (pearly nautilus) or shell-less mollusks (octopus): fossil forms generally have a flat, spiral shell with internal chambering and complex folded partitions; mouth surrounded by tentacles; locomotion by jet-propulsion system; well-developed brain and sense organs; marine only; increasingly abundant fossils from Early Cambrian onward.

Phylum Annelida: Segmented worms (for example, earthworm): 6 classes; marine and nonmarine; fossils consist of burrows, trails, tubes, impressions, and carbonized remains, also abundant teeth (scolecodonts); range from Precambrian(?) to Recent.

Phylum Onychophora: Onychophorians: related to arthropods and annelids; known as very rare fossils from the Cambrian onward.

Phylum Arthropoda: The most abundant phylum (estimated at a million species making up 80 percent of all known animals): diverse body form; practically all have chitinous external skeleton (exoskeleton), jointed appendages, segmented body; locomotion by burrowing, crawling, swimming, and flying; marine, freshwater, and terrestrial environments; fossils of marine forms abundant from Late Precambrian onward, land forms, such as insects, are less well represented.

Class Trilobita: Extinct segmented arthropods with a longitudinally trilobed exoskeleton consisting of head (cephalon), thorax, and tail (pygidium); simple and/or compound eyes; diverse marine habitats; many fossils from Early Cambrian to Late Permian.

Class Chelicerata: Segmented and unsegmented arthropods with pincerlike claws: terrestrial forms (spiders, scorpions, and mites) and marine forms (king crab and the extinct eurypterids); fair fossil record from Cambrian onward.

Class Crustacea: Thin- to thick-shelled arthropods (ostracods, barnacles, lobsters, and crabs), mostly with segmented bodies; also with antennae and other head appendages, compound eyes, and varied means of locomotion; marine, freshwater, and terrestrial; fossils from Cambrian onward.

Class Myriapoda: Centipedes and millipedes: terrestrial; the centipedes are represented by rare fossils from Carboniferous onward, millipedes from Silurian onward.

Class Insecta: Extremely diverse (29 orders) and numerous (over 850,000) species; typically wing-bearing arthropods, with three pairs of walking legs; food habits diverse; locomotion by burrowing, walking,

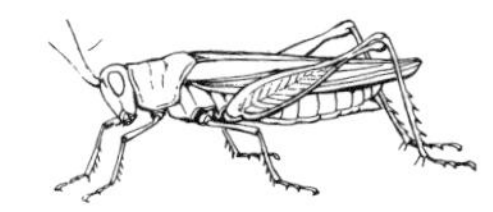

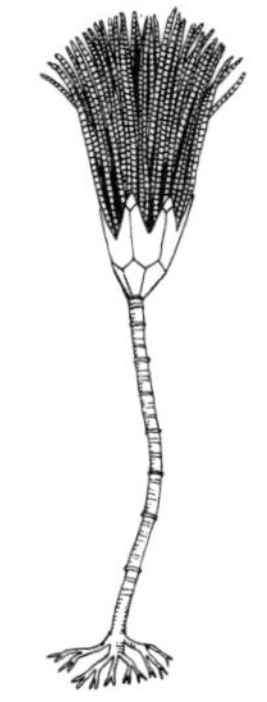
Crinoidea

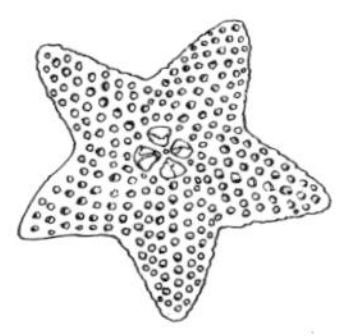
Stelleroida

Holothuroidea

Graptolithina

Conodont
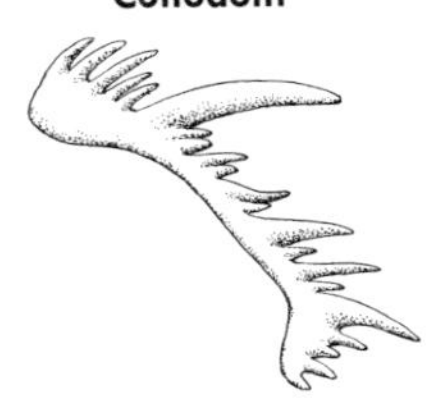

swimming, and flying; primarily terrestrial; rare as fossils except locally since Devonian.

Phylum Echinodermata: Echinoderms: internal skeleton (endoskeleton) of calcareous plates, some completely rigid (sea urchin), some flexible (starfish, sea cucumber); commonly with five-sided symmetry; food habits diverse; about half are sessile, the others move by burrowing or crawling; exclusively marine; common as fossils from Early Cambrian onward.

Class Cystoidea: Cystoids: stemmed, boxlike, rigid echinoderms with characteristic pore patterns; Early Ordovician to Late Devonian.

Class Blastoidea: Blastoids: boxlike stalked echinoderms with strong pentamerous symmetry, and 18 to 21 plates in the enveloping calyx (theca); Silurian to Permian.

Class Crinoidea: Crinoids: stalked (some secondarily stalkless), pentamerous echinoderms; complex structure of calcite plates; feed by filtering sea water; Cambrian(?) onward.

Class Stelleroidea: Starfish and brittle stars, typically with five radiating symmetrical arms: calcareous plates joined by flexible connective tissue allow free movement; scavengers and predators; starfish live near the shoreline, and brittle stars mainly in deep water; Ordovician onward.

Class Echinoidea: Sea urchins and kin: boxlike, many-plated endoskeleton, covered with movable spines for protection and burrowing; five-rayed symmetry expressed by pore and spine systems; Ordovician onward.

Class Holothuroidea: Sea cucumbers: free living, flexible, with rudimentary calcareous plates; Cambrian(?) onward.

NOTE: The following classes of echinoderms are known from rare fossil representatives and are not described in detail here: Eocrinoidea, Cambrian–Middle Ordovician; Paracrinoidea, Middle Ordovician; Parablastoidea, Middle Ordovician; Edrioblastoidea, Middle Ordovician; Helycoplacoidea, Lower Cambrian; Edrioasteroidea, Lower Cambrian–Mississippian; Ophiocistoidea, confined to Paleozoic.

Phylum Hemichordata: A diverse group of mostly colonial marine organisms showing some affinities with chordates. *Order Enteropneusta:* Wormlike, no fossils. *Order Pterobranchia:* Primitive hemichordates: very rare fossils beginning in the Early Ordovician.

Class Graptolithina: Colonial marine organisms with chitinous exoskeleton taking many forms: reproduction by budding, which yields branched, simple, or, rarely, encrusting colonies; both fixed and floating forms are known; Middle Cambrian to Late Mississippian.

NOTE: One of the most puzzling of organisms is the so-called conodont animal. This creature has left innumerable toothlike objects of varied form from Middle Cambrian to Middle Triassic. It was apparently bilaterally symmetrical, free living, and marine—beyond this, little is known. The general opinion is that it represents a distinct phylum; it is mentioned here because it is also thought to be related to the chordates.

Phylum Chordata: Highly organized bony animals and their primitive kin: extremely varied in most characteristics; flexible rod of cells (notochord) or segmented vertebral column; gill slits in embryonic and/or adult stages; food habits diverse; marine, freshwater, and terrestrial forms; fossils abundant from Ordovician onward.

Subphylum Urochordata: Small, soft-bodied, exclusively marine animals: no known fossils.

Subphylum Cephalochordata: Soft-bodied, elongate, primitive chordates with notochord: only two living genera; exclusively marine; no positive fossils.

Subphylum Vertebrata: Backboned animals with highly organized nervous system, diverse food habits, and varied means of locomotion; marine, freshwater, and terrestrial; many fossils from Ordovician onward.

Class Pisces: Fish: 35 orders: primarily egg-laying, coldblooded vertebrates occupying diverse habitats in marine, brackish, and freshwater environments; food habits varied; locomotion mainly by swimming; many fossil forms from Late Cambrian onward.

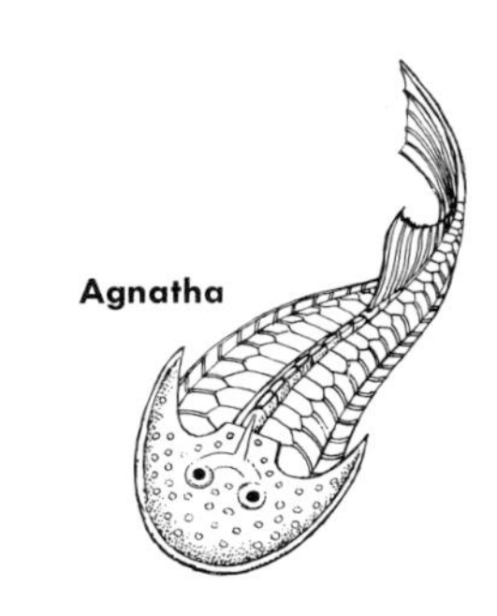

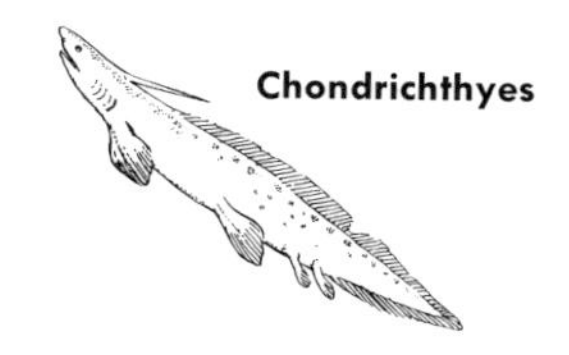

NOTE: In the extremely diverse and ancient group known broadly as fish, the following groups or classes are recognized: Agnatha, the jawless fish, Ordovician onward (represented today by lamprey and hagfish); Placodermi, early jawed fishes, mostly armored, Silurian to Pennsylvanian(?); Chondrichthyes, the cartilaginous fish, or sharks broadly considered, Devonian onward; Osteichthyes, the bony fish including most living types, Devonian onward.

Class Amphibia: Amphibians: 13 orders, water-dependent vertebrates with aquatic and terrestrial life stages; no marine forms positively known; rare fossils beginning with the Devonian.

Class Reptilia: Reptiles: 16 orders, most of which are extinct; aquatic, terrestrial, and aerial forms; mostly egg-laying and coldblooded; many fossils beginning with the Carboniferous.

Class Aves: Birds: 33 orders, mostly living: feathered, egg-laying, warmblooded; most can fly; rare fossils beginning with the Jurassic.

Class Mammalia: Mammals: 34 orders, about half of which are extinct; warmblooded, primarily live-bearing animals with a hairy covering and milk-producing organs; highly developed, with diverse food habits, varied modes of locomotion, including flight, and great range in size; gradually increasing fossil record beginning with the Triassic.

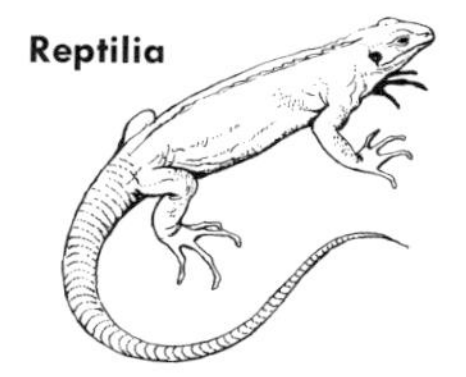

SUGGESTED READINGS

BARNETT, LINCOLN, *The World We Live In*. New York: Time, Inc., 1955.

FENTON, C. L., and M. A. FENTON, *The Fossil Book*. New York: Doubleday & Company, Inc., 1958.

GOLDRING, WINIFRED, *Handbook of Paleontology for Beginners and Amateurs*. Ithaca, N.Y.: Paleontological Research Institution. (Many editions.)

McALESTER, A. LEE, *The History of Life*. Englewood Cliffs, N.J.: Prentice-Hall, Inc., 1968.

MILLER, STANLEY L., and LESLIE E. ORGEL, *The Origins of the Life of Earth*. Englewood Cliffs, N.J.: Prentice-Hall, Inc., 1974.

SALTHE, STANLEY N., *Evolutionary Biology*. New York: Holt, Rinehart & Winston, Inc., 1972.

VALENTINE, JAMES W., *Evolutionary Paleoecology of the Marine Biosphere*. Englewood Cliffs, N.J.: Prentice-Hall, Inc., 1973.

Treatise on Invertebrate Paleontology, 24 parts. Copublishers: The University of Kansas Press and the Geological Society of America, Lawrence, Kansas, 1953—. A series on the major invertebrate animals by various authors. This is the best technical reference for identifying fossils.

16 The Precambrian

The generally accepted age of the earth is approximately 4.6 billion years. Within this lengthy period of time a number of events stand out significantly. However, many of the most ancient events and conditions are so far removed from the present that even their recognition is difficult. For example, it is not known whether the primitive earth was originally hot or cold. It is hot now and even partly molten, and it may have been entirely melted at least once in the past; but we do not know if a cool solid stage preceded the melting or not. The thermal history of the earth is one of the great unsolved problems of geology. If we understood this aspect, the rest would be relatively easy.

One event of particular significance is the appearance of a solid crust after the last molten stage. The earliest known rocks are 3.75 billion years old, but it is not certain that

Going-to-the-Sun Mountain in Glacier National Park, Montana, is carved from late Precambrian sedimentary rocks. (Montana Highway Commission.)

these are part of the original crust. Other notable events such as the appearance of water, of primitive life, of the atmosphere, and of the first continents are datable only in relatively semiquantitative ways. Later events become progressively easier to place in their proper time and setting. A major landmark by any standard is the appearance of abundant fossils about 600,000,000 years ago. This biologic event marks the end of the Precambrian interval; and it is the Precambrian interval that will be discussed in this chapter.

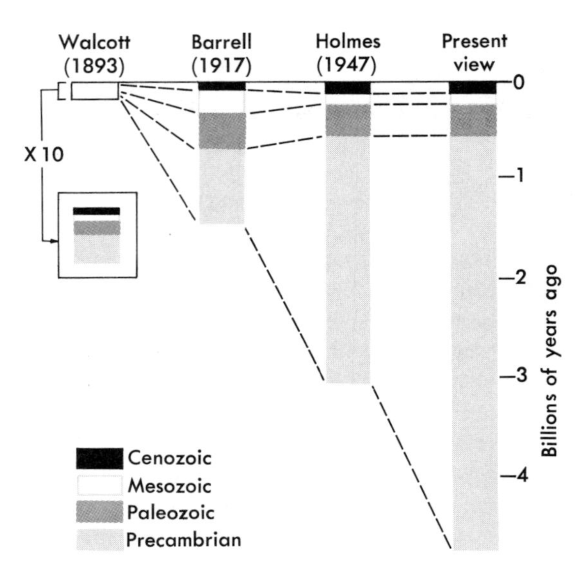

FIGURE 16-1 Changing view of the magnitude of Precambrian time in the twentieth century. [Adapted from H. L. James, "Problems of Stratigraphy and Correlation of Precambrian Rocks . . . ," American Journal of Science, 258–a (1960), 105.]

Naming and subdividing the Precambrian

Many methods have been suggested for subdividing the very lengthy Precambrian interval. Note that this period covers about seven-eighths of the total history of the earth—on the order of 3.5 to 4 billion years in duration. Unfortunately, the word *Postcambrian* cannot be used for time subsequent to the Precambrian because this would eliminate the Cambrian Period—a long and significant interval with abundant ancestral life forms. The word *Cryptozoic* (hidden life) is employed by many as a synonym for the Precambrian, and the companion word, *Phanerozoic* (evident or visible life), for all subsequent time. The Phanerozoic of this scheme is fairly well organized into the well-known geologic time scale with eras and periods, but the Cryptozoic has not been so divided. Instead, most geologists are using a rough threefold division: *Pre-Archean,* 4.6 to 3.8 billion years ago; *Archean,* 3.7 to 2.4 billion years ago; and *Proterozoic,* 2.4 to 0.6 billion years ago. Changing views of the duration of the Precambrian are shown by Figure 16-1.

In the absence of life forms and their evolutionary changes—which provide a satisfactory basis for subdividing the Phanerozoic—another basis for dealing with the Cryptozoic is being sought. Fortunately, a objective physical basis does exist in the form of widespread physical events that shape and build up the continents. The key to this classification is the fact that individual broad belts thousands of square kilometers in extent have similar rocks and structures throughout, and these belts plainly originated during rather specific intervals of time. Such individual geological regions are called *provinces.* The time periods in which provinces are formed are termed variously *orogenic* or *orogenetic cycles* or simply *orogens.* The word *orogenic* applies to mountains, so the term *mountain-building movement* is an approximate but incomplete synonym. This is because orogenies include more than the uplifting of mountains. Extrusion and intrusion of igneous rock, faulting, and metamorphism are all part of the process. Terminology is difficult; and some geologists would make mountain building the final phase of a much longer sequence of events—a sequence that they call the *geotectonic cycle.* This includes the formation of the rocks from which the mountains are made. Most major mountain chains have arisen from deep sedimentary accumulations in *geosynclines* (see Figure 16-2 for a graphic representation of the evolution of a geosyncline).

Geologists have found no very satisfactory fundamental explanation for the great geologic cycles. One possible explanation is that these cycles are related in some mysterious way to the rotation of the galaxy. The cosmic year, or time required for the solar system to complete one rotation, is approximately 275 million years. If this astronomical cycle does affect the earth, it is through powerful external forces we do not yet understand. Grand cycles may also be related to earth movements (plate tectonics) with or without astronomical influences. Thus there were no strong interactions between plates during the Cambrian Period, but there were extensive collisions during the Middle Ordovician to Devonian (Caledonian Revolution). Other great collisions took place in the late Paleozoic (Hercynian-Alleghenian Revolution) and in the Late Cretaceous–early Tertiary (Rocky Mountain Revolution).

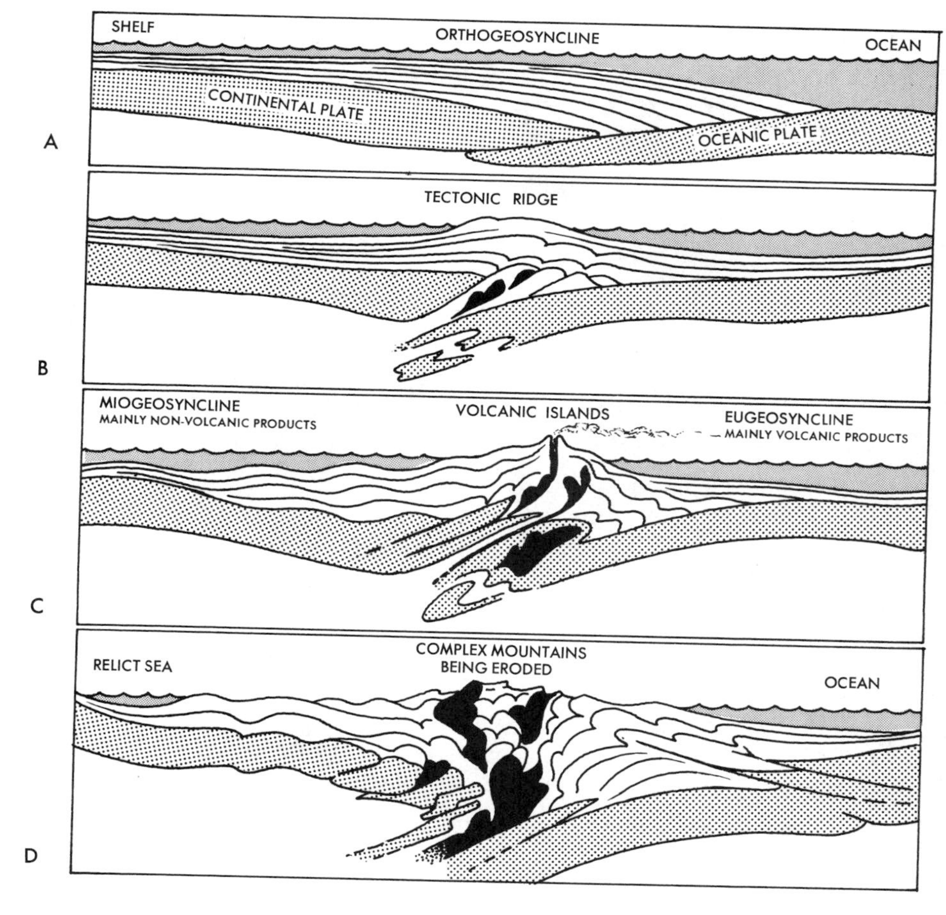

FIGURE 16-2 *Concept of the geosyncline. (A) Sediment derived from a continental area to the left is accumulating in great thickness at the margin of an ocean over the zone where the oceanic and continental plates join. This is similar to the configuration along the northern edge of the Gulf of Mexico. (B) The sediments are compressed as the oceanic plate begins to move under the continental plate; magma (shown in black) is generated at depth. (C) Compression continues, magma reaches the surface and volcanoes appear; metamorphism and faulting are intense; the outer or eugeosyncline receives chiefly volcanic debris from the median ridge. (D) Mountain building reaches a climax and ceases; the former geosyncline no longer exists but is represented by a mountain chain. Erosion may lower the mountains but most geosynclinal tracts once compressed are not again invaded by the ocean.*

Distribution of Precambrian rocks

All continents have cores or nuclei of ancient Precambrian rocks. Geologists refer to these Precambrian areas as *shields,* in allusion to their low-arching profiles—like a Roman shield in cross section. Each continent has one or more shield areas surrounded by belts of younger rock. Plainly the shields have been deeply eroded to furnish much of the fringing material. Shields are geologically very complicated, mostly highly metamorphosed, and penetrated by igneous bodies of great volume. Large subdivisions of a shield are called *cratons.*

Figure 16-3 shows the Precambrian outcrops of the world. In the Northern Hemisphere the major shield areas are the Canadian, Greenland, Baltic (in northwest Europe), Angarian or Siberian (in north-central Siberia), and Chinese or Korean (in eastern Asia). In the Southern Hemisphere are the Brazilian (sometimes subdivided in two or three pieces), Australian, and Antarctic shields. Africa, lying athwart the equator, is a special case; it is practically all Precambrian and may be referred to as a *platform* rather than a shield.

Much has been written about these shield areas, but most of it is highly technical and repetitious. For our purpose, brief summaries will suffice.

FIGURE 16-3 *Precambrian of the world. Black areas are exposed Precambrian rocks, ruled areas are covered Precambrian rocks, dashed line is border of Precambrian as presently known. (From The Geology of Continental Margins, Springer Verlag, with permission.)*

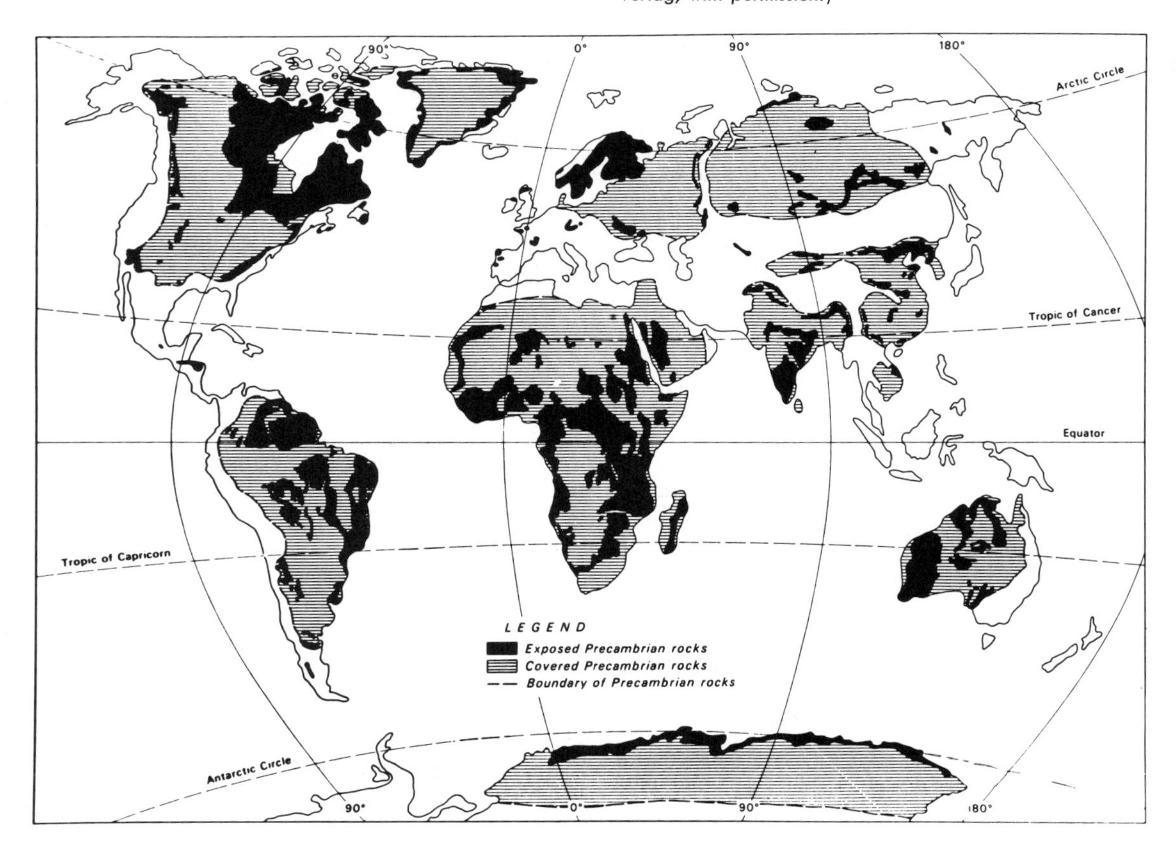

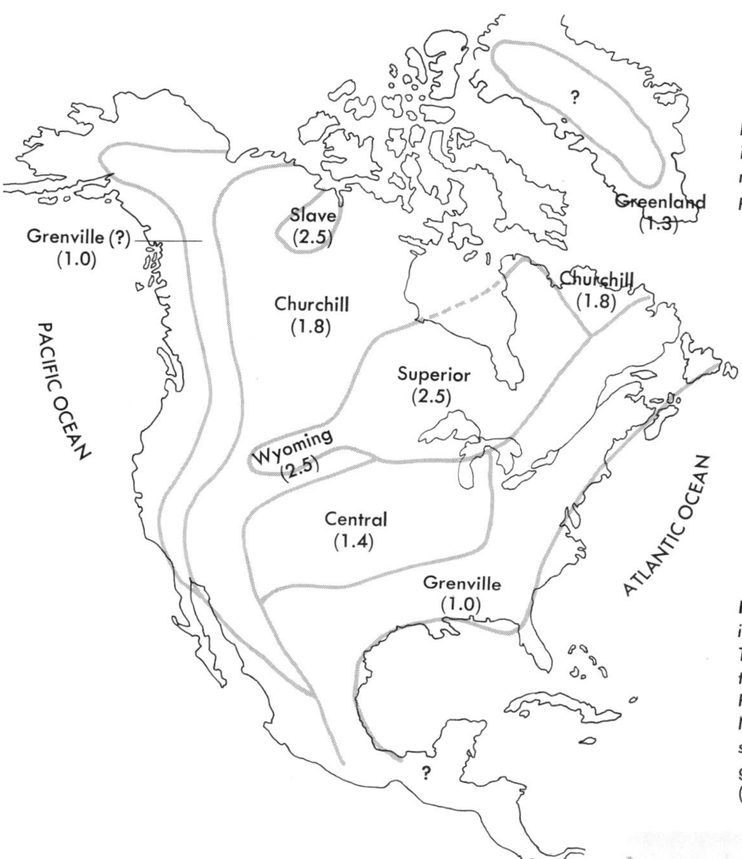

FIGURE 16-4 *(Left) Major geologic provinces of North America. The numbers in parentheses indicate the approximate age of the provinces in billions of years.*

FIGURE 16-5 *(Below) Looking down the inner gorge of the Grand Canyon in Arizona. The dark-colored, unstratified rock in which the river is running is the Vishnu Schist of Precambrian age, about 1.5 billion years old. Note light-colored veins in the walls. The stratified cliff above the Vishnu rests on a great unconformity and is of Cambrian age. (U.S. Bureau of Reclamation.)*

North America

In North America, Precambrian rocks are exposed chiefly in the *Canadian Shield,* an area of 7.2 million square kilometers if the nearby Greenland shield is included. The relationship of the shield area to other provinces is shown by Figure 16-4. The shield area is not centrally located but is displaced toward the northeast where it borders the Atlantic Ocean. Elsewhere around its edge the Precambrian rocks dip gently downward and disappear under a cover of younger sedimentary rocks. In what is called the *covered shield* of the north-central United States, Precambrian rocks are seen in uplifts such as the Black Hills and Ozark Mountains. They are also observed in the cores of the Rocky Mountain ranges and in deep canyons such as the Grand Canyon (Figure 16-5).

The oldest known rocks of North America come from a small area in southwestern Minnesota. These rocks are dated at 3.75 billion years. They lie within a much larger region stretching from Wyoming to Greenland, known as the *Superior-Wyoming Province.* Rocks of this province average about 2.5 billion years, and they are considered to have been produced by an intensive rock-making episode, the *Kenoran Orogeny.* Altered volcanic flows, granite bodies, and impure sediments make up most of this terrain.

Following the Kenoran Orogeny and presumably after the Superior-Wyoming Province had been eroded and partly surrounded by erosion products, a second intensive disturbance is recorded. A second generation of granitic masses was formed with the result that two new provinces, the Churchill to the north and the Central to the south, were created. The Churchill Province makes up most of the exposed Canadian Shield, whereas the Central Province occupies much of the north-central United States where it is mostly covered. The great disturbance that created these provinces, called the *Hudsonian Orogeny,* culminated roughly 1.8 billion years ago. A typical view of Precambrian terrane in the Canadian Shield is shown in Figure 16-6.

FIGURE 16-6 *Air view of rocks and structures in the Precambrian shield near Great Slave Lake, Canada. The great fault separates Precambrian granite on the left from Precambrian sediments on the right. The body of water is McDonald Lake. (Courtesy of Department of Mines and Technical Surveys, Canada.)*

A third great episode of rock formation and mountain building produced a new belt of continental rock that wraps about the eastern margins of North America from Labrador to Mexico. This is the *Grenville Orogeny,* and the belt of new rock is known as the *Grenvillian.* The ages of rocks gathered from this province average about 1 billion years.

Obviously there have been later orogenies and additions to the North American continent, such as those that produced the Appalachians and the Rocky Mountains. These too have enlarged the continent in the same way that the Precambrian orogenies did before them.

Europe

Most of northern Europe is underlain by Precambrian rocks. These rocks are found in many mountain ranges and make up the surface of the *Baltic Shield,* which is located chiefly in Sweden and Finland (Figure 16-7). The Precambrian lies under a thin blanket of sediments east and southeast of the Baltic Shield, an area known as the *Russian platform.* Precambrian rocks are also found in northern Scotland and neighboring islands and in various area of the European continent. Included are parts of the Alps, the Vosges, Brittany, the Massif Central of France, northwest Spain, Sardinia, and the Carpathians.

The European Precambrian has been intensively studied, and many concepts bearing on the origin of metamorphic and igneous

FIGURE 16-7 The Baltic Shield as seen in Finland. The region was heavily glaciated, and the Precambrian rocks are covered in many places by lakes, moraines, or forests. (Finnish Tourist Board.)

rocks have been developed by European geologists. Four orogenic periods are generally recognized in the Baltic Shield. The most intense and widespread effects occur in the third or *Karelidic* cycle, which culminated about 1.8 billion years ago. The youngest group of Precambrian rocks created after the Karelidic is little metamorphosed and contains vague fossil impressions. In northern Norway, beds of tillite resting on glaciated rock surfaces occur. Russian geologists, who know these late Proterozoic rocks chiefly through borings, call them the *Rhiphaean System;* in Norway they are known as the *Eocambrian,* or *Sparagmitian.*

Asia

Precambrian rocks are exposed at the surface or lie under shallow cover over a vast area of north-central Siberia. The central area in which the ancient rocks are actually visible is called the *Angara Shield;* a much wider area, thinly veneered with younger rocks, is the *Siberian platform.* Between the Siberian and Russian platforms lie the Ural Mountains and the west Siberian lowlands, which are sites of important basins and geosynclines of later date.

Precambrian rocks appear in the cores of many of the great ranges of southern Asia. There are also large disconnected patches in China, Korea, and Southeast Asia (Figure 16-8). Precambrian areas in India and the Arabian peninsula are best considered in connection with Africa with which they are geologically related.

FIGURE 16-8 *This section of the Great Wall of China in Hopeh Province crosses a rugged section of Precambrian rocks. (Photo courtesy of Frank H. Brown.)*

FIGURE 16-9 *Pão de Açúcar (Sugarloaf Mountain), an imposing monolith eroded from hard Precambrian rock, rises well above the massive formation that juts into the harbor area at Rio de Janeiro, Brazil. (Pan American Airways.)*

South America

About half of South America has Precambrian rocks at, or relatively near, the surface. Three large areas are recognized: the *Guiana Shield* bordered by the Orinoco River, the Atlantic Ocean, and the Amazon Valley; the *Central Brazilian Shield,* including most of the southern drainage of the Amazon; and the *Coastal Brazilian Shield* along the eastward bugle of the continent. Figure 16-9 shows a famous exposure of Precambrian rocks at Rio de Janeiro, Brazil.

The Precambrian rocks of South America are usually divided into three divisions: Early, Middle, and Late Precambrian. The oldest rocks are gneisses and schists about which little is known. The Middle Precambrian, a time of active deposition in geosynclinal belts, produced great thicknesses of sedimentary and igneous rock and ended with a period of granite formation and metamorphism. The Late Precambrian consists of less metamorphosed sediments such as quartzite, slate, phyllite, and conglomerate.

The South American shields have been mostly emergent since Precambrian time, but the Amazon Valley, which lies between two of these shields, has been repeatedly submerged by shallow seas since the early Paleozoic.

Africa

Africa has more extensive exposures of Precambrian rocks than any other continent. Madagascar is about half Precambrian (Figure 16-10). Africa is in fact a succession of shields and minor resistant blocks that have not yet been sorted out and named in a systematic way. Precambrian rocks make up much of the continent south of the Sahara and also the western bulge that fronts the Atlantic Ocean.

FIGURE 16-10 *Precambrian terrane in the Anosy region, Madagascar. Ancient rocks, chiefly gneisses and schists, appear frequently in rounded forms such as this. (Photo courtesy of UNESCO.)*

The *Arabian-Nubian Shield* is partly in Arabia and partly in Egypt, having been split by the relatively recent Red Sea rift. Some geologists refer to Africa as a platform and think of the entire continent as essentially Precambrian in nature.

The oldest rocks so far discovered are in Barberton Mountain Land of South Africa. In this area, piles of metamorphosed but still recognizable layered rocks having a thickness of about 19 km and ranging in age from 3 to 3.6 billion years have been found. One date of 4.1 billion years is recorded for a pebble of granite taken from a conglomerate within this section. This specimen, if correctly dated, is the oldest earth rock so far identified and may possibly be a fragment of the original crust.

Australia

Rocks of Precambrian age are found over wide areas of Australia, especially in the western half of the country. A striking feature is the great volume of sedimentary and volcanic rocks up to 2.3 billion years old that are relatively unmetamorphosed and are comparable in their unaltered appearance with rocks of much younger age elsewhere.

Australian geologists divide the Precambrian into the *Archaean* and *Proterozoic.* The Archaean, as defined by Australian geologists, includes rocks over 2.5 billion years old; it is complex and has not been subdivided except locally. The Proterozoic has been divided into three systems, from oldest to youngest: the *Nullaginian, Carpentarian,* and *Adelaidean.* These subdivisions include much basaltic lava, quartz-rich sediments, and cherty formations. These rocks occur in scattered patches with thicknesses up to 15,000 m or more. The Ade-

FIGURE 16-11 *Ayers Rock, an eroded remnant of Precambrian rock in the central desert region of Australia. Said to be the world's largest monolith, it is 335 m high and about 9 km in diameter at the base. (Quantas Airways Ltd.)*

laidean System is famous for the occurrence of many well-preserved primitive plant and animal fossils, which are described later in this chapter. Ayers Rock (Figure 16-11) is a famous Precambrian landmark in the heart of Australia.

India

About two-thirds of the surface of India consists of Precambrian rocks. The oldest series is chiefly gneiss intruded by granite of unknown thickness and age. Lying upon this is the Cuddapah System, a slightly deformed and slightly metamorphosed sequence of shale, conglomerate, limestone, and other ordinary sediments, which reaches a thickness of 7,000 m. Yet higher, in the central part of the peninsula, is the Vindhyan System, consisting of well-stratified, almost horizontal beds of sandstone, shale, and limestone. These beds can be identified as marine, fluvial, and estuarine in origin. Traces of primitive organisms have been discovered, and the rocks are considered to be very late Precambrian, with perhaps even some Cambrian at the top.

Antarctica

Although most of Antarctica is covered by ice and snow, enough of the bedrock is exposed around the margins and in protruding mountains to give a general idea of the geology. East Antarctica, the larger rounded part, is almost entirely of Precambrian age. Most of the exposures are dated at between 500 million and 1.5 billion years old; earlier Precambrian time is apparently not represented. There are many intrusive igneous rocks, and the effects of metamorphism and disturbances are much more intense than are shown by rocks of similar age in Australia. Figure 16-12 shows Precambrian rocks in the Transarctic Mountains.

FIGURE 16-12 *Exposure of Precambrian gneiss and schist, Transantarctic Mountains, Antarctica. (Institute of Polar Studies, photograph by John Gunner.)*

The Archean or Archaeozoic

The oldest rocks

It is a reasonable assumption that the outer crust of the earth solidified while the interior was still hot and molten. Geologists are interested in what the original crust may have been like. Was it spread in a continuous homogeneous layer, or was it disposed in islandlike patches? Was it basaltic in composition like the present-day ocean bottoms, or was it granitic like the continents? Any hope of finding patches of original crust has faded, but an intensive search is on to find the oldest rocks possible. Since the test of age is painstaking measurement of radioactive elements, a great amount of careful collecting and expensive analysis has been expended to reach the present state of knowledge.

It may be premature to designate specific rocks or any one area as being the oldest.

However, at present the oldest rocks on earth, dated at 3.75 billion years, are found near Granite Falls, Minnesota. Other areas of ancient rocks are known: South Africa, 3.4 to 3.6 billion years; Australia, 3.1 billion years; South America, 3.0 to 3.4 billion years; and Russia, rocks possibly as old as 3.8 billion years.

Although really old rocks are rare, many large areas of later Archean time (3.0 to 2.4 billion years) are known. These rocks are mostly enclosed within Precambrian material of later age from which they are significantly different. Most Archean rocks consist of intrusions of granite surrounded by belts of *greenstone,* a general name for greatly altered volcanic products, chiefly lava and pyroclastics. There are also great volumes of sedimentary rocks, and where these are not too highly metamorphosed, they are found to consist of conglomerate, arkose, breccia, and greywacke—this last being a general term for "dirty," unsorted, dark-colored mixtures of rock fragments derived from nearby sources and quickly buried without much sorting action.

Carbonate rocks (limestone and dolomite) are very rare in the Archean. Most Archean rocks, both igneous and sedimentary, are richer in sodium and poorer in potassium than later rocks. In fact the gradual increase in the K_2O/Na_2O ratio is one of the marked progressive changes of rock formation through time. Archean rocks show an abundance of dark, heavy igneous rocks of the gabbro-basalt class. The proportion of these iron-rich rocks falls off with time as varieties richer in silica are produced. The progressive changes in types of rocks produced through geologic time is illustrated by Figure 16-13.

Although it is difficult to relate Archean rocks to presently observed processes, many workers see similarities with what is going on now at growing continental margins. The Archean greenstone belts and granitic intrusions are geochemically similar to island arcs such as are forming along the western Pacific today. Since Archean belts or cratons are relatively small, it is concluded by some that the earliest landmasses were also small, and might best be called *protocontinents.* The Archean crust or lithosphere may have been one-third to one-half its present thickness and plate movements, if the movements really involved large tracts, were much livelier, and more localized than today. Relatively unmetamorphosed but still ancient deposits were laid down marginal to and between plates. Figure 16-14 is a view of sedimentary deposits in southwestern Rhodesia dated at more than 2.6 billion years old.

The appearance of water

The chemical compound H_2O exists in solid, liquid, or gaseous form, and by a peculiar coincidence all three states are possible on the surface of the earth at the present time. The total amount of water on the globe is calculated at about 330,000,000 cubic miles ($1{,}370 \times 10^6$ km^3). One of the most intriguing problems of geochemistry is the origin of this water and its role in the earlier stages of earth history. Clues to this problem are found not only in the rock record of the earth but also on neighboring planets and on the moon—all of which are waterless or nearly so. If we knew why Mars has no surface water, we might understand better why the earth is so abundantly supplied with it.

On the theory that a molten globe could not support oceans and streams, it is assumed that the water now in existence made an appearance during or after the cooling process. It was once supposed that all water was originally present in a thick and extensive canopy of clouds from which it might precipitate but could not accumulate until the crust had cooled and solidified. This theory has been replaced by a better one: that water emerged

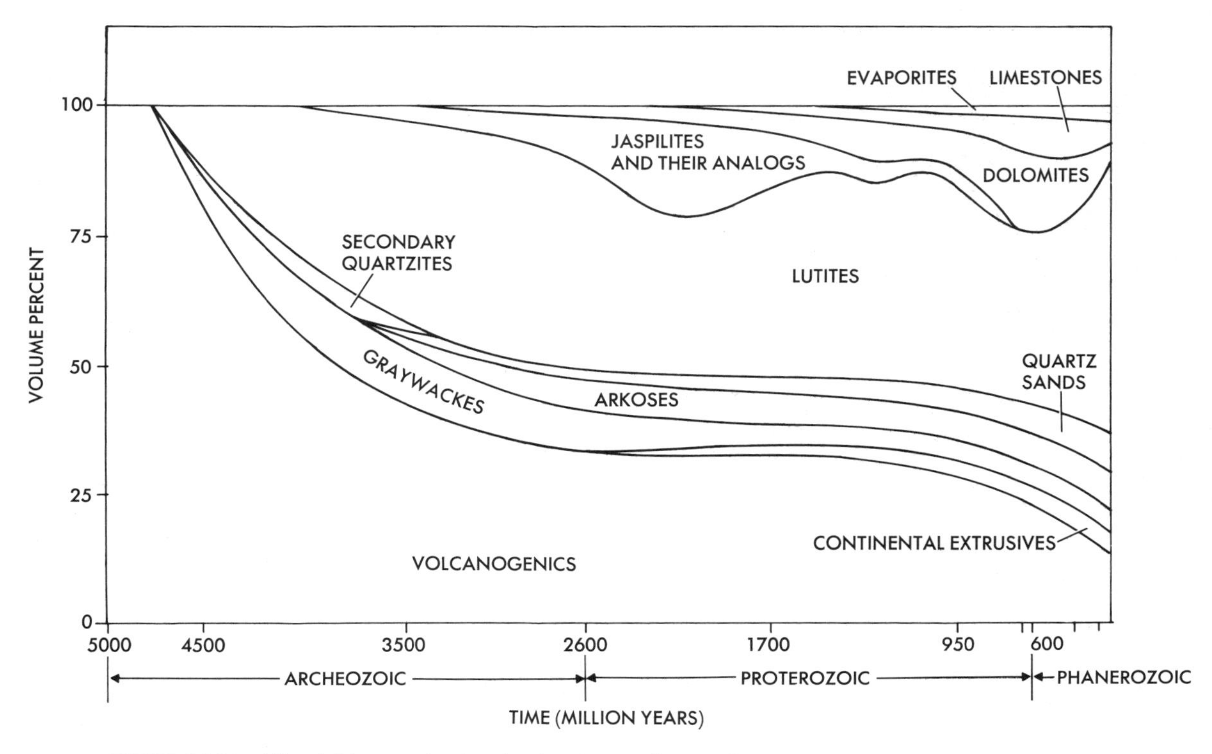

FIGURE 16-13 *(Above) Diagram showing the changing production of rock types from the beginning of the earth to present. Explanation of terms:* arkoses: *coarse, gritty sandstone derived with little change from granitic rocks;* continental extrusives: *volcanic lava, tuff, and ash produced on the continents, relatively high in silica;* dolomites: *magnesium-rich rocks, chiefly* $MgCO_3$, *generally associated with limestones;* graywackes: *unsorted "dirty" rocks, mostly sandstone with a high proportion of dark fragments together with quartz;* jaspilites: *siliceous sedimentary rocks usually rich in iron and chert;* limestones: *calcium-rich rocks that are chiefly calcium carbonate* $CaCo_3$; lutites: *fine-grained rocks such as mudstone, claystone, and fine siltstone;* quartz sands: *well-washed and relatively pure sand that is predominantly quartz* (SiO_2); secondary quartzites: *firmly cemented sedimentary rocks composed chiefly of silica sand with impurities;* volcanic products: *lava, tuff, ash, breccia, older formations are more mafic and make up much of so-called greenstones. (Modified from A. B. Ronov,* Common Tendencies in the Chemical Evolution of the Earth's Crust, Ocean, and Atmosphere: Geochemistry International, *No. 4, p. 713, 1964.)*

FIGURE 16-14 *(Right) Precambrian rocks dated at more than 2.6 billion years exposed in a limestone quarry, southwestern Rhodesia. These yielded the fossils shown in Figure 16-19. (NASA photo by K. A. Kenvolden.)*

FIGURE 16-15 Eruption of the volcano Kilauea, Hawaii, 1924. The great clouds emitted from the volcano consist largely of steam and water vapor. (U.S. Geological Survey.)

from the earth during and as a consequence of the cooling process. This event has been crudely compared with a stewing process; the correct term is *degassing,* borrowed from metallurgy and designating the expulsion of gas during the annealing process. Modern volcanic eruptions such as that illustrated by Figure 16-15 show that water is still being produced.

The early atmosphere

Geologists believe that the earth has possessed not one but several successive "atmospheres." In thinking about the origin of an atmosphere, bear in mind that what has been said about water holds true in large part for gases also. Volcanic eruptions obviously produce gaseous as well as solid products from within the earth.

It is generally believed that the early earth passed through a very hot if not entirely molten stage with surface temperatures of at least 2000°C. At this stage any gases clinging to the globe probably escaped into space, and for a time the earth was without an appreciable atmosphere. Later, as the surface cooled, water was produced, and an atmosphere accumulated partly from internal and partly from external sources.

Early in the history of the solar system we might expect many gaseous compounds of hydrogen since this is the most common element of the universe. Thus, water (H_2O),

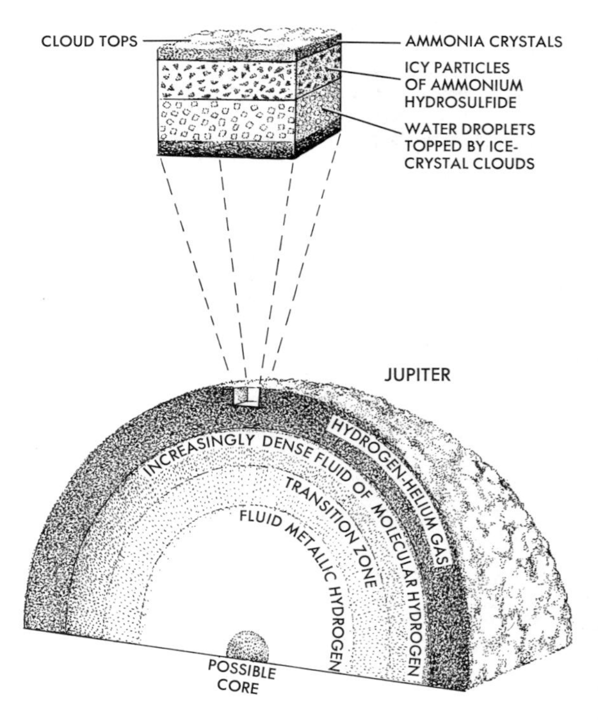

FIGURE 16-16 The giant planet Jupiter is predominantly hydrogen with complex outer cloud layers. Pioneer 10 and 11 spacecrafts passed near the planet in 1974 and 1975 and sent back data that reveal the surface and near-surface zones. What lies at depth is poorly understood.

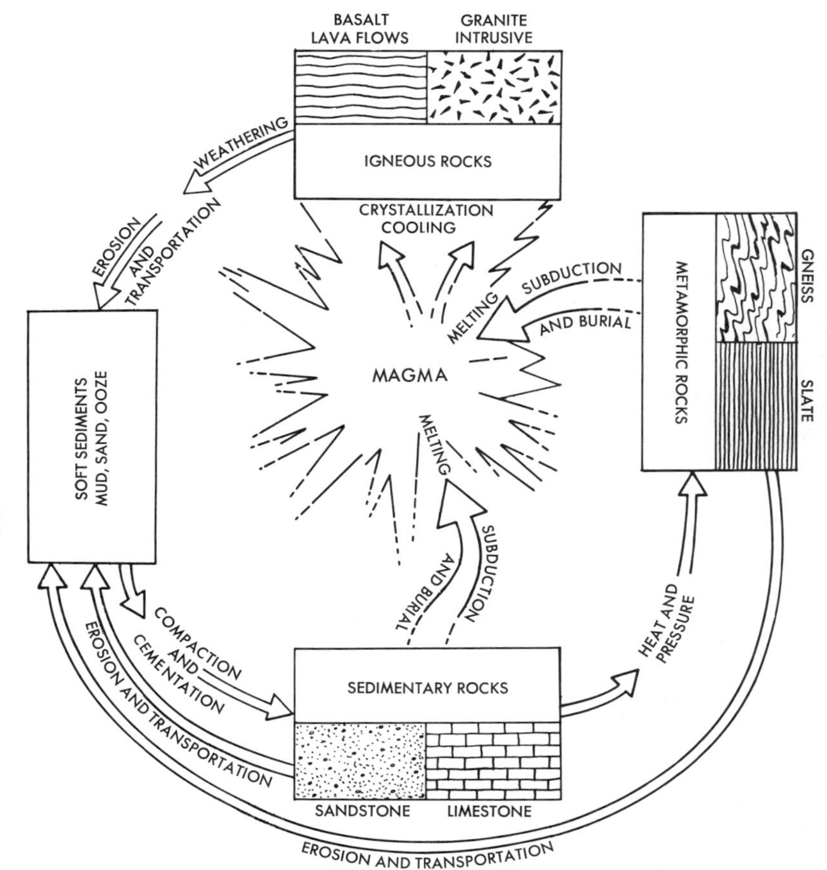

FIGURE 16-17 *The rock cycle showing nature and sequence of changes that convert rock material from one class to another.*

methane (CH_4), and ammonia (NH_3), all compounds of hydrogen, are assumed to have been major components of the earth's early atmosphere. These are still important in the major planets such as Saturn and Jupiter (Figure 16-16). At one time these gases may have been dispersed throughout a wide region around the primitive sun. The early atmosphere of the earth is commonly referred to as oxygen-poor or chemically reducing.

Early Proterozoic

As a general rule, the older a rock, the greater the changes that have affected it—that is, the more likely it is to have been altered and metamorphosed from its original condition. Agents that alter rocks are mainly heat and pressure; these agents convert sandstone to quartzite, limestone to marble, and shale to slate. As metamorphism intensifies, the affected rocks lose more and more of their original features so as to become gneiss and schist, which are probably the most common rock types of the Precambrian. At the ultimate extremes of metamorphism all constituents are melted and a second-generation igneous rock such as granite is created. Metamorphism alone cannot alter the chemical composition of a rock, but it can greatly affect its appearance. Refer to Figure 16-17 in connection with the topic of how the great groups of rocks have been produced and modified.

Two other terms and concepts should accompany metamorphism in thinking about Precambrian rocks. One of these is *differentiation*, the other *recycling*. *Differentiation* as applied to igneous rocks means the various chemical and physical actions that cause an originally homogeneous melted magma to separate into portions of different composition. Consider a mass of magma beneath the earth's surface that is gradually cooling and at the same time moving toward the surface. One simple type of differentiation controlled by gravity begins to operate: light fractions rise and heavy fractions settle. This alone seems capable of separating the crustal rocks into their best-known divisions, that is, the *Sial* or higher level and the *Sima* or lower level. (Sial from silicon and aluminum, and Sima from silicon and magnesium.) Other forms of differentiation, depending more on chemical reactions under different temperature and pressure relations, exist but need not be discussed here.

The other term to be considered is *recycling*. It applies to stratified rocks mostly and refers to what happens as sedimentary rocks are eroded, reworked by transportation, and redeposited a number of times. Thus the "dirty" rocks known as *wackies* are cleaned and sorted or winnowed into their various constituents, along with the production of "clean" sandstone and deposits of homogeneous fine-grained shale. Metamorphosed Precambrian graywacke is illustrated in Figure 16-18. With time there is a general reduction of particle size with less conglomerate and breccia and more siltstone or fine sandstone.

As the component grains of sedimentary rocks are reworked and recycled, they are said to become more mature. With the weathering

FIGURE 16-18 *Metamorphosed graywacke on the shore of Cross Lake, Manitoba. The age is determined by radioactive dating as about 2 billion years. (Photo courtesy Geological Survey of Canada.)*

of primary rocks there is also a gradual release of more soluble components, giving rise to the carbonates and evaporites, materials that are deposited directly from solution.

The Proterozoic, from about 2.5 to 0.5 billion years ago, was a time of metamorphism, differentiation, and recycling on a worldwide scale. The following trends are evident: (1) a decrease in rock types rich in iron and magnesium and a corresponding increase in the silica-rich varieties; (2) increasing production of limestone ($CaCO_3$) and dolomite ($MgCO_3$); and (3) thickening of the crust and widening of the continents through the differentiation and accretion of lighter siliceous material. In terms of actual igneous rocks produced, the trend during the Archaeozoic and Proterozoic is gabbro-basalt ⟶ diorite-andesite ⟶ granodiorite-dacite ⟶ quartz monozonite-granite-rhyolite. More and more silica was working upward and becoming manifest in less dense and lighter-colored rocks such as ordinary sandstone and siltstone.

Beginning of the fossil record

It may be helpful to recall that *fossils* are defined as *the remains or evidences of ancient organisms preserved in materials of the earth's crust.* This definition includes more than objects with definite form and shape. Disaggregated particles of carbonaceous material as well as chemical residues of materials known to have been synthesized by living things are genuine evidences of life and may be considered in the broad sense as fossil material. Petroleum is a mixture of organic liquid hydrocarbons and thus indicates the prior existence of life. Systematic search has been made in sediments for organic molecules such as phytane and pristane, which are related to chlorophyll. Starting with relatively young oil-rich rocks in which recognizable fossils are abundant, the examination has proceeded backward to sedimentary rocks about 3.6 billion years old. So far, rocks of all ages have yielded molecular fossils that suggest the presence of photosynthetic organisms. At no place in a long series of analyses has a point been discovered where the *residues* or products of living things are replaced by the more simple *predecessors* of life.

Oldest fossil organisms

Well-preserved remains of minute organisms have been found in sedimentary rocks billions of years old. Such preservation is possible mainly because of the physical properties of the rock called *chert,* which in pure form is silicon dioxide (SiO_2). This material precipitates rapidly as nodules or thin beds in sea water, hardens quickly, and becomes a very durable, noncompressible rock; therefore this mineral is ideal for the preservation of simple, semirigid organisms such as bacteria and fungi. When geologists suspect that a sample of chert contains microscopic remains, they cut it into thin sections and study it under high magnification.

African discoveries

The oldest known, structurally preserved objects that have been interpreted as organisms are from carbonaceous chert of the Onverwacht Series of the Barberton-Badplace region of the eastern Transvaal, South Africa. An igneous intrusion that cuts the chert beds is dated at 3.2 to 3.3 billion years, a minimum date for the age of the remains. There are algallike spheroids, filamentous structures, and carbonaceous fragments. That these are probably true fossils is indicated by their close association with chemical compounds not known to originate independently of living things.

(a)

(b)

FIGURE 16-19 *(a) Oldest known stromatolites, from the Bulawayan Group of southwestern Rhodesia. Age is more than 2.6 billion years—probably between 2.8 and 3.1 billion. Portion shown is about 9 cm long. (b) Stromatolites from the Precambrian rocks of Glacier National Park, Flathead County, Montana. The photo shows the weathered upper surface of a number of colonies. [(a) NASA, photograph by K. A. Kvenvolden; (b) U.S. Geological Survey.]*

In the same sequence of rocks but several hundred kilometers distant are found the earliest known *stromatolites* (Figure 16-19). These are distinctly layered accumulations of calcium carbonate having rounded, cabbage-like or branching fingerlike forms. They are not actual organisms but rather the deposits laid down by algae. That they are organic is shown by their resemblance to deposits created by living algae, by the discovery of scattered algal bodies within them, and by the absence of any other known mode of formation. Stromatolites are found in the Early Precambrian but do not become common until the Middle Precambrian. By Late Precambrian they are abundant enough to serve as specific guide fossils for rocks in which they occur.

An extensive assemblage, the first Precambrian microflora to be described, was reported from the Gunflint Iron Formation of Ontario, Canada, in 1954. The age as determined by radiometric means is about 2 billion years. The flora is diverse and includes threadlike, rodlike, and spheroidal species identical in form to present-day bacteria and blue-green algae. There are also abundant star-shaped and umbrella-shaped species of unknown affinities. Eight new genera with a large number of species have been named from the Gunflint Formation. All the forms are regarded as procaryotic (no organized nucleus), and none seem to be animals. Specimens from the Gunflint are shown in Figure 16-20. Extensive chemical analysis of the containing rocks has produced unmistakable hydrocarbons of organic origin as well as finely disseminated carbonaceous material.

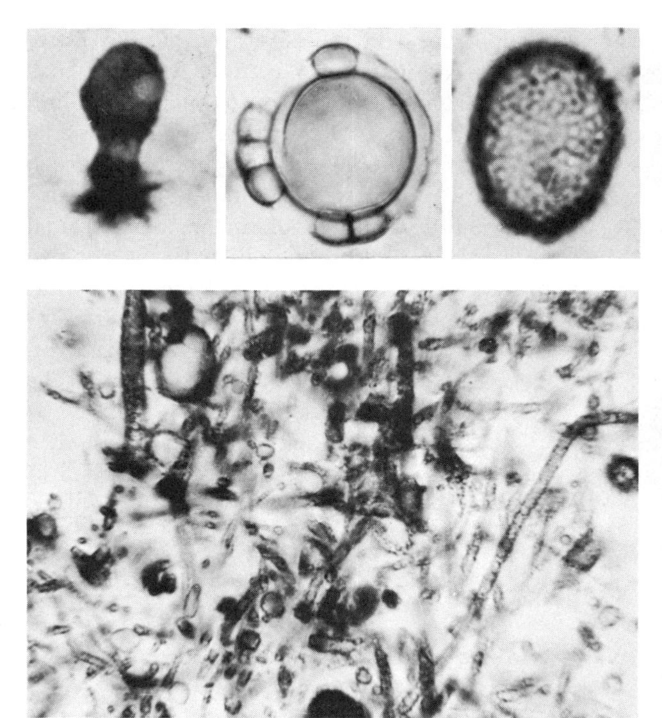

FIGURE 16-20 *Primitive algae and other microorganisms from the Gunflint chert near Schreiber, Ontario, Canada. All are highly magnified. The age is between 1.7 and 2.1 billion years. (Courtesy Elso S. Barghoorn.)*

FIGURE 16-21 *Microscopic plant fossils from the Late Precambrian Bitter Springs Formation, central Australia, approximately 1 billion years old. The fossils are preserved in chert, which can be cut into thin sections to reveal specimens such as these. Both spheroidal and filamentous forms are represented. The spheroidal forms (center) show what may be a nucleus. In each illustration the scale represents 10 μ. (Courtesy of Elso S. Barghoorn and J. W. Schopf.)*

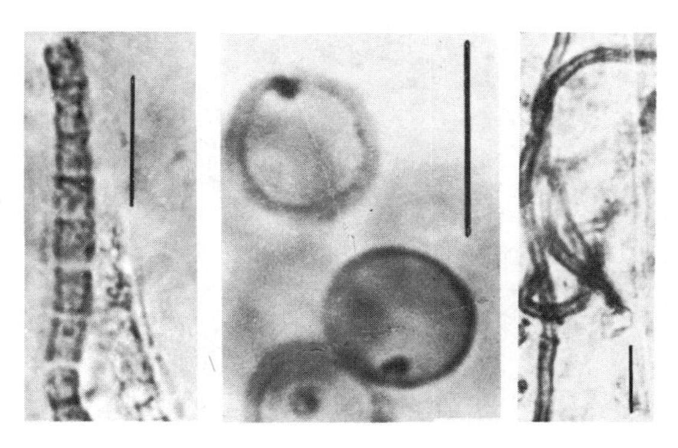

Other floras

The Beck Springs Dolomite of eastern California has yielded a Precambrian microflora younger than the Gunflint. The estimated age is 1.3 billion years. Another important and especially well-preserved assemblage of plants, found in the Late Precambrian Bitter Springs Formation of central Australia, has been determined to be approximately 1 billion years old. The fossils are preserved as organic residues in laminated black chert. Blue-green algae are abundant, and colonial bacteria, funguslike filamentous organisms, spheroidal green algae, and other cellular forms are also present (Figure 16-21). Fossil evidence indicates to some investigators that the first eucaryotic organisms (those with organized nuclei) developed from certain procaryotic cells about 1.5 billion years ago.

The Bitter Springs cherts have yielded 30 new species of plant microfossils; over half of these are blue-green algae, some of these show no significant differences from living forms of the same families.

The present atmosphere accumulates

The present atmosphere, which now surrounds the earth, was produced by processes that are probably unique in the solar system (Figure 16-22). Strange as it seems, living things may be directly responsible for eliminating primitive atmospheric compounds such as methane (CH_4); carbon dioxide (CO_2); and ammonia (NH_3) and replacing them with oxygen and nitrogen, which make up 20.95 percent and 78.09 percent of the atmosphere respectively (all other components make up less than 1 percent). The essential organic process is photosynthesis:

$$6CO_2 + 12H_2O \xrightarrow[\text{radiation}]{\text{solar}} C_6H_{12}O_6 + 6O_2 + 6H_2O$$

Note that photosynthesis breaks down both water and carbon dioxide to release part of the oxygen into the atmosphere in molecular form. Relevant reactions of oxygen in the atmosphere are:

$$2O_2 + CH_4 \longrightarrow CO_2 + 2H_2O \text{ (liquid)}$$
$$H_2 + \tfrac{1}{2}O_2 \longrightarrow H_2O \text{ (liquid)}$$
$$3O_2 + NH_3 \longrightarrow N_2 + 6H_2O \text{ (liquid)}$$

This last reaction may have produced the nitrogen of the present atmosphere. Other

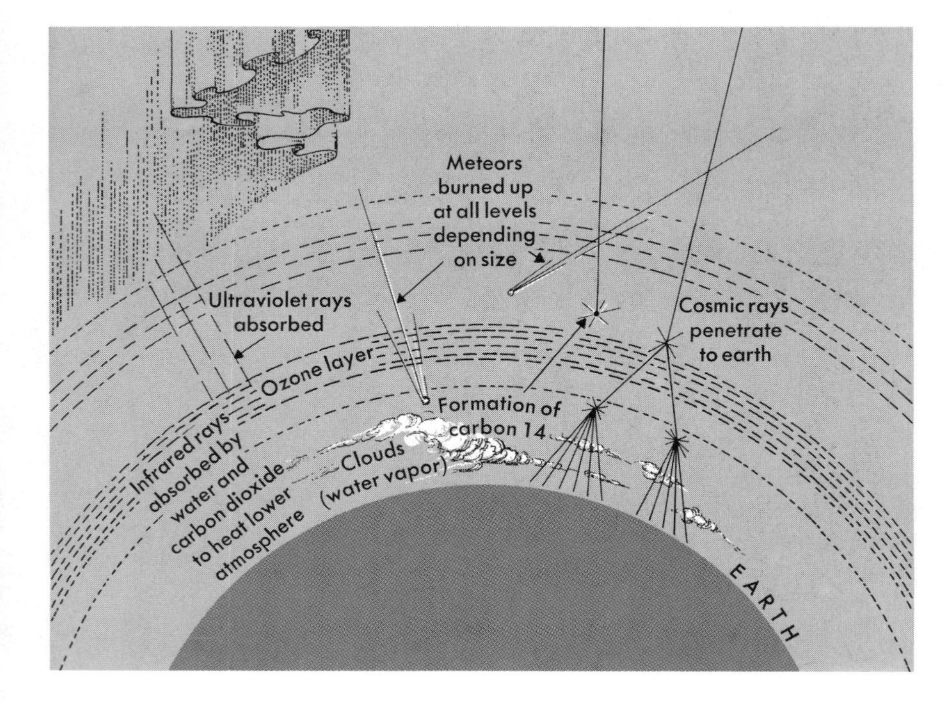

FIGURE 16-22 Some ways in which the atmosphere influences living things.

important reactions can be visualized. Possibly all surface carbon was once in the form of CH_4. But hydrogen can escape from the atmosphere, which would eventually leave carbon to react with oxygen and the resulting CO_2 in turn to react with calcium, so that $CaCO_3$ (calcite and limestone) could be deposited. This would explain why carbonate rocks are rare in the Archean but began to increase in volume during the Proterozoic.

Another reaction that some say was important in producing oxygen in the early atmosphere is the breaking up of water (*photolysis*) by ultraviolet radiation in the upper atmosphere:

$$2H_2O \longrightarrow 2H_2 + O_2$$

Early continents

An important aspect of the early history of the earth is the origin of continents. We know that the material of continents is essentially lightweight, frothlike material floating as if it were on a heavier substratum. Continental material has a specific gravity of 2.7; it rests on a heavier layer with a density of 3.3. The continental material is frequently referred to as *Sial* (from silicon and aluminum) or as *granitic* (from its typical light-colored igneous rock). The layer on which it rests is called *Sima* (from silicon and magnesium) or *basaltic* (from its major rock components). A most significant fact emerges: the continental material or Sial is the only component that does not occur in an earth-encasing shell. Why is this? In other words, why continents? If gravity alone were the agent responsible for stratification of the earth, it would seem that the continental material would also be a shell, even though a thin one, encasing the entire earth and supporting a universal ocean about 2,450 m deep. Something is working contrary to gravity to cause the lighter material to clump together in the form of large continental masses. A seemingly effective mechanism for the aggregation of the continents has been discovered in the form of continental drift and sea-floor spreading. By this process the lighter elements and minerals reach the surface and are then conveyed laterally to accumulate in discrete pancakelike masses quite distinct from, and elevated above, the ocean basins.

All this is fine in theory but does not answer the question of when and where the first continents appeared. All that the Archean rocks tell us is that there were patches of solid rock in existence. What remains of these today is certainly smaller than the original areas, considering the effects of subsequent melting and erosion. When the areas of Proterozoic rocks are added to the Archean, they make up the shield areas that are mostly large enough to be called continents. But it may not be safe to say that the shield areas were always in the solid aggregations that now exist. Some say they originated in smaller fragments and came together by a primitive form of continental drift. This is probably true, but it may not be justified to say that this process of accretion was truly plate tectonics as it now operates. The crust of the earth may have been thinner, even molten, so that solid blocks could actually traverse the surface of the earth like giant rafts.

It has been observed that if existing continents are reassembled to make up Pangaea, the ancient protocontinent of the early Paleozoic, the Precambrian rocks lie in a gigantic semicircle (see Figure 16-23). The Archean rocks constitute a core for this semicircle and appear to have once been continuous. This is the most ancient geographic configuration of the earth's surface that can be detected. It is proposed that this basic feature was created by the impact of a gigantic comet or asteroid about 4 billion years ago. This object might have contributed a great deal of exotic substance to the outer earth. The cataclysmic ef-

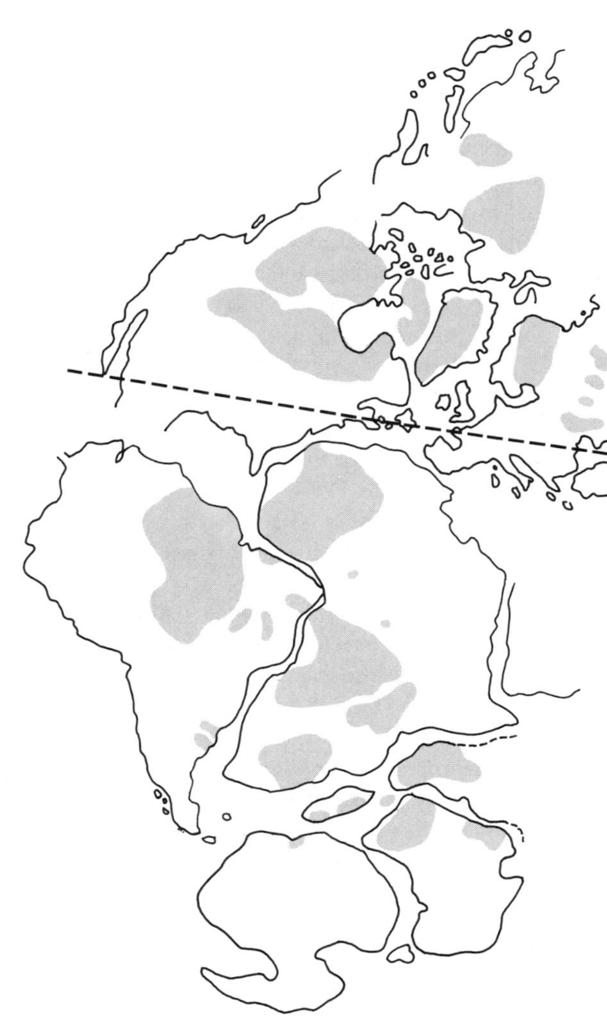

FIGURE 16-23 A predrift reconstruction showing the continental blocks (tinted areas) isolated in two restricted areas about 1.7 billion years ago. The blocks are transected and surrounded by belts of younger rocks.

fects of such a collision might include the piling up of crustal material on one side of the earth and its removal on the other, to create the water hemisphere and land hemisphere that characterize the globe. Is the earth still in the process of recovery from an event that took place over 4 billion years ago? We know that the Moon, Mars, and Mercury are geologically asymmetrical, apparently due to a period of giant impacts. Why not Earth? Scars of ancient impacts such as that shown in Figure 16-24 are fairly common in ancient rocks of the Canadian shield.

Rocks and environment of the late Proterozoic

Most rocks have been derived from previous rocks. Archaeozoic greenstone, granite, and graywacke formations were eroded and redeposited to make up later deposits of all ages starting with the Proterozoic. This recycling involved a variety of cleaning, sorting, and winnowing processes, and the resulting rocks tend to be both more homogeneous and diversified. Proterozoic rocks include much quartzite, the common erosional end product of siliceous rocks such as granite and arkose. At least 6,100 m of predominantly quartzite formations make up most of the Uinta Range in Utah (Figure 16-25). Precambrian processes tended to produce still finer grades of sediment such as siltstone and mudstone, the metamorphic equivalents of which are argillite, slate, and phyllite. A notable example is the Belt System of Montana, 18,300 m thick and mainly fine grained. Deposits such as these composed of reworked fragments are said to be *mature* or *submature.*

FIGURE 16-24 *(Right) Vertical air photo of the circular Clear-Water structures in Precambrian rocks near the east shore of Hudson Bay, Labrador. The dark area is water occupying circular depressions; the larger is about 36 km across. It is thought that these are ancient deeply eroded meteoric impact scars. (Photo courtesy of Geological Survey of Canada.)*

FIGURE 16-25 *(Below) The Green River flowing through Precambrian quartzite strata estimated to be over 1 billion years old. Near the eastern end of the Uinta Range, Utah-Colorado. (Photo courtesy of U.S. Bureau of Reclamation.)*

Well-sorted conglomerates abound in the Proterozoic. These were derived from older formations and reveal much about vanished mountains and source areas. Included among the conglomerates are great gold- and uranium-bearing formations. The valuable metals in most cases are plainly derived from older deposits. Also, in the older conglomerates, are fragments of pyrite (FeS) and uraninite (UO_2). These are not stable in the presence of free oxygen; pyrite "rusts" to form other iron minerals and UO_2 becomes UO_3. The presence of unoxidized minerals in surface deposits is considered strong supporting evidence for an oxygen-poor atmosphere before about 2 billion years ago (see Figure 16-26).

The carbonate rocks, limestone and dolomite, are rare but increasing in the Proterozoic. Not that they are entirely absent, since deposits of dolomite 2,000 m thick are known in the Mackenzie Mountains of Canada, and there are scattered thick carbonate formations in Africa. These are considered to be mainly chemical precipitates and not of biologic origin.

FIGURE 16-26 Grains of pyrite (white) thought to have been worn and rounded by water action. East Daggofontein Mines, East Rand, South Africa. Age is Middle Precambrian. (Photo courtesy of Desmond A. Pretorius.)

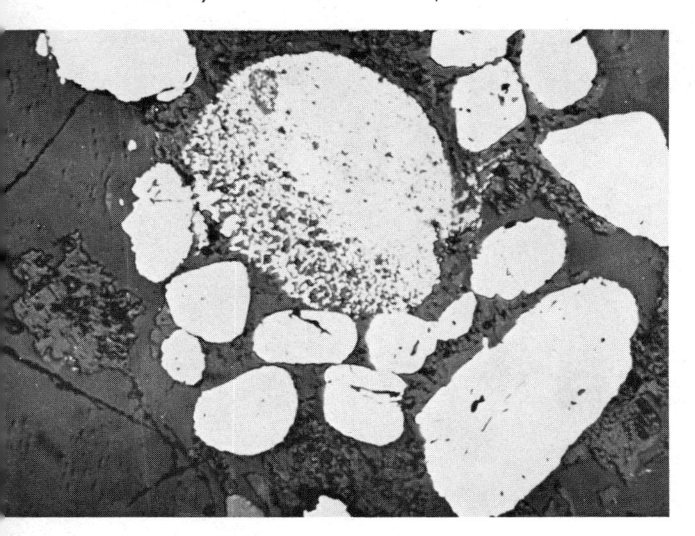

Precambrian resources

The Great Iron Age

Iron is the fifth most abundant element of the earth, so it is not surprising to find valuable concentrations in rocks of all ages (Figure 16-27). After a great deal of prospecting for minable deposits, it is now recognized that there was one period of time during which deposition of iron reached spectacular proportions. This period centered at about 2.2 billion years ago and might well be called the Great Iron Age. At this time iron was deposited and is now being mined in the Lake Superior area of the United States; Quebec-Labrador, Canada; northwestern Australia; Orissa and Bihar States, India; Sweden; Krivoyrog and Kurst areas, U.S.S.R.; Venezuela; Brazil; and many other places.

Iron-bearing sediments of the middle Proterozoic have distinctive properties. Mostly they are banded so that a typical sample shows alternations of iron-poor siliceous material (usually chert, SiO_2) and fine-grained, iron-rich minerals of various kinds, including oxides, silicates, and carbonates. The appearance and chemistry of these rocks are remarkably similar everywhere, and they are referred to as *banded iron formations* or *BIF* for convenient reference. The iron content is 15 percent or more and may be greatly increased by weathering and metamorphism. It is the enriched areas of the BIF that are sought after for commercial exploitation. The huge Lake Superior mines yield enriched ore with 50 to 60 percent iron that had been concentrated from original rock with 25 percent iron.

Banded iron formations are thought to have accumulated as sediments in extensive shallow lakes or seas. The iron appears to have been supplied from nearby lands made up of volcanic rocks. The very fine banding is thought by many geologists to be the result of

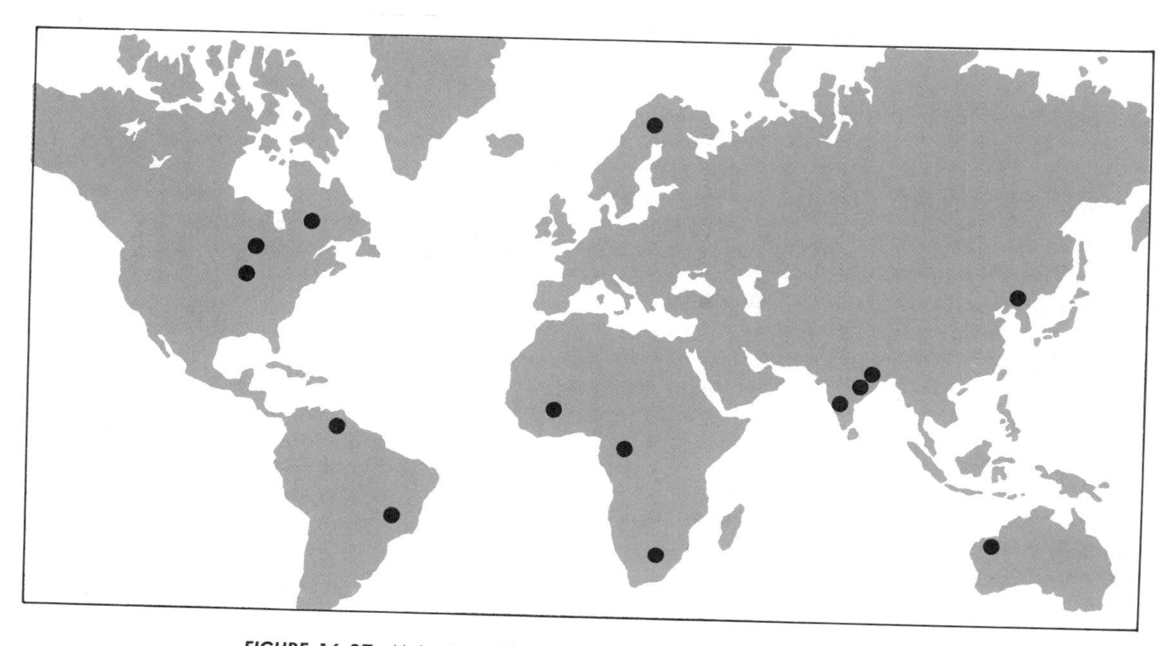

FIGURE 16-27 Major iron deposits of the earth. Most are in Precambrian rocks.

FIGURE 16-28 Banded iron formation, over two billion years old, Michipicoten district, Ontario. Alternating bands of iron-bearing minerals and chert are exposed. Pocketknife, upper center, gives scale. (Photo courtesy of Geological Survey of Canada.)

seasonal differences in temperature and water chemistry (Figure 16-28). The layers are, in other words, annual deposits or varves. Bacterial action may have been important, and stromatolites are present in many deposits.

Wider implications are that the BIF record a stage of the development of the earth when oxygen became abundant enough to react with iron in aqueous environments. In other words, by about 2200 million years ago, photosynthesis had released enough oxygen to react with iron and other metals on a global scale. Later, as oxygen-hungry elements were satisfied, the production of iron deposits declined.

Other ore deposits

In addition to iron, Precambrian rocks contain great reserves of other metallic ores. The Canadian Shield has a rich variety of ore deposits, most of which are in Canada, but a few are in the United States. Archean rocks are particularly rich in gold. The Slave Province in northwestern Canada, with rocks averaging 2.5 billion years old, has at least 1,000 known gold deposits. Other products of the Archean include silver, nickel, lead, and zinc, all of which show a preferred association with igneous rocks.

Later Precambrian rocks contain iron and also ores of other metals. Great deposits of uranium are found in conglomerates at Blind River, southern Ontario; copper ores, now mainly worked out, are found in the Lake Superior region; the world's greatest nickel deposit is at Sudbury, Ontario; and complex silver, lead, and zinc deposits occur at many localities. The Homestake Mine, most important gold producer of the United States, is in Precambrian rocks in the Black Hills (Figure 16-30), and the Iron Mountain iron deposit is in the Precambrian core of the Ozark Moun-

FIGURE 16-29 *Open-cut iron mine in the Precambrian rocks of the Mesabi Range, northeastern Minnesota. Note the overburden of waste material above the dark-colored iron ore. (Courtesy of M. A. Hanna Company and U.S. Bureau of Mines.)*

tains. These examples remind geologists that there are probably numerous other great mineral deposits under shallow cover of sediments in the United States and Canada.

The Baltic Shield of Norway, Sweden, Finland, and adjacent U.S.S.R. yields a variety of nonferrous (not iron or manganese) ores including those of copper, lead, zinc, and silver. Important reserves of copper, chromium, and asbestos are found in the Russian sector.

The shield areas of Asia, other than India, may not be particularly rich in mineral deposits, but much territory remains to be explored and developed.

India has, in addition to iron and manganese, gold at Kolar, asbestos at Travancore, and diamonds at a number of localities. Australia has several great gold camps in the Precambrian; Kalgoorlie in Western Australia is the most notable.

The Precambrian rocks of Africa constitute the world's greatest source of many important mineral products. The famous Copper Belt of Zambia contains an estimated 25 percent of the world's copper and much of its cobalt. Since 1905 South Africa has been the world's greatest gold producer; currently 75 percent of the world's output comes from here. Production centers at the Rand, and the gold is obtained from sedimentary rocks, chiefly conglomerate, at great depths. The same deposits contain enough uranium to make South Africa the world's third largest producer. South Africa is second in world production of platinum and is also important in vanadium, chrome, and titanium reserves. Zaire (formerly the Congo) is first in cobalt, and Niger is first in niobium.

Gold is widespread and there are important deposits, some worked since ancient times, in Ethiopia and Egypt. Diamonds are found in significant quantities in 12 African countries, and the African continent produces 67 percent of the world's supply. The geology of diamonds presents special problems. Diamonds are derived from deep-seated sources and may have been formed at one period and then were erupted to the surface by volcanic action at a later date. At Kimberley, South Africa, the diamonds are taken from the volcanic "pipes" that brought them to the surface (Figure 16-31). In most other diamond fields the stones have been reworked and deposited in river beds or beaches. Their great hardness makes them practically indestructible, and they may be carried hundreds of miles from their source.

FIGURE 16-30 *Mining gold-bearing quartz veins in the Homestake Mine, Lead, South Dakota. This is the most productive gold mine in the United States. The ore is in highly contorted Precambrian rocks. (U.S. Bureau of Mines.)*

The Precambrian rocks of South America, in addition to great deposits of iron and manganese, also have diamonds, chiefly in the Rasaima area of Venezuela. Gold is mined at Morrow Velho, Brazil. Prospecting in the jungle-covered regions is difficult, and other deposits will undoubtedly come to light.

FIGURE 16-31 The "Big Hole" of the Premier Diamond Mine, South Africa, created by removal of diamond-bearing rock. The diamond-rich rock occurs in the form of vertical cylindrical pipes that are thought to have been exit ways for deep-seated volcanic action during the Cretaceous Period. (Photo courtesy of U.S. Bureau of Mines.)

Animals appear

Animals, analogous to the amoeba, may have been in existence in early Precambrian time, but it is unlikely that such soft-bodied organisms could leave fossils. Note that plants have rigid cell walls and are therefore easily fossilized. Remains of chitinozoans generally considered to be animals with rigid cell-wells have recently been reported from rocks in the Grand Canyon estimated at about 850 million years old. Fossils of worm-like organisms have also been reported from Precambrian sediments of North Carolina; the estimated age is 620 million years.

Zoologists recognize as many as 11 phyla of living wormlike animals, but these are mostly soft-bodied and at best leave only elongate, winding trails and burrows. In spite of the poor Precambrian record, much can be inferred regarding these earliest of animal fossils. The worms possess a number of very significant adaptations: they have varied means of locomotion, a head end containing a brain and rudimentary sense organs, bilateral symmetry, and a means of taking food into the body where it is digested and from which it is excreted. These are all rather advanced characteristics not possessed by the majority of living things. In explaining the evidence of highly evolved characteristics at an early geologic date, paleontologists assume a long period of evolution in which these characteristics were produced and selected.

The most varied and distinctive trails and burrows so far discovered come from the Precambrian of Australia. These fossils consist of sinuous trails; meandering, closely spaced "grazing" patterns; evenly spaced grooves or indentations; and rows of ovoid structures interpreted as fecal material. All of these remains seem to have been created by elongate, wormlike animals, some of which may belong to annelids, ancestral mollusks, or coelenterates. Specimens from the Torrowangee Group of New South Wales are so consistently well defined that seven species have been described. The rocks are thought to be Late Precambrian in age, within the range of 1 to 1.5 billion years old.

Much more satisfactory than trails and burrows are fossils showing complete shapes. A considerable number of these are now known from Precambrian rocks at widely scattered spots on the globe. The most important locality thus far reported is the Ediacara Hills in southern Australia. Here the fossils occur as impressions in sandstone. Many hundreds of specimens have now been collected and have been classified into the following basic forms: rounded impressions with radiating grooves resembling jellyfish; stalklike fronds with grooved branches, similar to living sea pens; elongate, segmented, wormlike impressions with a horseshoe-shaped head and about 40 segments, which closely resemble annelids; flattened, nearly round, wormlike impressions; oval impressions with T-shaped grooves that seem to be unrelated to any living organisms; and curious circular impressions with bent arms that radiate outward, also unlike any known living thing. Among these forms there are unquestionably representatives of two living phyla, the coelenterates and the annelids; arthropods are probably present also. A great variety of trails and burrows and possible plant remains are also found in the Ediacara region.

Another locality with fossils similar to the Ediacara fauna is in southern Newfoundland (Figure 16-32). Casts and molds of a variety of organisms are described in terms of outline as spindle-shaped, leaf-shaped, round to lobate, and dendritelike. Some of these are definitely coelenterates, but most cannot yet be assigned to specific zoological groups. The age of this locality is in doubt, but the preponderance of evidence indicates that it is Late Precambrian.

FIGURE 16-32 *Unidentified fossil organisms from Late Precambrian rocks, southeastern Newfoundland. Impressions only are preserved. Both the branching and circular impressions are tentatively regarded as coelenterates, distantly related to modern coral and jellyfish. (Courtesy of S. B. Misra.)*

FIGURE 16-33 (Left) Varved sediments of the Gowganda Formation, Middle Precambrian, dated at about 2 billion years old. Note pebble embedded in the rock. From evidence such as this, the Gowganda is considered to record an early ice age. View at Espanola, Ontario. (Photo courtesy of Geological Survey of Canada.)

FIGURE 16-34 (Below) An exposure of some of the earth's oldest red beds. The Hakatai Formation of Late Precambrian age appears in the cliffs immediately above the Colorado River. Light-colored cliffs forming the point at the upper left are composed of the basal Cambrian Tapeats Sandstone. (Photo courtesy of Donald P. Elston.)

With the discovery of the abundant Australian and Newfoundland fossils, interest has been revived in previous less spectacular and hence unappreciated finds. Similar fossils has been found in England and South Africa, but these were so few in number and so poorly preserved that scientists were hesitant to place much weight on them.

Although doubt has been expressed about the correct age assignments of the early metazoan fossils just described, it is generally agreed that they are Late Precambrian. Certainly they are distinctly different from the trilobite-dominated faunas of the Cambrian. Evidence indicates that diversification into modern phyla was well under way, and further discoveries can be expected. Of significance is the fact that most, if not all, Precambrian animals so far discovered are soft-bodied and shell-less, a fact that may account for the general scarcity of remains.

Early ice ages and red beds

At least two ice ages have left their marks in the Precambrian record. The earlier of these is a little older than 2.0 billion years and is best recorded in the Huronian (early Proterozoic) of Canada (Figure 16-33). The second ice age was apparently more widespread; characteristic deposits have been found and authenticated on every continent except Antarctica. The most common rock type is *tillite,* the solidified equivalent of the loose heterogeneous material of glacial moraines. In addition, varved sediments, smoothed and striated surfaces, and abraded cobbles are common. This ice age may have been the greatest to have affected the earth. It is not surprising that no obvious explanations have been found for it—since we have none for the ice age we are currently experiencing either. The late Proterozoic glaciation is dated at about .7 billion years ago, and it may have set the stage in some way for the appearance of abundant fossils that came shortly after, at the beginning of the Cambrian.

Another widespread rock type of the late Proterozoic is "red beds." These are deposited in the presence of air and hence are said to be of terrestrial or dry-land origin. An excellent example is the Hakatai Formation in the Grand Canyon (Figure 16-34). The red color is mainly due to the pigment of oxidized iron-bearing minerals. Red beds are not of the same origin as the banded-iron formations that were water-deposited. The iron content of red beds is usually a fraction of a percent. Their significance is that they cannot form in an oxygen-free atmosphere. Their appearance about 1.0 billion years ago is taken to mean that enough oxygen had been set free by photosynthesis to impart a substantial and growing amount to the atmosphere.

SUMMARY The span of time from the formation of the earth to the appearance of abundant fossils is called the *Precambrian*—an interval of about 4 billion years. Rocks produced during this lengthy period are now mostly highly deformed and metamorphosed, and their study and correlation are difficult. Precambrian rocks are found in each of the large continental masses making up roughly centralized tracts called *shields.* A rough division of the Precambrian followed by most geologists places all time between 4.6 and 3.8 billion years in the Pre-Archean, between 3.7 and 3.4 billion years in the Archean, and between 2.4 to 0.6 billion years in the Proterozoic.

Few rocks of Pre-Archean time have been discovered. Any solid material of this interval was apparently destroyed by repeated meltings. During the Archean, solidification of igneous rocks and production of primitive types of sediment were under way. Water was produced early in the period, and sedimentary rocks began to accumulate. An atmosphere rich in CO_2, NH_5, CH_4, and H_2O prevailed, but there was little or no free oxygen. Traces of organic molecules indicate that life had probably appeared by about 3.2 billion years ago. Landmasses of unknown size and configuration, probably smaller than present continents, had formed by processes not greatly different from those seen today along active continental margins.

Fossils with definite structure and shape appear early in the Proterozoic, and there is evidence that photosynthesis was beginning to enrich the environment in free oxygen. This was a time of widespread formation of iron-rich sediments. Sedimentary rocks are mainly siliceous and are better sorted and more diversified than in the Archean. Carbonate rocks are relatively rare. Evidence is that the Proterozoic lands were in one or two supercontinents.

Fossils of many-celled animals appear near the close of the Proterozoic. Remains of three of the standard phyla (coelenterates, annelids, arthropods), have been identified. The atmosphere had apparently reached approximately its present composition, and continental red beds (requiring atmospheric oxygen) were beginning to form. A widespread ice age left effects on all major continents about .7 billion years ago. This may have set the stage for the appearance of abundant fossils at the beginning of the Cambrian about 100 million years later.

Precambrian rocks contain large, economically important deposits of metal-bearing ores including iron and gold in almost all continents. Especially notable is nickle, silver, and uranium in Canada, and copper, uranium, and diamonds in South Africa.

SUGGESTED READINGS

BRANCUZIO, C. J., and A. G. W. CAMERON, eds., *The Origin and Evolution of Atmospheres and Oceans.* New York: John Wiley & Sons, Inc., 1964.

BURK, CREIGHTON, and CHARLES DRAKE, eds., *The Geology of Continental Margins.* New York: Springer Verlag, 1974.

CLARK, T. H., and C. W. STERN, *The Geologic Evolution of North America.* New York: The Ronald Press Company, 1969.

KING, PHILIP B., *The Evolution of North America.* Princeton, N.J.: Princeton University Press, 1959.

RANKAMA, KALERVO, ed., *The Geologic Systems: The Precambrian.* New York: Interscience Publishers. Series includes: Vol. 1, Denmark, Norway, Sweden, and Finland (1963); Vol. 2, Spitsbergen and Bjoroya, British Isles, Greenland, and Canada (1965); Vol. 3, India, Ceylon, Seychelles Archipelago, Madagascar, and the Congo, Rwanda, and Burundi (1967); Vol. 4, Southeastern United States, South-central United States, Northwestern United States, Mexico (1970).

The Early Paleozoic Periods

This chapter deals with the combined Cambrian, Ordovician, and Silurian Periods with a total duration of about 175 million years (Cambrian, 75 million years; Ordovician, 60 to 70 million years; and Silurian, 30 to 45 million years). The obvious great disparity in the lengths of these periods gives good reason for not discussing each separately as equivalent entities. The next chapter treats the late Paleozoic (Devonian, Mississippian, Pennsylvanian, and Permian Periods) with a total duration of about 175 million years. The Mesozoic Era, consisting of the Triassic, Jurassic, and Cretaceous Periods, with a total length of about 165 million years, will be discussed in Chapter 19.

The three- and four-period intervals mentioned above are of nearly equal duration, and each displays a roughly similar or cyclic succession of events. The beginning of a major

17

Specimens of a trilobite (Olenellus garetti) from Lower Cambrian rocks of Cranbrook, British Columbia. (Courtesy of V. J. Okulitch.)

cycle, as typified by the Cambrian or Triassic period, is a time of relative quiet insofar as geologic activity is concerned. Mountain building, volcanic activity, glaciation, and climatic extremes are nonexistent or weakly expressed. The succeeding middle phase, typified by the Ordovician, Mississippian, and Jurassic, is a time of low-lying lands, extensive shallow seas, and a low ebb in mountain building. Each major cycle was terminated by geologic unrest and widespread mountain building. Thus the Devonian, Pennsylvanian-Permian, and Cretaceous–Early Tertiary are all marked by elevation of mountains, withdrawal of interior seas, and climatic extremes.

Like most historical cycles, these geologic rhythms are neither perfect nor universal. History repeats itself, but not exactly. At best, the great geological cycles constitute crude generalized patterns for earth history. Some geologists refer to times of great emergence as *land periods* and times of great submergence as *sea periods.* European and Russian geologists refer to the events of the early Paleozoic as *Caledonian,* those of the late Paleozoic as *Hercynian,* those of the Mesozoic as *Kimmerian,* and those of the Cenozoic as *Alpine.* These terms are applied not only to the mountain-building episodes but also to rocks formed during the same intervals.

The distribution of early Paleozoic landmasses and ocean basins is known in a general way, and information is rapidly accumulating that will make better reconstructions possible (Figure 17-1). The two great supercontinents of Gondwanaland and Laurasia were in evidence. Gondwanaland included what is now Africa, South America, Australia, and Antarctica. In addition it almost certainly included lands now in the Northern Hemisphere. India is the best known migrant from the southern aggregation, but there is growing evidence that other fragments such as Tibet, Tarim, and possibly North China, now all deeply embedded in central Asia, were also once parts of Gondwanaland.

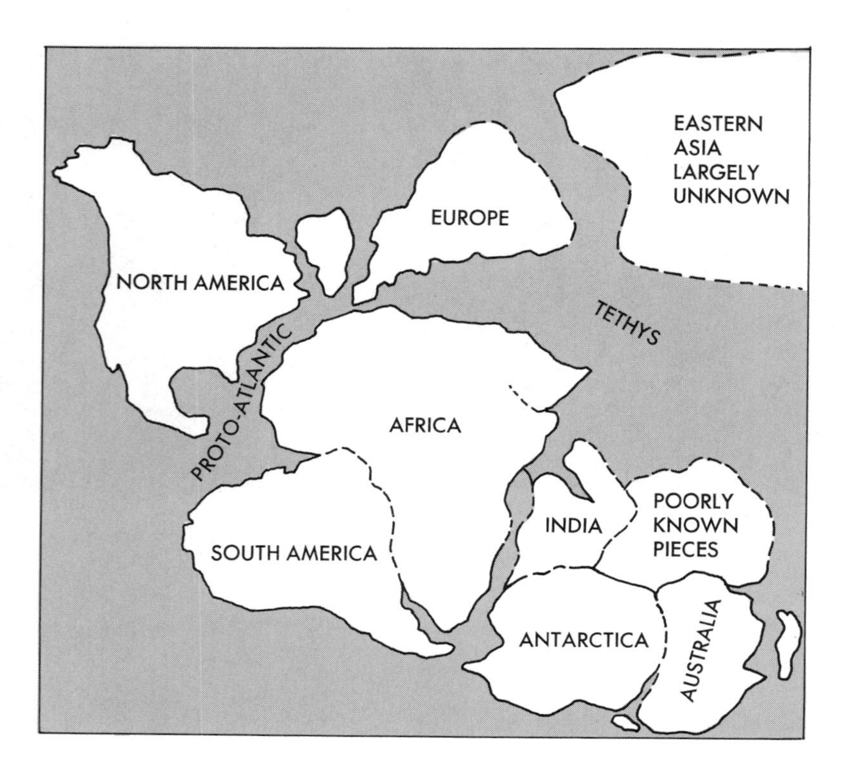

FIGURE 17-1 *Distribution and approximate positions of major landmasses in the Cambrian period. Semidiagrammatic only.*

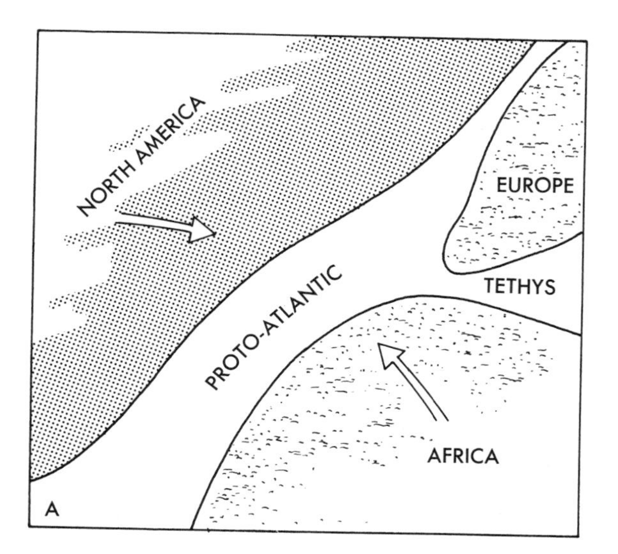

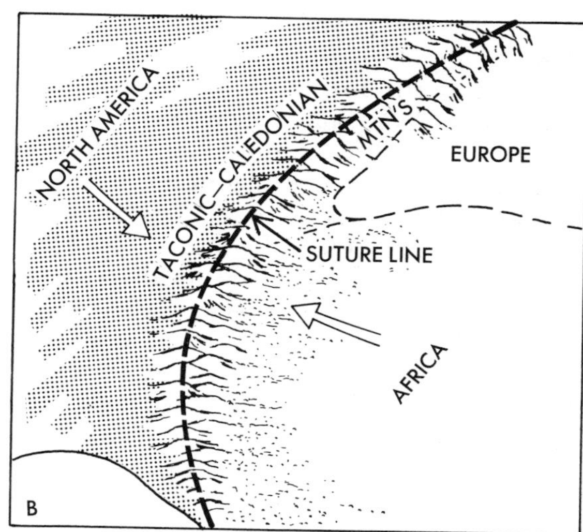

FIGURE 17-2 *Reconstructed history of the North Atlantic Ocean shown in a series of semidiagrammatic maps. (A) Early Paleozoic: an ocean of unknown width, the proto-Atlantic, separates North America and Europe-Africa. (B) Late Paleozoic: the proto-Atlantic is destroyed, and the landmasses are in contact; the collision has created the Caledonian mountains of Europe and the Acadian-Taconic mountains of North America. (C) Early Mesozoic: the landmasses are splitting apart along a rift zone that is parallel to but not in the same course as the previous suture line. (D) Present configuration of the North Atlantic: a detached slice of Europe is found in New England and southeast Canada; a sliver of North America remains in Ireland and Scotland. Fossil evidence supports this reconstruction.*

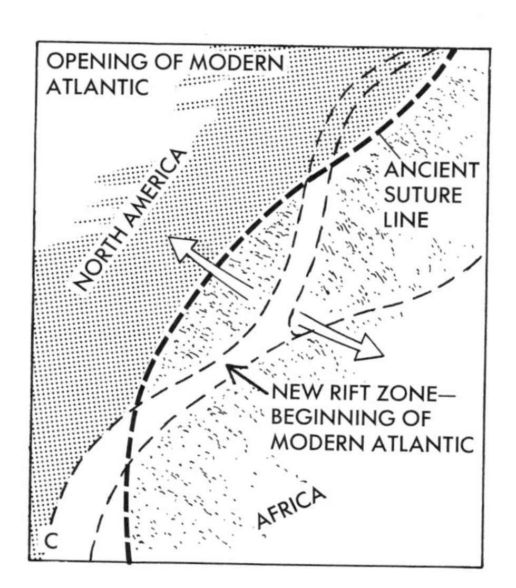

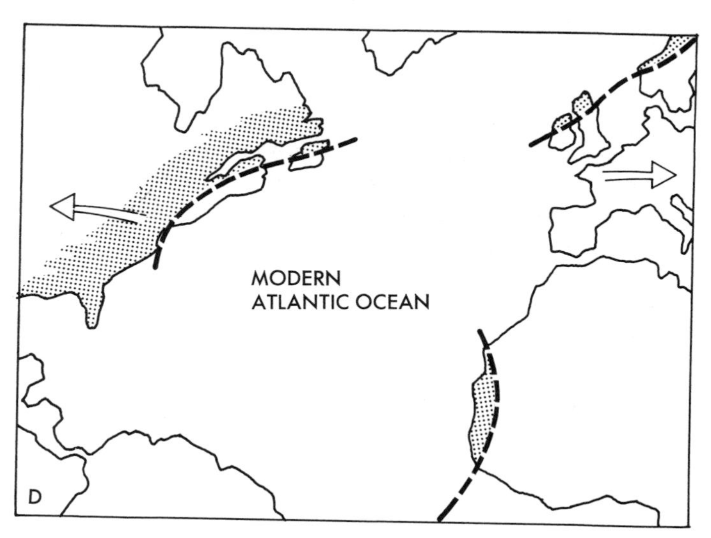

The Tethys seaway between Africa and the Baltic Shield area was of great but unknown width. In general it appears to have had a jawlike configuration—narrow to the west, widening eastward. Eurasia as a landmass was obviously smaller than now and was bisected by a north-south geosynclinal seaway in the position of the present Ural Mountains. There were other important waterways in the Northern Hemisphere. During early Paleozoic time, Europe and North America were separated by a body of water that geologists call the *proto-Atlantic* or more strictly the *proto-North Atlantic.* This seaway was almost surely much narrower than the present Atlantic and was destined to be entirely obliterated during later Paleozoic time as opposing landmasses moved together in a grand continental collision. The reconstructed history of the North Atlantic is shown by Figure 17-2. It is important to note that South America and Africa remained tightly welded together and did not separate until the Mesozoic. There was, in other words, no ancestral South Atlantic Ocean in the early Paleozoic. Most of Mexico and Central America were not yet in existence, so North America was separated from South America by open water, continuous with the proto-North Atlantic and with Tethys. Such an arrangement permitted marine organisms to migrate from eastern North America to western South America. Cambrian and Ordovician fossils of Argentina resemble those of the Baltic and Appalachian regions much further north. The proto-Atlantic and its continuation between North and South America received sediments that were later compressed to form the Appalachian Range and equivalents in the Ouachita and Wichita Ranges of Oklahoma and Arkansas.

No significant ice ages are known to have upset the earth during the early Paleozoic. The interval may be thought of as being climatologically less harsh than the preceding late Precambrian or the succeeding late Paleozoic. This is not to say that evidence of ice action is entirely lacking. During the Cambrian the South Pole appears to have been located in the proto-Atlantic off northwest Africa. During the Ordovician it moved onto land in the region of the western Sahara desert, where many evidences considered indicative of polar ice sheets are found. The opposite pole would then be in the open Pacific Ocean.

The Cambrian equator seems to have occupied a north-south position across western North America and diagonally across Eurasia. During the later Cambrian and Ordovician it rotated in a clockwise manner so that the trend became more easterly. The position of the equator helps to explain the variety and abundance of Cambrian and Ordovician life along the western and northern margins of North America and Siberia, which would have been under a near-vertical sun.

Disregarding minor interior seas, the Pacific Ocean with the great Tethys embayment was the major water body of the earth in early Paleozoic time. Circulating currents of the Pacific were able to carry warm water along extensive continuous coastlines, which in turn created rather uniform living conditions for marine animals. A recent tabulation shows that the number of families of shallow marine animals increasd from about 25 at the beginning of the Cambrian to about 150 at the close of the Silurian.

North America

The interior plains of North America have the most complete and extensive record of the Cambrian, Ordovician, and Silurian Periods of any continent. Seas of these periods spread far inland and at times covered practically all of the Canadian Shield. The fossils abundantly preserved in the troughs and shelf

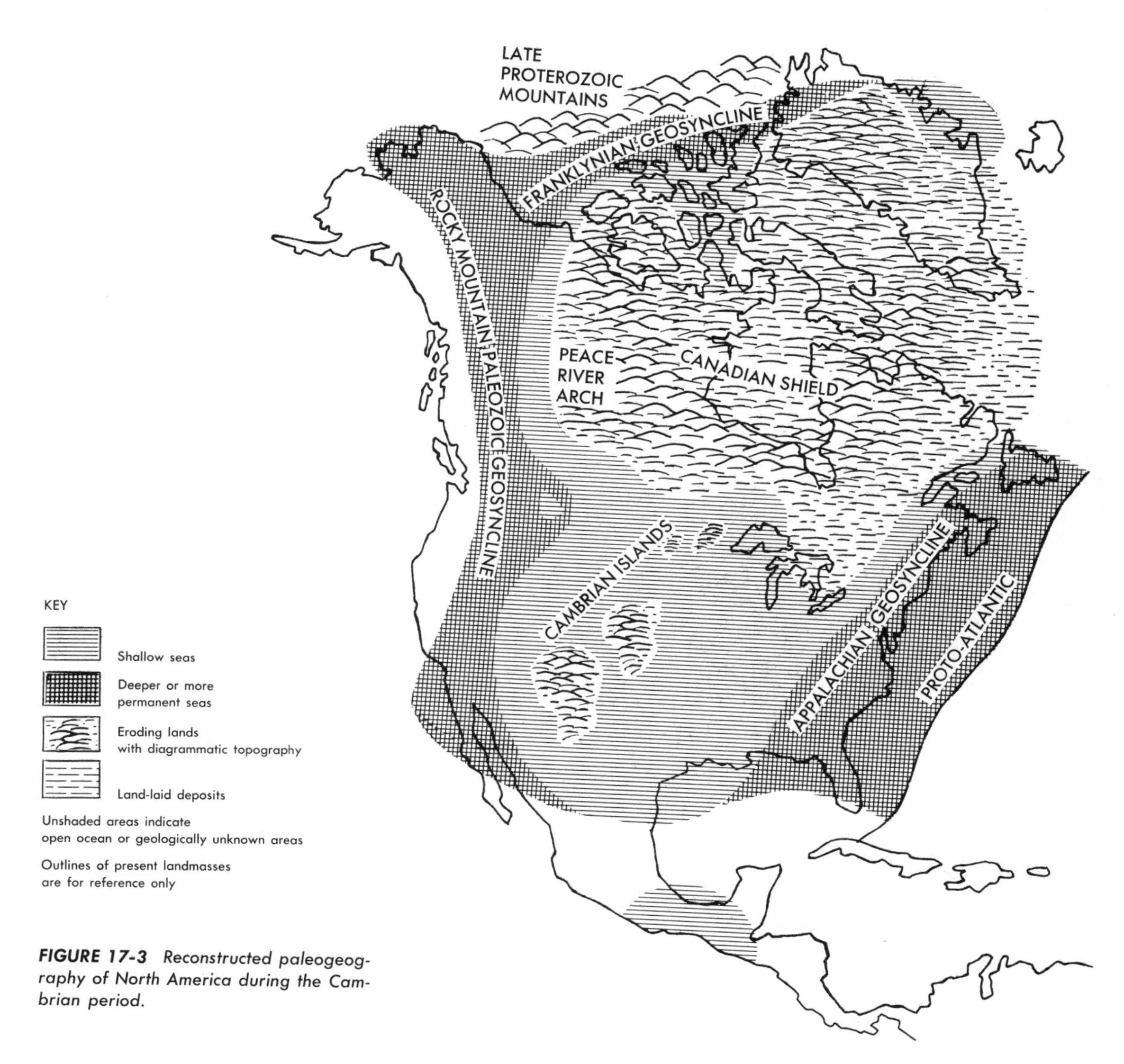

FIGURE 17-3 *Reconstructed paleogeography of North America during the Cambrian period.*

areas give a remarkably complete picture of early Paleozoic life.

Commencing with the Cambrian, the North American continent may be thought of as being divided by a sort of "backbone" into two great provinces—one to the southeast, the other to the northwest (see Figure 17-3). The backbone, or *Transcontinental Arch* as it is properly called, extends from the Mojave Desert area of California entirely across the United States to Lake Superior and into Labrador. Across its summit the geologic systems are thin and discontinuous; they thicken away from it in both directions.

Eastern North America

That portion of North America lying east and southeast of the Transcontinental Arch displays extensive outcrops of early Paleozoic rocks. The interior regions, chiefly the drainage area of the Mississippi River, are underlain

FIGURE 17-4 *(Above) Great unconformity between Late Ordovician marine strata exposed in the cliff and the much harder Precambrian granitic gneiss in foreground, Churchill River, Manitoba. (Photo courtesy of Geological Survey of Canada.)*

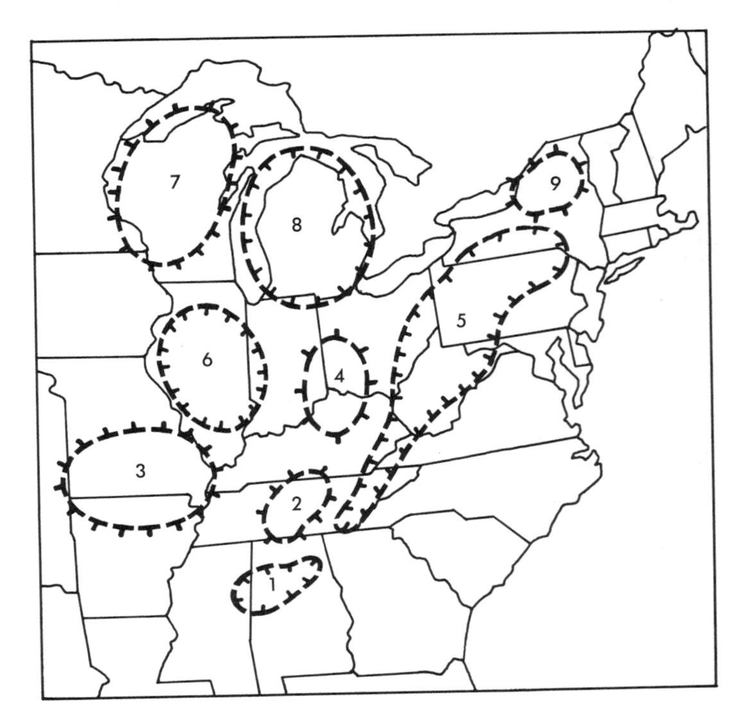

FIGURE 17-5 *(Right) Domes and basins of the eastern interior of the United States. These broad features were evident in early Paleozoic time and were accentuated later. Note ages of oldest rocks in the center of each structure.*

1. Black Warrior Basin—Pennsylvanian
2. Nashville Dome—Ordovician
3. Ozark Uplift—Precambrian
4. Cincinnatti Arch—Ordovician
5. Allegheny Basin—Permian
6. Illinois Basin—Pennsylvanian
7. Wisconsin Dome—Precambrian
8. Michigan Basin—Pennsylvanian
9. Adirondack Uplift—Precambrian

by rocks like those of the Canadian Shield with a thin cover of Paleozoic sediments (Fig. 17-4). A number of broad structural domes and basins surrounded by wide bands of Paleozoic sediments characterize this region. The major uplifts, along with the age of the oldest rock now exposed in their respective central areas, are as follows (see Figure 17-5): Ozark Mountains, Precambrian; Cincinnati Arch, Ordovician; Wisconsin Dome, Precambrian and Cambrian. The major basins with the age of youngest rocks now exposed in their central regions are: Illinois Basin, Pennsylvanian; Michigan Basin, Pennsylvanian; and Allegheny Basin, Permian. These domes and basins mostly began to form in Ordovician time and have been accentuated by later events, not by intensive mountain building.

Bordering the interior plains are mountain systems born of Paleozoic geosynclines. Along the eastern margin from Newfoundland to Georgia is the *Appalachian system.* The *Ouachita system,* now largely buried, arose from a geosyncline that lay along the southern margins of the continental nucleus. These troughs received sediments from the continental interior and from nearby landmasses, presumably neighboring continents.

The concept of the geosyncline as a great belt of downsinking and deposition came from studies by American geologists in the Appalachian region. The idea has proved a most useful one and has now been applied on all continents. The original pictures of a single sinking trough has had to be modified, and the geosyncline is now visualized as being divided into two parallel subtroughs. The outer or seaward trough is called the *eugeosyncline.* It contains chiefly highly siliceous material—such as volcanic products, course and poorly sorted, with few fossils. This belt appears primarily in New England where early Paleozoic siliceous and volcanic rocks reach a thickness of 5,000 meters.

By contrast, the inner, or *miogeosynclinal,* belt and shelf areas received mainly nonvolcanic shallow-water deposits such as limestone, dolomite, clean sandstone, and shale. There are many fossils, and coal and oil are present. In the Appalachian miogeosyncline, the Early Cambrian is chiefly sandstone. Later, in the Cambrian and continuing into the Ordovician, (Figure 17-6) came great thicknesses of dolomite and limestone. These calcareous rocks reach a thickness of 3,000 meters in Alabama and 1,800 meters in Oklahoma. The immense number of fossil shells and other organically precipitated carbonates attest to the importance of life in the building of rocks at this time. The early Paleozoic carbonate sheets effectively sealed off, or "armor-plated," much of the Precambrian crystalline outcrops so that sources of coarse sandy sediments were not available to erosion until laid bare in the subsequent mountain building and erosion of late Paleozoic time.

During the Cambrian there was surprisingly little volcanic activity in North America, and even the outer (eugeosynclinal) belt was inhabited by a variety of silt-loving trilobites. During the Ordovician, increasing amounts of dark-colored shale with fossil graptolites were laid down. In recognition of the almost worldwide association of dark shale and graptolites in the Ordovician and Silurian periods, the term *graptolite facies* has come into general use. This term contrasts with *shelly facies,* a term that designates contemporaneous shell-bearing limy rocks.

Lands bordering the North Atlantic began to show deep-seating unrest in mid-Ordovician time. Earth movements and volcanic activity mounted in intensity and culminated near the close of the period in the Taconic Disturbance, so named from effects in the Taconoic Range of Vermont and New York. Disturbances are recorded northward into Newfoundland (Figure 17-7).

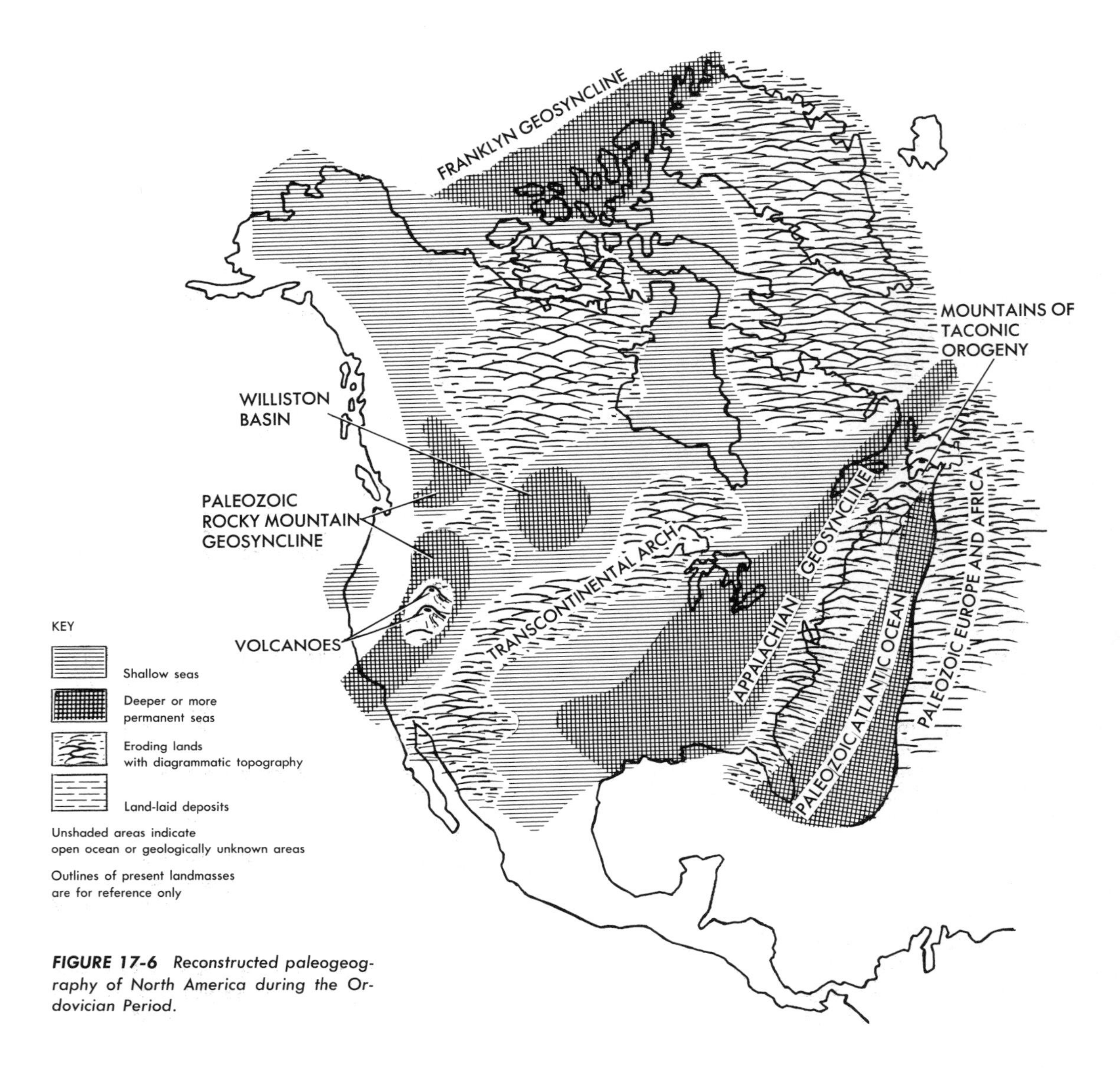

FIGURE 17-6 *Reconstructed paleogeography of North America during the Ordovician Period.*

Simultaneous effects in northwestern Europe are known as *Caledonian* from Caledonia, Scotland. The localization of intense geological effects at this time seems to be due to the collision of North America and Europe with the consequent closing and destruction of the ancient or proto-Atlantic ocean. The highlands raised on the borders of both continents became sources of coarse-grained sediment for Late Ordovician and Silurian formations. In New York and Pennsylvania these coarse sediments make up the ancient Queenston Delta.

By the end of the Silurian, the seas had cleared again and limestone was deposited. Not to be overlooked are the extensively salt-bearing beds of the eastern interior, which also accumulated near the close of the Silurian. Through a peculiar combination of landlocked basins and restricted seaways, a vast, almost "dead sea" type of environment came into being in western New York, Pennsylvania, Ohio, and Michigan. The greatest salt thick-

FIGURE 17-7 *Contorted Ordovician shale, Black Point, Port au Port, Newfoundland. The intense folding resulted from the Taconic Orogeny of Late Ordovician time. (Photo courtesy of Geological Survey of Canada.)*

FIGURE 17-8 Niagara Falls, the best-known exposure of Silurian rocks in North America. Diagram shows the formations at and below the falls. In this view, water and rock fall material cover formations below the Rochester Shale. (Photo courtesy of New York State Department of Commerce.)

ness, aggregating 480 meters, is in the center of the Michigan Basin. Salt from this formation is extensively mined at many places. The most famous exposure of Silurian rocks is at Niagara Falls; the rocks here are older than the salt-bearing series (Figure 17-8).

Western North America

During the early Paleozoic, the Pacific Ocean washed inland to the margins of the Transcontinental Arch and the Canadian Shield. There was no Rocky Mountain chain, but the Cordilleran Geosyncline, sometimes referred to as the Rocky Mountain Geosyncline, was a dominant feature. This great curving trough began in southern California, extended into Utah and Idaho, and then continued back toward the Pacific through Montana, Alberta, and Alaska. By contrast with eastern North America, there is practically no evidence of nearby landmasses on the oceanic side in early Paleozoic time. Essentially all of the earlier deposits of the trough came from sea water or from the wearing down of interior lands; and for the early Paleozoic, at least, the Rocky Mountain Geosyncline was essentially one-sided. Another significant difference from eastern North America is the almost complete lack of volcanic products in this western trough. Here the sedimentary history of the early Paleozoic was unbroken by strong mountain-building or volcanic effects. This is expectable as there was no westward movement of the continent at this time.

The Cambrian began with the encroachment of shallow seas at the northwest and southwest parts of the Rocky Mountain Geosyncline. The initial deposits were chiefly clean quartz sands washed from the continental interior. With the passage of time and the steady transgressions of the seas, the sediments became finer and more calcareous. The amount of limestone and dolomite deposited in the inner, or miogeosynclinal, belts is even greater than that of the eastern North America. The combined Cambrian-Ordovician carbonate section reaches a thickness of 5,000 meters in western Utah. Late in the Cambrian the seas had flooded inward until the Transcontinental Arch was reduced to a number of islands.

During the Ordovician the Rocky Mountain trough shows a clear division into an inner, or miogeosynclinal, carbonate belt and an outer, or eugeosynclinal, siliceous belt. In this period the east and west margins of the continent reached a high degree of symmetry and similarity. In both east and west the inner limestone facies passes into the outer or graptolite facies rather abruptly, but the structural complications are greater in New England than in Nevada. In general, throughout the western miogeosyncline, the early Ordovician is limy and the later part is dolomitic. Between them is a widespread sandstone or quartzite phase similar to the better-known formation called the *St. Peter Sandstone* of the midcontinent. Thin marine deposits record the eastward spread of shallow seas onto and occasionally across the Transcontinental Arch. The Williston Basin in North Dakota and south central Canada began to form in the Ordovician and received sediments of all the later Paleozoic periods.

The Silurian is a short period, and its deposits are correspondingly thin (Figure 17-9). A single sheet of dolomite in the deeper parts of the Rocky Mountain Geosyncline gives a relatively incomplete record of the time. That the seas did occasionally spread farther inland is known from small isolated remnants in Wyoming and Colorado, but it is certain that no very thick or extensive deposits were laid down.

Early Paleozoic rocks are widespread in northern Canada, including the Arctic Islands. Thin deposits on the mainland thicken northward into the Franklyn Geosyncline, where as much as 6,000 meters of early Paleozoic marine beds accumulated. Sediments are varied, but carbonates, as elsewhere in North America, are dominant. An outer belt of graptolitic shale correlates with the inner carbonates.

Eurasia

Geology as a science had its beginning in northwestern Europe, and an important phase of its formative period was the designation of the Cambrian, Ordovician, Silurian, and Devonian systems in Great Britain. From the type areas, rocks of these periods have been traced widely throughout Europe and Asia. The early Paleozoic systems form a somewhat coherent group, that were deposited in several narrow subsiding troughs and their shelving marginal seas. One of these, the Caledonian or Baltic Geosyncline, trended northeastward between Europe and North America and may be visualized as a narrow forerunner of the Atlantic Ocean. This seaway curved around the Baltic Shield area of Scandanavia and turned southerly to cross the Eurasian continent, essentially where the Ural Mountains now are.

Another geosynclinal belt extended the length of Eurasia from Spain to Malayia. This is called the *Mesogean Seaway*, the *Tethian Sea*, or simply *Tethys*. In it accumulated the sediments that were later elevated into the great ranges of the Alpine-Himalayan chains. Although the original shape and relation of Tethys to adjacent landmasses is now difficult to determine, it is obvious that the western reaches of the seaway lay between the stable African platform to the south and the Baltic Shield on the north. Tethys opened eastward beyond Africa into what is now the Indian Ocean, but India then lay south rather than north of this body of water. During the early

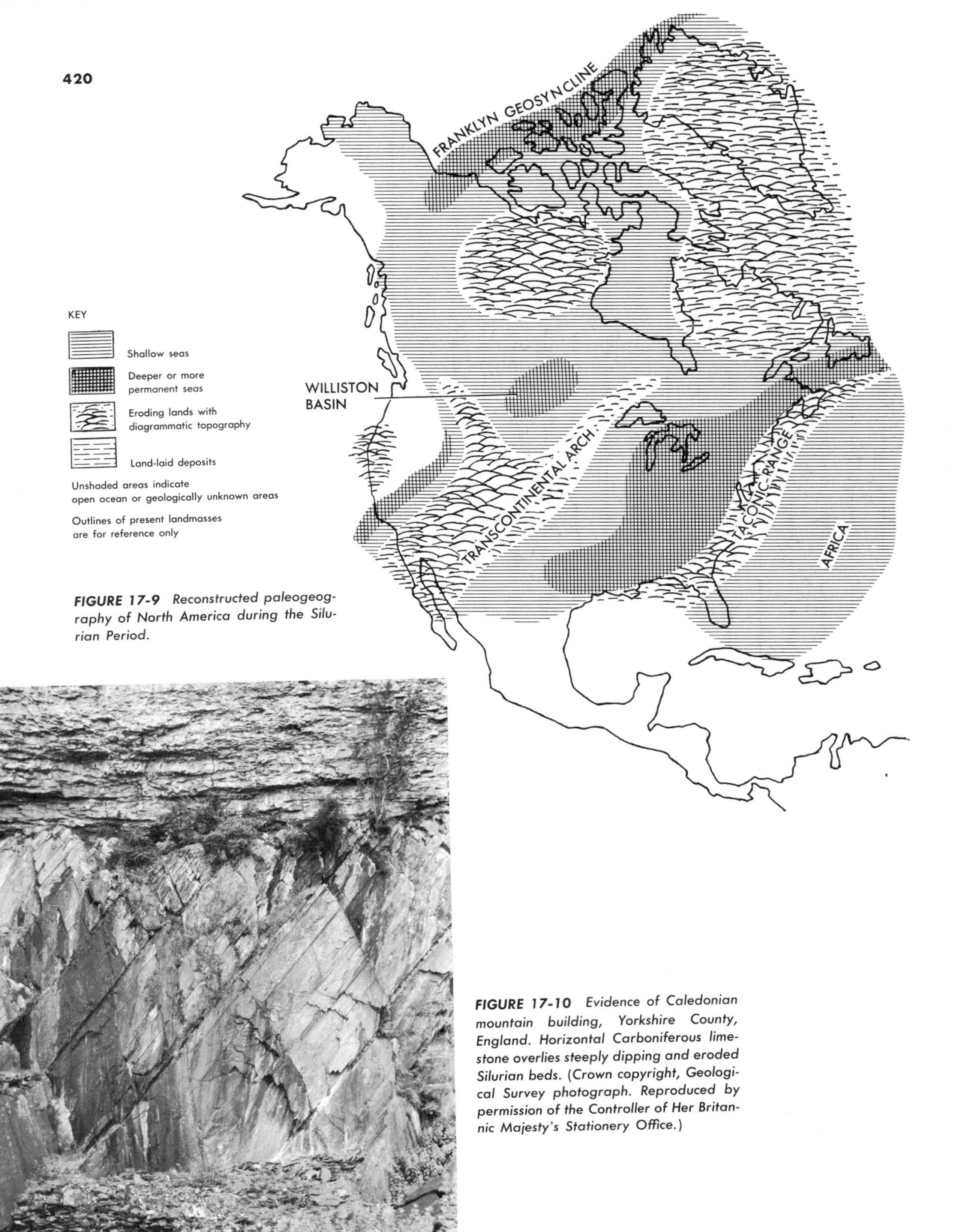

FIGURE 17-9 *Reconstructed paleogeography of North America during the Silurian Period.*

FIGURE 17-10 Evidence of Caledonian mountain building, Yorkshire County, England. Horizontal Carboniferous limestone overlies steeply dipping and eroded Silurian beds. (Crown copyright, Geological Survey photograph. Reproduced by permission of the Controller of Her Britannic Majesty's Stationery Office.)

Paleozoic, Tethys did not receive unusually thick deposits of sediment; but Cambrian, Ordovician, and Silurian formations are known across the entire distance from Spain to Southeast Asia. Early Paleozoic rocks of Africa north of the Sahara were deposited along the southern margin of the Tethian trough and furnish important reservoirs for a number of great oil fields. Of special interest is the occurrence of extensive glacial deposits of Ordovician age in the central Sahara (Fig. 17-11). In connection with other evidence these deposits are considered ample proof that the South Pole was located here about 450 million years ago.

Also well known is the Angarian Geosyncline, which occupied much of what is now southern Siberia. It trended essentially east-west, south of the Angaria Shield. Cambrian and Ordovician sediments in excess of 15,000 meters are reported from central Kazakhstan. The great ranges of central Asia constitute one of the most complex geologic regions on earth and display intersecting elements of the Uralian, Angarian, and Tethian geosynclines.

FIGURE 17-11 *Upper Ordovician glacial deposit, east of the Hoggar, Algeria. Large boulders are granite transported from source area to southwest of present locality. (Photo from Institut Algerienne du Petrole, courtesy Rhodes Fairbridge)*

Shallow seas that occasionally flooded outward from the Caledonian and Tethian geosynclines left important deposits upon the adjacent shield areas. Much of Russia west of the Urals was flooded during the early Paleozoic. The Cambrian here is described as claylike and remarkably fresh in appearance. The Cambrian of Siberia is thick, with many fossils, and enormous salt deposits that are among the oldest known. (Deposits of Late Precambrian age are also thick, and there is no distinct break at the base of the Cambrian. Russian geologists believe that the transitional series might well be considered as representing an additional period; for this they propose the name *Eocambrian.* Chinese geologists refer to these same rocks as *Sinian.*)

Ordovician and Silurian rocks are widespread in Eurasia, being represented in outcrops and surface exposures from their type areas in Great Britain eastward across Europe and Asia to Malaysia. To be noted is the very sparse record of early Paleozoic time in peninsular India, a not unexpected situation when it is realized that this landmass was then in the Southern Hemisphere, locked into the Gondwanaland landmass.

FIGURE 17-12 An important geologic boundary on Mt. Weaver, Antarctica. The geologist is standing on the contact of late Paleozoic glacial conglomerates (above) and granitic rocks of Cambrian or Ordovician age (below). (Courtesy Velon H. Minshew, Ohio State University Institute of Polar Studies.)

The southern continents

Compared with North America and Eurasia, the southern continents have a relatively incomplete and scattered sedimentary record of the Early Paleozoic periods. This deficiency is probably due to the relatively emergent and consolidated condition of the southern landmasses, which favored erosion rather than deposition.

The most complete representation of the Cambrian, Ordovician, and Silurian systems in the Southern Hemisphere is in Australia. In the eastern part of the continent there was steady deposition in the northerly-trending *Tasmanian,* or *Tasman, Geosyncline.* The types of fossils found in these beds indicate that migration of organisms to and from distant lands was possible. Cephalopods of the Australian Ordovician are much like those of western North America, and the graptolites of the Ordovician and Silurian are almost identical to those from Wales and the United States. The migration routes that permitted these similarities to exist are still obscure.

No Cambrian, Ordovician, or Silurian rocks have been positively identified in that part of Africa lying south of the equator. The few nondiagnostic fossils that have been found in pre-Devonian rocks do not resemble typical early Paleozoic forms from other places and may be of Precambrian age. South Africa was probably well above sea level in the early Paleozoic, just as it now is.

Although the early Paleozoic rock record of South America is much better than that of Africa, it is still less complete than that of North America. Fossiliferous Cambrian deposits are rare, but formations in Argentina have yielded Middle Cambrian trilobites almost identical with those of the Baltic region. By contrast, the Ordovician is fairly well represented along the entire Andean belt, in places, by deposits of great thickness. Graptolite-bearing shales are common, and many of the forms are related to those of eastern North America and the Baltic region of Europe.

Early Silurian seas spread across wide areas of South America, mainly east of the Andean belt into the Amazon trough and across northern Argentina. Devonian sediments are also represented in Brazil, particularly in the Parana Basin.

Although geologic investigations of Antarctica are incomplete, it is fairly well established that no Ordovician or Silurian sedimentary rocks exist there (Figure 17-12). The Cambrian, with *archaeocyathids* and other typical fossils, is present along the central mountain ranges.

FIGURE 17-13 *These rounded and banded structures known as Girvanella are thought to have been formed by algae. They are widespread in Cambrian rocks. Spheres are about 10 mm in diameter. (Courtesy R. E. Cohenour.)*

Life of the early Paleozoic

Plant life

By contrast with the animal kingdom, no striking changes in plant life seem to have taken place at the Precambrian-Cambrian boundary. The late Precambrian has been called the *Age of Stromatolites,* and although these organisms are present in later ages, they underwent a great decline during the Cambrian. It has been suggested the stromatolites were eaten out of existence by trilobites. Impressions of "seaweeds" are occasionally found in Cambrian rocks, as are abundant spores of water-living plants. The rounded bodies, called Girvanella found in Cambrian rocks are also thought to have had an algal origin (Figure 17-13). Organic material is abundant enough to constitute rich oil shale in the Cambrian of central Sweden. There is also a great deal of carbonaceous material in Ordovician formations, but it is mainly unidentifiable and must be of algae origin. Ordovician rocks produce much of the oil in the United States.

It is not until the Silurian that land plants are definitely known. An aggregation of plant fossils from Lower Silurian rocks of Maine contains several types that are preserved in an erect position and are probably very primitive land plants. These specimens lack true vascular (conductive) tissue but have tiny spines along the stems. Spores that may have been produced by this early land flora have been found in Lower Silurian rocks in New York State.

It has been suggested that plants could not thrive in the open atmosphere until later in the Paleozoic because of the strength and intensity of the ultraviolet rays of the sun, which at present are screened out by the ozone layer of the atmosphere. The ozone layer had to be built up from excess oxygen liberated by plants in the photosynthetic process. Plants were therefore forced to live partly shielded by shallow water until the protective ozone layer permitted both plants and animals to emerge on land.

The invertebrates

The appearance of abundant marine fossils

With the beginning of the Cambrian Period, fossils began to be preserved in great numbers. Notable was the marked increase in the number of animals with calcareous shells—chiefly *molluscs* and *echinoderms,* which use calcium carbonate almost exclusively; and the *brachiopods* and *arthropods,* which secrete mixtures of calcareous and phosphatic material. The increased use of calcium by shell-building animals seems to coincide with the formation of thick beds of limestone and dolomite. The prevalence of limy formations and shelled animals during the early Paleozoic may be related to a general warming of the ocean, for calcite is deposited and secreted more easily in warm water than in cold water.

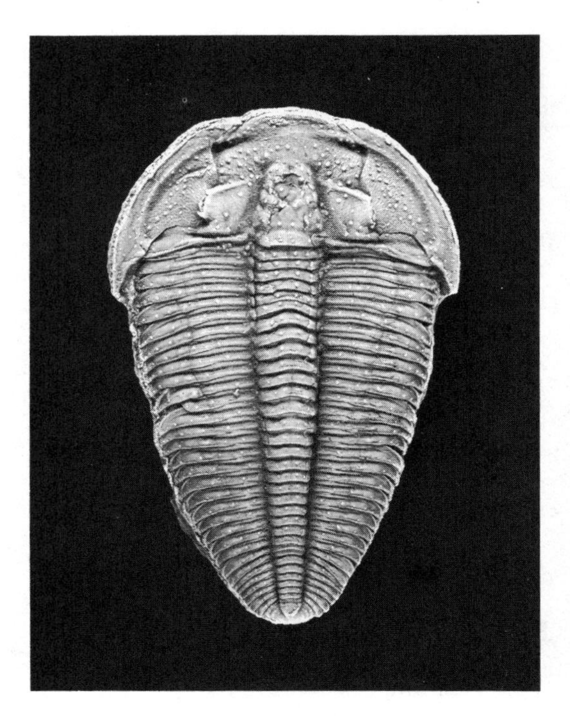

FIGURE 17-14 *A well-preserved Middle Cambrian trilobite (*Alokistocare harrisi*) from the House Range, western Utah. Specimen is about 2 centimeters long. (Photo courtesy of Richard A. Robison.)*

The Cambrian—age of trilobites

The *trilobites* were crawling or swimming arthropods with light, jointed skeletons (Figure 17-14). They are now extinct and have no close living relatives. Their remains are abundant in Cambrian rocks of all continents, and thousands of species have been described. The typical trilobite "shell," or *carapace,* is formed of a number of separate articulating pieces that tended to fall apart when the animal moulted or died. Perfect specimens are rare, but pieces constituting the heads and tails are so diagnostic that an expert can readily distinguish one species from another without the entire fossil. The trilobites were obviously adapted for life in shallow seas where they swam, floated, crawled, or burrowed, seeking food and protection. Of unusual significance is the evidence of very highly developed eyes, said by one investigator to be the "most sophisticated lenses ever produced by nature."

A second important group of Cambrian organisms is the *brachiopods.* These, in contrast to trilobites, are immobile and have a two-piece shell that closed to protect the animal inside. The earlier brachiopods resembled two shallow saucers fitted face to face and held together by a system of small muscles (Figure 17-15). They were oval, round, or tongue-shaped, and their shells were mostly phosphatic. The earlier forms are called *inarticulates* because their shells lacked definite hinge structures. More advanced brachiopods developed definite hinge structures and are called *articulates* (Figure 17-16). Almost all the brachiopods, living or extinct, are fixed permanently in one spot, usually by a sort of root or fleshy extension of the body. The trilobites could go

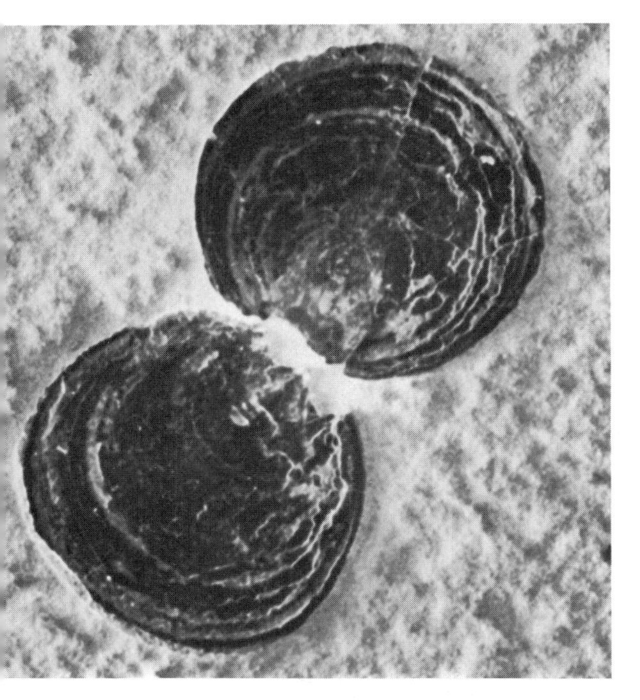

FIGURE 17-15 Acrothele, *a typical inarticulate brachiopod, from the Middle Cambrian. The small, two-piece shell is preserved as it lay open on the Cambrian seabed. Width of shells is about 1 cm.*

wherever food was most abundant, but the brachiopods had to depend on currents of water to bring food within reach of the feeble inflowing currents that were generated by ciliated coiled structures within their shells. Trilobites squandered some energy dragging their shells about; brachiopods conserved energy by remaining in one spot. But since both animals survived and prospered, each mode of life must have had its particular advantages.

A puzzling group of Cambrian organisms, the *Archaeocyatha* (ancient cups), were the earliest known reef-forming animals (Figure 17-17). They left conical calcareous fossils up to 10 cm long. The group is confined almost entirely to the Early Cambrian. So far as is known, this may be the first animal phylum to become extinct.

Trilobites make up about 60 percent of all Cambrian fossils; brachiopods constitute 10 to 20 percent; and the remainder includes archaeocyathids, protozoans, sponges, worms, graptolites, conodonts, gastropods, pelecypods, echinoderms such as cystoids, cephalopods, and arthropods other than trilobites. The trilobites and brachiopods so completely dominate the Cambrian fossil record that most other forms are interesting chiefly for the information they yield about the primitive beginnings of their respective lines. Among the groups not yet found in Cambrian rocks are *bone-bearing animals, bryozoans, true corals, starfish,* and *sea urchins.*

FIGURE 17-16 *Typical Orodvician brachiopods. (a)* Platystrophia *and (b)* Dinorthis. *The largest specimen measures about 3 centimeters across.*

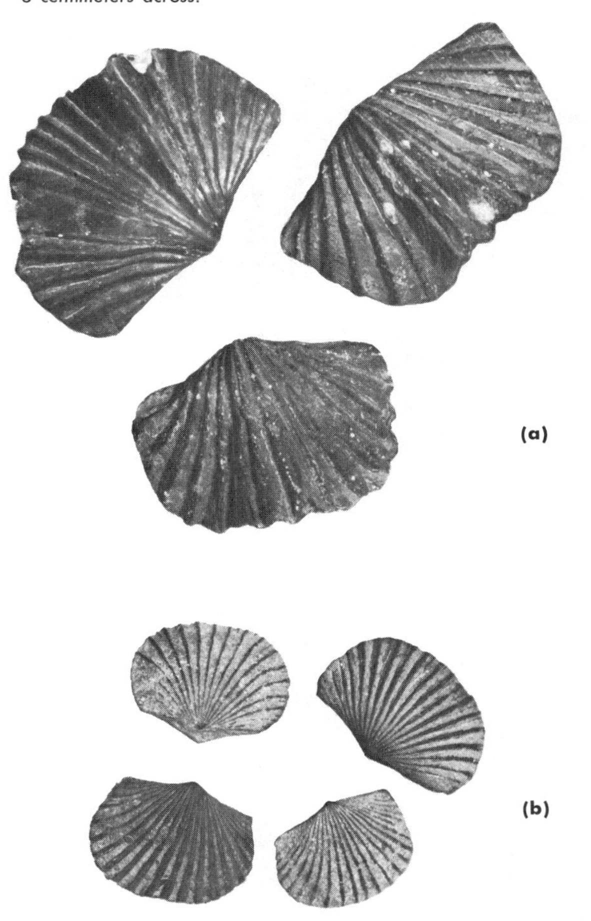

(a)

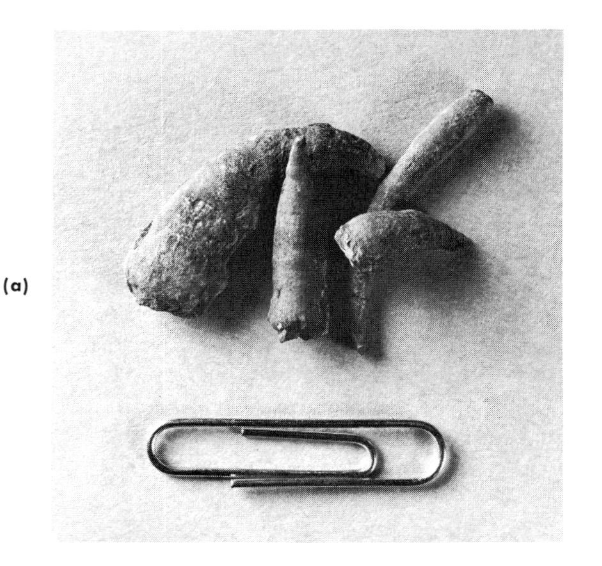

(b)

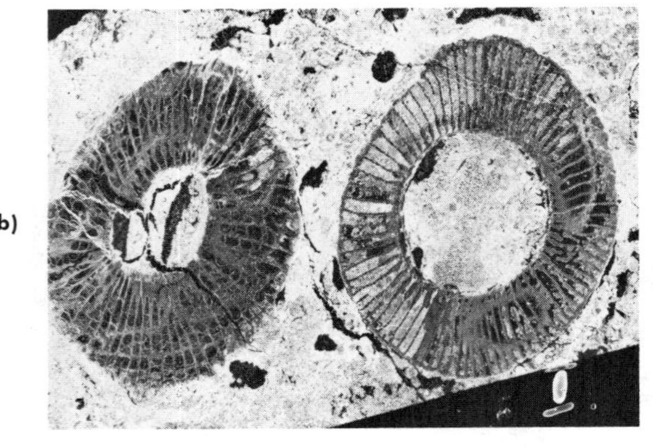

FIGURE 17-17 (a) *Archaeocyathids from Cambrian rocks of Nevada. The exact zoological classification of archaeocyathids is unknown; they show affinities both with corals and with sponges.* (b) *Cross-sectional views of two specimens in Lower Cambrian rocks of the Yukon Territory. Cross sections are about 1 centimeter across.* [(a) *Smithsonian Institution;* (b) *courtesy of V. J. Okulitch.*]

Ordovician faunas—all major phyla in existence

During the Ordovician Period, conditions continued to be favorable for marine invertebrates. New groups joined those that had appeared in the Cambrian, and by the close of the Ordovician all major animal phyla capable of leaving fossils were in existence. Calcite continued to be the chief construction material of shells and skeletons, and calcareous shells were locally so abundant that entire formations are composed of them. The first appearance of abundant oil and gas resources in Ordovician formations may indicate increasing organic productivity of the ocean waters during this period.

Trilobites reached the height of their development during the Early Ordovician and assumed a great variety of shapes and sizes. Ordovician trilobites tended to be either smooth and rounded or bristled with nodes and spines. A number of forms probably took up a free-floating existence at this time. The brachiopods also were numerous and varied. Ordovician brachiopods were mostly articulates, less than 3 cm across with shells of calcite. The majority lacked spines and ornaments other than simple ribs.

The *bryozoans,* which are unknown in Cambrian rocks, appeared early in the Ordovician and increased tremendously during that period. They are extremely small animals, all of which grow in composite masses or colonies of multitudes of single individuals (Figure 17-18). They construct colonies of calcite or other material in the form of twigs, branches, crusts, mounds, or networks. The bryozoans were the first group to exploit thoroughly the possibility of community existence. Existing specimens suggest that their food has always consisted of small particles strained from surrounding water.

The *graptolites* became common in the Ordovician. Some types were fixed like small shrubs to the sea bottom, and others floated freely in the upper levels or were attached to seaweeds. They are extinct, and little is known about their relationships to other animals. Their skeletons were composed of chitinous material and were light enough to float but too

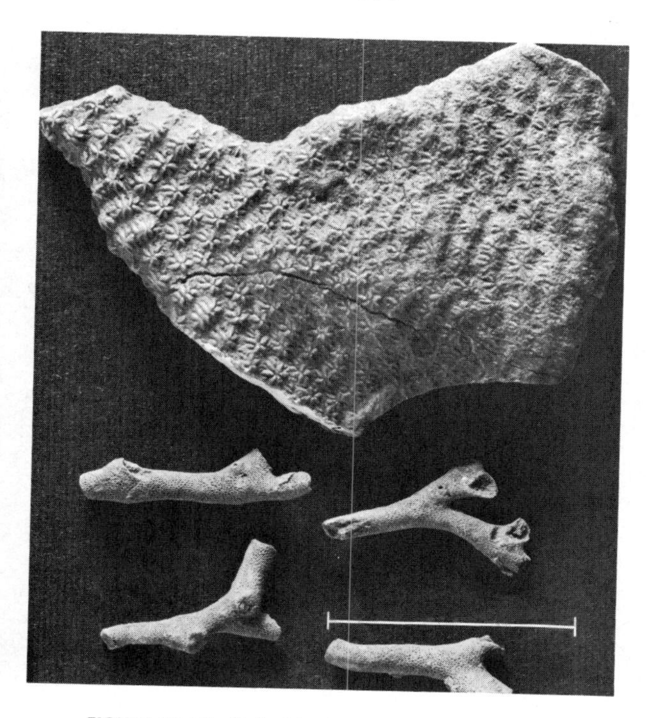

FIGURE 17-18 *Ordovician bryozoans. The large specimen at top is* Constellaria, *so named for the small, star-shaped elevations that cover the surface. The small, twiglike specimens are* Eridotrypa; *the small pores are the living chambers of extremely small, individual animals. The* Constellaria *specimen measures about 6 cm across.*

thin to afford much protection. Graptolites existed in many forms and went through distinctive evolutionary stages during the Ordovician and Silurian (see Figure 17-19 for a common genus).

The *cephalopods,* an important class of molluscs, also flourished during the Ordovician. Their variously shaped, chambered shells were buoyant enough to permit the animals to move about rapidly, and their keen senses probably made them the most advanced of all marine invertebrates.

The *crinoids* (stemmed or stalked echinoderms) also began to leave an abundant fossil record during the Ordovician. They have plantlike stems and roots and a flowerlike crown or head (Figure 17-20). The stem enabled the animals to keep the food-collecting devices well above the ocean bottom, but crinoids were delicate affairs and could easily be broken. Crinoid heads (correctly called *calyces*) are composed of many calcite plates that usually fall apart on the death of the animal so that complete specimens are very rare. The most common remains are the round, flat, disclike structures that make up the stem.

FIGURE 17-19 *Specimens of the Ordovician graptolite* Climacograptus *on a piece of shale from the Vinini Formation, central Nevada.*

FIGURE 17-20 A well-preserved crinoid (Eucalyptocrinites crassus) *of Silurian age. (American Museum of Natural History.)*

During the Ordovician, *corals* became increasingly common, both colonial as well as solitary types (see Figure 17-21 for a common Ordovician species). The corals capture food from the surrounding water with the aid of stinging cells and threadlike structures that they shoot at their prey.

The Silurian— heyday of the brachiopods

The Silurian was a relatively short period, and its life represents an orderly outgrowth from the Ordovician. Shallow seas still spread widely over the continental areas and provided the chief environments for life. Apparently few animals besides the graptolites were adapted to existence in the open oceans, and nonmarine or even brackish-water invertebrates are not positively known.

More families of brachiopods have been identified in the Silurian than in any other period. There was an increase of larger forms with complex internal structures and roughly five-sided outlines (*Pentamerus*). *Spirifieroid* forms with wing-shaped shells appeared suddenly, to launch a long and successful career. Trilobites were still abundant but had definitely passed their zenith, and few new genera were produced. Spiny and smooth forms were characteristic.

The graptolites of the Silurian were superficially more simple than their Ordovician predecessors. Most are classed as *monograptids* with living chambers (*theca*) ranged along one side of a single threadlike support. Corals proliferated and began to build extensive reefs; those of the Late Silurian of the east-central United States and of the Island of Bothnia in the Baltic are outstanding. Also contributing to the Silurian reef structures were great numbers of lime-secreting crinoids and bryozoans. A peculiar coral with a chainlike cross section, *Halysites* (Late Ordovician to Early Devonian), is very characteristic of Silurian rocks (Figure 17-22).

The mobile and predaceous cephalopods increased in importance; the *nautiloids* with simple, unfolded internal partitions and a variety of shell shapes represented this group. The rulers of Silurian seas were the *eurypterids*, or "water scorpions" (Figure 17-23), judging by their obviously fierce appearance and inferred predaceous habits. They were the largest animals of the time, some reaching a length of almost 2 meters. Another arthropod group, the *ostracods*, with small bivalved shells, was also increasing.

FIGURE 17-21 *Ordovician cup, or horn, corals. The illustration shows several individuals of the common genus* Lambeophyllum. *The fossils represent only the stony framework in which the soft, polyplike body of the animal was fixed. The opening of the top specimen measures about 2 centimeters across. Scale represents $2\frac{1}{2}$ centimeters.*

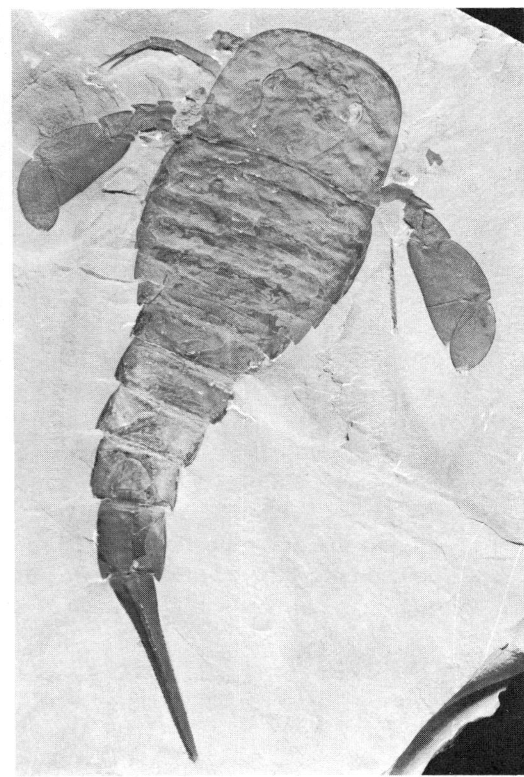

FIGURE 17-23 *A well-preserved eurypterid (*Eurypterus lacustris*) from Silurian rocks of western New York. Length is about 20 cm. (Buffalo Museum of Science.)*

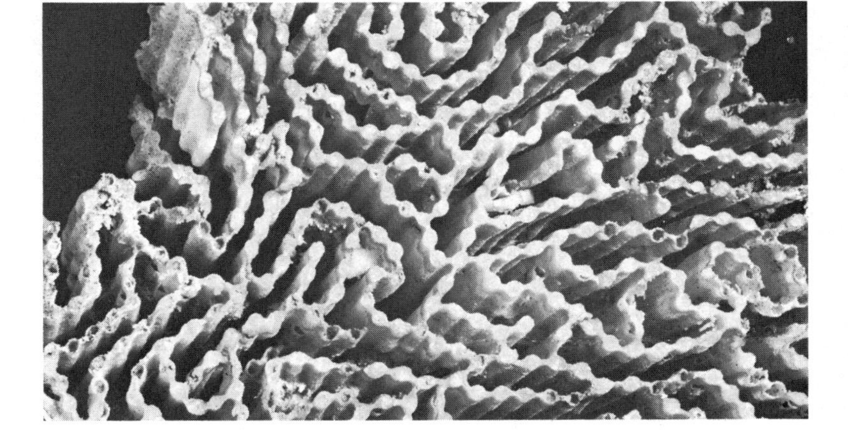

FIGURE 17-22 Halysites, *the chain coral. This specimen was etched from limestone of Middle Silurian age from western Utah.*

Vertebrates and possible kin

Although it is customary to consider vertebrates as more highly organized and specialized than invertebrates, it is impossible to make a fully accurate classification of all animals on the basis of whether or not they possess a backbone. Instead of the term *vertebrate*, it is more precise to use the term *chordate*, which designates not only the vertebrates but also their relatives that are not strictly invertebrates.

A few fragments of bone are known from Late Cambrian rocks and may represent the earliest vertebrates. We have already described the graptolites as invertebrates in the preceding section, but it is only fair to state that many investigators believe them to be primitive chordates, members of the subphylum Hemichordata to be exact. The conodonts are perhaps the most puzzling of all fossils for they are known almost entirely from small toothlike structures (Figure 17-24). A few poorly preserved specimens from the Pennsylvanian show vague carbonized outlines with the so-called teeth in the stomach region, not in the mouth.

FIGURE 17-24 *One of the earliest known conodont specimens (Proconodontus) from the Late Cambrian of Utah. The specimen is about 0.2 cm long. (Courtesy of James F. Miller.)*

If either the graptolite or conodont is a chordate, we can say that this phylum was present in the Cambrian. But if we ignore these problematic forms, the geologic history of vertebrates begins with the appearance of bone. This hard, durable, easily fossilized substance assumes many forms and is ideal for the construction of internal skeletons.

The earliest vertebrates

A few objects resembling true teeth and fragments of bone are known from Early Ordovician rocks and may represent the first vertebrates. The first really abundant remains occur in the Middle Ordovician Harding Sandstone of Colorado. This formation contains numerous fragments of bone and enamel, but no complete specimens have yet come to light. The bone fragments occur intermingled with shells of invertebrates and with tiny conodont "teeth," teeth that do not appear to belong to the same animals as the bones. It is possible that these early fishlike vertebrates may have lived in fresh water and that their remains were washed into the sea and mixed with ordinary marine forms.

So far, Late Ordovician rocks have yielded very little information about vertebrates, but impressive evidences of the group are found in Silurian rocks of northwestern Europe, where a number of strange fishlike forms are represented. Here, for the first time, the real nature of the earliest vertebrates is revealed. These forms were fishlike, mostly not over a few centimeters long. Some, known

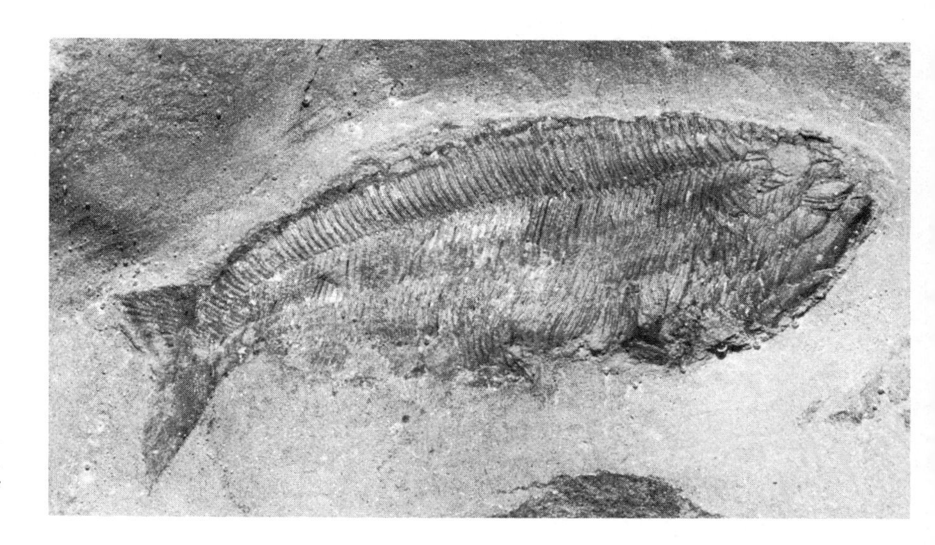

FIGURE 17-25 A complete specimen of Rhyncholepis, *a primitive fish, from Silurian rocks near Oslo, Norway. The length is about 12 cm. (Courtesy of A. Heintz.)*

as *ostracoderms,* were partly encased in bone; others were covered with queer scale patterns (Figure 17-25); and a few were apparently devoid of any protective covering except skin. Ordovician and Silurian fish seem to have been mainly mud-grubbers, for most of them had neither teeth nor jaws.

However obscure their beginnings may have been, these early fish possessed the potential for becoming masters of the sea. By slow degrees and through the evolution of many improvements, early vertebrates achieved a dominant position in the ocean that has never been successfully contested by the invertebrates. Jaws appear to have developed from gill supports that lay behind and near the mouth, while teeth may have been derived from modified toothlike structures that are similar to the tiny pointed scales that make the skin of modern sharks feel so rough. Marvelously well-reserved fossils of the brain cavity and nerve canals reveal that their nervous system and sense organs were already well developed by Silurian time, and there can be little doubt that certain other bodily systems were also relatively efficient and highly organized. In spite of the incomplete fossil record, it is evident that these vertebrates made many of their most important advances in early Paleozoic time.

Economic products of the early Paleozoic period

The study and evaluation of economic mineral deposits constitute an important subdivision of geology (see Chapter 22). For present purposes, however, we must confine our attention to the historical aspects of the subject—that is, to the times and manner in which mineral deposits were formed and their relation to other geologic events. If we disregard building materials and ordinary stone products, which are available almost everywhere, it is here convenient to consider mineral products as (1) those originating by igneous activity, (2) those originating by sedimentary processes and not affected by organic influences, and (3) those originating directly or indirectly through the influence of organisms.

The early Paleozoic is not particularly rich in metallic minerals because igneous action (which is usually connected with their deposition) was not strong. A few important sedimentary deposits with metallic minerals should be noted, however. Economic deposits of sedimentary iron occur in Ordovician and Silurian rocks in northern Norway, Great Britain, Germany, northwest Siberia, Spain, and the eastern United States. The Clinton Formation, which is the chief iron-producer in the

FIGURE 17-26 *Underground workings of large lead mine in Cambrian strata, near Bonneterre, Missouri. (Photo courtesy U.S. Bureau of Mines)*

Appalachian region, is a sedimentary deposit that extends hundreds of kilometers from New York State to Alabama. It is possible that certain bacteria were actively engaged in precipitating iron-rich deposits at this time. One particularly valuable type of ore deposit, yielding chiefly lead and zinc ores from flat-lying limestone deposits, is almost entirely confined to the Paleozoic. Great deposits of this type are found in Oklahoma, Kansas, Missouri, Kentucky, and Illinois (Figure 17-26).

Commercial deposits of copper are found in Late Cambrian and Ordovician sediments of Siberia, and there is phosphate-bearing ore in the Cambrian of central Asia.

Among the deposits associated with intrusive igneous rocks are platinum and gold in the Ural Range; copper, nickel, titanium, and chromium in areas of Caledonian disturbances in northern Norway; gold in New South Wales and Tasmania; and copper and gold in eastern North America. Important commercial deposits of salt are found in the Silurian of western New York and the Great Lakes region and in the Cambrian of central Asia.

Unlike the situation in the other continents, a great deal of oil and gas is derived from Ordovician rocks in North America. As a matter of fact, in the United States the Ordovician ranks third among the systems in known oil and gas yield and reserves. The largest Ordovician fields are in the midcontinent region. Plants had not yet become abundant enough, however, for the formation of significant coal beds.

SUMMARY

The Paleozoic Era began with the appearance of abundant fossils and a slow submergence of the continental margins. During the Cambrian, Ordovician, and Silurian Periods the major geosynclines were occupied almost continually by shallow seas, and the shield areas occasionally were flooded. The Cambrian was generally quiet, with volcanism and mountain building at a low ebb. Sediments from this period are chiefly fine-grained, and there are thick accumulations of limestone, dolomite, and shale.

During the Ordovician the eastern margin of North America was affected by the Taconic mountain-building disturbance. Other areas bordering the North Atlantic were in a state of geologic unrest during the Ordovician and Silurian. Major effects during the Silurian were in northwest Europe, where the previously downfolded Baltic Geosyncline was compressed and intruded by granites (Caledonian Revolution). The Southern Hemisphere and Pacific borderlands appear to have been geologically quiet.

The early Paleozoic was an age of marine invertebrates. All the major phyla were in existence, and the shallow continental seas were teeming with life. The vertebrates appear as fossils in the Ordovician, and fish multiplied in the seas during the Silurian. The first land vegetation appeared in the Silurian, but there were no forests of inland vegetation sufficient to leave coal beds.

SUGGESTED READINGS

HOLLAND, C. H., ed., *Cambrian of the New World.* New York: John Wiley & Sons, Inc., 1964.

LEVI-SETTI, RICCARDO, *Trilobites: A Photographic Atlas.* Chicago: University of Chicago Press, 1975.

PALMER, ALLISON R., "Search for the Cambrian World," *American Scientist, Vol. 62,* p. 216–224, 1974.

RAUP, DAVID M., and STEVEN M. STANLEY, *Principles of Paleontology.* San Francisco: W. H. Freeman and Co., 1971.

A Treatise on Invertebrate Paleontology, 24 parts. Lawrence, Kansas, and Boulder, Colorado. The University of Kansas Press and the Geological Society of America, 1957—. A series by various authors on the major invertebrate animals. Important references for early Paleozoic fossils are those on arthropods, brachiopods, bryozoans, graptolites, and echinoderms.

18 The Late Paleozoic Periods

The late Paleozoic comprises four periods: Devonian, 35–40 million years long; Mississippian, 40 million years long; Pennsylvanian, 50 million years long; and Permian, 45 million years long. A roughly cyclic progression of events is evident in these four periods, which is similar to that of the three periods of the early Paleozoic. The Late Devonian and Early Mississippian were relatively quiet geologically, with widespread shallow seas and few continental collisions or mountain-building movements. An exception to this generalization was the Ellesmerian Orogeny in Arctic North America. During the succeeding Pennsylvanian and Permian periods, there was accelerated mountain building, which seems to have brought on extremes of climate including widespread aridity and a great ice age. The late Paleozoic mountain-building episode is referred to by American geologists as the *Alleghe-*

Reconstruction of a coal-forming forest of the Pennsylvanian period. (Field Museum of Natural History, Chicago.)

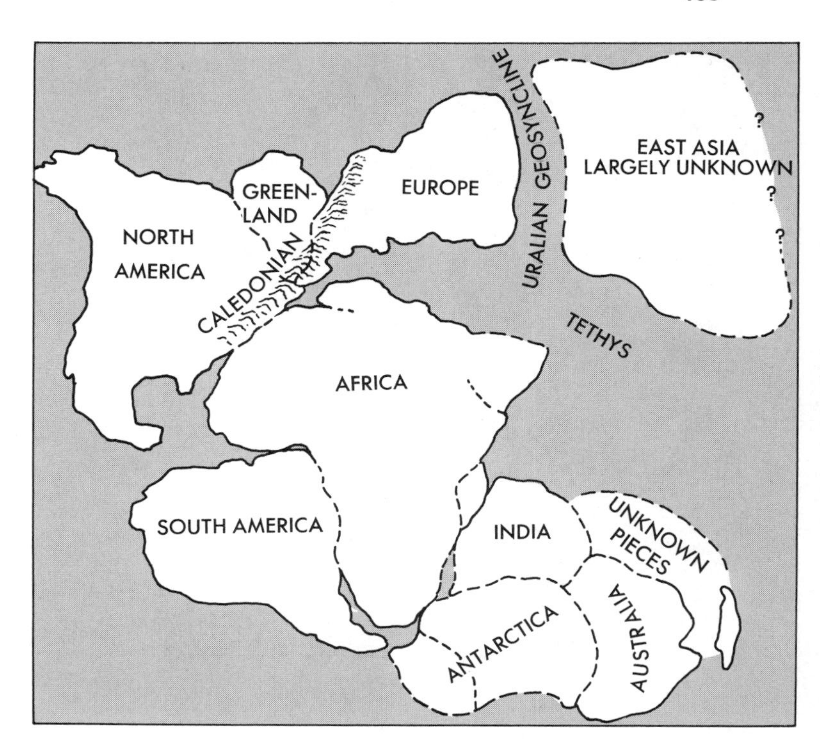

FIGURE 18-1 *Distribution of continental landmasses in the Middle Devonian. Semi-diagrammatic only.*

nian; in Europe the name *Hercynian* is used. During this major period of unrest many localized uplifts and depressions were created. These are practically all in the Northern Hemisphere and have mainly an east-west orientation.

During the entire late Paleozoic, the southern supercontinent, Gondwanaland, retained its compact form and was the greatest landmass on earth. On the other hand, several separate lands and intervening seaways existed in the Northern Hemisphere, and there were significant movements of component elements. Figure 18-1 shows the reconstructed global geography at the Middle Devonian and Figure 18-2 shows conditions in the late Permian. Chiefly to be noted is closure of the proto-Atlantic ocean, which was progressively obliterated during the late Ordovician to late Pennsylvanian as western Europe approached eastern North America. The Urlian Geosyncline between Europe and Asia remained an open seaway and was not destroyed until the Triassic period. The exact configuration and extent of land connections between Gondwanaland and Laurasia are poorly known, but there appears to have been intermittent obstructions within Tethys so that free east-west movement of water and marine organisms was sometimes hampered. In the Permian, Africa and Europe may have been united or nearly so. In any event there were extensive almost landlocked seas in northwestern Europe in which great salt deposits were laid down.

The late Paleozoic was marked by one of the earth's greatest ice ages. Effects were mainly in Gondwanaland, but there are also minor evidences in northeast Siberia. During this interval the South Pole moved relatively southward across Africa and into Antarctica

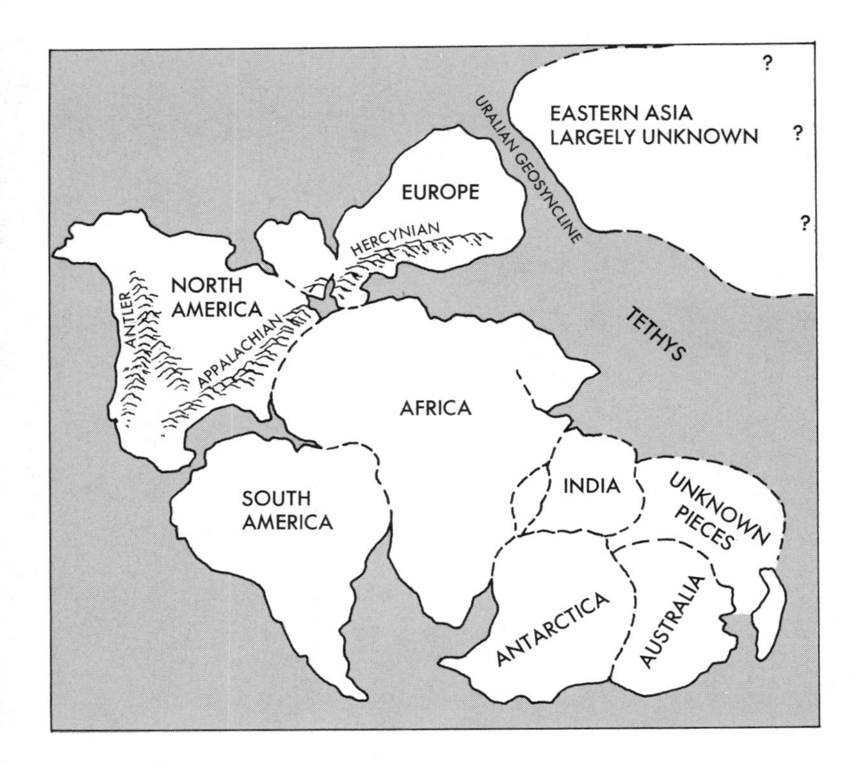

FIGURE 18-2 Distribution of continental landmasses in the Late Permian. Semidiagrammatic only.

and Australia, leaving a trail of glacial evidences in its wake. The equator lay diagonally across the area of the United States, from southern Arizona to Newfoundland. Since North America and Eurasia were then conjoined, the equator continued across Scotland, central Europe, central Asia, and Southeast Asia. As in the early Paleozoic, conditions were most favorable for life in the tropics and semitropics. The great Pennsylvanian coal forests of America and Europe were in this favorable zone, and the most complete record of continental animal life is likewise in the equatorial belt.

After about middle Mississippian time there was a decline in the abundance and variety of marine life. The period of greatest extermination of animal life came near the close of the Permian. The number of families of fossil animals of shallow marine environments declined from about 170 in the early Devonian to about 75 at the Permian–Triassic boundary.

North America

The late Paleozoic history of northeastern and eastern North America is best understood in terms of the large-scale reactions between Europe and North America. The two great landmasses had been moving together since sometime late in the Ordovician and had made contact in the north as early as the end of that period. During the Late Devonian another collision affected southeastern Canada and New England. This had been named the *Acadian Orogeny* long before plate tectonics had been thought of. Intense folding and metamorphism of older rocks and extrusion of lava and intrusion of granite were notable effects. The complex geology of the northern Appalachians dates from this time, and the area has never again been covered by the ocean.

The proto-Atlantic continued to close and was finally eliminated by a jawlike motion between Europe-Africa and North America. The middle and southern Appalachians were formed in the Pennsylvanian and Permian as the continental masses joined. Sediments of the proto-Atlantic sea bed were uplifted into significant mountains from which large rivers flowed westward into what is now the Appalachian region. The Catskill mountains preserve the deeply eroded remnants of a large delta deposit, 4,000 m thick, which was built up by a westward flowing river system whose source was beyond the present continental margin. Figure 18-3 shows the appearance of North America in the Devonian Period and Figure 18-4 in the Mississippian Period.

Geologists generally believe that the Appalachian trough turns rather abruptly to the northwest somewhere under the edges of the Gulf of Mexico. Perhaps it reappears again well within the continent, for the succession of rocks found in the Ouachita-Wichita uplifts in the Oklahoma-Arkansas region are similar to

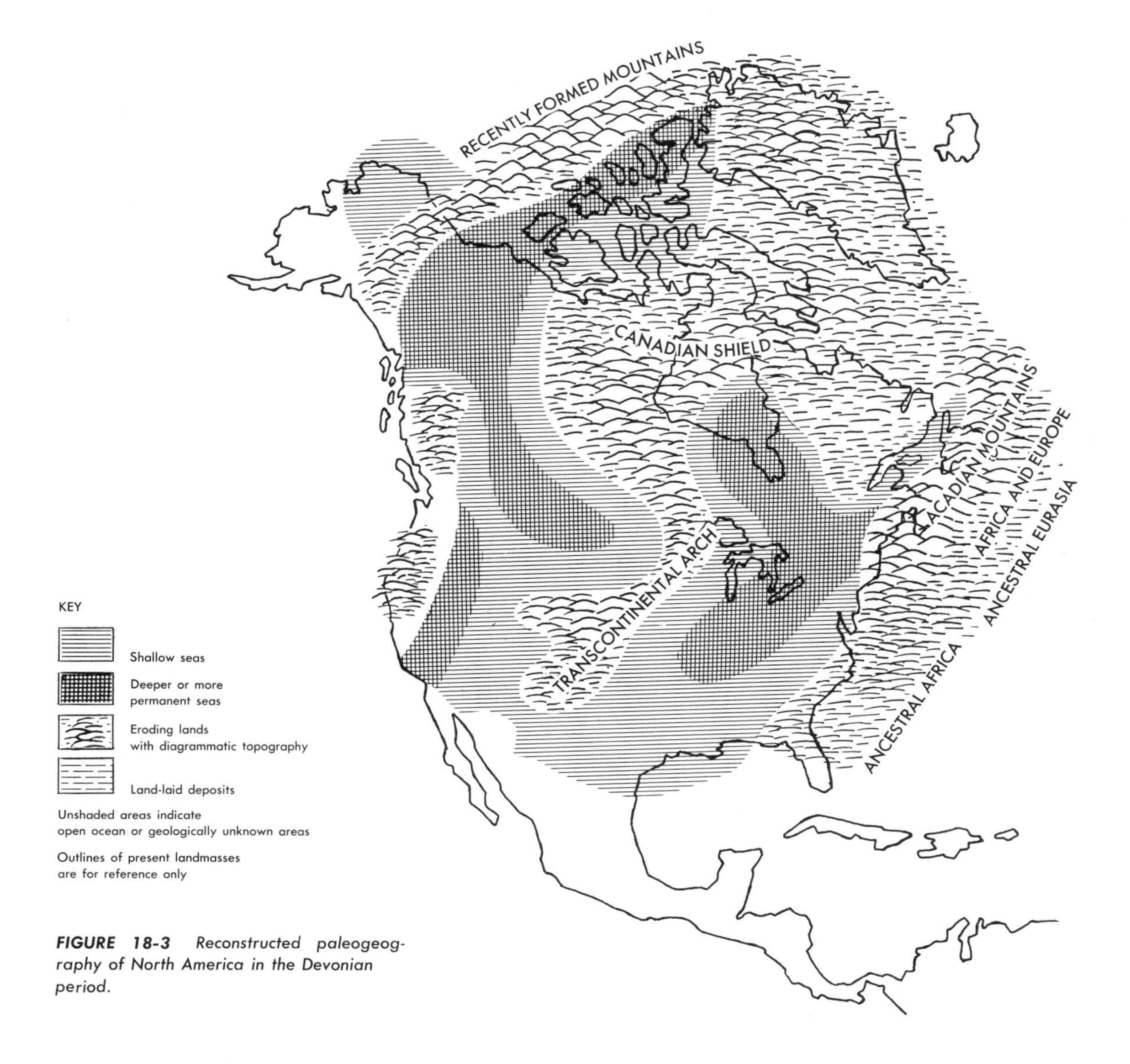

FIGURE 18-3 *Reconstructed paleogeography of North America in the Devonian period.*

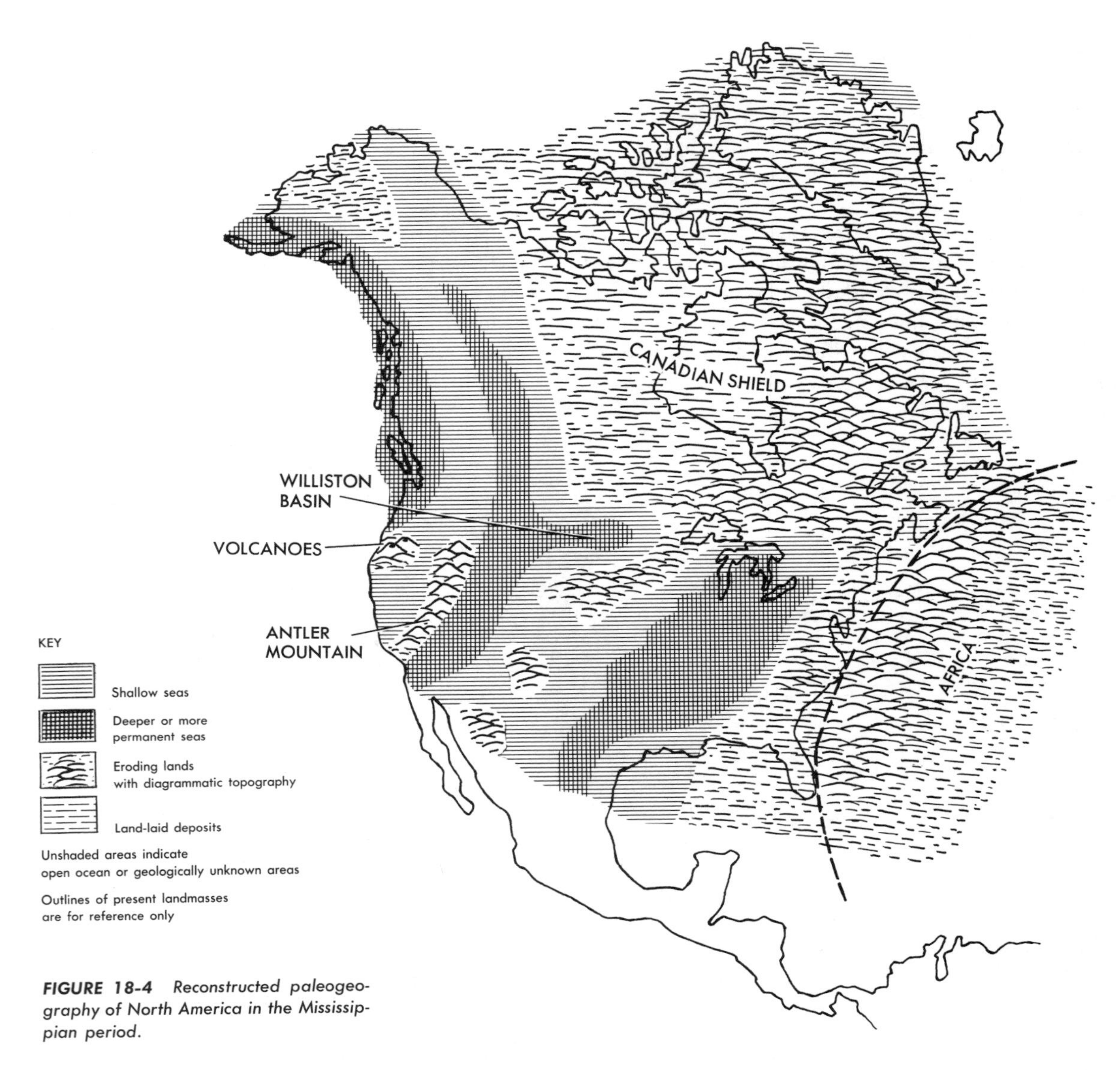

FIGURE 18-4 *Reconstructed paleogeography of North America in the Mississippian period.*

formations in the Appalachians. In Oklahoma and Arkansas we find evidences of intense orogeny in the late Mississippian and several times during the Pennsylvanian. Another great mountain chain, known as the *Ancestral Rockies,* was formed during the Pennsylvanian in the Utah-Colorado-New Mexico area. The Ancestral Rockies trend in the same general direction as the uplifts in Texas and Oklahoma that came into existence between the middle Mississippian and late Pennsylvanian. The entire group is broadly considered by most geologists to have been a product of the Alleghenian Revolution.

The late Paleozoic mountains of the southern and southwestern United States are difficult to relate to compression between Europe-Africa and North America. It is possible that South America may have been pressing against North America at this time, but the exact time and space relations of such reactions remain to be worked out.

The western margins of North America also became geologically active in the late Paleozoic. Intensive folding of previous geosynclinal sediments in western Nevada began in the late Devonian and continued into the Mississippian. This epoisode has been called the *Antler Orogeny.* It has been logically explained as the expression of a westward thrust of North America in response to the closure of the proto-Atlantic on the east. A second western disturbance called the *Sonoma Orogeny* affected a parallel zone along the Nevada-California border during Permian and Early Triassic time.

The overall effect of late Paleozoic disturbances was to create uplifts encircling most of North America. These uplifts and attendant depressions had important effects on the deposition of sediment in the interior regions. Late Devonian and Mississippian rocks of the Appalachian Geosyncline and most of the interior of North America reflect relatively quiet geologic conditions. Most of the interior of North America was covered by wide shallow seas in which limestone was the dominant sediment. The *Mississippian* gets its name from the general region near the junction of the Mississippi and Missouri rivers where the formations are for the most part very limy and favorable for cave formation (Mammoth Cave, Kentucky for example) (Figure 18-5).

The western and southwestern parts of the United States were occupied by a succession of shifting seaways and uplifts during the Late Paleozoic. Downsinking and deposition followed the northerly trend of the great Rocky Mountain Geosyncline, and there were substantial interior seas in the Williston Basin in the heart of the continent. A major change in the pattern of geologic development took place in the late Mississippian and Pennsylvanian as the Allegheny Orogeny reached a climax. Prior to this time, the Appalachian and Rocky Mountain geosynclines had been receiving most of the sediments. In the later Paleozoic periods a number of additional basins came into being in the southwestern part of the continent in connection with the Ancestral Rockies and Ouachita-Wichita Mountains. In these basins, Pennsylvanian and Permian deposits reaching 6,000 to 10,000 m in thickness were laid down. The paleogeography of the Pennsylvanian Period is depicted in Figure 18-6.

FIGURE 18-5 *Large quarry in Mississippian Salem Limestone near Oolitic, Lawrence County, Indiana. (Indiana Geological Survey.)*

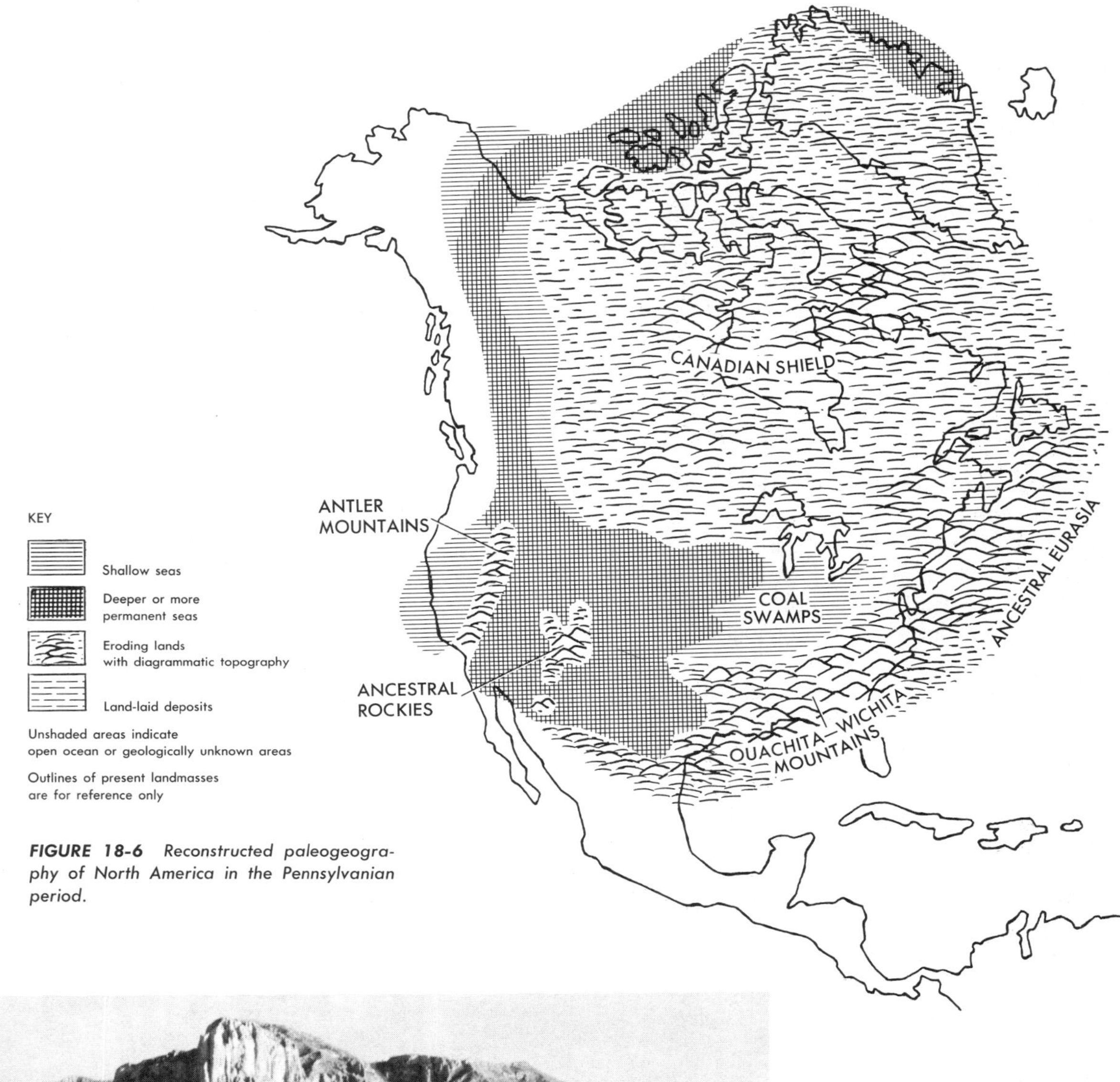

FIGURE 18-6 *Reconstructed paleogeography of North America in the Pennsylvanian period.*

FIGURE 18-7 *El Capitan and Guadalupe Peak, Culberson County, Texas. The cliffs are eroded in fossil reef material of Middle Permian age. (U.S. Geological Survey.)*

Recent exploration of the Arctic regions has shown an extensive rock record of late Paleozoic time. The Devonian reaches 3,000 m in thickness and is chiefly dolomite and shale. The Mississippian, Pennsylvanian, and Permian have lesser volume. Evaporite deposits (salt and gypsum) occur in the Pennsylvanian and Permian, but coal is not known. Great reefs were common (Figure 18-7).

The various orogenic movements that affected the borderlands of the Canadian Shield during the late Paleozoic touched off a distinctive change in sedimentation. Prior to this period, limestone, dolomite, and shale predominated; afterward, the sediments contained a higher proportion of sandstone and conglomerate, both in the marine and continental sections. It was a time of formation of gypsum, salt, and red beds. The Appalachian Geosyncline was filled with sediment and ceased to exist as an open seaway during the Pennsylvanian, and areas of Permian deposition follow a distinctly different pattern (Figure 18-8).

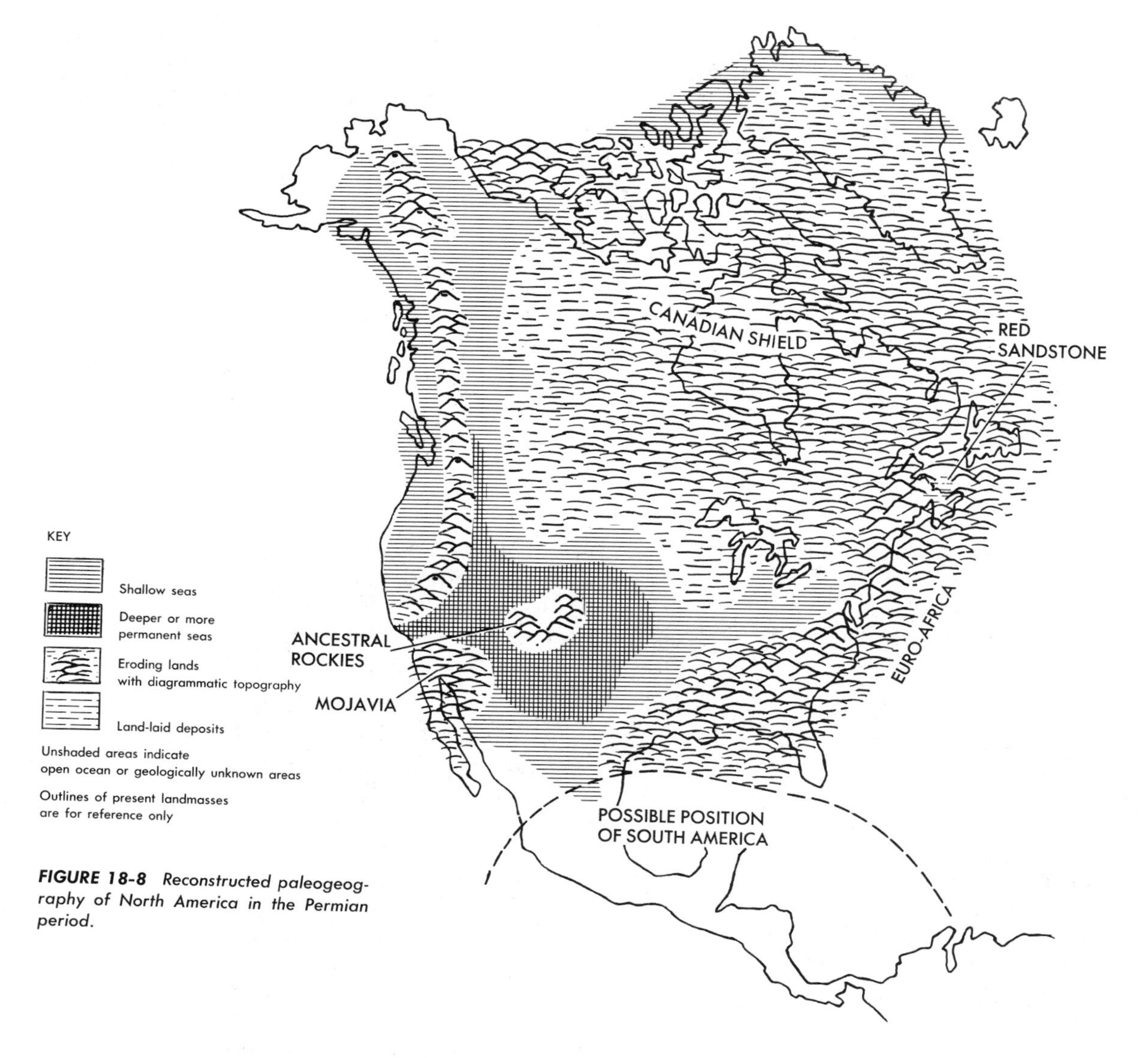

FIGURE 18-8 *Reconstructed paleogeography of North America in the Permian period.*

FIGURE 18-9 Aerial view of the Cape Peninsula, Union of South Africa. Stratified rocks of early Paleozoic age make up Table Mountain, which rises above Cape Town. (South African Information Service.)

The southern continents

The geologic records of each of the southern continents are remarkably parallel, a fact that lends weight to the theory that they were once united in one great landmass. Fossils are relatively few, but on the other hand, a variety of physical events and processes are clearly recorded that have no counterparts in northern lands.

FIGURE 18-10 Fossil cephalopods of both straight-shelled and coiled species exposed by natural weathering of Devonian limestone, Morocco, Sahara Desert, near Rissani. (Photo courtesy A. A. Ekdale.)

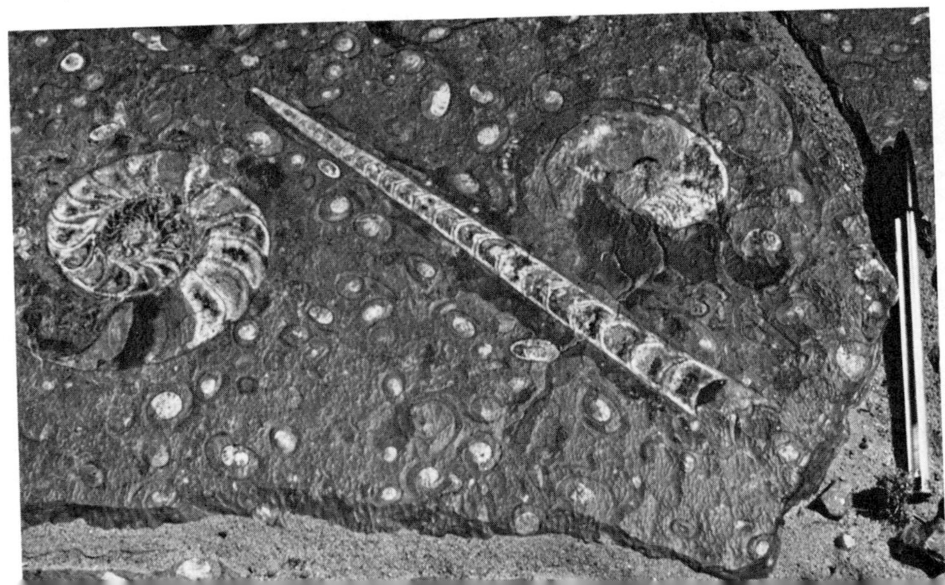

At the extreme southern tip of Africa a section of marine Devonian rocks about 2,100 m thick indicates temporary subsidence (Figure 18-9). About half of South America was covered by Devonian seas, including the present Andes and Amazon Basin. Some of the fossils are related to North America, some to other southern continents. Cephalopods are good guides to configuration of seaways at this time (Figure 18-10). Devonian rocks with brachiopods and primitive plants have been found in Antarctica, indicating shallow, warm seas at that time. Devonian formations are also widespread in Australia, and are 6,000 m thick in the Tasmanian Geosyncline.

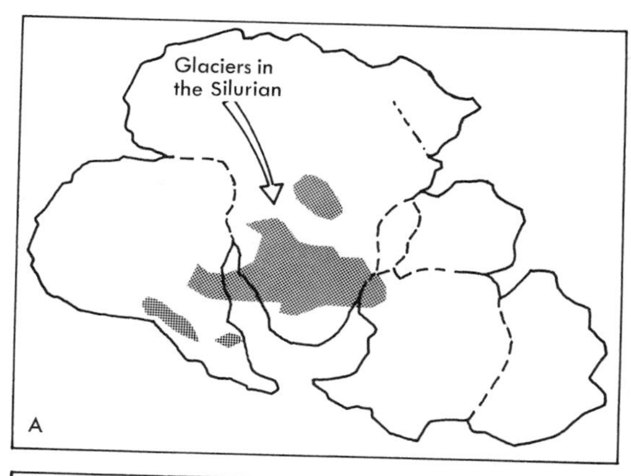

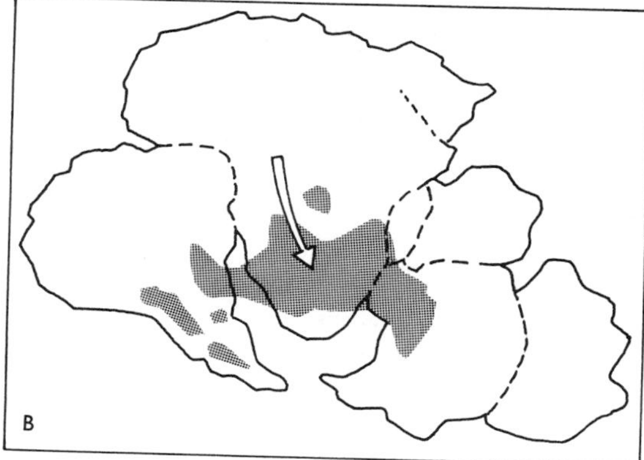

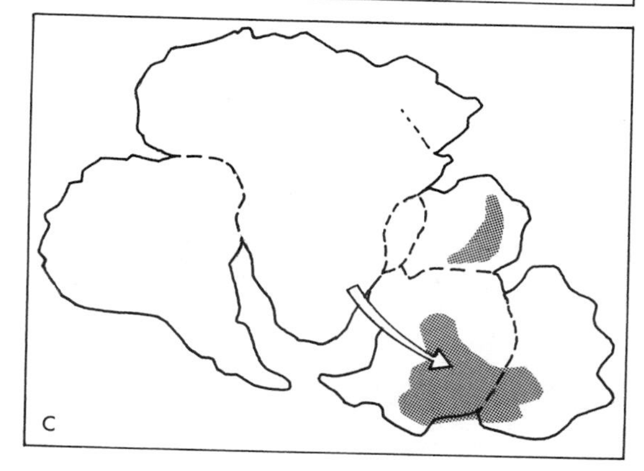

FIGURE 18-11 *Continental glaciers of the southern landmasses during the Late Paleozoic. (A) Mississippian (early Carboniferous). (B) Pennsylvanian (late Carboniferous). (C) Permian. Apparent path of the South Pole is shown by the large arrows. Drifting of the combined landmasses across the South Pole would explain the evidence.*

Ice age in the Southern Hemisphere

Abundant evidence proves that a great ice age gripped the Southern Hemisphere during the late Paleozoic. Much of the evidence of ice action is centered in what are now tropical and semitropical regions. Most geologists find it easier to believe that the continents bearing this evidence have been displaced to their present position from a former location near the South Pole, rather than that thick ice caps could have formed and spread in tropical or semitropical lands (see Chapter 2 and Figure 18-11).

Middle Devonian glacial deposits, dated by associated fossils, occur in western Argentina; and evidence of extensive ice sheets occurs in the Pennsylvanian of much of southern South America. In Australia, where marine beds with fossils are found interbedded with glacial deposits, it is possible to distinguish Permian and probable Pennsylvanian glaciations. The *Dwyka Tillite,* which covers many thousands of square kilometers and reaches 1,000 m in thickness in South Africa, is regarded by geologists as a product of large continental glaciers and has been dated as

Mississippian or possibly Devonian. It will be recalled that the South Pole was in the central Sahara during the Ordovician. Glacial centers apparently expanded as Gondwanaland passed under the pole, which was moving relatively southward in terms of present-day Africa. Maximum glacial effects during the Pennsylvanian were in South America and South Africa; later, in the Permian, the chief centers were in Australia and Antarctica (see Figures 18-12 and 18-13).

When the theory of plate tectonics is taken into account and the various elements of Gondwanaland are reassembled, the late Paleozoic ice age appears as a coherent climatological phenomenon with glacial evidences and ice movements consistent with a path that carried the pole across the length of Africa into Antarctica and Australia.

FIGURE 18-12 *Evidence of Carboniferous glaciation in Antarctica. The parallel groovings on the rock floor were formed by glacial ice moving across it. The glacial formation is the Buckeye tillite; locality, Wisconsin Range, Antarctica. (Courtesy V. H. Minshew.)*

FIGURE 18-13 *Rock surface polished and grooved by ice of the late Paleozoic glaciation of South Africa. View at Noostgedacht Farm, near Kimberly. (Courtesy George W. Bain.)*

The beginning of the Karroo series

The *Karroo,* a geographical region of the Cape Province of South Africa, gives its name to a sequence of rocks that contributes much to an understanding of the late Paleozoic–early Mesozic geologic history of the southern Hemisphere and of the land life of the time. The *Karroo Series,* which reaches a thickness of 6,000 m, was deposited almost entirely under continental conditions. Deposition started early in the Carboniferous and continued without interruption through the Permian and Triassic into the Jurassic. At the base is the Dwyka Tillite, the previously mentioned deposit of glacial origin.

Above the Dwyka lies the *Ecca Series,* which contains the best coal of Africa and a variety of freshwater fossils. Among these remains are the first fossil leaves of the well-known *Glossopteris flora.* Following the Ecca is the much thicker *Beaufort Series,* famous for its fossils of fish, amphibians, and varied reptiles.

The Permian–Triassic boundary lies within this series. The Mesozoic portion of the Karroo will be described in the next chapter.

Eurasia

Early in the Devonian Period, land-laid sediments began to fill the depressions between and around the Caledonian folded mountains that had formed in the Late Silurian. The area thus affected included much of northwestern Europe, and west-central Asia. Devonian land-laid sediments are preserved in several areas of Great Britain, where they are known as the *Old Red Sandstone.* Although the total area now covered by these beds is small, they are of interest because their study and classification had a significant influence on geological thought. Scattered throughout the Old Red series are beds containing brackish or freshwater fish, eurypterids, and fragments of land plants. The continental beds pass into thick marine formations in Belgium and West Germany, these are standards of comparison for the Devonian system.

Although European and Asiatic geologists do not recognize the Mississippian Period, they do treat the early Carboniferous as an equivalent time period. As in America this interval was one of relative quiet and much limestone was deposited (Figure 18-14). Shallow seas and swamps spread widely in Russia, and important coal deposits were laid down in the Moscow and Kama Basins. Not much Mississippian sediment exists in Siberia.

FIGURE 18-14 *Karst topography in Carboniferous limestone as seen along the Li Jiang River, Kwang Si Province, China. (Photo courtesy Frank H. Brown.)*

The Hercynian Orogeny

During the late Paleozoic a great system of mountains known as the *Hercynian Chain* was formed in Europe and Asia. It seems significant that the dominant trend of this system was east-west, the same direction taken by nearly contemporaneous structures in North America. The region affected included southern Wales, northern and central France, southern Germany, Bohemia, and parts of the USSR and central Asia (Figure 18-15). Volcanic outpourings and granitic intrusions accompanied the folding, and many important ore deposits were formed. This ancient mountain chain was subsequently deeply eroded; portions sank and were flooded by Tethys, and still later were incorporated in the great Alpine and Himalayan chains.

The Hercynian disturbance created many depressions in western Europe that became sites of coal forests in Pennsylvanian time. The Hercynian Orogeny may be thought of as having created the geologic framework of central Europe—a very diverse belt lying between the more ancient Baltic Shield to the north and the younger Alpine ranges to the south.

FIGURE 18-15 *Late Paleozoic mountains of N. America, Europe, and Africa showing probable relationship of landmasses. Inferred directions of drift shown by large arrows; Anc., Ancestral Rockies; Ant., Antler; Ap., Appalachian; H.-V., Hercynian-Variscian; Ou.-W., Ouachita-Wichita.*

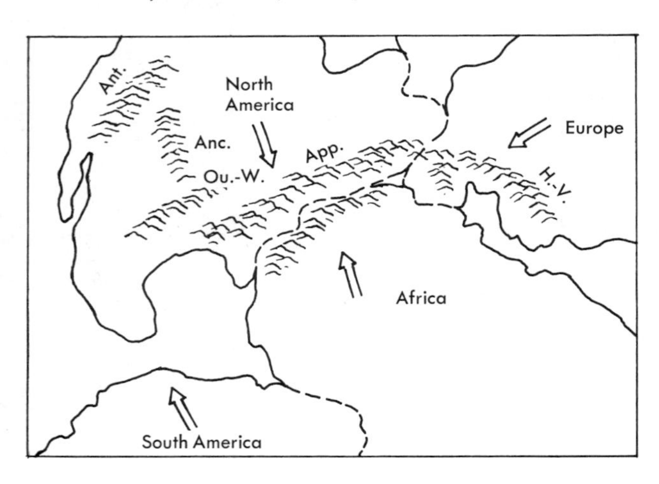

Beginning of the Great Salt Age

Used in its widest geological sense, the term *salt* includes not only the common household variety, NaCl, but also gypsum, $CaSO_4H_2O$; sylvite, KCl; magnesium chloride, MgCl; and many minor variates. All are included in the general term *evaporites*—meaning they originate by evaporation of the water of oceans, seas, or lakes. Evaporites are much more common than might be suspected because they are easily dissolved and rarely appear at the surface of the earth. Only by drilling and mining have they been discovered and their true magnitude made known.

Deposition of salt in nature is favored when evaporation exceeds inflow, a condition brought about chiefly in more or less enclosed basins having restricted connections with the ocean and little inflow of fresh water. The high temperatures associated with equatorial climates are also effective in bringing about maximum deposition. Through a combination of favorable factors, the period from the late Pennsylvanian to the early Cretaceous became what may be called the *Great Salt Age.* The semienclosed basins in which salt accumulated came into being between North America, South America, Africa, and Europe as those land masses began to split apart to create the Atlantic Ocean and certain embayments of Tethys. The narrow basins originating by continental splitting allowed water from the ocean to spill into areas where evaporation was exceptionally high. Figure 18-16 shows the distribution of salt deposits referred to above.

Early effects are to be found in the southwestern United States where the Paradox Basin of Utah and Colorado filled with salt during the late Pennsylvanian. Later, in the Permian, other great salt deposits were created in the Carlsbad Basin of New Mexico, Texas, Oklahoma, and Kansas. Another great Permian salt basin if that which underlies much of

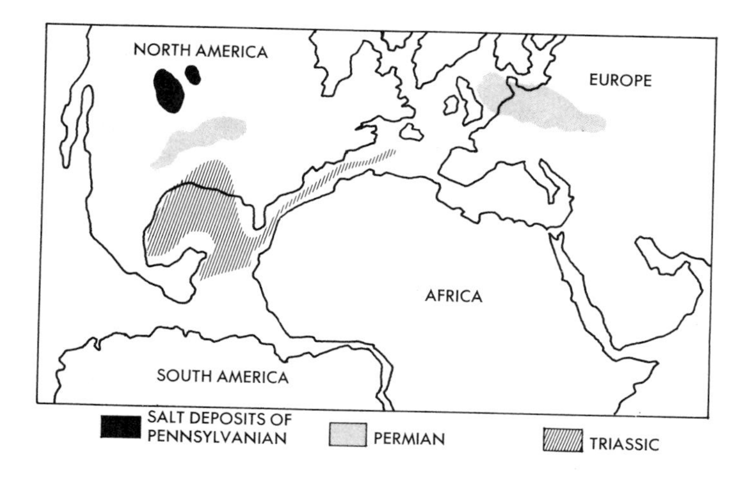

FIGURE 18-16 *Salt deposits of Pennsylvanian (black), Triassic (hatched) and Permian (grey) shown in relation to the major landmasses and possible seaways of the late Paleozoic-Early Mesozoic time.*

Germany, the North Sea, and parts of Great Britain. The parent brines for the European deposits were apparently derived through narrow connections with Tethys to the east and southeast.

Continuation of salt deposition during the Mesozoic will be described in the next chapter.

Life of the late Paleozoic

The first forests

The plant kingdom advanced and diversified tremendously during the late Paleozoic. The most important developments took place on the continents, where the spread of vegetation was amazingly rapid, considering how barren the long preceding periods had been. Our knowledge of early Devonian land plants is rather imperfect, although remains have been discovered in Europe, North America, China, and Australia. In general, the earliest known land plants lacked true roots and leaves, and bore spores at the ends of simple branching stems. A typical late Silurian–early Devonian plant is *Cooksonia*. A middle Devonian descendant is *Rhynia,* represented by silicified specimens from Rhynia, Scotland (Figure 18-17).

FIGURE 18-17 *Model of the Middle Devonian plant* Rhynia. *The simple, leafless stems are about 8 cm high. (Courtesy Field Museum of Natural History.)*

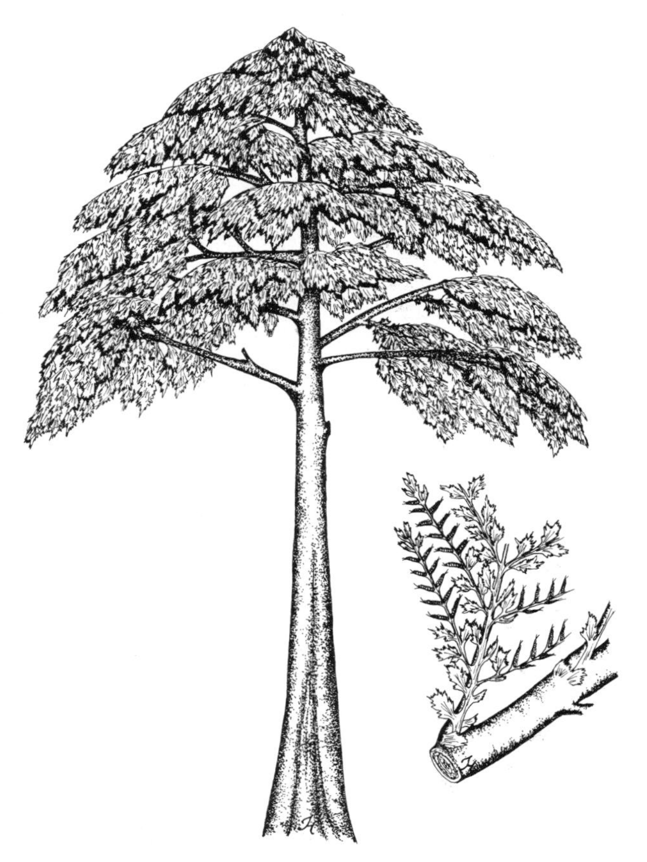

FIGURE 18-18 An artist's conception of the appearance of Archaeopteris, *dominant genus of Late Devonian floras. (Drawing by Naoma E. Hebbert.)*

Other advanced types of vegetation appeared in the middle Devonian. The rather sparse and tantalizing fossil record shows this to be a time of great diversification with a variety of advanced types broadly called *progymnosperms* appearing. These had spores suitable for land dispersal and increasing amounts of woody tissue.

There is a distinct break between the flora of middle and late Devonian times, marked by the disappearance of rhynia-type plants and the rise of a new group called the *Archaeopteris* flora. *Archaeopteris* had large fern-like leaves with clusters of spore-bearing organs (Figure 18-18). Fossil remains of the *Archaeopteris* flora have been found in such widely scattered areas as Russia, Ireland, Ellesmere Island to the north of Canada, and Australia. As a whole, the dominant late Devonian plants were larger than their mid-Devonian relatives and had branching root systems, stronger stems, and better reproductive systems. They were mainly lowland types and grew along shores and valley bottoms. There were ancestrial scouring rushes, or horsetails (the living *Equisetum* is a descendant), seed ferns (*Archaeopteris, Protopteridium*), and lycopods (spore-bearing plants) such as *Protolepidodendron* and *Archeosigillaria,* which would give rise to immense forests in the following period. The most primitive gymnosperm, *Callixylon,* is found in late Devonian rocks. By the end of the Devonian, plants had developed practically all of their basic characteristics: true roots, leaves, seeds, tree form, and complex vascular systems.

Late Devonian plants produced the earliest known *fossil forests.* Near Gilboa, New York, numerous stumps up to a meter in diameter have been uncovered.

The coal-forming swamps

The Carboniferous (combined Mississippian and Pennsylvanian) was ideal for swamp vegetation and became the great age of coal formation. The fossil record is exceptionally complete mainly because mining operations have yielded a wealth of specimens. Coal-forming vegetation included ferns (filicales), arborescent lycopods, herbaceous lycopods, sphenopsida, cordaites, and seed ferns. Perhaps the most important contributors to actual coal beds were the lycopods, *Lepidodendron* and *Sigillaria. Lepidodendron* had a slender trunk and a crown of forking branches that soared skyward to heights of over 15 m. The leaves were lance-shaped and arranged in spirals around the branches. When the leaves were shed, they left a characteristic pattern of scars on the branches, which remained to give the tree its popular name of *scale tree.* Only about 10 percent of the trunk of *Lepidodendron* was actual woody tissue; it was therefore not as sturdy as most living trees. In *Sigillaria* the leaf scars are in vertical rows and there is little or no branching. Figure 18-19 shows reconstructions of these common Carboniferous trees.

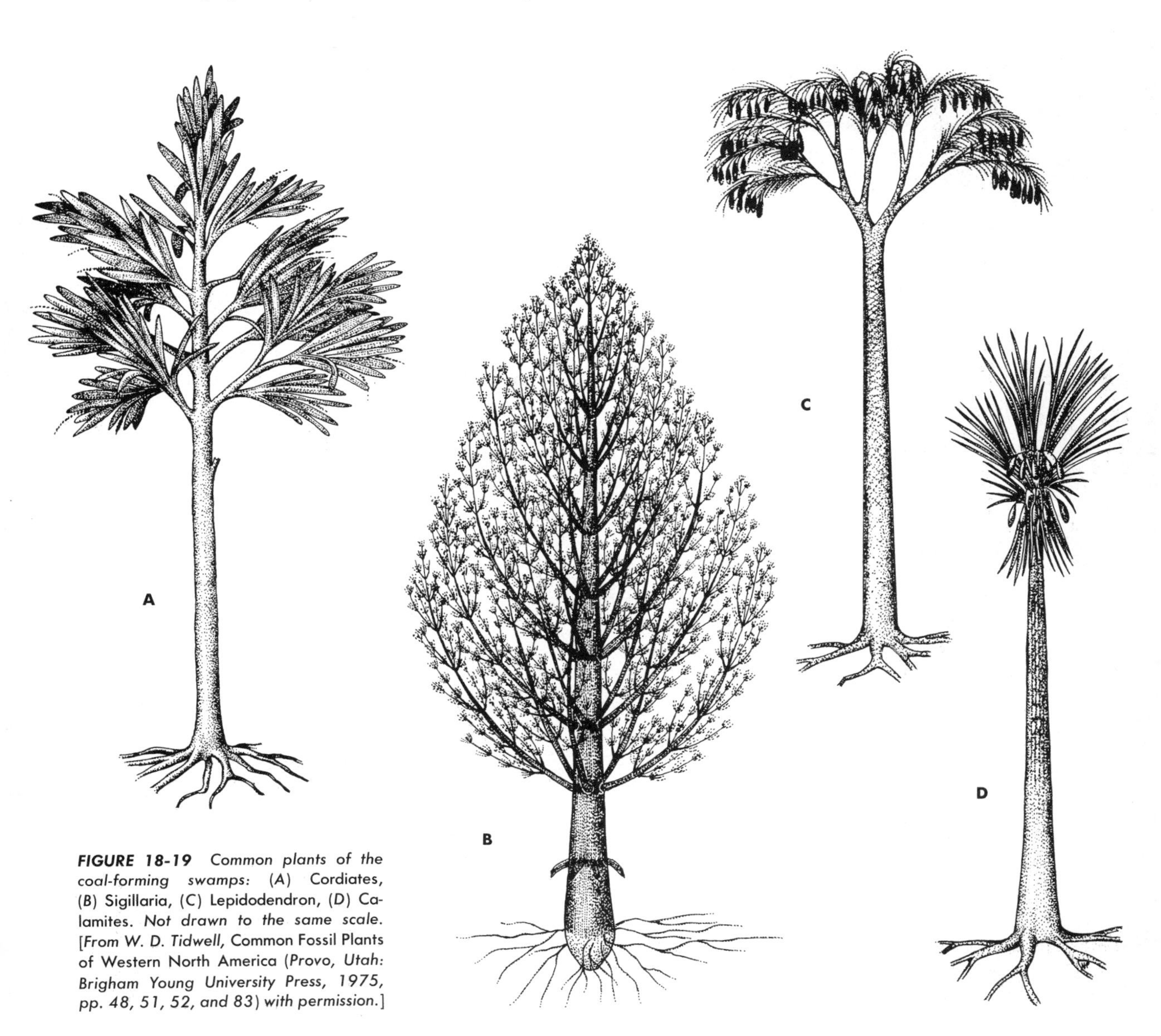

FIGURE 18-19 *Common plants of the coal-forming swamps: (A)* Cordiates, *(B)* Sigillaria, *(C)* Lepidodendron, *(D)* Calamites. *Not drawn to the same scale. [From W. D. Tidwell,* Common Fossil Plants of Western North America *(Provo, Utah: Brigham Young University Press, 1975, pp. 48, 51, 52, and 83) with permission.]*

Also inhabiting the coal forest were many jointed plants collectively called the *Arthrophyta* or *Sphenopsids*. The leaves of these plants radiate in whorls around the joints, as in the modern horsetail, *Equisetum*. Largest of the jointed plants was *Calamites*, which reached a height of over 30 m and a diameter of a meter.

In addition to the important coal-forming plants, there were minor Pennsylvanian forms destined to give rise to important vegetative groups in later periods. True ferns made up much of the undergrowth, and the *gymnosperms*, which were to dominate the Mesozoic, were represented by at least five orders, including the Pteridospermae, Cordaitales, and Bennettitales, which are extinct; the Ginkgoales, represented by one living species; and the Coniferales (conebearers), which make up an important segment of present-day vegetation. It is thought that these advanced types were evolving in dryer highland areas, while the more common coal-forming plants dominated the humid lowland basins.

Plants of the Permian period

The Permian was a time of increasing dryness, climatic diversity, and extensive glaciation in the Southern Hemisphere. A worldwide withdrawal of the seas seems to have begun in the late Pennsylvanian, and swamp vegetation was being replaced by upland and desert types. In a general way, the gymnosperms increased while the older groups declined during this interval. Some of the trees show definite annual rings, and leaves are coarser and thicker than in the early Pennsylvanian. An important group of the late Paleozoic plant world (frequently mentioned in connection with continental drift) is the so-called Glossopteris flora, named from the tongue-shaped leaves of the seed fern, *Glossopteris* (see Figure 18-20). Fossil remains of this flora are broadly distributed in South Africa, South America, Australia, and India, and also in Antarctica, within 500 km of the South Pole. This wide dispersal supports the theory of continental drift, for these plants could probably not have distributed themselves across the vast oceans that now separate their fossilized remains. It also seems impossible that lush vegetation could have lived within 500 km of the South Pole. The presence of the Glossopteris flora on all the southern continents and in India has long been cited as evidence for Gondwanaland and for continental drift.

Coal as fossil vegetation

Coal is the most important by-product of past life known to man. It has accumulated in prodigious quantities at many times and in many places when the right biologic and physical factors have happened to coincide. All evidence indicates that coal consists of altered plant remains, and every stage in this process of alteration has been thoroughly studied (see Figure 18-21). Commencing with peat, which today is still in the process of formation, the successive ranks of coal are *lignite, subbituminous, bituminous,* and *anthracite.* The rank depends on the conditions to which the plant remains were subjected after they were buried—the greater the pressure and heat, the higher the rank of the coal. Higher-ranking coals are denser and contain less moisture and volatile gases and have a higher heat value than low-ranking coals. Obviously, more than a heavy growth of vegetation is required for coal formation. The plant debris must be buried, compressed, and protected from erosion and from intensive metamorphism. The complete story of coal formation involves not only a study of the coal forests

FIGURE 18-20 *Slab of Permian siltstone with fossil leaves of Glossopteris browniana, Transantarctic Mountains, Antarctica. (Institute of Polar Studies, photograph by David H. Elliot.)*

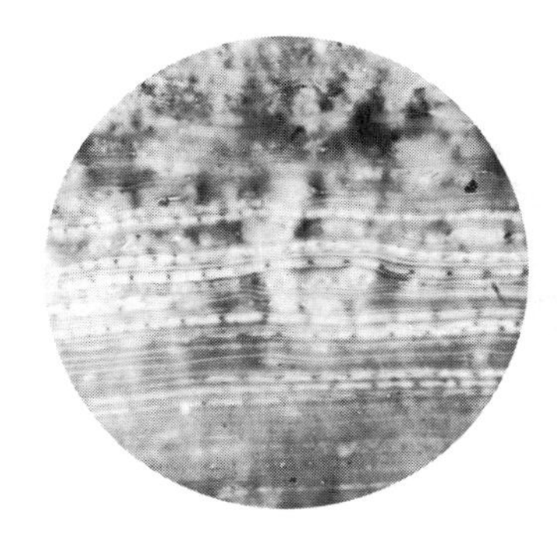

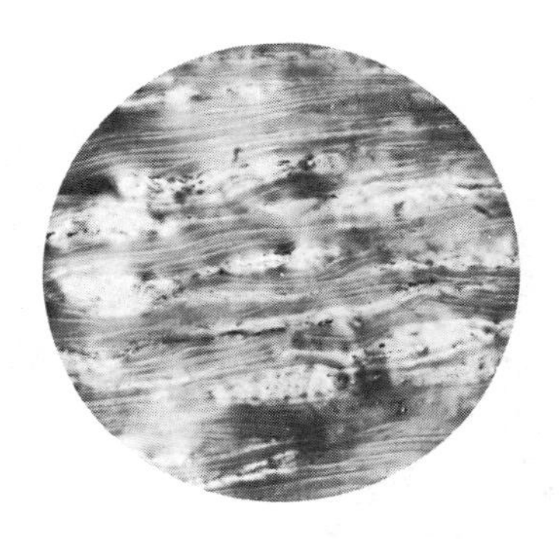

FIGURE 18-21 *Highly magnified, thin sections of coal showing remnants of cell structure in compressed and altered vegetative material. (U.S. Bureau of Mines.)*

themselves but also a consideration of the types of sediment in which the coal is buried and an understanding of the processes acting on the plant material after it has been buried.

A regular sequence of repeating or cyclic sediments of which coal is a normal member is called a *cyclothem.* When fully developed a cyclothem has the following units:

TOP

10. Shale with ironstone concretions.
9. Marine limestone.
8. Black shale with black limestone concretions or layers.
7. Impure, lenticular, marine limestone.
6. Shale.
5. Coal.
4. Underclay.
3. "Freshwater" limestone.
2. Sandy shale.
1. Sandstone, at the base and locally unconformable on underlying beds.

BOTTOM

A number of possible mechanisms have been proposed to account for the alternate flooding and emergence that produced the cyclothems. All theories agree that there must be a prolonged regional downsinking of coastal areas. One school of thought maintains that the downsinking is not steady, but alternately slows down and speeds up in giant pulsations. During the pauses coal swamps flourish; during more rapid downward movements the sea spreads inland, burying and compressing the vegetation of former swamps. An opposing school of thought attributes the cyclic effects to the rise and fall of sea level superimposed on the sinking lands. During emergent or low-water periods forests flourish; during flooding they are covered with water. Supporters of this idea point out that the period of coal formation in the Northern Hemisphere coincided with glaciation in the Southern Hemisphere and that alternate peri-

FIGURE 18-22 *Folded strata of lower part of Oquirrh Formation, Provo Canyon, Utah County, Utah. This well-exposed section illustrates the cyclic repetition of certain types of strata that characterizes Carboniferous formations. (Photo courtesy Walter P. Cottam.)*

ods of freezing (withdrawal of water from the sea) and melting (restoration of water to the sea) could have caused changes in the sea level. The whole problem is in an interesting state of speculation, and no one theory is accepted by all students of coal geology. What we do know is that a peculiar combination of apparently unrelated factors came into existence and operated over a period of many millions of years to produce a large part of the coal resources on which we depend for a great deal of our energy and many raw materials. An example of cyclic Pennsylvanian sediments, without coal, is shown by Figure 18-22.

Invertebrate life of the late Paleozoic

The invertebrates of late Paleozoic time show few spectacular advances and reveal fewer adaptations of evolutionary significance than either the plants or vertebrates of the same period. In general it was a time of intensive competition in the seas, with a consequent gradual elimination of less effective types and a flowering of more progressive ones.

Aquatic animals had already achieved their basic adaptations before the late Paleozoic and had settled down, as it were, to intensive competition involving mainly minor changes. As a rule, there were more relatively mobile creatures in the ocean than there had been in earlier periods. Legs, swimming organs, and fins were more in evidence, and more animals were able to crawl or burrow. Even sluggishly moving animals such as starfish, echinoids, and sea cucumbers gradually replaced the fixed members of their phylum typified by the cystoids and blastoids. Crawling molluscs were on the increase, and the cephalopods already had developed their coiled many-chambered, buoyant shells to high levels of mobility.

FIGURE 18-23 Spine-bearing brachiopods (productids) from Late Permian rocks of the Khisor Range, West Pakistan. Specimens have been etched from the enclosing limestone with their fragile spines almost intact. These samples are about 1 cm across. (Courtesy of Richard E. Grant, Smithsonian Institution.)

At the same time there is a great evidence of both defensive and offensive mechanisms and structures. Brachiopods were abundant, with a preponderance of thick-shelled and highly spiny types. As if to balance the scales, many species of sharks with heavy, flat teeth, well adapted for shell-crushing, cruised the seas. Slow-moving animals with weak skeletons, such as trilobites, inarticulate brachiopods, and graptolites, were either extinct or in decline as the more mobile or better-protected forms flourished.

Food-gathering and food-protecting techniques improved. Passive feeders, that depended on the vagaries of passing currents for their sustenance were gradually replaced by animals able to move about or at least to gather food by self-generated currents. Virtually all brachiopods of the late Paleozoic pos-

FIGURE 18-24 A remarkable slab of fossil crinoids from Mississippian rocks near Legrande, Iowa. Preservation of entire individuals such as these is very rare. (Courtesy Iowa State Department of History and Archives.)

FIGURE 18-25 Blastoid calyxes from Mississippian rocks of the midwestern United States. The large specimen and the five smaller ones at right belong to the genus Pentremites*; the two middle specimens are* Globoblastus*. The large specimen is about 2 cm long.*

sessed more powerful food-gathering mechanisms than their ancestors. Particularly efficient in this respect were the stalked crinoids, whose food-gathering arms branched and subdivided to cover wide areas of water (Figure 18-24). As the crinoids flourished, their relatives, such as cystoids and blastoids (Figure 18-25), with less efficient food-collecting systems, declined and disappeared, the former in the Devonian, the latter in the Permian.

Finally, we should mention the evolutionary advances made by the reef-building and colonial animals. These creatures were able to create interlocking structures of great size in which both plants and animals cooperated. The Permian reefs of Texas are examples.

(a)

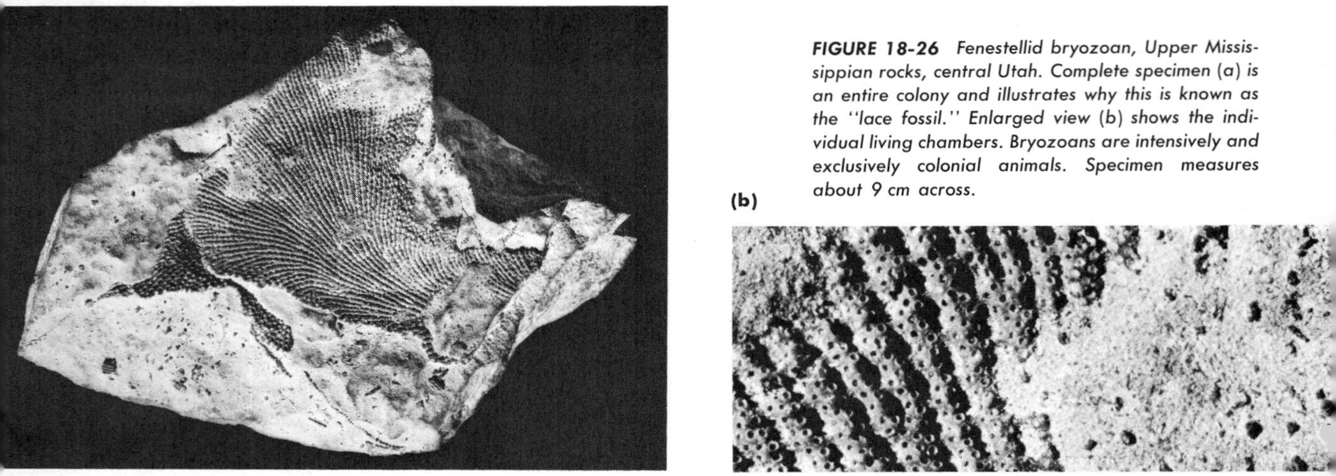

FIGURE 18-26 Fenestellid bryozoan, Upper Mississippian rocks, central Utah. Complete specimen (a) is an entire colony and illustrates why this is known as the "lace fossil." Enlarged view (b) shows the individual living chambers. Bryozoans are intensively and exclusively colonial animals. Specimen measures about 9 cm across.

(b)

Here, by presenting a sufficiently strong front to the open ocean, organisms could find food and living space.

Although late Paleozoic marine species were more advanced than their predecessors, they had still not achieved maximum efficiency or security, as we shall see from the pages that follow.

The late Paleozoic saw the rise and fall of many important groups of invertebrate animals. Important debuts included the *fusulines* in the Mississippian, cephalopods with moderately crinkled septa in the early Mississippian, cephalopods with highly crinkled septa in the middle Permian, land gastropods in the Pennsylvanian, spiny brachiopods in the late Devonian, and insects in the middle Devonian.

(a)

(b)

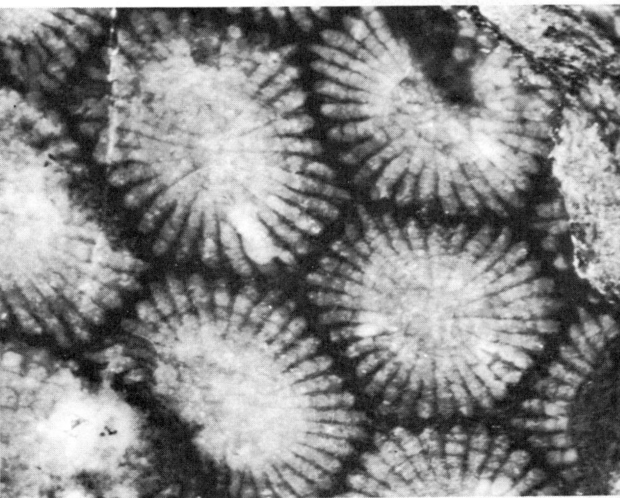

(c)

***FIGURE 18-27** Cut and polished specimens of three Devonian colonial corals.* (*a*) Hexagonaria, (*b*) Prismatophyllum, *and* (c) Billingsastrea. [(c) *Courtesy of W. A. Oliver, Jr.*]

The corals (Figure 18-27), bryozoans (Figure 18-26), brachiopods (Figures 18-23 and 18-28), gastropods, and pelecypods managed to maintain a relatively stable level of existence and were common and abundant during the late Paleozoic. These groups contribute the vast bulk of the marine fossils, and many thousands of species of each have been preserved. Within each group there were significant trends, marked by the extinction and replacement of individual families. Among the important coral genera were *Halysites* (Upper Ordovician into Lower Devonian), *Favosites* (Lower Ordovician–Permian), *Chaetetes* (Ordovician–Jurassic), and *Syringopora* (Silurian–Pennsylvanian). An important group of late Paleozoic bryozoans were the lacy forms, or *fenestellids* (Silurian–Permian), which left deli-

FIGURE 18-28 Spiriferoid brachiopods of late Paleozoic periods. Largest specimens are Paraspirifer from the Middle Devonian; medium-size specimens are Mucrospirifer, common in the Appalachian region; smaller specimens are Punctospirifer kentuckiensis of Pennsylvanian age, common in the midcontinent.

cate fossils representing colonies of many individual animals (Figure 18-26); also prevalent was the curious screw-shaped bryozoan, *Archimedes* (Mississippian–Permian). The most distinctive brachiopods during the late Paleozoic were *spirifers* (Ordovician–Jurassic), characterized by pointed, wide, and winged shells (Figure 18-28), and the very spiny *productids* (Devonian-Permian), some of which reached 30 cm in diameter. Pelecypods were represented by a variety of forms but were still subordinate to brachiopods in late Paleozoic seas. Gastropods of many types abounded locally but made no special advances, except for the evolution of certain air-breathing species that we have already mentioned.

The eurypterids (Ordovician–Permian), which have been aptly described as "sea scorpions," were the largest known arthropods of the late Paleozoic. Their jointed external skeletons ranged in size from a few cm to over 2 m. They reached the apex of their development in the Silurian and Devonian and gradually declined thereafter. We may assume that they competed with early fish, especially during the Ordovician and Silurian periods.

Beginning in the late Devonian, the coiled cephalopods with complex internal structures (the *ammonites*) began to be abundant and widespread. About ten worldwide cephalopod zones are recognized in the Devonian, seven in the Mississippian, six in the Pennsylvanian, and five in the Permian.

In North America the best guide fossils for marine Pennsylvanian and Permian rocks are the protozoans called *fusulines* (Figure 18-29). Their spindle-shaped skeletons superficially resemble grains of wheat, oats, rice, or rye. They have left their skeletons abundantly in limestone, sandstone, and shale, indicating a cosmopolitan type of existence. Although simple in outward appearance, their internal structure is very complex, consisting of numerous coils separated by curving partitions into a great many chambers. The internal details are characteristic for the different species, and since the fusulines evolved rapidly, we can distinguish the fossils of one interval or formation fairly easily from those of succeeding or preceding ones.

Because fusulines are generally quite small and tend to occur packed together in dense masses, they are frequently brought up in cores from wells and are very useful in correlating certain oil-bearing formations.

Late Paleozoic vertebrates

The Devonian Period is known as the *Age of Fishes,* and for the first time the fossil record reveals the existence of numerous and

(a)

FIGURE 18-29 *Fusolines. (a) A collection of wheat-shaped specimens of* Triticites. *(b) A highly magnified, thin section showing the complex internal structure. In the center is a specimen cut along the long axis; to the right of it is a specimen cut at right angles to the long axis to show the spiral coiling and chambers.*

(b)

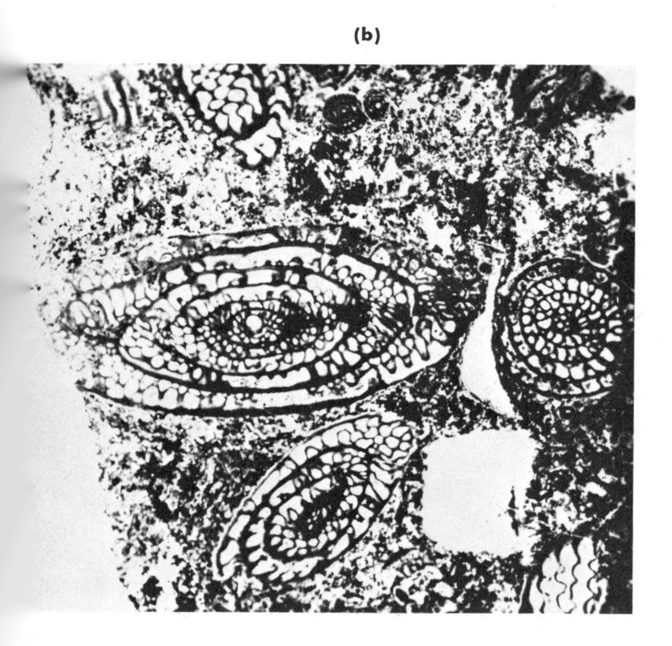

varied fish forms that represent a distinct evolutionary advance over the contemporary Paleozoic invertebrates.

As the fish diversified, competition began. Among those eliminated were the jawless *ostracoderms* (Figure 18-30), a few of which had lingered on from the Silurian. These creatures were the first large group of vertebrates to become extinct. Forms that enjoyed a temporary success and then disappeared before the close of the period were the *antiarchs* (spiny sharks) and the *arthrodires* (joint-necked fish). One of the arthrodires, *Dinichthys,* from the Late Devonian rocks of Ohio, was 9 m long, probably the largest animal of the time.

FIGURE 18-30 *An ostracoderm* (Cephalaspis lyelli) *from the Old Red Sandstone of Scotland. The crescent-shaped head shield is characteristic. Length about 17 cm. (NERC copyright. Photograph reproduced by permission of the Director, Institute of Geological Sciences, London.)*

Success of the fish

During the late Devonian two classes of fish began to establish their superiority: (1) the *bony fish,* or *Osteichthyes,* and (2) the *cartilage fish,* or *Chondrichthyes.* The bony fish, the most numerous, varied, and successful of the aquatic vertebrates, include the vast majority of living fish and many extinct forms. These fish are adapted for life in both salt and fresh water and have entered virtually all aquatic environments on earth. Their story is mainly a chapter in later geologic time.

FIGURE 18-31 *Well-preserved fossil of the Late Devonian crossopterygian fish* Eusthenopteron foordii. *This fish is near the line leading to the first land-living amphibians. (Courtesy of Erik Jarvik.)*

Included in the *Osteichthyes* class is a less numerous and rather unimpressive group, the *Choanichthyes,* or *nostril-bearing fish,* which can take in air through their nostrils as well as their mouths. This group includes the Devonian ancestors of the modern *lungfish,* and are characterized by peculiar teeth and the ability to survive dry periods by burrowing in the moist beds of streams or lakes. The *Choanichthyes* also include the *crossopterygians,* or *lobe-finned* fish (Figure 18-31), which are in the direct line of evolution from fish to land-living vertebrates and are now represented by the solitary "living fossil," *Latimeria.* In these fish the fin is a solid, muscular structure with a central axis of bones.

Out of the water

Late in the Devonian Period certain lobe-finned fishes established themselves on land, a step made possible by a great many structural and functional modifications. The lobe-fin, with its axis of internal bones, had to be converted into a walking limb; the lung was adapted to breathe air just as it now does in the lungfish; and the circulatory and excretory systems were modified along much the same lines as in present-day tadpoles. The earliest known amphibians, called *ichthyostegides,* of which the genus *Ichthyostega* (Figure 18-32) is a typical example, are found in late Devonian

FIGURE 18-32 *Drawing and reconstruction of the earliest known amphibian (Ichthyostega). The specimen, discovered in the uppermost Upper Devonian of East Greenland, is about 1 m long. (Courtesy of Erik Jarvik.)*

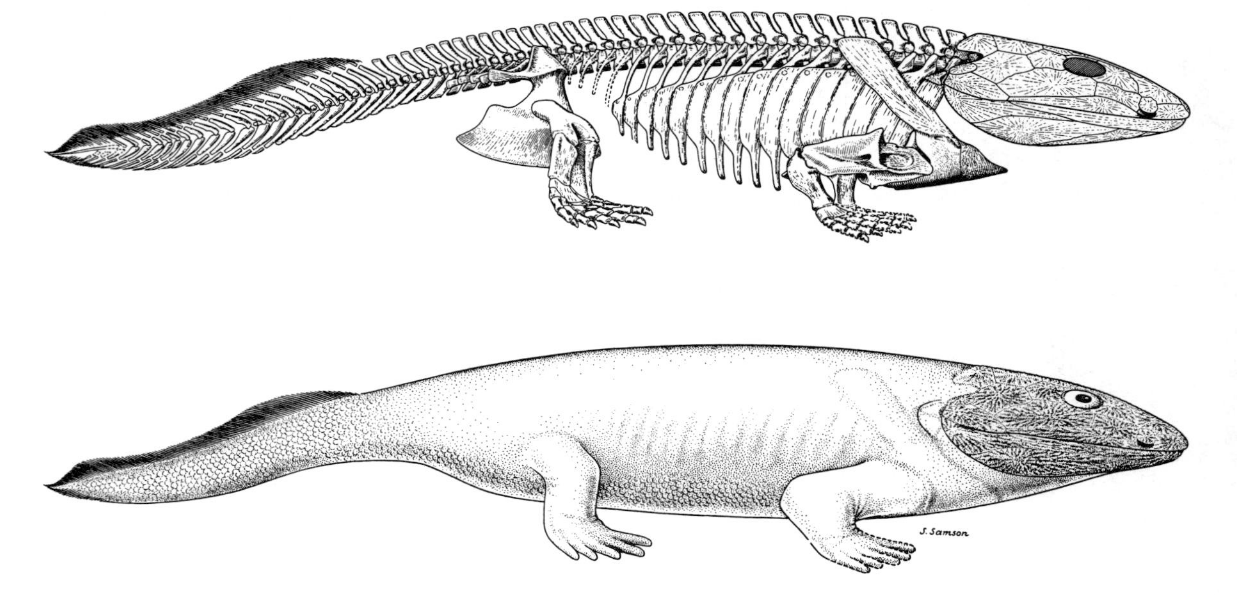

rocks of Greenland. *Ichthyostega* was about 60 cm long and possessed a strange mixture of newly acquired characteristics and older traits inherited from its fish ancestors. The legs were weak and the tail was long and had a fringe-like fin. Its fishlike skull can be compared bone for bone with the skulls of its fish relatives that never left the water. There were many pointed conical teeth with a peculiar structure that occurs also in the crossopterygian ancestors. This tooth structure, known as *labyrinthodont*, is characterized by deep infolding of the enamel. All in all, the fossil record of the skeletal changes that accompanied the important transition from water to land is quite complete.

Recent discoveries of footprints of four-legged animals have been made in late Devonian rocks of Victoria, southeastern Australia. These tracks have proportions appropriate to *Ichthyostega* and show that late Devonian land animals were probably widespread although rare.

We cannot be certain just why the crossopterygians invaded the land. Certainly competition in the water had become very severe, and the land offered food and protection to any animals that could live out of water. It seems logical that they made their first approach to land across the moist and sandy beaches where food, cast up by the tides, was generally available. Here, too, the opportunity or necessity of digging in the moist sand favored the development of stout limbs with toes. Apparently, then, the enticement of live food in the forest in the form of various arthropods eventually led the first pioneers to abandon the shores and to forsake the water completely.

Success of the amphibians

During the Mississippian and Pennsylvanian Periods, *amphibians* became the dominant land animals. The Mississippian, with extensive shallow seas, was not so favorable to their expansion as was the succeeding Pennsylvanian, when great swamps, mild climates, and luxuriant forests of coal-forming plants were widespread. Such surroundings were obviously ideally suited to the amphibian way of life, and most of our knowledge about Pennsylvanian vertebrates comes from specimens found in coal-bearing rocks. From these remains we can reconstruct many types of amphibians—some with lizardlike shapes, some resembling snakes, and some much larger, with contours resembling crocodiles and salamanders. We find not only the full-grown animals, but also many remains of immature or tadpole stages. And associated with the bones of the amphibians are the fossilized remains of the creatures they preyed on: spiders, centipedes, scorpions, and a variety of winged insects. In the world of the swamp forests, the amphibians ruled supreme.

Difficulties of land life

The gradual drying and cooling that characterized the Permian Period caused a corresponding decline in the number and variety of amphibians. Those that survived were mainly confined to water courses in dry regions, and we find their remains chiefly in red-bed types of sediments. These later amphibians were adapted to a large extent for life out of water, but they were awkward and inefficient by comparison with later land animals. As we look at their skeletons, we are impressed by the clumsy sprawling legs, heavy tails, and immense flattened heads (Figure 18-33). Their skeletons tell us that one of the amphibians' chief foes was the force of gravity. Merely to raise their bodies off the ground must have required intense physical effort, and really rapid locomotion was out of the question.

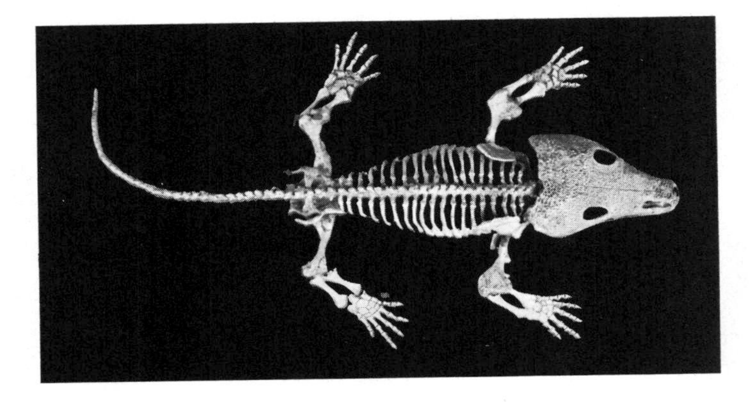

FIGURE 18-33 *(Left) Skeleton of Trematops, a sprawling, flat-headed amphibian from the Early Permian of Texas. (Field Museum of Natural History, Chicago.)*

FIGURE 18-34 *(Below) Restoration of Paracyclotosaurus, a 3-m-long amphibian from the Permian of New South Wales, Australia. (NERC copyright. Photograph reproduced by permission of the Director, Institute of Geological Sciences, London.)*

Many adaptive changes in the structure of the backbone strengthened it for its new role of supporting the weight of the amphibian's body. The changes involved mainly the development of interlocking devices and processes for muscle attachment. These variations in the structure of the backbone serve as a basis for classifying the amphibians. The ancestral amphibians had rather deep, fishlike heads, about as wide as they were high. With the passage of time the skull gradually flattened until the entire head was many times wider than it was thick. This curious adaptation eventually enabled the animal to open its mouth by raising its upper jaws and skull while its lower jaw lay flat on the surface on which the animal rested. The usual method of chewing by lowering and raising the lower jaw is obviously inefficient if an animal must raise its entire body in order to permit the lower jaw to operate. Since virtually all groups of late amphibians developed broad, flat skulls, we suppose that this same principle of conservation of energy was acting on them all.

The term *stegocephalian* ("roof-headed") is applied to the larger flattened amphibians of the Pennsylvanian, Permian, and Triassic Periods. A reconstruction is given in Figure 18-34. Although this type of amphibian lingered on into the Triassic, it was already declining at the close of the Paleozoic. Frogs and toads are products of a later time, and paleontologists have found no precise ancestors among Paleozoic fossils.

The coming of the reptiles

All evidence indicates that *reptiles* developed from amphibians some time during the Mississippian or Pennsylvanian. A few isolated bones that some geologists assign to reptiles have been found in Mississippian deposits, but the first complete and authentic reptile skeletons were taken from rocks of Pennsylvanian age. The Permian red beds of Texas have yielded a number of well-preserved skeletons of a small animal that stands

FIGURE 18-35 Reconstructed scene depicting the ancestral reptile Petrolacosaurus. Specimen is from the rocks of Late Pennsylvanian age near Garnett, Kansas. Fossils of the other plants and animals depicted in the illustration were found in the same deposit. (From F. E. Peabody, courtesy of the University of Kansas.)

almost midway between reptiles and amphibians. This creature, called *Seymouria,* although not the exact reptile ancestor, does indicate what the actual ancestor may have been like.

Another important link in the history of the reptiles was discovered in late Pennsylvanian rocks near Garnett, Kansas. Here, the complete 60 cm skeleton of a small, lizardlike reptile was found in natural association with a large number of other plant and animal fossils. This animal, called *Petrolacosaurus,* shows clearly the improved adaptations to land life that the reptiles had by this time achieved (Figure 18-35).

We must say something here about the one unique and basic feature that separates reptiles from amphibians—that is, the ability to produce an egg that can be laid and hatched out of water. Just how did the reptiles develop this capability? On this point the fossil record is silent, but it is significant that a fossil egg, the oldest known, has been discovered in Permian rocks of Texas. The ability to produce an egg complete with food, water, and oxygen supply encased in a protective shell entails many basic adjustments that involve intricate chemical and mechanical processes.

Long before the Permian Period ended, the amphibians had sharply declined and reptiles were spreading over the continental areas. Remains of late Paleozoic vertebrates are scattered over all the great continents, but the most complete and continuous record is found in the Karroo Basin in South Africa. The most characteristic animals of the Karroo belong to the *Therapsida,* or *mammallike reptiles.* This group, which includes animals of various sizes, developed adaptations similar to those of later mammals. Some had well-differentiated teeth, the limbs were "pulled in" and not sprawling, there were fewer ribs, the tail was smaller, and the whole skeleton was light and well constructed. These were all reptiles, however; the transition to mammals took place during the Triassic, and it took many millions of years before the reptiles yielded their supremacy to the mammals. Tracks as well as bones of mammal-like reptiles have been found (Figure 18-36).

FIGURE 18-36 Tracks of Permian pelycosaurs (mammallike reptiles) preserved in sandstone, Tunisia. (Photo courtesy D. W. Boyd.)

End of the Paleozoic—time of the great dying

The closing stages of the Paleozoic were marked by unusually rigorous conditions for living things. A great ice age affected much of the Southern Hemisphere, and the chemistry of the ocean was significantly altered by wholesale removal of salt and phosphate. Great deserts with little or no life were widespread, and thousands of square kilometers on several continents were submerged in lava. The combined effect of these events was to create a crisis in the history of life—one so great that paleontologists refer to it as the *Time of the Great Dying.*

At or near the close of the Permian Period, nearly half of the known families of animals disappeared. This included 75 percent of the amphibian families and 80 percent of reptile families. The marine realm was also seriously affected. One survey shows that the number of families of creatures living in continental shelves and capable of leaving fossils was reduced from about 165 in the Middle Devonian to about 75 at the close of the Permian. Groups that were entirely eliminated were the trilobites, eurypterids, fusulinids, and productid brachiopods. Many bryozoan, crinoid, and coral families also disappered. In pinpointing the exact time of transition from Paleozoic to Mesozoic time, evidence from all groups is considered, with great weight being placed on certain species of conodonts and ammonites (Figure 18-37).

It is difficult to understand what factors caused the late Paleozoic exterminations, but the underlying cause was almost certainly radical changes in the earth's physical features that adversely influenced the climates of the lands and the temperature and chemistry of the oceans.

Economic geology of the late Paleozoic

Mineral deposits of all types were formed during the late Paleozoic periods. Especially significant are the vast stores of fossil fuels: coal, oil, and gas. Coal of Mississippian age is mined in Russia; and Pennsylvanian coal supports heavy industry in the eastern United States, Great Britain, and western Europe. Permian beds are the chief source of coal in India, Australia, South Africa, and China, and are also important in Russia. Figure 18-38 shows the distribution of coal deposits in the United States.

Oil and gas are plentiful in late Paleozoic rocks, especially in North America. The Devonian is a major source of oil in Canada; and important Mississippian, Pennsylvanian, and Permian pools are found in interior parts of the United States. It has been estimated that reserves of Paleozoic oil account for about 10 percent of the world's total, which indicates just how extensive the later Mesozoic and Cenozoic contributions have been.

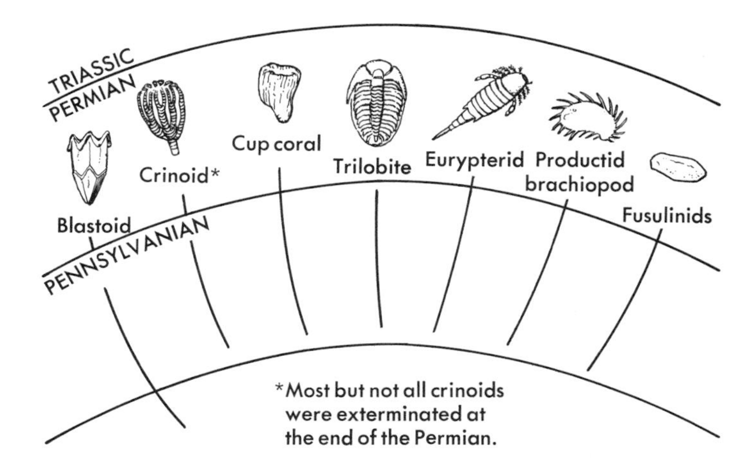

FIGURE 18-37 Animals exterminated at or near the close of the Paleozoic Era.

The extensive mountain-building activities and granitic intrusions of the late Paleozoic, brought many important metalliferous deposits into existence throughout the affected areas. Tin, lead, zinc, copper, and silver deposits were formed in western Europe, and the iron, copper, nickel, and chromite ores of the Urals date from this period. Large deposits of sedimentary copper occur in lower Carboniferous rocks of south-central U.S.S.R. Gold, lead, silver, and other metals were deposited in central Asia, along with tin and tungsten ores in the Malaya-Burma region, and tin, zinc, antimony, and mercury deposits in China. Important ores of various metals also were deposited in Australia and New Zealand.

By contrast, there are apparently very few metal deposits of this age in the Western Hemisphere.

The late Paleozoic was a time of extensive salt deposition, especially in the northern continents. Ordinary table salt, halite, is found in the Pennsylvanian of Colorado and Utah,

FIGURE 18-38 Coal-bearing regions of the contiguous 48 states. Geologic age of the deposits is shown by the legend. Alaska also has large reserves; Hawaii has none. (U.S. Bureau of Mines.)

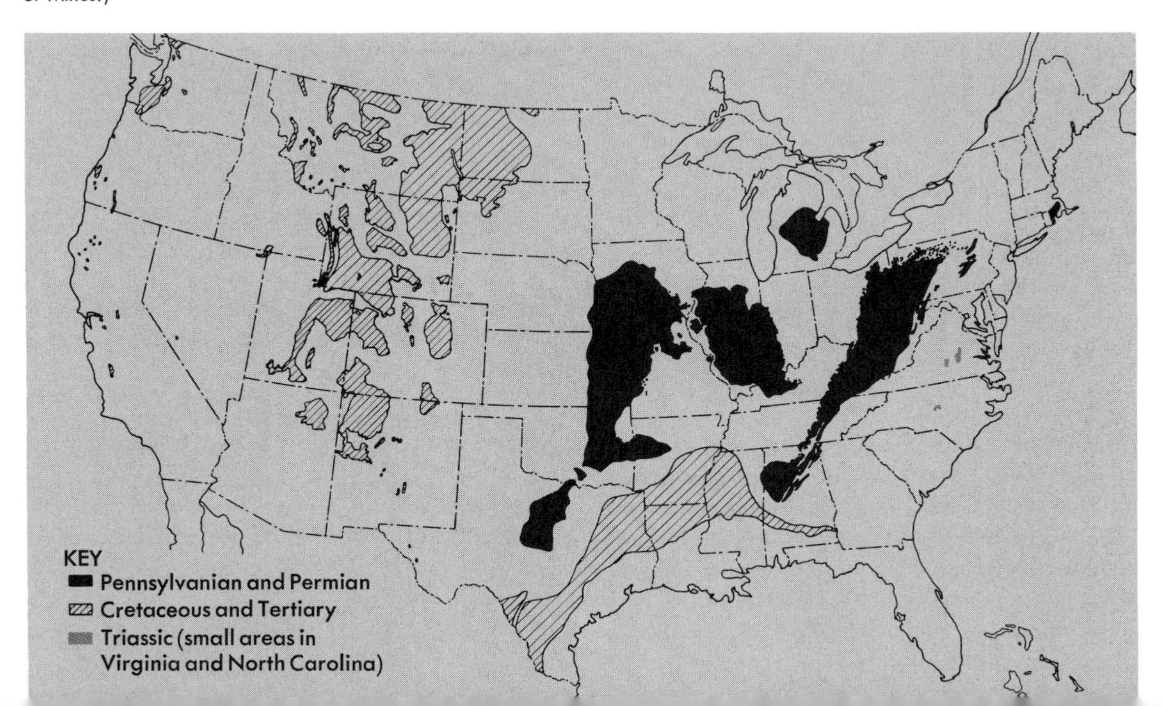

and in the Permian of Texas, New Mexico, and Kansas. It is accompanied by the more valuable potash salts in these western localities. Salts of various kinds are also mined from Permian rocks in Germany and Russia.

Important reserves of phosphate, which were laid down in late Permian (*Phosphoria Formation*) rocks in Idaho and adjacent states, now support a flourishing mineral-fertilizer industry.

SUMMARY

Sediments of late Paleozoic age occur on all the continents, including Greenland and Antarctica. The northern continents were flooded by extensive shallow seas in the Devonian and Mississippian periods. In the Pennsylvanian and Permian there is a high proportion of land-laid sediments, including extensive coal beds. The record of marine sedimentation for the Southern Hemisphere is fragmentary except in Australia. Signs of glacial action in the Pennsylvanian and Permian have been found in Africa, Antarctica, Australia, South America, and India. This feature, taken in connection with similarities in fossil forms and igneous activity, is considered by many geologists to indicate that all southern continents were then joined in one super landmass called *Gondwanaland*.

The late Paleozoic was a time of evolution and expansion of land life. Forests appeared in the Devonian and became dominant features of the landscape in succeeding periods. Amphibians appeared in the Devonian and were common in the coal swamps. Reptiles appeared in the late Mississippian or early Pennsylvanian and had become adapted to life in the continental interiors by the end of the Permian.

The closing stages of the Paleozoic were marked by varied topographic conditions, climatic extremes, and exterminations, especially of marine animals such as trilobites, fusulines, and productid brachiopods. Vegetation and land animals were less affected.

SUGGESTED READINGS

DAPPLES, EDWARD C., and M. E. HOPKINS, *Environments of Coal Deposition,* Special Paper 114. Geological Society of America, 1969.

McKEE, EDWIN D., RAYMOND C. GUTSCHINK, *History of the Redwall Limestone of Northern Arizona,* Memoir 114. Geological Society of America, 1969.

MUIR-WOOD, HELEN, and G. ARTHUR COOPER, *Morphology, Classification and Life Habits of the Productoidea (Brachiopoda),* Memoir 81. Geological Society of America, 1960.

Treatise on Invertebrate Paleontology, 24 parts. Lawrence, Kansas, and Boulder, Colorado: The University of Kansas Press and the Geological Society of America © 1957 by the University of Kansas Press and the Geological Society of America. A series by various authors on the major invertebrate animals. Important volumes for late Paleozoic fossils are those on brachiopods, corals, protozoans, echinoderms, and protozoans.

The Mesozoic Era

The Mesozoic or "time of middle life" lasted for an estimated 160 million years and is divided into three periods: Triassic, 30–35 million years; Jurassic, 55–60 million years; and Cretaceous, about 70 million years.

At the beginning of the Triassic Period the northern lands were apparently still closely united in the great Laurasian supercontinent, as were the southern lands that made up Gondwanaland (Figure 19-1). Between the two major landmasses lay the Tethys seaway or ocean, but the exact width and configuration of this waterway are difficult to reconstruct. The present Atlantic Ocean quite certainly did not exist, and the Pacific was correspondingly much wider than now. The Uralian seaway, which separated Europe and Asia, was closed to form the Ural Mountains in late Triassic time. In contrast to the Atlantic, which closed and reopened again, the

19

Reconstructed landscape of the Cretaceous period near the close of the Age of Reptiles. Tyrannosaurus rex *in center; duck-billed dinosaur,* Anatosaurus, *right rear;* Triceratops, *left rear;* Ankylosaurus, *the armored dinosaur, front right. (Yale Peabody Museum.)*

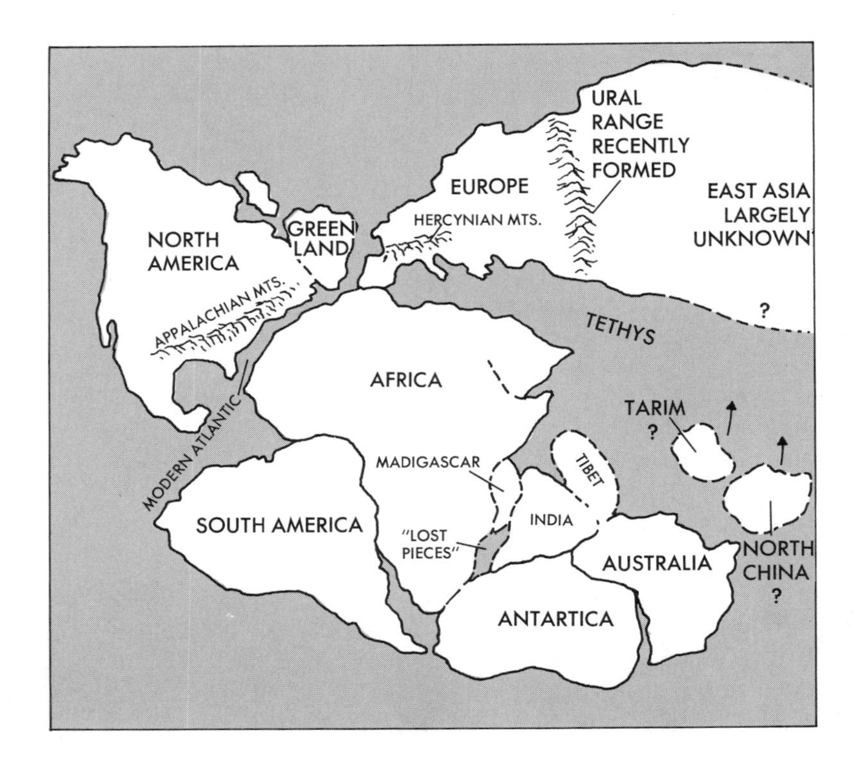

FIGURE 19-1 *Distribution of major landmasses in the Middle Triassic.*

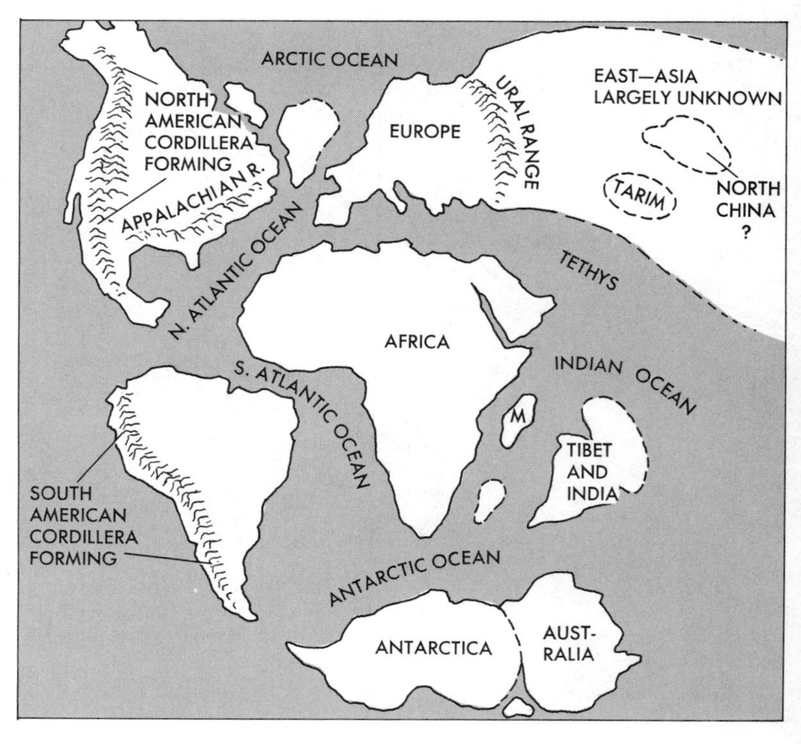

FIGURE 19-2 *Distribution of major landmasses near close of the Cretaceous period.*

Uralian Geosyncline remained closed. All major landmasses were generally well above sea level, and remnants of late Paleozoic highlands still were being eroded to supply sediment to interior basins and alluvial plains. The Triassic was a time of only minor mountain building, but volcanic action was locally strong.

Late in the Triassic the supercontinents began to split apart. Rifts appeared in the area joining Africa with North America. Some of this splintered material was carried away with Africa, while some remained with North America. North Africa apparently separated from North America at about the Jurassic–Cretaceous transition some 135 million years ago. South America and Antarctica were in contact until late Cretaceous–early Cenozoic, and both were attached to Africa throughout the Triassic. India probably broke away from a position between Africa and Australia during the late Jurassic–Early Cretaceous. Least understood are the relations of other possible fragments that once bordered Australia and India. There is increasing evidence that Tibet and Tiram may have been part of the southern continental mass that broke away before India did and drifted northward early in the Mesozoic. Figure 19-2 shows the reconstruction of major features near the close of the Cretaceous Period.

Conditions for life during the Mesozoic ranged between very unfavorable and favorable. This era began with one great crisis in the biologic world and ended with another. Practically all forms of life had suffered a decline at the end of the Paleozoic. Recovery was slow, and it was not until the Jurassic that conditions were more or less stable again. Life was prolific during the lengthy Cretaceous Period, but near its end conditions again worsened and other exterminations took place.

The history of Mesozoic life reflects climatic influences that were in turn determined by geographic configurations of the lands and seas. During the early Mesozoic the equator ran diagonally across North America at the latitude of the southern United States or northern Mexico. With the passage of time it shifted gradually southward until in the late Cretaceous it was nearly at its present position. The North Pole moved out of a locality corresponding to northeastern Asia into the Arctic Ocean, and the South Pole approached Antarctica. Shallow seas spread gradually inward during the Jurassic and by the Cretaceous had inundated large areas in one of the greatest floods of all time. This condition was favorable for the poleward dispersal of tropical heat by shallow marine currents so that warm-loving organisms could exist much farther north than now—dinosaurs in Canada for example.

Note that the great displacements of New World continents were mainly parallel to climatic zones rather than transverse to them, so that the movements did not affect climate as much as might be expected.

Eastern North America

The Triassic and Jurassic are poorly represented in the eastern half of North America. Nonmarine red beds, mostly of Triassic age, are found in a dozen downfaulted basins from Nova Scotia to South Carolina. These red formations, known collectively as the *Newark Series,* consist of shale, sandstone, and conglomerate, with great sheets of dark lava and intrusions of dolerite. The famous Palisades of the Hudson is one of these (Figure 19-3). Freshwater fish and dinosaur tracks are common fossils in the Newark Series. These rocks are fairly well known because they crop out near centers of population. Many similar basins are known to exist under deep cover along the Atlantic shelf.

FIGURE 19-3 The Palisades of the Hudson River near Alpine, N. J. The vertical cliff is columnar diabase of Triassic age. (U.S. Geological Survey.)

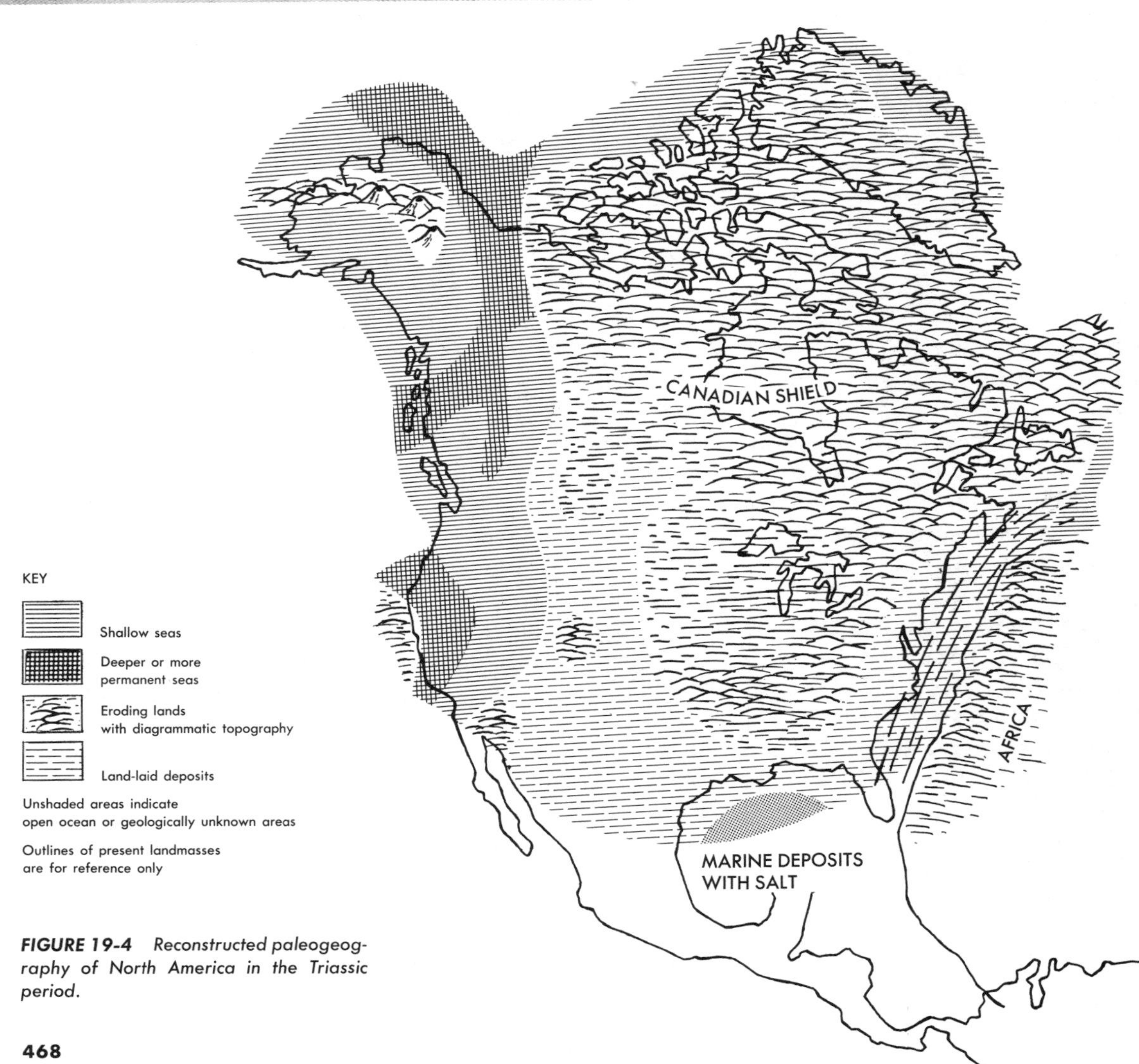

FIGURE 19-4 Reconstructed paleogeography of North America in the Triassic period.

Except for an unknown amount of the upper Newark Series, no Jurassic rocks are present at the surface east of the Mississippi River. However, thick Jurassic formations, some with oil pools, have been found by drilling in states marginal to the Gulf of Mexico. The Cretaceous is well represented as a curving marginal belt along the Atlantic and Gulf Coasts; this belt lies under the Atlantic from Long Island northward, but is 959 km wide in the Mississippi valley. The Cretaceous is mixed marine and nonmarine, and the marine facies die out landward in such a way as to prove that there was an Atlantic Ocean at this time and that the east coast of North America had achieved essentially its present form.

The Appalachian Mountains, the Canadian Shield, and other central areas were undergoing erosion throughout the Mesozoic. At the beginning of the Triassic, the Appalachians must have been an imposing mountain range; by the Cretaceous Period, nothing remained

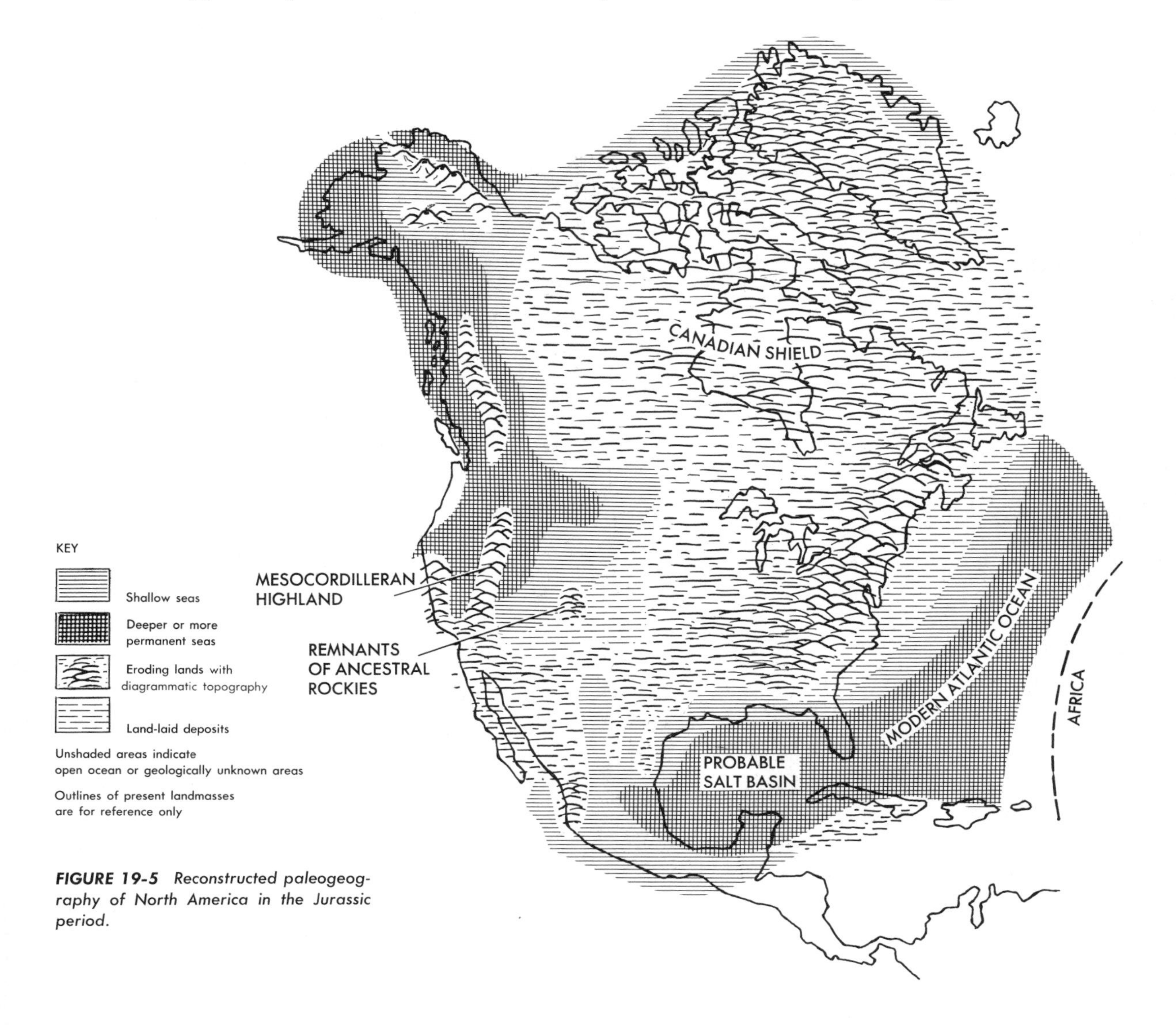

FIGURE 19-5 *Reconstructed paleogeography of North America in the Jurassic period.*

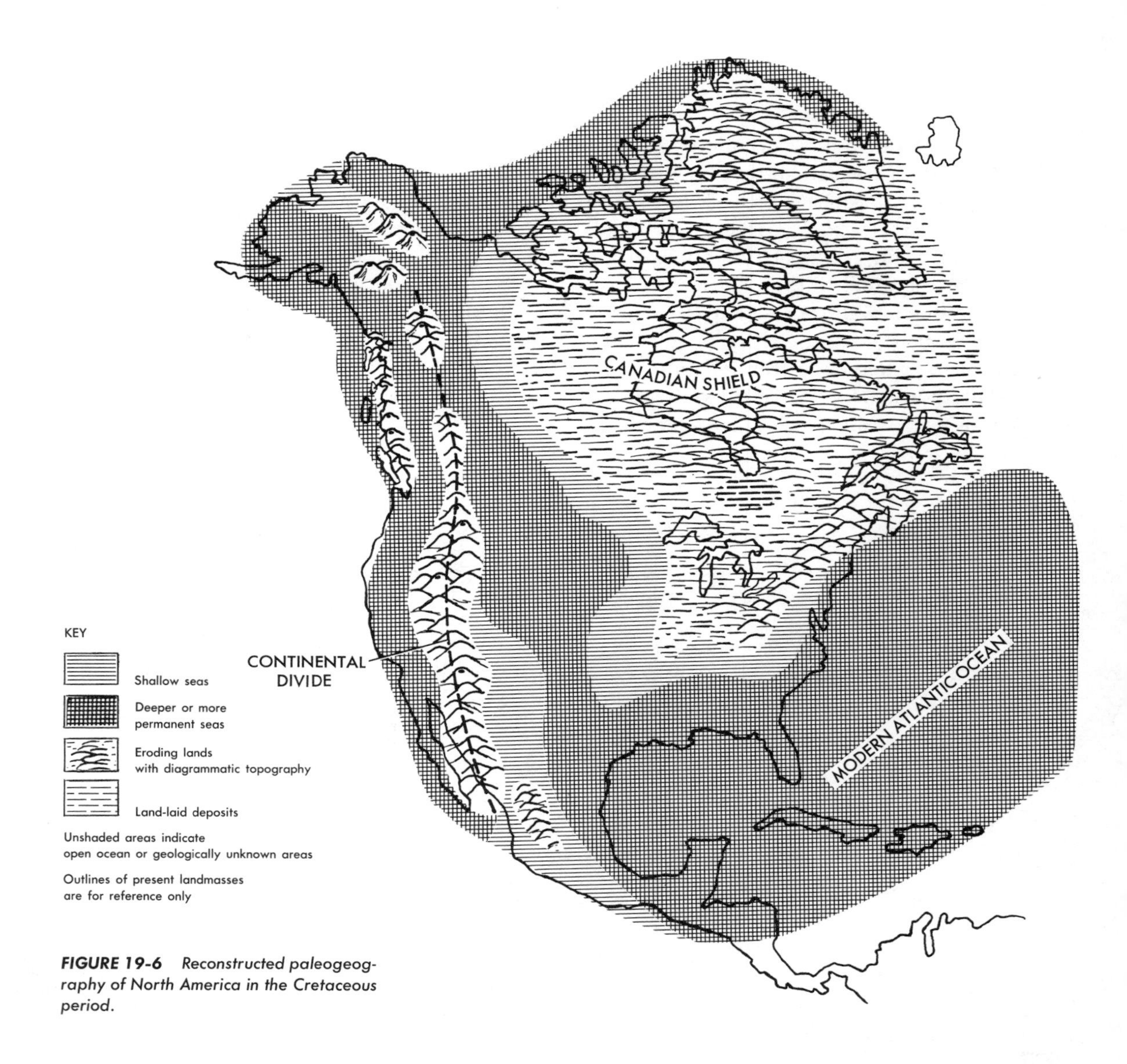

FIGURE 19-6 *Reconstructed paleogeography of North America in the Cretaceous period.*

but the planed-off, deeply eroded stumps. Most of the material removed in the Triassic and Jurassic was carried beyond the present margins of the Atlantic and Gulf coasts and may in fact lie in deep trenches under the continental shelf. Material eroded in the Cretaceous is still near at hand in existing exposed formations. The paleogeography of North America during the Triassic, Jurassic and Cretaceous Periods is shown by the maps of Figures 19-4, 19-5, and 19-6.

Western and northern North America

During the Mesozoic, North America bodily moved westward to interact strongly with the Pacific lithospheric plates. During this process a number of shallow marine invasions washed over the western portions of the continent to leave interfingering marine and nonmarine deposits. Also during the Mesozoic a strip of land up to 600 km wide was formed by sedimentation and igneous action and was welded to the western margin of the preexisting continental mass.

In early Triassic time the seas spread inward from the Pacific to central Utah and Wyoming, while broad low flood plains extended beyond into the central part of the continent. This was the last time marine waters crossed the western United States from the Pacific because a narrow uplift, the *Mesocordilleran Highland,* arose from the geosyncline in the Middle Triassic to block out the western seas.

For a while the interior protected area east of the highland was the site of accumulation of desert sands on a large scale (Figure 19-7). Eventually with continued downsinking, the seas entered a strait to the north in Canada or Alaska and again flowed into the west-central parts of the continent. During the Jurassic Period this interior seaway was occupied by at least two successive shallow embayments up to 1,200 km wide. These failed, however, to connect with other embayments that spread inward from the ancestral Gulf of Mexico and from California. One of the last events of the Jurassic was the accumulation of the Morrison Formation, which covers about 195,000 km^2 in the Rocky Mountains and western plains. The Morrison is composed of material worn from the Mesocordilleran Highland and spread eastward by sluggish, meandering rivers. Buried in the Morrison are skeletons of the gigantic dinosaurs and other late Jurassic life forms. This formation also supplies much of the uranium ores of the United States.

During the Cretaceous, shallow seas spread inward on all sides in the greatest flood

FIGURE 19-7 *Erosion forms carved from continental formations of Late Jurassic age near Lake Powell, south-central Utah. (U.S. Bureau of Reclamation.)*

of the Mesozoic. Although the Mesocordilleran Highland was not submerged, it was surrounded like an island, as seas from the north connected with seas from the south across the central United States and Canada. In this wide Cretaceous seaway many hundreds of cubic kilometers of silty mud accumulated. On the shallow shelving deltas and flood plains along the western borders of the seaway, successive swamps with abundant vegetation gave rise to immense coal beds.

The Cretaceous came to a close as the central sea was crowded out by its own deposits (Figure 19-8). Swamps and lakes marked the final phases, and the stage was set for the great Rocky Mountain Revolution, which affected much of the Pacific borderlands.

FIGURE 19-8 *Cliffs of sandstone of Late Cretaceous age, Mesa Verde National Park, Colorado. In cliffs such as these are dozens of dwellings of the ancient inhabitants of the region. (Photo courtesy National Park Service.)*

Northern Canada has a Mesozoic history much like that of the western United States. The Triassic exists in minor volume with nommarine red beds and marine shale. The Jurassic has beds of both marine and nonmarine origin, with little or no limestone. The Cretaceous is also well represented by varied sediments. Notable is the presence of basalt flows ranging up to the Cretaceous in age.

Greenland has richly fossiliferous Triassic rocks, a thick and varied Jurassic section, and Cretaceous beds showing alternate retreats and transgressions of the sea. A very large basalt field of late Cretaceous or early Tertiary age is found in east-central Greenland.

Eurasia

Rocks of Mesozoic age are widespread at the surface and under thin cover in both Europe and Asia. For the most part these were produced in shallow seas that spread widely across continental interiors. Locally, in the Tethys Geosyncline, great thicknesses of deep-water sediments accumulated. Remnants of late Paleozoic uplifts, referred to as the *Hercynian Mountains* or *Paleozoic Alps* of Europe, remained as low barriers between Tethys to the south and the low-lying continental shelves and platforms to the north. The north-trending Uralian Geosyncline, important seaway of the Paleozoic, was filled with sediment and ceased to exist in the early Mesozoic. A new geosyncline in Northeastern Siberia, as yet not well studied, began to form in the Triassic and received great thicknesses of Mesozoic rocks.

Historically the Triassic, Jurassic, and Cretaceous were first studied and named in western Europe and Great Britain. The Triassic is here represented mainly by continental sediments mostly of red color. In Germany, Triassic rocks have been a source of copper ores for hundreds of years. The Jurassic of Europe is varied and extremely fossiliferous. The Cretaceous, as the name implies, is predominantly chalk (light-colored, poorly compacted limestone); typically exposed in the famous White Cliffs of Dover (Figure 19-9). Very important oil and gas fields are being developed in Mesozoic rocks, chiefly Jurassic, beneath the North Sea.

The rock record of the Mesozoic in eastern Europe, Russia, and Siberia is not greatly different than that of western Europe. The Triassic was a time of volcanic activity (basaltic lava flows) and erosion across much of Siberia, and only in the northeast portion is the record

FIGURE 19-9 *Cliffs of chalk of Cretaceous age on the coast of Selwick's Bay, Yorkshire, England. Glacial deposits overlie the chalk beds. (NERC copyright. Photograph reproduced by permission of the Director, Institute of Geological Sciences, London.)*

fairly complete. It may be that the North Pole was located over water in what is now northeastern Siberia, but the organic record there is very sparse. With the Jurassic the geologic record of northern Eurasia improves. The rocks are chiefly continental in origin with important coal beds. During the mid-Cretaceous there was great flooding of shallow interior areas. Much of Russia and western Siberia were under water and received chalky and marly sediments. Later in the period as the seas withdrew, continental beds rich in coal and with abundant dinosaur remains accumulated. By this time the North Pole had reached essentially its present position in the Arctic Ocean.

The shallow-marine or continental environments that generally prevailed across northern Asia gave way southward to deeper water conditions in the Mediterranean Geosyncline. Here great thickness of Triassic, Jurassic, and Cretaceous rocks were deposited.

FIGURE 19-10 Dinosaur footprints preserved in rocks of Late Jurassic age near Pica oasis, northern Chile. The sediments have been steeply tilted by forces that created the Andes Mountains. (Courtesy Carlos Galli.)

The Cretaceous inundation was one of the most extensive ever to affect southern Asia. Mesozoic rocks make up entire mountain ranges in central Asia, and cover extensive regions of China and India as well.

Over most of northern Asia between the Ural Mountains and the east coast, there is no lengthy interruption between Permian and Triassic rocks, all of which are of marine origin. There was, however, a break between the early and middle Triassic, after which time most of northern Asia was free of marine invasions and received a great variety of land-laid deposits including windblown sand.

Southern continents

Mesozoic rocks are widespread in the Southern Hemisphere, and each of the great continents has representative formations of Triassic, Jurassic, and Cretaceous age (Figure 19-10). These rocks reveal a dramatic record of the breaking apart of the Gondwanaland supercontinent and the resultant climatic and biologic changes. At the beginning of the Mesozoic, southern lands were united, and the

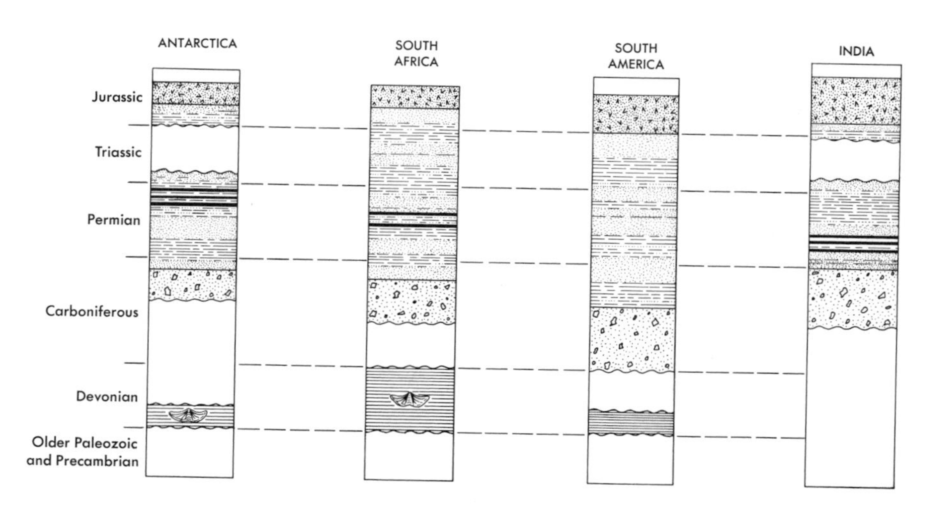

***FIGURE 19-11** Late Paleozoic and early Mesozoic rock units of the chief land masses of the Gondwanaland assemblage. Shows the broad similarities which prove parallel development and probably close geographic proximity. Sections from 3,000 to 7,000 meters thick are represented.*

rock and fossil record is rather uniform throughout (Figure 19-11). In the mid-Mesozoic, fracturing was under way, the modern Atlantic was beginning to form, and the plants and animals could no longer pass freely from one land to another. Finally, as the era drew to a close, practically all connections were broken, and the continents had reached almost their present position. The wholesale reorganization of major geographic features that took place in the Mesozoic was greater than that of any similar period of geologic time.

Completion of the Karroo series

The most complete record of Triassic continental conditions is found in the Karroo System, the beginnings of which were in the late Paleozoic and have been mentioned in the preceding chapter. Although the name comes from Karroo Basin in South Africa, what is now seen in this region is only a part of a much greater body of sediment. Basic to an understanding of the Karroo and related rocks is the fact that fragments now existing in such widely separated areas as South Africa, Antarctica, South America, India, and Madagascar were once in near proximity.

In the Karroo itself the passage from Permian to Triassic is transitional and is marked by the evolutionary appearance of certain distinctive reptiles. The Triassic part of the Karroo is sandstone and shale, mainly red in color. Here are found the earliest known mammals and dinosaurs. Fish, crustaceans, and plants also occur. The closing stages of Karroo deposition in the Jurassic were marked by strong volcanic action. Both lava (basalt) and intrusive rocks (dolerite sills) were produced on a large scale. These are well displayed in the spectacular Drakensburg Mountains of Natal (Figure 19-12).

The African sequence of Triassic and Jurassic rock types was repeated in Antarctica and South America: red, bone-bearing, land-laid deposits at the base with increasing signs of volcanic activity higher up. Flows of dark lava covered nearly 2 million square kilometers of southern Brazil. Antarctica consists of an ancient shield area called *East Antarctica* and a western folded-mountain area called *West Antarctica,* which is in many ways a continuation of the Andes Range. West Antarctica has red Triassic rocks with fossils like those of Africa overlain and intruded by masses of dark volcanic rock dated as Jurassic.

Australia is similar to and yet different from other Gondwanaland pieces. There are great continental deposits of Triassic and Jurassic age, but little or no signs of volcanic activity. The Jurassic was a time of extensive swamps and bodies of fresh water, one of which, Lake Wallon, covered at least 700,000 km^2 and received deposits over 1,500 m thick. India, a final major piece of the Gondwanaland puzzle, is now attached to Asia, far from its original position between Africa and Australia. It betrays its Gondwanaland connections in having continental deposits of Triassic and Jurassic age capped with thick outpourings of lava. These are known as the *Deccan,* and are dated as chiefly Cretaceous.

The fossil record of Mesozoic time in the Southern Hemisphere is most informative. A small aquatic reptile, *Mesosauras,* is known only from Early Permian rocks of opposing

FIGURE 19-12 *Edge of the Drakensburg Plateau, Natal, South Africa where over 1,500 m of lava of Jurassic age is exposed. These outpourings terminated the Karroo Series. (Courtesy South African Information Service.)*

regions of South Africa and South America. To reach its present resting places it would have had to navigate hundreds of miles of salt water, a feat so unlikely that paleontologists generally favor the view that its homeland was once in one piece. A major discovery bearing on the Gondwanaland puzzle was the finding of a distinctive assemblage of reptiles and amphibians in the Triassic of Antarctica. A key member of this fauna is the reptile *Lystrosaurus,* which had already been discovered in South Africa, India, and China (Figure 19-13). Dating of this and associated fossils reinforces the idea of a great breakup in the Jurassic followed by rapid drifting of landmasses to present localities.

It is the discovery of *Lystrosauras* in China that is most intriguing. In all fairness, how can the existence of *Lystrosaurus* in Antarctica be taken as proof that this land was attached to South Africa and its presence in China be more or less ignored? The solution to this problem may be in some very radical rethinking about continental drift, which would derive Tarim, north China, central Siberia, and Tibet from a position in Gondwanaland adjacent to northern Australia and northern India. These Asiatic blocks could well have been detached from the Gondwanaland mass and been moved rapidly northward to become welded to and incorporated into the great Asian complex.

The history of India is better understood. From an original position between Africa and Antarctica, it became detached in the middle Mesozoic, and after a rapid transit across the Tethys seaway, collided with and partly underthrust southern Asia. The collision was a complex event with the sea floor and much marginal sediment passing under southern Asia first, and the continental masses making forceable contact later. The suture between Asia and India is not at the base of the mountains but higher and deeper in the ranges.

FIGURE 19-13 *Restoration of the Triassic mammallike reptile* Lystrosaurus. *Remains of this animal have been found in China, South Africa, and, most recently, in Antarctica. Its presence in these widely separated localities is taken as convincing evidence of the breakup of Gondwanaland in post-Triassic time. (Courtesy of Edwin H. Colbert.)*

The ocean basins

Ocean water as such has a history almost as long as that of the earth itself, but the basins in which the waters reside are of later age, chiefly products of Mesozoic sea-floor spreading and continental drift. The Pacific may be regarded as the mother of oceans because other oceans have formed at its expense and are filled with water drawn from the Pacific Basin. Note that previous chapters have had little to report regarding the ocean basins or the rocks making up their floors. It is almost literally true that the history of oceans began only about 180 million years ago. Not that there were no oceans or basins before this time, it is simply that the rocks that made up the more ancient oceanic floors have been literally destroyed by the process of sea-floor spreading.

To be specific, we can now say with fair certainty that most of the rocks underlying the Pacific, Atlantic, Indian, Antarctic, and Arctic Oceans are of Cretaceous and Tertiary age.

Relatively small areas of Jurassic age are found in the Pacific southeast of the Philippines and north of New Guinea, and in the Atlantic just east of the Bahamas. If there is anything older, it is insignificant.

Although the early Mesozoic Pacific sea floor has disappeared, it is possible to construct some Cretaceous configurations (Figure 19-14). Most of the Pacific basin is now made up of the great Pacific Plate that is moving northwestward to destruction under the Ryuko, Japanese, Kurile, and Aleutian island arc systems. It has been preceded by two or three other large plates. A small remnant of one of these, the *Farallen Plate,* is still to be seen off the Oregon-Washington coast. It was overridden by North America in the late Mesozoic and Tertiary. Indirect evidence in the form of magnetic anomalies is interpreted to mean that another large plate, the *Kula Plate,* has been subducted under the Kurile and Japanese arcs mostly during the Cretaceous Period. Note that not only has the Kula Plate been consumed but also the ridge between it and the Pacific Plate. It is supposed that once a ridge has been subducted, all spreading ceases, and the ridge is engulfed along with its products.

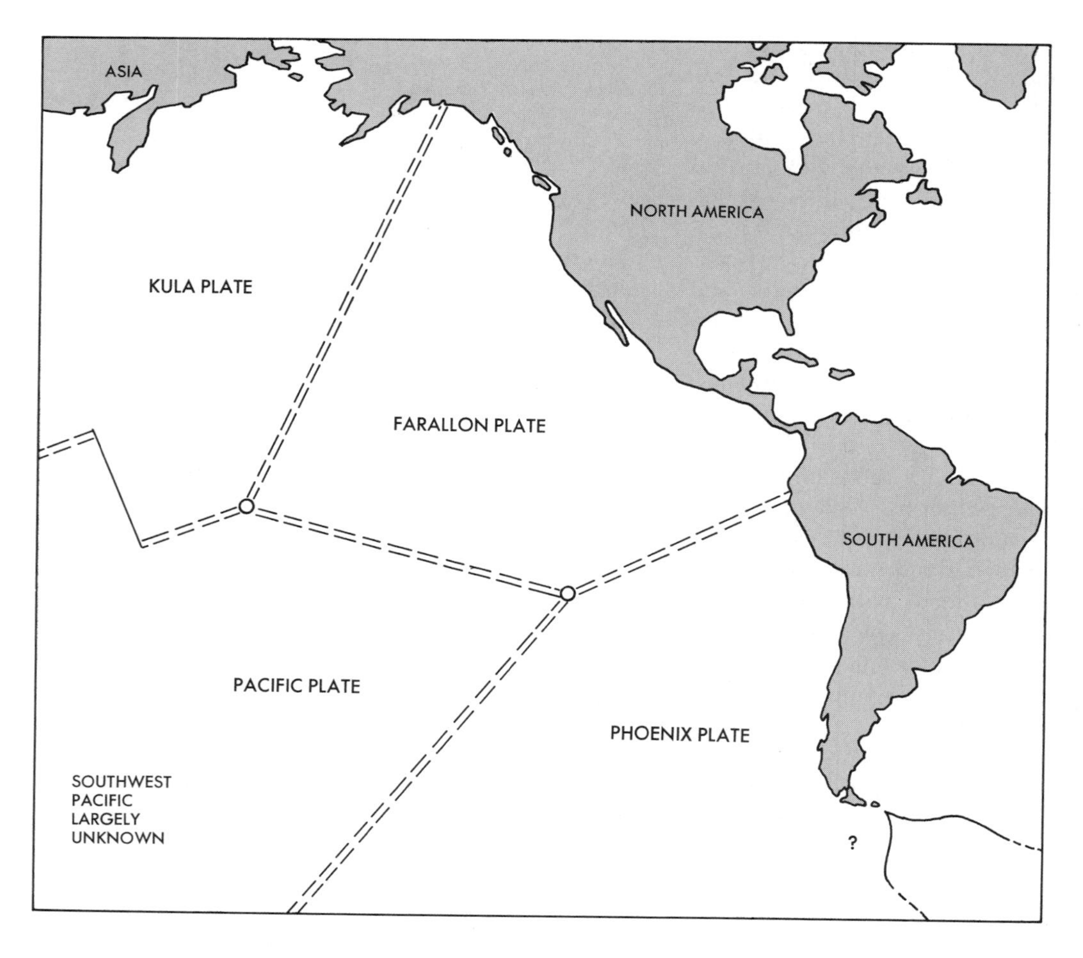

FIGURE 19-14 *Reconstructed configuration of plates in the Pacific Basin in middle Cretaceous time about 110 million years ago. The Kula, Farallon, and Phoenix Plates have been almost entirely destroyed; see Figure 20.7. The exact configuration and courses of the plate boundaries cannot be known, and the outline of the surrounding continents, shown as they are today, were then different in detail.*

FIGURE 19-15 *Artist Chesley Bonestell's conception of the topography of the Pacific Ocean bottom. If the water should be removed, the flat-topped guyots, or seamounts, would be revealed. The steep front and fringing deposits of a coral reef are in the foreground. (Courtesy of E. L. Hamilton.)*

Continuation of the Great Salt Age

Conditions favorable to deposition of massive salt deposits, as described in the previous chapter (p. 446), continued through much of the Mesozoic. These deposits seem to be confined chiefly to newly opened arms of the Atlantic Ocean. In Late Triassic time, massive deposition of salt took place in and marginal to, the Gulf of Mexico. The salt from the buried deposits has risen to the surface in salt domes (Figure 19-16). At roughly the same time, salt was deposited in downfaulted belts of northwestern Africa and offshore Newfoundland—two areas then in near proximity. Later during late early Cretaceous time, when the splitting of Africa and South America was well under way, salt was deposited in narrow troughs near the mouth of the Niger delta and offshore Brazil, also then closely joined.

The East and West Indies

For practical purposes, the history of the island-dotted West and East Indies begins in the late Mesozoic. The oldest known rocks of the East Indies are middle Paleozoic and are seen only in limited outcrops on larger islands.

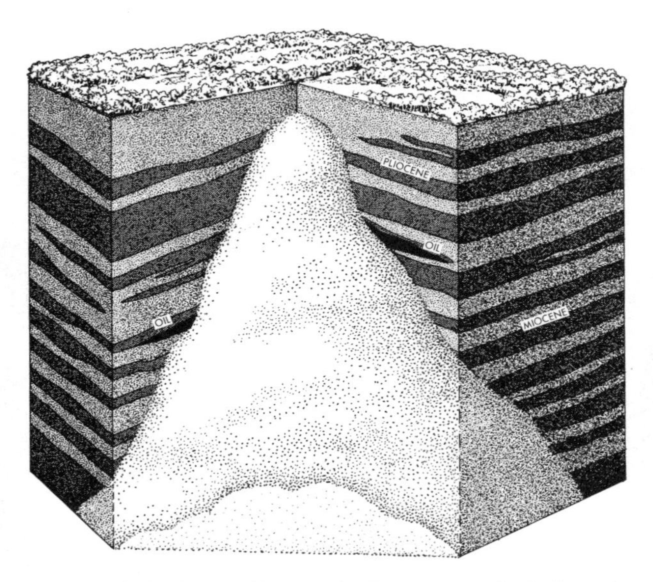

FIGURE 19-16 Diagram showing structure of salt domes in the Gulf Coast region of the United States. Salt rising from Early Mesozoic deposits at depths of several miles has penetrated into younger sediments to create favorable sites for oil and gas accumulation.

Important Permian deposits are found on Timor. About one-fourth of the surface outcrops on the islands are Cretaceous; the rest are Cenozoic. Sedimentary and volcanic rocks are intermingled and structure is complex. Geologic activity in the form of volcanism and earthquakes is intense. This is to be expected, since the Java-Sumatra chain marks a subduction zone where destruction of the northeasterly moving Indian-Australian Plate is going on, and there are also actively moving plates to the north and northeast.

The West Indies with its island arcs and great faults presents many problems. Only a few questionable outcrops are older than Jurassic. The common sedimentary rock is limestone of Cretaceous age, which is found in great quantities on various islands and in neighboring borderlands such as the Florida Platform, the Bahama Islands, and the Yucatan Peninsula.

Life of the Mesozoic

Plants

The Mesozoic was a time of transition and change in the plant kingdom. Later Cretaceous vegetation bears little resemblance to that of the early Triassic, truly revolutionary changes having occurred during the interval between these periods. By comparison, the changes between late Cretaceous and present-day vegetation are relatively minor. Emphasizing the distinct nature of the early Mesozoic floras is the fact that plants such as *seed ferns, lycopods,* and *horsetails*—all common in Carboniferous coal forests—are represented in the Mesozoic only by small, insignificant species.

Ferns prospered during the Mesozoic, and their delicate remains are particularly common in continental rocks of Triassic and Jurassic age. Insofar as our understanding of past climates is concerned, it seems significant that the types of fern that lived during the Mesozoic in temperate latitudes are now restricted mainly to tropical countries. As a rule, these Mesozoic ferns are more closely related to species now living than they are to Paleozoic types.

Another characteristic group of Mesozoic plants are the *cycads* (Figure 19-17). The term *cycad* includes not only true cycads, or *Cycadales,* but also certain cycadlike plants known as the *Cycadeoidales,* or *Bennettitales,* which were extinct by late Cretaceous time and are thus referred to as the "fossil cycads." The *Cycadeoidales* have a fossil record ranging from Permian into late Cretaceous. They are gymnosperms that bore true flowers with spore-bearing stamens and a seed-bearing female organ quite similar to flowers of more recent plants. The true cycads, which are now represented by 9 genera confined chiefly to tropical regions, did not become evident as fossils until the Triassic. They are so charac-

FIGURE 19-17 A well-preserved frond of a cycad tree from the Mesozoic rocks of Great Britain. Length about 46 cm. (NERC copyright. Photograph reproduced by permission of the Director, Institute of Geological Sciences, London.)

FIGURE 19-18 A portion of a cycad trunk (Cycadeoidea marylica), about 33 cm high, found in Early Cretaceous sediments of Anne Arundel County, Maryland. (Smithsonian Institution.)

teristic of the Mesozoic that this era is sometimes called the *Age of Cycads.* The group was very cosmopolitan, ranging across all the great continents. Cycads have reproductive parts similar to living conifers, and there were small, medium, and large species. The typical cycad fossil is a silicified, keg-shaped trunk with a pattern of diamond-shaped indentations marking the position of the leaf bases (Figure 19-18). The foliage, which is seldom found attached to a trunk, is typically palmlike. Another gymnosperm group, which is now represented by one species, *Ginkgo biloba* (the Maidenhair tree), was represented by at least 16 genera in the Mesozoic. Its remains have been found in Mesozoic rocks of North America, Europe, Central America, Malaya, South Africa, and Australia. *Ginkgo* may be the oldest living genus of seed plant and is not known to exist in a wild state today.

The early Mesozoic marked a high point in the history of primitive cone-bearing trees. Dominating many landscapes were the descendants of primitive conifers, such as *Walchia* and *Lebachia,* which were minor inhabitants of the Carboniferous forests. The cool and dry climate of the Permian and Triassic favored the spread of conifers throughout the world. Especially prominent during the early Mesozoic were members of the family *Araucariaceae,* which now survives only in the Southern Hemisphere. Petrified trunks of

FIGURE 19-19 *Broken sections of fossil logs in the Chinle Formation of Late Triassic age in Petrified Forest National Park, Arizona. (U.S. National Park Service.)*

araucarians account for most of the fossil remains in the famous Petrified Forest in Arizona (Figure 19-19). Here lie trunks as much as 1.5 m in diameter and over 30 m long. The first true pines appeared in the late Jurassic and spread widely during the Cretaceous, as their fossilized remains—cones, needles, and wood—indicate. The *sequoias* appeared during the Jurassic and had become very common by the Cretaceous. Conifers declined in the Cretaceous but have held a rather stable position since then.

A distinct and relatively sudden change in the vegetation of the earth took place during the middle Cretaceous. This has been called the *Angiosperm Revolution.* Earlier, during the Triassic and Jurassic, the most abundant plants had been ferns, various types of cone-bearing plants, and the cycads and their relatives. After the Middle Cretaceous, the chief plants were members of the great group known as the *flowering plants,* or *angiosperms.* There are about 175,000 species of these flowering plants now living, and at least 30,000 fossil species have been found. Angiosperms have flourished in all climates and include trees, shrubs, and herbs.

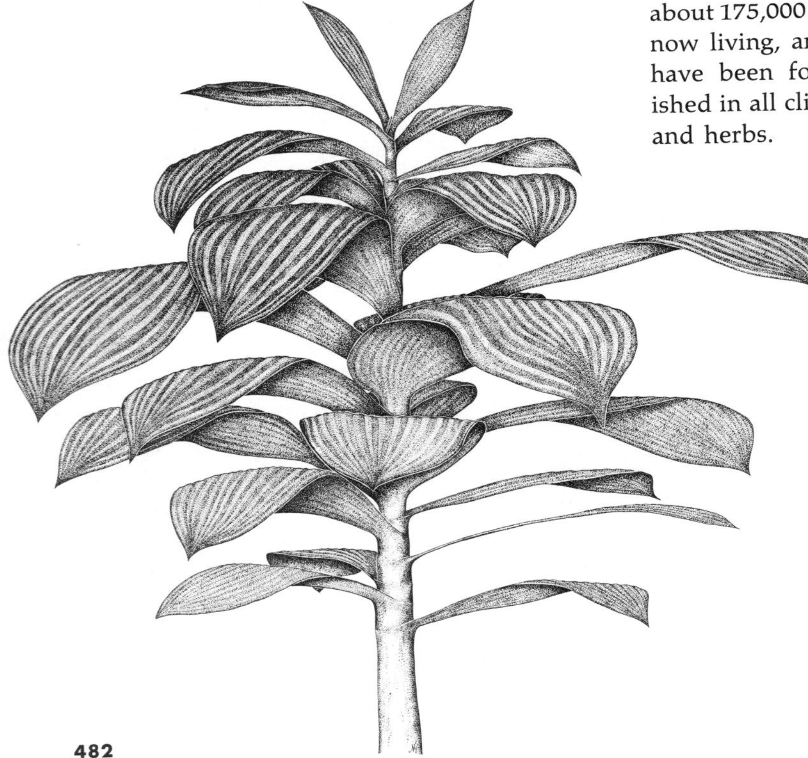

FIGURE 19-20 *Artist's reconstruction of the unusual plant* Sanmiguelia, *which may be an early angiosperm. (Reconstruction and photo by W. D. Tidwell.)*

The origin of the angiosperms is an unsolved problem. They apparently developed first in a restricted unknown area and became almost worldwide by the close of the Cretaceous Period. A peculiar palmlike plant, *Sanmiguelia,* found in Upper Triassic rocks of southwestern Colorado, may be the oldest known angiosperm (Figure 19-20). Leaves of the magnolia, sassafras, fig, and willow are common in Upper Cretaceous rocks. Forests of angiosperms contributed to the formation of Cretaceous coal, and pollen grains of this group are useful in understanding climatic conditions and also in correlating sedimentary formations.

Angiosperms are important not only because they have become the dominant form of plant life on earth but also because of the influence they exert on animal life. With these flowering plants came a variety of grains, nuts, and fruits, which furnish a food supply that ensures the survival of the plant embryo. The relatively concentrated nature of the food supply and the small size of most seeds, nuts, and fruits make them ideal fare for small animals and, unfortunately, for insects as well. It is unlikely that the large Mesozoic reptiles ate food of this sort; they probably fed on large quantities of coarse herbage and on succulent water vegetation.

No less important than the origin and spread of the angiosperms on land was the expansion of lower forms of plant life in marine and fresh water. The small aquatic algae known as *diatoms* are not positively known before the Jurassic, but they quickly increased in importance and are now a major food source for animal life in the sea. Diatoms secrete skeletons of silica that accumulate to form oceanic ooze or the sediment *diatomite.*

The chalk beds characteristic of the Jurassic, Cretaceous, and younger ages are composed largely of extremely small fossils called *coccoliths,* which are secreted by very simple floating algae (Figure 19-21).

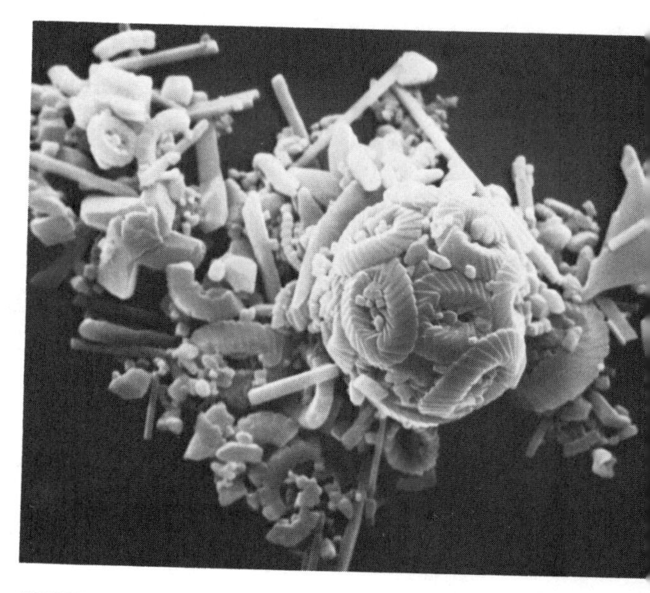

FIGURE 19-21 *Chalk-making fossils of Late Cretaceous age obtained by drilling on the Shatsky Rise, northwest Pacific. The oval object in the center is a coccosphere consisting of a hollow sphere of button-shaped structures called coccoliths. The structure, which falls apart rather easily, is secreted by a single-celled planktonic marine alga. Fragments of other calcareous fossils are also seen in the photograph. The intact coccosphere is 10 μ in diameter. (Photo courtesy Peter Roth.)*

Nonmarine invertebrates

Invertebrates living on dry land and in fresh water were fairly well established by the Mesozoic. Streams and lakes were well stocked with clams, snails, arthropods such as *ostracods* and *branchiopods,* and small, inconspicuous sponges. All these fed chiefly on water plants with which their remains are usually associated. Air-breathing snails are also found.

The history of insects is particularly important, especially as it relates to the flowering plants. The plants furnish food for the insects in the form of nectar and pollen, in return for which insects serve the vital function of fertilization by transferring pollen from flower to flower. This association undoubtedly evolved gradually over the ages so that many of the complicated behavior patterns of insects and the myriad forms of flowering plants are the outcome of a long process of adaptation. The insects chiefly involved in this insect-plant dependency are bees, butterflies, beetles, wasps, and flies.

The ascendancy of insects and flowering plants during the Mesozoic was greatly aided by the cooperative relationship that developed between them. The benefits that accrued to birds, mammals, and ultimately to man are incidental to the process. Most fruits, vegetables, ornamental plants, and industrial plants are insect-pollinated.

The insects, besides being man's chief competitor for food and living space on land, are important in other ways. Although we cannot trace by fossil remains the intricate stages of insect adaptations, we should examine some of their more significant achievements. Insects have developed a number of social orders that, although composed of individuals, function as a unit. This type of social arrangement combines the benefits of individual action and of group solidarity. Nowhere in nature has this ideal been approached so closely as in the case of the bees or ants. Ants not only harvest the natural vegetation, but some species also plant and cultivate special crops. The earliest known ant is a well-preserved individual in amber from Upper Cretaceous rocks of New Jersey (Figure 19-22). The adaptations of the bees to plants and to one another are becoming well known. Apparently their small size is the only thing that has prevented insects from becoming unquestioned rulers of the earth. Their evolution has been rapid, but once a group reaches a certain level, it tends to stagnate. Many fossil insects found preserved in amber 50 million years old are practically indistinguishable from living forms. They apparently completed all essential phases of their evolution at least that long ago.

FIGURE 19-22 *The earliest known fossil ant, discovered in Late Cretaceous deposits of New Jersey. The specimen is preserved in amber. Besides being the first known ant and the first clear indication of the social level of insect organization, it is also an almost perfect link between wasps and ants. Note stinger at end of abdomen. (Courtesy of Frank M. Carpenter.)*

Marine invertebrates

Triassic marine invertebrates show the deadly effects of the late Paleozoic exterminations. Apparently the early Triassic seas continued to be unusually cool or otherwise inhospitable to such animals as corals, sponges, bryozoans, and protozoans. Other groups that were greatly reduced and relegated permanently to minor positions were crinoids and brachiopods, which are generally rare in Mesozoic rocks. Two groups that eventually recovered a great deal of lost ground were the protozoans and bryozoans, but the types that became common during the late Mesozoic and Cenozoic were quite different from their Paleozoic predecessors.

The mollusks weathered the late Paleozoic without great losses and became the most important shelled invertebrates of the Mesozoic seas. This group includes the coiled, single-chambered gastropods, the bivalved (two-shelled) pelecypods, typified by the clams and oysters (Figure 19-23), and the coiled many-chambered cephalopods. All mollusks gradually became more varied and numerous as the Mesozoic progressed. Freshwater clams and gastropods were plentiful, and even air-breathing snails were locally abundant. Of special importance are the large conical or twisted shells of a group of pelecypods known as *rudistids,* some of which reached a length of over 1 m and a diameter of more than 0.5 m. Fossil rudistids are distributed in Cretaceous rocks in a world-circling belt along Tethys and in warmer regions of the Western Hemisphere.

The most spectacular, varied, and successful marine invertebrates of the Mesozoic were the cephalopods (Figure 19-24). Because they were able to swim and crawl and because the shells of some species could float after the animals had died, their remains have been scattered and preserved in many localities. Cephalopod shells make striking fossils, and they are usually beautifully preserved. The coiled varieties range from a few centimeters across to more than 1 m in diameter. Also included in the cephalopod class are the *belemnites,* squidlike forms whose internal skeletons have the form of a solid, stony, cylindrical object shaped like a cigar. Fossilized belemnite skeletons are particularly abundant, resist erosion, and have been known for centuries as "thunderbolts."

The ammonities (cephalopods whose shells have complex suture patterns) are unequaled as guide fossils for marine Triassic, Jurassic and Cretaceous rocks. Individual species ranged widely, lived short lives, were independent of ocean-bottom conditions, and, of

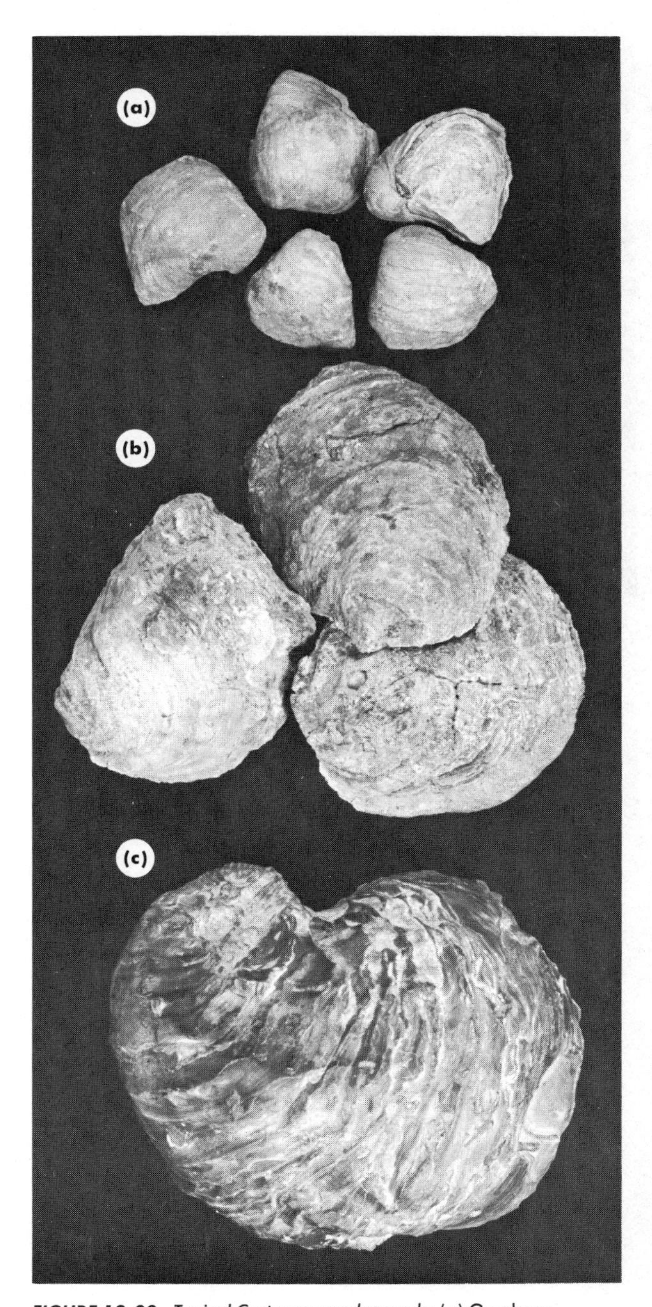

FIGURE 19-23 *Typical Cretaceous pelecypods:* (a) Gryphaea, (b) Ostrea (oyster), *and* (c) Exogyra. *In these forms the shells are of irregular shape and one valve is larger than the other. The* Exogyra *specimen measures about 10 cm across.*

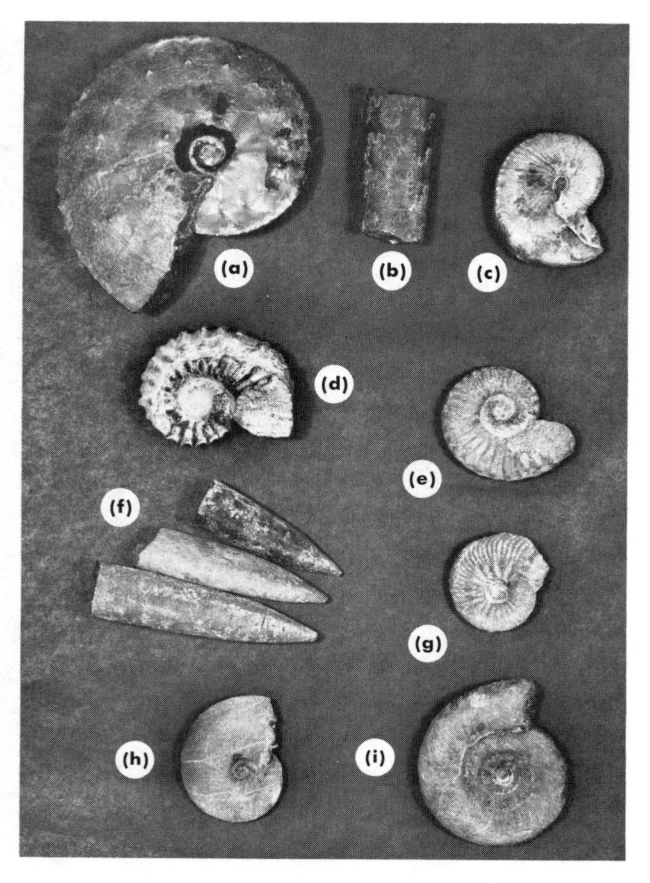

FIGURE 19-24 *Some typical Mesozoic cephalopods: Late Cretaceous:* (*a*) Plancenticeras, (*b*) Baculites, *and* (*c*) Scaphites; *Early Cretaceous:* (*d*) Cheloniceras *and* (*e*) Dufrenoya; *Jurassic:* (*f*) Pachyteuthis *and* (*g*) Cardioceras; *Triassic:* (*h*) Meekoceras *and* (*i*) Columbites. *The* Placenticeras *specimen measures about 8 cm across.*

course, are usually well preserved. Ammonites successfully weathered the critical Permian–Triassic transition, and descendants of Paleozoic forms blossomed in Mesozoic seas over the entire world. Although the group suffered another serious setback at the close of the Triassic and again near the close of the Jurassic, the surviving members recovered rapidly and each time succeeded in repopulating the seas. Only at the close of the Cretaceous did the race suffer final extinction.

During the Mesozoic the arthropod phylum evolved and spread in the oceans, as well as on land and in the air. The living space vacated by the trilobites was taken over at least in part by a variety of crustaceans. The arthropod group that includes the familiar shrimps, crabs, crayfish, and lobsters appeared in the Triassic (Figure 19-25). Over 8,000 living species of this very successful group have been identified. The first true crabs especially adapted for life along the beaches and shallow offshore areas appeared during the Jurassic, along with the earliest barnacles.

Mesozoic echinoderms were represented by a variety of forms, including starfish (Figure 19-26), sea urchins, crinoids, and sea cucumbers. It is significant that the majority of these creatures were capable of some degree of locomotion.

FIGURE 19-25 *Lobsterlike crustacean,* Eryon, *from the Upper Jurassic Solnhofen limestone, Bavaria. About 15 cm long. (Courtesy Field Museum of Natural History.)*

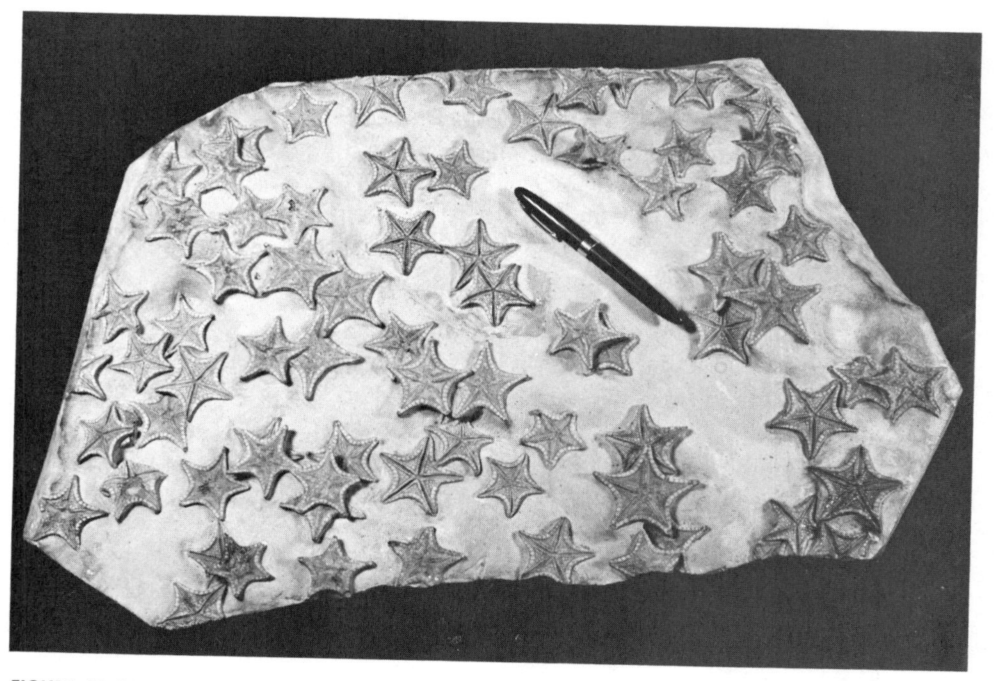

FIGURE 19-26 (Above) Numerous specimens of the starfish Austinaster from the Cretaceous Austin Formation of Austin, Texas. (Texas Memorial Museum.)

FIGURE 19-27 (Below) A thick-scaled fish (Lepidotus) from the Early Jurassic of Germany. The scales are relatively thick, bony, and inflexible. Specimen is about 70 cm long. (NERC copyright. Photograph reproduced by permission of the Director, Institute of Geological Sciences, London.)

Protozoans were at a very low ebb in the Triassic but increased in the Jurassic and reached spectacular numbers in the Cretaceous. A recent survey shows 9 families in the Triassic, 24 in the Jurassic, and 36 in the Cretaceous. An important development of the Cretaceous was the appearance and spread of floating forms such as *Globigerina,* a lime-secreting form living in open oceans.

Vertebrates

The Mesozoic was a time of abundant land life, popularly known as the *Age of Reptiles* because of the dominant position this group achieved on land, in the air, and in the seas. Although mammals and birds appeared during the Mesozoic, they were not particularly abundant. Fish continued to evolve and became better adapted to their special ways of life, but the amphibians declined to a level of comparative insignificance and have never recovered.

Fish and amphibians

Fossil fish are common in many Mesozoic formations of both freshwater and marine origin. The most important advances were made by the bony fish, especially the *actinopterygians* (*ray-finned fish*). During the early stages of the Mesozoic most fish belonged to the *Chondrostei,* a group characterized by heavy diamond-shaped scales, asymmetrical tails, somewhat scaly fins, and considerable cartilage in their skeletons (Figure 19-28). Toward the middle of the era these fish were gradually replaced by the *Holostei,* with lighter scales, more symmetrical tails, fins supported by flexible rays, and relatively more bone in their skeletons. This group was in turn displaced during the Cretaceous by the *Teleostei,*

FIGURE 19-28 Graveyard of Buettneria, a giant amphibian, found in Triassic beds near Santa Fe, New Mexico. *The massed remains of many specimens probably record the drying up of a pond or lake. The skulls are about 60 cm long.* (Smithsonian Institution.)

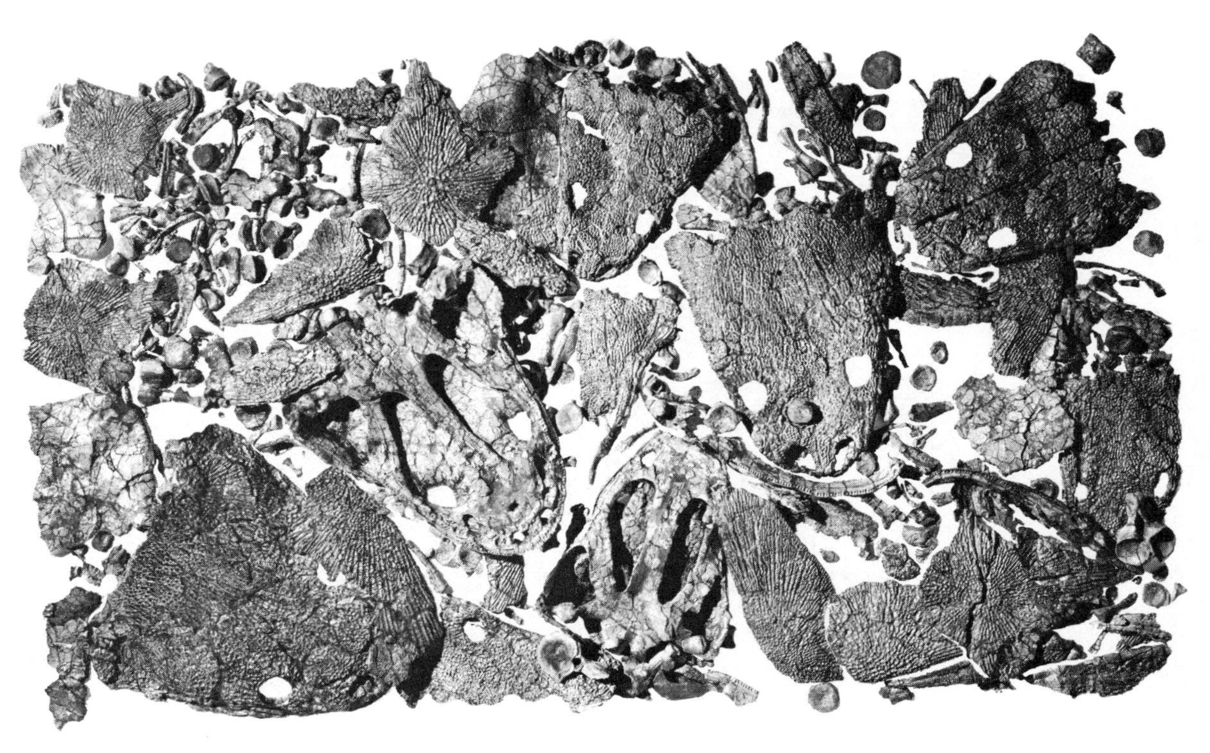

flexible-scaled fish with completely bony skeletons and powerful, well-formed fins and tails; this group dominates present-day seas and rivers. The sharks were poorly represented during the early Mesozoic but gradually expanded during the Jurassic and Cretaceous to regain ground lost during the late Paleozoic. The crossopterygians and lungfish held a subordinate position in out-of-the-way surroundings, just as they now do.

As mentioned above, the Mesozoic was a period of decline for the amphibians. All the large, flatheaded stegocephalians became extinct early in the Triassic, and the group thereafter is represented by the familiar toads, frogs, salamanders, and the legless, wormlike apoda. Figure 19-28 shows a graveyard of stegocephalians from New Mexico. The first true frogs are found in Triassic rocks of Madagascar, but ancestral forms from which they could have evolved have been discovered in the preceding period. The first known salamanders come from the Late Jurassic Morrison Formation.

Reptiles

Although the amphibians as a group never achieved complete sway over the land, their descendants, the reptiles, became rulers of the earth. About a dozen orders of reptiles were present during the late Triassic and early Jurassic, but only four orders exist today. Their basic adaptations permitted them to expand into hitherto unoccupied territory and even to reenter the ocean in competition with the fish.

The few paragraphs that we can devote to the story of the Mesozoic reptiles obviously cannot do justice to the dramatic spread and remarkable achievements of this group. The reptile family tree is illustrated by Figure 19-29. It may help our perspective to remember that the reptiles were dominant for 125 million years, compared with the approximately 65 million years of the Cenozoic (*Age of Mammals*), and the few thousand years of man's reign.

The Triassic Period was marked by the decline and disappearance of certain of the older orders of reptiles and by the appearance of new and vigorous stocks. Among the Permian holdovers that disappeared in the Triassic were the *protosaurs* and certain mammal-like reptiles, the *therapsids* (Figure 19-30). The exact nature of these animals need not concern us here, but the newly evolved groups merit more detailed attention. The first turtles appeared during the Triassic and quickly achieved their typical form. The *plesiosaurs* and *ichthyosaurs*, adapted to life in the ocean, also appeared at this time, along with the first crocodiles. Most notable from the viewpoint of evolutionary progress were the *thecodonts*, lightly built, relatively small reptiles adapted to land life. As a group, they are confined to the Triassic, but they gave rise to a remarkable array of descendants that dominated the Jurassic and Cretaceous Periods. Crocodiles, dinosaurs, flying reptiles (*pterosaurs*, or *pterodactyls*), and birds all sprang from thecondont ancestors. One short-lived line of thecondonts, the *phytosaurs*, resembled the crocodile in size, shape, and mode of life, so nearly in fact that they are called the "ecological ancestors" of the crocodiles, meaning that they occupied a similar place in nature but are distantly related.

Dinosaurs

The term *dinosaur* is popularly applied to members of two orders of reptiles—*Saurischia* and *Ornithischia*—both of which achieved gigantic size and occupied similar environments. Dinosaurs of the saurischian order first appeared during the Middle Triassic as certain slender two-legged forms. A typical genus is *Coelophysis*, a carnivore found in Late Triassic

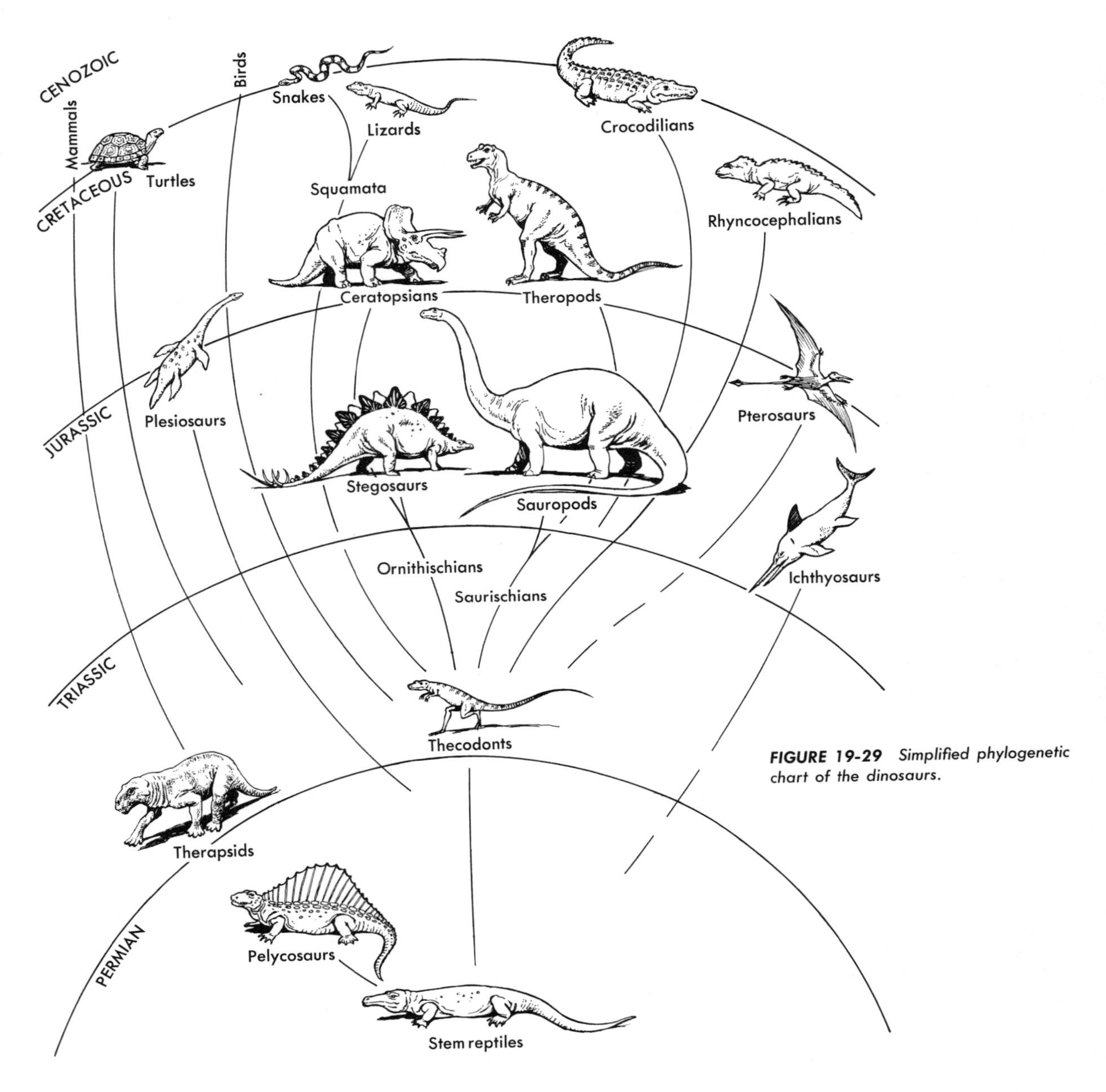

FIGURE 19-29 Simplified phylogenetic chart of the dinosaurs.

rocks of New Mexico and adapted for life on dry uplands. *Coelophysis* was the forebear of a host of saurischian dinosaurs that adapted to many different terrestrial environments. Some increased in size and seem to have reverted to the four-footed stance, eventually giving rise to the gigantic, long-tailed, long-necked *sauropods.* The sauropods were most common in the Jurassic but existed in restricted localities until the very end of the Cretaceous. Among the well-known sauropods are the following: (1) a primitive form, *Plateosaurus,* from the Late Triassic of Germany; (2) *Diplodocus,* a slender, whip-tailed type from the Jurassic Morrison

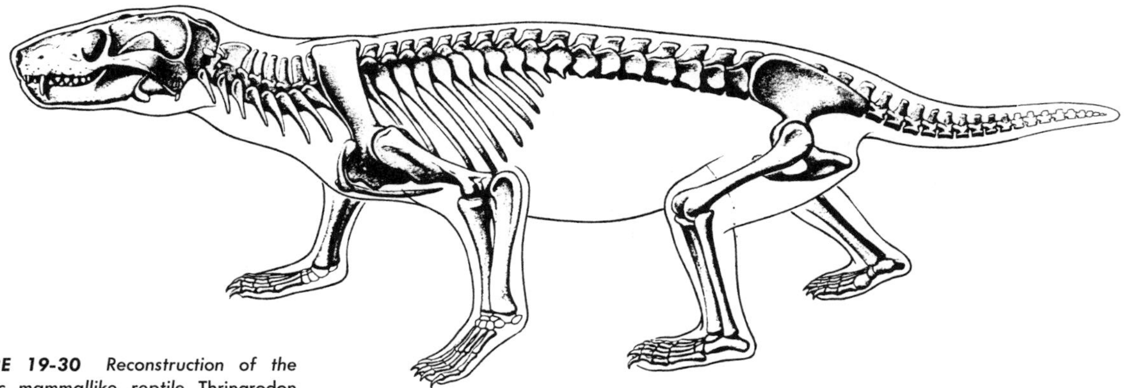

FIGURE 19-30 Reconstruction of the Triassic mammallike reptile Thrinarodon liorhinus. *It is a member of the group known as cynodonts. Remains on which the reconstruction are based are from South Africa. Note that the lower jaw is composed of several bones (reptilelike), but the teeth are differentiated (mammallike). Length about 45 cm. (Courtesy of Peabody Museum, Yale University.)*

Formation in the western United States; (3) *Brontosaurus,* a heavier contemporary of *Diplodocus;* and (4) *Brachiosaurus,* the long-necked but short-tailed form, perhaps the greatest land animal of all time.

Those saurischians that retained their two-legged pose and carnivorous habits continued to live side by side with the sauropods and perhaps eventually began to prey on them. Carnivorous saurischians include the well-known *Allosaurus,* from the Morrison Formation; *Tyrannosaurus rex,* the greatest land carnivore of all time; and *Tarbosaurus,* which lived during the Late Cretaceous in the region of the Mongolian Gobi Desert.

The earliest known member of the second order of dinosaurs, the *Ornithischia,* thus far discovered is from late Triassic rocks of Cape Province, South Africa. This animal, *Lycorhinus,* had a skull about 10 cm long and a strange mixture of tooth types, perhaps indicating a varied diet. The descendants of the Triassic ornithischians became numerous and varied during the Jurassic and Cretaceous. Here again there are two-legged types, but few members of the group lost the use of their forelimbs to the extend that the saurischian bipeds did. All post-Triassic ornithischians appear to have been herbivores, and the group invaded environments ranging from arid deserts to marshes and lagoons. The suborder *Ornithopoda* ("bird foot") included many large, water-loving plant-eaters with duck-billed

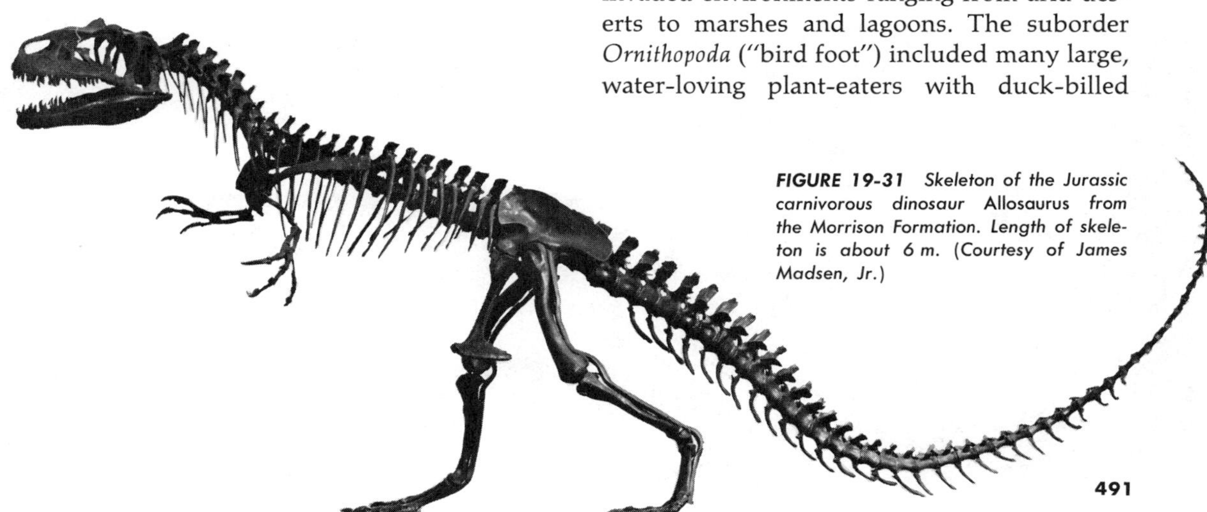

FIGURE 19-31 Skeleton of the Jurassic carnivorous dinosaur Allosaurus *from the Morrison Formation. Length of skeleton is about 6 m. (Courtesy of James Madsen, Jr.)*

faces. That they existed on tough woody vegetation is indicated by the numerous closely packed teeth that were set in their strong jaws. The duck-billed dinosaurs were most plentiful during the Cretaceous. The suborder *Stegosauria* included heavy four-legged herbivores armed with various types of spikes, plates, and bony protuberances; they were exclusively of Jurassic age. The suborder *Ankylosauria* included heavily armored types known only from the Cretaceous. A fourth suborder, *Ceratopsia,* included large four-legged species that had tremendous skulls and a sort of collar or frill that extended backward across the neck and shoulders. Most of the later ceratopsians had one or more forward-pointing horns. This group is known only from the Cretaceous. Representative ornithischian dinosaurs are illustrated by Figure 19-32.

A list of the inhabitants of any of the larger continents during the late Triassic, Jurassic, or Cretaceous invariably includes a great number of dinosaurs. These animals spread widely, and even the sauropods reached Australia, a fact suggesting that this landmass remained in contact with the other continents as late as the Cretaceous. Generally, the various types of dinosaurs occur together in fossil "boneyards," for the carnivorous types appear to have preyed on, and lived among, the herbivores. The most widely distributed forms are the sauropods, but we must

FIGURE 19-32 *Duck bills and neck frills typify Cretaceous dinosaurs. (a) Lambeosaurus, a duck-billed specimen, has a peculiar skull structure that may correlate with underwater feeding habits. (b) Triceratops, the giant horned dinosaur, is a Late Cretaceous ceratopsian from North America. (c) Protoceratops is an Early Cretaceous ceratopsian. This skeleton is from the Gobi Desert. [(a) and (c) courtesy Field Museum of Natural History, Chicago; (b) courtesy Smithsonian Institution.]*

(b)

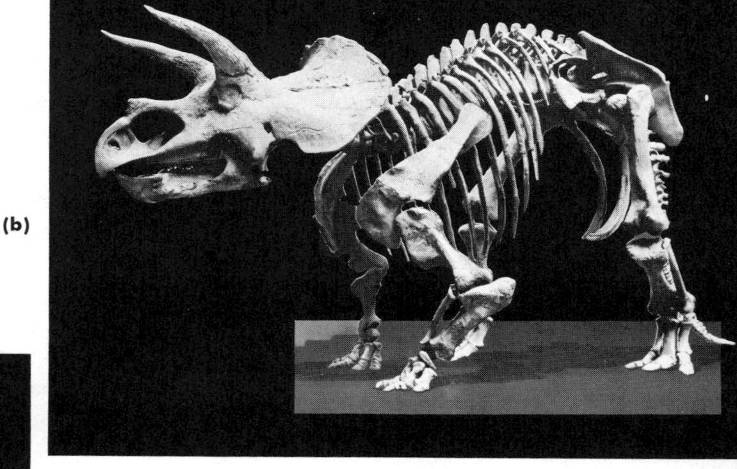

(a)

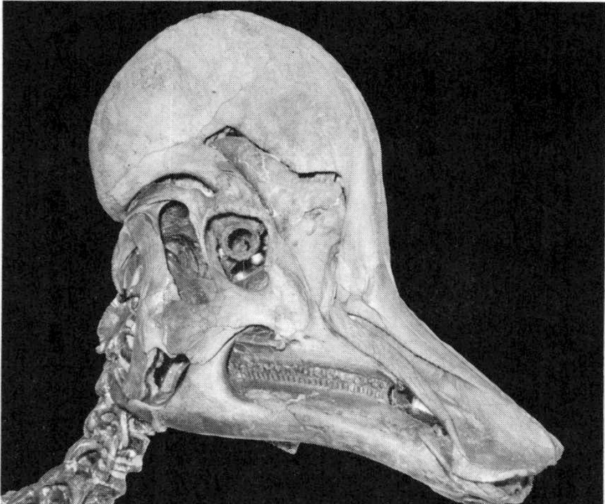

(c)

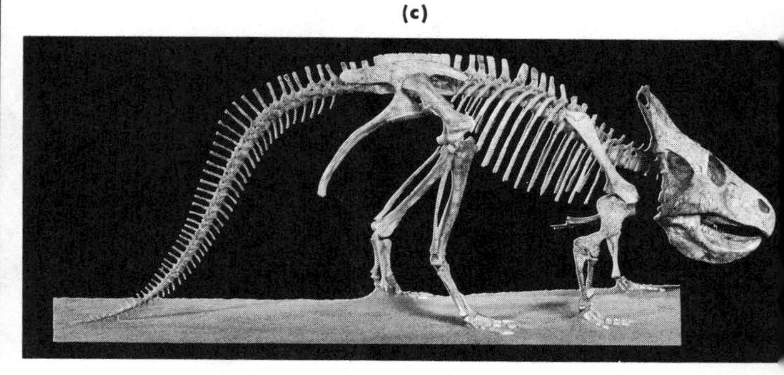

FIGURE 19-33 *An ichthyosaur (Stenopterygius quadricissus) with the outline of the body preserved. The specimen is from Lower Jurassic deposits near Holzmaden, Germany, and is nearly 3 m long. (Field Museum of Natural History, Chicago.)*

remember that their huge bones fossilized readily and when exposed can scarcely escape being discovered. The dinosaurs disappeared at the close of the Cretaceous and as a group are therefore strictly confined to the Mesozoic Era.

Sea monsters

During the Mesozoic a number of reptilian groups adapted to life in the seas. These were as follows: (1) fishlike forms, appropriately called *ichthyosaurs* (Figure 19-33), (2) paddle-swimmers, mostly with short tails and long necks, the *plesiosaurs:* (3) gigantic long-tailed marine lizards, the *mosasaurs,* and (4) large sea turtles such as *Archelon.* Of these, only the sea turtles survive today; the rest became extinct at or before the close of the Cretaceous, 65 million years ago.

The ichthyosaurs were the first large reptiles to invade the seas, and they became the most fishlike in their adaptations. The earliest members of the group thus far identified are of Triassic age; they became very common during the Jurassic and Cretaceous. They had fishlike limbs, a forked tail, and unusually large eyes, indicating that they may have frequented the dimly lighted depths of the ocean. Some were toothless, whereas others had jaws lined with many sharp teeth; stomach contents, which have been found petrified inside the body of some specimens, indicate that the animals lived on a varied diet, chiefly fish and cephalopods. The largest specimens reached a maximum length of a little over 4 m. All evidence indicates that the ichthyosaurs hatched their young internally and brought them forth into the sea fully formed.

The plesiosaurs appeared in the Triassic and had vanished by the close of the Cretaceous. There were two chief lines of evolution. One, typified by *Alasmosaurus,* had long necks and small heads; the other, including the remarkable *Kronosaurus,* had short necks and large heads. The plesiosaurs had wide oarlike flippers, or paddles, and evidently preyed on fish. Because they could not move as quickly as the contemporaneous ichthyosaurs, they probably depended heavily on the flexibility of their long necks as an aid in capturing prey. The number of vertebrae in the long-necked group increased steadily during their evolutionary history. The last form had 76 neck vertebrae, which added up to over half the animal's total length.

The mosasaurs, 7 m in length, appeared in the late Cretaceous but had vanished by the close of the period. With their long, flat tails and sinuous bodies, they were undoubtedly excellent swimmers, and their numerous conical teeth were well suited for capturing large active prey.

Archelon, the giant sea turtle found in the Cretaceous chalk beds of Kansas, had a flattened body and was over 3 m long from its nose to the tip of its tail. The bones of the shell were very small, and the covering leatherlike. The estimated weight is 3 metric tons.

Crocodiles, snakes, and lizards

The earliest known crocodile, *Proterochampsa,* was discovered in Triassic rocks of Argentina in 1958. Crocodiles have succeeded in a modest way through the Mesozoic periods and into the present time.

The first lizards made their debut during the Jurassic, but except for the large marine forms, they left few remains in Mesozoic rocks. The snakes, which are the last major group of reptiles to appear, are first found in Cretaceous rocks. They are obviously highly modified lizards, with which they are grouped into a single order, *Squamata.* Lizards and snakes have thrived both in wet and in dry areas.

Flying reptiles

For many millions of years in the late Paleozoic, the only animals capable of flight were the insects. This aerial monopoly was broken in the Triassic when vertebrates took to the air. A gliding reptile from the Newark Group of New Jersey is the oldest known aerial vertebrate. It has been named *Icarosaurus siefkeri* and is a highly modified lizard. In the succeeding Jurassic, the feathered birds and leathery-winged pterosaurs became airborne and were highly successful (Figure 19-34). The pterosaurs appear in the fossil record slightly before the birds and may have outnumbered them in the Jurassic and Early Cretaceous. The remains of both pterosaurs and birds are rare because both had light and delicate skeletons and were apt to live and die in forests under conditions unfavorable for fossilization.

The pterosaurs ranged from sparrow size to giants with a wingspread of over 15 m. In general, Jurassic forms were smaller than their Cretaceous descendants and had many small teeth, long tails, and relatively heavier skeletons. *Rhamphorhynchus,* represented by well-preserved skeletons from the Solnhofen quarries in Bavaria, is typical of the group. Later species, such as the well-known Cretaceous form, *Pteranodon,* were generally larger and toothless and had short tails and delicate bones. A recently discovered and yet unnamed pterosaur from the late Cretaceous of west Texas is estimated to have had a wingspan of 15 m. The remains of pterosaurs were preserved only under very special conditions, usually in extremely fine-grained sediments that accumulated in quiet, shallow water. Paleontologists generally agree that the pterosaurs were poor flyers, for their remains show no signs of attachments for powerful wing muscles. The discovery of well-pre served footprints with impressions of both the hind feet and front feet has led some investigators to speculate that pterosaurs could walk when necessary and could take to the air after gathering speed on the ground. The wing membrane was stretched between an elongated "little finger" and the sides of the body, and it is evident that damage to such a structure would be difficult to repair and would perhaps permanently disable the animal.

Birds

Birds appear in the fossil record during the late Jurassic. They seem to be rather closely related to crocodiles and, surprisingly enough, to dinosaurs. Two well-preserved skeletons and parts of two others, all with imprints of feathers, have been found in the Solnhofen quarries. These remains are of two genera, *Archaeopteryx* and *Archaeornis;* these were creatures about the size of a crow. In anatomy, *Archaeopteryx* stands about midway

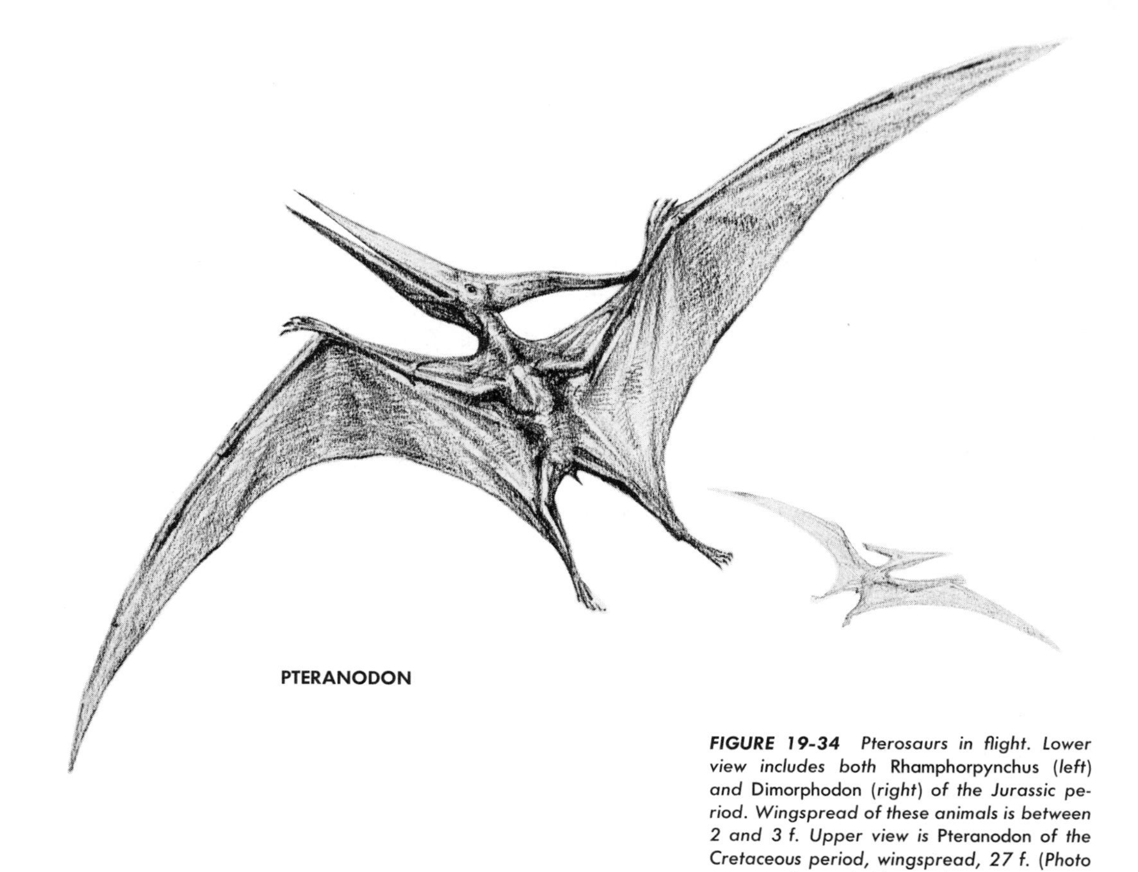

FIGURE 19-34 Pterosaurs in flight. Lower view includes both Rhamphorpynchus *(left) and* Dimorphodon *(right) of the Jurassic period. Wingspread of these animals is between 2 and 3 f. Upper view is* Pteranodon *of the Cretaceous period, wingspread, 27 f. (Photo courtesy Carnegie Museum.)*

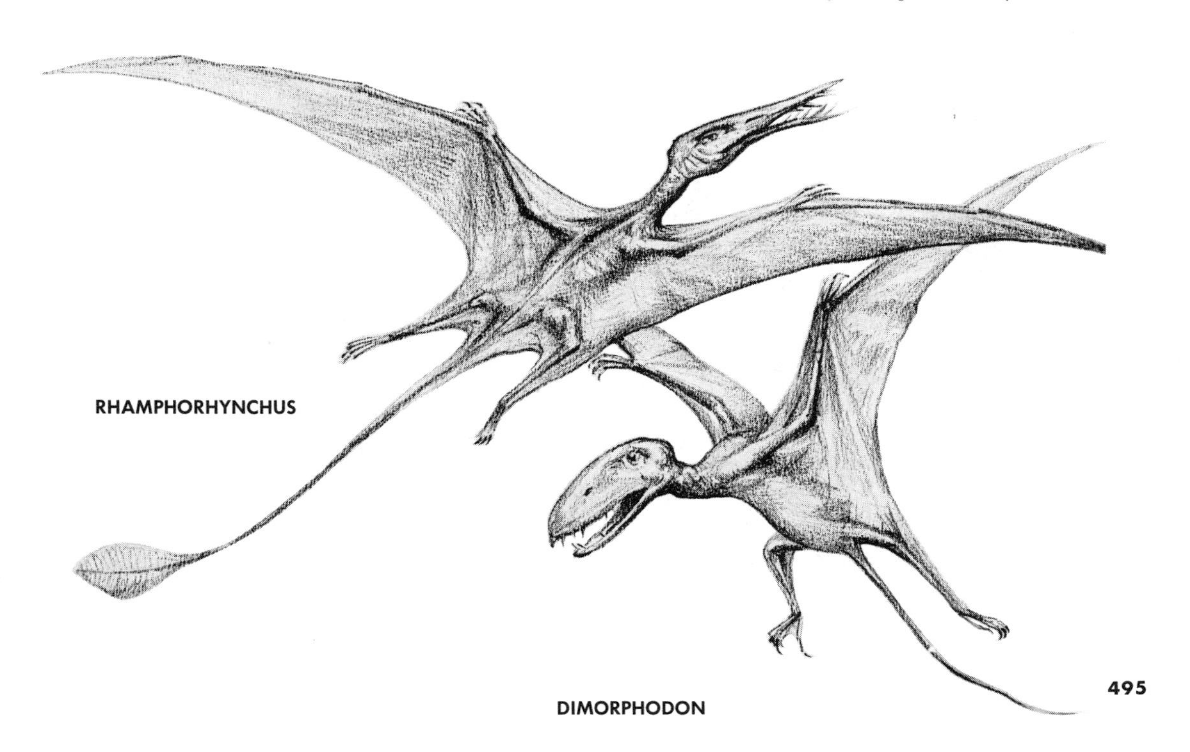

between thecodont reptiles and later birds (Figure 19-35). The Mesozoic birds had entered many new environments, but their fossils are nevertheless rare. *Hesperornis,* a bird from the Cretaceous chalk beds of Kansas, was about 2 m long and had a good set of teeth but no trace of wings. It was probably a powerful diver and swimmer.

A comparison of wing structure plainly reveals one reason why birds succeeded and pterosaurs failed. The pterosaur's wing, formed of membrane, was difficult to repair, and was adapted chiefly for gliding. But a bird's wing permits true flight and is composed of many individual feathers, each of which can be renewed and replaced as needed to keep the wing in a state of constant efficiency.

FIGURE 19-35 *Restoration of Archaeopteryx constructed at the British Museum of Natural History. (NERC copyright. Photograph reproduced by permission of the Director, Institute of Geological Sciences, London.)*

Mesozoic mammals

The fossils of Mesozoic mammals are of greatest evolutionary importance even though they are insignificant in quantity when compared with the remains of contemporary reptiles. Bones of early mammals are extremely small and generally scattered. At a few places, however, they are abundant; and thousands of individual bones and teeth have been taken from concentrated pockets and fissures in South Wales. This locality is of very late Triassic age and is the most prolific and significant source of early mammals so far discovered.

The most important vertebrate found in the South Wales deposits is *Morganucodon,* currently judged to be the earliest mammal. It is less than 10 cm long and has many primitive characteristics that suggest relationship with the modern *monotremes* (platypus and others). Species of *Morganucodon* have also been discovered in such widely scattered areas as China and South Africa. Contemporaries of *Morganucondon* had also achieved many mammalian characteristics. Of these so-called mammallike reptiles, the *tritylodonts* are best known. They have been found in many localities, including a site with many complete skeletons near Kayenta, Arizona.

Fossils of Jurassic mammals have been found both in Europe and in North America, but again remains are relatively rare and consist only of partial skulls, jaws, and teeth (Figure 19-36). All are rodentlike, about the size of mice and rats or smaller. Several distinct orders are represented. The teeth are variously modified and differentiated into incisors, canines, and molars; the animals seem to have subsisted on a high-energy diet of insects, seeds, and fruit. The typical Jurassic mammal inhabited the undergrowth or the branches of trees and was ever on the alert for the large predatory reptiles with which it could not hope to compete in open combat.

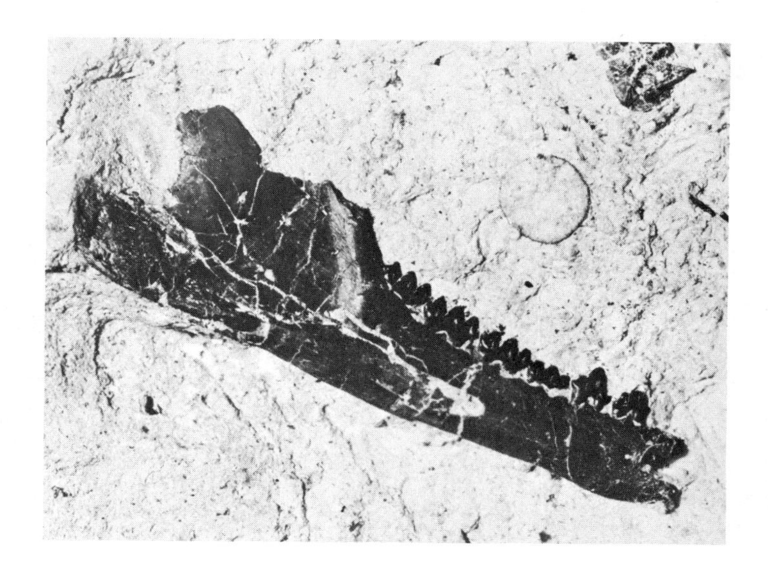

FIGURE 19-36 Jaw and teeth of Triconodon, a primitive mammal from the Late Jurassic of England. The specimen is about 2.5 cm long. (NERC copyright. Photograph reproduced by permission of the Director, Institute of Geological Sciences, London.)

Mammals continued to be scarce during the Cretaceous, and some of the Jurassic forms even disappeared. The best-known late Mesozoic remains are from the Upper Cretaceous of Mongolia and the western United States. Early in the period true *marsupials* (pouched mammals) appeared. *Placentals* (advanced mammals) came in the late Cretaceous. All in all, the mammals were quite rare during Mesozoic time, but they had already developed the potential to replace the reptiles when the time arrived.

Close of the Mesozoic

No single dramatic event marks the end of the Mesozoic era. Although a number of gradual geological changes occurred near the Mesozoic–Cenozoic time bundary, geologists have found no evidence of a rapid, simultaneous, worldwide physical alteration. In general, the oceans were withdrawing during this interval, converting former sea bottoms into swamps, flood plains, and tidal flats. At places the withdrawal was accompanied or caused by mountain building.

In the biological sense, however, there were a number of startling changes near the end of the Mesozoic Era. Almost all fossiliferous areas indicate that large and important groups were being exterminated both on the land and in the oceans. Dinosaurs and pterodactyls vanished from the lands, and in the ocean the last plesiosaurs, mosasaurs, and ichthyosaurs disappeared. Of even greater significance to paleontologists was the extinction of the ammonite cephalopods. Although all these groups did not disappear simultaneously everywhere, their extinction does provide a guide to the end of the Mesozoic era. In practice, the line between the Cretaceous and Tertiary is placed above the highest, or last, occurrence of dinosaurs, ammonites, and other key forms.

Economic products of the Mesozoic

Mineral resources of Mesozoic age are critically important to our modern industrialized civilization. On an average, Mesozoic rocks are richer in oil than either Paleozoic or Cenozoic rocks. According to a recent survey,

almost 20 percent of all the world's oil fields tap Mesozoic rocks, and these rocks contain 52.7 percent of the world's known reserves. The prolific Middle East fields draw oil chiefly from Jurassic and Cretaceous rocks. Other areas of Mesozoic production are the Rocky Mountains, the upper Gulf Coast, western Venezuela, and Argentina. Recently, large oil and gas accumulations have been discovered in underwater Cretaceous formations in the Bass Strait between Australia and Tasmania. Large reserves have also been discovered in Triassic rocks of northern Alaska and under the North Sea.

__FIGURE 19-37__ A bed of Cretaceous coal, 10 m thick, being mined by underground methods at Castle Gate, Utah. (Photo courtesy Utah State Historical Society.)

Coal is found in rocks of each Mesozoic period: Triassic coal occurs in the eastern United States, South Africa, and China; important Jurassic coal deposits are found in China and Siberia; and Cretaceous coal is so abundant in the United States and Canada that the system ranks second only to the Pennsylvanian in amount of total reserves (Figure 19-37). Cretaceous coal is also important economically in northeast Asia.

Metallic mineral products were formed in many places in connection with Mesozoic igneous activity and sedimentation. Copper occurs in scattered deposits in Triassic sandstone and shale in Germany, Russia, and the United States. Jurassic rocks contain important reserves of sedimentary iron ore in England and Alsace-Lorraine. Cretaceous sedimentary iron ore is mined in Siberia. Both Triassic and Jurassic rocks in the western United States contain important reserves of uranium. The famous Mother Lode, source of much of California gold, was formed in the late Jurassic or the Cretaceous. The salt plugs, or domes, in the Gulf Coast area, which yield sulfur in addition to salt, arose from Triassic beds thousands of meters beneath the surface. Salt is also found in Lower Cretaceous formations of Brazil and Gabon. The diamond-bearing pipes of South Africa are probably of Cretaceous age, and those of Siberia are thought to belong to the Triassic and Jurassic. The diamonds they contain, however, may have been formed much earlier, and of course many stones were freed from their original matrix and redeposited in rocks that were formed much later.

SUMMARY

The Mesozoic Era dawned with the lands relatively high and with few seas in the interiors of the continents. The Triassic Period was geologically rather quiet, and large volumes of red continental sediments were laid down on the land areas. The succeeding Jurassic Period was without extensive mountain building; the oceans spread more widely over the continents; and a variety of marine and nonmarine sediments accumulated. The Triassic and Jurassic were marked by widespread intrusion and extrusion of basic igneous rocks, especially in the Southern Hemisphere. Proponents of continental drift maintain that the breakup of Gondwanaland was accomplished chiefly in the mid-Mesozoic.

During the Cretaceous, interior seas spread across the continents in one of the greatest floods of all times. Commencing in middle Cretaceous time, the margins of the Pacific became geologically active; igneous masses were intruded, and older geosynclinal deposits were folded and uplifted. The era closed with the continents being uplifted and eroded.

Mesozoic life was in transition. Plants became better adjusted to dry-land environments and to seasonal changes. Pines, ferns, and cycads were dominant during most of the era. During the middle Cretaceous the flowering plants appeared in considerable numbers. Marine organisms slowly recovered from the critical late Paleozoic. Pelecypods, gastropods, and coiled cephalopods multiplied, and toward the end of the era, floating organisms of various kinds increased. On land, the reptiles, represented by dinosaurs, were dominant vertebrates. Mammals appeared in the Triassic and birds in the Jurassic, but neither group succeeded in becoming dominant during the Mesozoic.

Many important deposits of coal and oil were formed in the Mesozoic. The great oil fields of the Middle East produce from Jurassic and Cretaceous formations and the coal reserves in Cretaceous rocks of western North America are among the greatest in the world.

SUGGESTED READINGS

ARKELL, W. J., *Jurassic Geology of the World.* New York: Hafner Publishing Co., Inc., 1956.

AUGUSTA, J., and Z. BURIAN, *Prehistoric Reptiles and Birds,* trans. M. Schierl. London: Hamlyn, 1961.

AUGUSTA, J., and Z. BURIAN, *Prehistoric Sea Monsters,* trans. M. Schierl. London: Hamlyn, 1964.

BREED, CAROL S., and WILLIAM J. BREED, eds., "Investigations in the Triassic Chinle Formation," *Museum of Northern Arizona, Bullentin 47* (1972).

COLBERT, E. H., *Dinosaurs, Their Discovery and Their World.* New York: E. P. Dutton & Co. Inc., 1961.

HALLAM, A., *Jurassic Environments.* Cambridge, England: Cambridge University Press, 1975.

KURTÉN, BJÖRN, *The Age of Dinosaurs.* New York: McGraw-Hill Book Company, World University Library, 1968.

ROMER, A. S., *The Vertebrate Story,* 4th revised and enlarged edition. Chicago and London: University of Chicago Press, 1959.

Treatise on Invertebrate Paleontology, 24 parts. Lawrence, Kansas, and Boulder, Colorado: The University of Kansas Press and the Geological Society of America. © 1957 by the University of Kansas Press and the Geological Society of America. A series by various authors on the major invertebrate animals. Important volumes for Mesozoic fossils are those on ammonite cephalopods, pelecypods, gastropods, echinoderms, and protozoans.

WOOD, PETER, DORA JANE HAMBLIN, and JONATHAN NORTON LEONARD, *Life Before Man.* New York: Time, Inc., 1972. (A Time-Life Book.)

The Tertiary Period

This chapter deals with the Tertiary Period, which together with the Quaternary Period makes up the Cenozoic Era. The Tertiary has a total duration of about 65 million years and came to an end about 1.8 million years ago. The generally accepted subdivisions are as follows: Paleocene, Eocene, Oligocene, Miocene, and Pliocene Epochs. However, not all geologists agree with this scheme of subdividing the time scale; Europeans generally prefer to divide the Cenozoic into the Paleogene Period (Paleocene, Eocene, and Oligocene) and Neogene Period (Miocene, Pliocene and Pleistocene). The important Old World mountain-building episode known as the *Alpine Revolution* culminated near the end of the Oligocene and serves as a natural division between the Paleogene and Neogene. The term *Nummulitic Period* is also used in Europe for the combined Paleocene, Eocene, and Oli-

20

Peaks of the Grand Teton Range, Wyoming. The central higher portions are ancient Precambrian rocks. The range was uplifted during the Rocky Mountain Revolution early in the Tertiary period. (Courtesy of Hal Rumel.)

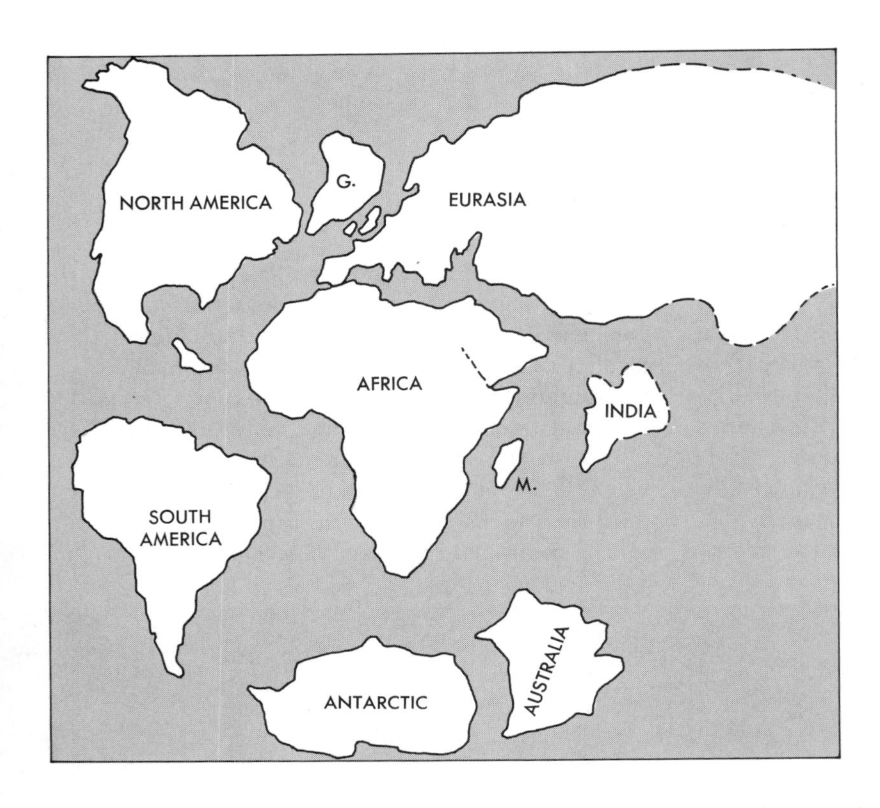

FIGURE 20-1 *Distribution and approximate positions of major landmasses in the Eocene Epoch, about 50 million years ago. Semidiagrammatic, no exact scale or directions implied.*

gocene Epochs in recognition that nummulites (large disc-shaped fossil protozoans) are common in this interval. However, for obvious reasons, the Cenozoic is recognized by all concerned as the *Age of Mammals.* Future usage will dictate which of the competing subdivisions will eventually gain universal acceptance.

The great continental landmasses and oceans, together with larger islands and seas that are familiar geographic features today, were either already well marked or roughly outlined at the beginning of the Tertiary Period (Figure 20-1). In striking comparison with the preceding Cretaceous Period, all the continents were relatively high and dry with practically no interior seas. A general withdrawal of water into the ocean basins is indicated.

Continental separation that had been going on at a lively pace during the late Mesozoic continued without interruption into the early Tertiary. The Americas were separated from Europe and Africa except for a connection across the North Atlantic where Greenland formed a passable land bridge. In the Southern Hemisphere, Australia and Antarctica had lost contact with all other southern lands but remained joined to each other until the Eocene. Antarctica soon drifted southward into a polar position, and Australia commenced a northward trip that is still continuing today.

Important reactions took place between the southern and northern landmasses, practically obliterating the ancient Tethyan seaway and converting the contained sediments into the great Alpine and Himalayan ranges, which dominate so much of the earth's geography today. Details are immensely complex but are explainable chiefly as a northward movement of Africa and other southern blocks toward Eurasian lands. The Alps represent great sheets and slices of Tethyan sediment driven northward out of the sea and into central Eu-

FIGURE 20-2 *Space photo of India showing the snow-covered Himalayan Range to the north, and the wide band of foothills and the plains of the peninsular section to the south. The rivers that traverse the view are all tributaries of the Ganges. Mt. Everest appears with a large triangular shadow near the upper right corner. This view encompasses the complex structure produced by the collision of the older, more solid shield area to the south with the geosynclinal sedimentary rocks making up the ranges. (Photo courtesy of NASA.)*

rope. The narrowing or shortening of distances across the former seaway has been calculated to have equaled many hundreds of kilometers. The Arabian Peninsula, which was orginally part of Africa, split apart from the parent continent to create the Red Sea in the Oligocene and Miocene and at the same time pushed up the Zagros ranges and became part of the Eurasian continent. Most spectacular of all was the collision of India in the Eocene and Oligocene giving rise to the Himalayan ranges and Tibetan Plateau (Figure 20-2). All these dramatic movements reached a culmination in the middle Tertiary to create one of the earth's greatest disturbances, the *Alpine Revolution.*

As shown by fossil vegetation, Tertiary climatic zones were much like the present (Figure 20-3). The North Pole had reached essentially a landlocked position in the Arctic sea at the beginning of the Tertiary and seems only to have migrated a short distance from west to east over the past 60 million years. The position of the equator with respect to the continents was not significantly different from what it is today. Climatic reactions in the southern polar and temperate regions are not as simple. The South Pole was on Antarctica in the Eocene, and there could have been glaciers on its higher elevations at that time. But Australia was in near proximity and created an obstacle to circum-Antarctica circulation until about 30 million years ago, at which time the present oceanic current system and climatic regime began.

FIGURE 20-3 *The "palm line" marks the northern and southern limits of the natural occurrence of palms. This map shows this boundary as it exists today and as it existed in the early Tertiary. Note parallelism with the present equator.*

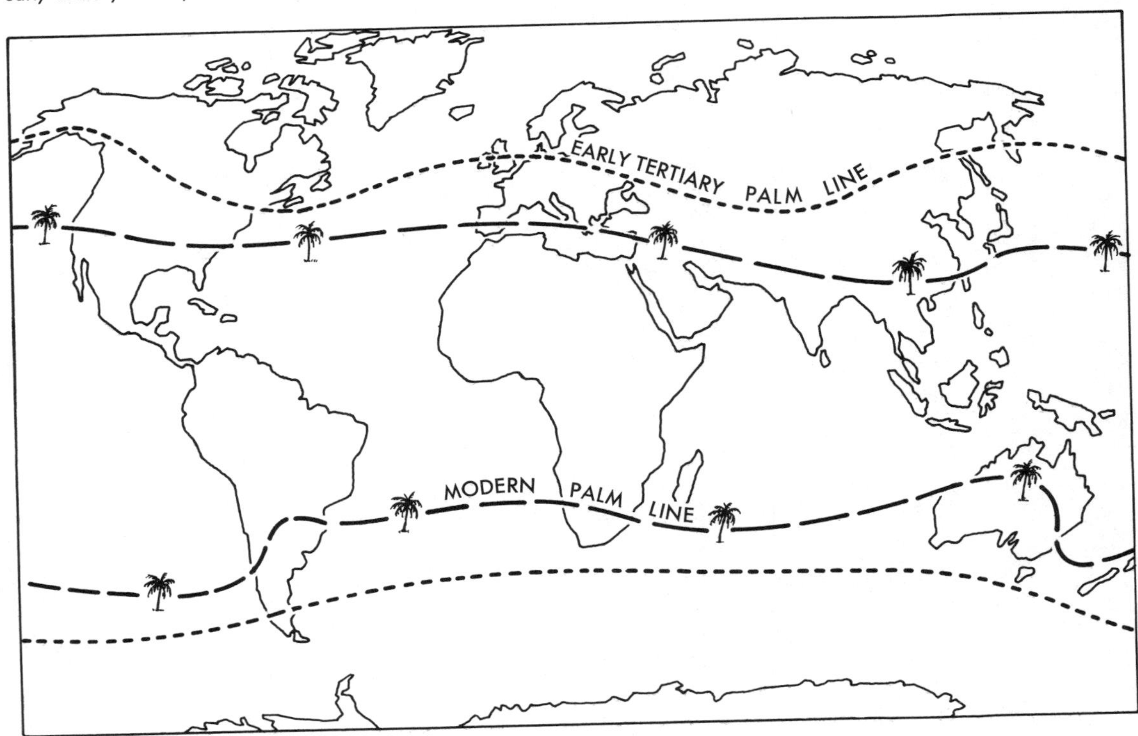

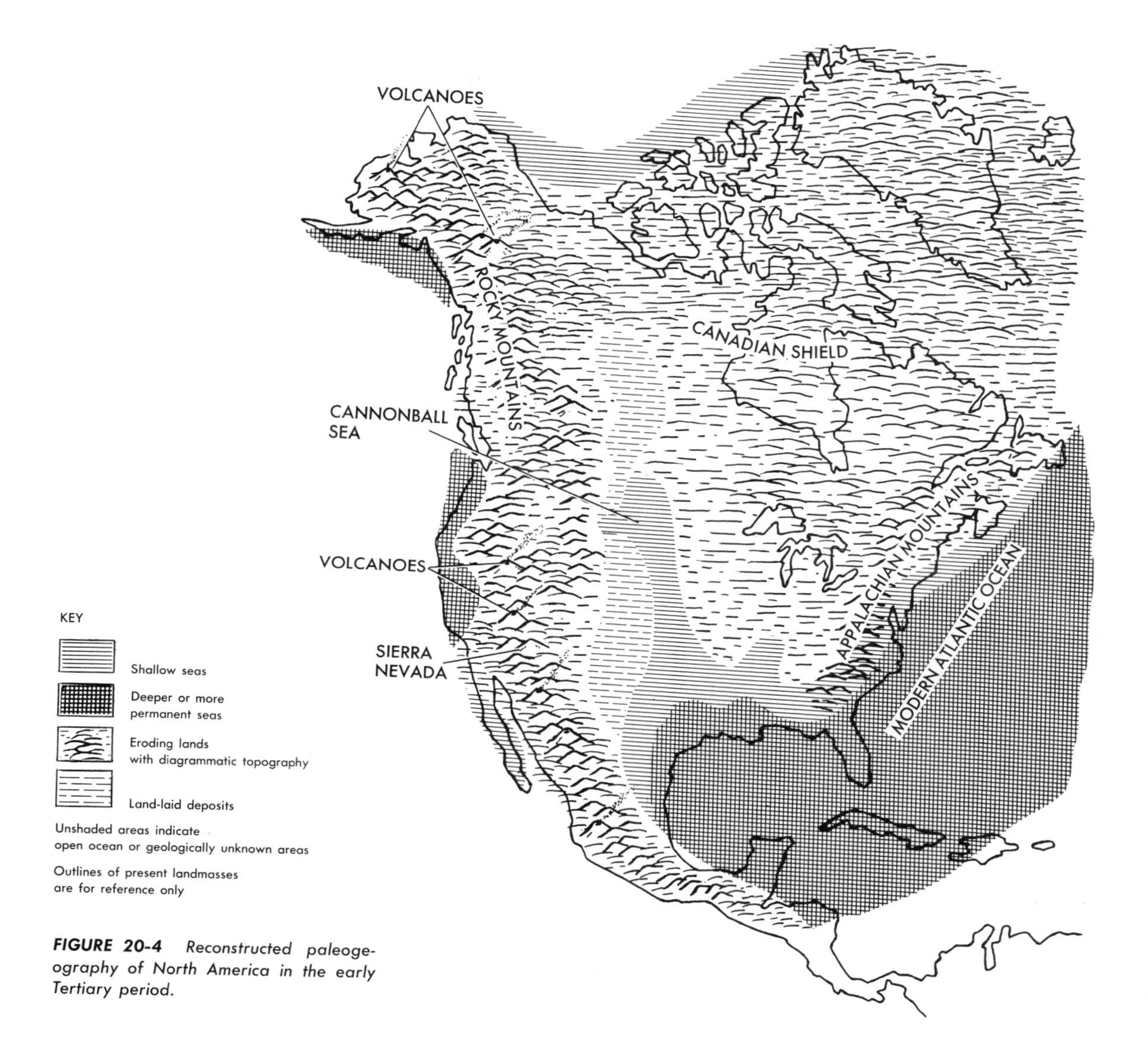

FIGURE 20-4 *Reconstructed paleogeography of North America in the early Tertiary period.*

North America

North America had achieved roughly its present size, shape, and relative position by the beginning of the Tertiary 65 million years ago (Figure 20-4). Its separation from Eurasia was not entirely complete however, as it was still attached to the Old World landmass by way of Greenland. About 65 million years ago, rifts appeared east of Greenland and west of Norway, heralding the opening of the present North Atlantic Ocean. Greenland remained attached to North America until about 35 million years ago when Davis Strait and Baffin Bay appeared. The North Atlantic deepened and widened through a complex series of movements, and the Mid-Atlantic Ridge came to occupy its present position about 16 million years ago. At this time Iceland appeared, chiefly by volcanic outpourings. It is important to note that North America and Eurasia were joined also at the region of the Bering Straits during most of the Tertiary. The Arctic Ocean was thus surrounded by an unbroken circumpolar landmass until about 20 million years ago, at which time warm water from the Atlantic penetrated to the polar regions with important climatic effects.

Eastern North America

A relatively narrow belt of Tertiary sediments fringes the Atlantic and Gulf coasts of North America. The visible part of the belt begins at Cape Cod, widens southward to include all of Florida, and extends up to 320 km wide in the lower Mississippi Valley. The interior regions, including the stumps of the Appalachian Range, supplied most of the sediments. Because the coastal plain has been little deformed and because its sediments can be related directly to their source areas, we know that the Tertiary was relatively quiet and uneventful insofar as the eastern part of North America is concerned.

As would be expected, the Tertiary rocks thicken and contain more and more marine elements as they are traced seaward. Coarse sands and silts characterize the northern outcrops, but limy rocks reflecting warmer seas increase southward. Florida has many limy formations with masses of shells and even coral reefs (Figure 20-5). The oldest formation seen at the surface in Florida is of Eocene age.

Not to be overlooked are the Bahama Banks, which resemble Florida but contain even more calcareous material. A well drilled on Andros Island directly east of the Florida Keys passed through a continuous sequence of nearly 4,500 m of Tertiary and Cretaceous calcareous sediments, which suggests that the area has been subsiding for a long time and that deposition of calcareous sediment kept pace with the downsinking.

FIGURE 20-5 *The Key Largo Limestone as exposed in a quarry on Windley's Key, southern Florida. A large colony of the coral Montastrea annularis can be seen in position of growth. The entire rock is composed of calcareous organic debris. (Photo courtesy of James I. Jones)*

FIGURE 20-6 (Above) Columbia River Basalt, in Franklin and Whitman counties, Washington. Successive flows are exposed as cliffs in the canyon walls. (U.S. Geological Survey.)

Western North America

The Tertiary brought extensive changes along the Pacific Coast of North America, and the Coast Ranges are essentially a product of this interval. At the beginning of the Tertiary, the Pacific extended inward over the site of the present Cascade Range and to the foothills of the Sierra Nevada. Great outpourings of lava, especially in the Miocene, spread widely in Washington, Idaho, and Oregon (Figure 20-6).

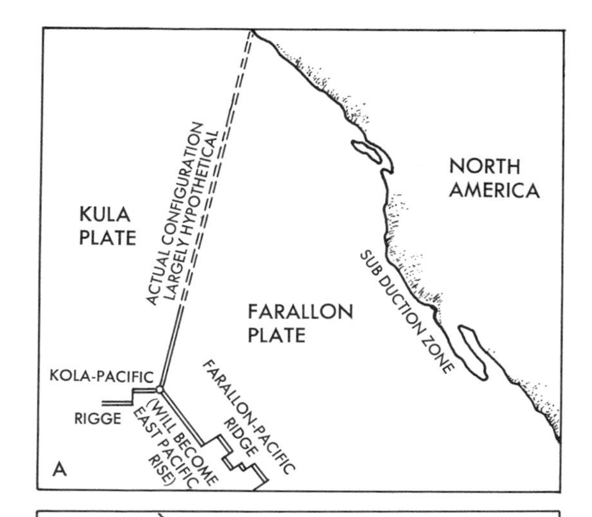

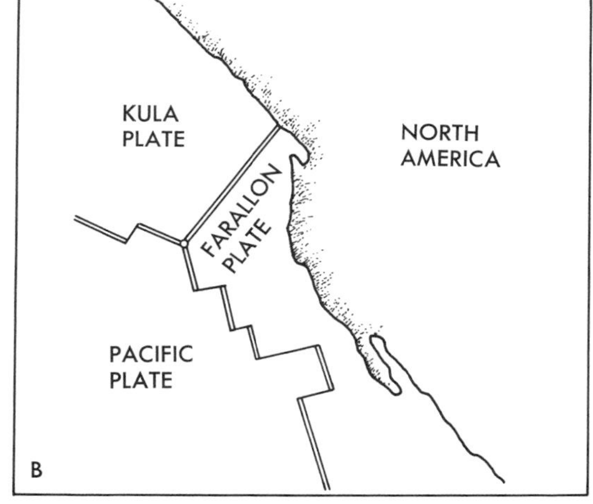

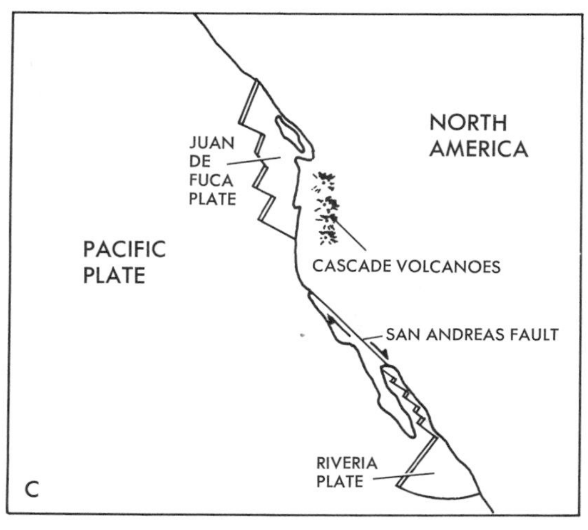

FIGURE 20-7 (Right) Plate tectonic history of the eastern Pacific Basin and North America. (A) 60 million years ago near Cretaceous-Tertiary transition: North America is chiefly overriding the Farallon Plate; Kula Plate is moving northwesterly; Pacific Plate appears in the central Pacific. (B) 40 million years ago, Late Eocene: much of Farallon and Kula plates has been consumed; Pacific Plate enlarging. (C) present time: the Juan de Fuca and Rivera plates, disconnected descendants of the Farallon Plate, are nearing extinction; Kula Plate subducted chiefly under Aleutian Islands; Pacific Plate makes up practically all of east Pacific Basin.

The Columbia River basalts cover at least 500,000 km^2 and are 3,000 m thick. Twenty-three separate flows, one above another, have been counted, and a total of about 150,000 km^3 of lava is estimated to have been produced in the Columbia Plateau.

Westward movement of the continent that had been going on since the early Mesozoic continued into the middle Tertiary. During this time there was continued subduction of the sea floor with consequent addition of material that now makes up much of the Coast Ranges. At about 25 million years ago a radical change in plate motions took place. The oceanic and continental plates no longer moved toward each other; instead, an oblique motion began with North America moving southeasterly and the Pacific Plate northwesterly. This was the origin of the San Andreas fault system, which is thus given an age of no more than 25 to 30 million years. The postulated sequence of events is illustrated by Figure 20-7.

The Rocky Mountains continued to grow during the Early Tertiary. The Great Basin, with its peculiar north-south trending ranges, began to form in the Oligocene as great numbers of parallel faults cut the crust. The area between the Sierra Nevada and the Wasatch Range subsided unevenly to create basins for the lakes that appeared in the more humid phases of the Pleistocene Ice Age.

Tertiary deposits are widespread in Alaska but rare in northern Canada and the Arctic islands. In Alaska, coal beds and fossil plants are common, but no Tertiary vertebrates have yet been found, hinting that climates were not as favorable as they were farther south.

Continental sediments

The most complete sequence of Tertiary land deposits known is found in the western United States. Here conditions were especially favorable for the production and preservation of large amounts of sediment. Deposits of the Paleocene Epoch accumulated mainly in the interior basins lying between the ranges created by the *Laramide Orogeny* (Figure 20-8). The Paleocene is well represented by formations in the *Big Horn Basin* of Wyoming and the *San Juan Basin* of New Mexico. There are also many thousands of square kilometers of Paleocene rocks in the plains section of Montana and the Dakotas. The rocks are chiefly sandstone and shale, with immense reserves of low-grade coal. The fossil vegetation and animal life indicate that a generally humid, semitropical climate prevailed over the western interior of the United States at this time. Apparently the whole region was at least a thousand meters lower than it is now. Paleocene sediments that now lie at relatively high elevations in the Rockies accumulated originally only slightly above sea level.

The Eocene is represented by formations in at least a dozen separate Rocky Mountain basins, but there are practically no sediments of this age in the plains area. The sediments that filled the individual basins were obviously derived from the adjacent uplifted ranges and show considerable variety. Among the most common rock types are thick conglomerates and coarse sandstones laid down by powerful streams. An unusual feature of the Eocene was a system of large interior lakes that formed in the adjoining parts of Colorado, Utah, and Wyoming. This water body, called the *Green River Lake System,* covered over 100,000 km^2. Its waters teemed with plant and animal life whose remains have become incorporated in the fine-grained bottom sediments. These organic-rich deposits constitute the world's greatest single oil reserve. If all the contained oil could be recovered, it would exceed all other known reserves in the world. The delicate fossils of insects, leaves, and fish found in the Green River Formation are world famous.

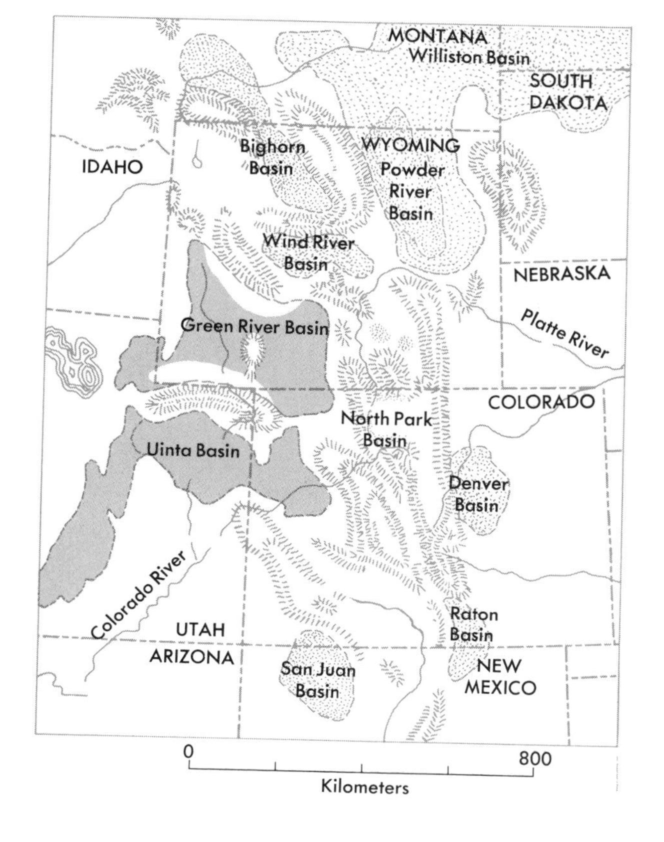

FIGURE 20-8 *(Left) Basins with Tertiary sediments within and near the Rocky Mountains. From these basins come most of the fossil remains that are the basis for reconstructing the early Tertiary life of North America. Areas occupied by lakes for significant periods of time are indicated by light grey. [Adapted from P. B. King, The Evolution of North America (Princeton, N.J.: Princeton University Press, 1959). Reprinted by permission.]*

During the period of lake formation, the western United States remained near sea level, and a subtropical or humid climate prevailed. Erosion and sedimentation were the dominant geologic processes, and the mountain ranges were worn down and buried in their own debris. The many excellent fossils of primitive mammals and other organisms that occur in the Eocene of the Rocky Mountains provide us with a near-perfect picture of life of the time.

FIGURE 20-9 *(Below) The San Juan Mountains, Colorado. These scenic mountains, sometimes called the "Alps of the United States" are built up chiefly of volcanic rocks erupted in Middle Tertiary time. (Colorado Department of Public Relations.)*

With the coming of the Oligocene Epoch, the geologic quiet was broken by volcanic outbursts that blanketed large areas of the American West with a thick covering of ashes and dust. Explosive volcanoes erupted in many places and were particularly active in the Yellowstone Park area of Wyoming and the San Juan Mountains of Colorado (Figure 20-9). Although the volcanic deposits were originally hundreds of meters thick, they have since been eroded away to a point where the topography that existed before the eruptions is again exposed. There is relatively little Oligocene sediment preserved within the Rocky Mountains proper. The most complete record of this epoch is in the foothills and plains areas of Colorado, Wyoming, Nebraska, and South Dakota. Here are the famous White River beds, from which come the most perfect and complete fossil mammals to be found anywhere in the world (Figure 20-10). The fine preservation of these remains is attributable to the frequent overflowing of sediment-laden rivers in the area, which buried not only scattered bones but also the complete skeletons of animals drowned on the flood plains. From the White River badlands have come remains of at least 150 species of fossil animals, including types ancestral to forms still alive and others that are completely extinct. The mammalian life of the western plains of North America at this time must have resembled that seen today only in central Africa. The Oligocene climate was evidently somewhat cooler and drier than the preceding Eocene.

The succeeding Miocene Epoch witnessed a general uplift over a large part of North America. The entire Rocky Mountain area was raised bodily over 1,000 m, and the climate became drier and cooler. Miocene deposits are spread out over thousands of square kilometers between the Rocky Mountains and the Mississippi River. Here are found ancient stream channels, lake beds, flood plains, and soils composed of sediments derived from the rejuvenated mountain ranges to the west. Mammal remains are abundant, but the warmth-loving semitropical forms are no longer in evidence. The animal life suggests that extensive grassy plains were spreading over the area, replacing forested sections.

Pliocene deposits are found over much of the same area as are those of the Miocene. The chief source of sediment was still the Rocky Mountain ranges, and the place of deposition was the western plains. Animal and plant re-

FIGURE 20-10 *Badlands National Monument, South Dakota. The barren, rugged outcrops are eroded in river-laid deposits of Oligocene age. Fossil mammals are abundant in this formation. (U.S. National Park Service.)*

FIGURE 20-11 *Lake Bikal, U.S.S.R., largest freshwater lake in the world, over 1750 m deep, occupies a trench formed by Tertiary faulting. The area is one of active earthquakes. (Photo courtesy of Kenneth L. Cook.)*

mains and the types of sediment deposited indicate that climates were cooler and drier than they were during the Miocene. Mammals adapted to moist, warm climates were no longer represented. Some had migrated elsewhere, and some had become extinct. The stage was set for the great Pleistocene Ice Age, which we shall consider at some length in the following chapter.

Eurasia

Although the wide and lengthy Tethys seaway was still in existence at the beginning of the Tertiary, the downsinking and sedimentation phases of its history were drawing to a close. Mountain range after mountain range began to appear in Europe and Asia as compressive and volcanic processes affected the area. Basins came into being among the ranges, one of these containing Lake Bikal, Figure 20-11.

The closing of the seaway and the severe compression of its contained sediments can be explained by the northward movement of the African block toward Europe and by the collision of India with southern Asia.

The closing of the European segment of Tethys evidently resulted from a scissorlike motion that began at the west end and progressed eastward. Thus the Pyrenees had emerged at the close of the Eocene, as had the Atlas Mountains of northwest Africa. The most important foldings of the European Alps and the Carpathians took place in the course of the Oligocene. In Asia the Himalayan area began to be affected early in the Tertiary, with the most intense phase coming in the middle Miocene. In the Middle East, mountain building commenced in the Cretaceous and continued into the Pliocene, with diminishing marine deposition and corresponding nonmarine accumulations. The entire *Alpine Revolution,* as it is generally called, seems to have culminated in the Miocene, and the total effect was to create the highest mountain ranges from the area of deepest sediments (Figure 20-12). A number of residual seas in the late Miocene and Pliocene encircled the previously formed mountains and received extensive deposits of

FIGURE 20-12 *Limestone and sandstone of Tertiary age deformed by the Alpine Orogeny, Hindu-kush Mountains, Afghanistan. (Photo courtesy of Laurence H. Lattman.)*

FIGURE 20-13 *A geologic cross section showing the internal structure of the Helvetian Alps in northwestern Switzerland. Great sheets of Jurassic and Cretaceous strata have been pushed northward upon Tertiary deposits that are seen at the left end of the section and are buried at great depths below the range. Length of section 30 km; "Om" designates sea level; true vertical scale. (From Rudolf Trümpy: Die helvelischen Decken der Ostschweiz, Eclogae Geologicae Helvetiae, Basel. Juni, 1969, Table 1. Reproduced with permission.)*

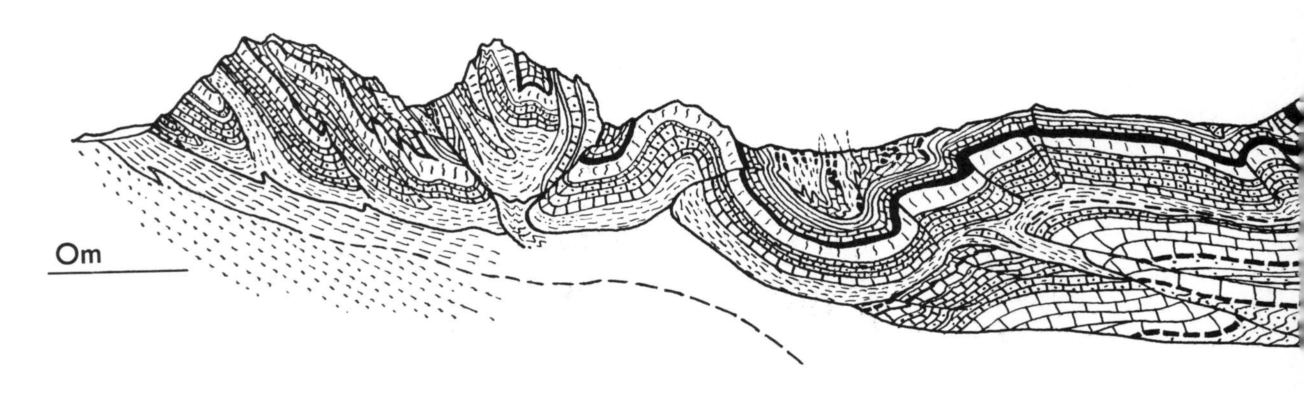

sediment from the newly formed ranges. These deposits later became involved in secondary movements and foldings. A cross-section of part of the much-studied Alps is given as Figure 20-13.

Rising ranges cut off large areas from the sea, and these became lagoons with brackish and freshwater deposits. As the Himalayan range rose, it shed debris southward into a great subsiding trough that was developing upon the newly arrived Indian block. This deposit, the *Siwalik Series,* is 4,500 to 6,000 m thick and yields a key record of life and environments from Late Miocene to Early Pleistocene time. Here was found the ancestral hominid, *Ramapithecus,* 12 to 14 million years old.

Much of the actual folding and mountain building of the Tethyan belt took place at or below sea level. Later effects served to elevate the mountains as well as the surrounding areas on a broad (*epeirogenic*) scale, so that the seas were permanently expelled and broad tracts bordering the ranges became dry land. This effect came chiefly in the Pliocene and early Pleistocene. The Tibetan Plateau, highest in the world, originated as the area was lifted by the underthrust of the Indian block. The Mediterranean Sea is not technically regarded as a remnant of Tethys, although it lies along the same trend and covers some of the same territory. Tethys was destroyed by mountain building, and the Mediterranean was opened by later movements related to the formation of the Atlantic Ocean.

Most of Europe between the Tethys seaway and the Baltic Shield was occupied by a series of shallow Tertiary seas or lagoons. There was a discontinuous barrier between the northern lagoons and Tethys in the form of the eroded remnants of the Hercynian ranges. The famous succession of deposits in the Paris Basin, where Lyell conceived the subdivisions of the Tertiary, represents deposition in shallow embayments from the Atlantic and contrasts sharply with contemporaneous deposits in Tethys. Southeastern Britain, France, northern Germany, and the Netherlands have extensive Tertiary deposits.

Eastward, the Tertiary seas were less extensive. They did not reach Moscow but covered much of southern Russia. For a short period in the Oligocene a narrow seaway lying east of the Urals extended from the Arctic Ocean to Tethys. This was the last flooding of

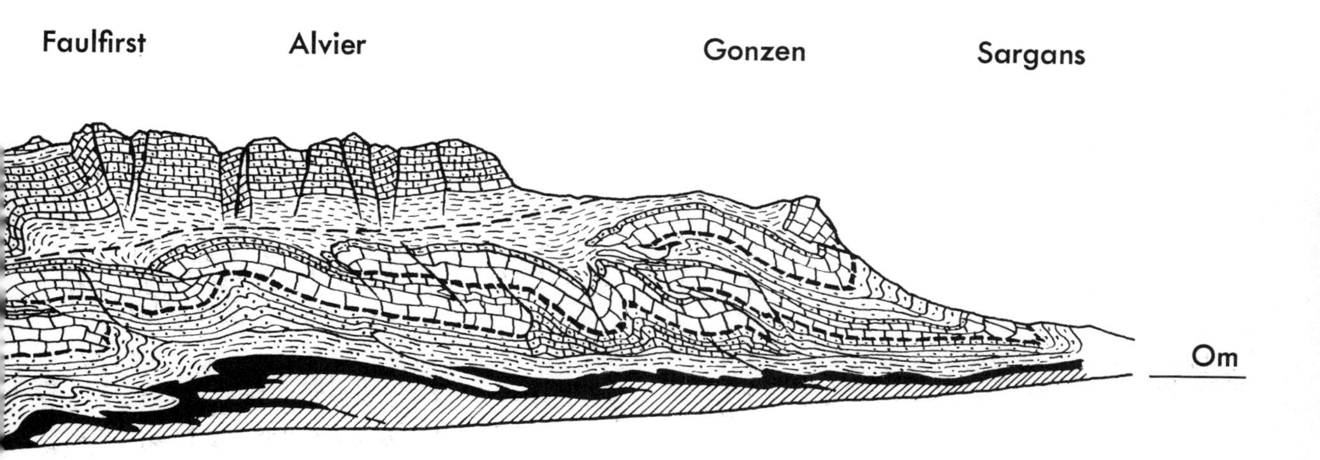

Eurasia north of Tethys. Most of Siberia and China was free of marine transgressions and received instead a variety of continental rocks including coarse conglomerates. Many large basins came into existence and trapped sediments from adjacent ranges. In these basins are well-preserved fossils of contemporary mammals and plant life. The largest land mammal known, *Indricotherium,* a rhinoceros, lived during the Oligocene in central Asia.

In extreme northeastern Siberia the Tertiary is again represented by marine rocks, and there are lava flows and granitic intrusions appropriate to the circum-Pacific zone. Deposits of valuable metals and petroleum are found in Tertiary rocks of northeastern Asia.

Southern hemisphere

The Tertiary history of southern lands is read mainly in sediments of interior lakes and basins and in narrow bands of marine sediment along continental margins. Only in the East Indies and New Zealand were thick geosynclinal accumulations laid down. Africa, as a sort of central heartland of the ancestral Gondwanaland supercontinent, remained relatively stable during the Tertiary. Extensive sloping shelves are few, and the transition from continent to ocean is unusually abrupt—witness the almost total lack of interior waterways or other indentations on Africa's coastline. A number of extensive interior basins received Tertiary deposits. The Congo Basin and its great river system began in the late Pliocene (Figure 20-14). Chad Basin in and adjacent to the southern Sahara received late Tertiary and Quaternary lake deposits with volcanic components. Much of Africa south of the Congo, especially the Kalahari Desert, is covered by the remarkable Kalahari beds, which are mainly loose or partly consolidated red sands with a wide age range.

Most important of African continental formations are those filling the long, narrow rift valleys. The rift valleys were formed by massive fracturing associated with an incipient

FIGURE 20-14 *Lake Kivu in the Congo Basin, which is filled mainly with Tertiary sediments. Thick vegetation and deep soil cover hinder geologic explorations in regions such as this. (UNESCO, photograph by Basal Zarov.)*

FIGURE 20-15 Craters on the island of Java, Indonesia. Volcanic activity and crustal unrest have characterized the southwestern Pacific for many geological periods. (Courtesy Indonesian Information Office.)

breakup of East Africa. The Red Sea and Gulf of Aden were penetrated by seas in the middle Tertiary and seem to have widened and deepened subsequently. Within the continent, large lakes such as the present-day Albert, Nyasa, and Tanganyika occupy rifted depressions, and even larger water bodies are recorded for the past. Olduvai Gorge, the Afar Triangle, and Lake Rudolf, famous hunting grounds for ancient man, are associated with the rift valleys.

All epochs of the Tertiary are represented by a wide belt of deposits on the Mediterranean coast of Africa. An embayment several hundred kilometers long entered Nigeria in the early Tertiary. Elsewhere the marginal band is relatively narrow. Many narrow depressions have been located during exploration for oil off the east and west coasts of Africa. These are logically considered as having resulted as other great landmasses split off from Africa.

The East Indies as they now exist are essentially products of Tertiary geologic events. About three-fourths of the exposed island surfaces are sediment or volcanic material of Cenozoic Age. The sedimentary rocks are chiefly marine limestone rich in corals, algae, and foraminifera. Extensive oil fields are found in Mesozoic and Cenozoic rocks of many of the larger islands. Tremendous volcanic accumulations consisting of flows, tuffs, and breccias are widespread. Many of these solidified below sea level. The East Indies today are a veritable crossroads of geologic unrest as shown by the intensive earthquake and volcanic activity (Figure 20-15). What is happening today is a continuation of what has been going on in this region for the past 150 million years.

Australia has marine, nonmarine, and volcanic rocks of Tertiary age. Outcrops are concentrated chiefly in the southeast quarter of the continent and occupy a relatively small fraction of the total area. Marine deposits are chiefly of Oligocene and Miocene age and are marginal or beneath the shallow shelves. Many scattered lake and river deposits are known, some containing fossil plants and important coal beds. The record of vertebrate life is surprisingly scanty and casts little light on the peculiar marsupial fauna's early history.

Geologically, Australia comprises a single unit with its large islands, Tasmania to the south and New Guinea to the north. The whole aggregation has been moving north and northeasterly since early Tertiary time and is currently approaching southeast Asia, where even greater collision effects may be expected a few million years from now.

South America and Antarctica

A fairly complete and varied rock record of the Tertiary Period is found in South America. All epochs of the period are represented in continental deposits east of the Andes in much the same relation as similar beds are present east of the Rockies in the United States. Numerous mammalian fossils are found in the Argentina Tertiary. The Andes and associated ranges contain much volcanic material of Tertiary age; in fact, much of the southern tip of South America is a product of Tertiary activity. As ever-increasing quantities of debris were supplied by the rising Andes, the belt of sediment widened eastward. The Pliocene deposits reach several thousand meters in thickness and extend from Venezuela to Patagonia with a great lobe down the Amazon Valley. Figure 20-16 is a view in the central Andes of eastern Bolivia.

Marine deposition was strong in the Miocene when areas near the mouth of the Amazon and east-central Argentina were inundated. Many cubic miles of sediment were swept into the Atlantic along the Argentine coast, where they make up an unusually wide continental slope and shelf.

West Antarctica has a geologic history much like that of southern South America, especially of the Andean chain. Volcanic eruptions broke forth intermittently, and there was

FIGURE 20-16 *Andes Mountains as viewed from the east in eastern Bolivia. Mount Mururata on the skyline is 5,800 m high; it is about 4,000 m from the bottom of the canyon to the summit. (Courtesy of Eugene Callaghan.)*

intensive faulting and folding. The most significant sedimentary formations are chiefly found in the Antarctic Peninsula, which points fingerlike toward South America. Fossils there are mainly of plants and marine mollusca. Mild climates are suggested by the plants, mollusca, and brachiopods. No land vertebrates have yet been found. There is evidence of close connections with South America early in the Tertiary, but similarities diminish through later time. The breaking of an originally continuous ridge to form the Scotia Arc is clearly another expression of plate tectonics.

Life of the Tertiary

In a general but unmistakable way the life of the Tertiary was modified in response to new and rigorous conditions. As shown by the preceding review of the physical environments, the continents became relatively high and rugged while the interior and marginal seas withdrew.

Extremes of heat and cold and great variations in precipitation have characterized world climates during the last 60 million years. It seems safe to say that there has been greater diversity in living conditions during this interval than during any other equivalent span of the earth's history. The rigorous environments of the Tertiary Period presented many challenging opportunities to living things. Although the great continental glaciers and the rainless deserts were almost devoid of life, the extent to which animals and plants have invaded other unfavorable habitats, such as high mountains, caves, the depths of the ocean, and the ice-free polar regions, is truly remarkable. In general, the story of Tertiary life is one of response to climatic extremes and topographic diversity, with a premium on adaptations that could overcome severe cold and seasonal changes.

Plants

All major groups of plants are represented by fossils in Tertiary rocks. Even soft and unsubstantial forms, such as mosses, fungi, molds, and bacteria, have left actual remains or indirect evidence of their existence at this time. The dwindling descendants of earlier important groups, such as club mosses, horsetails, ferns, ginkgoes, and cycads, also continued to leave a few scattered fossils. Cone-bearing plants, including sequoia, pine, juniper, cypress, fir, and cedar, are locally well represented.

The angiosperms, or flowering plants, dominate the plant world and are preserved in great profusion. This large group had become well established in the Cretaceous, and all modern families appear to have evolved by Miocene time. Remains include trunks, branches, leaves, flowers, and pollen grains. Where conditions were favorable, Tertiary plants accumulated to form coal, lignite, and peat.

Grass is the most important angiosperm of the Tertiary (Figure 20-17). The first fossil grass seeds appear late in the Cretaceous; the family now includes about 500 genera embracing 5,000 species, and there are scarcely any areas where grass will not grow.

Among the important items of food furnished by the grass family are rice, wheat, barley, oats, rye, corn, and millet. Other important members of the group include bamboo, useful as a food and as a building material; sugar cane, a giant grass that supplies half the world's sugar; and sorghum, which provides grain, forage, straw, and molasses. Although all these plants have been modified by man, ancestors of these plants provided food for plains-living mammals throughout the Tertiary epochs. The grazing habits of many large mammals must have evolved in response to the availability of grass. The grains also

FIGURE 20-17 Grassland. This remnant of undisturbed prairie in Pottawatomie County, Kansas, illustrates the type of vegetation that has dominated the western plains through millions of years of the late Cenozoic era. (U.S. National Park Service.)

FIGURE 20-18 Fossil ginkgo leaf on a slab of Miocene shale from southwestern Montana compared with the leaves from a living tree. As far as can be determined, there are no ginkgo trees growing in a wild state at the present time, but fossil remains are common in Mesozoic and Cenozoic rocks.

sustained many rodents, birds, and insects. It is difficult to appreciate how important this one plant group was to Tertiary life.

During the Tertiary, plant groups migrated extensively in response to climatic changes. In general, these migrations were marked by a withdrawal of tropical and semitropical vegetation toward the equator as the climate of the continents grew cooler and drier. The uplifting of mountain ranges such as the Alpine-Himalayan chain and the American Cordilleras profoundly affected this migration. In regions of seasonal change only those forms of vegetation that could shed their leaves and live through cold seasons in a resting stage were able to survive. In the seas and oceans, plant life also continued to evolve. The diatoms (Figure 20-20) expanded in importance and have been the basic food source in the marine world ever since the middle Mesozoic. The first known freshwater diatoms are of Cretaceous age, and there are now an estimated 4,000 nonmarine compared with 1,000 marine species.

(a)

(b)

(c)

(d)

FIGURE 20-19 (Above) Fossil nuts and seeds from the Clarno Formation of Oligocene age, Washington: (a) Bursericarpum (elephant tree); (b) unidentified Vitaceae (grape); (c) Meliosma (no modern counterpart); and (d) Juglandaceae (walnut). All except the last named are about 0.5 cm in diameter. (Specimens collected and photographed by Thomas J. Bones.)

Marine invertebrates

The chief marine invertebrates of the Tertiary, arranged roughly in order of decreasing practical importance to paleontologists, are foraminiferal protozoans, pelecypods, gastropods (Figure 20-21), corals, bryozoans, crustaceans, and sea urchins. Of course, any such arrangement is somewhat arbitrary, and it does not apply in all places for all subdivisions. Missing from the seas at the dawn of the Tertiary were ammonities, rudistids, *Inoceramus* pelecypods, and several types of oysters. Although most of the modern families of pelecypods and gastropods appeared in the Mesozoic, their great expansion occurred during the Tertiary. They are locally so varied and abundant that they serve as a basis for correlating the marine formations.

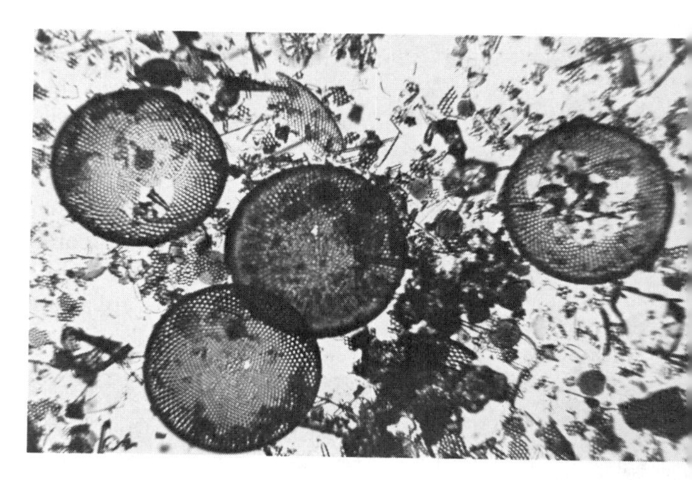

FIGURE 20-20 (Above) A highly enlarged view of diatomaceous ooze collected from the ocean bottom northwest of Honshu, Japan. (Scripps Institute of Oceanography, photograph by Taro Kanaya.)

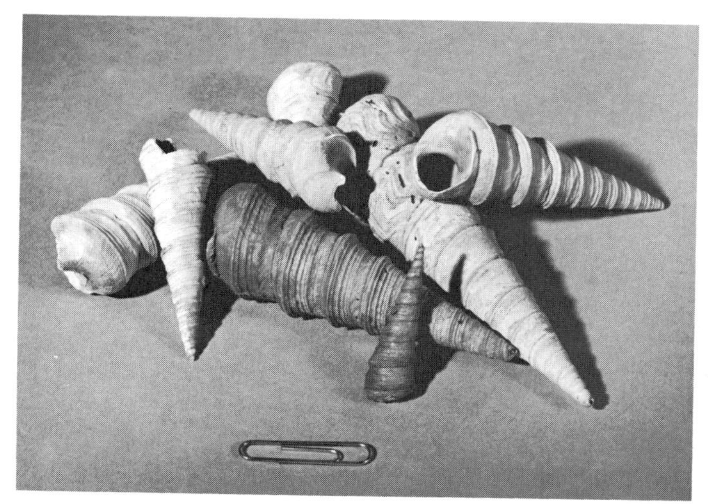

FIGURE 20-21 (Left) Fossil shells of Turitella, a very abundant gastropod of the Tertiary period (Smithsonian Institution.)

All the shelled cephalopods except the ancestors of the pearly nautilus had disappeared by the Paleocene, but the shell-less types continued to be well represented by squids, cuttlefish, and the octopus. These forms leave few fossils, and their race histories are poorly known. Squids and their relatives became plentiful in the open ocean where they competed directly with the fish.

The importance of the protozoans during the Tertiary is indicated by (1) the number of species that appeared, (2) the extent to which they contributed their shells and skeletons to the formation of shallow-water sediments and deep-sea oozes, and (3) their practical value as guide fossils in correlating oil-bearing rocks. These unicellular, calcareous-shelled organisms have left innumerable remains in all Tertiary epochs. Many forms have been discovered, but the most important guide fossils are the large varieties commonly called *nummulites* (Figure 20-22). This group derives its name from the Greek word for "coin," which refers to their flat disclike shape. These "coin" fossils are common in Egypt and other Mediterranean lands. Some nummulites are more than 2 cm across, and they occur locally in such profusion that they make up large masses of rock. The nummulites lived mainly in warm waters and are highly characteristic of the Eocene and Oligocene of the Tethys seaway. They swarmed over the coral and algal reefs and contributed their shells to the building of thick limestone formations. They are also found in deposits laid down in the warmer waters of the Western Hemisphere, chiefly adjacent to the Gulf of Mexico and the Caribbean Sea.

Many other protozoans have been found useful as guides for both local and worldwide correlations. They have been especially useful in correlating oil-producing areas of the Gulf Coast, California, Venezuela, the East Indies, and the Near East. As many as 30 zones for each of the Tertiary epochs have been established locally on the basis of protozoans.

No new major groups of invertebrates appeared, and it may even be that Tertiary marine life was somewhat less prolific than it was during the Cretaceous. The relative uplift of the landmasses during the Tertiary stimulated erosion and flooded the offshore waters with heavy loads of mud and silt. Such organisms as sponges, corals, brachiopods, bryozoans, crinoids, and other fixed forms cannot endure muddy water, and even pelecypods, gastropods, starfish, and sea urchins do not thrive in it. Unfavorable living conditions on continental shelves and the elimination of interior seas may have forced many species into deeper waters and certainly cut into their living space.

FIGURE 20-22 *Nummulites from the Eocene Giza Limestone of Egypt.*

FIGURE 20-23 *Fossil fish from the Green River Formation near Kemmerer, Wyoming. Literally millions of freshwater fish were buried and preserved in the fine-grained deposits of an extensive lake that covered parts of Wyoming, Utah, and Colorado during the Eocene epoch. The larger of the two specimens is about 40 cm long.*

Vertebrates

The Cenozoic is commonly referred to as the *Age of Mammals* in recognition of the group that contributed the dominant land-living animals during the era. But in giving due credit to the mammals, we should not overlook the fact that birds also progressed during this same period and became adapted to a great variety of environments. Likewise, in the seas the modernized bony fish achieved a supremacy equal to that of the mammals on land. We would therefore be more correct in characterizing the Tertiary as the *Age of Mammals, Birds, and Teleost Fish.* The reptiles had been reduced to relative insignificance at the end of the Mesozoic, and joined the amphibians among the ranks of the dispossessed. The story of Tertiary vertebrates then is essentially a tale of the warmblooded vertebrates. (This statement takes on particular significance when we realize that the period in which they thrived was one of relatively cold and changeable climates.)

Mammals

Conditions at the beginning of the Tertiary were ideal for mammals, and they were quick to seize their opportunities. Their ancient reptilian enemies were gone, food was abundant, and living space was almost unlimited. Mammals are characterized by such obvious physical traits as warm blood, hairy covering, efficient reproductive systems, milk glands, strong teeth, and sturdy skeletons. But more important than any of these in the struggle for survival was the gradual enlargement of the mammalian brain, which became an organ capable of storing and retaining impressions that could be used in directing subsequent intelligent action.

Although it is obviously impossible to find out a great deal about the mental processes and behavior of extinct mammals, we do have an important source of information in the form of fossil brain casts. The brain cavity inside a mammalian skull is easily filled with sediment, which may harden to preserve a perfect cast of the brain and its important nerve connections. The study of fossil skulls shows that the mammalian brains gradually increased in both absolute and relative size during the Tertiary. If the mass or complexity of the brain is a reliable guide, the most stupid mammal was a mental giant compared with his reptilian predecessors. Even the earliest mammals were relatively "brainy" in the sense that their brains were large in proportion to their bodies. As individual species grew ever larger, their brains also expanded but at a slower rate. The higher primates, including man, constitute the only known group in which the rate of brain growth kept pace with or exceeded the rate of body growth.

Fossil casts reveal another very important fact about the brain—namely, the phenomenal increase in size and importance of the cerebrum, which fills most of the cranial cavity of the mammalian skull. The growth or expansion of the cerebrum resulted mainly from the increase in its outer layer, or *cortex,* commonly known as the *gray matter.* A diagrammatic representation of the evolution of the brain is given by Figure 20-24.

We shall put off for the time being a discussion of the role of the brain in the evolution of man, but we should repeat that intelligent activity is the key to the success of mammals generally. Intelligent activity implies a degree of freedom of action—a mammal's behavior is not stereotyped and is not driven by purely reflex action. Sometimes the animal knows that he is going to do—apparently after having chosen between or among alternatives—before he does it.

Aided by superior brains, mammals were able to compete successfully with animals that were much stronger than they, to escape unfavorable environments, and to survive during periods of danger and stress. Circumstances forced the mammals to "live by their wits" during the Mesozoic. During the Tertiary, as they came to compete with one

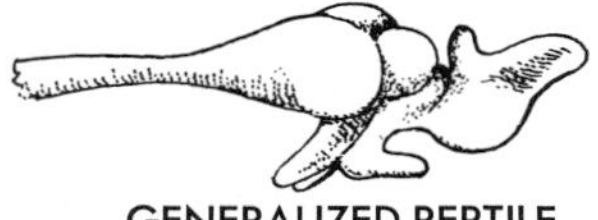

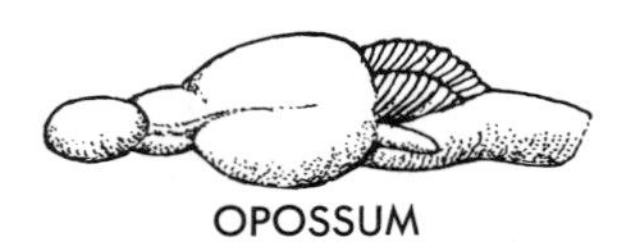

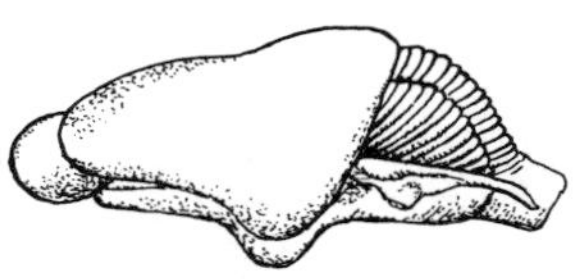

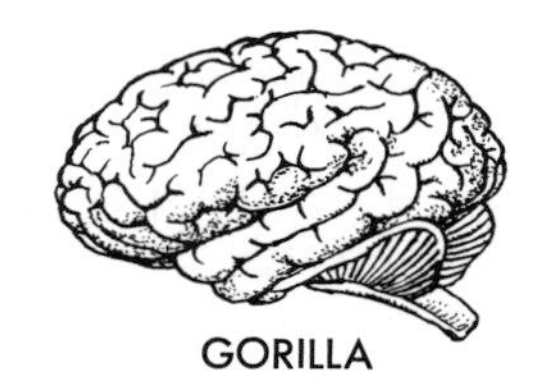

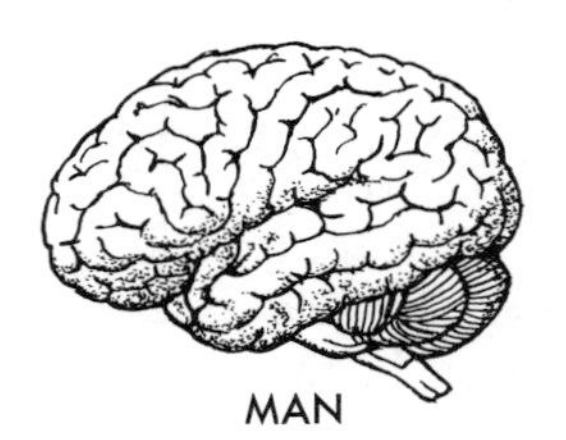

FIGURE 20-24 *The brain, from fish to man. The diagram shows the increase in size of the cerebrum and the complex foldings that characterize the more advanced forms of life. Not intended to show true relative sizes.*

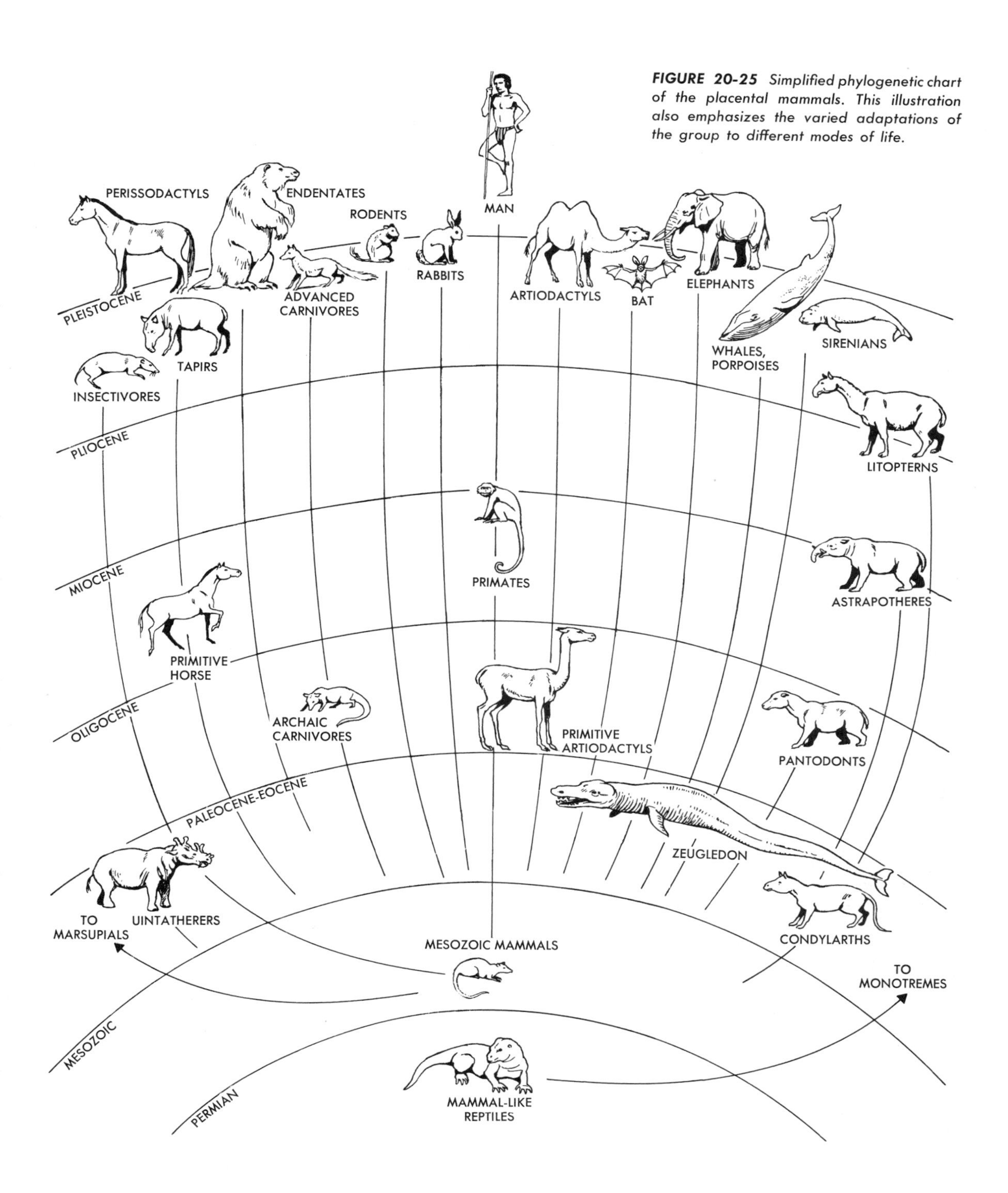

FIGURE 20-25 *Simplified phylogenetic chart of the placental mammals. This illustration also emphasizes the varied adaptations of the group to different modes of life.*

another and with an unfriendly environment, they became even more intelligent, adaptable and varied (Figure 20-25).

In general, mammals of the Paleocene were primitive, unspecialized, and relatively small (Figure 20-26). Of the 15 orders of mammals known from Paleocene rocks, only 6 have been found in the preceding Cretaceous and only 6 are alive today. The orders in existence before the Paleocene included the *multituberculates, marsupials, condylarths, primates,* and *insectivores.* Among the primitive Paleocene forms, we perceive possible ancestors of many succeeding families including the first known horse, *Hyracotherium.* However, the differences by which we distinguish many living groups of carnivores and herbivores were not yet clearly evident. The Paleocene life of Europe and North America was similar, indicating that a land bridge probably connected the two continents.

Nature continued to favor the mammals during the Eocene epoch. The descendants of the archaic Mesozoic and Paleocene forms lived side by side with the newly evolved and more progressive ancestors of modern families. Ten new orders were added to the 15 that had carried over from the Paleocene. The multituberculates, an order of mammals that originated in the mid-Mesozoic, disappeared at the end of the Eocene, after having lived through at least two geologic periods. They may have been pushed aside by the more efficient rodents. There were a number of clumsy primitive herbivores such as *Coryphodon* and *Phenacodus,* which had teeth adapted for eating vegetation but also some carnivorous characteristics such as claws, long tails, and short limbs. Among the important groups that took on recognizable characteristics during the Eocene were the carnivores and the hoofed animals. For the first time, in fact, we find true hoofed animals. Included here are odd-toed forms (*perissodactyls*), including ancestral horses, tapirs, and rhinoceroses. The even-toed hoofed animals (*artiodactyls*) spread tremendously during the late Eocene; among those appearing on the scene were the ancestors of the deer, pig, and camel, although all were small and very much alike. Rabbits were present in a recognizable form, and other small mammals such as rodents, insectivores, and primates were gaining in importance. There were bats in the air and whales in the

FIGURE 20-26 *Scene in western Colorado during the Paleocene epoch. The animal is Barylambda, a large-hoofed mammal that has no living descendants. This picture of a well-watered, semitropical landscape is based on study of the sediments and fossil vegetation associated with the vertebrate remains. (Field Museum of Natural History, Chicago.)*

FIGURE 20-27 *Middle Eocene life of the Wyoming area.* Uintatherium, *a large, six-horned, saber-toothed herbivore, dominates the scene; lower right, the primitive tapir* Helaletes; *lower center right,* Stylinodon, *a gnawing, toothed mammal; center,* Trogosus, *another gnawer; lower left,* Hyrachyus, *a fleet-footed rhinoceros. The tropical vegetation is reconstructed from fossil evidence. (Mural by Jay H. Matternes, courtesy of Smithsonian Institution.)*

ocean, but the history of both of these groups is obscure. A reconstruction of Eocene life in what is now Wyoming is presented by Figure 20-27.

South America was now completely separated from the other continents and supported a number of odd herbivorous groups. The Eocene climate was temperate and equable, and the lands were generally lower than at present. Migration between North America and Europe appears to have been possible during the early Eocene but not late in the epoch.

With the close of the Eocene came a transitional period during which the archaic mammals were rapidly eliminated and modernized forms began to spread. A few apparently indestructible forms, such as the opossum, moles, and shrews, continued to survive and were joined by rodents such as beavers, rats, and mice. New families appeared among the artiodactyls and perissodactyls: horses, deer, tapirs, rhinoceroses, camels, and antelopes, as well as true cats and dogs were all now clearly recognizable.

Among other forms that disappeared during the Oligocene were the gigantic *titanotheres,* typified by the *Brontops,* which was as large as an elephant but with weak teeth and a small brain. Another common Oligocene mammal was the *oreodont,* a grazing animal about the size of a modern sheep.

The climates of the northern landmasses were becoming cooler and drier, and the semitropical forests were retreating southward during the Oligocene. The warmth-loving primates that had been plentiful in North America during the Paleocene and Eocene now disappeared from the continent. There was a marked increase in the number of grazing animals adapted to subsist on grass. Figure 20-28 is a reconstruction of life in the High Plains area during the Oligocene.

FIGURE 20-28 *Mammals of the early Oligocene of the South Dakota-Nebraska area. Center and upper right,* Brontotherium, *a titanothere; lower right,* Merycoidodon, *a sheeplike grazing animal; lower center,* Protapiras, *an ancestral tapir; center left,* Hyracodon, *a three-toed rhinoceros. (Mural by Jay H. Matternes, courtesy of Smithsonian Institution.)*

During the Miocene, there was general uplift of the northern continents and important mountain building along the Tethys belt. With the uplift came cooler and drier climates and corresponding changes in the plant and animal world. Grasslands expanded, forests retreated, and the time grew favorable for animals that could exist on open plains. It is not surprising that the hoofed animals, able to travel widely over rough ground, should multiply considerably. There were many species of three-toed horses, rhinoceroses, giant pigs, camels, ancestral deer, primitive antelopes, mastodons, and the last survivors of the oreodonts and chalcotheres. Also making their first appearance were large panther- and tiger-like cats, the first bears and raccoons, and a host of lesser flesh-eaters, such as weasels, wolverines, skunks, and otters. Life in what is now Nebraska is shown as it existed during the Miocene by Figure 20-29.

Uplift had brought Eurasia into contact with North America via the Bering Straits, and Africa was connected with Europe and Asia. Not all types of mammals found and crossed the available land bridges however. South American life continued to evolve in isolation from the rest of the world, and Miocene deposits of the southern continent contain abundant fossils of the edentates, ground sloths, armadillos, and other bizarre native herbivores. In many ways the Miocene was a high point in mammalian evolution. Conditions were ideal for land life, and wholesale exterminations had not yet begun.

The Pliocene Epoch has been called the *autumn of the Cenozoic,* for its climate heralded the coming of the ice and cold of the Pleistocene. Pliocene mammals were clearly adapted to cool conditions and seasonal changes. Warmth-loving mammals, such as the primates, which formerly had been widespread, now lived only in restricted tropical areas. The open plains of the northern continents were inhabited by numerous highly specialized forms, many of which were giants of their races. A variety of bears, dogs, cats, wolves, antelopes, camels, horses, and mastodons roamed North America and Eurasia. There appears to have been intermittent migration between the northern landmasses and Africa, but South America still remained cut off from the rest of the world until near the end of the epoch. When the Panama land bridge between the two Americas was elevated above sea level a lively exchange of land mammals commenced. The competition was disastrous for the less adaptable South American forms, many of which were exterminated.

The Pliocene epoch on the whole produced few new types of mammals. The time of rapid expansion and diversification had passed, and an increasingly harsh and inhospitable environment was beginning to eliminate the stupid and inefficient. This weeding-out process would continue during the Pleistocene. Figure 20-30 is a reconstructed scene of life in the southern High Plains during the Pliocene.

FIGURE 20-29 *Mammals of the early Miocene of Nebraska. Foreground,* Parahippus, *a three-toed horse; center right,* Dinohyus, *a giant piglike mammal; near center left,* Stenomylus, *a small camel; center left,* Promercychoerus, *an oreodont; under the tree,* Diceratherium, *a rhinoceros; upper right,* Oxydactylus, *a long-legged camel;* Syndyoceras, *a horned artiodactyl, is near horses. (Mural by Jay H. Matternes, courtesy of Smithsonian Institution.)*

FIGURE 20-30 *Mammals of the early Pliocene of the southern High Plains. Center and center left,* Amebelodon, *the shovel-toothed mastodon; lower left,* Teleoceros, *a short-legged rhinoceros; lower right,* Hipparion, *a characteristic three-toed horse. (Mural by Jay H. Matternes, courtesy of Smithsonian Institution.)*

FIGURE 20-31 Thick seam of coal exposed by strip mining at Wyodack, Wyoming. Coal of this field is of early Tertiary age. (Photo U.S. Bureau of Mines.)

Economic products of the Tertiary

Although it was a relatively short period, many important mineral deposits were formed during the Tertiary, including coal, oil, and gas in widely scattered areas of the world and rich deposits of metallic minerals in the Western Hemisphere.

According to a recent statistical survey, 50 percent of the world's oil fields tap Tertiary rocks and are responsible for 38.2 percent of the world's total oil reserves. The Oligocene and Miocene are especially prolific in the rich Middle East fields. Other Tertiary oil and gas fields have been discovered along the Gulf Coast and in California, Venezuela, Colombia, Russia, the East Indies, and the Bass Strait, between Australia and Tasmania. We should also mention here the great Eocene oil-shale deposits in the western United States. Large reserves of coal and lignite occur in Paleocene rocks of Wyoming, Montana and the Dakotas (Figure 20-31). There are also important deposits of Oligocene brown coal in Germany and France.

Although geologists cannot always positively determine when deep-seated ore deposits were formed, most of the metal-bearing ores of western North America and South America are considered to be products of Tertiary activity. Deposits of mercury, gold, silver, lead, zinc, and copper are widespread in and adjacent to the great Rocky Mountain-Andes Cordilleras. These deposits are associated mainly with intrusive igneous rocks, and many, out of the hundreds that have been discovered, are bonanzas of the richest type (Figure 20-32). Similar Tertiary deposits occur around the western margin of the Pacific Ocean and in Japan, the Philippines, the East Indies, and Australia. Almost everywhere these Tertiary deposits have been more or less eroded, and native metals such as gold, platinum, and tin have been released to form placer deposits.

FIGURE 20-32 One of the richest hills on earth in a famous silver-mining district, Cerro de Potosi, in east-central Bolivia. The deposits are in an intrusion of Tertiary age and have been worked for almost 400 years. (Courtesy of Eugene Callaghan.)

A variety of metallic deposits—gold, silver, lead, zinc, copper, mercury, and other rarer metals—came into being in southern Europe during the Alpine Orogeny. Similar deposits were also formed in the mountain ranges of southern Asia.

Nonmetallic minerals such as clay, diatomaceous earth, gypsum, salt, phosphate rock, and building stone are mined in great quantities from Tertiary rocks at many scattered localities.

SUMMARY

During the Tertiary period the continents were elevated, topography was rugged, and rigorous climatic conditions prevailed. The Alpine-Himalayan chain was elevated, and disturbances of various kinds affected the margins of the Pacific Ocean. The East Indies and West Indies also underwent many modifications. Sediments produced during the period chiefly reflected the accelerated erosion of the continents—coarse conglomerates, sandstone, and shale were produced in abundance, and fine muds and limestone were rare.

The fringing continental shelves were considerably enlarged, and the rate of marine deposition also appears to have speeded up. All types of reefs expanded, especially the ones attached to Pacific volcanic isles.

Few new major groups of plants and animals appeared during the Tertiary, but many species evolved in response to changed conditions. Mammals were the dominant form of land life. Birds and teleost fish continued to evolve in their respective environments and became highly specialized. Plants migrated widely, and there was a general withdrawal of warmth-loving vegetation toward the equator. Grass became the dominant vegetative form in temperate regions. Marine life was somewhat restricted, but all forms that could adapt to life in the open ocean increased.

Many important raw materials came into existence: oil, gas, and coal were formed in connection with Tertiary sedimentation, and many deposits of metallic minerals were formed in association with igneous activity.

SUGGESTED READINGS

LAROCQUE, AURELE, *Molluscan Faunas of the Flagstaff Formation of Central Utah,* Memoir 78. Geological Society of America, 1960.

SCOTT, W. B., *A History of Land Mammals in the Western Hemisphere*. New York: The Macmillan Company, 1937.

Treatise on Invertebrate Paleontology, 24 parts. Lawrence, Kansas, and Boulder, Colorado: The University of Kansas Press and the Geological Society of America/A series by various authors on the major invertebrate animals. Important volumes for Tertiary fossils are those on gastropods, pelecypods, echinoderms, and protozoans.

TRUMPY, RUDOLPH, *Paleotectonic Evolution of the Central and Western Alps.* Geological Society of America, 1960; reprinted 1962.

Man and the Great Ice Age

At first glance the two topics combined in the above chapter title may seem to bear no close relationship to each other. There is a connection, however, in that modern man is essentially a product of the Ice Age, and we cannot understand his evolutionary development apart from the subjects of climatic change and glacial chronology.

Early man lived close to nature and was largely at the mercy of the elements. Climate and weather controlled his comings and goings and dictated the nature of his food, clothing, and shelter. His chief enemies were the large mammals with which he shared the earth. The ice-age environment was harsh and inhospitable, and life was full of challenges. Since constant vigilance was necessary merely to stay alive, there were definite advantages to be gained by the exercise of intelligence, by inventiveness, and by adaptability. Man, perhaps more than any other creature, was shaped by the rigors of climatic change.

21

Mount Russell and icefall tributary to Yentna Glacier, from the east. (U.S. National Park Service.)

FIGURE 21-1 A glacier-filled valley in the Swiss Alps. (Swiss National Tourist Office.)

Concept of the Pleistocene Ice Age

The effects of the last glacial period, or Great Ice Age, were first recognized in Europe. The living glaciers of the Alps (Figure 21-1) provide vivid examples of ice in action, and the effects created by them led to intelligent interpretations of the analogous but more widespread continental ice caps. An early student of the Ice Age was Louis Agassiz, who is known also for his contributions in zoology. He traveled extensively in Eurpoe and North America and was largely responsible for establishing the Ice Age as a fact of geology.

Once geologists, geographers, and biologists were convinced of the validity of the glacial theory, they searched widely for proof of its past extent. When the exact outlines of the ice masses were carefully charted, it became clear that nearly half of Europe had been submerged by ice that radiated from the Scandanavian highlands to cover 4,300,000 km^2. This sheet pushed across the North Sea and joined with local ice masses to cover all of Great Britain except a narrow strip along the southern part of the island. The Alps supported much larger glaciers than at present, and these flowed onto adjacent lowlands but did not merge with the main sheet coming from the north.

About 4,100,000 km^2 of northern Asia were covered by another icecap that moved from a center in northwestern Siberia. This sheet and ice from the Ural Mountains eventually coalesced with the Scandinavian ice in the vicinity of Moscow. The higher ranges of Asia supported glaciers even larger than those of the Alps, and ice streams from the Himalayan ranges descended to within 900 m of sea level.

The North American continental ice sheet covered no less than 11,700,000 km^2. Unbroken ice blanketed nearly all of Canada and the United States as far south as the Ohio and Missouri rivers. Large independent mountain glaciers came into being in the Rocky Mountains as far south as New Mexico and also in the Sierra Nevada. The Greenland

FIGURE 21-2 *Glacial effects, Stefansson Island, Northwest Territories. The elongate hills, called drumlins, indicate the ice was flowing northward (to the right) into Viscount Mellville Sound. (Photo courtesy Geological Survey of Canada.)*

icecap was much larger than it now is; in fact, solid ice connected it with the Canadian islands (Figure 21-2). Floating pack ice covered the Atlantic as far south as Iceland, and the surface of the entire Arctic ocean was probably frozen solid.

The great icecap of Antarctica may be regarded as a reminder of the Ice Age. It is even now almost as thick and extensive as it could possibly be because it reaches the ocean almost everywhere and shows no clear evidence of being on the decline; it may in fact be

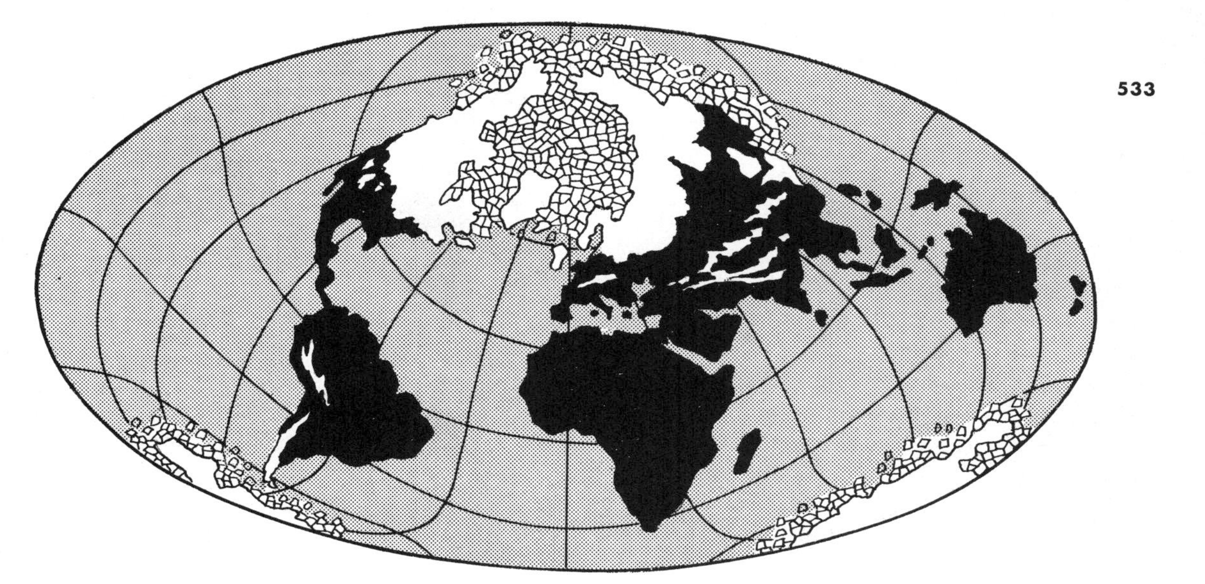

FIGURE 21-3 *The world of the Pleistocene Ice Age. Major continental and mountain glaciers shown in white, and areas of oceanic pack ice in broken patterns.*

FIGURE 21-4 *Glacial pavement showing deep grooves and striations cut by moving ice, Wasatch Range, Utah. (Photo Robert Rogers.)*

expanding. Elsewhere in the Southern Hemisphere glaciers were present in mountainous areas. South American glaciers were confined chiefly to the Andes, but along the narrow southern tip, ice spread from the highlands eastward onto the Argentine Pampas and westward into the Pacific Ocean. Africa had no extensive ice masses, but the glaciers of Mt. Kenya descended 1,600 m below the limits of the present shrunken remnants.

The glaciers of New Zealand descended below present sea level, and Tasmania supported a sizable icecap where none exists today. Mountainous islands in the open ocean, such as New Guinea and Hawaii, also had large glaciers.

It is estimated that 27 percent of the land surface of the earth, a total of 39 million square kilometers, was covered by ice during the last glacial stage (Figure 21-3). Evidence of mountain glaciers are widespread and unmistakable (Figure 21-4).

Multiple glaciations and subdivisions of the Pleistocene

The reality of a great Pleistocene Ice Age is now accepted by all students of the past, but they disagree about its duration in years and the number of subdivisions that should be recognized. It became known rather early that the Ice Age was not one long, unbroken cold period. Evidences of successive retreats and advances of both continental and mountain glaciers can be seen in countless places. Periods of ice formation (technically called *glaciations*) alternated with ice-free periods (called *interglaciations*) on a grand, possibly worldwide, scale. The interglaciations were as long or longer than the glaciations; as a matter of fact, it is generally thought that the earth is now in an interglacial stage because the length of time since the disappearance of the continental ice sheet from the United States is shorter than the time assigned to any of the previous major interglacial stages.

Glaciations are recognizable by ice-deposited materials, varved clays, scratched and polished rock surfaces, and fossils of cold-adapted organisms. Interglacials are indicated by old soil horizons, deep chemical weathering of rocks in place, some types of windblown deposits, and fossils of warmth-loving organisms.

In both Europe and America the Pleistocene has usually been subdivided into four glaciations and three intervening interglaciations. These subdivisions are shown in Figure 21-5. Many minor retreats and advances of the ice are recognized, and details differ from place to place. Much effort is being expended in correlating the record of the Ice Age that is preserved in the ocean with that of the lands. Deep-sea cores from the Atlantic are thought to indicate a greater number of possible ice advances than do the lands. Events and deposits far from the actual ice sheets are even more difficult to correlate with the advances and retreats of the ice. In such areas the current practice is to refer correlations to the early, middle, and late Pleistocene. Continued detailed studies of fossils and the application of various dating methods will eventually provide a reliable framework for identifying the subdivisions of the Pleistocene.

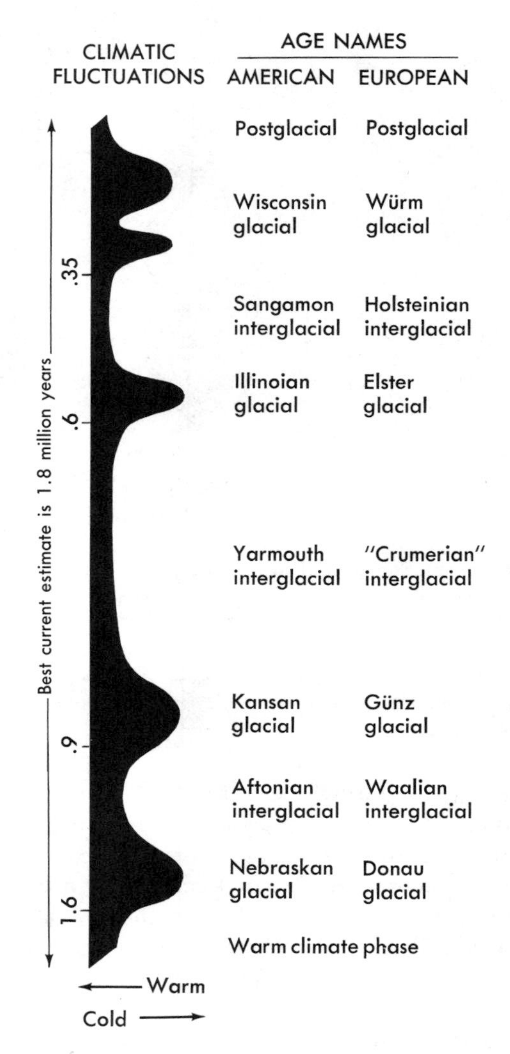

FIGURE 21-5 *Terminology of the subdivisions and correlation of the Pleistocene as commonly accepted in North America and Europe. Because of the intensified study of glacial geology, the terminology and subdivisions are subject to constant revision.*

Duration of the Pleistocene

The Pleistocene is generally regarded as having commenced with the first great ice advance. In theory this might appear to be a satisfactory clear-cut milepost, but in practice it is not easy to determine, especially in areas hundreds or thousands of kilometers from major ice masses. At least for the present, it is agreed that fossils give the best evidence for fixing the base or beginning of the Pleistocene in both glacial and nonglacial areas. For land areas the best guides are the mammals, while for the oceans small floating organisms are best.

It is one thing to determine the beginning of Pleistocene deposition in sediments, and it is another thing to assign an age in years to these deposits. Until rather recently it was customary to date the commencement of the Pleistocene at about 1,000,000 years. Now, with the advent of more reliable radiometric methods, the evidence is that this estimate should be increased to about 1,800,000 years. Clearly a great deal of work must be done before the recorded events of the Pleistocene are unified by a good dating system. The information included in Figure 21-5 is the best currently available.

Beyond the ice

There is more to an ice age than just ice. Glaciers are an essential feature, it is true, but attendant climatic influences go far beyond the icecaps. Because of the effects of the ocean and atmosphere in transporting heat and cold, it is impossible to extensively alter the weather or climate of one area without affecting other areas as well. We see a good example of this in present-day Antarctica, which is called the "weather factory of the world."

Wind and storm patterns

Many attempts have been made to reconstruct the weather and climate of an ice age. It may seem paradoxical that it is the weather that brings on glaciation, and yet the weather itself is, in turn, profoundly influenced by the ice sheets once they are formed. We may ask if glaciers are a cause or a result of an ice age. For the present we must leave this question unanswered and attempt to portray the climatic conditions of the earth during a time when the glaciers were near maximum size.

Since cold air sinks and warm air rises, we would expect the great ice sheets to be centers of dispersal of outflowing cold air. Opposing the colder and heavier air from the glaciers would be warm air from the equatorial regions. The belts or zones where cold and warm air meet would be the places of storms and high precipitation. These zones would in general follow the edge of the icecap and would be pushed ahead of a continental glacier as it expands. Another effect of glaciation would be to compress and intensify the climatic zones between the ice and the equatorial regions. This shifting and intensification of storm belts must have resulted in increased precipitation in areas that were previously relatively dry if not actually desertlike.

Lakes and rivers

The Ice Age brought great changes to lakes and river systems of the world. Lakes expanded in arid and semiarid latitudes. In the western United States, Lake Bonneville (Figure 21-6) and Lake Lahontan appeared in the Great Basin. In the Sahara, Lake Chad expanded to cover thousands of square kilometers. Deeper water bodies such as the Caspian Sea and the Dead Sea also oscillated in volume with glacial conditions.

FIGURE 21-6 Ancient beaches of Pleistocene Lake Bonneville in the Terrace Mountains near Great Salt Lake, Utah. Water stood at the highest levels about 20,000 years ago. (Courtesy of Peter B. Stifel.)

FIGURE 21-7 Farmland near Larimore, North Dakota. The fertile soil that supports this agricultural area has developed on marginal deposits of the last major advance of the continental glacier. (U.S. Department of Agriculture, photograph by B. C. McLean.)

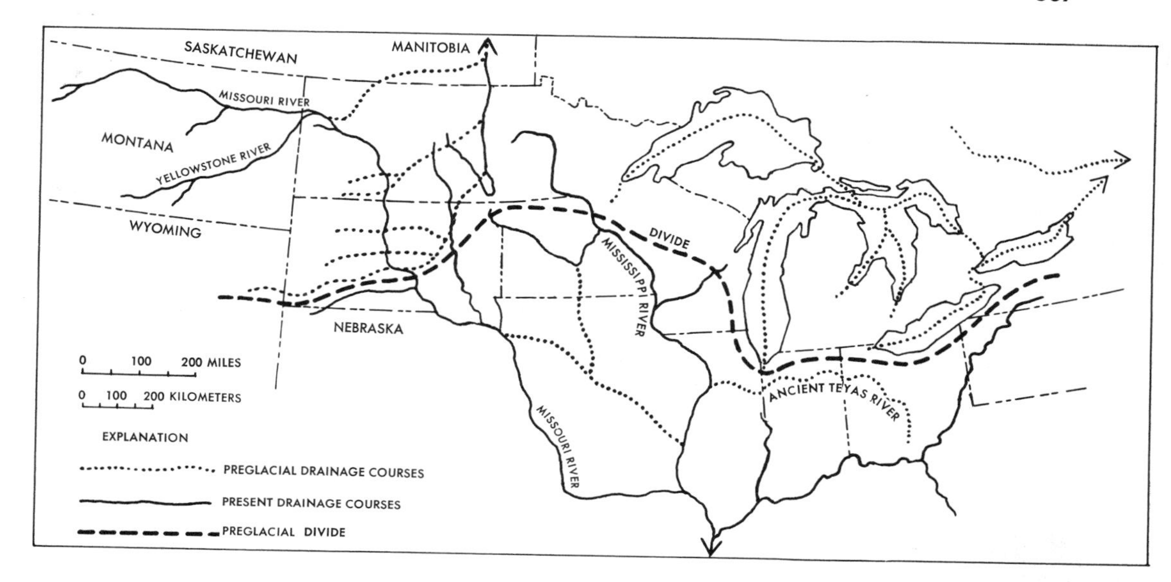

FIGURE 21-8 *Drainage changes caused by the North American ice sheets. Present rivers shown as solid lines, preglacial courses are dotted. Note position of former divides.*

Other lakes were formed by direct ice action. In these the water was held in check by glaciers that dammed up the normal drainage ways. Such lakes would invariably disappear as the ice melted. The greatest known ice-dammed body of water was Lake Agassiz, which inundated over 260,000 km^2 in north-central Canada and adjacent parts of the United States (Figure 21-7). Since the water was held in check mainly by the Canadian icecap, when this disappeared so did the lake. Other lakes, numbering in the tens of thousands, occupy depressions formed by continental ice sheets. The characteristic appearance of much of Canada and the Baltic region of Europe is due to abundant glacial lakes.

Mountain glaciers left lakes of two kinds: (1) those occupying basins gouged out at the head and along the upper courses of the ice streams and (2) those dammed up by the moraines at their lower ends.

Ice ages affected rivers in many ways. Some systems were completely obliterated by being overridden by ice. Some were forced to empty into entirely different bodies of water from what they did before the ice age. All of the upper Missouri drainage flowed to Hudson Bay in preglacial times (Figure 21-8); now it flows into the Gulf of Mexico. In a few cases, rivers were actually reversed, as in the famous Finger Lake region of New York.

The oceans

The ocean is the great thermal regulator of the earth. Water warms and cools slowly, and as it circulates in the massive currents of the ocean, it carries warmth to northern regions and cools the equatorial regions. A peculiar physical property of water is that it expands as it freezes; thus ice will float and not accumulate on the bottoms of water bodies. Water cooled to the freezing point is nevertheless heavier than warm water, so that it sinks in polar regions and flows toward the equator beneath the warmer higher layers. Oceanic circulation is very complicated, but the tendency is always to equalize the temperature, pressure, and salinity throughout.

The bearing of the above facts on our present subject is that a lengthy period of glaciation will tend to cool the oceans. Cold water derived from polar icecaps and cooled by radiation in the open ocean will tend to sink to lower levels, and so, given ample time, the ocean will have drastic effects upon sea life and may drive many species to migration or actually to extinction. The impact of cold oceans on the weather and climate of the continents will be somewhat delayed, but inevitably the temperature and precipitation must be lessened.

FIGURE 21-9 *Extent of the continental shelves, shown in black, about the margins of North and South America. The shallow, submerged shelves have a combined area of 3,550,000 square km for North America and 1,570,000 square km for South America. (Courtesy of U.S. Geological Survey, reproduced from Bulletin 1067.)*

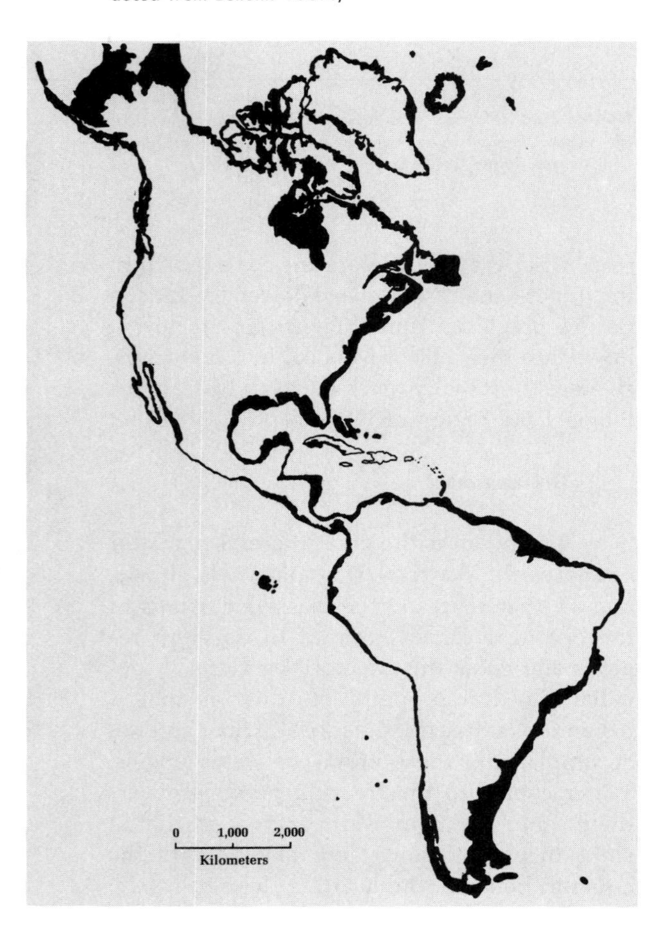

The Pleistocene Ice Age brought drastic worldwide changes in sea level. As glaciers accumulated on land, the level of the oceans fell; as glaciers melted, sea level rose. It has been calculated that 44,000,000 km^3 of water, or about 5 percent of all the water on earth, was locked in glaciers at the height of the last Ice Age. At the period of maximum withdrawal, the ocean was 105 m below the present levels (Figure 21-9). Obviously, from the persistence of the Greenland and Antarctic icecaps, all the ice did not melt. In fact, if all existing ice should melt, there would be an additional sea level rise of 60 m.

Remember that changes of level due to the addition or withdrawal of water are superimposed upon other effects, such as those due to the rise and fall of landmasses, movements in the ocean basins, sedimentation, and changes in the shape or motion of the earth. It is known for instance, that the submergence of the New Jersey coast has been about 1.5 m per thousand years for the past 2,600 years, before which it was about 3.0 m per thousand years. An independent study of the Connecticut coast shows a submergence of 2.7 m in the last 3,000 years and 10.0 m in the last 7,000 years. If continued, this submergence could flood most of the major cities of the world; New York, Tokyo, Paris, Berlin, and many others are within a hundred meters or so of sea level. Marine installations and port facilities everywhere would have to be abandoned or rebuilt. A reverse trend of falling sea level might appear to be less dangerous, but it would be an

FIGURE 21-10 *Curves from three localities on the Atlantic and Gulf of Mexico coastline showing the rise of sea level in the last few thousand years. Carbon-14 dates from organic material buried in submerged shoreline sediments provide the points on the curves. [After David Scholl and Minze Stuiver, Recent Submergence of Southern Florida: A Comparison with Adjacent Coasts and Other Eustatic Data: Geological Society of America Bulletin, 78 (1967), 449.]*

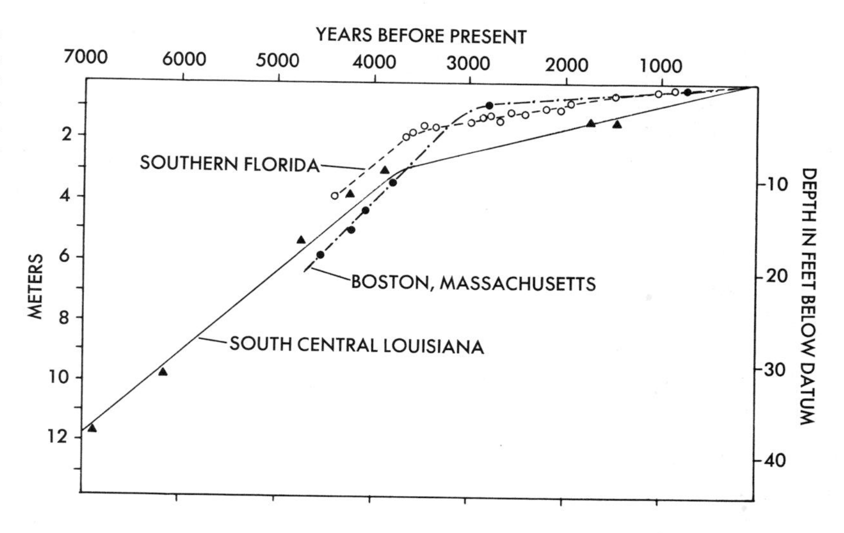

FIGURE 21-11 *Great Britain and nearby parts of continental Europe, showing the wide tracts that lie less than 180 m below sea level. During the Ice Age there was free communication between the islands and the Continent, and plants and animals migrated in both directions. A large river, of which the Rhine and the Thames were tributaries, drained into the North Sea. The last separation of Great Britain from the mainland took place about 7,000 years ago. Present lands are shown in light grey, chief rivers and water over 180 m deep in dark grey, and shallowly submerged land in white.*

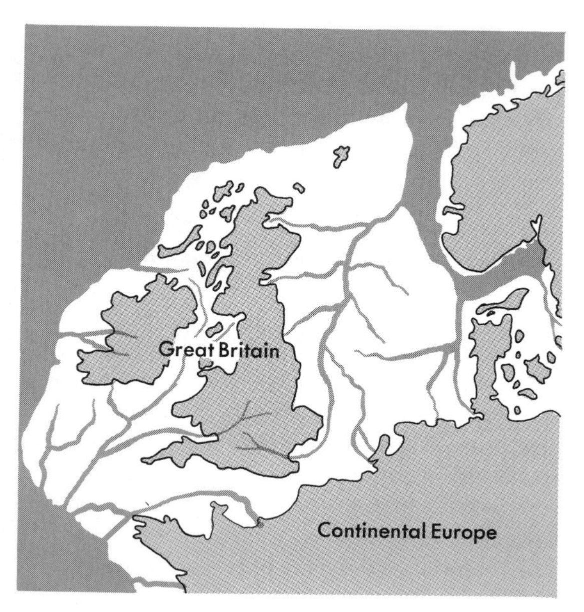

ominous sign that glaciers were again building up to devastate the lands. Figure 21-10 shows in graphic form the rise of sea level along the eastern shores of the United States.

Although we are individually not likely to be greatly disturbed by either rising or falling sea level, the effects have been drastic and far-reaching in the past. Great tracts of low-lying land equal in total area to a good-sized continent were alternately laid bare and inundated with each glacial-interglacial interval. By any standard this territory was and is choice living space for plants and animals. Its possession alternated between denizens of the ocean and the inhabitants of the land. As far as land animals, including man, were concerned, these tracts had to be abandoned as seas rose and recolonized as they retreated. Also affected were many shallow banks and islands, which were not only valuable in themselves but were also migration routes between major landmasses. A sea-level drop of 90 m united the Old and New World by converting the Bearing Straits to dry land; Ireland joined England and both became part of Europe (Figure 21-11). A shallowly submerged ridge also appeared across the mid-Mediterranean to join Italy with Africa by way of Sicily. Literally thousands of East Indian islands became elevations on a vast low-lying plain extending southeastward from mainland Asia. Examples are endless; and examination of any good map showing elevations of land and depths of water illustrates what can happen and, in fact, has happened with relatively minor rises and falls of sea level.

Effects of the Ice Age on plants and animals

The extensive changes of climate and physical geography that came with the Ice Age were responsible for correspondingly great changes in the biological realm. Although no radically different forms of life, except modern man, appeared during the Pleistocene, there were notable evolutionary changes at all levels. The really significant impact of the Ice Age occurred through the enforced *migration, mixing,* and *isolation* that developed as ice sheets contracted and expanded. New living space became available as low-lying lands and islands were laid bare of ice or water. Corresponding contractions resulted as the sea level rose during interglaciations. Migration routes were opened and closed on a scale unknown under nonglacial conditions. Tropical regions, far from actual ice, were affected by changes in precipitation and by variations in the volume and capacity of rivers. Even the depths of the ocean were stirred by influx of cold water or by changes in current directions.

FIGURE 21-12 *Inhospitable Antarctic terrain offers little opportunity for colonization by any type of life. (Official photograph, U.S. Navy.)*

Each advance of glacial ice totally depopulated millions of square kilometers of the earth's surface. Present day Antarctica is still in the grip of the ice (Figure 21-12). Although most plant and animal species were able to escape extermination by retreating before the ice, their living space was drastically reduced. To put it another way, the total food production of glaciated lands was greatly curtailed, and fewer individuals, though not necessarily fewer species, were able to exist. Organisms were not forced to adapt to climatic conditions for which they were entirely unsuited; rather, the climatic zones themselves were greatly compressed. Thus, the tundra belt, now perhaps a thousand kilometers wide in North America, was narrowed to a few tens of kilometers adjacent to the expanded ice fields. During the glacial advances, for example, species that are now found only in Canada were found in the United States, together with those that are now here.

Enforced migrations of mammals

It is not unusual to find fossils of cold- and warmth-loving animals alternating with one another in the sediments and cave deposits of glaciated areas. Thus, in central and western Europe the glacial deposits contain woolly rhinoceroses, mammoths, lemmings, reindeer, arctic foxes, and moose—forms now extinct or confined to more northern lands. The interglacial deposits of the same area contained fossils of lions, rhinoceroses, hippopotami, and hyenas, now characteristic of African climates.

The island of Malta, now isolated in the Mediterranean, has yielded reindeers, arctic foxes, mammoths, bisons, horses, and wolves, recording not only much cooler climates but also connections with the European mainland. Similar less spectacular faunal changes occurred in North America. Reindeer and woolly mammoths reached southern New England, and moose lived in New Jersey during the glacial stages. During warmer periods, sea cows, now found in coastal waters off Florida, ranged as far north as New Jersey, and the tapir and peccary roamed Pennsylvania. Ground sloth remains have been found as far north as Alaska. Elephants were isolated on the Channel Islands off the coast of California.

The history of the musk ox during the Ice Age is instructive. Its bones have been found in Iowa, Nebraska, and Minnesota, recording a time when tundra conditions prevailed in the north-central United States. Today, the musk ox lives only in the far northern reaches of continental Canada, the Arctic Archipelago, and northern Greenland (Figure 21-13). The animal has evidently moved with its customary environment as the environment has shifted with the ice front across a distance of about 3,000 km.

FIGURE 21-13 *The musk ox, a survivor from the Ice Age. Scattered remains prove that the musk ox migrated widely as the ice sheets advanced and retreated in the Northern Hemisphere. (Courtesy Field Museum of Natural History.)*

Displacements of Pleistocene vegetation

Pleistocene plants belong mostly to still-living species, but there have been great displacements and mingling of plant groups. Indeed, the present distribution of plants is difficult to explain without reference to the Ice Age. In Europe the forests of hardwood that characterized the Tertiary were considerably reduced by the advancing Scandinavian and Alpine ice sheets. In North America, where the mountain ranges trend north-south, the hardwood forests advanced and retreated across open lowlands and were not imprisoned as they were in Europe. As glaciers retreated, the arctic floras followed the margin of the ice and also ascended mountains where favorable cool conditions still prevailed. Thus, once-continuous plant populations became more and more widely separated, one group retreating northward and the other ascending available mountains. Some plants that now grow along the highest slopes of the White Mountains in New Hampshire and on Labrador have disappeared from the intervening lowlands. On Greenland species grow that are found elsewhere only in the Alps and Himalayas. The great vegetational flux of the Pleistocene is still going on, and is of tremendous importance to man.

The study of pollen grains has thrown considerable light on climatic fluctuations and plant migration during the Pleistocene. Pollen is produced in great quantities by many plants, and is so resistant to decay that it remains recognizable for extremely long periods. Experts can identify the grains produced by specific plants, and are thus able to reconstruct the general composition of the vegetation of particular times and places (see Figure 21-14).

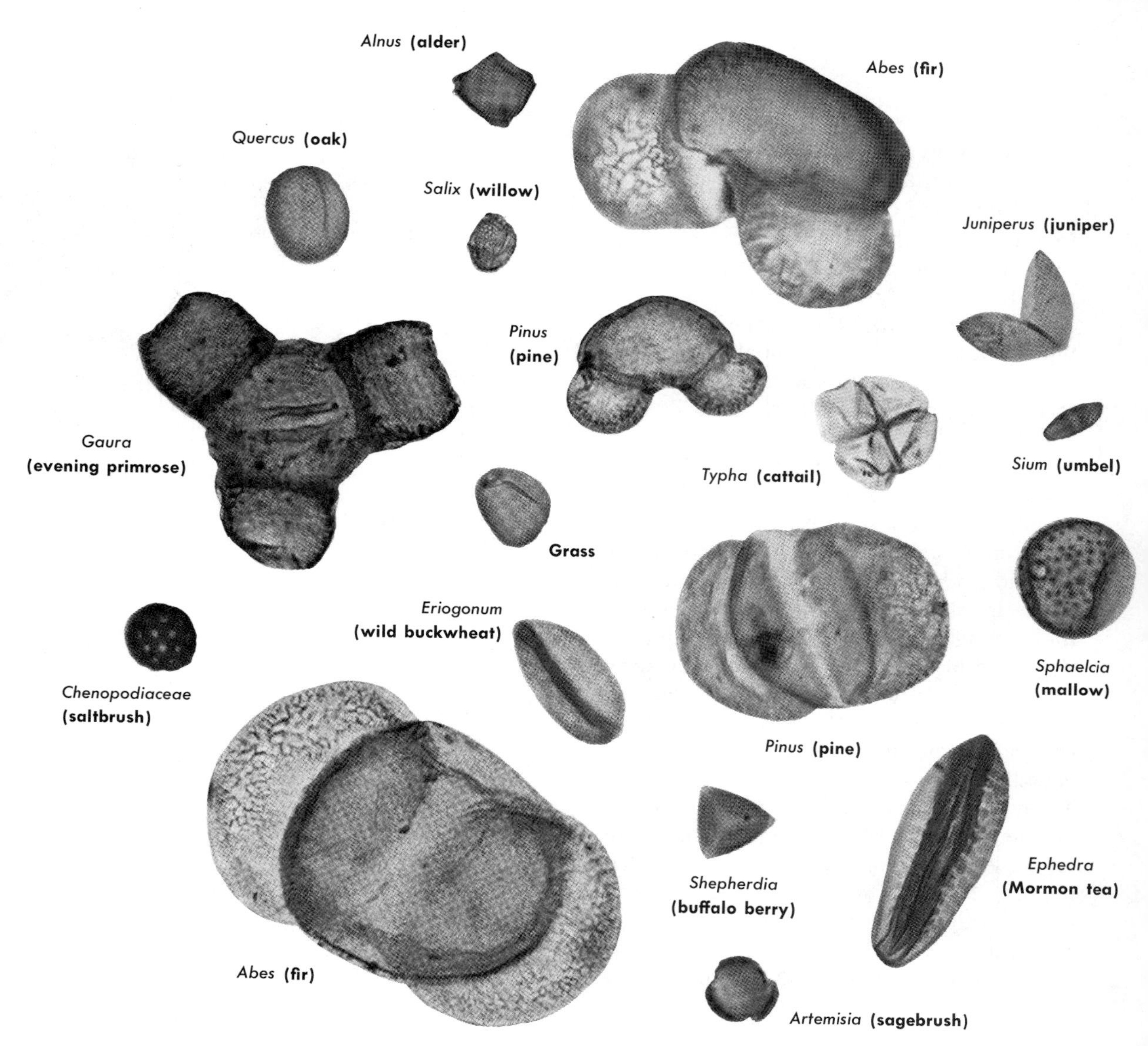

FIGURE 21-14 *Pollen grains from Pleistocene deposits of Las Vegas Valley, Nevada. This assemblage is indicative of semiarid conditions. (Courtesy of P. J. Mehringer, Jr., and the Nevada State Museum.)*

Since plants are very sensitive to environmental factors such as temperature and moisture conditions, the analysis of pollen offers perhaps the best means of discovering what the climates of the past were really like.

Such species as birch, spruce, and fir indicate cold and moist conditions; pine signifies warmth and dryness. Oak, alder, and hemlock suggest warm, moist surroundings; and an absence of tree pollen coupled with an increase of pollen from arctic herbs would indicate a tundra environment. Pollens of grass and drought-resistant shrubs are a sign of dryness. Even the pollens of cultivated plants such as corn yield important clues about the agricultural and food habits of early man.

Fate of the giant mammals

The Pliocene and Pleistocene were characterized by giant mammals (Figure 21-15). Almost every group of mammals produced one or several colossal members. These are abundantly preserved on all continents, occasionally in great "graveyards" such as the La Brea Tar Pits in Los Angeles, California, or the Big Bone Lick in Kentucky. Some of the fossils are so recent that their unpetrified, dried, mummified, frozen, or embalmed remains are still to be found.

Among the most common and well-known giants of the Pleistocene were the *mammoths* and the *mastodons.* The Imperial Mammoth attained an average height of 4 m at the shoulders and had great curving tusks reaching up to 4 m in length. Mammoth and mastodon bones are surprisingly abundant; over 100 mastodon skeletons have been recorded from New York State alone (Figure 21-16). Another famous Pleistocene animal is the saber-toothed "tiger," *Smilodon,* of which hundreds of individuals left their bones in the La Brea Tar Pits. There were true lions much larger than the present "king of beasts," and bears more massive than the grizzly. The giant beaver, large as the black bear, could topple the largest trees, and there were other rodents of proportionally large size. Several kinds of bison roamed the American West, one species with a horn spread of over 2 m. Large camels, pigs, and dogs lived in North America, together with the huge ground sloth, heavy as an elephant, which reared clumsily on its hind legs to browse on foliage 6 m above the ground.

FIGURE 21-15 Some giant mammals, birds, and reptiles of the Pleistocene. The bar near each figure represents 1 m.

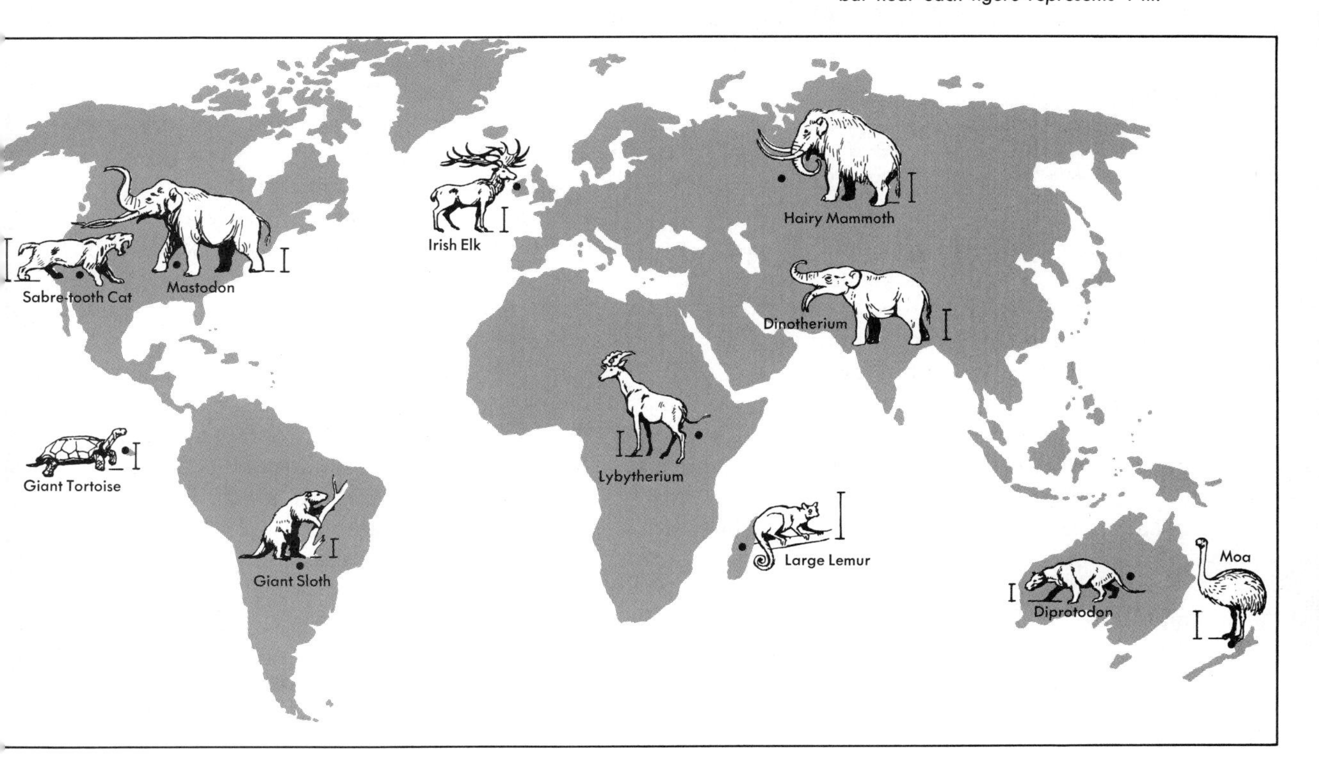

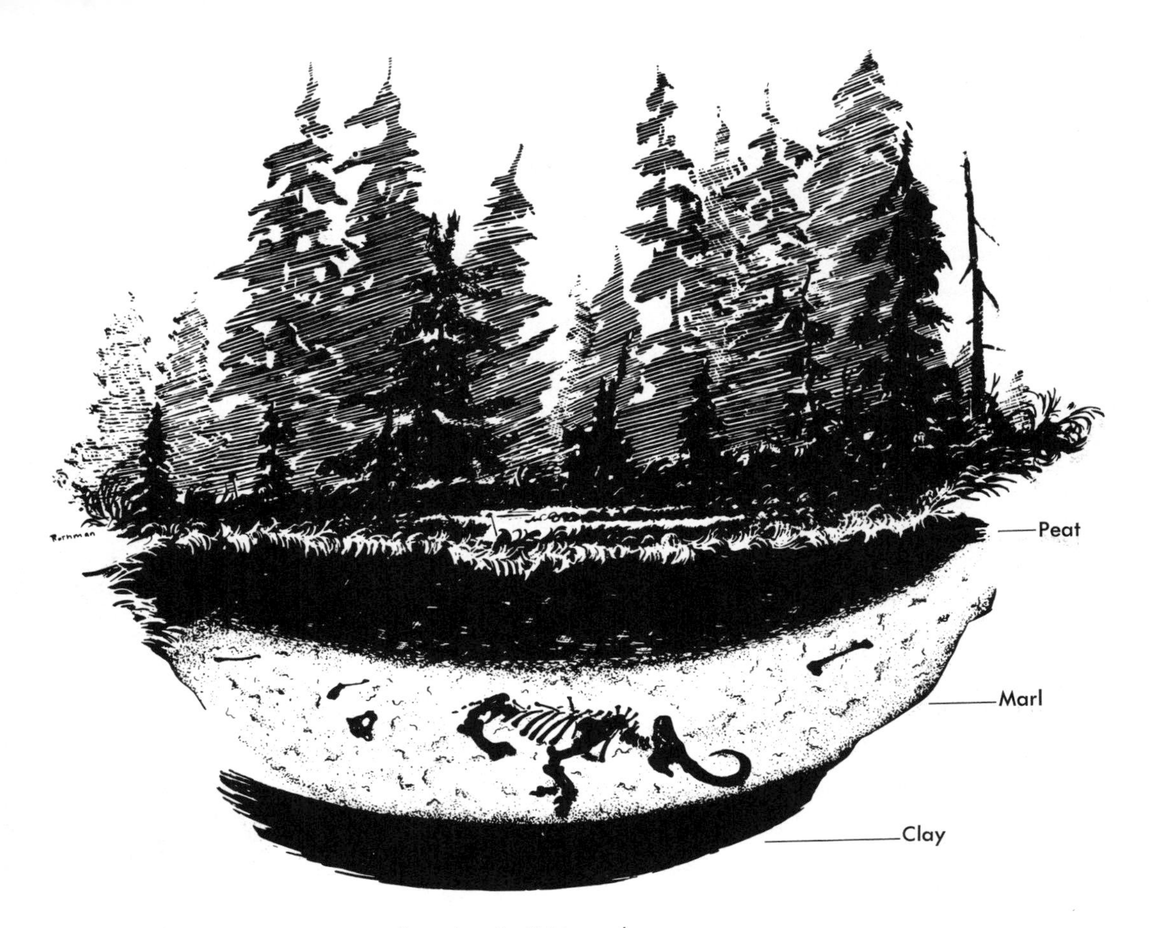

FIGURE 21-16 Diagrammatic illustration of a Pleistocene bog deposit with fossil mammoth skeleton. This mode of preservation is common in the glaciated portions of the United States. (New York State Museum and Science Service.)

Other continents also had their giants. In Africa there were pigs as big as a present-day rhinoceros, sheep that stood 2 m at the shoulders, giant baboons larger than the gorilla, and ostrich relatives over 4 m tall. From South America come fossils of the giant anteater, the glyptodon, and the sloth, together with rodents big as calves. Here also were abundant large, flightless, flesh-eating birds, up to $2\frac{1}{2}$ m tall and with 38-cm beaks. Even Australia had giant kangaroos and other marsupials.

Most of the giant Pleistocene land animals can be traced to smaller ancestors in preceding epochs of the Tertiary. The gradual increase in size is a tendency that is observed many times in the history of land life and presents no particular problems. It is puzzling, however, to find that most of the giants survived the repeated glacial onslaughts, only to disappear within the last few thousand years. This condition is particularly true of North America, where glaciation was both extensive and intensive. A recent survey gives the following estimated dates of extermination of some key forms in North America: saber-tooth cat, 14,000 years ago; woolly mammoths, 10,500 years ago; ground sloth, 9,500 years ago; native North American horse, 8,000 years ago; Columbian mammoth, 7,800 years ago; and mastodon, 6,000 years ago.

Man has been blamed for the destruction of many giant mammals, but such a charge is difficult to prove and can scarcely be true for those animals that disappeared before his arrival. However, for later species, available evidence suggests that man accomplished by indirect means what he could not do with the primitive weapons then at his disposal. His probable ally was fire, which, by accident or design, he applied to forest and prairie with catastrophic results. We know that burning out forests to clear agricultural plots was a common practice among earlier inhabitants of northern Europe, and certain American Indians also were arsonists when necessary.

The North American "big game" of today is but a pitiful remnant of small and medium-sized mammals left over from the Ice Age. And Africa today suggests, but does not duplicate, the teeming life that was characteristic of other continents during the Tertiary and Pleistocene.

Man in an Ice Age setting

The history of modern man and of his immediate predecessors is mainly a story of the Ice Age. It is still debated whether true man existed in the Pliocene. That man as such should emerge when he did is not entirely coincidental, for the challenge of the environment placed a premium on just such characteristics as were needed to put the "final touches" on the human race. Figure 21-17 illustrates the evolutionary history of primates leading into the Pleistocene and Recent.

FIGURE 21-17 *Evolutionary history of the primates.*

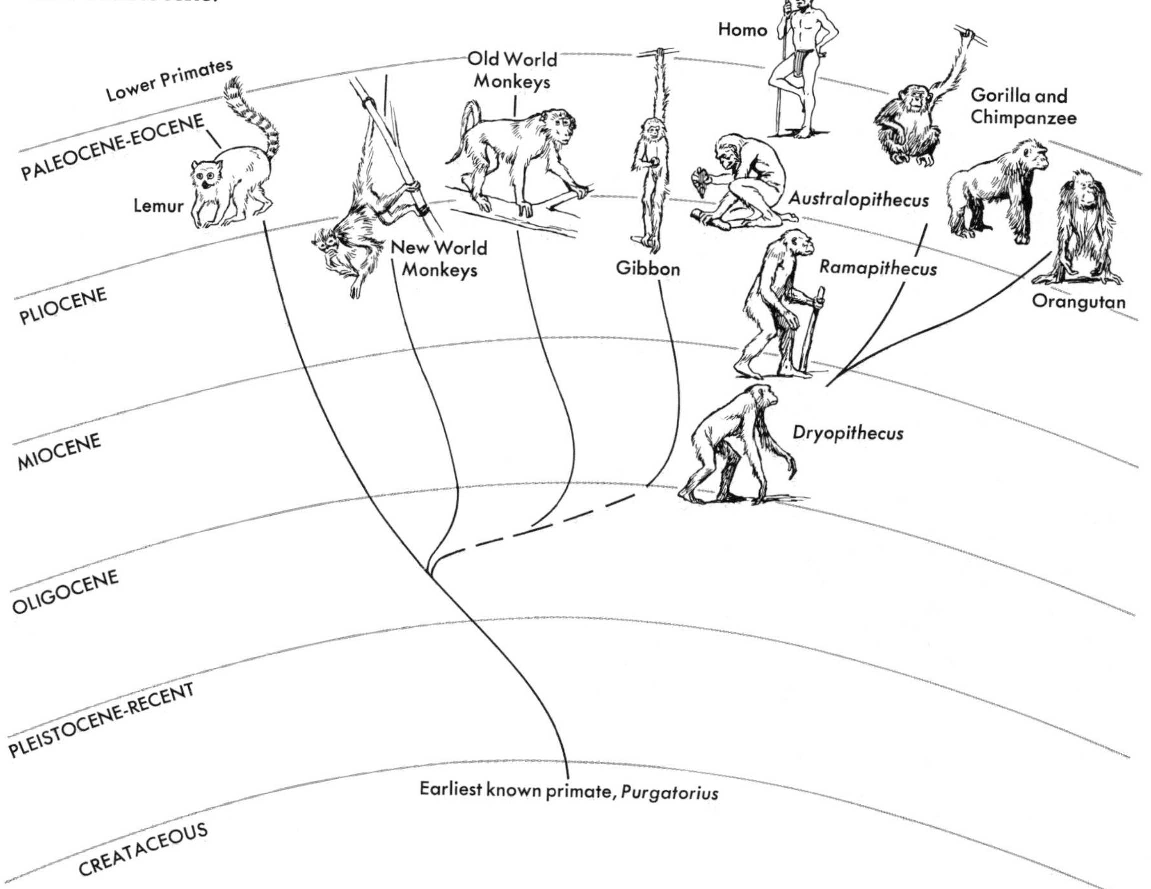

The Pleistocene and Pliocene were a time of accelerated climatic change and unsettled conditions. In the plant world there was a general retreat and destruction of junglelike tropical and subtropical forests. Grasslands and tundra expanded at the expense of trees; large grazing animals with herd instincts were favored over browsers and tree dwellers. There was therefore more opportunity for life on the ground and in the open. Man himself has followed this line of development. The scarcity of such forest foods as monkeys and apes found desirable forced man and his immediate ancestors to adapt to other sources. The shift to an increasingly carnivorous diet was natural. At first only small, easily killed creatures were taken—lizards, birds, tortoises, and fish. The enticement of larger game led to the development of weapons and traps. More than artificial aids were needed; raw courage, skill, and daring became matters of survival. The need for cooperative effort and family groupings is also obvious.

Climatic changes insured that no place or mode of existence became permanent. Cold taught man the comforts of clothing and benefits of fire. Life in caves, where available, fostered the accumulation of material culture and community existence. Competition on many fronts weeded out the weak and inefficient and drove the survivors to further efforts, which at length reached the plane of conscious and purposeful planning and premeditated action. Some of man's chief advances are briefly described in the following pages.

The Australopithecines

It is now fairly certain that the family of man (Hominidae), of which we are the sole surviving members, had a beginning at least 2,500,000 years ago and possibly even more that 5,500,000 years ago. An extension of the Age of Man has been one of the more significant results of research during the past decade. Centers of discovery and research have been South and East Africa, which may be the veritable cradle of mankind. The crucial problem of dating has been greatly lessened by application of new techniques, particularly the potassium-argon method. Datable volcanic materials are fortunately associated with most of the East African discoveries but not with those of South Africa.

FIGURE 21-18 Australopithecus africanus transvaalensis, *discovered at Sterkfontein, South Africa. This skull furnished the first certain proof of the small brain size of the australopithecines. Its probable age is in excess of 2 million years. (Wenner-Gren Foundation.)*

FIGURE 21-19 Exposure of the Pliocene Shungura Formation in southwest Ethiopia. The ridges are carved from westward-dipping river deposits; prominent bands are volcanic tuff deposits from which ages ranging from $1\frac{1}{2}$ to 3 million years have been obtained. The site has yielded many remains of Australopithecus. (Photo courtesy of F. H. Brown.)

Fossils of manlike creatures began to turn up in early Pleistocene deposits of South Africa in 1924. These were variously termed "ape-men," "man-apes," and "half-men," but the more precise name *Australopithecines* (from the generic name *Australopithecus* meaning "southern ape") soon became established. Specimens were taken from Swartkrans, Sterkfontein (Figure 21-18), and Makapansgat in South Africa; collecting still continues at these sites.

Recently (since 1959) many additional remains referred to as *Australopithecus* have been found in East Africa, specifically at Olduvai Gorge and Lake Natron (Tanzania), East Rudolf (Kenya), and Omo and the Afar Triangle (Ethiopia). Exploration and discoveries are still going on in these areas, and there is an understandable tendency on the part of individual investigators to adopt cautious and conservative viewpoints in dealing with their many new finds. Past experience has shown that the urge to rush into print with new names and bold assertions about the position of any new fossil in the family tree is likely to lead to error and embarrassment.

The Australopithecines are represented by dozens of specimens, mostly teeth, but including skulls and other skeletal elements. From the material at hand it is apparent that the group was extremely variable and may include at least three species or varieties. The following names have been proposed: *Australopithecus africanus* (relatively small), *Australopithecus robustus* (large), and *Australopithecus boisei* (largest) (Figure 21-20). The calculated body weights are respectively 32 kg, 40.5 kg and 47.5 kg. Compared with modern man, the

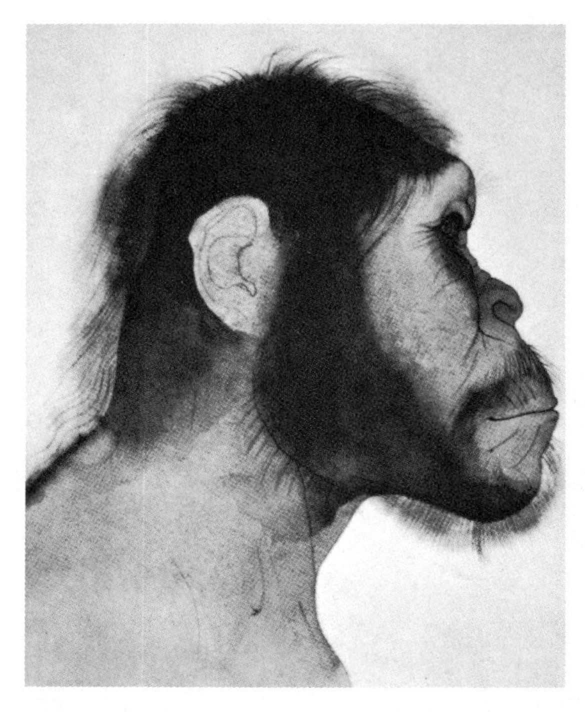

FIGURE 21-20 *Reconstruction of* Australopithecus, *a hominid that lived in East Africa during a period of several million years. There seems to have been much variation among australopithecines from place to place and with the passage of time. (NERC copyright. Photograph reproduced by permission of the Director, Institute of Geological Sciences, London.)*

Australopithecines were nevertheless rather small (an average Australian aborigine is 57 kg, and a chimpanzee is 45 kg). They walked erect, a pose probably acquired 3 to 4 million years ago, and stood between 1 and 1.2 m tall.

Some of the Australopithecines made and used tools of primitive sorts, including crude choppers 2.5 million years ago and more advanced hand axes 1.5 million years ago. No evidence of the use of fire has come to light.

Another branch is found

In 1972 the Afar Depression of Ethiopia began to be explored for hominid fossils. Finds have been relatively abundant, but as of now these have not been fully studied and described. Teeth and jaws found in 1974 and dated at over 3.25 million years have been determined as being more like *Homo* than *Australopithecus,* showing that man may have set a separate course much earlier than previously supposed.

More important is the finding of a yet-unnamed type that is probably neither *Homo* nor *Australopithecus.* Remains pertain to one individual and include vertebrae, pelvis, limb bones, teeth, and skull parts. About 40 percent of the skeleton is represented, making this the most complete hominid yet found in Africa.

FIGURE 21-21 *Fragment of a human humerus from northwestern Kenya, Africa. Dated at 2.5 million years old, this very diagnostic bone adds its weight to the story of primate evolution. (Photo courtesy of Harvard University News Office.)*

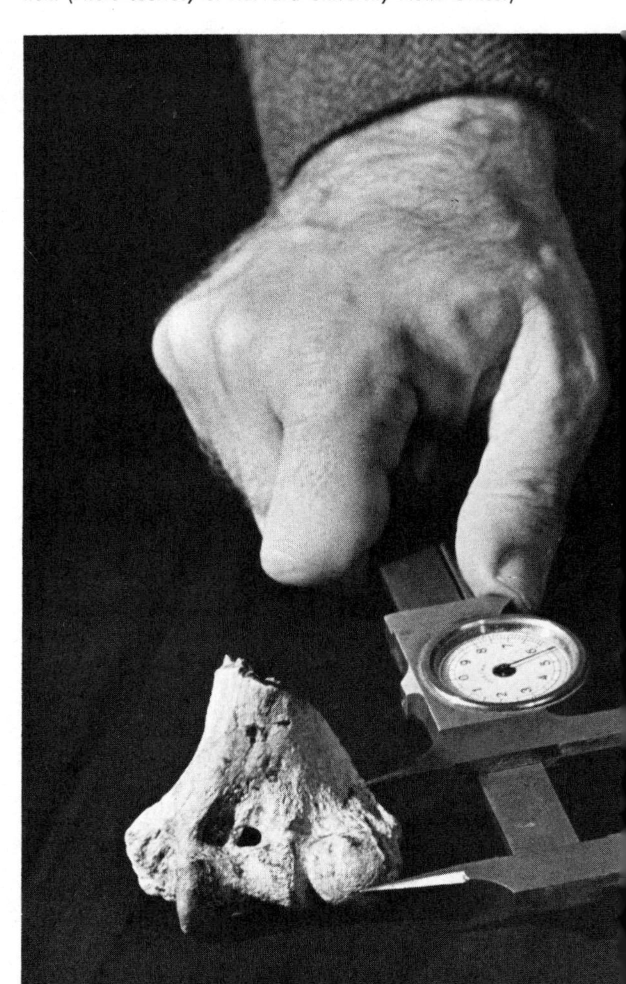

The individual represented is a female, full-grown and about 20 years old. She walked erect and was not over 1 m tall. The enclosing sediments suggest a marshy or lakeside environment. A date between 3.01 and 3.25 million years has been determined.

Anthropologists, confronted with a fairly complete skeleton for the first time, are uncertain what to do with this specimen. Current opinion is that since it is neither *Homo* nor *Australopithecus,* it may belong to a line that gave rise to both. It is surprising, although not impossible for ancestral types to continue living among their descendants. Figure 21-22 gives a tentative family tree based chiefly on African discoveries.

Homo, true man emerges

A textbook writer must proceed with extraordinary caution in dealing with the early history of the human family. This is because of a current lack of unified opinions among the experts and the unusual rapidity with which new finds are coming to light. What seems to be emerging at the present time is an opinion that one branch of the Australopithecine lineage, most probably *africanus,* is on the direct line to modern man. This group possesses more traits compatible with modern man than do the other two, which apparently became extinct, possibly 1,500,000 years ago.

That *Australopithecus africanus* is close to the transition of preman to man is indicated by the fact that some scholars would designate him as *Homo africanus.* In other words, he might be transferred from one branch of the family tree to another. However, near the place of divergence of these two groups, this is not really a great change. Be this as it may, the Australopithecines were living at the same time and place as creatures that are generally admitted as belonging to the genus *Homo.* Remains of these first men were described by the famous paleonanthropologist, Louis S. B. Leakey, in 1964, under the name *Homo habilis.* Compared with the *Australopithecus* species, *habilis* has smaller cheek teeth and larger front teeth, a relatively larger brain, and a skeleton more like that of modern man. In many ways *habilis* is a satisfactory transition between *Australopithecus* and modern man. His weight is estimated at 43 kg and his cranial capacity at 725 cm^3 (the brain of *africanus* is estimated at 450 cm^3, that of a gorilla is 550 cm^3).

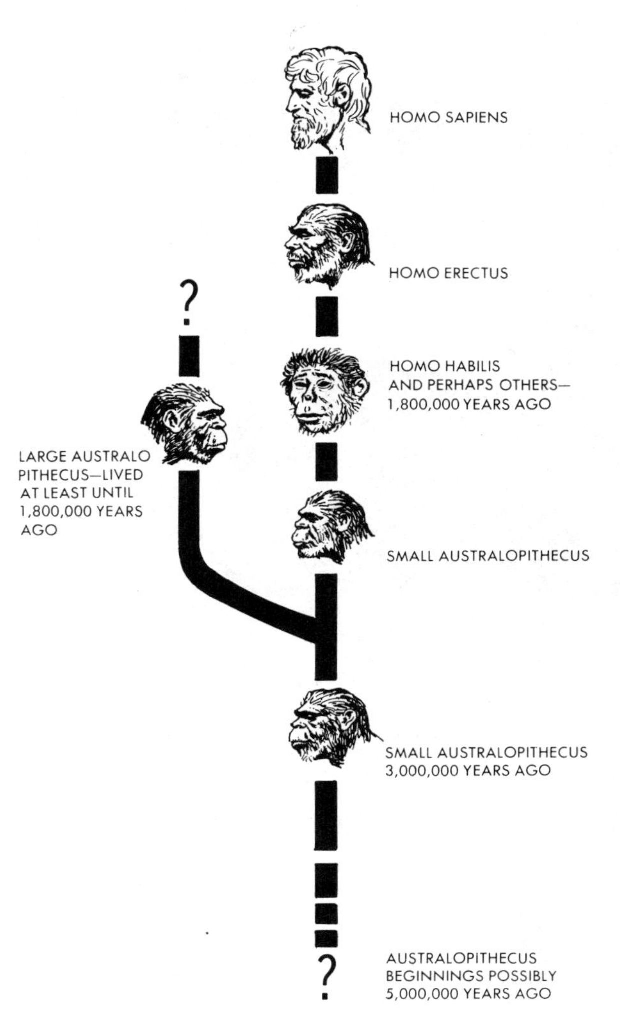

***FIGURE 21-22** Tentative view of the hominid lineage as constructed in 1976. New finds are constantly coming to light that may alter family trees such as this.*

FIGURE 21-23 The banks of the Solo River, Java, where the first discovery of the Java man (Homo erectus) was made in 1890. (Courtesy of George H. Hansen.)

"Homo erectus"— man of the middle Pleistocene

One of the most famous fossils ever discovered, the so-called **Java "ape-man,"** long known as *Pithecanthropus erectus,* was unearthed in 1890 in the banks of the Solo River near Trinil, Java (Figure 21-23). Because of its obviously intermediate status between man and ape, this find gave rise to intensive and often bitter debate, but its study laid the foundations for modern ideas about the origin of man.

The original discovery consisted of a skull cap, thighbone, and a few teeth. Additional material from Java found 40 to 50 years later included skull and skeletal material from a number of individuals. Olduvai Gorge has also contributed a good cranium of the same type; the discovery comes from a higher level than that which contains *Australopithecus* and *Homo habilis.* No artifacts have been found with the Java specimen, but some are associated with the Olduvai remains. Equally important to an understanding of this group is the so-called **Peking man,** formerly known as *Sinanthropus erectus* and before that as *Sinanthropus pekinensis.* Remains of over 50 individuals were found, chiefly at a locality near the village of Chou-kou-tien, 50 km southwest of Peking (Figure 21-24). Charcoal layers, suggesting the use of fire, and crude stone implements are associated with the bones.

After comparing all remains of the Pithecanthropines, paleontologists have made a satisfactory reconstruction. The cranial capacity is 700 to 800 cm^3, the brain case is low, the forehead flat, and the eyebrows heavy and protruding. The upper jaw is very large and thrusts forward, the lower jaw is also heavy and lacks a chin. The teeth are large, but the canines project only slightly beyond the level of the other teeth. *Homo erectus* is almost certainly a descendant of *Homo habilis,* the transition from one species to another having taken place near the beginning of the middle Pleistocene.

Java has yielded the so-called **Solo man,** whom some consider intermediate between *Pithecanthropus* and Neanderthal man. With the 11 skulls attributed to Solo man were a number of skillfully formed artifacts.

The earlier Pithecanthropines appear to have overlapped in time and territory with the Australopithecines, and they may have survived to within 400,000 years of the present. Their distinctly hominid features would permit them to be the ancestors not only of Neanderthal man but also of modern *Homo sapiens.* Thus, they constitute a distinct and essential stage in evolution, just as was suspected when the first "ape-man" was found. Figure 21-25 is one of a great many restorations of *Homo erectus.*

FIGURE 21-24 *Chou-kou-tien, site of discovery of Peking man (Homo erectus). Excavations were mainly in the 1930s. Note that the material removed occupied an open fissure between solid bedrock walls. Remains were mixed with broken rubble, some of which still remains above the opening. (Photo courtesy of F. H. Brown.)*

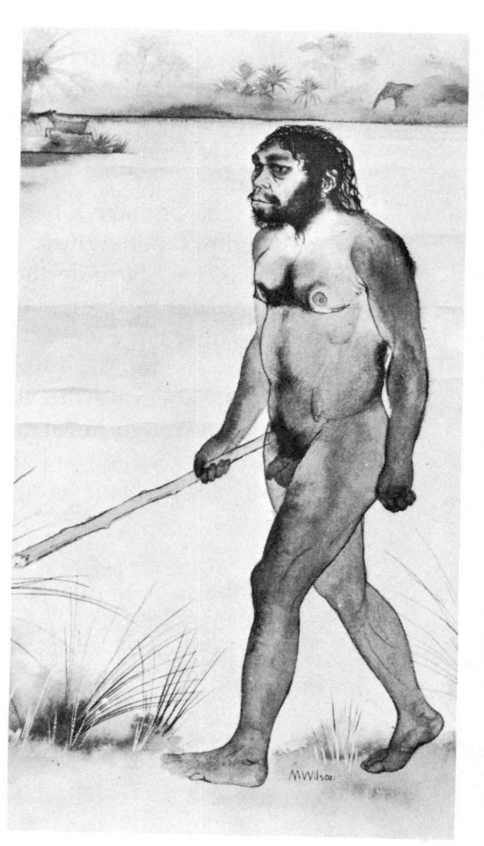

FIGURE 21-25 Reconstruction of Homo erectus. *(NERC copyright. Photograph reproduced by permission of the Director, Institute of Geological Sciences, London.)*

Neanderthal man

Second in interest only to *Homo sapiens sapiens,* or modern man, is Neanderthal man, *Homo sapiens neanderthalensis.* His remains are widespread; specimens are known from Germany, where the first discoveries were made, and also from France, Belgium, Hungary, Rumania, Gilbraltar, Malta, Greece, Czechoslovakia, Iraq, central and north Africa, and China. Over 100 individuals are represented. Stone tools and weapons manufactured by him have also been collected in great numbers.

The popular concept of Neanderthal man is that he was hairy, brutish, and crude, that he dwelled in caves, that he was inferior to modern man, and that he was pushed aside and superseded in the struggle for existence. This picture is not entirely accurate. Serious mistakes were made in assembling the earlier finds, which resulted in a stooped brutish appearance that contrasts unfavorably with modern man.

Now that we have a better idea of him, we may write a better description than the original workers. Neanderthal man had a stocky body with a deep chest, very strong limbs, and rather large feet. This suggests that he was sturdy and tough with great endurance. The skull as shown by the classic or typical individuals was rough-hewn and thick-boned. The chin was receding and the brow ridges heavy and barlike. The brain, as the cranial capacity of 1200 to 1600 cm^3 indicates, was just as large as, or perhaps even larger than, modern man's. This description must be modified to include remains of nonclassical types in which the skull is much less rugged and the brow ridges much less prominent. Figure 21-26 is a more up to date reconstruction and Figure 21-27 is a comparison with other fossil men.

FIGURE 21-26 *Reconstruction of a Neanderthal family group of about 50,000 years ago. (NERC copyright. Photograph reproduced by permission of the Director, Institute of Geological Sciences, London.)*

FIGURE 21-27 *Skull of Neanderthal man (a) compared with Pithecanthropus (b) and Cro-Magnon (c). (American Museum of Natural History.)*

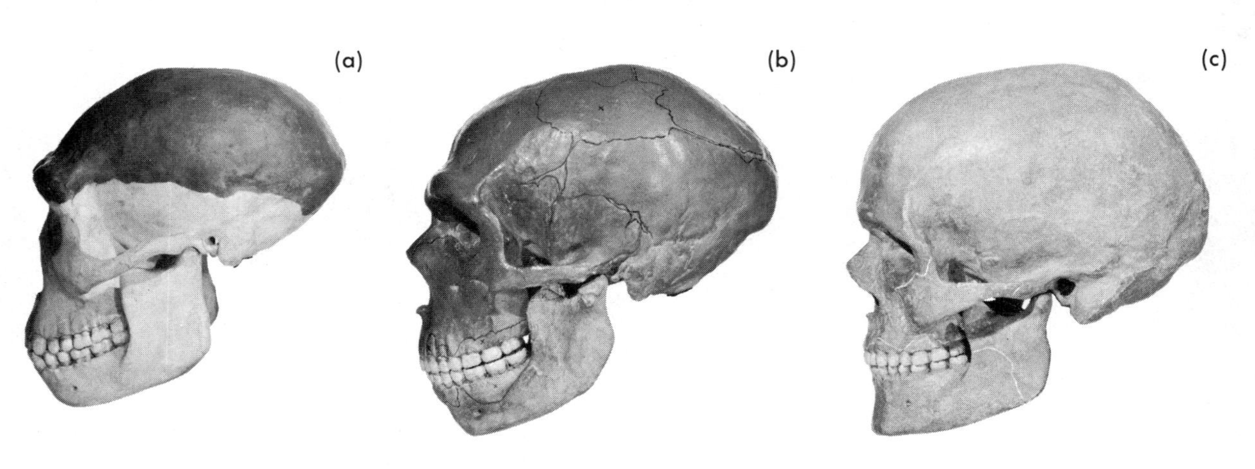

Many theories have grown up around Neanderthal man. Although the first discoveries were dismissed as freaky or diseased examples of modern man, it soon became apparent that there had been on earth a numerous race quite unlike ourselves. Neanderthal man has been variously regarded as a direct ancestor of modern man, a hybrid between modern man and the pithecanthropines, and early contemporary of modern man, and finally as a branch of humanity that sprang from the *sapiens* stock, developed distinctive characters, and then disappeared.

Theories had to be modified as successive discoveries came to light. Thus at Mount Carmel in Israel, remains of ten individuals, discovered in the 1930s, show a mixture of traits bridging the gap in appearance between true Neanderthals and modern man. Some believed this represents a hybrid situation. Most helpful in settling the true position of Neanderthal man has been the application of improved dating techniques. It is generally believed that most if not all of the Neanderthal race lived between 45,000 and 100,000 years ago. In fact some authorities contended that all hominids living in this period should be classed as Neanderthals. This simplistic view proved to be unsatisfactory as it is evident that a very wide variation existed in the human population of this period. The problem is difficult because the best method of close dating, that of carbon-14, is not applicable beyond about 50,000 years and so cannot be helpful in most Neanderthal problems.

What emerges is that specimens classed as Neanderthals include individuals that appear to be in the line to modern *Homo sapiens* as well as others leading away from him toward extinction. Disregarding the problem of classification, we may turn briefly to what is known about human life in the depths of the Great Ice Age winter. Many Pleistocene human fossils were found in close association with glacial deposits of the Old World. This may be a fortuitous thing, and it is not likely to be a true indication of the distribution of Pleistocene man. In other words, most humans may actually have lived near the glacier in preference to other localities. There are valid reasons for this: for example, partly wooded or grassy regions produce more animal food in contrast to jungles, deserts, or coniferous forests. Game to a large extent follows grass, and grass in turn is a product of the temperate regions into which glaciers moved. On the basis of the evidence we must conclude that man lived in small bands, say 30 to 100 individuals, and that these individuals were basically nomadic in habits, finding the necessities of life where and when they could.

There can be no doubt that ice age man was a hunter of big game. Bison, mammoths, woolly rhinos, horses, and other large herd-living mammals abounded, and their very abundance invited the development of hunting skills. More than meat was derived: hides, sinews, bones, and antlers were put to use. This is not to say that plant food was ignored, but we know that most Neanderthal teeth are little worn—a sign that abrasive plant food was not common fare. The most durable and interesting product of Neanderthal man was stone work. From thousands of spear points, hand axes, scrapers, awls, knives, and other artifacts much has been deduced about the habits and accomplishments of Neanderthal man. Arranged in order of their appearance or invention, the stone artifacts shows evidence of slow but steady improvement. Neanderthal artisans did more than merely shape natural stones. They learned to strike precisely shaped flakes from suitable pieces of flint and to process the flakes into a variety of tools and weapons.

Shelter, like food, was a major problem. Neanderthal man was a cave dweller at times it is true, but caves are not plentiful enough to house more than a very few in comfort. Evidence has been found in Molodova, in the Soviet Union, of artificial shelters built out in the open. Although traces of actual clothing were not found, it is assumed from the evidence of certain stone tools that hides of animals were fashioned into warm protective shoes, capes, and other apparel. Fire was a common possession of the Neanderthals generally; its use in warming and cooking in cold climates needs no special explanation.

FIGURE 21-28 *Skull fragments of Swanscombe man. The large brain and smooth skull leave no doubt that the remains pertain to Homo sapiens. (Wenner-Gren Foundation.)*

True man advances

It has become increasingly evident that Neanderthal man, at least as he is represented by the majority of his remains, cannot be directly ancestral to modern man. He is too late in time to hold this position in the family tree. There are, however, a number of rare and tantalizing discoveries that could be direct predecessors of *Homo sapiens sapiens.*

First to be noted is an almost perfect jaw of large size with 16 teeth intact, referred to as *Heidelberg man.* This very incomplete fossil discovered near Heidelberg, Germany, in 1907 is dated at an early stage of the Ice Age, possibly at 450,000 years. No tools are associated, and the true position in the human lineage is debatable.

A second important link is *Steinheim man,* represented by a crushed and incomplete skull found in 1933 in a gravel pit near Steinheim, Germany. The geological setting and the associated mammalian remains give a fairly reliable date in the second interglacial, 200,000 to 400,000 years ago. This clearly makes Steinheim man older than Neanderthal. Steinheim man appears to have had a smaller brain than Neanderthal, but the reconstructed vocal tract is "essentially modern."

From the valley of the Thames River near Kent, England, comes *Swanscombe man,* parts of which were discovered in 1935, 1936, and 1955. The brain case is fairly complete but lacks the frontal bone with its important brow ridges (Figure 21-28). The cranial capacity is about 1,350 cm^3. Well-formed hand axes are associated, along with a warm interglacial fauna. The age is fairly well fixed at about 250,000 years.

Other remains stand in the poorly known lineage between the Pithecanthropines and modern man. One is *Vertesszöllös man,* discovered near Budapest and represented by only an occipital bone. The age is estimated at between 400,000 and 500,000 years. Another is the *Es skuhl V individual* found in a cave on Mt. Carmel, Israel, and dated at about 40,000 to 50,000 years. Some refer to him as a Neanderthal, some as a "Neanderthal contemporary."

Another borderline individual, *Rhodesian man,* was found at the Broken Hill Mine in Zambia, Africa, in 1929. He also shows intermediate characteristics, particularly in the matter of the brow ridges, which are less prominent than Neanderthal but more prominent than modern man.

FIGURE 21-29 A site at Dordogne, central France, that was occupied by Cro-Magnon man. (Service Commercial Monuments Historiques.)

Modern man arrives

In spite of the many significant discoveries of the past few decades, the exact time and place of origin of modern man are still uncertain. True, some former candidates for his ancestry have been eliminated and their places taken by better ones, but specimens unquestionably ancestral to *Homo sapiens sapiens* have so far eluded the scholars. As usual we need new, or at least more complete, fossils to settle this important question.

The first and best-known representative of *Homo sapiens sapiens* is *Cro-Magnon man,* who takes his name from a French rock-shelter where his remains were first discovered in 1868. Cro-Magnon remains are widespread in western and central Europe, and over 100 individual specimens have been recovered. Figure 21-29 is a famous site in central France occupied by Cro-Magnon people. The race dates to the waning stages of the last glacial advance, from 35,000 to 8,000 years B.C.

The average Cro-Magnon was rather large and massively built; many of the men were 6 feet tall or more. There are no traces of Neanderthaloid characteristics in the typical Cro-Magnon skeleton. The forehead and skull vault are high, the brow ridges are small, the nose narrow and prominent, the chin highly developed. The brain capacity averaged 1,700 to 1,800 cm^3, probably more than the average of living men.

Cro-Magnon was progressive and adaptable. During his period of existence, he developed at least five distinct successive cultures and became an expert at shaping stone tools and weapons. By 8000 B.C., he had learned to fashion bone and antler and to make awls, saws, needles, and delicately fashioned weapons. He apparently did not learn to domesticate animals, but he was a mighty hunter, as bones of his prey and his cave art indicate (Figure 21-30). Use of fire and clothing enabled him to withstand the elements and to establish more or less permanent settlements.

FIGURE 21-30 An example of Cro-Magnon art from Les Eyzies (Dordogne), France. Note the superimposed animal figures.

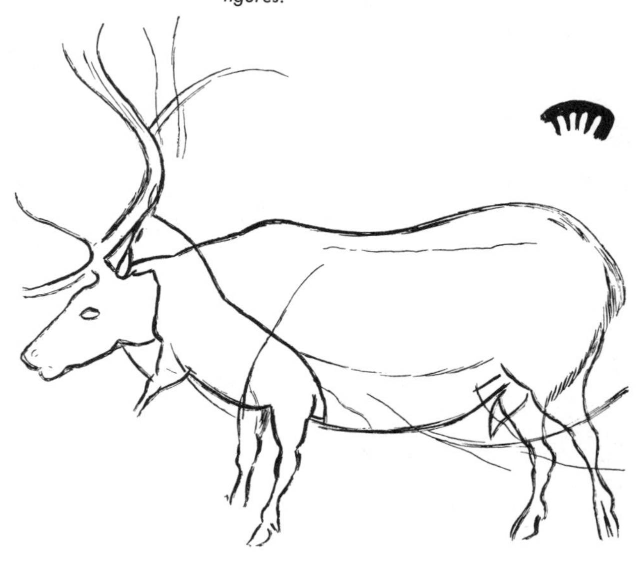

Man in the New World

The history of primates in the New World is discontinuous and incomplete. North America, lying farther from the equator than any other large continent, became too cold or otherwise unsuited for primates by the middle Tertiary, and there are no known fossil remains of the group between the early Oligocene and the late Pleistocene. Even in South America, where we might expect the record to be more abundant, primate evolution seemingly did not progress past the monkey stage. As far as primates are concerned, the New World was effectively separated from the Old World until late in the Pleistocene, when modern man successfully migrated from one hemisphere to the other. Paleontologists generally agree that America's oldest human inhabitants probably reached North America from Asia via the Bering Straits area, which occasionally formed a dry-land bridge between the two continents. Most evidence is interpreted as indicating that man populated the New World within the last 20,000 years. There are, however, a few specimens judged to be older than this, some even possibly as old as 100,000 years. Part of the uncertainty arises because there are no thoroughly reliable dating methods in the range just beyond carbon-14, which is about 50,000 years.

Anthropologists are most interested in the period beginning with the arrival of man and ending about 5,000 years ago. During this interval, ancient Americans lived in association with many large mammals that are now extinct and pursued a nomadic existence governed by climatic conditions and the distribution of food sources. It seems unrealistic to call

FIGURE 21-31 *Discovery sites of important Old World fossil men and premen mentioned in the text.*

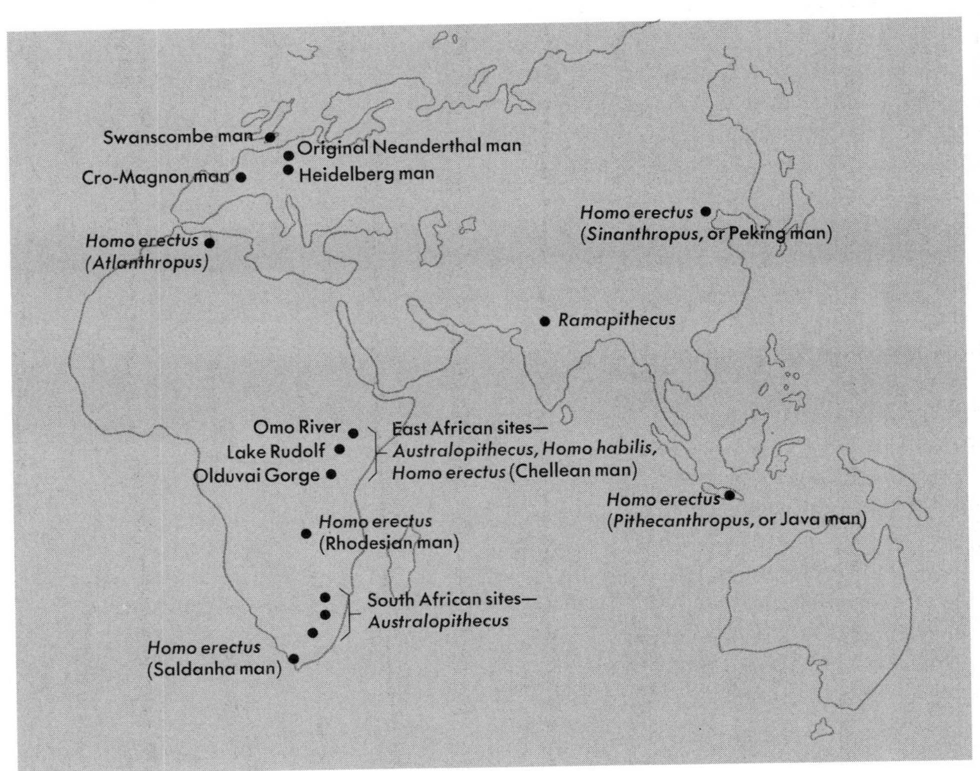

FIGURE 21-32 *Folsom man hunted the giant bison by killing it with a spear-thrower or driving it over a cliff. (Courtesy of Alfred M. Bailey, Denver Museum of Natural History.)*

these early people *Indians,* for they may not have been ancestors of the inhabitants Columbus encountered in the fifteenth century. Following the anthropologists, we will call these earliest Americans *paleo-Indians.*

Paleo-Indians are known to have hunted a variety of now-extinct mammals because many stone projectile points and bones of these animals have been found mingled together. Among the aminals we know that were positively associated with early man in the Americas are several types of extinct bison, the ground sloth, the extinct horse, the mammoth, the camel, the mastodon, and various antelopes. The most common source of meat appears to have been the bison, of which many varieties, mostly larger than the type with which we are familiar, have been found (Figure 21-32).

America's oldest humans knew how to use fire, and their stone work is distinctively unlike anything that was produced in Europe. They created a characteristic type of projectile point with wide, shallow grooves or channels on one or both faces. This artifact, best typified by the well-known *Folsom point,* is not the only distinctive American contribution, however, for other types of arrow and spear points, both older and younger than the Folsom, have also been discovered (Figure 21-33).

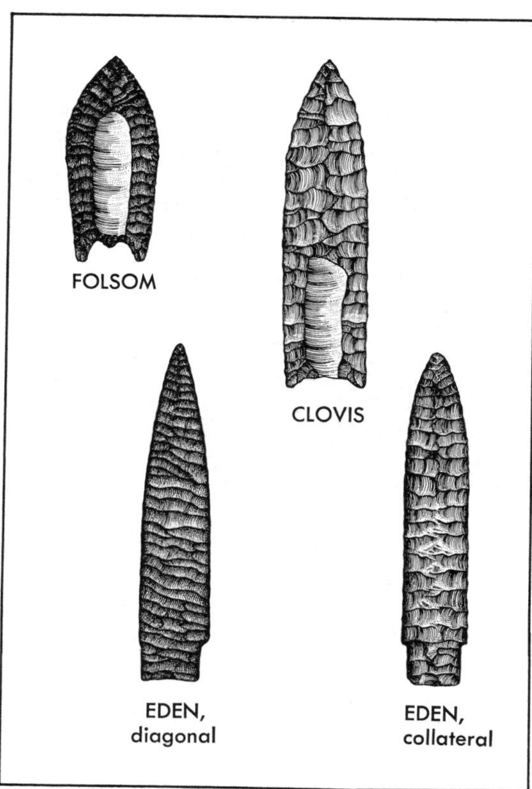

FIGURE 21-33 *Distinctive projectile points of early North Americans. (Drawing by Jeffry B. Hulen.)*

Modern Indians and their direct ancestors are comparatively well known. They were associated with animals that still exist, and they lived under climatic conditions much like those of the present.

SUMMARY

During the last l to 3 million years there have been drastic climatic changes that caused the advance and retreat of continental glaciers over at least one-fourth of the land area. Simultaneous effects included rise and fall of sea level, cooling of the oceans, filling of interior basins by freshwater lakes, permanent freezing of rock and soil, and heightening of wind action. All parts of the earth were directly or indirectly affected.

The Ice Age caused great disturbances in the biologic realm. Plants and animals were forced to migrate with changing climates, and there was an overall reduction of organic productivity of the lands. There were notable exterminations, especially among the larger mammals, but the greatest extinctions seem to have occurred after the last glaciation was over.

Man is in a sense a product of the Ice Age. The first erect, primates (the Australopithecines and *Homo habilis*) lived 3 to 2 million years ago in East and South Africa. During successive glacial and interglacial stages, more advanced hominids appeared in the Old World. The Pithecanthropines (*Homo erectus,* including the Java "ape-man" and the "Peking man") inhabited Southeast Asia and parts of Africa in the middle Pleistocene. From the middle Pleistocene of Europe came Heidelberg man, Steinheim man, Swanscombe man, and Vertesszöllös man, all near the *Homo sapiens* ancestral line.

Glacially adapted Neanderthal man lived in Europe, Asia, and Africa during the late Pleistocene and left relatively abundant cultural remains. Modern man (*Homo sapiens sapiens*) appeared in the second interglacial stage and was contemporary with, rather than subsequent to, Neanderthal man. Cro-Magnon man is the best-known fossil representative of modern man.

Man colonized the New World at a relatively late date. In the Americas he lived with and hunted a variety of giant mammals that are now extinct.

SUGGESTED READINGS

BATES, MARSTON, *Man in Nature*. Englewood Cliffs, N.J.: Prentice-Hall, Inc., 1964.

FLINT, RICHARD FOSTER, *Glacial and Quaternary Geology*. New York: John Wiley & Sons, Inc., 1971.

HOWELL, F. CLARK, and Editors of Time-Life Books, *Early Man*. New York: Time-Life Books, 1965.

PILBEAM, DAVID, *The Ascent of Man—An Introduction to Human Evolution*. New York: The Macmillan Company, 1972.

SCHUMM, STANLEY A., and WILLIAM C. BRADLEY, *United States Contributions to Quaternary Research*, Special Paper 123. Geological Society of America, 1969.

SIMONS, ELWYN L., *Primate Evolution—An Introduction to Man's Place in Nature*. New York: The Macmillan Company, 1972.

Minerals and Energy Resources

All living things depend on the physical resources of the earth for their existence. Man is no exception; indeed, he has, more than any other organism, learned to exploit the materials of the earth. His ability to find and use mineral and energy sources has allowed him to control and modify his environment; yet at the same time he has been dependent on a continuing and usually expanding supply of energy and mineral materials—a supply that is finite and exhaustible (Table 22-1).

A study of the earth's economic resources involves geology; but it also involves the technology of exploration, extraction, and fabrication, as well as the economics of man's social systems. Space here demands that we focus our attention on the geologic aspects of natural resources, but some understanding of the technologic and human aspects is desirable. This chapter also brings together many things we have learned and applies them to real situations of vital importance to man.

22

Offshore mobile rig at Dunlin field in the North Sea. (Courtesy of Exxon Company, U.S.A.)

Table 22-1 Amounts of New Energy and Mineral Materials Required Annually for Each U.S. Citizen

MATERIAL	AMOUNT USED ANNUALLY (in kilograms)
Petroleum	3,600
Coal	2,318
Natural gas	2,115
Uranium	0.045
Iron, steel	585
Aluminum	29
Copper	11
Zinc	7
Lead	7
Other metals	16
Stone	4,163
Sand and gravel	3,825
Cement	360
Clays	248
Salt	203
Other nonmetals	540
Total	18,027 (about 40,000 lbs)

SOURCE: Data from U.S. Bureau of Mines, 1976.

Table 22-2 Potential U.S. Domestic Resources for the Minimum Anticipated Demand, 1968 to 2000

10 times	*2 to 10 times*
Gypsum	Aluminum
Sulfur	Iron
	Molybdenum
$\frac{3}{4}$ to 2 times	Phosphate
	Titanium
Copper	Zinc
Gold	
Lead	*$\frac{1}{3}$ to $\frac{3}{4}$ times*
Manganese	
Nickel	Tungsten
Silver	
$\frac{1}{10}$ to $\frac{1}{3}$ times	*$\frac{1}{10}$ times*
Asbestos	Chromium
Mercury	

SOURCE: Information from U.S. Geological Survey, 1976. Includes reserves currently of economic value plus those that might be used with favorable price conditions or improved technology.

Ore deposits

Minerals of use to man are grouped into two broad categories: (1) *metals,* such as aluminum, copper, gold, silver, iron, tin, platinum, chromium, nickel, lead, and zinc, and (2) *nonmetals,* such as diamonds, salt, limestone, cement, sulfur, and asbestos. When minerals are present so that they can be worked at a profit, they are called *ore deposits.*

Ore deposits are extremely rare and of limited extent. Only a few geologic environments favor their formation. The favored places of occurrence are unevenly distributed throughout the world. Except for the U.S.S.R., there is no nation that is completely self-sufficient in all industrial minerals.

Once an ore deposit is mined, the deposit is depleted. The minerals cannot be grown again in the time span of man's probable existence. The supply of ore minerals is finite, and many important industrial minerals are already in short supply (Table 22-2). In the last 30 years, more metal has been mined than in all of preceding history.

Although there are many low-grade ore deposits that will be mined as scarcities arise and as prices increase, this solution will not suffice for long. Considerable energy must be expended in working low-grade ores, and energy resources are also finite and have become extremely costly. We must begin to make more intelligent use of our ore deposits and to conserve them.

A convenient way of discussing ore deposits is to classify them on the basis of the geologic processes that have created them: igneous activity, weathering, sedimentation, and the formation of the original rock masses (Table 22-3).

Concentration by igneous activity

In Chapter 4 we discussed the formation of igneous rocks from a mixture of elements in a solution called a *magma.* Some magmas, however, also contain elements that, because of the size of their ions, do not combine readily with the common rock-forming minerals. If one of the early-formed minerals is much denser than the parent magma, it may sink rapidly and form a concentration by *magmatic segregation* on the floor of the magma chamber. Sometimes magmatic segregation minerals are formed by settling of late-crystallizing but dense metalliferous parts of the magma.

Diamond

Diamonds are familiar to us as a precious gem, the gem of innocence. But they also are widely used in industry as an abrasive, for they are the hardest mineral known. Diamonds are formed where carbon is trapped

Table 22-3 Modes of Occurrence of Mineral Deposits Throughout the World

PROCESS, MANNER OF FORMATION	EXAMPLES
Magmatic Segregation	
A. By settling of early-formed minerals during magma crystallization.	Layers of chromite, magnetite, and platinum-rich pyroxenite in Bushveldt intrusion, South Africa.
B. By settling of late-crystallizing but dense metalliferous parts of magma, which either crystallize in the interstices of older silicate minerals or are injected along faults and fissures of the wall rocks.	Copper-nickel deposits of Norway and most of those of Sudbury, Canada.
C. By direct magmatic crystallization.	Diamond deposits of South Africa.
Hydrothermal Deposits	
By filling fissures in and replacing both wall rocks and the consolidated outer part of an intrusion by minerals whose components were derived from a cooling magma.	Copper deposits of Montana and Utah; lead deposits of Idaho and Missouri; zinc deposits of Mississippi valley district; silver deposits of Nevada; gold deposits of California and South Dakota.
Contact Metamorphic	
By replacement of the wall rocks of an intrusive by minerals whose components were derived from the magma.	Magnetite deposits of Iron Springs, Utah; some copper deposits of Morenci, Arizona.
Chemical Weathering	
By weathering that causes leaching out of valueless minerals, thereby concentrating valuable minerals.	Iron ores of Cuba; bauxite ores of Arkansas, Jamaica, and Surinam.
Sedimentary	
A. By deposition of rocks in which the detrital grains of valuable minerals are concentrated because of superior hardness or density.	Placer gold deposits of California, Siberia, Alaska; probably of Rand, South Africa; diamond placers of Southwest Africa.
B. By deposition of rocks unusually rich in particular elements.	Iron deposits of Alabama, and some in Minnesota, Michigan, and Labrador; phosphate deposits in Florida, Montana, Wyoming, and Utah.
C. By evaporation of saline waters, leading to successive precipitation of valuable salts.	Salt and potash deposits of New Mexico, Utah, Michigan, Ohio, New York, Germany, and Canada.

under extremely high pressures, pressures greater than those normally encountered in the earth's crust. Most diamonds are found in volcanic necks where the cooling magma formed garnet, peridotite, eclogite, and other mafic rocks. Laboratory experiments indicate that these rocks are stable only at great depths (150 km or more), where temperatures do not exceed 1100°C because diamonds convert to graphite at higher temperatures. Diamond-bearing volcanic necks are believed to have originated as local molten pockets in the upper mantle, which then moved into the overlying crust where they are now exposed on or near the earth's surface.

FIGURE 22-1 *Most metals are associated with a particular temperature range within the overall temperature pattern of ore deposits. Pegmatites are coarse-grained igneous rocks, some of which contain unusual minerals. (This diagram is in part courtesy of A. F. Buddington.)*

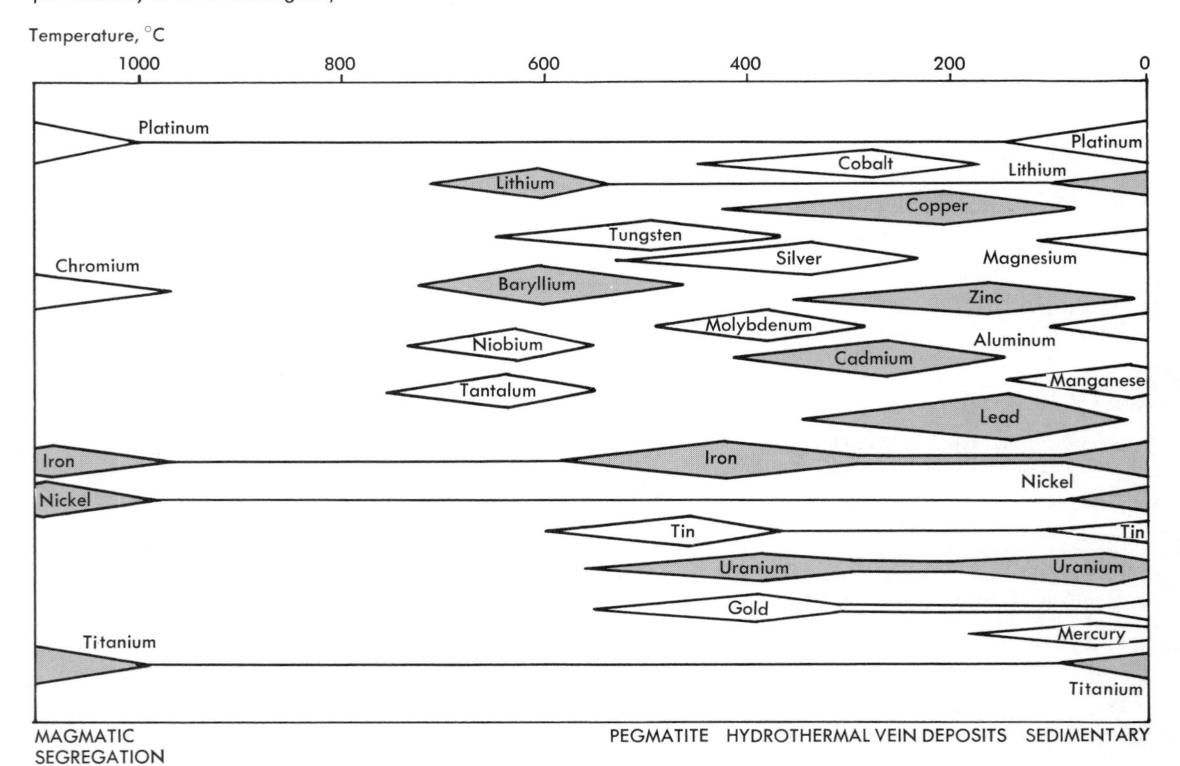

Chromium, nickel, and platinum

Deposits of chromium, nickel, and platinum are found in formations of simatic rocks all around the world. Apparently all three of these minerals somehow became concentrated in the cooling magma; and since they were denser than the rock-forming minerals, they settled out during crystallization.

Although the chromium and nickel that make up these deposits have combined with other elements to form compounds, the platinum is present in an uncombined state. The chief source of chromium is *chromite,* $(MgFe)_2CrO_4$, which is associated closely with ultrabasic igneous rocks such as dunite (mostly olivine) and peridotite. Apparently chromite is a magmatic segregation deposit (see Table 22-3 and Figure 22-1), where the chromite sank to the bottom of magma chambers because of its denseness. Most of the chromium produced in the United States comes from Shasta County, California. The leading world producers of chromite are South Africa, Russia, Turkey, Southern Rhodesia, and the Philippines.

Chromium is used chiefly to form an alloy with steel that has extreme hardness, toughness, and resistance to chemical attack. It is also used for plating hardware, plumbing fixtures, and automobile accessories. More than 60 percent of the chromite consumed in the United States goes into metallurgical uses of this sort. The remainder is used in the construction of furnaces and other equipment where heat resistance is required, and in various other chemical processes.

Although nickel is a relatively rare element in the earth's crust, it is extremely important in modern industry. It is used in the manufacture of a strong, tough alloy known as *nickel steel* (2.5 to 3.5 percent nickel) and in the preparation of *monel metal* (68 percent nickel) and *nichrome* (35 to 85 percent nickel). It is also used in various plating processes, and it forms 25 percent of the U.S. five-cent coin. Finally, its low expansion tendency makes it an ideal metal for watch springs and other delicate instruments.

The most important deposits of nickel are in Canada at Sudbury, Ontario, and Thompson Lake, Manitoba, where the nickel is in the mineral *pentlandite* $(Fe, Ni)_9S_8$. Magmatic segregation is responsible for these deposits, and the dense, sulfide liquid sank in a late event to the bottom of magma chambers where crystallization of the pentlandite took place.

Platinum also is present in association with the nickel deposits at Sudbury. The value of platinum in industry results from its high melting point, 1,755° C, and its resistance to chemical attack. These properties make it especially useful in laboratory equipment such as crucibles, dishes, and spoons, and for the contact points of bells, magnetos, and induction coils. Platinum also finds special uses in the manufacture of jewelry, in dentistry, and in photography.

Gold

Gold is a rare element used principally in coins and jewelry. It is normally present in the uncombined state in sialic igneous rocks, particularly in those that are rich in quartz. About one-half of the gold that is mined in California, the leading gold-producing state, comes from the Mother Lode, a series of hydrothermal veins lying along the western slope of the Sierra Nevada. The rest comes from placer deposits.

Copper

Copper, which is second only to iron among the important industrial metals, is present in the continental crust in very small amounts. The average crustal concentration is 60 parts per million (ppm), and deposits are unevenly distributed. The United States is the largest producer of copper.

Technologic advances have lead to the mining of many low-grade (as low as $\frac{1}{2}$ percent) copper ores (porphyry copper). The major copper mineral, *chalcopyrite*, $CuFeS_2$, is present as small blebs or veinlets throughout the ore. More than 50 percent of the world's copper is produced from porphyry copper deposits.

All of the porphyry copper is found on the inner side of present or past convergent plate boundaries and is within or adjacent to subduction-related, intrusive igneous rocks. This has led to the suggestion that the copper (and associated metals) originated in oceanic crust; then subduction of the oceanic crust took place, and magma was generated and rose toward the surface. Hydrothermal solutions then concentrated and moved the metals from the cooling magmas to their depositional sites.

Tin

The only important ore of tin is *cassiterite*, SnO_2. A good bit of tin is present in original hydrothermal deposits, but 75 percent of the world's supply comes from placers. Tin is used principally as a coating on steel to form *tin plate* for food containers.

There are almost no tin deposits in North America. Most of the world's production is from two narrow belts: one along the Malayan Peninsula, then southeast to Java and Indonesia; the second along the eastern side of the high Andes in Bolivia and Peru.

Silver

Native silver has been deposited from hydrothermal solutions, as has another ore of silver—*argentite*, Ag_2S, or silver sulfide. Argentite as an ore may also be of secondary origin when it has been concentrated by weathering processes. For centuries, silver was used almost exclusively in jewelry and coins. By 1940, however, it had become an important industrial metal. It is extensively used in photography, in laboratory and electrical goods, in medical and dental work, and as an alloy in bearings, solders, and brazing compounds.

Lead and zinc

Lead and zinc deposits have been created mainly by the process of *metasomatism* (see Chapter 7), in which hydrothermal solutions and magmatic gases have replaced some of the original components of the rock surrounding a magma. Limestone exposed to igneous activity is particularly susceptible to metasomatism.

In the famous Mississippi valley deposits of Missouri, Oklahoma, Kansas, and Wisconsin, the hydrothermal lead-zinc deposits are present mainly as replacement bodies in limestone of many ages. Metal-rich solutions apparently dissolved the limestone and slowly deposited the principal ores, *galena* (PbS) and *sphalerite* (ZnS). The deposits generally are far from any obvious igneous activity. Many geologists therefore now believe that the hydrothermal solutions were metamorphic rather than magmatic in origin.

Lead is used in the manufacture of bullets, cable coverings, foil, pipes, storage batteries, weights, and as a gasoline additive. It forms an alloy with tin to make solder, with antimony to make type metal, and with bismuth and tin to make metals that melt at low temperatures. Large quantities of lead are used every year in the preparation of paint pigment.

Zinc is used chiefly for galvanizing iron, as an alloy with copper in making brass, and in the manufacture of batteries.

Concentration by weathering

So far, we have discussed deposits that were originally created by igneous activity in about the same form in which they now appear. But other important deposits have accumulated through the action of weathering on preexisting rocks. There are three important weathering processes in the formation of ore deposits.

1. The chemical alteration of compounds from which desired elements cannot otherwise be extracted economically.

Aluminum, although it is one of the most common elements in the earth's crust, almost always is present in feldspars and other silicates from which it cannot be extracted economically by any process now known. However, under tropical conditions, weathering breaks the feldspars down into clay minerals; they in turn become hydrous oxides of aluminum and iron. The soils produced by this activity are *laterites* and the aluminum ore is *bauxite*.

At least one-half of the world's bauxites are developed from carbonate rocks that contained small amounts of clay minerals and iron oxides. In tropical climates, limestone and dolomite dissolve rapidly, and the clay residues remain. Bauxite develops from the clays, as a product of chemical weathering.

The principal deposits of bauxite in the United States are near Little Rock, Arkansas, where the deposit developed in nepheline syenite, a coarse-grained, igneous rock containing feldspars and *nepheline* ($NaAlSiO_4$) but no free quartz. During the leaching process the nepheline and feldspars altered to form clays. The clays then were leached to form *gibbsite* (H_3AlO_3), the primary mineral of the Arkansas bauxites. No more than one-half of the original rock remained after the leaching process was completed.

Aluminum is a very light, strong metal used extensively in the manufacture of cooking utensils, furniture, household appliances, automobiles, airplanes, railway cars, and machinery. It is becoming increasingly popular as an insulating material in buildings.

2. The removal of undesired components, by leaching or percolation of water, leaving the desired compounds more concentrated than they were originally.

Iron, which accounts for more than 95 percent of all metals consumed, has been concentrated by this weathering process in many areas of the world—in the extremely important deposits around Lake Superior, for example. In Minnesota and Michigan, iron-bearing formations underlie thousands of square kilometers; but for years, only where percolating ground water had removed enough silica from the parent rock was the iron (in the form of Fe_2O_3) sufficiently concentrated to make mining practicable.

3. The solution and redeposition of desired elements in useful concentrations, a process sometimes called secondary or supergene enrichment.

In some regions, igneous activity has built up original deposits such as copper, but not in large enough concentrations to be worked. Sometimes, though, ground water has dissolved the copper and has carried it down to be deposited in an enriched zone. At Bingham Canyon, Utah (see Figure 22-2), there is a spectacular open-pit mining operation that recovers at a profit an ore containing as little as .04 percent of metal. The ore is silicic porphyry that contains finely disseminated primary sulfides. The benches of the mine range from 12 to 15 m in height and are not less than 20 m in width. The operation covers an area of 62 km^2 including the waste dumps, and during 1975 about 470,000 tons of rock per day were moved. At Morenci, Arizona, the impoverished zone is as deep as 65 m, but beneath it the enriched zone extends about 300 m farther down. Underlying the enriched zone is the unaltered bedrock, which is mostly too low grade to be worth mining.

Concentration by sedimentary processes

Some of the geologic agents described in Chapters 10, 11, and 12 pick up the products of weathering and ultimately deposit them below base level. Both mechanical and chemical processes are involved in the transportation and deposition of the weathered rock (see Figure 22-1).

Flowing water moves great quantities of mineral material along the channels of streams, particularly in mountainous regions. The heavier minerals—harder for the water to transport, yet resistant to chemical decay—tend to accumulate in the channel basins, in a deposit called a *placer* (rhymes with "passer"). Gold is exceptionally well adapted to placer deposition. Weathering breaks it from the rocks and veins where it originally crystallized from hydrothermal solutions, but its mallea-

FIGURE 22-2 *The Bingham Mine near Salt Lake City, Utah. The open-cut workings are descending into a large deposit of low-grade copper ore that originated in association with an igneous intrusion of Middle Tertiary age. Surrounding rocks that contain much of the ore are Pennsylvanian and Permian in age. (Courtesy Kennecott Copper Corporation.)*

bility prevents it from being finely pulverized. Moreover, its high specific gravity (ranging from 15 to 19, depending on the percentage of impurities present) causes it to settle readily from agitated mixtures of water, sand, and lighter materials. The gold discovered in 1848 on the western slopes of the Sierra Nevada in California was concentrated in placers so rich that great fortunes were made by panning it out by hand (Figure 22-3). Less concentrated placer deposits can now be worked by modern hydraulic giants, which wash away the barren material that overlies the pay dirt and sluice the gold-bearing gravels into boxes where the gold is trapped.

Even deposits of gold in gravel below ground water or ocean level can be worked by specially designed dredges that recover at a profit gold so thinly dispersed that there are only a few cents worth in each cubic meter. Almost 70 percent of the world's annual production of gold is mined from the Witwatersrand District of South Africa, on a plateau standing 1,800 m above sea level and about 1,100 km northeast of Cape Town. The deposits in this rich area are in conglomerates, themselves formed from ancient placer deposits according to some geologists. Other geologists believe that permeable channels in the original rock were invaded by gold-bearing hydrothermal solutions. A third interpretation is that the gold was originally deposited as detrital particles, but was later dissolved and reprecipitated after short distances of transport.

Nearly 75 percent of the world's tin production comes from placer deposits. Most of the ore is *cassiterite* (Figure 22-4), or *tin dioxide* (SnO_2) (which has a specific gravity of 7).

The world's largest deposits of iron ore originated initially by chemical precipitation in sediments, perhaps as colloidal products of the life processes of iron-secreting bacteria. But most of the deposits could not be worked

FIGURE 22-3 *Panning stream gravel in the Northern Cascades, Washington, for gold, silver, copper, lead, and zinc. (Courtesy U.S. Geological Survey.)*

FIGURE 22-4 *Cassiterite from Cornwall, England. Waterworn pebbles from a placer deposit. (Photo by Benjamin M. Shaub.)*

FIGURE 22-5 *Iron ore is loaded into railroad cars that will carry it to a nearby plant for processing before shipment to the steel mills, Chisholm, Minnesota. (Courtesy U.S. Steel Corporation.)*

commercially until weathering processes had increased the concentrations by secondary enrichment.

About nine-tenths of the iron ore in the United States is present in *hematite* (Fe_2O_3). In the Lake Superior District, the zone that has still not been leached by weathering consists of a mineral assemblage called *taconite*, containing chert with *hematite*, *magnetite* (Fe_3O_4), *siderite* ($FeCO_3$), and *hydrous iron silicates*. The iron content of taconite averages only about 25 percent. But in the zone that has been leached, most of the iron has been oxidized to hematite, which produces ores of from 50 to 60 percent iron. Recently, however, commercial methods have been developed for recovering iron even from the taconite of this district. Thus great volumes of original unleached rock have been added to our iron reserve (Figure 22-5).

The newest reserves of iron ore on the North American continent, near 55° N, 67° W, on the Labrador-Quebec border, also consist of deposits enriched by weathering. Other deposits are being worked in sedimentary formations of Silurian age, known as the *Clinton beds*, which outcrop across Wisconsin and New York, and along the southern Appalachians. These beds are also being mined extensively in Alabama, in the Birmingham District (Figure 22-6). The primary unleached ores from the Clinton beds are often high in $CaCO_3$ and contain 35 to 40 percent iron. But after the $CaCO_3$ has been leached out by weathering, they may contain as much as 50 percent iron.

The known reserves of iron ore available in the leached and enriched ores are very large, but are miniscule compared with the amount present in the unaltered iron formations. Estimates in 1965 by the U.S. Geological Survey showed that taconite reserves in the Lake Superior region exceeded 10^{11} tons of Fe. Thus depletion of iron ores in the United States should not become a serious problem for many centuries.

Concentration during rock formation

Many rock materials are valuable in their original condition, as originally produced by rock-forming processes, without undergoing any additional enrichment or concentration. Stone, of course, has been used for several thousand years as a building material. But its importance has grown tremendously during the last half century with the discovery of new techniques for removing it from the ground by blasting and for crushing it into usable sizes. Every mode of transportation in the modern world depends in some degree on crushed or broken stone; for example, crushed or broken stone provides the basis for countless kilometers of modern highways, ballast for railways, bases for landing fields, and jetty stones for harbor facilities. Each year, 950 million tons of sand and gravel, valued at $1.1 billion, are mined in the United States.

Other rocks have commercial value because of their chemical properties. *Limestone*, for example, is used to neutralize acids in the processing of sugar, to correct the acidity of soil, and to supply calcium to plants. Limestone that contains small amounts of impurities serves as the raw material in the manufacture of cement; the impurities give cement its characteristic hardness. The type known as *Portland cement* consists of 75 percent calcium carbonate (limestone), 13 percent silica, and 5 percent aluminum oxide, along with the silica and alumina that are normally present in clays. Some manufacturers add the right percentage of impurities to the limestone; others use limestone deposits called *cement rock*, in which the impurities are present naturally.

FIGURE 22-6 *Outcrop of Clinton iron ore near Birmingham, Alabama. The ore is the thick bottom layer. (Photo by Aloia Studio; courtesy Republic Steel Corporation.)*

Phosphate rock is a popular term used for sedimentary rocks that contain high percentages of phosphate, commonly with the mineral *apatite*, $Ca_5(PO_4)_3(OH,F)$. Phosphorus is a vital element in the physiological processes of vertebrates, including man, and is essential to plant growth. Phosphate rock is extremely important as a source of agricultural fertilizer. The Rocky Mountain states have phosphate reserves estimated at 6 billion metric tons in the Phosphoria-Park City Formation of Permian age.

More than 90 percent of the world's phosphate rock production is from marine sedimentary beds like those found in the Rocky Mountain area. Although the origin of these beds is uncertain, the best interpretation is that the upwelling of cold, phosphate-rich ocean water led to the precipitation of apatite in the accumulating sediment when these waters were carried into shallow marine basins.

All living things contain phosphate in the most basic parts of the cell chemistry. The famous DNA double-helix molecule that carries the genetic information and the ATP molecule that is the universal energy carrier are both phosphorus compounds. The mineral apatite constitutes the bulk of our bones and teeth. How, in our distant past, did our bones, our energy carrier, and our genes all come to be made of such a scarce element?

It is possible that phosphorus is the only element that is suitable for constructing a living cell. Still another possibility is that during an early period in organic evolution, phosphorus was locally more abundant or more soluble, and under these conditions organisms came to use phosphorus and continued to use it after it became scarce. A third possibility is related to the fact that some biological molecules stick tightly to apatite, but other molecules flow on through the apatite crystals. On the surfaces of apatite crystals, primitive organic molecules may have played some role in getting living cells organized. However, so far all of the explanations for our dependence on phosphorus are speculative, and the true answer is quite unknown.

Organisms require nitrogen because proteins are nitrogen compounds. As protein sources become rarer, a supply of nitrogen is essential. Nitrogen compounds, such as ammonia (NH_3), are made by reacting nitrogen from the atmosphere with hydrogen. The cheapest hydrogen source is associated with petroleum refining; nitrogen for fertilizers and chemicals is thus indirectly related to the production of oil.

Sulfur is another very important element to man, especially because of its use as a fertilizer. However, sulfur is available in sedimentary deposits as the mineral *gypsum* ($CaSO_4 \cdot H_2O$), as *free sulfur*, and in some natural gases as *hydrogen sulfide* (H_2S). Sulfur is also present in sulfide minerals, such as *pyrite* (FeS_2). Our supply of sulfur (see Table 22-2) is very large indeed.

Asbestos is the name applied to certain minerals that form soft, silky, flexible fibers in metamorphic rocks. The most common asbestos is *chrysotile*, $Mg_3(Si_2O_5)(OH)_4$, a variety of the mineral *serpentine*. The longer fibers are woven into yarn for use in brake linings and heat-resistant tapes and cloth. Asbestos materials are extremely versatile, for they withstand fire, insulate against heat and sound, are light in weight, can be made into pliable fabrics, and resist soil, corrosion, and vermin. However, there is evidence that asbestos fibers can cause cancer if inhaled into the lungs, which sometimes happens in certain kinds of work. The United States, the largest user of asbestos, imports up to 90 percent of its needs from Canada where there is an important belt of serpentine in Quebec.

Salt, NaCl (the mineral halite), essential to life, and fortunately one of the most abundant substances in the world, is derived commercially both from sea water and from rocks that were formed by the natural evaporation of sea water. Salt is produced by 99 countries on a regular basis. The primary use of salt is in the chemical industries, but it is also valuable in the preparation and transportation of foods, in various manufacturing processes, and in the treatment of icy highways.

Under the pressure of a few thousand kilometers of sediment, rock salt flows plastically. In some regions it has been pushed up into overlying beds in great plugs known as *salt domes.* Although the details of shape and history vary a good bit from one dome to another, all salt domes tend to have a cylindrical shape with a top diameter of about 1.5 km. They may rise to the surface, or they may get no nearer than a few thousand meters. Some salt domes have forced their way upward from the original salt bed through 6,000 m of overlying beds. Traps for oil and gas are often formed by salt domes as they rise.

Sources of energy

Man used more energy in the 30-year period from 1941 to 1970 than he consumed in all history prior to 1940. In 1970, the equivalent of 87 million barrels of crude oil per day (about 12.4 million metric tons) were consumed. By the year 2000, world energy demand is expected to reach 400 million barrels daily (57.1 million metric tons), representing nearly a fivefold increase.

In 1970 almost all of the energy used was derived from crude oil (43 percent), natural gas (33.3 percent), and coal (19.4 percent). Hydropower furnished about 3 percent of the energy used and nuclear power less than 1 percent.

Projections through 1990 (Figure 22-8) indicate that in the United States oil will continue to be the most important source of energy. In fact, our reliance on energy from oil will increase to about 50 percent by 1980. It is hoped that nuclear power will develop rapidly and contribute 20 percent or more of total energy needs by 1990. However, environmental concern about safety and pollution is strong and locating of new plants is difficult.

Production of natural gas in the United States peaked in 1972 and is now on a downward curve. It is not expected that this trend can be reversed, but declining production will be offset to some degree by Alaskan and Arc-

FIGURE 22-7 *Cut and polished quartz gemstones. The 20-mm stone in the center on the left is agate, a popular cryptocrystalline variety of quartz. Carnelian (reddish chalcedony) is the stone to the right of the agate. Surrounding the two center stones, clockwise from the northwest corner, are amethyst (purple or violet quartz), opal matrix (hydrated silica), rose quartz, tiger's-eye (quartz that is pseudomorphous after crocidolite), citrine (yellow quartz), and smoky quartz (smoky yellow to dark brown or black in color).*

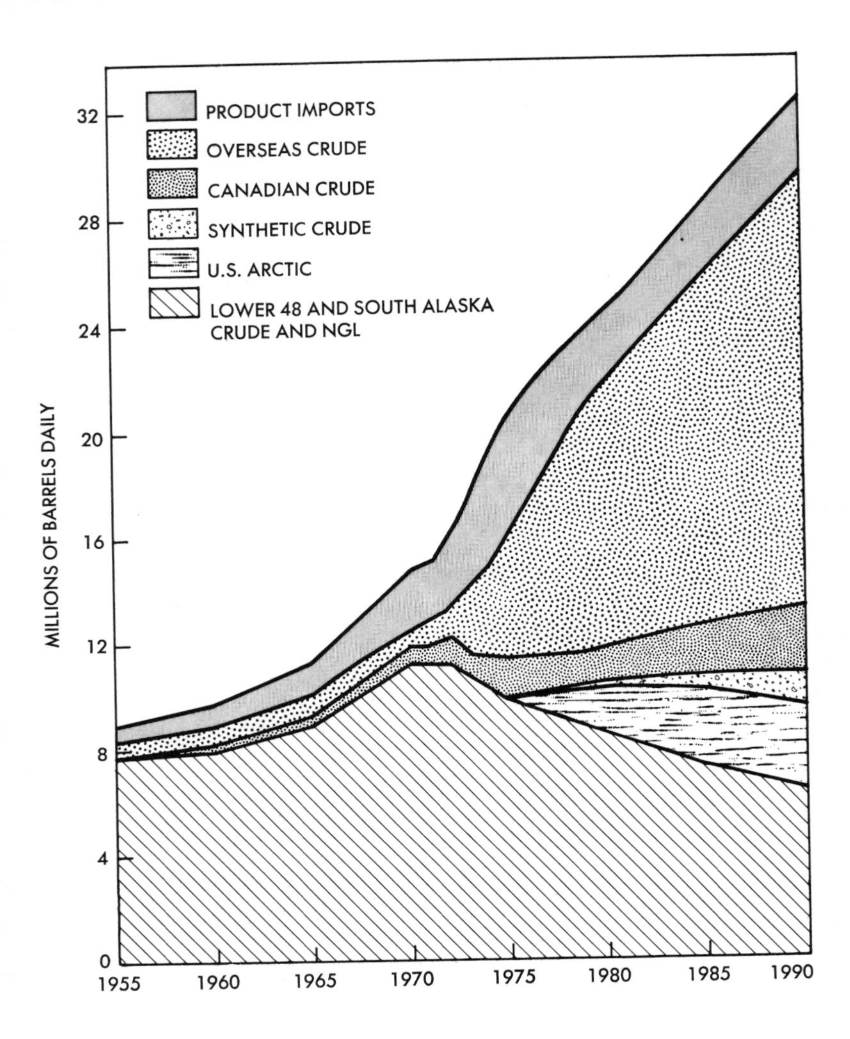

FIGURE 22-8 *Projected U.S. petroleum supply through 1990. (Estimate courtesy of Shell Oil Company.)*

tic gas, Canadian imports, gas derived from coal and oil, and imported liquefied natural gas.

Coal is our most abundant source of fossil fuel, and at present rates of consumption, we have supplies enough to last several hundred years. Only in Asia, including the European part of the Soviet Union, are coal resources larger. However, coal will now be consumed at a much greater rate in the United States because of the high cost of imported crude oil.

Hydropower potential is limited by the scarcity of natural sites and by the high costs of new projects. It is expected that the share of total energy furnished by hydropower will decline slightly during the next 15 years.

Consumption of energy and mineral resources is closely related to a nation's gross national product (GNP). Standard of living and GNP also are closely related. We are dependent on energy for our standard of living, and during the early 1970s total energy consumption was doubling every 14 years. With 6 percent of the world's population, the United States consumes about 35 percent of the world's energy.

Energy demands are increasing throughout the world, but energy is no longer cheap, and new large resources are very difficult to find. The energy resources of the earth, including coal, oil, and gas; oil shale; oil-impregnated sandstone; water power; and nuclear, geothermal, and solar energy, are discussed in sections that follow.

Coal, oil, and gas

The three fuels—coal, oil, and gas—represent energy that has been concentrated by the decay of organic material and are often referred to as *fossil fuels.* During the decay process, the least combustible components are driven off, leaving behind the highly combustible elements carbon, hydrogen, and oxygen.

When these organic fuels are burned, large quantities of stored chemical energy are released in the form of heat energy. This heat may either be used directly or be converted into other forms, such as electrical energy.

Coal

Coal is the end product of vegetable matter that accumulated in the swamplands of the earth millions of years ago (Figure 22-9). Geologic conditions favorable for coal formation are much like those of the present Florida Everglades. Because of its slow rate of formation, coal is essentially a nonrenewable resource. To produce one meter of coal requires several centuries of plant growth and accumulation in a tropical setting.

Sedimentary rocks of diverse ages contain coal beds. The major deposits in the United States are in beds deposited during the Pennsylvanian and Cretaceous Periods. The important Cretaceous coal-bearing sequences of western Wyoming, eastern Utah, and western Colorado were deposited in freshwater environments marginal to a large inland sea about 100 million years ago. Periodic episodes of swamp conditions then migrated eastward with time as the sea gradually withdrew.

Varieties of coal (Table 22-4) are distinguished on the basis of their carbon content, which increases the longer the material has undergone decay. The plant matter from which coal has developed contains about 50 percent carbon. Peat, the first stage in the decay process, contains about 60 percent; anthracite, the final stage, contains 95 to 98 percent. Although carbon is by all odds the most important element in coal, as many as 72 elements have been found in some coal deposits. Over 1 percent of the ash formed by the bituminous coals of West Virginia consists of sodium, potassium, calcium, aluminum, silicon,

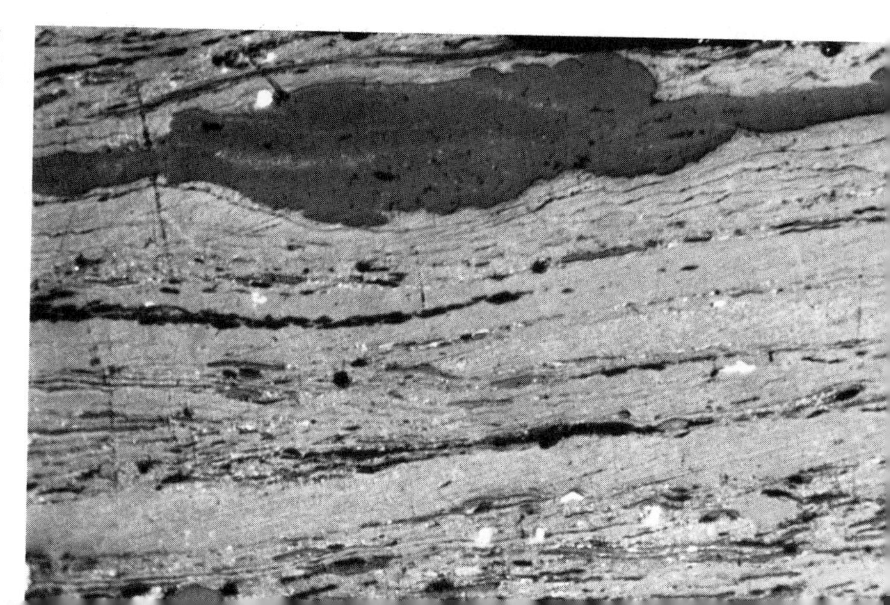

FIGURE 22-9 Photomicrograph of a particle of bituminous coal, Christian County, Illinois. (Courtesy Illinois State Geological Survey.)

Table 22-4 Characteristics of Coal of Various Ranks

TYPE OF COAL	APPEARANCE	CHARACTERISTICS
Lignite	Brown to brownish black	Weathers rapidly; plant residues apparent.
Bituminous coal	Black, dense, brittle	Does not weather easily; plant structures can be seen with microscope; burns with short blue flame.
Anthracite coal	Black, hard, glossy luster	Very hard and brittle; burns with almost no smoke.

iron, and titanium. And there are 26 metals present in concentrations ranging down to .01 percent, including lithium, rubidium, chromium, cobalt, copper, gallium, germanium, lanthanum, nickel, tungsten, and zirconium.

Coal is also important as the source of coke used in the steel industry. The coke is burned in blast furnaces, where it supplies carbon, and the carbon then combines with the oxygen of iron ores to free the metallic iron.

Most of the world's coal resources are found in China and the Soviet Union (Table 22-5). Only minor amounts are present in the Southern Hemisphere. The United States has significant reserves, estimated at about 1,400 billion metric tons. Of this total, 25 percent is economically recoverable by present strip-mining techniques. More than one-half of the present coal mined comes from strip mining (Figure 22-10), which creates several unfavorable environmental consequences, consequences that can be reduced significantly if care is taken.

Mine safety is frequently a difficult problem in all coal mining operations and must be carefully watched (Figure 22-11).

Oil and gas

Oil and gas are the remains of living matter that has been reduced by decay to a state in which carbon and hydrogen are the principal elements. These elements are combined in a great variety of ways to form molecules of substances called *hydrocarbons.* The distinguishing feature of the molecule of each hydrocarbon is the number of carbon atoms it contains. One carbon atom combined with four hydrogen atoms, for example, forms a molecule of a gas called *methane,* CH_4. And two carbon atoms combined with six hydrogen atoms form a molecule of a gas called *ethane,* C_2H_6.

Table 22-5 World Coal Resources in Billions (10^9) of Metric Tons

	IDENTIFIED	PREDICTED	TOTAL POTENTIAL
Asia	6,370	3,640	10,010
North America	1,565	2,621	4,186
Europe	564	189	753
Africa	73	146	219
Oceania	55	64	119
South and Central America	18	9	27
Totals	8,645	6,669	15,314

FIGURE 22-10 *Two ledges in a coal seam, 20 m thick, at Amax Coal Company strip mine in Campbell County, Wyoming. (Courtesy Illinois State Geological Survey.)*

FIGURE 22-11 *Operations to stabilize coal mine for safe working conditions. (Courtesy Illinois State Geological Survey.)*

Natural deposits of oil contain many kinds of hydrocarbons mixed together. These hydrocarbons are separated by an industrial process called *fractional distillation,* based on the principle that light molecules are volatilized more readily than heavy molecules. As early as 600 B.C., Nebuchadnezzar, king of Babylon, was building roads that consisted of stones set in asphalt. Asphalt is the hydrocarbon left behind where natural oil has seeped to the surface and lost its lighter components by evaporation.

Source beds. Most petroleum (from the Latin *petra,* "rock," and *oleum,* "oil"; hence "rock oil") and natural gas have originated from organic remains formerly deposited in marine or lacustrine sedimentary environments. A modern example of such an environment is the Black Sea. Here the water circulates slowly, and the bottom sediments contain as much as 35 percent organic matter, in contrast to the 2.5 percent that is normal for marine sediments. When the putrefaction of the organic remains takes place in an environment of this sort, the product is a slimy black mud known as *sapropel* (from Greek *sapros,* "rotton," and *pelagos,* "sea"). Petroleum and natural gas are believed to develop from the sapropel through a series of transformations not unlike the stages in coal's development from peat.

Three conditions are required for the development of a deposit of petroleum or natural gas: (1) source beds where the hydrocarbons form, (2) a relatively porous and permeable reservoir bed into which they migrate, and (3) a trap at some position in the reservoir bed where they are entrapped.

The most important source beds are dark, organic-rich, finely bedded, fine-grained sedimentary rocks, mainly claystone, silty claystone, and micrite. Source beds for petroleum were predominantly deposited in shallow marine and lacustrine environments. In contrast, many source beds for natural gas were deposited in fluvial and deltaic environments.

Location of reservoir beds. Just where one will find a reservoir of petroleum or natural gas depends on laws that govern the migration of these substances to reservoir rocks. Much remains to be understood about these laws, but several empirical relationships have been established.

Gravity seems to explain the location of many occurrences. According to the *gravitational theory,* if oil, gas, and water are present in a reservoir bed, the oil and gas, being lighter than water, will rise to the top, with the gas uppermost. If the reservoir is trapped in a dome or an anticline capped by an impermeable bed, the oil and gas will accumulate along the crest of the anticline or dome (Figures 22-12, 22-13, and 22-14). This *anticlinal theory* of accumulation, one aspect of the gravitational theory, has proved to be a valuable guide to exploration, and has led to a substan-

FIGURE 22-12 Symmetrical anticlinal trap for oil and gas separated from water by differences in specific gravity. The oil and gas move upward above the water associated with them in a permeable reservoir rock (shown here as a sandstone) until they encounter an impermeable shale folded into an anticline and can rise no further. There they accumulate, with gas above the oil.

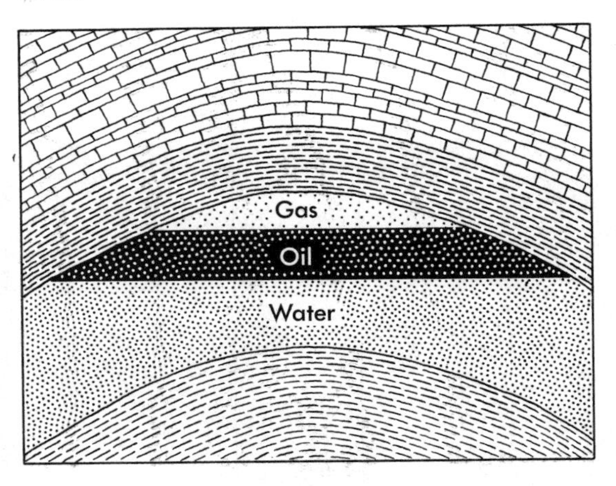

FIGURE 22-13 The dream of all petroleum geologists, a beautifully exposed anticline that can be drilled. (Courtesy U.S. Geological Survey.)

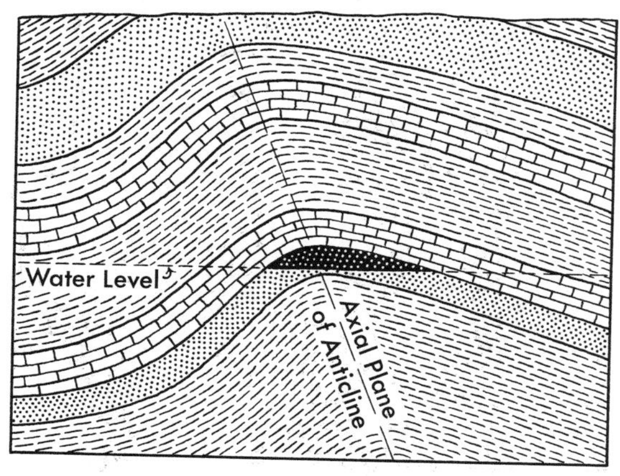

FIGURE 22-14 (Above) Distorted anticlinal trap for oil. If the anticlinal fold is slightly steeper on one side than on the other, its crest at the surface (outcrop of the axial plane) will not be vertically above its crest at the depth where oil and gas have accumulated.

tial volume of production. A corollary of the gravitational theory is that if no water is present, the oil will gather in the trough of a syncline with the gas above it.

Another structure that is important in the gravitational theory is the *stratigraphic trap*, formed when oil and gas in the presence of water are impeded by a zone of reduced permeability as they migrate upward (Figure 22-15). This situation may develop, for example, along old shorelines or in ancient sand bars, where facies change horizontally from sandstone to claystone. Or the upward progress of the oil and gas through a permeable reservoir bed may be blocked by an impermeable bed at an unconformity, or at a fault.

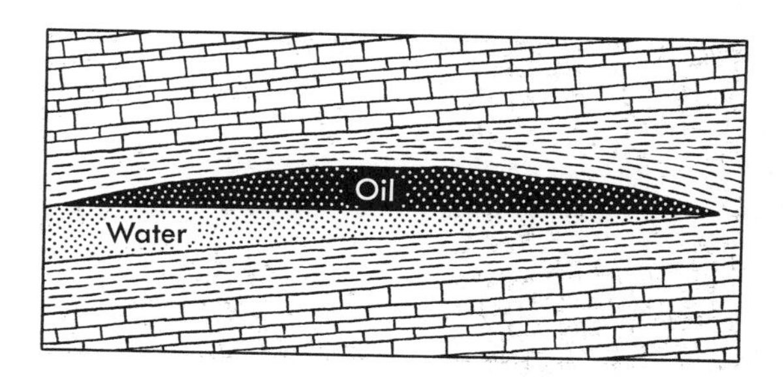

FIGURE 22-15 (Right) Stratigraphic trap for oil. If old shorelines or sand bars developed conditions in which sand graded into clay, the rock equivalents of these deposits will show permeability for the sand and none for the clay. If oil migrating upward in the sand is trapped by a sudden change to impermeable shale, without structural deformation, a stratigraphic trap develops.

Economic impact of giant petroleum fields. Nearly 22,050 oil and gas fields have been discovered in the world, 16,100 of them in the United States and Canada. Between 70,000 and 80,000 oil and gas fields will ultimately be discovered, more than 52,000 of them outside the United States and Canada. According to Meyerhoff (1976)[1], most of these fields (about 36,000) will be found in the Soviet Union. These estimates emphasize the growing shortage of petroleum resources in North America, exclusive of Mexico.

The daily production of the leading 15 petroleum-producing nations during 1974 is shown in Table 22-6. Of the total daily production, about three-fourths came from giant petroleum fields. A *giant oil field* is one that contains a minimum of 68 million metric tons of recoverable oil; a *giant gas field* contains a minimum of 86 billion cubic meters of recoverable gas. A map showing the distribution of the giant petroleum fields of the world is given in Figure 22-16, and a test well drilling for oil and gas in Saudi Arabia is shown in Figure 22-17.

The difference between a giant petroleum field and an ordinary one is the size of the accumulation. Size of a field is directly related to the sizes of the reservoir and trap and the amount of petroleum generated in the source beds. If the trap is large, reservoirs are thick, and large amounts of petroleum are available to move from source beds through carrier beds to the trap, a gigantic field will result.

In the United States and Canada, most of the giant fields already have been discovered, and petroleum in them has been largely depleted. Therefore, strenuous efforts must be made to conserve energy and to develop much more fully alternate forms of energy.

[1]Meyerhoff, A. A., 1976, Economic impact and geopolitical implications of giant petroleum fields: American Scientist, v. 64, pp. 536–541.

Water power

The energy of moving water is all-important in the transportation of material from the continents to the ocean basins and in the sculpturing of the world's landscape. For over 2,000 years man has captured and turned some of this energy to his own ends. Water power in recent years has been converted to electricity by large dams whose water flows through power-generating turbines.

It has been estimated that the world's potential water power supplied by rivers may be almost 3 million megawatts, of which less than 10 percent is now being utilized. Such a potential is three to four times the world's present electricity demand. However, to achieve the full potential, virtually every stream and river in the world would have to be dammed. And since demand for electricity doubles every few decades, this would consume the potential resource in several decades or less.

Table 22-6 World Crude Oil Production in Fifteen Leading Nations, 1974

NATION	METRIC TONS/DAY (1,000s)	PERCENT OF WORLD PRODUCTION
U.S.S.R.	1,320	16.4
United States	1,252	15.6
Saudi Arabia	1,173	14.6
Iran	865	10.8
Venezuela	425	5.3
Kuwait	325	4.0
Nigeria	322	4.0
Iraq	266	3.3
Canada	241	3.0
United Arab Emirates	241	3.0
Libya	217	2.7
Indonesia	196	2.4
China	186	2.3
Algeria	146	1.8
Mexico	93	1.2
Fifteen-nation total	7,267	90.4
World total	8,030	100.0

SOURCE: Data from American Petroleum Institute, 1976.

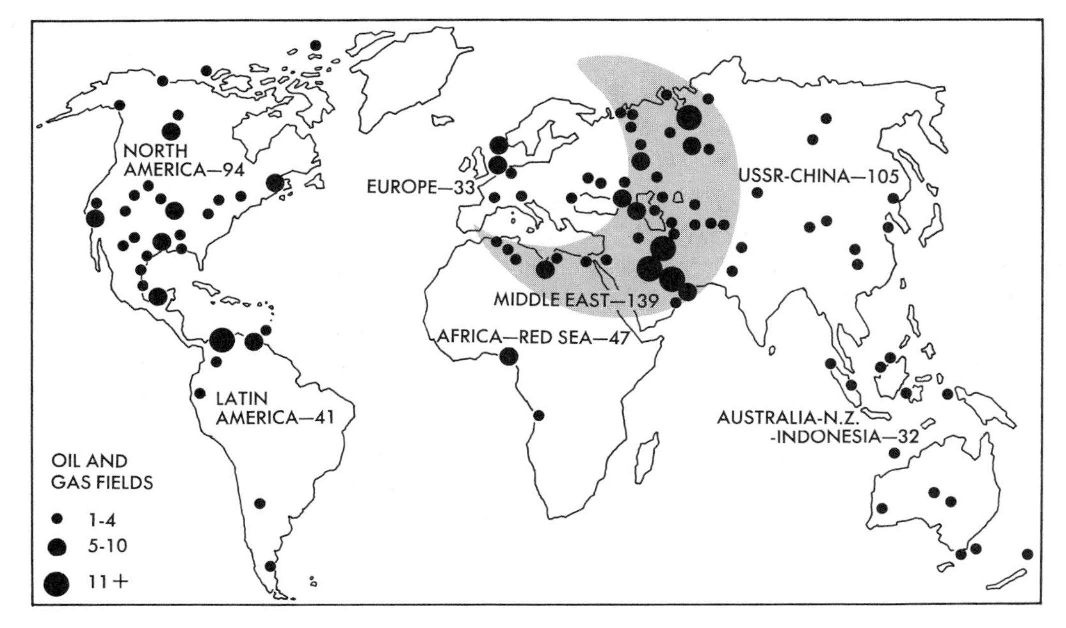

FIGURE 22-16 *(Above) Distribution of giant petroleum fields of the world. More than 60 percent of the giant fields are found in the crescent-shaped area, which also contains nearly 68 percent of the world's known oil and gas reserves. (From Meyerhoff, A. A., 1976, Economic impact and geopolitical implications of giant petroleum fields: American Scientist, v. 64, p. 536–541.)*

FIGURE 22-17 *(Below) A test well drilling for oil and gas in Saudi Arabia. Tertiary beds are prolific producers of petroleum in the general vicinity of the Persian Gulf. (Courtesy Arabian American Oil Company.)*

Nuclear energy

Nuclear energy is expected to be our greatest source of energy in the future (Figure 22-18). Coal, oil, and gas yield only the chemical energy stored in the electrons of atoms; atomic fuels release the much greater energy that is locked in atomic nuclei.

In the *fission process* of nuclear energy generation, neutrons bombard uranium-235 atoms and break them down into other isotopes. The breakdown triggers a tremendous energy release that can be used to generate steam to drive turbines that produce electricity. A single gram of U^{235} releases energy equal to 2 metric tons of oil.

Uranium deposits have been formed by igneous activity, and they are present in igneous rocks, pegmatite dikes, and vein deposits. The primary ore of uranium is the mineral *uraninite*, (UO_2), an oxide sometimes called *pitchblende*. Another oxide, containing smaller amounts of uranium, is the soft, yellow mineral *carnotite*, found in sandstone of the Colorado Plateau. Carnotite constitutes an important source of uranium in the United States. Uranium in this form has gone through several steps, including solution from igneous rocks, transportation in underground solutions, and redeposition.

Most of the world's supply of uranium comes from primary deposits at Great Bear Lake, Canada, in the Congo, and in Czechoslovakia. Nearly all uranium ores in the United States are in freshwater or marine sandstone, which averages 0.2 percent uranium. Uraninite and carnotite in the pore spaces of sandstone

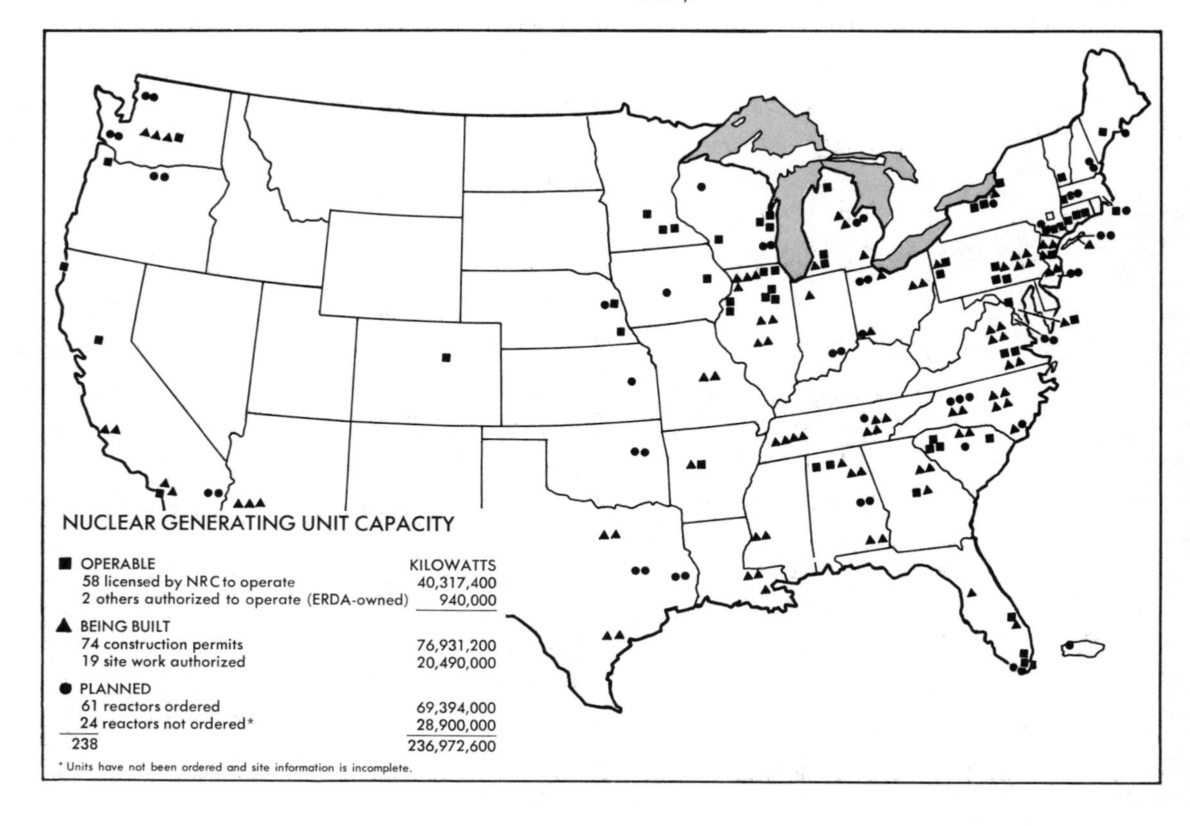

FIGURE 22-18 *Location of nuclear power reactors and nuclear generating unit capacity in the United States. (Courtesy ERDA).*

are the chief sources of occurrence. Many fossil logs, completely replaced by carnotite, and worth a fortune, have been discovered in the Colorado Plateau region. Most of the uranium in the United States is in Wyoming, New Mexico, Colorado, Utah, and Texas.

FIGURE 22-19 *Polished slab of rich oil shale showing thin laminations from the Green River Formation, Colorado. Width of specimen is about 15 cm. (Photo by Rex D. Cole.)*

Oil shale

Oil shale is a fine-grained sedimentary rock—containing organic matter derived mainly from aquatic organisms or from waxy spores and pollen grains—that is largely distillable into artificial petroleum. The largest known deposits of oil shale are in lake beds of the Green River Formation in Colorado and Utah (Figures 22-19 and 22-20). Within this area there are beds at least 3 m in thickness that would yield 76.5 liters per metric ton in sufficient quantity to provide a total of 84.3 billion metric tons of petroleum. In the United States, shale oil resources probably exceed 286 billion metric tons of petroleum according to the U.S. Geological Survey, but there is no present production.

Several factors have inhibited commercial development of these extensive petroleum reserves. The most serious of these is the need to develop a technology that is competitive or nearly competitive with crude oil and natural gas. Although there has been some experimentation, a technology has not yet been perfected for commercial production. The cost of constructing a plant to produce 7,143 metric tons per day is estimated to be at least $800 million and probably closer to $1 billion.

FIGURE 22-20 *Remote and desolate outcrop of oil shale beds in the Green River Formation of Colorado. Rich oil shale (Mahogany Ledge) is present above hammer. (Photo by Rex D. Cole.)*

Two types of processing are being considered: (1) surface processing and (2) in-place processing. *Surface processing,* which includes mining, crushing, conveying of shale to the plant, and disposing of the spent shale after extraction of the petroleum, is the more advanced technology. However, strip mining of oil shale would present many problems. Every cubic meter of oil shale that was processed would produce 1.3 m^3 of waste rock. Also to produce 160 liters of oil, it would be necessary to use two to three times as much water, a resource in short supply in western Colorado and eastern Utah.

In-place processing, which would entail the extraction of the petroleum from the underground bed without handling of the rock itself, would eliminate the mining and spent-shale disposal problems. However, in-place experiments have been discouraging and have not led to presently promising methods. At the moment, surface processing is much closer to being commercial than in-place processing.

Oil-impregnated sandstone

Oil-impregnated sandstone (tar sands) at Athabasca in Alberta, Canada, contain 40 billion metric tons of oil, which is as much as all the oil resources in North America. Many plans for mining the tar sands and extracting the oil have been proposed, and there presently is a plant in operation. However, this venture has not been profitable, and production has not been large. Because of the large size of this energy resource, large-scale development will undoubtedly take place in the future when economic factors are more favorable and technology is more advanced.

In the United States, there are large tar sand deposits at several localities, but there has been no development of them. More than 95 percent of all known reserves in the United States are in Utah. Geologists in other prospective states have not looked as diligently for their tar sands.

A tar sand deposit is a dead oil field. When erosion cuts down to an oil field, or oil migrates almost to the earth's surface, the smaller and lighter hydrocarbon molecules easily evaporate, leaving dark heavy tar behind. Many such deposits existed in the past, but have since been removed by cycles of erosion.

Geothermal energy

Geothermal energy—steam, hot water, and hot rock—may in the future be developed into an important source of energy. Although the amount now produced is small, electric power is produced from geothermal energy at the Geysers in northern California (Figure 22-21) and from steam from the Larderello region of Italy. Electric power is produced commercially from geothermal energy in seven countries.

The earth's heat is produced primarily through the decay of radioactive minerals within the earth's interior. Where ground water encounters hot rock, steam or hot water will form. Hot springs are present throughout the world but are concentrated in areas of recent volcanism and at the boundaries of major crustal plates.

Wells drilled in geothermal areas commonly produce a mixture of steam and hot water. This mixture must be separated before the steam can be used to generate electricity. Disposal of the water is presently a problem, but means will likely be found to use the water for agricultural or municipal use or to generate power.

Proven and potential sources of hot water are present in parts of all of the western United States. In the future, such hot water may be used to generate power by systems in which the heat is transferred to a secondary

FIGURE 22-21 *A Union Oil Company drilling rig encounters shows of geothermal steam coming from the special muffler system on the right of the rig; photographed at the Geysers in northern California.*

fluid, which is then passed through power turbines. A geothermal power plant of this type is already operating in Russia.

Hot rock is a much larger geothermal resource but far more difficult to develop than steam or hot water. Dry geothermal resources are not now economically competitive with other energy sources, but the potential size of the resource means that much study will continue in hopes of developing low cost methods of development.

One estimate places the world's potential geothermal resources at about equal to 203 billion metric tons of oil, with 5 to 10 percent of this available in the United States.

Solar energy

Solar radiation is the most abundant form of energy available to man. The solar energy arriving on 0.5 percent of the land area of the United States is more than the projected total energy needs of the country for the year 2000. Despite its abundance, however, solar energy is a very minor energy source because the technology is not yet available to make it commercially productive. Nevertheless, some scientists believe that within five years, solar-powered systems for heating and cooling homes could be commercially available at prices competitive with gas or oil furnaces and electric air conditioners.

Of the proposed uses of solar energy, heating and cooling for homes and for low-rise commercial buildings are the most advanced techniques presently developed and will constitute the first significant use of solar energy. Solar water heaters are in use in Florida and in several countries. A few homes are equipped with solar heating systems, and preliminary development of cooling systems has begun.

Acceptance of solar power may be slow, even with elimination of some of the technical problems, because the building industry is slow to adopt new techniques. Solar heating systems will increase building costs, although later savings in fuel costs may offset initial installation costs. However, the growing shortages and increased costs of oil and gas will lend impetus to development of the necessary technology for economical utilization of solar energy in homes and buildings.

A more difficult technical challenge is the generation of electricity with heat from solar energy. One group of engineers favors small generating units located where the electricity will be used; another group favors large central solar-thermal facilities. The two concepts differ philosophically and technically, and either system will be difficult to implement commercially. However, both groups believe that the cost of solar-thermal plants will not be more than two or three times what fossil-fueled or nuclear-generating plants cost now, and that rising fuel costs will eventually favor solar-thermal plants whose fuel is essentially free.

Small vapor turbines that use heat from solar collectors to generate electricity have already been built. A miniature solar power plant is operating in Senegal. Experimental solar engines have been developed. And preliminary efforts to develop large central power plants are now under way. Solar energy is a dream whose day may almost be here.

Solar cells, which convert sunlight directly into electricity, are the main source of power for space satellites. However, solar cells are not now competitive with other means of generating electricity for use on earth.

Other energy sources

About 1 billion metric tons of inorganic mineral wastes and more than 1.8 billion metric tons of organic wastes are generated each year in the United States. Experimental studies indicate that the solid organic wastes can be converted into synthetic fuels, but these amounts would be small compared with our yearly consumption of crude oil and natural gas unless the wastes can be supplemented with plant material (algae) grown for the purpose. However, it has been estimated that it will be at least 25 years before large-scale production of algae could begin to contribute as much as 1 percent of our energy supply.

Scientists also know how to extract hydrogen from water for use as a fuel, but the cost is great and it would be difficult to adapt automobiles to hydrogen fuel. Experimental studies will continue, however, because hydrogen can be extracted from water and it burns without polluting the air.

SUMMARY

An ore deposit is a mineral deposit that can be mined at a profit. Metals come from rocks that can be broken down atom by atom to produce the metallic elements. Nonmetal deposits generally are used with little processing after extraction from the earth.

The origin of metal deposits is related to the cooling of magmas, contact metamorphism, weathering of preexisting rocks, or the deposition of sedimentary rocks. New metal deposits are found on the basis of an understanding of the origin and occurrence of other deposits and the application of exploration techniques to surface outcrops and the subsurface. Patterns of metal use range from more than 10^8 tons of iron per year to about 0.1 ton of osmium per year. Ore deposits are a nonrenewable resource; the supply of high-grade ore is limited.

Through 1990 oil will continue to be the most important source of energy. The origin of oil commences with the accumulation of organic-rich sediment in a reducing environment. The organic matter is buried by new sediment and is changed to hydrocarbon-rich liquid and gas. These fluids and gases migrate through rock until they are trapped in reservoir beds (sandstone, limestone, dolomite) beneath an impermeable barrier or until they escape to the surface.

Coal is our most abundant source of fossil fuel, and at present rates of consumption we have supplies to last several hundred years. Coal is part of a series that goes from plant material to lignite to bituminous coal to anthracite coal. Much coal is high in sulfur and is difficult to produce and process, but coal will be consumed at a great rate in the United States because of the high cost of imported crude oil and because of declining reserves of oil and natural gas.

Nuclear energy is expected to be our greatest source of energy in the future. At present reactors burn uranium 235. If uranium becomes too difficult to find, breeder reactors using uranium 238 will replace current nuclear reactors.

Commercial development of our vast oil shale reserves has been inhibited by the lack of a technology that is competitive or nearly competitive with crude oil and natural gas. Geothermal energy is of increasing interest as a source of energy. The development of solar energy has not received the attention it deserves.

SUGGESTED READINGS

BATEMAN, ALAN M., *Economic Mineral Deposits,* 2nd ed. New York: John Wiley & Sons, Inc., 1950.

BATES, ROBERT L., *Geology of the Industrial Rocks and Minerals.* New York: Dover Publications, Inc., 1969.

HAMMOND, ALLEN L., WILLIAM D. METZ, and THOMAS H. MAUGH, II, *Energy and the Future.* Washington, D.C.: American Association for the Advancement of Science, 1973.

LANDES, KENNETH K., *Petroleum Geology of the United States.* New York: Wiley-Interscience, 1970.

LAPORTE, LEO F., *Encounter with the Earth.* San Francisco: Canfield Press, 1975.

LEVORSEN, A. I., *Geology of Petroleum,* 2nd ed. San Francisco: W. H. Freeman and Co., 1967.

PARK, CHARLES F., JR., and ROY A. MACDIARMID, *Ore Deposits.* San Francisco: W. H. Freeman and Co., 1964.

PICARD, M. DANE, *Grit and Clay.* New York: Elsevier Scientific Publishing Company, 1975.

SHELL OIL COMPANY, *The National Energy Outlook.* Houston, Texas: Shell Oil Company, 1973.

SKINNER, B. J., *Earth Resources.* Englewood Cliffs, N.J.: Prentice-Hall, Inc., 1969.

WEBSTER, ROBERT, *Gems: Their Sources, Description and Identification.* London: Archor Books, 1970.

WILLIAMSON, I. A., *Coal Mining Geology.* London: Oxford University Press, 1967.

Lessons from the Earth

Imagine a microscopic but intelligent super-mite, which finds itself in the center of a lettuce head in the Great Valley of California. Of course it doesn't know where it is, who it is, or where it came from. But, being possessed of an insatiable curiosity and an appetite to match, it sets about exploring and devouring its way through the leaves of its lettuce head. It gets to know a great deal about its home and in the process meets other bugs less super than it is. Some of these it tolerates, some it avoids, some it eats, and some it destroys for pleasure. Occasionally the lettuce home is inundated by water, suddenly shaken, or permeated by strange almost deadly odors; in its ignorance, super-mite doesn't know much about rain, cultivation, or insecticides.

One day super-mite eats its way through the outside leaves and breaks forth into sunlight. It is a shocking experience. Super-mite

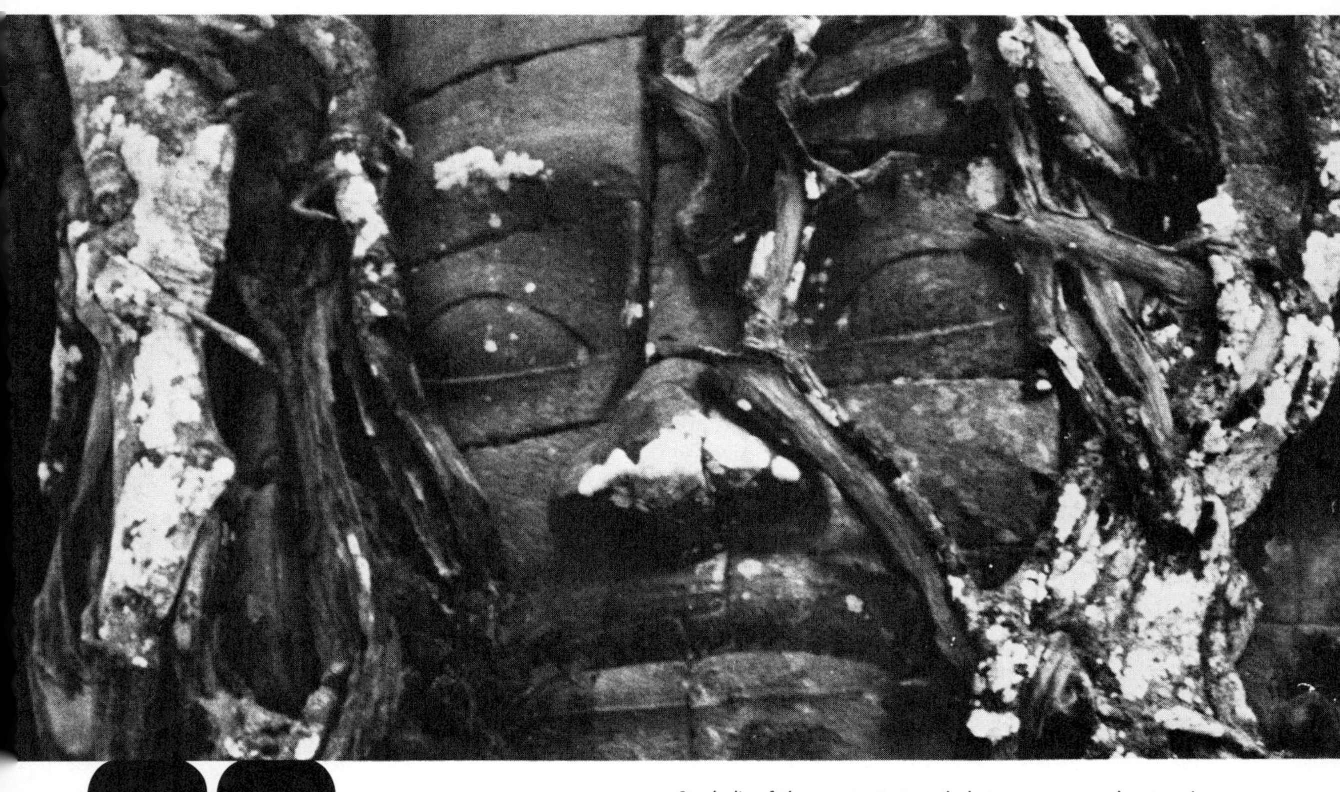

23

Symbolic of the constant struggle between man and nature is this ancient sculptured face overgrown by roots of trees; Ankor, Cambodia. (Photo courtesy of UNESCO.)

contemplates the situation and is deathly afraid of falling off its world to a fate it is not at all sure of. It begins to perceive, however, that its lettuce head is only one of many, and of course it wonders if they too are inhabited by beings like itself. Little does it realize that it is in California and not in New York or Florida.

Unfortunately super-mite is not allowed time to assess its position or to fully understand the meaning of what it sees. A harvester observes that the lettuce head has been rendered somewhat unfit by borings . . .

Finish this fable as you will.

All men share the same earth and depend on its resources for their enjoyment and survival. But not all men view their earthly environment in the same way. Those who live close to nature and who understand the basic ways in which the earth contributes to their existence are inclined to see themselves as caretakers and tenants. Others who do not understand the limitations of water, soil, and minerals may proceed to exploit the earth in short-sighted and unwise ways. A few religions seem to regard the earth as something inferior and insignificant to man—a fallen and degraded planet due to come to an end while men pass on to other spheres. Finally, there are many who enjoy the earth aesthetically, others who study it scientifically, and still others who seek to understand it philosophically. No matter which of these attitudes may dominate a person's view of things, it is obvious that a deeper knowledge of geology is helpful if not essential.

Almost every ideology proposed by philosophers, religionists, politicians, writers, and scientists has its roots in some peculiar views of nature. But history teaches that past concepts of the earth have been mostly erroneous, and insofar as these have been accepted as guides for action, they have led, if not to ruin and disaster, at least to pain and distress. Nothing has contributed more to the fall of nations than man's failure to cooperate with nature. Now, with heightened understanding and educated hindsight we should be in a position to assess our true relations to the earth, its resources, and other living things. Today's assessment is gloomy but not hopeless. What we do know is that decisions and adjustments can not be put off much longer.

Who should be better equipped to understand the earth than those who consider it the subject of their science? Geologists have been studying and mapping the earth and its resources in a systematic way for almost a century and a half. What have they learned that bears on the present predicament of humanity? What are they doing to correct ancient abuses or to stave off future disasters?

On many specific mundane problems, it is not unusual to find geologists arrayed against one another. For example, in some circumstances the task of a geologist is to find a mineral commodity such as oil and then devise means of extracting it at least expense, with minimum attention given to the environment or to future consequences. On the other hand, geologists are sometimes hired to see that proper conservation methods are employed and to help preserve ecological and aesthetic values.

Over and above day-to-day problems, however, there are many basic concepts that unite those who study the earth. Therefore, it would seem to be a major contribution of geologists to present their well verified findings as guides for intelligent action by those who must make decisions about the earth-bound resources that support human life.

Nature of geological science

Geology is not usually regarded as one of the basic sciences. It is said to be a derived science because its subject matter is dealt with according to fundamental laws, principles, and techniques of physics and chemistry. However, geology resembles biology both in its dependence on the basic sciences and in having many time-related aspects. Both biology and geology deal with evolutionary or developmental processes—biology more with that which is learned from currently living organisms, geology with what took place in the past. The science of ancient life (paleontology) is a borderline field that overlaps broadly both biology and geology. Geophysics unites physics and geology, while mineralogy and petrology combine geology and chemistry.

Other comparisons and contrasts emerge. Geology and biology are usually considered to be observational sciences, whereas chemistry and physics are experimental. It is best not to think of this as a hard and fast distinction, however. Both observation and experiment must be brought to bear whenever they are appropriate and as the solution of specific problems requires. What is important is that much of physics and chemistry is amenable to laboratory study; whereas much of biology and geology must be provided by observation—without much hope that what is seen in its natural setting can be duplicated or understood through indoor experiments. For example, no one has been very successful in duplicating even the most simple rocks and ores in the laboratory, or in engineering the genetic changes that bring viable animal and plant species into existence. What is obviously lacking here is the time factor. Both rocks and species are the outcome of long-continued processes that cannot be comprehended and controlled by a single observer or be fully duplicated in a laboratory.

Long-term changes that are of basic importance to geology are essentially evolutionary and historical in nature. Perhaps half of what geologists deal with is embraced under the term *historical geology.* Again a partial comparison with the basic sciences of chemistry and physics is called for. Chemistry and physics are classed by some as nonhistorical because their subject matter is not bound by limitations of space and time. On the other hand, in geology a student is interested primarily in answering the question: What happened? Geologists are frequently and not inappropriately compared with detectives in that they must reconstruct a past event from clues that may be few and far between. Historians of all types deal with events that are bound by temporal and spatial limitations. The events that are studied happened once in the past and can never be exactly duplicated. Furthermore, the past is uncontrollable and yet variable and of so vast a scope that laboratory methods are limited (though not entirely impossible). Therefore the historical scientist must rely heavily on careful and precise observations, and on intuitions and judgments derived from such observations. Unlike the physicist who commonly moves from controlled and regulated causes to anticipated results, the geologist frequently moves backward from observed results to possible causes.

Historians, including all who study the prehistoric past, must deal with the best material that they can discover. Evidence ranges from direct eyewitness testimony, through written records and monuments, to "things" of the prehistoric past such as rocks and fossils. The historical geologist, since he has no human documents, must rely on what the legal profession might call "circumstantial evidence." As with all sciences, the information or evidence of geology has accumulated step by step and bit by bit from a state of ignorance and superstition to present levels of sophistication. The past decade or so of geology has been considered truly revolutionary—on a par with the so-called Darwinian revolution of the mid-eighteen hundreds. Consider some of the crucial steps leading to the present concept of plate tectonics described in Chapter Two.

In summary, geology is a field of knowledge that interacts with many other sciences. By its nature it is derived and not basic, largely but not entirely observational, and historical rather than nonhistorical. Its subject matter has been gathered bit by bit through the efforts of many workers everywhere. The present is a time of unusually fruitful synthesis—geology like biology is now essentially global.

Energy

The earth is an energy-charged planet, and it is the manifestations of energy that give variety and excitement to the science of geology. Compare the static and lifeless Moon with the dynamic, life-bearing Earth—the differences are due basically to how energy is produced and utilized on the two bodies. Energy sources of the earth include solar radiation, electrical fields and discharges, cosmic rays, natural radioactivity, mechanical movement and impacts (including gravity effects), volcanoes and hot springs, chemical reactions (including combustion), and diffuse heat flow from deep-seated mostly unknown sources. Geologists have investigated the origin and effects of all these energy sources, not only as these have influenced the development of the earth and its life but also as to how they might be put to work by man.

The earth has apparently passed through a molten stage and may still retain a large quantity of its primitive heat. The source of heat for melting the earth and keeping it hot is generally believed to be supplied by radioactive isotopes of uranium, thorium, and potassium. It is entirely conceivable that radioactive energy has been responsible for major physical alterations of the earth, but this energy is not suitable for supporting life. Man is presently working out ways to control and concentrate atomic energy, and this form of energy may contribute a large fraction of his future needs. The natural subterranean heat of the earth, known as *geothermal energy*, may also be harnessed as a power source.

Life has evolved in the presence of solar radiation and has come to depend on it for survival. The process of photosynthesis was incorporated into plant cells at a very early stage, even before the earliest fossils, and is the basis of practically all subsequent activity of life. Plants, through photosynthesis, are able to put together energy-rich starches, sugars, and oils. These are stored in special structures to be utilized in time of darkness, drought, or cold. Such food stores encouraged the development of the animal kingdom. At first it may have been only a simple osmosis process whereby one cell extracted specific compounds from another, but the means by which animals are parasitic upon plants has reached unbelievable heights of sophistication. It is certainly no coincidence that the first plants were simple one-celled algae and that similar types have always been the basic food supply of the seas. The first animal fossils were simple food-sifters or mud-grubbers able to exist on contemporary lowly plants.

The whole succeeding story of animal life, as revealed by fossils, is a see-saw contest between eating and being eaten. Animals that prey upon other animals soon put in an appearance. Teeth, as specialized structures for tearing, rending, piercing, grinding, crushing, and gnawing, became standard equipment among predaceous animals. Better teeth were countered by thick shells, sharp spines, hard armor, tough hides, and above all a means of rapid escape.

By the middle Paleozoic the oceans were essentially full of a variety of competing forms, and a few hardy pioneers crossed the beaches onto land. The land-based food chain also depended on plants that were already there, and the ancient struggle to obtain food stores continued. Primary carnivores in the form of air-breathing arthropods furnished an essential link, which enticed the first land-living vertibrates into new environments. In short order, fearsome predators such as dinosaurs arose to command the land. Almost inevitably it seems, intelligent action, as an outcome of evolutionary advancements of the nervous system, began to supplant unthinking brute force, and man arose to occupy the top of the food pyramid.

Reducing this already simplified story to its aphoristic essentials, we can say that nothing comes or can succeed unless preceded by an adequate food (energy) source. No living or dead organism can be understood without reference to its position in the food chain—Where did it get its energy and for what contemporaries did it furnish energy? In the economy of nature, energy is the one universal medium of exchange.

This is probably the best place to give proper perspective to the fossil fuels, those most important byproducts of the past that man has put to use and currently depends on for energy to support his civilization. Fossil fuels represent bottlenecks in the energy cycle of the earth. In a perfectly balanced situation, as much energy would leave the surface of the earth as reached it. In the biologic world, the amount of each year's new growth would be exactly offset by death and decay so that nothing would be left over from year to year.

Equilibrium was not always in effect; at many times and places growth exceeded decay. Not that organisms did not die; they did die but were incompletely decomposed so that an energy-rich residue in the form of hydrocarbon compounds remained and was sealed in the earth. Thus, came beds of coal, oil-rich shale, asphalt sands, and the parent materials of petroleum and natural gas. A fraction of the energy of ancient sunlight, captured by photosynthesis millions of years ago, has in effect been held over to the present time. Once used this store is gone forever. It is a one-time resource, never to be replaced or renewed.

At least 60 percent of all geologists are engaged in searching for fossil fuels, mainly petroleum. Much of what is known about the earth is a byproduct of this search.

Change

Much of the science of geology deals with change; in fact, geology is almost literally the science of change. Whereas the ordinary view is that the earth is static or changeless, a geologist looks at everything as being in the process of flux and transition. On the principle that every effect must have a cause, the subject of geology invites us to seek to detect past causes from present effects and to predict future effects from current causes. Many important changes are seen to be irreversible and cumulative in one way or another, and for these we can extrapolate with fair certainty both backward and forward. Radioactivity, a process that transmutes elements and generates heat in the earth, diminishes the parent material on which it feeds and increases the daughter products that it generates. Organic evolution, like radioactivity, is also irreversible; genetic material cannot be restored to exact previous configurations, and evolution never reverses itself to reduce a complex creature to a previous less complex stage. Every species—once and only once.

Many other geologic changes are seen to be more or less cyclic in nature. Cycles may be so extremely long that their presence is not evident even in the recorded history of the earth, or they may be so short that millions are compressed within a second's duration. Some cycles are almost perfect, while others are crude and irregular. As a practical example we are certain that some sort of cyclic activity was responsible for the repetitious deposition of valuable coal beds one above another. The same rhythmic repetition is seen in the accumulation of salt or gypsum or limestone. The knowledge that such cyclic deposits exist is of use in exploration and exploitation, even though their cause is not known. Perhaps of greater importance are the repetitive and probably cyclical changes of climate that affected the earth during the Great Ice Age. The fact that gigantic ice sheets formed simultaneously at least four times over much of North America and Europe during the past 2 million years gives cause to consider whether the ordeal is over or not. The problem of what the climate of coming decades, centuries, and millennia will be like is certainly of paramount importance to the human race. Evidence is good that we are still in the Ice Age and that sooner or later the glaciers will again form and devastate millions of square kilometers of the earth's surface. It is small consolation that this may be far in the future. The drought that is currently affecting large areas of the globe has spread with devastating suddenness. One wonders if such disasters could have been foretold and their effects softened if greater attention had been paid to climatic records preserved in sediments of the arid and semiarid lands that are now affected. Here, as with other problems, the remedy may come too little and too late.

Equilibrium

The Earth is frequently described as being alive. Certainly it is a very active and dynamic planet by comparison with neighboring members of the solar family. Winds and convection currents stir the atmosphere to create a constant state of flux, which is dramatically displayed in the shifting patterns of the daily weather map. Currents, waves, and tides agitate the oceans, rivers, and lakes as these water bodies respond to gravity, atmospheric influences, and the rotation of the earth. Ongoing reactions among the liquid, gaseous, and solid constituents accomplish the weathering and erosion of rocks and eventually destroy entire mountain ranges. Even the solid "crust" of the earth is slowly shifting as it shears and buckles under the force of immense internal stresses. There is good evidence that even the core of the earth is stirred by gigantic currents. The presence of life forms on the surface is but another expression of the dynamic character of the planet Earth.

Underlying all this seemingly random restless movement is the universal drive toward equilibrium. In the final analysis, the action is always toward a condition of stability, rest, or repose. Water runs downhill, and winds blow from areas of high pressure to areas of low pressure, so as to achieve a state of stability. The sediment carried along a river bed or seashore comes to rest only when it can no longer be moved. Mineral particles such as sand may become part of a solid rock that will remain solid until it is no longer in equilibrium with prevailing conditions. As rocks are subject to heat, they pass through various stages of metamorphism and may melt or even volatilize. But the process always ends when a stage of equilibrium is reached.

The idea that nature works toward a condition of rest has been called the *principle of least action.* This comes near to being a universal law. The great medieval genius, Leonardo da Vinci, wrote that "every action of nature is made along the shortest way possible." In more formal English, the principle has been restated thus: "A system tends to change so as to minimize an external disturbance." In everyday human terms we may think of nature as being lazy, always doing what is forced upon it and no more. The principle of least action appears in science under several names. Chemists commonly refer to it as the theorem of LeChatelier. Biologists may not realize that the maxim of survival of the fittest reflects the same idea. And the economists' law of supply and demand is yet another expression of this great truth. It is clearly beyond the limits of this summary to discuss all these expressions, but our subject justifies more than mere assertion that the principle of least action is illustrated by organic evolution. Individual organisms or entire communities are continually being subject to external stresses such as changes of climate, food shortages, newly arrived competition, or destruction of their habitats. In all cases life reacts to "minimize the disturbance." The common reaction of living things is to migrate, which is probably more correctly visualized as moving from area to area in order to remain with favorable conditions. If migration is not possible, the many obvious adaptations to heat or cold, less food, crowding, or competition come into play. The search for security, comfort, or survival is everywhere. That the search is not always successful is evident from the deaths of individuals and the extermination of species. Those who have successfully adjusted to the stresses of existence, or have minimized disturbances to tolerable levels, are alive to inhabit the present world. Man is no exception—he illustrates the principle of least effort when he puts on or takes off just the right amount of clothing to be comfortable.

Nature is currently being greatly disturbed by a very upsetting influence, namely, man. Does the principle of least action apply? Is nature reacting to minimize the disturbance? Evidence is clear that man is in no way immune from this universal law. His soil will erode until it can erode no more, until the falling raindrops strike only bare rock and nothing more can be removed. The animal species on which man preys will diminish until they can diminish no more, even if some are forced into extinction. Water bodies will absorb what pollution they must and no more, even though they may or may not pass the point of being fit or useful to man. And when worst comes to worst, man himself will have to go if he gets too far out of equilibrium with his resources.

For what consolation it may offer, we should note that nature is neither vengeful nor malicious. Soil, water, and living things will react only to the stresses placed upon them. This means that they may recover and be salvaged. Species not persecuted need not become extinct, and not all of the human race must be eliminated to satisfy the least action principle.

Survival of the fittest

The phrase "survival of the fittest" arose from Charles Darwin's explanation of how organic evolution operates. He pointed out that competition is a universal fact of nature and that it is chiefly by struggle that the less fit are eliminated, leaving the more fit to survive and take possession of the world. This way of looking at things has been condemned as cruel, inhuman, and anti-Christian. Its critics usually focus attention on such brutal and bloody face-to-face examples as the wolf slaughtering the defenseless lamb, the hawk striking the dove, and the armed soldier shooting down the defenseless civilian. Despots have used Darwin's theory as an excuse for attacking their weaker neighbors, while individuals without number have translated it into a "might makes right" philosophy of life. Those so inclined have found in the concept of survival of the fittest an excuse for oppressing or even eliminating their supposedly less advanced neighbors, and for the forceable removal of the sick, crippled, aged, less intelligent, or unproductive members of human society. After setting one's self or one's group up as being superior (more fit), the theory of ridding the world of anyone who doesn't measure up under these criteria becomes much easier to accept.

The perversion of the evolutionary concept of survival of the fittest to the degradation or detriment of man has been cause for blaming Darwin and his theory for most of the ills of man's society. Thus, evolution has been said to oppose the basic Christian principles of love, mercy, and concern for the individual. Although we obviously cannot pursue the extensive topic of moral conduct, we should not leave the subject without attempting to put the process of organic evolution in a more realistic context.

In a world without man, such concepts as cruelty, war, hate, and oppression are meaningless. Nature is apparently mindless, being neither malicious nor benevolent in the human sense. The lion in killing the zebra is in no way reprehensible; it kills to live, no more, no less. Consider the billions of years of prehistoric evolution during which untold numbers of species came into being, lived for many generations, and passed from the scene despite valiant efforts and clever tricks to survive. Has the outcome of this long process of elimination, which we now observe and are a part of, been good or bad for humanity? It would be a very cynical person indeed who would say that the earth as man originally found it was bad or undesirable. Indeed, it would be difficult to imagine a more suitable, well-balanced environment in which to live. If we can accept the world as being good, as it was before man became a significant factor in it, then it must be that the concept of the survival of the fittest is likewise good. We are more than the beneficiaries of the process; we are, in a sense, the products of the process.

What of the unfit? If the present is inhabited by those beings that nature has found to be better fitted than their contemporaries, will not the future belong to the most fit of those now existing? The ancient standard of fitness has not been abandoned just because man has entered the scene. Survival is basically what it has always been—the preservation of the individual to the end that the group (race, species) is perpetuated. If anything is evident from the lengthy panorama of past life, it is that the survival of the individual is of no consequence unless the race is preserved. Man is different from dinosaurs chiefly in that he can make a conscious choice as to the relative importance of self-preservation and group preservation. The choice is not an easy one. We may say that man as a species is the most successful animal ever to appear. He rules, commands, and dominates all other living things. Contemporary animals live or die at his good pleasure; he eliminates some and preserves others without restraint. It seems pointless to ask if man is a biologic success—the answer according to the rules of the ancient game of survival must be yes. But if we ask if man will continue to survive, we cannot be so positive. Perhaps we are uncertain because we have not asked the right questions. Is man *fit* to survive? Optimists will say yes, pessimists, no. Let us put it in yet another way: What type of man is more likely to survive? This question is intended to imply that here and there among mankind generally, there are individuals more likely to perpetuate the race than those with whom they may associate. But even this may not be the best question we could ask because it lacks sufficient realism in that it does not take account of the fact that men live, act, and die in groups. Therefore, our question might best be: What type of society is most likely to survive? The answer to such a question focuses on the thought that some nations, some sects, some ethnic groups might be expected to have greater chances of survival than other groups.

We are reminded of the old saying that Nothing succeeds like success. The obvious paraphrase is that Nothing survives like survival. But to repeat aphorisms is an easy way out and really gives no answer at all. There are reasons why things have survived in the past, and these same reasons will hold in the future if we know how to apply them. Seldom can we know or detect all the factors of success, but one thing is certain, *these factors have survival value.* They somehow give their possessors a reproductive edge over their contemporaries, whether these be of the same or of different species. We note with certainty the survival value of skeletons of bones, of light-sensitive organs, of circulating blood, of internal fertilization, of brains, or the upright posture, and many others. Man possesses all of these admirable traits and many more.

It may be argued that man as a physical entity could be improved upon here and there, but the question is: Improved for what? Would the suggested improvements be for the prolongation of the individual life and for personal selfish pleasures, or would they be for the more effective propagation of the race as a whole?

Natural selection, if we may personify it, managed things very well, even though it took a lot of time and was an expensive process in terms of life and energy. Darwin referred to the process as a natural selection and went further by pointing out that the basis of selection is the relative ability to survive not only under prevailing but also under future conditions. The best that the past has to contribute to one's chances for survival is never entirely sufficient in itself. Something must be added or achieved by the individual. The real question that every person must answer for himself and for his group is: Do I and my group have survival value?

Search for an ethic

From its earliest inception to the present, the science of geology has had to win its way against entrenched popular opinion. Opposition has been almost entirely from those religious sects with literal views of the Judeo-Christian scriptures. The first two chapters of the Bible presented an account of the origin and early history of the "heavens and the earth and all that in them are," which devout Christians through the centuries had, until fairly recently, considered to be accurate and sufficient.

A literal interpretation of Genesis would seem to be that the sun, moon, stars, earth, plants, animals and man were produced during a single week of ordinary 24-hour days about 6,000 years ago. Furthermore, the creation was accomplished by God in a miraculous, supernatural manner that cannot be understood or in any way duplicated by man. Churchmen considered it to be impious and improper to inquire into or question the manner of creation, and it was heretical to entertain the thought that the creation had not been accomplished exactly as described in the Bible.

The first direct confrontation between science and organized Christian religion involved the position, shape and movements of the earth rather than its age and origin. In 1633 Galileo was tried and condemned as a heretic for teaching that the earth is round and instead of being the center of the universe is but one of a number of relatively small bodies circling the large central sun. Galileo was right, and the observational method of discovering truth about the natural world proved superior to literal scriptural interpretations. This episode is of greatest significance in the intellectual development of Western thought. Practically everyone once believed in a flat, stationary earth. The evidence that this might not be true came in scattered bits of direct and indirect information and not in one grand conclusive revelation. Some thinkers became convinced of a round earth rather easily; others waited for more evidence; while still others held out until the circumnavigation of the globe was accomplished by Magellan in 1519–1522. (There is still a Flat Earth Society.) Some ecclesiastical authorities had come out strongly in favor of a flat earth and had supported their stand with scriptural quotations. They found themselves in the embarrassing position of having to admit grave errors, not only in interpretation of the facts that Galileo had tried to explain to them but also in interpretation of the scriptures of which they were supposedly the ultimate interpreters.

Fully as difficult to justify is the manner in which Galileo and other scientists were treated for daring to express heretical ideas of any sort. As a consequence of the events of this period, organized religion suffered an immense loss of prestige and influence, and the veracity of literal scriptures came increasingly into question. Two hundred and twenty-four years after Galileo, a second scientific concept emerged that sparked an even greater intellectual upheaval. This was the Darwinian theory of organic evolution, put forth in what is said to be the most influential nontheological book ever written, *The Origin of Species,* issued in 1857. The Darwinian theory appeared more threatening to Christian sects than anything previously put forth by science, for it seemed to attack the divine nature and origin of man. Darwin proposed that all species, including man, had been derived through normal reproductive processes from previous species backward through time to one or a few simple beginnings. Each species is a natural product, not a special handiwork of God called forth in perfection on the day of creation. That new species have appeared and old ones disappeared is clearly indicated by the fossil record. The emergence of new species, Darwin explained, had to come through the modification of previous ones.

Organic evolution has been most difficult for many Christians to accept. If species make themselves or come into being through some blind mechanical process, there seems to be no need of a creator God. And the thought of man, supposedly the crowning product of creation, being no better than a beast, perhaps even the descendant of an ape, is almost intolerable.

Anti-evolution feeling came to a climax in the so-called "monkey trial" held in Dayton, Tennessee, in 1925. John Scopes was brought into court for having taught the theory of evolution in his classroom in defiance of the law. Although Scopes was judged guilty and paid a small fine, the trial has been regarded as a triumph of science over bigoted religion. One thing was proven beyond doubt, that it is unwise if not impossible to insist on strict literal interpretations of Genesis in the face of scientific facts.

The creationist-evolutionist battle is by no means over. Latest events center on the teaching of evolution in public schools to the exclusion of the Genesis account. The General Assembly of the State of Tennessee passed an act in April, 1973, which prohibited the use of any biology textbook used for teaching in the public schools that

> expresses an opinion of, or relates to a theory about origins of creation of man and his world . . . unless it specifically states that it is a theory . . . and is not represented to be scientific fact. Any textbook so used . . . shall give, in the same textbook and under the same subject, commensurate attention to, and an equal amount of emphasis on, the origins and creation of man and his world, as . . . is recorded in other theories, including . . . the Genesis account in the Bible. . . .

It seems most unfortunate that the facts and well-founded theories of science must be judged in moral terms. Worse yet, scientists themselves are accused of unworthy motives and of organized attempts to undermine religion and faith in God. What is there that is intrinsically wrong about belief in organic evolution, in an ancient earth, or in an origin of man's physical body from lower forms of life? Many worthy Christians have successfully reconciled such concepts with their religious beliefs. Perhaps a lesson is to be learned from the past. The futility and even absurdity of trying to prove and maintain strictly literal interpretations of scriptures are evident in the trials of Galileo and Scopes. But such actions do not destroy scripture. Certainly more than one interpretation of the cryptic language of Genesis is possible.

SUMMARY

As home for man, the earth has been viewed in a variety of ways. Primitive people were understandably ignorant of its structure, place in the universe, and the extent of its resources. Scientific investigation has dispelled this ignorance and replaced it with facts and well-founded theories. Not only curiosity but also dire necessity have driven mankind to learn everything possible about the earth and its resources. The science of geology in particular has to do with the earth in all its aspects. Geology is a derived science based on physics and chemistry but having strong historical and observational bias. The origin of life and evolution of species are essential parts of geology and relate it to all aspects of biology.

Aside from the mere statistical cataloging of facts and figures and the making of maps, geologists have accumulated information from which certain very basic generalizations are possible, as noted below.

Energy relations

The earth has developed within the constraints of its energy supply from external and internal sources. In compliance with the second law of thermodynamics, the energy of the sun-earth system is diminishing, but energy from the sun has been sufficient to drive the earth's living systems to increasing levels of complexity. No species has appeared or succeeded without being preceded by an adequate energy (food) source.

Change

The earth and its systems are not static, changeless, or eternal. Everything is in a state of flux and transition. Some processes such as radioactive disintegration and organic evolution are one-way processes whose past and future states can be known with some accuracy. Other changes such as mountain building or climatic changes may be cyclic or repetitive.

Equilibrium

All natural processes tend toward a state of rest, repose, or equilibrium; that is, "systems tend to change so as to minimize external disturbances." The everyday processes of erosion and transportation of sediment, of metamorphism of rocks, and even of the movement of massive surface plates and interior of the earth obey the rule of least action. Living things in searching for security, comfort, and survival also act to minimize upsetting external conditions.

Survival of the fittest

The 3-billion-year history of life has been one of constant progress if this be defined as more complete utilization of earth's space and energy resources, greater efficiency in reproduction, and improvement of nervous tissue. Today's survivors are few compared with those exterminated. Success depends on *survival value*, the true measure of adaptation.

Conflict with religion

Literal translation of Judeo-Christian scriptures is in conflict with the findings of geology. It is unfortunate that this has led to ill-feeling, distrust, and confusion. Both sides may have to make adjustments if science and theology are to be unified for the good of man.

The Elements—Mass and Energy

APPENDIX A

ELECTRONIC CONFIGURATION OF THE FIRST 30 ELEMENTS

(These elements constitute 99.6 per cent of the earth's crust)

Atomic Number (Protons)	*Name of Element*	*Symbol*	*Number of Electrons in*										*Mass Number (Protons + Neutrons) of Stable Isotopes in Order of Abundance*	*Parts Per Million in Earth's Crust*
			1-shell	*2-shell*		*3-shell*			*4-shell*					
			s	*s*	*p*	*s*	*p*	*d*	*s*	*p*	*d*	*f*		
1	Hydrogen	H	1										1, 2	1,400.
2	Helium	He	2	Inert Gas									4, 3	.003
3	Lithium	Li	2	1									7, 6	65.
4	Beryllium	Be	2	2									9	<.001
5	Boron	B	2	2	1								11, 10	3.
6	Carbon	C	2	2	2								12, 13	320.
7	Nitrogen	N	2	2	3								14, 15	46.
8	Oxygen	O	2	2	4								16, 18	466,000.
9	Fluorine	F	2	2	5								19	300.
10	Neon	Ne	2	2	6	Inert Gas							20, 22, 21	<.001
11	Sodium	Na	2	2	6	1							23	28,300.
12	Magnesium	Mg	2	2	6	2							24, 25, 26	20,900.
13	Aluminum	Al	2	2	6	2	1						27	81,300.
14	Silicon	Si	2	2	6	2	2						28, 29, 30	277,200.
15	Phosphorus	P	2	2	6	2	3						31	1,180.
16	Sulfur	S	2	2	6	2	4						32, 34, 33, 36	520.
17	Chlorine	Cl	2	2	6	2	5						35, 37	314.
18	Argon	A	2	2	6	2	6	Inert Gas					40, 36, 38	.04
19	Potassium	K	2	2	6	2	6		1				39, 41	25,900.
20	Calcium	Ca	2	2	6	2	6		2				40, 42, 43, 44, 46, 48	36,300.
21	Scandium	Sc	2	2	6	2	6	1	2	Transition Elements			45	5.
22	Titanium	Ti	2	2	6	2	6	2	2				48, 46, 47, 49, 50	4,400.
23	Vanadium	V	2	2	6	2	6	3	2				51	150.
24	Chromium	Cr	2	2	6	2	6	5	1				52, 53, 50, 54	200.
25	Manganese	Mn	2	2	6	2	6	5	2				55	1,000.
26	Iron	Fe	2	2	6	2	6	6	2				56, 54, 57, 58	50,000.
27	Cobalt	Co	2	2	6	2	6	7	2				59	23.
28	Nickel	Ni	2	2	6	2	6	8	2				58, 60, 62, 61, 64	80.
29	Copper	Cu	2	2	6	2	6	10	1				63, 65	70.
30	Zinc	Zn	2	2	6	2	6	10	2				64, 66, 68, 67, 70	132.
														996,108. (99.61 per cent)

Atomic Number (Protons)	Name of Element	Symbol	Number of Electrons in																						
			1-shell	2-shell		3-shell			4-shell				5-shell				6-shell				7-shell				
			s	*s*	*p*	*s*	*p*	*d*	*s*	*p*	*d*	*f*	*s*	*p*	*d*	*f*	*s*	*p*	*d*	*f*	*s*	*p*	*d*	*f*	
31	Gallium	Ga	2	2	6	2	6	10	2	1															
32	Germanium	Ge	2	2	6	2	6	10	2	2															
33	Arsenic	As	2	2	6	2	6	10	2	3															
34	Selenium	Se	2	2	6	2	6	10	2	4															
35	Bromine	Br	2	2	6	2	6	10	2	5															
36	Krypton	Kr	2	2	6	2	6	10	2	6															Inert Gas
37	Rubidium	Rb	2	2	6	2	6	10	2	6			1												
38	Strontium	Sr	2	2	6	2	6	10	2	6			2												
39	Yttrium	Y	2	2	6	2	6	10	2	6	1		2												Transition Elements
40	Zirconium	Zr	2	2	6	2	6	10	2	6	2		2												Transition Elements
41	Niobium (Columbium	Nb Cb)	2	2	6	2	6	10	2	6	4		1												Transition Elements
42	Molybdenum	Mo	2	2	6	2	6	10	2	6	5		1												Transition Elements
43	Technetium	Tc	2	2	6	2	6	10	2	6	6		1												Transition Elements
44	Ruthenium	Ru	2	2	6	2	6	10	2	6	7		1												Transition Elements
45	Rhodium	Rh	2	2	6	2	6	10	2	6	8		1												Transition Elements
46	Palladium	Pd	2	2	6	2	6	10	2	6	10														
47	Silver	Ag	2	2	6	2	6	10	2	6	10		1												
48	Cadmium	Cd	2	2	6	2	6	10	2	6	10		2												
49	Indium	In	2	2	6	2	6	10	2	6	10		2	1											
50	Tin	Sn	2	2	6	2	6	10	2	6	10		2	2											
51	Antimony	Sb	2	2	6	2	6	10	2	6	10		2	3											
52	Tellurium	Te	2	2	6	2	6	10	2	6	10		2	4											
53	Iodine	I	2	2	6	2	6	10	2	6	10		2	5											
54	Xenon	Xe	2	2	6	2	6	10	2	6	10		2	6											Inert Gas
55	Cesium	Cs	2	2	6	2	6	10	2	6	10		2	6			1								
56	Barium	Ba	2	2	6	2	6	10	2	6	10		2	6			2								
57	Lanthanum	La	2	2	6	2	6	10	2	6	10		2	6	1		2								"Rare Earths"; Transition Elements
58	Cerium	Ce	2	2	6	2	6	10	2	6	10	1	2	6	1		2								"Rare Earths"; Transition Elements
59	Praseodymium	Pr	2	2	6	2	6	10	2	6	10	2	2	6	1		2								"Rare Earths"; Transition Elements
60	Neodymium	Nd	2	2	6	2	6	10	2	6	10	3	2	6	1		2								"Rare Earths"; Transition Elements
61	Promethium	Pm	2	2	6	2	6	10	2	6	10	4	2	6	1		2								"Rare Earths"; Transition Elements
62	Samarium	Sm	2	2	6	2	6	10	2	6	10	5	2	6	1		2								"Rare Earths"; Transition Elements
63	Europium	Eu	2	2	6	2	6	10	2	6	10	6	2	6	1		2								"Rare Earths"; Transition Elements
64	Gadolinium	Gd	2	2	6	2	6	10	2	6	10	7	2	6	1		2								"Rare Earths"; Transition Elements
65	Terbium	Tb	2	2	6	2	6	10	2	6	10	8	2	6	1		2								"Rare Earths"; Transition Elements
66	Dysprosium	Dy	2	2	6	2	6	10	2	6	10	9	2	6	1		2								"Rare Earths"; Transition Elements
67	Holmium	Ho	2	2	6	2	6	10	2	6	10	10	2	6	1		2								"Rare Earths"; Transition Elements
68	Erbium	Er	2	2	6	2	6	10	2	6	10	11	2	6	1		2								"Rare Earths"; Transition Elements
69	Thulium	Tm	2	2	6	2	6	10	2	6	10	12	2	6	1		2								"Rare Earths"; Transition Elements
70	Ytterbium	Yb	2	2	6	2	6	10	2	6	10	13	2	6	1		2								"Rare Earths"; Transition Elements
71	Lutetium	Lu	2	2	6	2	6	10	2	6	10	14	2	6	1		2								"Rare Earths"; Transition Elements
72	Hafnium	Hf	2	2	6	2	6	10	2	6	10	14	2	6	2		2								Transition Elements
73	Tantalum	Ta	2	2	6	2	6	10	2	6	10	14	2	6	3		2								Transition Elements
74	Wolfram (Tungsten)	W	2	2	6	2	6	10	2	6	10	14	2	6	4		2								Transition Elements
75	Rhenium	Re	2	2	6	2	6	10	2	6	10	14	2	6	5		2								Transition Elements
76	Osmium	Os	2	2	6	2	6	10	2	6	10	14	2	6	6		2								Transition Elements
77	Iridium	Ir	2	2	6	2	6	10	2	6	10	14	2	6	7		2								Transition Elements
78	Platinum	Pt	2	2	6	2	6	10	2	6	10	14	2	6	8		2								Transition Elements
79	Gold	Au	2	2	6	2	6	10	2	6	10	14	2	6	10		1								Transition Elements
80	Mercury	Hg	2	2	6	2	6	10	2	6	10	14	2	6	10		2								
81	Thallium	Tl	2	2	6	2	6	10	2	6	10	14	2	6	10		2	1							
82	Lead	Pb	2	2	6	2	6	10	2	6	10	14	2	6	10		2	2							
83	Bismuth	Bi	2	2	6	2	6	10	2	6	10	14	2	6	10		2	3							

Atomic Number (Protons)	Name of Element	Symbol	1-shell s	2-shell s	2-shell p	3-shell s	3-shell p	3-shell d	4-shell s	4-shell p	4-shell d	4-shell f	5-shell s	5-shell p	5-shell d	5-shell f	6-shell s	6-shell p	6-shell d	6-shell f	7-shell s	7-shell p	7-shell d	7-shell f	
84	Polonium	Po	2	2	6	2	6	10	2	6	10	14	2	6	10		2	4							
85	Astatine	At	2	2	6	2	6	10	2	6	10	14	2	6	10		2	5							
86	Radon	Rn	2	2	6	2	6	10	2	6	10	14	2	6	10		2	6 Inert Gas							
87	Francium	Fr	2	2	6	2	6	10	2	6	10	14	2	6	10		2	6			1				
88	Radium	Ra	2	2	6	2	6	10	2	6	10	14	2	6	10		2	6			2				
89	Actinium	Ac	2	2	6	2	6	10	2	6	10	14	2	6	10		2	6	1		2				Transition Elements
90	Thorium	Th	2	2	6	2	6	10	2	6	10	14	2	6	10	1	2	6	1		2				
91	Protactinium	Pa	2	2	6	2	6	10	2	6	10	14	2	6	10	2	2	6	1		2				
92	Uranium	U	2	2	6	2	6	10	2	6	10	14	2	6	10	3	2	6	1		2				
93	Neptunium	Np	2	2	6	2	6	10	2	6	10	14	2	6	10	4	2	6	1		2				
94	Plutonium	Pu	2	2	6	2	6	10	2	6	10	14	2	6	10	5	2	6	1		2				
95	Americium	Am	2	2	6	2	6	10	2	6	10	14	2	6	10	6	2	6	1		2				
96	Curium	Cm	2	2	6	2	6	10	2	6	10	14	2	6	10	7	2	6	1		2				
97	Berkelium	Bk	2	2	6	2	6	10	2	6	10	14	2	6	10	8	2	6	1		2				
98	Californium	Cf	2	2	6	2	6	10	2	6	10	14	2	6	10	9	2	6	1		2				
99	Einsteinium	En	2	2	6	2	6	10	2	6	10	14	2	6	10	10	2	6	1		2				
100	Fermium	Fm	2	2	6	2	6	10	2	6	10	14	2	6	10	11	2	6	1		2				
101	Mendelevium	Me	2	2	6	2	6	10	2	6	10	14	2	6	10	12	2	6	1		2				
102	Nobelium	No	2	2	6	2	6	10	2	6	10	14	2	6	10	13	2	6	1		2				

(Number of Electrons in the shells; elements 89–102 bracketed as Transition Elements.)

ALPHABETICAL LIST OF THE ELEMENTS

Element	Symbol	Atomic Number	Element	Symbol	Atomic Number	Element	Symbol	Atomic Number
Actinium	Ac	89	Gold	Au	79	Promethium	Pm	61
Aluminum	Al	13	Hafnium	Hf	72	Protactinium	Pa	91
Americium	Am	95	Helium	He	2	Radium	Ra	88
Antimony	Sb	51	Holmium	Ho	67	Radon	Rn	86
Argon	A	18	Hydrogen	H	1	Rhenium	Re	75
Arsenic	As	33	Indium	In	49	Rhodium	Rh	45
Astatine	At	85	Iodine	I	53	Rubidium	Rb	37
Barium	Ba	56	Iridium	Ir	77	Ruthenium	Ru	44
Berkelium	Bk	97	Iron	Fe	26	Samarium	Sm	62
Beryllium	Be	4	Krypton	Kr	36	Scandium	Sc	21
Bismuth	Bi	83	Lanthanum	La	57	Selenium	Se	34
Boron	B	5	Lead	Pb	82	Silicon	Si	14
Bromine	Br	35	Lithium	Li	3	Silver	Ag	47
Cadmium	Cd	48	Lutetium	Lu	71	Sodium	Na	11
Calcium	Ca	20	Magnesium	Mg	12	Strontium	Sr	38
Californium	Cf	98	Manganese	Mn	25	Sulfur	S	16
Carbon	C	6	Mendelevium	Me	101	Tantalum	Ta	73
Cerium	Ce	58	Mercury	Hg	80	Technetium	Tc	43
Cesium	Cs	55	Molybdenum	Mo	42	Tellurium	Te	52
Chlorine	Cl	17	Neodymium	Nd	60	Terbium	Tb	65
Chromium	Cr	24	Neon	Ne	10	Thallium	Tl	81
Cobalt	Co	27	Neptunium	Np	93	Thorium	Th	90
Columbium (or Niobium	Cb Nb)	41	Nickel	Ni	28	Thulium	Tm	69
Copper	Cu	29	Niobium (or Columbium	Nb Cb)	41	Tin	Sn	50
Curium	Cm	96	Nitrogen	N	7	Titanium	Ti	22
Dysprosium	Dy	66	Nobelium	No	102	Tungsten (or Wolfram)	W	74
Einsteinium	En	99	Osmium	Os	76	Uranium	U	92
Erbium	Er	68	Oxygen	O	8	Vanadium	V	23
Europium	Eu	63	Palladium	Pd	46	Wolfram (or Tungsten)	W	74
Fermium	Fm	100	Phosphorus	P	15	Xenon	Xe	54
Fluorine	F	9	Platinum	Pt	78	Ytterbium	Yb	70
Francium	Fr	87	Plutonium	Pu	94	Yttrium	Y	39
Gadolinium	Gd	64	Polonium	Po	84	Zinc	Zn	30
Gallium	Ga	31	Potassium	K	19	Zirconium	Zr	40
Germanium	Ge	32	Praseodymium	Pr	59			

The Metric System

APPENDIX B

Conversion Tables (Approximate Values)

METRIC TO ENGLISH			ENGLISH TO METRIC		
When You Know	*Multiply by*	*To Get*	*When You Know*	*Multiply by*	*To Get*
centimeters	0.39	inches	inches	2.54	centimeters
meters	3.28	feet	feet	0.30	meters
meters	1.09	yards	yards	0.91	meters
kilometers	0.62	miles	miles	1.61	kilometers
square centimeters	0.15	square inches	square inches	6.45	square centimeters
square meters	11	square feet	square feet	0.09	square meters
square meters	1.20	square yards	square yards	0.84	square meters
hectares	2.47	acres	acres	0.40	hectares
square kilometers	0.38	square miles	square miles	2.6	square kilometers
cubic centimeters	0.06	cubic inches	cubic inches	16.4	cubic centimeters
cubic meters	0.37	cubic feet	cubic feet	0.27	cubic meters
cubic meters	0.13	cubic yards	cubic yards	0.76	cubic meters
cubic kilometers	0.24	cubic miles	cubic miles	4.19	cubic kilometers
grams	0.04	ounces	ounces	28.3	grams
kilograms	2.20	pounds	pounds	0.45	kilograms
tons	1.1	tons	tons	0.9	tons
milliliters	0.033	ounces	fluid ounces	30	milliliters
liters	1.06	quarts	quarts	0.95	liters
liters	0.26	gallons	gallons	3.8	liters

Energy

1 barrel of crude oil = 42 gallons
7 barrels of crude oil = 1 metric ton = 40 million BTUs
1 metric ton of coal = 28 million BTUs
1 gram U_{235} = 2.7 metric tons of coal = 13.7 barrels of crude oil
1 BTU (British Thermal Unit) = 252 calories = 0.0002931 kilowatt-hour
1 kilowatt-hour = 860,421 calories 3412 BTU

Minerals

APPENDIX C

MINERALS

Many of the most common minerals may be identified in hand specimens by their physical properties. Among the characteristics useful for this purpose are (1) hardness, (2) specific gravity, (3) streak (sometimes color), (4) shape (that is, crystal form, cleavage, and fracture), and (5) response to light as indicated by luster and transparency.

Hardness

The hardness of a mineral is determined by scratching the smooth surface of one mineral with the edge of another. In making a hardness test, be sure that the mineral being tested is actually scratched. Sometimes particles simply rub off the specimen, suggesting that it has been scratched, even though it has not been.

Ten common minerals have been arranged in the Mohs scale of relative hardness.

Mohs Scale of Hardness

Softest	1	Talc	
	2	Gypsum	
	3	Calcite	$2\frac{1}{2}$ Fingernail 3 Copper coin
	4	Fluorite	
	5	Apatite	
	6	Orthoclase	$5\frac{1}{2}$–6 Knife blade or plate glass $6\frac{1}{2}$–7 Steel file
	7	Quartz	
	8	Topaz	
	9	Corundum	
Hardest	10	Diamond	

Each of these minerals will scratch all those lower in number on the scale and will be scratched by all those higher. In other words, this is a *relative scale.* In terms of absolute hardness, the steps are nearly, though not quite, uniform up to 9. Number 7 is 7 times as hard as 1, and number 9 is 9 times as hard as 1. But number 10 is about 40 times as hard as 1.

Luster

Luster is the way a mineral looks in reflected light. There are several kinds of luster.

Metallic, the luster of metals.
Adamantine, the luster of diamonds.
Vitreous, the luster of a broken edge of glass.
Resinous, the luster of yellow resin.
Pearly, the luster of pearl.
Silky, the luster of silk.

Fracture

Many minerals that do not exhibit cleavage (see Chapter 4) do break, or fracture, in a distinctive manner. Some of the types of fracture are:

Conchoidal: along smooth, curved surfaces like the surface of a shell (*conch*). Commonly observed in glass and quartz.

Fibrous or **splintery:** along surfaces roughened by splinters or fibers.

Uneven or **irregular:** along rough, irregular surfaces.

Hackly: along a jagged, irregular surface with sharp edges.

Mineral	Chemical Composition and Name	Specific Gravity	Streak	Hardness	Cleavage or Fracture	Luster
Actinolite (An asbestos; an amphibole)	$Ca_2(Mg,Fe)_5Si_8O_2(OH)_2$ Calcium iron silicate	3.0–3.3	Colorless	5–6	*See* Amphibole	Vitreous
Albite	(*See* Feldspars)					
Amphibole	(*See* Hornblende)				Perfect prismatic at 56° and 124°, often yielding a splintery surface	
Andalusite	Al_2SiO_5 Aluminum silicate	3.16	Colorless	$7\frac{1}{2}$	Not prominent	Vitreous
Anhydrite	$CaSO_4$ Anhydrous calcium sulfate	2.89–2.98	Colorless	$3–3\frac{1}{2}$	3 directions at right angles to form rectangular blocks	Vitreous; pearly
Anorthite	(*See* Feldspars)					
Apatite	$Ca_5(F,Cl)(PO_4)_3$ Calcium fluophosphate	3.15–3.2	White	5	Poor cleavage, one direction; conchoidal fracture	Glassy
Asbestos	(*See* Actinolite, Chrysotile, Serpentine)					
Augite (A pyroxene)	$Ca(Mg,Fe,Al)(Al,Si_2O_6)$ Ferromagnesian silicate	3.2–3.4	Greenish gray	5–6	Perfect prismatic along two planes at nearly right angles to each other, often yielding a splintery surface	Vitreous
Azurite	$Cu_3(CO_3)_2(OH)_2$ Blue copper carbonate	3.77	Pale blue	4	Fibrous	Vitreous to dull, earthy
Bauxite	Hydrous aluminum oxides of indefinite composition; not a mineral	2–3	Colorless	1–3	Uneven fracture	Dull to ear
Biotite (Black mica)	$K(Mg,Fe)_3AlSi_3O_{10}(OH)_2$ Ferromagnesian silicate	2.8–3.2	Colorless	$2\frac{1}{2}–3$	Perfect in one direction into thin, elastic, transparent, smoky sheets	Pearly, glassy
Bornite (Peacock ore; purple copper ore)	Cu_5FeS_4 Copper iron sulfide	5.06–5.08	Grayish black	3	Uneven fracture	Metallic
Calcite	$CaCO_3$ Calcium carbonate	2.72	Colorless	3	Perfect in 3 directions at 75° to form unique rhombohedral fragments	Vitreous

Color	*Transparency*	*Form*	*Other Properties*
hite to ht green	Transparent to translucent	Slender crystals, usually fibrous	A common ferromagnesian metamorphic mineral.
			A group of silicates with tetrahedra in double chains; hornblende is the most important; contrast with pyroxene.
esh-red, reddish own, olive-green	Transparent to translucent	Usually in coarse, nearly square prisms; cross section may show black cross	Found in schists formed by the middle-grade metamorphism of aluminous shales and slates. The variety *chiastolite* has carbonaceous inclusions in the pattern of a cross.
hite; may have nt gray, blue, red tinge	Transparent to translucent	Commonly in massive fine aggregates not showing cleavage; crystals rare	Found in limestones and in beds associated with salt deposits; heavier than calcite, harder than gypsum.
een, brown, l	Translucent to transparent	Massive, granular	Widely disseminated as an accessory mineral in all types of rocks; unimportant source of fertilizer; a transparent variety is a gem, but too soft for general use.
		A general term applied to certain fibrous minerals that display similar physical characteristics although they differ in composition. The most common asbestos mineral is *chrysotile*, a variety of serpentine.	
rk green to ck	Translucent only on thin edges	Short, stubby crystals with 4- or 8-sided cross section; often in granular crystalline masses	An important igneous rock-forming mineral found chiefly in simatic rocks.
ense azure e	Opaque	Crystals complex in habit and distorted; sometimes in radiating spherical groups	An ore of copper; a gem mineral; effervesces with HCl.
low, brown, y, white	Opaque	In rounded grains; or earthy, clay-like masses	An ore of aluminum; produced under subtropical to tropical climatic conditions by prolonged weathering of aluminum-bearing rocks; a component of *laterites;* clay odor when wet.
ck, brown, k green	Transparent, translucent	Usually in irregular foliated masses; crystals rare	Constructed around tetrahedral sheets; a common and important rock-forming mineral in both igneous and metamorphic rocks.
wnish bronze fresh fracture; ckly tarnishes variegated ple and blue, l finally black	Opaque	Usually massive; rarely in rough cubic crystals	An important ore of copper.
ally white or orless; may be ed gray, red, en, blue, yellow	Transparent to opaque	Usually in crystals or coarse to fine granular aggregates; also compact, earthy; crystals extremely varied—over 300 different forms	A very common rock mineral, occurring in masses as limestone and marble; effervesces freely in cold dilute hydrochloric acid.

Mineral	Chemical Composition and Name	Specific Gravity	Streak	Hardness	Cleavage or Fracture	Luster
Carnotite	$K_2(UO_2)_2(VO_4)_2$ Potassium uranyl vanadate	4		Very soft	Uneven fracture	Earthy
Cassiterite (Tin stone)	SnO_2 Tin oxide	6.8–7.1	White to light brown	6–7	Conchoidal fracture	Adamantine submetallic and dull
Chalcedony	(*See* Quartz)					
Chalcocite (Copper glance)	Cu_2S Copper sulfide	5.5–5.8	Grayish black	2½–3	Conchoidal fracture	Metallic
Chalcopyrite (Copper pyrites; yellow copper ore; fool's gold)	$CuFeS_2$ Copper iron sulfide	4.1–4.3	Greenish black; also greenish powder in groove when scratched	3½–4	Uneven fracture	Metallic
Chlorite	$(Mg,Fe)_5(Al,Fe'')_2 Si_3O_{10}(OH)_8$ Hydrous ferromagnesian aluminum silicate	2.6–2.9	Colorless	2–2½	Perfect in 1 direction like micas, but into inelastic flakes	Vitreous to pearly
Chromite	$FeCr_2O_4$ Iron chromium oxide	4.6	Dark brown	5½	Uneven fracture	Metallic to submetallic pitchy
Chrysotile (Serpentine asbestos)	(*See* Serpentine)					
Clay	(*See* Kaolinite)					
Corundum (Ruby, sapphire)	Al_2O_3 Aluminum oxide	4.02	Colorless	9	Basal or rhombohedral parting	Adamantine to vitreous
Diamond	C Carbon	3.5	Colorless	10	Octahedral cleavage	Adamantine greasy
Dolomite	$CaMg(CO_3)_2$ Calcium magnesium carbonate	2.85	Colorless	3½–4	Perfect in 3 directions at 73°45′	Vitreous or pearly
Emery	(*See* Corundum)					
Epidote	$Ca_2(Al,Fe)_3(SiO_4)_3(OH)$ Hydrous calcium aluminum iron silicate	3.35–3.45	Colorless	6–7	Good in 1 direction	Vitreous

Color	*Transparency*	*Form*	*Other Properties*
lliant canary low	Opaque	Earthy powder	An ore of vanadium and uranium.
own or black; ely yellow or ite	Translucent; rarely transparent	Commonly massive granular	The principal ore of tin.
ny lead-gray; nishes to dull ck	Opaque	Commonly fine-grained and massive; crystals rare; small, tabular with hexagonal outline	One of the most important ore minerals of copper; occurs principally as a result of secondary sulfide enrichment.
ass-yellow; tar- shes to bronze or descence, but ore slowly than rnite or chalco- e	Opaque	Usually massive	An ore of copper; distinguished from pyrite by being softer than steel while pyrite is harder than steel; distinguished from gold by being brittle while gold is not; known as "fool's gold," a term also applied to pyrite.
een of various ades	Transparent to translucent	Foliated massive, or in aggregates of minute scales	A common metamorphic mineral characteristic of low-grade metamorphism.
n-black to ownish black	Subtranslucent	Massive, granular to compact	The only ore of chromium; a common constituent of peridotites and serpentines derived from them; one of the first minerals to crystallize from a cooling magma.
own, pink, or ie; may be iite, gray, green, by-red, sapphire- ie	Transparent to translucent	Barrel-shaped crystals; sometimes deep horizontal striations; coarse or fine granular	Common as an accessory mineral in metamorphic rocks such as marble, mica schist, gneiss; occurs in gem form as *ruby* and *sapphire;* the abrasive emery is black granular corundum mixed with magnetite, hematite, or the magnesian aluminum oxide *spinel.*
olorless or pale llow; may be d, orange, green, ie, black	Transparent	Octahedral crystals, flattened, elongated, with curved faces	Gem and abrasive; 95 per cent of natural diamond production is from South Africa; abrasive diamonds have been made in commercial quantities in the laboratory in the United States.
nk, flesh; may be ite, gray, green, own, black	Transparent to opaque	Rhombohedral crystals with curved faces; coarse-grained cleavable masses, or fine-grained compact	Occurs chiefly in rock masses of dolomitic limestone and marble, or as the principal constituent of the rock named for it; distinguished from limestone by its less vigorous action with cold hydrochloric acid (the powder dissolves with effervescence, large pieces only if the acid is hot).
stachio-green, llowish to ackish green	Transparent to translucent	Prismatic crystals striated parallel to length; usually coarse to fine granular; also fibrous	A metamorphic mineral often associated with chlorite; derived from metamorphism of impure limestone; characteristic of contact metamorphic zones in limestone.

Mineral	*Chemical Composition and Name*	*Specific Gravity*	*Streak*	*Hardness*	*Cleavage or Fracture*	*Luster*
Feldspars	Alumino silicates	2.55–2.75		6	Good in 2 directions at or near 90°	
Orthoclase	$K(AlSi_3O_8)$ Potassic feldspar	2.57	White	6	Good in 2 directions at or near 90°	Vitreous
Plagioclase	Soda-lime feldspars, a continuous series varying in composition from pure albite to pure anorthite					
Albite	$Na(AlSi_3O_8)$ Sodic feldspar	2.62	Colorless	6	Good in 2 directions at 93°34′	Vitreous to pearly
Anorthite	$Ca(Al_2Si_2O_8)$	2.76	Colorless	6	Good in 2 directions at 94°12′	Vitreous to pearly
Fluorite	CaF_2 Calcium fluoride	3.18	Colorless	4	Good in 4 directions parallel to the faces of an octahedron	Vitreous
Galena	PbS Lead sulfide	7.4–7.6	Lead-gray	2½	Good in 3 directions parallel to the faces of a cube	Metallic
Garnet	$R''_3R'''_2(SiO_4)_3$ R″ may be Calcium, Magnesium, Iron, or Manganese. R‴ may be Aluminum, Iron, Titanium, or Chromium. Ferromagnesian silicates	3.5–4.3	Colorless	6½–7½	Uneven fracture	Vitreous to resinous
Graphite (Plumbago; black lead)	C Carbon	2.3	Black	1–2	Good in one direction; folia flexible but not elastic	Metallic or earthy
Gypsum	$CaSO_4 \cdot 2H_2O$ Hydrous calcium sulfate	2.32	Colorless	2	Good cleavage in one direction yielding flexible but inelastic flakes; fibrous fracture in another direction; conchoidal fracture in a third direction	Vitreous, pearly, silk
Halite (Rock salt; common salt)	NaCl Sodium chloride	2.16	Colorless	2½	Perfect cubic cleavage	Glassy to dull
Hematite	Fe_2O_3 Iron oxide	5.26	Light to dark Indian-red; becomes black on heating	5½–6½	Uneven fracture	Metallic
Hornblende (An amphibole)	Complex ferromagnesian silicate of Ca, Na, Mg, Ti, and Al	3.2	Colorless	5–6	Perfect prismatic at 56° and 124°	Vitreous; fibrous vari often silky

Color	*Transparency*	*Form*	*Other Properties*
			The most common igneous rock-forming group of minerals; weather to clay minerals.
hite, gray, sh pink	Translucent to opaque	Prismatic crystals; most abundantly in rocks as formless grains	Characteristic of sialic rocks.
			Important rock-forming minerals; characteristic of simatic rocks.
lorless, white, ay	Transparent to translucent	Tabular crystals; striations caused by twinning	Opalescent variety, *moonstone.*
lorless, white, ay, green, low, red	Transparent to translucent	Lath-like or platy grains, tabular crystals; striations caused by twinning; lath-like or platy grains	A unique and beautiful play of colors is common on plagioclase feldspars intermediate between albite and anorthite in composition, as with *andesine* (70 to 50 per cent albite) and *labradorite* (50 to 30 per cent albite).
riable; light en, yellow, ish green, rple, etc.	Transparent to translucent	Well-formed interlocking cubes; also massive, coarse or fine grains	Some varieties fluoresce; a common, widely distributed mineral in dolomites and limestone; an accessory mineral in igneous rocks; used as a flux in making steel.
ad-gray	Opaque	Cube-shaped crystals; also in granular masses	The principal ore of lead; so commonly associated with silver that it is also an ore of silver.
d, brown, llow, white, en, black	Transparent to translucent	Usually in 12- or 24-sided crystals; also massive granular, coarse or fine	Common and widely distributed, particularly in metamorphic rocks; brownish red variety *almandite,* $Fe_3Al_2(SiO_4)_3$, used to define one of the zones of middle-grade metamorphism; striking in schists.
ack to el-gray	Opaque	Foliated or scaly masses common; may be radiated or granular	Feels greasy; common in metamorphic rocks such as marble, schists, and gneisses.
lorless, white, ay; with impuri-s, yellow, red, own	Transparent to translucent	Crystals prismatic, tabular, diamond-shaped; also in granular, fibrous, or earthy masses	A common mineral widely distributed in sedimentary rocks, often as thick beds; *satin spar* is a fibrous gypsum with silky luster; *selenite* is a variety which yields broad, colorless, transparent folia; *alabaster* is a fine-grained massive variety.
lorless or white; pure: yellow, l, blue, purple	Transparent to translucent	Cubic crystals; massive granular	Salty taste; permits ready passage of heat rays (i.e., diathermanous); a very common mineral in sedimentary rocks; interstratified in rocks of all ages to form a true rock mass.
ddish brown to ck	Opaque	Crystals tabular; botryoidal; micaceous and foliated; massive	The most important ore of iron; red earthy variety known as *red ocher;* botryoidal form known as *kidney ore,* micaceous form *specular;* widely distributed in rocks of all types and ages.
rk green to ck	Translucent on thin edges	Long, prismatic crystals; fibrous; coarse- to fine-grained masses	Distinguished from augite by cleavage; a common, rock-forming mineral which occurs in both igneous and metamorphic rocks.

Mineral	*Chemical Composition and Name*	*Specific Gravity*	*Streak*	*Hardness*	*Cleavage or Fracture*	*Luster*
Kaolinite (Clay)	$Al_2Si_2O_5(OH)_4$ Hydrous aluminum silicate	2.6	Colorless	2–2½	None	Dull earthy
Kyanite	Al_2SiO_5 Aluminum silicate	3.56–3.66	Colorless	5 along, 7 across crystals	Good in one direction	Vitreous to pearly
Limonite (Brown hematite; bog iron ore; rust)	Hydrous iron oxides; not a mineral	3.6–4	Yellow-brown	5–5½ (Finely divided, apparent H as low as 1)	None	Vitreous
Magnetite	Fe_3O_4 Iron oxide	5.18	Black	6	Some octahedral parting	Metallic
Mica	(*See* Biotite and Muscovite)					
Muscovite (White mica; potassic mica; common mica)	$KAl_3Si_3O_{10}(OH)_2$ Nonferromagnesian silicate	2.76–3.1	Colorless	2–2½	Good cleavage in one direction, giving thin, very flexible and elastic folia	Vitreous, silky, pearly
Olivine (Peridot)	$(Mg,Fe)_2SiO_4$ Ferromagnesian silicate	3.27–3.37	Pale green, white	6½–7	Conchoidal fracture	Vitreous
Pyrite (Iron pyrites; fool's gold)	FeS_2 Iron sulfide	5.02	Greenish or brownish black	6–6½	Uneven fracture	Metallic
Pyroxene	(*See* Augite)					
Quartz (Silica)	SiO_2 Silicon oxide but structurally a silicate, with tetrahedra sharing oxygens in 3 dimensions	2.65	Colorless	7	Conchoidal fracture	Vitreous, greasy, splendent

Color	*Transparency*	*Form*	*Other Properties*
ite	Opaque	Claylike masses	Usually unctuous and plastic; other clay minerals similar in composition and physical properties, but different in atomic structure, are *illite* and *montmorillonite;* derived from the weathering of the feldspars.
e; may be ite, gray, en, streaked	Transparent to translucent	In bladed aggregates	Characteristic of middle-grade metamorphism; compare with andalusite, which has the same composition and is formed under similar conditions, but has a different crystal habit; contrast with sillimanite, which has the same composition but different crystal habit and forms at highest metamorphic temperatures.
rk brown to ck	Opaque	Amorphous; mammillary to stalactitic masses; concretionary, nodular, earthy	Always of secondary origin from alteration or solution of iron minerals; mixed with fine clay, it is a pigment, *yellow ocher.*
n-black	Opaque	Usually massive granular, coarse or fine in grain	Strongly magnetic; may act as a natural magnet, known as *lodestone;* an important ore of iron; found in black sands on the seashore; mixed with corundum, it is a component of *emery.*
in: colorless; ck: light yellow, wn, green, red	Thin: transparent; thick: translucent	Mostly in thin flakes	Widespread and very common rock-forming mineral; characteristic of sialic rocks; also very common in metamorphic rocks such as gneiss and schist; the principal component of some mica schists; sometimes used for stove doors, lanterns, etc., as transparent *isinglass;* used chiefly as an insulating material.
ve to grayish en, brown	Transparent to translucent	Usually in imbedded grains or granular masses	A common rock-forming mineral found primarily in simatic rocks; the principal component of peridotite; actually, a series grading from *forsterite,* Mg_2SiO_4, to *fayalite,* Fe_2SiO_4; the most common olivines are richer in magnesium than in iron; the clear green variety *peridot* is sometimes used as a gem.
ass-yellow	Opaque	Cubic crystals with striated faces; also massive	The most common of the sulfides; used as a source of sulfur in the manufacture of sulfuric acid; distinguished from chalcopyrite by its paler color and greater hardness; from gold by its brittleness and hardness.
			A group of silicates with tetrahedra in single chains; augite is the most important; contrast with amphibole.
lorless or white en pure; any lor from impuri- s	Transparent to translucent	Prismatic crystals with faces striated at right angles to long dimension; also massive forms of great variety	An important constituent of sialic rocks; coarsely crystalline varieties: *rock crystal, amethyst* (purple), *rose quartz, smoky quartz, citrine* (yellow), *milky quartz, cat's eye;* cryptocrystalline varieties: *chalcedony, carnelian* (red chalcedony), *chrysoprase* (apple-green chalcedony), *heliotrope* or *bloodstone* (green chalcedony with small red spots), *agate* (alternating layers of chalcedony and opal); granular varieties: *flint* (dull to dark brown), *chert* (like flint but lighter in color), *jasper* (red from hematite inclusions), *prase* (like jasper, but dull green).

Mineral	Chemical Composition and Name	Specific Gravity	Streak	Hardness	Cleavage or Fracture	Luster
Serpentine	$Mg_3Si_2O_5(OH)_4$ Hydrous magnesium silicate	2.2–2.65	Colorless	2–5	Conchoidal fracture	Greasy, wa[illegible] or silky
Siderite (Spathic iron; chalybite)	$FeCO_3$ Iron carbonate	3.85	Colorless	$3\frac{1}{2}$–4	Perfect rhombohedral cleavage	Vitreous
Silica	(*See* Quartz)					
Sillimanite (Fibrolite)	Al_2SiO_5 Aluminum silicate	3.23	Colorless	6–7	Good cleavage in 1 direction	Vitreous
Sphalerite (Zinc blende; black jack)	ZnS Zinc sulfide	3.9–4.1	White to yellow and brown	$3\frac{1}{2}$–4	Perfect cleavage in 6 directions at 120°	Resinous
Staurolite	$Fe''Al_5Si_2O_{12}(OH)$ Iron aluminum silicate	3.65–3.75	Colorless	7–$7\frac{1}{2}$	Not prominent	Fresh: resi[illegible]ous, vitreo[illegible] altered: du[illegible] to earthy
Talc (Soapstone; steatite)	$Mg_3Si_4O_{10}(OH)_2$ Hydrous magnesium silicate	2.7–2.8	White	1	Good cleavage in 1 direction, gives thin folia, flexible but not elastic	Pearly to greasy
Topaz	$Al_2SiO_4(F,OH)_2$ Aluminum fluosilicate	3.4–3.6	Colorless	8	Good in 1 direction	Vitreous
Tourmaline	Complex silicate of boron and aluminum, with sodium, calcium, fluorine, iron, lithium, or magnesium	3–3.25	Colorless	7–$7\frac{1}{2}$	Not prominent; black variety fractures like coal	Vitreous to resinous
Uraninite (Pitchblende)	Complex oxide of uranium with small amounts of lead, radium, thorium, yttrium, nitrogen, helium, and argon	9–9.7	Brownish black	$5\frac{1}{2}$	Not prominent	Submetallic pitchy
Wollastonite	$CaSiO_3$ Calcium silicate	2.8–2.9	Colorless	5–$5\frac{1}{2}$	Good cleavage in 2 directions at 84° and 96°	Vitreous or pearly on cleavage surfaces

Reference: Cornelius Hurlbut, Jr., *Dana's Manual of Mineralogy,* 16th ed. New York: John Wiley and Sons, Inc., 1952.

Color	*Transparency*	*Form*	*Other Properties*
ıriegated shades green	Translucent	Platy or fibrous	Platy variety, *antigorite;* fibrous variety, *chrysotile,* an asbestos; an alteration product of magnesium silicates such as olivine, augite, and hornblende; common and widely distributed.
ght to dark own	Transparent to translucent	Granular, compact, earthy	An ore of iron; an accessory mineral in taconite.
own, pale green, hite	Transparent to translucent	Long, slender crystals without distinct terminations; often in parallel groups; frequently fibrous	Relatively rare, but important as a mineral characteristic of high-grade metamorphism; contrast with andalusite and kyanite, which have the same composition but form under conditions of middle-grade metamorphism.
ıre: white, een; with iron: llow to brown d black; red	Transparent to translucent	Usually massive; crystals many-sided, distorted	A common mineral; the most important ore of zinc; the red variety is called *ruby zinc;* streak lighter than corresponding mineral color.
ed-brown to ownish black	Translucent	Usually in crystals, prismatic, twinned to form a cross; rarely massive	A common accessory mineral in schists and slates; characteristic of middle-grade metamorphism; associated with garnet, kyanite, sillimanite, tourmaline.
ray, white, ver-white, apple-een	Translucent	Foliated, massive	Of secondary origin, formed by the alteration of magnesium silicates such as olivine, augite, and hornblende; most characteristically found in metamorphic rocks.
raw-yellow, ine-yellow, pink, uish, greenish	Transparent to translucent	Usually in prismatic crystals, often with striations in direction of greatest length	Represents 8 on Mohs scale of hardness; a gem stone.
aried: black, rown; red, pink, een, blue, yellow	Translucent	Usually in crystals; common: with cross section of spherical triangle	Gem stone; an accessory mineral in pegmatites, also in metamorphic rocks such as gneisses, schists, marbles.
lack	Opaque	Usually massive and botryoidal (i.e. like a bunch of grapes)	An ore of uranium and radium; the mineral in which helium and radium were first discovered.
olorless, hite or gray	Translucent	Commonly massive, fibrous, or compact	A common contact metamorphic mineral in limestones.

Glossary

Å Abbreviation for angstrom, a unit of length, 10^{-8} cm.

Abiogenesis The theory that living things can originate from nonliving matter, or, more commonly, the evolution of nonliving organic molecules. Compare *biogenesis.*

Ablation As applied to glacier ice, the process by which ice below the snow line is wasted by evaporation and melting.

Abrasion Erosion of rock material by friction of solid particles moved by water, ice, wind, or gravity.

Absolute Date A date expressed in terms of years and related to the present by a reliable system of reckoning.

Absolute Time Geologic time measured in terms of years. Compare with *relative time.*

Adaptation Any specific characteristic of an organism that is of value to its survival in a given situation. The term may also apply to the process by which beneficial characteristics are achieved or accumulated.

Adaptive Radiation The complex evolutionary changes by which any basic stock of organisms is able to enter and exploit a number of environments.

Aeolian Pertaining to wind. Designates rocks and soils whose constituents have been carried and laid down by atmospheric currents. It also applies to erosive and other geologic effects accomplished by wind.

Aftershock An earthquake that follows a larger earthquake and originates at or near the focus of the larger earthquake. Generally, major shallow earthquakes are followed by many aftershocks. These decrease in number as time goes on but may continue for many days or even months.

Agate A variety of chalcedony with alternating layers of chalcedony and opal.

Age A period of earth history of unspecified duration characterized by a dominant or important life form, i.e., the Age of Fishes. Also, a time when a particular event occurred, such as the Ice Age. Age may also refer to the position of anything in the geologic time scale; if possible, it may be expressed in years.

A-horizon The soil zone immediately below the surface, from which soluble material and fine-grained particles have been moved downward by water seeping into the soil. Varying amounts of organic matter give the A-horizon a gray to black color.

Algae Any member of a numerous group of simple plants usually classified as a subdivision of the *Thallophyta.* Algae contain chlorophyll and are capable of photosynthesis. They may appear blue-green, green, red, or brown and are classified accordingly. Forms capable of secreting calcium carbonate are important rock builders.

Alluvial Fan The land counterpart of a delta. An assemblage of sediments marking the place where a stream moves from a steep gradient to a flatter gradient and suddenly loses its transporting power. Typical of arid and semiarid climates, but not confined to them.

Alpha Particle A helium atom lacking electrons and therefore having a double positive charge.

Alpine Glacier A glacier confined to a stream valley. Usually fed from a cirque. Also called *valley glacier* or *mountain glacier.*

Amino Acid Any of a large group of nitrogenous organic compounds that serve as structural units in the proteins and are essential to all forms of life.

Ammonite Any member of an extinct group of marine molluscs, whose fossil remains are especially important as index fossils in the Permian period and the Mesozoic era. Typically, they possess a coiled, many-chambered shell with complex crenulations along the edges of the septa between the chambers.

Amorphous A state of matter in which there is no orderly arrangement of atoms.

Amphibole Group Ferromagnesian silicates with a double chain of silicon-oxygen tetrahedra. Common example: hornblende. Contrast with *pyroxene group.*

Amphibolite A faintly foliated metamorphic rock developed during the regional metamorphism of simatic rocks. Composed mainly of hornblende and plagioclase feldspars.

Amphibolite Facies An assemblage of minerals formed at moderate to high pressures between 850°F and 1300°F (450°C and 700°C) during regional metamorphism.

Andesite A fine-grained igneous rock with no quartz or orthoclase, composed of about 75 percent plagioclase feldspars and the balance ferromagnesian silicates. Important as lavas, possibly derived by fractional crystallization from basaltic magma. Widely characteristic of mountain-making processes around the borders of the Pacific Ocean. Confined to continental sectors.

Andesite Line A map line designating the petrographic boundary of the Pacific Ocean. Extrusive rocks on the Pacific side of the line are basaltic and on the other side andesitic.

Angiosperm Any member of the advanced group of plants that carries its seeds in a closed ovary and has floral reproductive structures.

Angstrom A unit of length, equal to one hundred-millionth of a centimeter, 10^{-8} cm. Abbreviation, Å.

Angular Momentum A vector quantity, the product of mass times radius of orbit times velocity. The energy of motion of the solar system.

Angular Unconformity An unconformity or break between two series of rock layers such that rocks of the lower series meet rocks of the upper series at an angle; in other words, the two series are not parallel.

Anorthosite A variety of igneous rock formed at depth and composed of

90 percent or more of the feldspar mineral anorthite.

Antecedent Stream A stream that maintains after uplift the same course it originally followed prior to uplift.

Anthracite Hard coal; compact, dense, very black with a hardness of 2.0 to 2.5 and a specific gravity of 1.4 to 1.8. Anthracite has 80 to 90 percent fixed carbon and burns with a short flame and great heat.

Anthropology The study of man, especially his physical nature and the ways in which he has modified or been modified by external forces.

Anticlinal Theory The theory that water, petroleum, and natural gas accumulate in up-arched strata in the order named (water lowest), provided the structure contains reservoir rocks in proper relation to source beds and capped by an impervious barrier.

Anticline A configuration of folded, stratified rocks in which the rocks dip in two directions away from a crest, as the principal rafters of a common gable roof dip away from the ridgepole. The reverse of a *syncline.* The "ridgepole," or crest, is called the *axis.*

Apparent Polar Wander The apparent migration of the pole with respect to one continent as inferred from paleomagnetic studies.

Aquifer A permeable material through which ground water moves.

Aragonite Calcium carbonate ($CaCO_3$) with crystals in the orthorhombic system. As a constituent of some shells it is less stable than calcite.

Archaeocyathid Any of a group of extinct spongelike animals restricted to the Early Cambrian.

Archeo- Combining form meaning "ancient"; from Gr. *archaios,* "ancient."

Arête A narrow, saw-toothed ridge formed by cirques developing from opposite sides into the ridge.

Arkose A detrital sedimentary rock formed by the cementation of individual grains of sand size and predominantly composed of quartz and feldspar. Derived from the disintegration of granite.

Arroyo Flat-floored, vertically walled channel of an intermittent stream typical of semiarid climates. Often applied to such features of southwestern United States. Synonymous with *wadi* and *wash.*

Artesian Water Water that is under pressure when tapped by a well and is able to rise above the level at which it is first encountered. It may or may not flow out at ground level.

Arthropod Any member of the great animal phylum that is characterized by jointed appendages, a bilaterally symmetrical body, and usually an external chitinous skeleton.

Artifact Anything of a material nature produced by human skill.

Artiodactyl Any of the even-toed hoofed mammals.

Asbestos A general term applied to certain fibrous minerals that display similar physical characteristics although they differ in composition. Some asbestos has fibers long enough to be spun into fabrics with great resistance to heat, such as those used for automobile brake linings. Types with shorter fibers are compressed into insulating boards, shingles, etc. The most common asbestos mineral (95 percent of U.S. production) is chrysotile, a variety of serpentine, a metamorphic mineral.

Asexual Reproduction Reproduction by one individual independent of others.

Ash In geology, the finest rock material from volcanic explosions.

Asphalt A brown to black solid or semisolid bituminous substance. Occurs in nature but is also obtained as a residue from the refining of certain hydrocarbons (then known as "artificial asphalt").

Asteroid A minor planet; one of many small bodies ranging from a few hundred kilometers to less than one kilometer. Most are in orbits between Mars and Jupiter.

Asthenosphere A world-circling zone of soft, hot, plastic material extending roughly 100 to 400 km below the earth's surface. This mobile layer separates the lithosphere above from the mesosphere below, and on it the brittle lithospheric plates shift slowly in a complex interlocking pattern.

Astrogeology Application of the principles of geology, geophysics, and geochemistry to the study of aggregations in the solar system, excluding the earth.

Asymmetric Fold A fold in which one limb dips more steeply than the other.

Atoll A roughly circular, elliptical, or horseshoe-shaped island or ring of islands of reef origin, composed of coral and algal rock and sand and rimming a lagoon in which there are no islands of noncoral origin.

Atom A combination of protons, neutrons, and electrons. Ninety-two kinds are found in nature; 125 kinds are now known.

Atomic Energy Energy associated with the nucleus of an atom. The energy is released when the nucleus is split, or is derived from mass that is lost when a nucleus is fused together.

Atomic Mass The nucleus of an atom contains 99.95 percent of its mass. The total number of protons and neutrons in the nucleus is called the *mass number.*

Atomic Number The number of positive charges on the nucleus of an atom; the number of protons in the nucleus.

Atomic Size The radius of an atom (average distance from the center to the outermost electron of the neutral atom). Commonly expressed in angstroms.

Axial Plane A plane through a rock fold that includes the axis and divides the fold as symmetrically as possible.

Axis The ridge, or place of sharpest folding, of an anticline or syncline.

Backset Beds Inclined layers of sand developed on the gentler dune slope to the windward. These beds may constitute a large part of the total volume of a dune, especially if there is enough vegetation to trap most of the sand before it can cross over to the slip face.

Badland Highly eroded, barren area with extensive exposure of the bedrock or subsoil; mostly slopes or deep gullies.

Barchan A crescent-shaped dune with wings or horns pointing downwind. Has a gentle windward slope and steep lee slope inside the horns. About 30 m in height and 300 m wide from horn to horn. Moves with the wind at about 8 to 16 m per year across a flat, hard surface where a limited supply of sand is available.

Barrier Island A low, sandy island near the shore and parallel to it, on a gently sloping offshore bottom.

Barrier Reef A reef that is separated from a landmass by a lagoon of varying width and depth opening to the sea through passes in the reef.

Basalt A fine-grained igneous rock dominated by dark-colored minerals, consisting of over 50 percent plagioclase feldspars and the balance ferromagnesian silicates. Basalts and andesites represent about 98 percent of all extrusive rocks.

Base Level (1) For a *stream,* a level below which it cannot erode. There may be temporary base levels along a stream's course, such as those established by lakes, or resistant layers of rock. Ultimate base level for a stream is sea level. (2) For a *region,* a plane extending inland from sea level and sloping gently upward from the sea. Erosion of the land progresses toward this plane but seldom, if ever, quite reaches it.

Basement Complex Undifferentiated rocks underlying the oldest identifiable rocks in any region. Usually sialic, crystalline, metamorphosed. Often, but not necessarily, Precambrian.

Basic Widely, but loosely, applied to rocks with a relatively low content of silica and a correspondingly high content of minerals rich in iron, lime, or magnesia, such as amphibole, pyroxene, and olivine. The content of silica in so-called basic rocks is on the order of 45 to 52 percent. Because basic, as applied to rocks, has no direct relation to *base* in the chemical sense, it is being replaced by the term *subsilicic.*

Basin A depression in the land surface. In geology, it is an area in which the rocks dip toward a central spot. Basins tend to be accentuated by continued downsinking and thus receive thicker deposits of sediment than surrounding areas. An example is the Michigan Basin.

Batholith A very large mass of intrusive rock, generally composed of granite or granodiorite, which in most cases cuts across the invaded rocks and shows no direct evidence of having a floor of older solid rock. A surface exposure exceeding about 100 sq km has been suggested as a lower size limit.

Bauxite The chief ore of commercial aluminum. A mixture of hydrous aluminum oxides.

Bay Barrier A sandy beach, built up across the mouth of a bay, so that the bay is no longer connected to the main body of water.

Bed "Bed" and "layer" refer to any tabular body of rock lying in a position essentially parallel to the surface or surfaces on or against which it was formed, whether these be a surface of weathering and erosion, planes of stratification, or inclined fractures.

Bedding (1) A collective term used to signify the existence of beds or layers in sedimentary rocks. (2) Sometimes synonymous with *bedding plane.*

Bedding Plane The surface that separates a layer of stratified rock from an overlying or underlying layer.

Bed Load Material in movement along a stream bottom, or, if wind is the moving agency, along the surface. Contrast with material carried in suspension or solution.

Bedrock The more or less solid undisturbed rock in place either at the surface or beneath superficial deposits of gravel, sand, or soil.

Beheaded Stream The lower section of a stream that has lost its upper portion through *stream piracy.*

Belemnite A general name for any ancient squidlike cephalopod with a pointed, cylindrical, internal skeleton of solid calcium carbonate.

Belt of Soil Moisture Subdivision of zone of aeration. Belt from which water may be used by plants or withdrawn by soil evaporation. Some of the water passes down into the intermediate belt, where it may be held by molecular attraction against the influence of gravity.

Bentonite A rock composed of clay minerals and derived from the alteration of volcanic tuff or ash. The color range of fresh material is from white to light green, or light blue. On exposure the color may darken to yellow, red, or brown.

Bergschrund The gap or crevasse between glacier ice and the headwall of a cirque.

Berms In the terminology of coastlines, berms are storm-built beach features that resemble small terraces; on their seaward edges are low ridges built up by storm waves.

B-horizon The soil zone of accumulation that lies below the A-horizon. Here is deposited some of the material that has moved downward from the A-horizon.

"Big Bang" Theory A theory that the universe began with a great explosion and has been expanding ever since.

Binding Energy The amount of energy that must be supplied to break an atomic nucleus into its component fundamental particles. It is equivalent to the mass that disappears when fundamental particles combine to form a nucleus.

Biochemical Rock A sedimentary rock made up of deposits resulting directly or indirectly from the life processes of organisms.

Biogenesis The origin of life from preexisting life, or, as now commonly used, the origin of life without regard to process.

Biosphere All earth life considered together, usually as an interdependent system.

Biota The plants (flora) and animals (fauna) of a particular time and place.

Biotite "Black mica," ranging in color from dark brown to green. A rock-forming ferromagnesian silicate mineral with its tetrahedra arranged in sheets.

Bituminous Coal A compact, brittle coal of a gray-black to velvet-black color. It burns with a yellow flame and gives off a strong bituminous odor. Generally, there are no traces of organic structures visible to the eye.

Bivalve An invertebrate animal whose shell is divided into two equal or subequal parts, or valves. Brachiopods, pelecypods, and ostracods are examples.

Blastoid Any of a large group of extinct marine echinoderms that possess a stem and a budlike head, or theca. They range from the Ordovician to the Permian.

Blowout A basin, scooped out of soft, unconsolidated deposits by the process of deflation. Ranges from a few meters to several kilometers in diameter.

Body Wave Push-pull, or shake, earthquake wave that travels through the body of a medium, as distinguished from waves that travel along a free surface.

Bond See *covalent bond; ionic bond.*

Borderland An actual or hypothetical landmass occupying a position on or near the edge of a continent and supplying sediment to a geosyncline or site of deposition on the continent.

Bottomset Bed Layer of fine sediment deposited in a body of standing water beyond the advancing edge of a growing delta. The delta eventually builds up on top of the bottomset beds.

Boulder Size A volume greater than that of a sphere with a diameter of 256 mm.

Boulder Train A series of glacier erratics from the same bedrock source, usually with some property that permits easy identification. Arranged across the country in the shape of a fan with the apex at the source and widening in the direction of glacier movement.

Bowen's Reaction Series A series of minerals for which any early-formed phase tends to react with the melt that remains, to yield a new mineral further along in the series. Thus, early-formed crystals of olivine react with remaining liquids to form augite crystals; these in turn may further react with the liquid then remaining to form hornblende. See also *continuous reaction series* and *discontinuous reaction series.*

Brachiopod A type of shelled marine invertebrate now comparatively rare but abundant in earlier periods of earth history. Brachiopods are common fossils in rocks of Paleozoic age. They have a bivalve shell and are symmetrical with regard to a plane passing through the beak and the middle of the front margin.

Braided Stream A complex tangle of converging and diverging stream channels separated by sand bars or islands. Characteristic of flood plains where the amount of debris is large in relation to the discharge.

Breccia A rock consisting of consolidated angular rock fragments larger than sand grains. It is similar to conglomerate except that most of the fragments are angular, with sharp edges and unworn corners.

Brine Liquid with high salt content resulting from evaporation; also, ground water with an unusually high concentration of salts.

Brown Clay An extremely fine-grained deposit characteristic of some deep ocean basins, particularly those of the Pacific.

Bryozoans Aquatic invertebrate animals that individually average less than 1 mm in length but that construct large colonial structures that have been preserved as fossils in rocks of all ages from Late Cambrian upward. Only lime-secreting varieties are common as fossils, and bryozoan limestone or marl is widespread. In older textbooks the bryozoans are combined with the brachiopods to constitute the phylum *Molluscoidea;* most recent students treat the bryozoans as a distinct phylum.

Burrow A tabular or cylindrical hole in sediment or rock made by an organism. It may be filled or unfilled and have any orientation with respect to bedding.

Calcareous Pertaining to material containing a relatively large amount of calcium carbonate.

Calcite A common, rock-forming, carbonate mineral whose chemical formula is $CaCO_3$. It has a hardness of 3 and a specific gravity of 2.7. A rock with much calcite is said to be *calcareous.*

Caldera A roughly circular, steep-sided volcanic basin with a diameter at least three or four times its depth. Commonly at the summit of a volcano. Contrast with *crater.*

Caliche A whitish accumulation of calcium carbonate in the soil profile.

Calving As applied to glaciers, the process by which a glacier that terminates in a body of water breaks away in large blocks. Such blocks form the icebergs of polar seas.

Capacity The amount of material that a transporting agency such as a stream, a glacier, or the wind can carry under a particular set of conditions.

Capillary Fringe Belt above zone of saturation in which underground water is lifted against gravity by surface tension in passages of capillary size.

Carbohydrate A compound of carbon, hydrogen, and oxygen. Carbohydrates are the chief products of the life process in plants.

Carbonaceous Containing carbon. In geology, containing coal in well-defined beds or small disseminated particles of carbon mingled with inorganic constituents.

Carbonate Any compound formed when carbon dioxide contained in water combines with the oxides of calcium, magnesium, potassium, sodium, and iron. Among the common carbonates are dolomite, siderite, and calcite.

Carbon-14 Radioactive isotope of carbon, ${}_6C^{14}$, with a half life of 5,730 years. Used to date events back to about 50,000 years ago.

Carbon-14 Dating A method of obtaining approximate age limits of carbon-bearing materials based on the radioactive disintegration of carbon-14, which enters living materials along with other nonradioactive isotopes of carbon. Because the half-life is only 5,730 years, the method is not very useful in dating materials over 50,000 years old.

Carbonization The process of converting a substance into a residue of carbon by removing other ingredients, as in the charring of wood, the natural formation of anthracite, and the fossilization of leaves and other plant organs.

Carbon-ratio A number obtained by dividing the amount of fixed carbon in a coal by the sum of fixed carbon and volatile matter, and multiplying by 100. This is the same as the percentage of fixed carbon, assuming no moisture or ash.

Cassiterite A mineral; tin dioxide, SnO_2. Ore of tin with a specific gravity of 7. Nearly 75 percent of the world's tin production is from placer deposits, mostly in the form of cassiterite.

Cast A natural or artificial reproduction of an object showing its outward shape. Natural casts are common as fossils.

Catastrophism The belief that the past history of the earth and of living things has been interrupted or greatly influenced by natural catastrophes occurring on a worldwide or very extensive scale.

Cellulose The most abundant carbohydrate, $C_6H_{10}O_5$, with a chain structure like that of the paraffin hydrocarbons. With lignin, an important constituent of plant material, from which coal is formed.

Cement The material that binds the particles of a consolidated sedimentary rock together. Various substances may act as cement, the most common being silica, calcium carbonate, and various iron oxides.

Cementation The process by which a binding agent is precipitated in the spaces between the individual particles of an unconsolidated deposit. The most common cementing agents are calcite, dolomite, and quartz. Others include iron oxide, opal, chalcedony, anhydrite, and pyrite.

Cement Rock A clayey limestone used in the manufacture of hydraulic cement. Contains lime, silica, and alumina in varying proportions.

Central Vent An opening in the earth's crust, roughly circular, from which magmatic products are extruded. A volcano is an accumulation of material around a central vent.

Cephalopod Any member of a large class of molluscs whose head and mouth are circled with muscular tentacles. Water is drawn in and expelled through a siphon. The eyes are well developed and the whole animal is highly organized for rapid, intelligent action. Many fossil remains have been discovered in rocks of Cambrian age and younger. Modern representatives include the squid, octopus, and pearly nautilus.

Chalcedony A general name applied to fibrous crypto-crystalline silica, and sometimes specifically to the brown translucent variety with a waxy luster. Deposited from aqueous solutions and frequently found lining or filling cavities in rocks. *Agate* is a variety with alternating layers of chalcedony and opal.

Chalk A variety of limestone made up in part of biochemically derived calcite, in the form of the skeletons or skeletal fragments of microscopic oceanic plants and animals, which are mixed with very fine-grained calcite deposits of either biochemical or inorganic chemical origin.

Chemical Energy Energy released or absorbed when atoms form compounds. Generally becomes available when atoms have lost or gained electrons, and often appears in the form of heat.

Chemical Rock In the terminology of sedimentary rocks, a chemical rock is composed chiefly of material deposited by chemical precipitation, either organic or inorganic. Compare with *detrital sedimentary rock.* Chemical sedimentary rocks may have either a clastic or nonclastic (usually crystalline) texture.

Chemical Weathering The weathering of rock material by chemical processes that transform the original material into new chemical combinations. Thus, chemical weathering of orthoclase produces clay, some silica, and a soluble salt of potassium.

Chert A very dense, usually light-colored siliceous rock usually found associated with limestone, either in the form of nodular or concretionary masses or as distinct beds.

Chitin A horny organic substance, chemical composition $C_{32}H_{54}N_4O_{21}$. It is present in the skeletons and protective coverings of most arthropods and in some sponges, coelenterates, and worms.

Chlorite A family of tetrahedral sheet silicates of iron, magnesium, and aluminum characteristic of low-grade metamorphism. Green color, with cleavage like that of mica, except that small scales of chlorite are not elastic whereas those of mica are.

Chondrite The largest class of meteorites, identified by the presence of small spherical bodies called *chondrules.*

Chondrule A small spherical body, about 1 mm in diameter, found in the class of meteorites called *chondrites.* Chondrules have the same composition as certain basic rocks. The origin is not understood.

C-horizon The soil zone that contains partially disintegrated and decomposed parent material. It lies directly under the B-horizon and grades downward into unweathered material.

Chute, or Chute Cutoff As applied to stream flow, the term *chute* refers to a new route taken by a stream when its main flow is diverted to the inside of a bend along a trough between low ridges formed by deposition on the inside of the bend where water velocities were reduced. Compare with *neck cutoff.*

Cinder Rough, slaglike fragment from a millimeter to several centimeters across, formed from magma blown into the air during an eruption.

Cinder Cone Built exclusively or in large part of pyroclastic ejecta dominated by cinders. Parasitic to a major volcano, it seldom exceeds 500 m in height. Slopes up 30° to 40°. Example: Parícutin.

Cirque A steep-walled hollow in a mountainside at high elevation, formed by ice-plucking and frost action, and shaped like a half-bowl or half-amphitheater. Serves as principal gathering ground for the ice of a valley glacier.

Class In connection with living things a major subdivision of a phylum; it in turn is made of one or more orders. In connection with rocks, a subdivision based on the relative proportions of certain standard minerals.

Clastic Rocks Include those deposits that are made up of fragments of preexisting rocks or of the solid products that are formed during the chemical weathering of such older rocks.

Clastic Texture Texture shown by sedimentary rocks formed from deposits of mineral and rock fragments.

Clay Minerals Finely crystalline, hydrous silicates that form as a result of the weathering of such silicate minerals as feldspar, pyroxene, and amphibole. The most common clay minerals belong to the kaolinite, montmorillonite, and illite groups.

Clay Size A volume less than that of a sphere with diameter of $\frac{1}{256}$ mm (.004 mm, or .00015 in.).

Cleavage (1) *Mineral cleavage.* A property possessed by many minerals of breaking in certain preferred directions along smooth plane surfaces. The planes of cleavage are governed by the atomic pattern, and represent directions in which atomic bonds are relatively weak. (2) *Rock cleavage.* A property possessed by certain rocks of breaking with relative ease along parallel planes or nearly parallel surfaces. Rock cleavage is designated as *slaty, phyllitic, schistose,* and *gneissic.*

Coal A general name for combustible, solid, black or brownish black, carbonaceous materials formed through the partial decomposition of vegetable debris. Its formation is distinctly traceable through a series of gradational steps starting with peat and passing through lignite, bituminous coal, and anthracite to a final theoretical limit of nearly pure carbon. It is usually distinctly stratified and is found in association with ordinary sedimentary rocks such as shale and sandstone and, more rarely, limestone.

Cobble Size A volume greater than that of a sphere with a diameter of 64 mm (2.5 in.), and less than that of a sphere with a diameter of 256 mm (10 in.).

Coccolith Any of various microscopic structural elements, usually of buttonlike shape, that make up the outer skeleton of the floating protistid organisms called *coccolithophores.*

Coelenterate A member of the animal phylum Coelenterata, characterized by a hollow body cavity, radial symmetry, and stinging cells; includes jellyfish, corals, and sea anemones.

Col A pass through a mountain ridge. Created by the enlargement of two cirques on opposite sides of the ridge until their headwalls meet and are broken down.

Colloidal Size Between two-tenths of a micron and one micron (.0002 mm to .001 mm, or 8×10^{-6} in. to 4×10^{-5} in.).

Colonial Animal An individual that lives in close association with others of the same species. Usually it cannot exist as a separate individual.

Column A column or post of dripstone joining the floor and roof of a cave; the result of joining of a stalactite and a stalagmite.

Columnar Jointing A pattern of jointing that blocks out columns of rock. Characteristic of tabular basalt flows or sills.

Columnar Section A geologic illustration that shows in a graphic manner, and by use of conventional symbols for rock types, the successive rock units that occur throughout a given area or at a specific locality.

Comet A small body of icy and dusty matter orbiting the sun.

Compaction Reduction in pore space between individual grains as a result of pressure of overlying sediments or pressures resulting from earth movement.

Competence The maximum size of particle that a transporting agency, such as a stream, a glacier, or the wind, can move.

Composite Volcanic Cone A cone composed of interbedded lava flows and pyroclastic material. Characterized by slopes of close to 30° at the summit, reducing progressively to 5° near the base. Example: Mayon.

Compound A combination of the atoms or ions of different elements. The mechanism by which they are combined is called a *bond.*

Conchoidal Fracture A mineral's habit of breaking, in which the fracture produces curved surfaces like the interior of a shell (*conch*). Typical of glass and quartz.

Concordant Pluton An intrusive igneous body with contacts parallel to the layering or foliation surfaces of the rocks into which it was intruded.

Concretion An accumulation of mineral matter that forms around a center or axis of deposition after a sedimentary deposit has been laid down. Cementation consolidates the deposit as a whole, but the concretion is a body within the host rock that represents a local concentration of cementing material. The enclosing rock is less firmly cemented than the concretion. Commonly spheroidal or disk-shaped, and composed of such cementing agents as calcite, dolomite, iron oxide, or silica.

Conformity The mutual relationships between sedimentary beds laid down in orderly sequence with little or no evidence of time lapses and, specifically, without any evidence that the lower beds were folded, tilted, or eroded before the higher beds were laid down.

Conglomerate The consolidated equivalent of gravel. The constituent rock and mineral fragments may be of varied composition and range widely in size. The matrix of finer material between the larger fragments may be sand, silt, or any of the common natural cementing materials such as calcium carbonate, silica, clay, or iron oxide. The rock fragments are rounded and smoothed from transportation by water or from wave action.

Conodont Any small, toothlike fossil of phosphatic composition. Conodonts range through rocks of Cambrian to Triassic age. Their origin is uncertain. They have been variously assigned to vertebrates and to several invertebrate phyla.

Connate Water Water that was trapped in a sedimentary deposit at the time the deposit was laid down.

Consequent Stream A stream following a course that is a direct consequence of the original slope of the surface on which it developed.

Consolidation In geology, any or all of the processes whereby loose, soft, or liquid earth materials become firm and coherent. Any action that increases the solidity, firmness, and hardness of earth materials is important in consolidation.

Contact The surface, in many cases irregular, that constitutes the junction of two bodies of rock.

Contact Metamorphism Metamorphism at or very near the contact between magma and rock during intrusion.

Continental Crust The portion of the earth's crust composed of two layers: first layer, sialic rock, 16 to 24 km thick; second layer, simatic rock, 16 to 24 km thick.

Continental Deposits Deposits laid down on land or in bodies of water not connected with the ocean. The term is applicable whether the landmass is a true continent or only an island. The continental environment embraces fluviatile, lacustrine, glacial, and aeolian conditions.

Continental Drift The process, considered by some to be theoretical and by others to be a fact, whereby one or more large landmasses split apart and "drifted" laterally to form the present-day continents.

Continental Glacier A large ice sheet that completely covers a large section of a continent, covering mountains and plains in an unbroken expanse.

Continental Shelf The gently sloping belt of shallowly submerged land that fringes the continents. It may be broad or narrow. The slope is roughly 1 in 540 and the break of slope into deeper water is generally at a depth of about 180 m. The geology of the continental shelves is similar to the geology of the adjacent emergent land.

Continental Slope Portion of the ocean floor extending from about 180 m (100 fathoms), at the seaward edge of the continental shelves, to the ocean deeps. Continental slopes are steepest in their upper portion, and commonly extend more than 3,600 m (2,000 fathoms) downward.

Continuous Reaction Series That branch of Bowen's reaction series (*q.v.*) comprising the plagioclase feldspars, in which reaction of early-formed crystals with later liquids takes place continuously—that is, without abrupt phase changes.

Convection A mechanism by which material moves because its density is different from that of surrounding material. The density differences are frequently brought about by heating.

Convection Current A closed circulation of material sometimes developed during convection. Convection currents normally develop in pairs; each pair is called a *convection cell.*

Convergence In terms of living things, convergence is the gradual process by which two or more originally unlike organisms become more and more similar in form, function, or reactions.

Coquina A coarse-grained, porous, friable variety of clastic limestone made up chiefly of fragments of shells.

Coral A general name for any of a large group of marine invertebrate organisms that belong to the phylum Coelenterata, which are common in modern seas and have left an abundant fossil record in all periods later than the Cambrian. The term *coral* is commonly applied to the calcareous skeletal remains. As found in the fossil state, coral consists almost exclusively of calcium carbonate.

Coral Reef A reef, usually very large, and made up chiefly of fragments of corals, coral sands, and the solid limestone resulting from consolidation.

Core A cylindrical piece of material cut and brought to the surface by special types of rock-cutting bits during the process of drilling. Also, the innermost part of the earth, surrounded by the mantle.

Core Drilling Drilling with a hollow bit and barrel, which cut out and recover a solid core of the rock penetrated.

Coriolis Effect The tendency of any moving body, on or starting from the surface of the earth, to continue in the direction in which the earth's rotation propels it. The direction in which the body moves because of this tendency, combined with the direction in which it is aimed, determines the ultimate course of the body relative to the earth's surface. In the Northern Hemisphere, the coriolis effect causes a moving body to veer or try to veer to the right of its direction of forward motion; in the Southern Hemisphere, to the left. The magnitude of the effect is proportional to the velocity of a body's motion. This effect causes cyclonic-storm wind circulation to be counterclockwise in the Northern Hemisphere and clockwise in the Southern, and determines the final course of ocean currents relative to trade winds.

Correlation The process of determining the position or time of occurrence of one geologic phenomenon in relation to others. Usually, and in the narrowest sense, it means determining the equivalence of geologic formations in separated areas through a comparison and study of fossil remains or lithologic peculiarities. In a wider sense, it applies to the cause-and-effect relationships of all geologic events in time and space and to the establishment of these phenomena in a logical and complete chronological system, such as the geologic time scale.

Cosmic Ray A very high-speed subatomic particle that reaches the earth from outer space.

Cosmology The science that deals with the universe, its parts, and the laws governing its operation.

Cosmopolitan As applied to fossil organisms, the term *cosmopolitan* implies a widespread geographic distribution.

Covalent Bond A bond in which atoms combine by the sharing of their electrons.

Crater A roughly circular, steep-sided volcanic basin with a diameter less than three times its depth. Commonly at the summit of a volcano. Contrast with *caldera.*

Craton A stable area of the earth's continental crust that has not been deformed for a long time period. Includes both shields and platforms.

Creep As applied to soils and surficial material, slow downward movement of a plastic type. As applied to elastic solids, slow permanent yielding to stresses that are less than the yield point if applied for a short time only.

Creodont One of the groups of early, primitive, carnivorous mammals included in the suborder Creodonta. They flourished early in the Tertiary period.

Crevasse (1) A deep crevice or fissure in glacier ice. (2) A breach in a natural levee.

Crinoid An exclusively marine invertebrate animal belonging to the phylum Echinodermata. Fossil crinoids are found in Late Cambrian and younger rocks. Typically, they are attached by a jointed stem, and their shape suggests a lilylike plant; hence the name "sea lily," by which they are commonly known. Crinoids were especially abundant in Devonian and Mississippian time, declined at the end of the Paleozoic era, and achieved a secondary maximum in the middle of the Mesozoic era. About 650 species are still in existence.

Cross-cutting Relationships, Law of A rock is younger than any rock across which it cuts.

Crossopterygian A type of fish considered to be ancestral to land vertebrates and characterized especially by a stout, muscular fin with a bony axis.

Cross Section A geologic diagram or actual field exposure showing the geologic formations and structures transected by a given plane. Cross-section diagrams are commonly used in conjunction with geologic maps and contribute to an understanding of the subsurface geology. The formations, faults, veins, and so forth are shown by conventional symbols or colors, and the scale is adapted to the size of the features present. Unless otherwise noted, cross sections are drawn in a vertical plane.

Crust In a general sense, the crust of the earth is the outermost shell from 32 to 48 km thick that encloses the weaker, less well-known central part of the earth. The term *crust* is frequently used to mean the outermost part of the earth in which relatively low velocities of earthquake waves prevail above the first major discontinuity, the so-called Mohorovičić discontinuity.

Cryptocrystalline A state of matter in which there is actually an orderly arrangement of atoms characteristic of crystals, but in which the units are so small (that is, the material is so fine-grained) that the crystalline nature cannot be determined with the aid of an ordinary microscope.

Crystal A solid with orderly atomic arrangement. May or may not develop external faces that give it crystal form.

Crystal Form The geometrical form taken by a mineral, giving an external expression to the orderly internal arrangement of atoms.

Crystalline Structure The orderly arrangement of atoms in a crystal. Also called *crystal structure.*

Crystallization The process through which crystals separate from a fluid, viscous, or dispersed state.

Cuesta A ridge with one steep and one gentle face formed by the outcrop of a resistant gently dipping bed.

Curie Temperature The temperature above which ordinarily magnetic material loses its magnetism. On cooling below this temperature, the material regains its magnetism. Example: iron loses its magnetism above 1,400°F (760°C) and regains it as it cools below this temperature. This is its Curie temperature.

Current Ripple Marks Ripple marks, asymmetric in form, formed by air or water moving more or less continuously in one direction.

Cutoff See *chute cutoff; neck cutoff.*

Cycad A plant of the class Gymnospermae, having a short, pithy trunk, a thin, woody covering marked by numerous leaf-base scars, and palmlike leaves and cones.

Cycle of Erosion A qualitative description of river valleys and regions passing through the stages of youth, maturity, and old age with respect to the amount of erosion that has been effected.

Cyclothem A succession of beds deposited during a single sedimentary cycle of the type that prevailed during the Pennsylvanian period. The orderly repetition of a sequence of various kinds of strata in a series of cyclothems reflects a similar repetition of conditions of deposition over fairly wide areas of shallow sea and adjacent low-lying land areas.

Cystoid Any member of a class of echinoderms with a box- or cystlike body constructed of numerous plates that may be arranged regularly or irregularly. The plates may be perforated by many pores, and the creature may have a stem and short food-gathering appendages. The known geologic range is from the Ordovician to the Devonian.

Daughter Element Any element of a radioactive series between the parent and the end product.

Debris Slide A small, rapid movement of largely unconsolidated material that slides or rolls downward to produce an irregular topography.

Decomposition Synonymous with *chemical weathering.*

Deep Focus Earthquake focus deeper than 200 miles (322 km). The greatest depth of focus known is 435 miles (700 km).

Deep-sea Trenches See *island arc deeps.*

Deflation The erosive process in which the wind carries off unconsolidated material.

Deformation of Rocks Any change in the original shape or volume of rock masses. Produced by mountain-building forces. Folding, faulting, and plastic flow are common modes of rock deformation.

Delta A plain underlain by an assemblage of sediments that accumulate where a stream flows into a body of standing water and its velocity and transporting power are suddenly reduced. Originally so named because many deltas are roughly triangular in plan, like the Greek letter *delta* (Δ), with the apex pointing upstream.

Dendrite A branching or fernlike deposit of mineral matter, usually manganese oxide, along a fracture surface.

Dendritic Pattern An arrangement of stream courses that, on a map or viewed from the air, resembles the branching habit of certain trees, such as the oaks or maples.

Dendrochronology The science of dating and correlating that involves matching growth rings of trees or other vegetation.

Density A number that measures the concentration of matter, expressed as the mass per unit volume. (Mass equals weight divided by acceleration of gravity.)

Density Current A current due to differences in the density of sea water from place to place caused by changes in temperature and variations in salinity or the amount of material held in suspension. Also called turbidity current.

Deposit Anything laid down. A natural accumulation of mineral matter in the form of solidified rock, unconsolidated material, useful ores, or organic materials such as coal and oil.

Depositional Remanent Magnetism Magnetism resulting from the tendency of magnetic particles such as magnetite to orient themselves in the earth's magnetic field as they are deposited. Their orientation is maintained as the soft sediments are lithified and thus records the earth's field when the particles were laid down. Abbreviation, DRM.

Derived Fossils or rock fragments that have been removed by erosion or some other process from their original sites and have been redeposited in later formations are said to be *derived.*

Desert Pavement A thin layer of rock fragments, usually of pebble size, left as a cover in desert regions after the wind has removed the finer material. They may be very extensive and one pebble thick.

Desiccation Loss of water from pore spaces of sediments through compaction or through evaporation caused by exposure to air.

Detrital Sedimentary Rocks Rocks formed from accumulations of minerals and rocks derived either from erosion of previously existing rock or from the weathered products of these rocks.

Detritus Any material worn or broken from rocks by mechanical means. The composition and dimensions are extremely variable. The deposits produced by accumulation of detritus constitute the *detrital sediments.*

Diabase A basic igneous rock of the basalt-gabbro series in which the essential minerals are plagioclase and augite, with the plagioclase in long, narrow, lath-shaped crystals oriented in all directions and the augite filling the interstices.

Diagenesis The physical and chemical changes undergone by sediments during lithification and compaction, exclusive of erosion and metamorphism.

Diamond A mineral composed of the element carbon; the hardest substance known. Used as a gem and industrially in cutting tools.

Diastem A minor or obscure break in sedimentary rocks that involves only a very minor time loss. A short interval.

Diastrophism The process or processes that deform the earth's crust.

Diatom A microscopic aquatic plant (one of the algae) that secretes a siliceous skeleton, or test. A sediment made of diatoms is called *diatomite.*

Diatomaceous Ooze A siliceous deep-sea ooze made up of the cell walls of one-celled marine algae known as *diatoms.*

Differential Weathering The process by which different sections of a rock mass weather at different rates. Caused chiefly by variations in composition of the rock itself but also by differences in intensity of weathering from one section to another in the same rock. The result is usually that harder materials stand higher or protrude above softer materials.

Differentiation In the study of igneous rocks, the various processes that produce rocks with a wide range of composition from an original homogeneous parent magma.

Dike A sheetlike body of igneous rock that fills a fissure in older rocks that it entered while in a molten condition. Dikes occur in all types of material—igneous, metamorphic, and sedimentary; if in sedimentary rocks or bedded volcanic rocks, dikes "cut" the formations or transect the beds at an angle.

Dinosaur Any of a large number of extinct reptiles, usually of large size and belonging to either the Saurischia or the Ornithischia. The dinosaurs were confined to the Mesozoic Era and were characterized by diapsid skull structure, three bones uniting to form the hip joint, and other peculiar structural features.

Diorite A coarse-grained igneous rock with the composition of andesite (no quartz or orthoclase), composed of about 75 percent plagioclase feldspars and the balance ferromagnesian silicates.

Dip (1) The acute angle that a rock surface makes with a horizontal plane. The direction of the dip is always perpendicular to the strike. (2) See *magnetic declination.*

Dipole Any object that is oppositely charged at two points. Most commonly refers to a molecule that has concentrations of positive or negative charge at two different points.

Dipole Magnetic Field The portion of the earth's magnetic field that can best be described by a dipole passing through the earth's center and inclined to the earth's axis of rotation. See also *nondipole magnetic field* and *external magnetic field.*

Discharge With reference to stream flow, the quantity of water that passes a given point in unit time. Usually measured in cubic meters per second, abbreviated cms.

Disconformity A break in the orderly sequence of stratified rocks above and below which the beds are parallel. The break is usually indicated by erosion channels with sand or conglomerate, which indicate a lapse of time or absence of part of the rock sequence.

Discontinuity (within the earth's interior) Sudden or rapid changes with depth in one or more of the physical properties of the materials constituting the earth, as evidenced by seismic data.

Discontinuous Reaction Series That branch of Bowen's reaction series (*q.v.*) including the minerals olivine, augite, hornblende, and biotite, for which each change in the series represents an abrupt phase change.

Discordant Pluton An intrusive igneous body with boundaries that cut across surfaces of layering or foliation in the rocks into which it has been intruded.

Disintegration Synonymous with *mechanical weathering.*

Dispersal The spread of a species from its point of origin into other territory where its existence is possible.

Distillation The process of creating fossils by eliminating the liquid or gaseous constituents from an organic substance so that only a carbonaceous residue remains.

Distributary Channel or Stream A river branch that flows away from a main stream and does not rejoin it. Characteristic of deltas and alluvial fans.

Divide Line separating two drainage basins.

Dolerite (1) Loosely, any dark igneous rock whose constituents cannot be easily determined megascopically. (2) Any coarse basalt. (3) Any rock of the composition of basalt regardless of grain size. (4) A rock of the diorite or gabbro clan with uniform medium to small grains.

Dolomite A mineral composed of the carbonate of calcium and magnesium, $CaMg(CO_3)_2$. Also used as a rock name for formations composed largely of the mineral dolomite.

Dome An upfold in which the strata dip downward in all directions from a central point or area. It is the reverse of a basin.

Drainage Basin The area from which a given stream and its tributaries receive their water.

Drift Any material laid down directly by ice, or deposited in lakes, oceans, or streams as a result of glacial activity. Unstratified glacial drift is called *till* and forms *moraines.* Stratified glacial drift forms *outwash plains, eskers, kames,* and *varves.*

Drill Hole An artificial hole cut or drilled in the earth to explore for valuable minerals or to secure scientific data.

Dripstone Calcium carbonate deposited from solution by underground water entering a cave in the zone of aeration. Sometimes called *travertine.*

DRM See *depositional remanent magnetism.*

Drumlin A smooth, streamlined hill composed of till. Its long axis is oriented in the direction of ice movement. The blunt nose points upstream, and a gentler slope tails off downstream with reference to the ice movement. In height, drumlins range from 7.5 m to 60 m, with the average somewhat less than 30 m. Most drumlins are between 0.4 and 0.8 km in length. The length is commonly several times the width. Diagnostic characteristics are the shape and the composition of unstratified glacial drift, in contrast to kames, which are of random shapes and stratified glacial drift.

Dune A mound or ridge of sand piled by wind.

Dust Size A volume less than that of a sphere with a diameter of $\frac{1}{16}$ mm (.06 mm or .0025 in.). Used in reference to particles carried in suspension by wind.

Earth (1) The solid matter of the globe as contrasted with water and air. (2) The loose or softer material composing part of the surface of the globe as distinguished from firm rock. The word is rather indefinite in this sense, meaning about the same as, but not technically synonymous with, the term *soil.*

Earthflow A combination of slump and mudflow.

Earthquake Waves in the earth generated when rocks break after being distorted beyond their strength.

Earthquake Sounds Sounds in air of audible frequencies, as generated by earthquake waves.

Echinoderm Any member of the phylum Echinodermata. The chief characteristics are radial symmetry and a spiny skin. Common living examples are the starfish, sand dollar, and sea urchin. Extinct forms include blastoids and cystoids.

Echinoid Any of a number of marine invertebrate animals belonging to the class Echinoidea of the phylum Echinodermata. Recent forms are variously known as sea urchins, sand dollars, and sea porcupines. They are abundant in present-day seas and have left a fossil record extending back to the Ordovician period. Certain forms are useful guide fossils for some formations of Mesozoic and Cenozoic age.

Echo Sounder An oceanographic instrument that emits sound pulses into the water and measures water depth by the travel time.

Ecliptic The apparent path of the sun in the heavens; the plane of the planet's orbit.

Eclogite A dense igneous rock with the composition of basalt, formed at high temperature and pressure.

Ecology The study of the relations between organisms and environment. *Paleoecology* is the same study applied to past conditions.

Edentate Any member of the mammalian order Edentata, a group characterized chiefly by degenerate teeth. Living examples include the armadillo and the sloth.

Elastic Deformation A nonpermanent deformation after which the body returns to its original shape or volume when the deforming force is removed.

Elastic Energy The energy stored within a solid during elastic deformation, and released during elastic rebound.

Elasticity A property of materials that defines the extent to which they resist small deformation from which they recover completely when the deforming force is removed. Elasticity = stress/strain.

Elastic Rebound The recovery of elastic strain when a material breaks or when the deforming force is removed.

Elastic Solid A solid that yields to applied force by changing shape or volume, or both, but returns to its original condition when the force is removed. The amount of yield is proportional to the force.

Electrical Energy The energy of moving electrons.

Electric Charge A property of matter resulting from an imbalance between the number of protons and the number of electrons in a given piece of matter. The electron has a negative charge, the proton a positive charge. Like charges repel each other; unlike attract.

Electric Log A record of the electrical responses of the geological materials encountered in a drill hole. Electric logs are useful in locating changes in composition and in making local correlations.

Electron A fundamental particle of matter, the most elementary negative electrical charge. Its mass is .00055 unit.

Electron Microscope An instrument utilizing a beam of electrons capable of producing magnifications on the order of 100,000×.

Element A unique combination of protons, neutrons, and electrons that cannot be broken down by ordinary chemical methods. The fundamental properties of an element are determined by its number of protons. Each element is assigned a number that corresponds to its number of protons. Combinations containing from 1 through 102 protons are now known.

End Moraine A ridge or belt of till marking the farthest advance of a glacier. Sometimes called *terminal moraine.*

Energy The capacity for producing motion. Energy holds matter together. It can become mass, or can be derived from mass. It takes such forms as kinetic, potential, heat, chemical, electrical, and atomic energy, and can be changed from one of these forms to another.

Energy Level The distance from an atomic nucleus at which electrons can have orbits. May be thought of as a shell surrounding the nucleus.

Entrenched Meander A meander cut into underlying bedrock when regional uplift allows the originally meandering stream to resume downward cutting.

Environment of Deposition The physical, chemical, and biological conditions at the site where sediment accumulates.

Epeirogenic Pertaining to or designating the deformation of broad tracts of the earth's crust; contrasts with *orogenic.*

Epicenter The point on the earth's surface directly above the focus of an earthquake.

Epicontinental Resting on a continent, as an epicontinental sea.

Epoch A unit of geologic time; subdivision of a period. Some geologists restrict the term to the equivalent of a rock series, such as the Eocene Epoch or Series of the Tertiary Period or System.

Era One of the major divisions of geologic time, including one or more periods. The eras usually recognized are the Archeozoic, Proterozoic, Paleozoic, Mesozoic, and Cenozoic.

Erg A unit of energy, the capacity for doing work. The energy expended when a force of one dyne acts through a distance of one centimeter.

Erosion The wearing away and removal of materials of the earth's crust by natural means. As usually employed, the term includes weathering, solution, corrosion, and transportation. The agents that accomplish the transportation and cause most of the wear are running water, waves, moving ice, and wind currents. Most writers include under the term all the mechanical and chemical agents of weathering that loosen rock fragments before they are acted on by the transporting agents; a few authorities prefer to include only the destructive effects of the transporting agents.

Erosional Flood Plain A flood plain that has been created by lateral erosion and the gradual retreat of the valley walls.

Erosional Unconformity A break in the continuity of deposition of a rock series that is made manifest by evidences of erosion. The strata above and below the break may be parallel, with no evidence of folding of the lower beds during the lapse in sedimentation.

Erratic A rock fragment, usually large, that has been transported from a distant source, especially by the action of glacial ice. In the terminology of glaciation, an erratic is a stone or boulder carried by ice to a place where it rests on or near bedrock of different composition.

Esker A widening ridge of stratified glacial drift, steep-sided, 3 to 30 m in height, and from a fraction of a km to over 160 km in length.

Eucaryote A type of living cell having a nucleus enclosed within a nuclear membrane and with well-defined chromosomes and plastids. Eucaryotic cells reproduce by miotic division in which genetic material is segregated among successive cells or descendants.

Eugeosyncline A geosyncline in which abundant volcanic products occur with clastic sediments, usually the outer seaward part of an orthogeosyncline.

Eurypterid An extinct type of anthropod with pincerlike claws and 13 abdominal segments.

Eustatic Change of Sea Level A change in sea level produced entirely by an increase or a decrease in the amount of water in the oceans, hence worldwide.

Evaporation The process by which a liquid becomes a vapor at a temperature below its boiling point.

Evaporite A rock composed of minerals that have been precipitated from solutions concentrated by the evaporation of solvents. Examples: rock salt, gypsum, anhydrite.

Evolution The unfolding or development of an organism so that it becomes more perfectly or completely adapted to the environments that become available to it with the passage of time. The implication of organic evolution is that all life has been derived from one or a few simple beginnings.

Exfoliation The process by which plates of rock are stripped from a larger rock mass by physical forces.

Exoskeleton A hard, outer skeleton or protective covering to which muscles are attached. The integument of the arthropod is a typical exoskeleton.

Exposure An unobscured outcrop of either solid rock or unconsolidated superficial material. In one sense or another, the term embraces all earth materials appearing at the surface that are not hidden by vegetation, water, or the works of man.

External Magnetic Field A component of the earth's field originating from activity above the earth's surface. Small when compared with the dipole and nondipole components of the field, which originate beneath the surface.

Extrusive Rock A rock that has solidified from material poured or thrown out upon the earth's surface by volcanic action.

Facies A term with many shades of meaning and hence difficult to define. In general, the term *facies* designates the aspect or appearance of a mass of earth material different in one or several respects from surrounding material. The features by which facies are named and recognized are usually selected more or less arbitrarily and may be lithologic (lithofacies) or biologic (biofacies). The two usages seem to be evident. The first applies only within a specific rock or time unit. A facies within the specific interval then designates some particular or general feature by which a part differs from other parts deposited at the same time and usually in physical continuity. The second usage applies to certain features not confined to specific intervals. Thus, a black shale facies might be of almost any age.

Facies Changes Lateral or vertical changes in the lithologic or paleontological characteristics of contemporaneous deposits. Since facies relationships are usually complex, the exact features selected for mapping or discussion should be clearly designated.

Fault A break in materials of the earth's crust in which there has been movement parallel with the surface along which the break occurs. A fault occurs when rocks are strained past the breaking point and yield along a crack or series of cracks so that corresponding points on the two sides are distinctly offset. One side may rise or sink or move laterally with respect to the other side.

Fault-block Mountain A mountain bounded by one or more faults.

Fauna The aggregation of animal species characteristic of a certain locality, region, or environment. The animals found fossilized in certain geologic formations or occurring in specified time intervals of the past may be referred to as *fossil faunas.*

Faunal Succession The observed sequence of life forms through past ages. The total aspect of life at any one period is different from that of preceding and succeeding periods. Faunal succession implies but does not prove evolution.

Feldspars Silicate minerals composed of silicon-oxygen and aluminum-oxygen tetrahedra linked together in three-dimensional networks with positive ions fitted into the interstices of the negatively charged framework of tetrahedra. Classed as aluminosilicates. When the positive ion is K^+, the mineral is orthoclase; when it is Na^+, the mineral is albite; when it is Ca^{2+}, the mineral is anorthite.

Felsite Light-colored igneous rock that is poor in iron and magnesium and rich in feldspars and quartz.

Ferromagnesian Silicate A silicate in which the positive ions are dominated by iron, magnesium, or both.

Fiord A glacially deepened valley that is now flooded by the sea to form a long, narrow, steep-walled inlet.

Firn Granular ice formed by the recrystallization of snow. Intermediate between snow and glacier ice. Sometimes called *névé.*

First Motion On a seismogram, the direction of ground motion at the beginning of the arrival of a *P* wave. Upward ground motion indicates a compression; downward motion, a dilatation.

Fissility A property of splitting along closely spaced parallel planes more or less parallel to the bedding. Its presence distinguishes shale from mudstone.

Fission Tracks Minute tubes formed in certain minerals such as mica by fragments of radioactive elements that have undergone fission in place.

Fissure Eruption Extrusion of lava from a fissure in the earth's crust.

Flint A dense, hard, siliceous rock composed of very finely crystalline and amorphous silica.

Flood Basalt Basalt poured out from fissures in floods that tend to form great plateaus. Sometimes called *plateau basalt.*

Flood Plain A strip of relatively smooth land bordering a stream, built of sediment carried by the stream and dropped in the slack water beyond the influence of the swiftest current. It is called a *living flood plain* if it is overflowed in times of high water but a *fossil flood plain* if it is beyond the reach of the highest flood.

Flood Plain of Aggradation A flood plain formed by the building up of the valley floor by sedimentation.

Flora The assemblage of plants of a given geologic formation, environment, region, or time interval.

Fluvial Pertaining to streams or stream action.

Focus The source of a given set of earthquake waves.

Fold A bend, flexure, or wrinkle in rock produced when the rock was in a plastic state.

Foliation A layering in some rocks caused by parallel alignment of minerals. A textural feature of some metamorphic rocks. Produces rock cleavage.

Footwall One of the blocks of rock involved in fault movement. The one that would be under the feet of a person standing in a tunnel along or across the fault. Opposite the hanging wall.

Foraminifera An important order of one-celled animals (protozoa), which have left an extensive fossil record in rocks of Ordovician and younger age. They are almost all marine and have durable shells, or tests, capable of fossilization. Being small, their remains are readily recovered from well cores and cuttings and have become very important in correlating oil-bearing rocks. Thousands of fossil species have been discovered, and they are especially useful as guide fossils in rocks of late Paleozoic, Cretaceous, and Tertiary age.

Foreset Beds Inclined layers of sediment deposited on the advancing edge of a growing delta or along the lee slope of an advancing sand dune.

Foreshock A relatively small earthquake that precedes a larger earthquake by a few days or weeks and originates at or near the focus of the larger earthquake.

Formation The formation is the fundamental unit in the local classification of rocks. The larger units, groups, and series may be regarded as assemblages of formations and the smaller units as subdivisions of formations. The discrimination of sedimentary formations is based on the local sequence of rocks, lines of separation being drawn at points in the stratigraphic column where lithologic characters change or where there are significant breaks in the continuity of sedimentation or other evidences of important geologic events.

Fossil Originally, any rock, mineral, or other object dug out of the earth. Now restricted to any evidence of the existence or nature of an organism that lived in ancient times and that has been preserved in materials of the earth's crust by natural means. The term is not restricted to petrified remains—i.e., those of a stony nature—and includes, besides actual remains, such indirect evidences as tracks and trails. Fossils are, with few exceptions, prehistoric, but no age limit in terms of years can be set. Fossils are useful in studying the evolution of present life forms and in determining the relative ages of rock strata. The term also is frequently, but loosely, used in connection with ancient inorganic objects and markings, such as fossil ripple marks or rain prints.

Fossil Assemblage All the fossil organisms that can be found or collected from a single bed or formation and that are assumed to have lived at the same time.

Fossil Fuels Organic remains (once living matter) used to produce heat or power by combustion. Include petroleum, natural gas, and coal.

Fractional Crystallization The separation of a cooling magma into components by successive formation of crystals at progressively lower temperatures.

Fractional Distillation The recovery, one or more at a time, of fractions of a complex liquid, each of which has a different density.

Fractionation A process whereby crystals that formed early from a magma have time to settle appreciably before the temperature drops much further. They are effectively removed from the environment in which they formed.

Fracture As a mineral characteristic, the way in which a mineral breaks when it does not have cleavage. May be conchoidal (shell-shaped), fibrous, hackly, or uneven.

Fracture Cleavage A system of joints spaced a fraction of a centimeter apart.

Fracture Zone A very long narrow belt of displacement on the deep sea floor that separates tracts of different depth. Many fracture zones cross and displace the midoceanic ridge.

Fringing Reef A reef attached directly to a landmass.

Front In connection with concepts of granitization, the limit to which diffusing ions of a given type are carried. The *simatic front*, for example, is the limit to which diffusing ions carried the calcium, iron, and magnesium that they removed from the rocks in their paths. The *granitic front* is the limit to which diffusing ions deposited granitic elements.

Frost Action Process of mechanical weathering caused by repeated cycles of freezing and thawing. Expansion of water during the freezing cycle provides the energy for the process.

Frost Heaving The heaving of unconsolidated deposits as lenses of ice grow below the surface by acquiring capillary water from below.

Fumarole A small vent in the ground from which volcanic gases and heated ground water emerge, but no lava.

Fusuline, Fusulinid Any of an important group of extinct, marine, one-celled animals (class Sarcodina, phylum Protozoa) that have left an extensive fossil record from Late Paleozoic time. Owing to their small size, they are easily recovered from well cuttings and have proved of great value in correlating oil-bearing rocks.

G Symbol for rigidity modulus.

Gabbro A coarse-grained igneous rock with the composition of basalt.

Galaxy A portion of space in which stars, dust, gas, and matter in general are concentrated.

Galena A mineral; lead sulfide, PbS. The principal ore of lead.

Garnet A family of silicates of iron, magnesium, aluminum, calcium, manganese, and chromium, which are built around independent tetrahedra and appear commonly as distinctive 12-sided fully developed crystals. Characteristic of metamorphic rocks. Generally cannot be distinguished from one another without chemical analysis.

Gas (1) A state of matter that has neither independent shape nor volume, can be compressed readily, and tends to expand indefinitely. (2) In geology, the word *gas* is sometimes used to refer to *natural gas*, the gaseous hydrocarbons that occur in rocks, and are dominated by methane. Compare with use of the word *oil* to refer to *petroleum*.

Gastropod Any member of a large and important class of molluscs that typically possess a coiled, single-chambered shell. Marine, freshwater, and terrestrial forms exist, and the group has left fossil representatives in Cambrian and all younger rocks. The gastropods are extremely numerous at present and have been important throughout the Cenozoic era. Snails are the best-known representatives.

Gene An hereditary determiner, located in a chromosome.

Genus (pl. **genera**) A group of closely related species of organisms.

Geochemistry All geologically oriented study involving chemical change. Attention is given to the distribution, amounts, and reactions of the chemical elements in minerals, ores, rocks, soils, water, and the atmosphere.

Geochronology The study and classification of time in relation to the history of the earth.

Geode A roughly spherical, hollow, or partially hollow, accumulation of mineral matter from a few cm to more than over 0.3 m in diameter. An outer layer of chalcedony is lined with crystals that project inward toward the hollow center. The crystals, often perfectly formed, are usually quartz, although calcite and dolomite are also found and, more rarely, other minerals. Geodes are most commonly found in limestone, and more rarely in shale.

Geographic Poles The points on the earth's surface marked by the ends of the earth's axis of rotation.

Geologic Age The time of existence of a fossil organism or the occurrence or duration of a particular event as stated in terms of the conventional geological time scale. Any event not datable in terms of years is usually assigned a relative geologic age.

Geologic Column (1) A chronologic arrangement of rock units in columnar form with the oldest units at the bottom and the youngest at the top. (2) A diagram showing the subdivisions of part or all of geologic time or the rock formations of a particular locality.

Geologic Hazard Any condition of a geologic nature existing naturally or created by man that either actually or potentially presents a risk to human life or property. Examples are landslides, earthquakes, flooding, mine disasters, and erosion.

Geologic Map A map showing geologic features such as distribution, age, and nature of rock units; structural features such as faults, folds, and joints; and associated works of man such as mines and roads. Surficial deposits may be shown separately from bedrock.

Geologic Section Any succession of rock units found exposed or revealed underground. Also the graphic or written description of the natural sequence.

Geologic Time The segment of time that elapsed before written history began. Although no precise limits can be set, the term implies extremely long duration or remoteness in the past.

Geologic Time Scale A chronologic sequence of units of earth time.

Geology The science that treats of the origin, composition, structure, and history of the earth, especially as revealed by the rocks, and of processes by which changes in the rocks are brought about. Included is the study of the origin and evolution of living organisms, especially in prehistoric times. There are many subdivisions of the science, of which the following are important: historical geology, physical geology, economic geology, structural geology, mineralogy, mining geology, physiography, geomorphology, petrography, petrology, vulcanology, stratigraphic geology, and paleontology.

Geomagnetic Poles The dipole best approximating the earth's observed field is one inclined 11.5° from the axis of rotation. The points at which the ends of this imaginary magnetic axis intersect the earth's surface are known as the geomagnetic poles.

Geomorphology The science of the surficial features of the earth; deals with the classification, description, origin, and relation of landforms to the underlying geologic structure and nature of the material making up the surface.

Geophysical Prospecting Mapping rock structures by methods of experimental physics. Includes measuring magnetic fields, the force of gravity, electrical properties, seismic wave paths and velocities, radioactivity, and heat flow.

Geophysics Broadly, the physics of the earth, including the fields of meteorology, hydrology, oceanography, seismology, vulcanology, magnetism, and geodesy. In the more popular and practical sense, the term implies the application of electrical, thermal, magnetic, gravimetric, and seismic methods to the search for petroleum, metals, and underground supplies of water.

Geosyncline Literally, a great, elongate downfold in the earth's crust. In general, the surface dimensions must be measured in terms of scores of kilometers, and the thickness of accumulated rocks must be on the order of 9,000 to 12,000 m. A typical geosyncline comes into being through long-continued, gradual subsidence with simultaneous filling by shallow-water sediments. Geosynclines usually originate between or adjacent to the more solid shield or platform areas of the globe. They may become, with suitable structural evolution, the sites of large-scale deformation, and it is recognized that many major mountain systems are formed of compressed geosynclinal sediments.

Geothermal Power Power generated by utilizing the heat energy of the crust, especially in volcanic areas.

Geyser A special type of thermal spring that intermittently ejects its water with considerable force.

Glacial Drift As used today, the term *glacial drift* embraces all rock material in transport by glacial ice, all deposits made by glacial ice, and all deposits predominantly of glacial origin laid down in the sea or in bodies of glacial meltwater, whether rafted in icebergs or transported in the water itself. It includes till and scattered rock fragments.

Glaciation A period of intensive ice action; also, the geologic work accomplished by ice masses.

Glacier A mass of ice, formed by the recrystallization of snow, that flows forward, or has flowed at some time in the past, under the influence of gravity. By convention we exclude icebergs from this definition even though they are large fragments broken from the seaward end of glaciers.

Glacier Ice A unique form of ice developed by the compression and recrystallization of snow, and consisting of interlocking crystals.

Glass A form of matter that exhibits the properties of a solid but has the atomic arrangements, or lack of order, of a liquid.

Globigerina Ooze A deep-sea calcareous ooze in which limy shells of minute one-celled animals called *Globigerina* abound.

Glossopteris Flora A Late Paleozoic assemblage of fossil plants named for the seed fern *Glossopteris*, one of the plants in the flora. Widespread in South America, South Africa, Australia, India, and Antarctica.

Gneiss A banded metamorphic rock with alternating layers of unlike minerals. Usually, equigranular minerals alternate with tabular minerals.

Gneissic Cleavage Rock cleavage in which the surfaces of easy breaking, if developed at all, are from a few hundredths of an inch to a cm or more apart.

Gondwanaland (or **Gondwana**) A hypothetical continent formed by the union of South America, Africa, Australia, India, and Antarctica. This landmass is thought to have broken into its present fragments in the Mesozoic era.

Gradation Leveling of the land. This is constantly being brought about by the forces of gravity and by such agents of erosion as water at the surface and underground, and wind, glacier ice, and waves.

Grade A term used to designate the extent to which metamorphism has advanced. Found in such combinations as high-grade or low-grade metamorphism. Compare with *rank*.

Graded Bedding The type of bedding shown by a sedimentary deposit when particles become progressively finer from bottom to top.

Gradient Slope of a stream bed.

Granite A coarse-grained igneous rock dominated by light-colored minerals, consisting of about 50 percent orthoclase, 25 percent quartz, and the balance plagioclase feldspars and ferromagnesian silicates. Granites and granodiorites constitute 95 percent of all intrusive rocks.

Granitic Having the general character of granite, especially the structure of interlocking crystals. The mineral composition may or may not be the same as true granite.

Granitization A special type of metasomatism by which solutions of magmatic origin move through solid rocks, change ions with them, and convert them into rocks that achieve granitic character without having passed through a magmatic stage.

Granodiorite A coarse-grained igneous rock intermediate in composition between granite and diorite.

Graphic Structure An intimate intergrowth of potassic feldspar and quartz with the long axes of quartz crystals lining up parallel to a feldspar axis. The quartz part is dark and the feldspar is light in color, so the pattern suggests Egyptian hieroglyphs. Commonly found in pegmatites.

Graphite A mineral composed entirely of carbon. "Black lead." Very soft because of its crystalline structure, in contrast to diamond, which has the same composition but is the hardest substance known.

Graptolite Any of a large number of extinct marine invertebrates that occur as fossils from late in the Cambrian period to the Mississippian period. Their zoological affinities are obscure, but they have been recently assigned to the phylum Protochordata and are thus distantly related to vertebrates. They are especially useful as guide fossils in Ordovician rocks.

Gravel Loose, or unconsolidated, coarse granular material larger than sand grains, resulting from erosion of rock by natural agencies. The lower size limit is usually set at 2 mm.

Gravity Anomaly Difference between observed value of gravity and computed value.

Gravity Fault A fault in which the hanging wall appears to have moved downward relative to the footwall. Also called *normal fault.*

Gravity Meter An instrument for measuring the force of gravity. Also called *gravimeter.*

Gravity Prospecting Mapping the force of gravity at different places to determine differences in specific gravity of rock masses, and, through this, the distribution of masses of different specific gravity. Done with a gravity meter (gravimeter).

Graywacke A variety of sandstone generally characterized by its hardness, dark color, and angular grains of quartz, feldspar, and small rock fragments set in a matrix of clay-sized particles.

Greenschist A schist characterized by green color. The product of metamorphism of simatic rocks. The green color is imparted by the mineral chlorite.

Groundmass The finely crystalline or glassy portion of a porphyry.

Ground Moraine Till deposited from a glacier as a veneer over the landscape and forming a gently rolling surface.

Ground Water Underground water within the zone of saturation.

Ground-water Table The upper surface of the zone of saturation for underground water. It is an irregular surface with a slope or shape determined by the quantity of ground water and the permeability of the earth materials. In general, it is highest beneath hills and lowest beneath valleys. Also referred to as *water table.*

Group A unit of stratigraphic classification. A local or provincial subdivision of a system based on lithologic features. A group is usually less than a standard series and contains two or more formations.

Guide Fossil Any fossil that has actual, potential, or supposed value in identifying the age of the rocks in which it is found. Also called *index fossil.*

Guyot A flat-topped submarine mountain whose summit is supposed to have been exposed to wave action and to have been planed away to the surface of the ocean. Same as *tablemount.*

Gypsum Hydrous calcium sulphate, $CaSO_4 \cdot 2H_2O$. A soft, common mineral in sedimentary rocks, where it sometimes occurs in thick beds interstratified with limestones and shales. Sometimes occurs as a layer under a bed of rock salt, since it is one of the first minerals to crystallize on the evaporation of sea water. Alabaster is a fine-grained massive variety of gypsum.

H Symbol for mineral hardness.

Half-life Time needed for one-half of the nuclei in a sample of a radioactive element to decay.

Halite A mineral; rock salt, or common salt, NaCl. Occurs widely disseminated, or in extensive beds and irregular masses, precipitated from sea water and interstratified with rocks of other types as a true sedimentary rock.

Hanging Valley A valley that has a greater elevation than the valley to which it is tributary, at the point of their junction. Often (but not always) created by a deepening of the main valley by a glacier. The hanging valley may or may not be glaciated.

Hanging Wall One of the blocks involved in fault movement. The one that would be hanging overhead for a person standing in a tunnel along or across the fault. Opposite the footwall.

Hardness A mineral's resistance to scratching on a smooth surface. The Mohs scale of relative hardness consists of ten minerals. Each of these will scratch all those below it in the scale and will be scratched by all those above it: (1) talc, (2) gypsum, (3) calcite, (4) fluorite, (5) apatite, (6) orthoclase, (7) quartz, (8) topaz, (9) corundum, (10) diamond.

Hard Water Water that contains sufficient dissolved calcium and magnesium to cause a carbonate scale to form when the water is boiled or to prevent the sudsing of soap.

Head Difference in elevation between intake and discharge points for a liquid. In geology, most commonly of interest in connection with the movement of underground water.

Heat Energy A special manifestation of kinetic energy in atoms. The temperature of a substance depends on the average kinetic energy of its component particles. When heat is added to a substance, the average kinetic energy increases.

Hematite Iron oxide, Fe_2O_3. The principal ore mineral for about nine-tenths of the commercial iron produced in the United States. Characteristic red color when powdered. The name is derived from the Greek word meaning "blood."

Hiatus A break or gap in the geologic record, as when rocks of a particular age are missing. The hiatus of an unconformity refers to the time interval not represented by rocks or to rocks missing by comparison with other areas.

Historical Geology The study of the history and development of the earth, including the life forms that have inhabited it, and the sum of that knowledge. Historical geology encompasses what astronomy and geophysics can tell of the earth's origin, the paleontologic evidence of the nature of ancient life and its development through geologic time, and the relations developed by stratigraphy, structural geology, and other branches of geology that place the events of earth history in a sequential order.

Horizon A surface of contact or an imaginary plane without actual thickness that marks a certain level in stratified rocks.

Horn A spire of bedrock left where cirques have eaten into a mountain from more than two sides around a central area. Example: Matterhorn of the Swiss Alps.

Hornblende A rock-forming ferromagnesian silicate mineral with double chains of silicon-oxygen tetrahedra. An amphibole.

Hornfels Dense, granular metamorphic rock. Since this term is commonly applied to the metamorphic equivalent of any fine-grained rock, its composition is variable.

Hornfels Facies An assemblage of minerals formed at temperatures greater than 1,300°F (704°C) during contact metamorphism.

Hot Spot (thermal center) A place at which volcanic products and other manifestations occur as a result of a rising column or plume of heated material from deep in the mantle.

Hot Spring A spring that brings hot water to the surface. A *thermal spring.* Water temperature usually 15°F (−9°C) or more above mean air temperature.

Hydration A chemical reaction, usually in weathering, that adds water or OH^- to a mineral structure.

Hydraulic Gradient Head of underground water divided by the distance of travel between two points. If the head is 3 m for two points 30 m apart, the hydraulic gradient is .1 or 10 percent. When head and distance of flow are the same, the hydraulic gradient is 100 percent.

Hydrocarbon A compound of hydrogen and carbon that burns in air to form water and oxides of carbon. There are many hydrocarbons. The simplest, methane, is the chief component of natural gas. Petroleum is a complex mixture of hydrocarbons.

Hydrologic Cycle The general pattern of movement of water from the sea by evaporation to the atmosphere, by precipitation onto the land, and by movement under the influence of gravity back to the sea again.

Hydrothermal Solution A hot, watery solution that usually emanates from a magma in the late stages of cooling. Frequently contains and deposits in economically workable concentrations minor elements that, because of incommensurate ionic radii or electronic charges, have not been able to fit into the atomic structures of the common minerals of igneous rocks.

Icecap A localized *ice sheet.*

Ice Sheet A broad, moundlike mass of glacier ice of considerable extent with a tendency to spread radially under its own weight. Localized ice sheets are sometimes called *icecaps.*

Ichthyosaur Literally "fish lizard"; any of the extinct, aquatic, fishlike reptiles belonging to the order Ichthyosauria.

Igneous Pertaining to or having the nature of fire. As used in geology to distinguish one of the three great classes of rocks, the name is a misnomer, for there is actually no fire involved; it should be interpreted to mean high temperatures.

Igneous Rocks Rocks formed by solidification of hot mobile rock material (magma), including those formed and cooled at great depths (plutonic rocks) that are crystalline throughout and those that have poured out on the earth's surface in the liquid state or have been blown as fragments into the air (volcanic rocks).

Illite A clay mineral family of hydrous aluminous silicates. Structure similar to that of montmorillonite, but with aluminum substituted for 10 to 15 percent of the silicon, which destroys montmorillonite's property of expanding with the addition of water because weak bonds are replaced by strong potassium-ion links. Structurally, illite is intermediate between montmorillonite and muscovite. Montmorillonite converts to illite in sediments, while illite converts to muscovite under conditions of low-grade metamorphism. Illite is the commonest clay mineral in clayey rocks and recent marine sediments, and is present in many soils.

Index Fossil See *guide fossil.*

Indigenous Said of an organism or rock body native to, or originating in, a specific place.

Induced Magnetism In the terminology of rock magnetism, one of the components of the rock's natural remanent magnetism. It is parallel to the earth's present field and results from it.

Infiltration The soaking into the ground of water on the surface.

Insectivore A member of the order of placental animals, the *Insectivoria.* Living examples include the shrews and hedgehog.

Intensity (of an earthquake) A number related to the effects of earthquake waves on man, structures, and the earth's surface at a particular place. Contrast with *magnitude,* which is a number related to the total energy released by an earthquake.

Intermediate Belt Subdivision of zone of aeration. The belt that lies between the belt of soil moisture and the capillary fringe.

Intermediate Focus Earthquake focus

between depths of 60 to 300 km (40 to 200 miles).

Intermittent Stream A stream that carries water only part of the time.

Intrusive Rock A rock that has solidified from a mass of molten material within the earth's crust but did not reach the surface.

Invertebrate An animal without a backbone; pertaining to such an animal or animals.

Ion An electrically unbalanced form of an atom, or group of atoms, produced by the gain or loss of electrons.

Ionic Bond A bond in which ions are held together by the electrical attraction of opposite charges.

Ionic Radius The average distance from the center to the outermost electron of an ion. Commonly expressed in angstroms.

Iron Formation A sedimentary deposit, typically thin-bedded or laminated, containing at least 15 percent iron of sedimentary origin. Usually of Precambrian age.

Island Arc A group of islands having an arclike pattern. Most island arcs lie near the continental masses, but inasmuch as they rise from the deep ocean floors, they are not a part of the continents proper.

Island Arc Deeps Arcuate trenches bordering some of the continents. Some reach depths of 9,000 m or more below the surface of the sea. Also called *deep-sea trenches* or *trenches.*

Isograd A line or curved surface connecting rocks that have undergone an equivalent degree of metamorphism.

Isolation A term used in biology to designate any process or condition whereby a group of individuals is cut off and separated for a considerable length of time from other areas or groups. The situation need not arise from actual geographical factors. Animals may become isolated as a result of food preferences or because of purely psychological reactions.

Isoseismic Line A line connecting all points on the surface of the earth where the intensity of shaking produced by earthquake waves is the same.

Isostasy The ideal condition of balance that would be attained by earth materials of differing densities if gravity were the only force governing their heights relative to each other.

Isotope Alternate form of an element. The fundamental properties of the element, and its place in the table of elements, are determined by the number of protons in its nucleus. Variations in the number of neutrons in the nucleus produce isotopes.

Jasper Granular cryptocrystalline silica usually colored red by hematite inclusions.

Jet or Shooting Flow A type of flow, related to turbulent flow, occurring when a stream reaches high velocity along a sharply inclined stretch, or over a waterfall, and the water moves in plunging, jetlike surges.

Joint A break in a rock mass where there has been no relative movement of rock on opposite sides of the break.

Juvenile Water Water brought to the surface or added to underground supplies from magma.

Kame A steep-sided hill of stratified glacial drift. Distinguished from a drumlin by lack of unique shape and by stratification.

Kame Terrace Stratified glacial drift deposited between a wasting glacier and an adjacent valley wall. When the ice melts, this material stands as a terrace along the valley wall.

Kaolinite A clay mineral; a hydrous aluminous silicate, $Al_4Si_4O_{10}(OH)_8$. Structure consists of one sheet of silicon-oxygen tetrahedra, each tetrahedron sharing three oxygens to give a ratio of Si_4O_{10}, linked with one sheet of aluminum and hydroxyl. The composition of pure kaolinite does not vary as it does for the other clay minerals, montmorillonite and illite, in which ready addition or substitution of ions takes place.

Karst Topography Irregular topography characterized by sinkholes, streamless valleys, and streams that disappear into the underground, all developed by the action of surface and underground water in soluble rock such as limestone.

Kerogen A mixture of organic substances found in many fine-grained sedimentary rocks and a major constituent of oil shale.

Kettle A depression in the ground surface formed by the melting of a block of ice buried or partially buried by glacial drift, either outwash, or till.

Key Bed A well-defined and easily recognizable bed that serves to facilitate correlation in geologic work. The term is also applied to the horizon or bed on which elevations are taken or to which elevations are finally reduced in making a structure contour map. The term is used interchangeably with *key horizon.*

Kinetic Energy Energy of movement. The amount of kinetic energy possessed by an object or particle depends on its mass and speed.

L Symbol for earthquake surface waves.

Labyrinthodont Pertaining to a peculiar tooth structure characterized by deep infolding of the enamel. This type of tooth is possessed by extinct amphibians and related ancestral fish.

Laccolith A concordant pluton that has domed up the strata into which it was intruded.

Lake A considerable body of inland water or an expanded part of a river.

Laminar Flow Mechanism by which a fluid such as water moves slowly along a smooth channel, or through a tube with smooth walls, with fluid particles following straight-line paths parallel to the channel or walls. Contrast with *turbulent flow.*

Land Bridge A land area, usually narrow and subject to submergence, that connects landmasses and serves as a route of dispersal for land plants and animals.

Land Form The term *land form* is applied by physiographers to each of the multitudinous features that taken together make up the surface of the earth. It includes all broad features such as plains, plateaus, and mountains, and also all the minor features such as hills, valleys, slopes, canyons, arroyos, and alluvial fans.

Most of these features are the products of erosion, but the term also includes all forms that result from sedimentation and from movements within the crust of the earth.

Landslide A general term for relatively rapid mass movement, such as slump, rock slide, debris slide, mudflow, and earthflow.

Large Waves Earthquake surface waves. Also called *L waves.*

Latent Heat of Fusion The number of calories per unit volume that must be added to a material at the melting point to complete the process of melting. These calories do not raise the temperature.

Lateral Moraine A ridge of till along the edge of a valley glacier. Composed largely of material fallen on the glacier from valley walls.

Laterite Tropical soil rich in hydroxides of aluminum and iron formed under conditions of good drainage.

Laurasia A hypothetical landmass composed of Asia, North America, and other minor landmasses of the Northern Hemisphere.

Lava A general name for molten rock poured out on the surface of the earth by volcanoes and for the same material that has cooled and solidified as solid rock.

Law of Superposition The general law that states that if undisturbed, any sequence of sedimentary rocks will have the oldest beds at the base and the youngest at the top.

Levee (natural) Bank of sand and silt built by a river during floods, where suspended load is deposited in greatest quantity close to the river. The process of developing natural levees tends to raise river banks above the level of the surrounding flood plains. A break in a natural levee is sometimes called a *crevasse.*

Lichen A plant composed of a fungus and an alga living in symbiotic relationship.

Lignite A low-grade coal with about 70 percent carbon and 20 percent oxygen. Intermediate between peat and bituminous coal.

Limb In geology, one of the two parts of an anticline or syncline on either side of the axis.

Limestone A sedimentary rock composed largely of the mineral calcite, $CaCO_3$, which has been formed by either organic or inorganic processes. Most limestones have a clastic texture, but nonclastic, particularly crystalline, textures are common. The carbonate rocks, limestone and dolomite, constitute about 22 percent of the sedimentary rocks exposed above sea level.

Limonite Iron oxide with no fixed composition or atomic structure. Always of secondary origin and not a true mineral. Is encountered as ordinary rust, or the coloring material of yellow clays and soils.

Lithification The process by which unconsolidated rock-forming materials are converted into a consolidated or coherent state.

Lithology The study of stones or rocks, especially those of sedimentary origin. Also, the description of the total physical characteristics of specified samples or formations.

Lithosphere The solid outer shell of the earth from 20 to 50 km thick. It includes the upper part of the mantle and the crust. It is divided into six major and many minor slablike plates that are shifting about on the plastic asthenosphere in a complex interlocking pattern.

Littoral Pertaining to the near-shore environment.

Living Fossil A term applied to any organism with a long geologic history, usually one that has outlived the forms with which it was once associated.

Load The amount of material that a transporting agency, such as a stream, a glacier, or the wind, is actually carrying at a given time.

Lode An unusually large vein or set of veins containing ore minerals.

Loess An unconsolidated, unstratified aggregation of small, angular mineral fragments, usually buff in color. Generally believed to be wind-deposited. Characteristically able to stand on very steep to vertical slopes.

Log A record of the earth materials passed through in digging or drilling a test pit or well. It may contain, in addition, notes regarding geologic structure, water conditions, casing used, and so on. Special types of logs are electric, caliper, radioactivity, sample, and so forth.

Longitudinal Dune A long ridge of sand oriented in the general direction of wind movement. A small one is less than 3 m in height and 60 m in length. Very large ones are called *seif dunes.*

Longshore Current A current that moves parallel to a shoreline and is formed from the momentum of breaking waves that approach the shore obliquely.

Mafic Pertaining to dark-colored igneous rocks that are mostly relatively rich in magnesium and iron.

Mafic Mineral A dark-colored mineral rich in iron and magnesium, especially pyroxene, amphibole, or olivine.

Magma Hot mobile rock material generated within the earth, from which igneous rock results by cooling and crystallization. It is usually conceived of as a pasty or liquid material, or a mush of crystals together with a noteworthy amount of liquid phase having the composition of silicate melt.

Magma Chamber A magma-filled cavity within the earth.

Magnetic Declination The angle of divergence between a geographic meridian and a magnetic meridian. It is measured in degrees east and west of geographic north.

Magnetic Inclination The angle that the magnetic needle makes with the surface of the earth. Also called *dip* of the magnetic needle.

Magnetic Pole The north magnetic pole is the point on the earth's surface where the north-seeking end of a magnetic needle free to move in space points directly down. At the south magnetic pole the same needle points directly up. These poles are also known as *dip poles.*

Magnetic Reversal An unexplained but verified phenomenon in which the south magnetic pole becomes the north magnetic pole and vice versa.

Magnetite A mineral; iron oxide, Fe_3O_4. Black, strongly magnetic. An important ore of iron.

Magnetometer An instrument for measuring either one orthogonal component or the entire intensity of the earth's magnetic field at various points.

Magnitude (of an earthquake) A number related to the total energy released by an earthquake. Contrast with *intensity*, which is a number related to the effects of earthquake waves at a particular place.

Mammal A vertebrate animal characterized by warm blood, a covering of hair, live birth (two egg-laying exceptions), and the ability to suckle its young.

Mantle In the geophysical sense, the part of the earth between the surface and the core, excluding the part above the Mohorovičić discontinuity. The term is occasionally used without this exclusion (see *crust*). In a more general sense, mantle refers to the loose material at or near the surface, above bedrock. In zoology, the mantle is the membrane lining the respiratory cavity of molluscs or brachiopods. It also secretes the shell substance.

Marble Metamorphic rock of granular texture, no rock cleavage, and composed of calcite or dolomite or both.

Mare Dark low-lying lunar plain, filled to an undetermined depth with mafic volcanic rocks (plural, *maria*).

Marl Sedimentary material that is mostly soft and clayey, containing shells or other calcareous matter.

Marsh Gas Methane, CH_4, the simplest paraffin hydrocarbon. The dominant component of natural gas.

Marsupial Any of the group of mammals that lack a placenta and have an abdominal pouch in which the immature young remain for some time after birth. Examples are the kangaroo and opossum.

Mass A number that measures the quantity of matter. It is obtained on the earth's surface by dividing the weight of a body by the acceleration due to gravity.

Mass Movement Surface movement of earth materials induced by gravity.

Mass Number Number of protons and neutrons in the nucleus of an atom.

Mass Spectrometer An instrument for separating ions of different mass but equal charge (mainly isotopes) and measuring their relative quantities.

Mass Unit One-sixteenth the mass of the oxygen atom. Approximately the mass of the hydrogen atom.

Matter Anything that occupies space. Usually defined by describing its states and properties: solid, liquid, or gaseous; possesses mass, inertia, color, density, melting point, hardness, crystal form, mechanical strength, or chemical properties. Composed of atoms.

Meander (1) A turn or sharp bend in a stream's course. (2) To turn, or bend sharply. Applied to stream courses in geological usage.

Meander Belt The zone along a valley floor that encloses a meandering river.

Mechanical Weathering The process by which rock is broken down into smaller and smaller fragments as the result of energy developed by physical forces. Also known as *disintegration.*

Medial Moraine A ridge of till formed by the junction of two lateral moraines when two valley glaciers join to form a single ice stream.

Mega- Combining form meaning "great" or "large" (*mega*fossil).

Member A subdivision of a geologic formation that is identified by lithologic characteristics such as color, hardness, composition, and similar features and has considerable geographic extent. Members may receive formal names.

Mesosphere The solid interior of the earth below the hot and plastic asthenosphere. Also the layer of the atmosphere above the stratosphere.

Meta- Combining form meaning "changed" or "altered."

Metal A substance that is fusible and opaque, is a good conductor of electricity, and has a characteristic luster. Examples: gold, silver, aluminum. Over three-fourths of the elements are metals.

Metamorphic Facies An assemblage of minerals that reached equilibrium during metamorphism under a specific range of temperature.

Metamorphic Rocks (1) "Changed-form rocks." Any rock that has been changed in texture or composition by heat, pressure, or chemically active fluids after its original formation. (2) One of the three great groups of rocks. Metamorphic rocks are formed from original igneous or sedimentary rocks through alterations produced by pressure, heat, or the infiltration of other materials at depths below the surface zones of weathering and cementation. Rocks that have undergone only slight changes are not usually considered metamorphic; for practical purposes, the term is best applied to rocks in which transformation has been almost complete or at least has produced characteristics that are more prominent than those of the original rock.

Metamorphic Zone An area subjected to metamorphism and characterized by a certain metamorphic mineral that formed during the process.

Metamorphism A process whereby rocks undergo physical or chemical changes, or both, to achieve equilibrium with conditions other than those under which they were originally formed. Weathering is arbitrarily excluded from the meaning of the term. The agents of metamorphism are heat, pressure, and chemically active fluids.

Metasomatism A process whereby rocks are altered when volatiles exchange ions with them.

Metazoan Any animal in which more than one kind of cell makes up the tissues or organs.

Meteoric Water Ground water derived primarily from precipitation.

Meteorite A mass of mineral or rock matter coming to the earth from space.

Methane The simplest paraffin hydrocarbon, CH_4. The principal constituent of natural gas. Sometimes called *marsh gas.*

Micas A group of silicate minerals characterized by perfect sheet or scale cleavage resulting from their atomic pattern, in which silicon-oxygen tetrahedra are linked in sheets. Biotite is the ferromagnesian black mica. Muscovite is the potassic white mica.

Micro- Combining form meaning "small."

Microfossil Any fossil too small to be

studied without magnification. Includes single organisms, fragments of complete organisms, or colonies of many organisms.

Micropaleontology The branch of paleontology dealing with fossils so small that they require magnification for identification and study.

Microseism A small shaking. Specifically limited in technical usage to earth waves generated by sources other than earthquakes, and most frequently to waves with periods of from a second to about 9 seconds from sources associated with atmospheric storms.

Middle Designates a portion of time between early and late; also a division of rocks, usually stratified, between comparable upper and lower divisions. When capitalized, the term designates a major division of a geologic period (time) or system (rocks).

Midoceanic Ridge A relatively narrow submarine mountain range extending through the center of the major ocean basins. It is composed predominantly of basalt derived from deep in the mantle through a narrow zone that splits to become new sea floor.

Migmatite A mixed rock produced by an intimate interfingering of magma and an invaded rock.

Mineral A naturally occurring solid element or compound, exclusive of biologically formed carbon components. It has a definite composition, or range of composition, and an orderly internal arrangement of atoms known as *crystalline structure,* which gives it unique physical and chemical properties, including a tendency to assume certain geometrical forms known as *crystals.*

Mineral Deposit A local accumulation or concentration of mineral substances or of a single mineral, either metallic or nonmetallic, that is of economic or potentially economic value.

Mobile Belt A belt or tract of the earth's crust, usually long and relatively narrow, that displays evidence of greater geologic activity such as geosynclines, folds, faults, and volcanic activity. Contrasts with stable block.

Mohorovičić Discontinuity A level of major change in the interior of the earth. It is found just beneath the crust at depths ranging from 8 to 32 km.

Mold An impression of the exterior or interior of an object from which it is possible to obtain a cast or reproduction of its outward shape.

Molecule The smallest unit of a compound that displays the properties of that compound.

Mollusc Any member of the numerous group of animals constituting the phylum Mollusca. In general, they are soft-bodied and are protected by a calcareous shell of their own making. There are marine, freshwater, and terrestrial forms, and the range of the phylum is from the Early Cambrian to present.

Monadnock A hill left as a residual of erosion, standing above the level of a peneplain.

Monel Metal Steel containing 68 percent nickel.

Montmorillonite A clay mineral family; a hydrous aluminous silicate with a structural sandwich of one ionic sheet of aluminum and hydroxyl between two (Si_4O_{10}) sheets. These sandwiches are piled on each other with water between them, and with nothing but weak bonds to hold them together. As a result, additional water can enter the lattice readily. This causes the mineral to swell appreciably and further weakens the attraction between structural sandwiches. Consequently, a lump of montmorillonite in a bucket of water slumps rapidly into a loose, incoherent mass. Compare with the other clay minerals, *kaolinite* and *illite.*

Moraine A general term applied to certain landforms composed of till.

Mosasaur A large extinct marine lizard commonly found in Upper Cretaceous rocks. Mosasaurs averaged about 4.5 to 6 m in length.

Mountain Any part of a landmass that projects conspicuously above its surroundings.

Mountain Chain A series or group of connected mountains having a well-defined trend or direction.

Mountain Glacier Synonymous with *alpine glacier.*

Mountain Range A series of more or less parallel ridges, all of which were formed within a single geosyncline or on its borders.

Mud Cracks Cracks caused by the shrinkage of a drying deposit of silt or clay under surface conditions.

Mudflow Flow of a well-mixed mass of rock, earth, and water that behaves like a fluid and flows down slopes with a consistency similar to that of newly mixed concrete.

Mudstone Fine-grained, detrital sedimentary rock made up of silt and clay-sized particles. Distinguished from shale by lack of fissility.

Multituberculate An extinct mammal of the order Multituberculata, which existed in the Late Mesozoic and Early Tertiary. Their teeth are characterized by numerous cusps, and their habits and appearance were evidently somewhat like those of modern rodents.

Muscovite "White mica." A nonferromagnesian rock-forming silicate mineral with its tetrahedra arranged in sheets. Sometimes called *potassic mica.*

Mutation An inherited change stemming from modification of the hereditary material in the reproductive cells. The change may be slight or great.

M.Y. Abbreviation for million years.

Nannofossil A very small microfossil, usually less than 100 microns in diameter.

Native State State in which an element occurs uncombined in nature. Usually applied to the metals, as in native copper, native gold, etc.

Natural Gas Gaseous hydrocarbons that occur in rocks. Dominated by methane.

Natural Remanent Magnetism The magnetism of a rock. May or may not coincide with present magnetic field of the earth. Abbreviation, NRM.

Natural Selection The complex process whereby organisms are eliminated or preserved according to their fitness or adaptation to their surroundings, especially to changes in the environment.

Nautiloid Any of a large group of marine invertebrate organisms constituting a division of the class Cephalopoda of the phylum Mollusca. Typically, they have a straight, curved, or coiled, many-chambered shell. The edges of the septa between chambers have a straight or curved pattern and are not acutely angular or crenulated as in the ammonoids. They range from the Cambrian to the present with a maximum development in the Silurian.

Nebula Faintly luminous object or appearance seen in the heavens. Some nebulae within the Milky Way are masses of gas and dust; others, outside of the local galaxy, are clusters of stars.

Neck Cutoff The breakthrough of a river across the narrow neck separating two meanders, where downstream migration of one has been slowed and the next meander upstream has overtaken it. Compare with *chute cutoff.*

Neutron A proton and an electron combined and behaving like a fundamental particle of matter. Electrically neutral, with a mass of 1.00896 units. If isolated, it decays to form a proton and an electron.

Névé Granular ice formed by the recrystallization of snow. Intermediate between snow and glacier ice. Sometimes called *firn.*

Nickel Steel Steel containing 2.5 to 3.5 percent nickel.

Nivation Erosion beneath and around the edges of a snowbank.

Nodule An irregular, knobby-surfaced body of mineral that differs in composition from the rock in which it is formed. Silica in the form of chert or flint is the major component of nodules. They are commonly found in limestone and dolomite.

Nonconformity A type of unconformity in which an older, eroded sequence of rocks meets a younger, overlying sequence at an angle. Tilting and erosion of the lower sequence before deposition of the higher beds are implied. Some geologists use nonconformity only for cases where the older rock is of plutonic origin; both usages are evidently correct.

Nondipole Magnetic Field That portion of the earth's magnetic field remaining after the dipole field and the external field are removed.

Nonferromagnesians Silicate minerals that do not contain iron or magnesium.

Nonmetal An element that is not a metal, such as oxygen, carbon, sulfur, phosphorus, and boron.

Normal Fault A fault in which the hanging wall appears to have moved downward relative to the footwall. Opposite of a thrust fault. Also called *gravity fault.*

Nova Literally a "new" star but more accurately one that increases suddenly in size and brilliance. After expending tremendous energy for a short while, it fades to obscurity.

NRM See *natural remanent magnetism.*

Nucleic Acid An organic acid characteristic of the nucleus of living cells; DNA (deoxyribonucleic acid) and RNA (ribonucleic acid) are best-known examples.

Nucleus (atomic) The protons and neutrons constituting the central part of an atom.

Nuée Ardente (pl. **nuées ardentes)** "Hot cloud." A French term applied to a highly heated mass of gas-charged lava ejected more or less horizontally from a vent or pocket at the summit of a volcano, onto an outer slope down which it moves swiftly, however slight the incline, because of its extreme mobility.

Nummulite Any of a large group of foraminiferal protozoans having coinlike shells. They are common in the Early Tertiary rocks of the warmer regions of the earth.

Obsidian Glassy equivalent of granite.

Oceanic Crust Portion of the earth's crust composed of one layer of simatic rock 32 to 48 km thick.

Offlap The arrangement of nonconformable sedimentary units in a depositional basin whereby the shoreward edge of each succeedingly younger unit is farther offshore than the unit on which it lies.

Oil In geology, refers to petroleum (*q.v.*).

Oil Shale Shale containing such a proportion of hydrocarbons as to be capable of yielding petroleum on slow distillation.

Olivine A rock-forming ferromagnesian silicate mineral that crystallizes early from a magma and weathers readily at the earth's surface. Its crystal structure is based on isolated SiO_4 ions and positive ions of iron or magnesium, or both. General formula: $(Mg,Fe)_2SiO_4$.

Ontogeny The sequential changes or events in the development of an individual organism.

Oölites Spheroidal grains of sand size, usually composed of calcium carbonate, $CaCO_3$, and thought to have originated by inorganic precipitation. Some limestones are made up largely of oölites.

Ooze Ooze in a sedimentary sense is any soupy deposit covering the bottom of any water body. Specifically, the term relates to more or less calcareous or siliceous deposits that cover extensive areas of the deep-ocean bottom. The marine oozes contain in greater or less quantities the shells of small organisms whose presence in quantities of 25 percent or more leads to differentiation into varieties based on their presence. Thus, there are the *Globigerina,* Pteropod, Radiolarian, and Diatom oozes. The percentage of the shells of these organisms may range from zero to nearly 100. Other constituents of the oozes are minerals of a wide range and various other kinds of organic matter.

Opal Amorphous silica, with varying amounts of water. A mineral gel.

Order of Crystallization The chronological sequence in which crystallization of the various minerals of an assemblage takes place.

Ore A natural deposit in which a valuable metallic element is present in high enough concentrations to make mining economically feasible.

Ore Mineral The mineral of an ore that contains the useful element.

Organic Pertaining to or derived from life or from an organism. Chemically, an organic compound is one in which hydrogen or nitrogen is directly united with carbon.

Orogenic Pertaining to or designating an orogeny or mountain-building disturbance; contrasts with *epeirogenic.*

Orogeny The process by which great

elongate chains and ranges of mountains are formed. Although the process or processes are not well understood, many orogenic movements appear to start with the downwarping of a large trough in the earth's crust, which is filled with sediments. The trough and its included sediments are then mashed, and the width of the belt is greatly shortened by folding and faulting. Igneous activity generally accompanies or follows deformation, and many of the largest bodies of intrusive igneous rocks lie within orogenic belts. The episode of deformation by which a specific system of mountains comes into being may be called an *orogeny*. The word thus seems to signify not only a process but also an event.

Orthoclase The feldspar in which K^+ is the diagnostic positive ion; $K(AlSi_3O_8)$.

Orthogenesis Evolution or development along definite lines as the result of a supposed directing influence.

Osteology The study of bone as such and the ways it is organized into skeletons of bony animals.

Ostracod Any of a great number of small aquatic invertebrates of the phylum Arthropoda (class Crustacea, subclass Ostracoda). Typically, their bodies are small, segmented, and encased in a bivalved, horny, or calcareous shell. There are both marine and nonmarine species, and they range from the Ordovician to the present. They are valuable guide fossils for many marine formations and are of special importance in correlating nonmarine continental rocks.

Ostracoderm Any of a large group of fishlike jawless chordates with the head encased in bony plates or scales.

Outcrop Part of a body of rock that appears bare and exposed at the surface of the ground. In a more general sense, the term also applies to areas where the rock formation occurs next beneath the soil, even though it is not exposed.

Outwash Material carried from a glacier by meltwater. Laid down in stratified deposits.

Outwash Plain Flat or gently sloping surface underlain by outwash.

Overburden Barren rock or soil overlying a mineral deposit or the upper part of any designated sequence of rock that compresses the material below.

Overthrust Fault In a general sense, any reverse fault with low dip; more specifically, a low-angle fault on which the mass above has demonstrably moved or been pushed over a relatively stable mass below the fault. Usually a reverse fault having a dip of less than 20 degrees and a displacement measured in kilometers.

Overturned Fold A fold in which at least one limb is overturned—that is, has rotated through more than 90°.

Oxbow An abandoned meander, caused by a neck cutoff.

Oxbow Lake An abandoned meander isolated from the main stream channel by deposition, and filled with water.

Oxide Mineral A mineral formed by the direct union of an element with oxygen. Examples: ice, corundum, hematite, magnetite, cassiterite.

Ozone A molecule consisting of three atoms of oxygen. Ozone is very reactive and is highly concentrated in a layer about 25 km above the earth's surface.

P Symbol for earthquake primary waves.

Pahoehoe A basaltic lava flow with a glassy, smooth and undulating, or ropy surface.

Paired Terraces Terraces that face each other across a stream at the same elevation.

Paleo- A combining form denoting the attribute of great age or remoteness in the past; e.g., *paleobotany*, the study of fossil plants.

Paleoecology The study of ancient ecology or the relations of fossils to their environment.

Paleocurrent An ancient current (usually of water), the direction of which may be inferred from study of sedimentary structures or textures.

Paleocurrent Map A map showing current directions inferred from crossbedding, ripple marks, or other primary sedimentary structures.

Paleogeographic Map A map that shows the reconstructed geographic features of some specified period in the ancient past.

Paleogeography The study of ancient geography.

Paleolithic Pertaining to the earliest stage in use of stone by mankind; the Old Stone Age.

Paleomagnetism The study of the earth's magnetic field as it has existed during geologic time.

Paleontology A study of the plant and animal life of past periods. It is based on the fossil remains found in the earth.

Pangaea A hypothetical continent from which all others are postulated to have originated through a process of fragmentation and drifting.

Parabolic Dune A dune with a long, scoop-shaped form that, when perfectly developed, exhibits a parabolic shape in plan, with the horns pointing upwind. Contrast *barchan*, in which the horns point downwind. Characteristically covered with sparse vegetation, and often found in coastal belts.

Paraconformity An obscure or uncertain unconformity above and below which the beds are parallel and there is little physical evidence of a long lapse in deposition.

Peat Partially reduced plant or wood material containing approximately 60 percent carbon and 30 percent oxygen. An intermediate material in the process of coal formation.

Pebble Size A volume greater than that of a sphere with a diameter of 4 mm and less than a sphere of 64 mm.

Pedalfer A soil characterized by the accumulation of iron salts or iron and aluminum salts in the B-horizon. Varieties of pedalfers include red and yellow soils of the southeastern United States, and podsols of the northeastern quarter of the United States.

Pediment In geology, a broad, smooth erosional surface developed at the expense of a highland mass in an arid climate. Underlain by beveled rock, and covered by a veneer of gravel and rock debris. The final stage of a cycle of erosion in a dry climate.

Pedocal A soil characterized by an accumulation of calcium carbonate in its profile. Characteristic of low rainfall. Varieties include black and chestnut soils of the northern plains states, and the red and gray desert soils of the drier western states.

Pedology The science that treats of soils—their origin, character, and utilization.

Pegmatite A small pluton of exceptionally coarse texture, with crystals up to 12 m in length, commonly formed at the margin of a batholith and characterized by graphic structure. Nearly 90 percent of all pegmatites are simple pegmatites of quartz, orthoclase, and unimportant percentages of micas. The others are extremely rare ferromagnesian pegmatites and complex pegmatites. Complex pegmatites have as their major components the sialic minerals of simple pegmatites, but they also contain various rare minerals.

Pelagic Deposit Material formed in the deep ocean and deposited there. Example: ooze.

Peneplain An extensive, nearly flat surface developed by subaerial erosion, and close to base level, toward which the streams of the region are reducing it. Originally defined as forming in a humid climate.

Perched Water Table The top of a zone of saturation that bottoms on an impermeable horizon above the level of the general water table in the area. Is generally near the surface, and frequently supplies a hillside spring.

Peridotite A coarse-grained igneous rock dominated by dark-colored minerals, consisting of about 75 percent ferromagnesian silicates and the balance plagioclase feldspars.

Period The fundamental unit of the standard geologic time scale; the time during which a standard system of rocks was formed. Examples are the Devonian, Cretaceous, and Tertiary periods.

Permafrost Permanently frozen ground, or more correctly, ground that remains below freezing temperatures for two or more years.

Permeability For a rock or an earth material, the ability to transmit fluids. Permeability for underground water is sometimes expressed numerically as the number of gallons per day that will flow through a cross section of 0.09 sq m, at 60°F (15°C), under a hydraulic gradient of 100 percent. Permeability is equal to velocity of flow divided by hydraulic gradient.

Petrifaction The process of petrifying, or changing into stone; conversion of organic matter, including shells, bones, and the like, into stone or a substance of stony hardness. Petrifaction is produced by the infiltration of water containing dissolved mineral matter such as calcium carbonate, silica, and so on, which replaces the organic material particle by particle, sometimes with original structure retained.

Petrify To convert organic material such as wood or bone into stone.

Petrographic Microscope An optical microscope designed for examining thin sections of rocks and equipped with polarizing filters on opposite sides of the specimen.

Petroleum A complex mixture of hydrocarbons, accumulated in rocks, and dominated by paraffins and cycloparaffins. Crude petroleums are classified as *paraffin-base* if the residue left after volatile components have been removed consists principally of a mixture of paraffin hydrocarbons; as *asphalt-base* if the residue is primarily cycloparaffins.

Petroleum Resources The total quantity of oil in the earth's crust that may possibly be recovered.

Petrology The science of the origin, occurrence, structure, and history of rocks.

Phase In physical chemistry, a homogeneous, physically distinct portion of matter in a system that is not homogeneous, as in the three phases—ice, water, and aqueous vapor.

Phenocryst A crystal significantly larger than the crystals of surrounding minerals.

Phosphate Rock A sedimentary rock containing calcium phosphate.

Photogeology The study of geology from photographs, usually taken from aircraft or spacecraft.

Photosynthesis The process by which carbohydrates are compounded from carbon dioxide and water in the presence of sunlight and chlorophyll. The synthesis of carbohydrates by green plants in the presence of sunlight.

Phreatic Zone The zone of soil and rock in which pores are completely filled with ground water. Also, termed the *saturated zone.*

Phyllite A clayey metamorphic rock with rock cleavage intermediate between slate and schist. Commonly formed by the regional metamorphism of shale or tuff. Micas characteristically impart a pronounced sheen to rock cleavage surfaces. Has phyllitic cleavage.

Phylogeny The history of any race or group. *Phylogenetic* is the adjective.

Phylum A large group of plants or animals; a major division of a kingdom. Usually divided into subphyla or classes.

Physical Geology The branch of geology that deals with the nature and properties of material composing the earth, distribution of materials throughout the globe, the processes by which they are formed, altered, transported, and distorted, and the nature and development of landscape.

Phytosaur An extinct reptile somewhat like a crocodile in appearance that is included in the order *Thecodonta.* Phytosaurs are characteristic of the Late Triassic.

Piedmont Glacier A glacier formed by the coalescence of valley glaciers and spreading over plains at the foot of the mountains from which the valley glaciers came.

Pillow Lava Lava in its fresh or hardened form that shows pillowlike or baglike forms. The structure results when molten rock encounters water.

Placental Pertaining to or possessing the embryonic organ known as the *placenta,* which attaches the embryo to the uterine wall.

Placer A deposit of sand or gravel, usually of river or beach origin, containing particles of gold or other valuable minerals.

Plagioclase Feldspars Albite and anorthite.

Planetesimal Literally, "little planet," meaning a small body in space that behaves like a small planet in following an orbit around the sun. Planetesimals may be somewhat hypothetical entities but are required in certain theories of earth origin.

Planetology The study of the condensed matter of the solar system including planets, satellites, asteroids, meteorites, and interplanetary material.

Plankton Floating organisms of seas or lakes.

Plastic Deformation Permanent change in shape or volume that does not involve failure by rupture, and that, once started, continues without increase in the deforming force.

Plastic Solid A solid that undergoes change of shape continuously and indefinitely after the stress applied to it passes a critical point.

Plate (lithospheric) One of the large, relatively thin and rigid blocks or segments that make up the lithosphere of the earth. It is thick enough to include the crust and upper mantle (50 to 250 km).

Plateau Basalt Basalt poured out from fissures in floods that tend to form great plateaus. Sometimes called *flood basalt.*

Plate Tectonics The concept that the crust and outer mantle of the earth are divided into large plates or tabular blocks that are capable of mutual reactions that produce earthquake belts, mountain ranges, and other major features.

Playa The flat-floored center of an undrained desert basin.

Playa Lake A temporary lake formed in a playa.

Pleochroic Halo Minute, concentric spherical zones of darkening or coloring that form around inclusions of radioactive minerals in biotite, chlorite, and a few other minerals. About .008 cm in diameter.

Plesiosaur Any of the extinct marine reptiles characterized by a long, flexible neck and flattened body which make up the order *Plesiosauria.*

Plunge The acute angle that the axis of a folded rock mass makes with a horizontal plane.

Pluton A body of igneous rock that is formed beneath the surface of the earth by consolidation from magma. Sometimes extended to include bodies formed beneath the surface of the earth by the metasomatic replacement of older rock.

Plutonic Igneous Rock A rock formed by slow crystallization, which yields coarse texture. Once believed to be typical of crystallization at great depth, but this is not a necessary condition.

Pluvial Lake A lake formed during a pluvial period (*q.v.*).

Pluvial Period A period of increased rainfall and decreased evaporation that prevailed in nonglaciated areas during the time of ice advance elsewhere.

Podsol An ashy gray or gray-brown soil of the pedalfer group. This highly bleached soil, low in iron and lime, is formed under moist and cool conditions.

Point Bar A crescent-shaped accumulation of sand and gravel deposited on the inside of a meander bend.

Polar Compound A compound, such as water, with a molecule that behaves like a small bar magnet with a positive charge on one end and a negative charge on the other.

Polar Wandering A movement of the magnetic poles during past geologic time in leading to the present positions.

Polymorph A mineral that shares the same chemical composition with another physically distinct mineral; calcite and aragonite are polymorphs of calcium carbonate.

Porosity The percentage of open space or interstices in a rock or other earth material. Compare with *permeability.*

Porphyritic A textural term for igneous rocks in which larger crystals, called *phenocrysts,* are set in a finer groundmass, which may be crystalline or glassy, or both.

Porphyry An igneous rock containing a considerable proportion, say 25 percent or more by volume, of large crystals or phenocrysts set in a finer groundmass of small crystals or glass, or both.

Porphyry Copper Deposit A porphyritic igneous rock with a low content of copper sulfides in abundant veins. Can be mined at a profit.

Portland Cement A hydraulic cement consisting of compounds of silica, lime, and alumina.

Positive Area A relatively large tract or segment of the earth's crust that has tended to rise over fairly long periods with respect to adjacent areas.

Potassic Feldspar Orthoclase, $K(AlSi_3O_8)$.

Potential Energy Stored energy waiting to be used. The energy that a piece of matter possesses because of its position or because of the arrangement of its parts.

Pothole A hole ground in the solid rock of a stream channel by sands, gravels, and boulders caught in an eddy of turbulent flow and swirled for a long time over one spot.

Prairie Soils Transitional soils between pedalfers and pedocals.

Precambrian Pertaining to or designating all rocks formed prior to the Cambrian period. The term *Precambrian* is used as a noun as well as an adjective by some geologists.

Precipitation The discharge of water, in the form of rain, snow, hail, sleet, fog, or dew, on a land or water surface. Also, the process of separating mineral constituents from a solution by evaporation (halite, anhydrite) or from magma to form igneous rocks.

Prehistoric Time The interval of time preceding the historic period or the invention of writing.

Pressure Force per unit area applied to the outside of a body.

Primary Waves Earthquake body waves that travel fastest and advance by a push-pull mechanism. Also known as longitudinal, compressional, or *P* waves.

Primate Any member of the placental mammal order *Primates,* characterized by large brains, prehensile hands, and five digits on hands and feet.

Primitive Pertaining to the origin or

beginning of anything; also, rudimentary.

Procaryote An organism consisting of cells that lack nuclear walls, well-defined chromosomes, or organelles. They cannot reproduce by miotic cell division.

Productid Any of a great number of extinct, spine-bearing brachiopods that were common in the Late Paleozoic.

Proton A fundamental particle of matter with a positive electrical charge of 1 unit (equal in amount but opposite in effect to the charge of an electron), and with a mass of 1.00758 units.

Protore The original rock, too poor in mineral values to constitute an ore, from which desired elements have been leached and redeposited as an ore. The process of leaching and redeposition of desired elements is sometimes called *supergene enrichment*, or *secondary sulfide enrichment*.

Protozoan A single-celled organism.

Pseudofossil An object of inorganic but natural origin that might resemble or be mistaken for a fossil.

Pterodactyl Any member of the reptilian order *Pterosauria*, or flying reptiles.

Pteropod Ooze A calcareous deep-sea ooze dominated by the remains of minute molluscs of the group Pteropoda.

Pumice Pyroclastic rock filled with gas-bubble holes. Cellular in texture, with many open compartments sealed from one another, it is usually buoyant enough to float on water.

Push-pull Wave A wave that advances by alternate compression and rarefaction of a medium, causing a particle in its path to move forward and backward along the direction of the wave's advance. In connection with waves in the earth, also known as *primary wave, compressional wave, longitudinal wave*, or *P* wave.

Pyroclastic Rock Fragmental rock blown out by volcanic explosion and deposited from the air. Includes bomb, block, cinder, ash, tuff, and pumice.

Pyroxene Group Ferromagnesian silicates with a single chain of silicon-oxygen tetrahedra. Common example: augite. Compare with *amphibole group* (example: hornblende), which has a double chain of tetrahedra.

Quartz A silicate mineral, SiO_2, composed exclusively of silicon-oxygen tetrahedra with all oxygens joined together in a three-dimensional network. Crystal form is a six-sided prism tapering at the end, with the prism faces striated transversely. An important rock-forming mineral.

Quartzite Metamorphic rock commonly formed by the metamorphism of sandstone and composed of quartz. Has no rock cleavage. Breaks through sand grains as contrasted to sandstone, which breaks around the grains.

Radial Adaptation or Adaptive Radiation The general process whereby a related group of organisms becomes adapted for life in all suitable environments that become available to it.

Radial Drainage An arrangement of stream courses in which the streams radiate outward in all directions from a central zone.

Radioactive Disintegration The change an element or isotope undergoes during radioactivity.

Radioactive Isotope A variety of any element that has a different atomic weight from other isotopes of the same element and is radioactive. Carbon-14 is a radioactive isotope of carbon.

Radioactivity The spontaneous breakdown of an atomic nucleus, with emission of radiant energy.

Radiocarbon Radioactive carbon, mostly carbon-14, but also carbon-10 and carbon-11.

Radiogenic Formed by or resulting from radioactive processes, such as radiogenic heat.

Radiolarian Ooze A siliceous deep-sea ooze dominated by the delicate and complex hard parts of minute marine protozoa called *Radiolaria*.

Radiometric Pertaining to measurements based on radioactive processes.

Radiometric Dating Determination of age of geologic materials by measuring the quantities of radioactive elements and their products.

Range A range of mountains; also the area over which an organism exists; also the duration in time from origin to extinction of a designated species or other group.

Rank A term used to designate the extent to which metamorphism has advanced. Compare with *grade*. Rank is more commonly employed in designating the stage of metamorphism of coal.

Reaction Series See *Bowen's reaction series*.

Recessional Moraine A ridge or belt of till marking a period of moraine formation, probably in a period of temporary stability or slight readvance, during the general wastage of a glacier and recession of its front.

Rectangular Pattern An arrangement of stream courses in which tributaries flow into larger streams at angles approaching 90°.

Recumbent Fold A fold in which the axial plane is more or less horizontal.

Red Beds Red sedimentary rocks of any age.

Reef An aggregation of organisms with hard parts that live or have lived at or near the surface of a body of water, usually marine, and that build up a mound or ridgelike elevation. Reefs are considered sedimentary accumulations.

Refraction (wave) The departure of a wave from its original direction of travel at the interface with a material of different index of refraction (light) or seismic wave velocity.

Refractory A mineral or compound that resists the action of heat and of chemical reagents.

Regional Metamorphism Metamorphism occurring over tens or scores of kilometers.

Regolith Soil, alluvium, and rock fragments lying upon bedrock.

Regression The large-scale withdrawal of sea water from a land surface. Also the succession of sedimentary deposits that results.

Rejuvenation A change in conditions of erosion that causes a stream to begin more active erosion and a new cycle.

Relative Age The age of a given geologic feature, form, or structure stated in terms of comparison with its immediate surroundings, that is, not stated in terms of years or centuries.

Relative Dating The placement of an object or event in its proper chronological order in relation to other things or events without reference to its actual age in terms of years.

Relative Time Dating of events by means of their place in a chronologic order of occurrence rather than in terms of years. Compare with *absolute time.*

Replacement The process whereby the substance of a rock, mineral, ore, or organic fragment is slowly removed by solution, and material of a different composition is deposited in its place.

Reverse Fault A fault in which the hanging wall appears to have moved upward relative to the footwall. Also called *thrust fault.* Contrast with *normal* or *gravity fault.*

Revolution A time of intense geologic disturbance usually manifested by widespread mountain building.

Reworked A fragment of rock or a fossil removed by natural means from its place of origin and deposited in recognizable form in a younger deposit.

Rhyolite A fine-grained igneous rock with the composition of granite.

Richter Magnitude Scale A measure of earthquake size, determined by taking the common logarithm (base *lp*) of the largest ground motion observed during the arrival of a *P* wave or seismic surface wave and applying a standard correction for distance to the epicenter.

Rift Zone A system of fractures in the earth's crust. Often associated with extrusion of lava.

Rigidity Resistance to elastic shear.

Rigidity Modulus The number that expresses a material's rigidity. For example, the number of kilograms per square centimeter necessary to cause a specified change of shape. Represented by the symbol G.

Ring Dike An arcuate, rarely circular, dike with steep dip.

Ripple Mark The undulating surface sculpture of ridges and troughs produced in noncoherent granular materials such as loose sand by the wind, by currents of water, and by agitation of water in wave action.

Roche Moutonnée (pl. **roches moutonnées**) A sheep-shaped knob of rock that has been rounded by the action of glacier ice. Usually only a few meters in height, length, and breadth. A gentle slope faces upstream with reference to the ice movement. A steeper slope attributed to plucking action of the ice represents the downstream side.

Rock In the popular and also in an engineering sense, the term *rock* refers to any hard, solid matter derived from the earth. In a strictly geological sense, a rock is any naturally formed aggregate or mass of mineral matter, whether or not coherent, constituting an essential and appreciable part of the earth's crust. A few rocks are made up of a single mineral, as a very pure limestone. Two or more minerals usually are mixed together to form a rock.

Rock Cycle A concept of the sequences through which earth materials may pass when subjected to geological processes.

Rock Flour Finely divided rock material pulverized by a glacier and carried by streams fed by melting ice.

Rock Flow The movement of solid rock when it is in a plastic state.

Rock-forming Silicate Minerals Minerals built around a framework of silicon-oxygen tetrahedra. Olivine, augite, hornblende, biotite, muscovite, orthoclase, albite, anorthite, quartz.

Rock Glacier A tongue of rock waste found in the valleys of certain mountainous regions. Characteristically lobate and marked by a series of arcuate, rounded ridges that give it the aspect of having flowed as a viscous mass.

Rock Slide Sudden and rapid slide of bedrock along planes of weakness.

Rossi-Forel Scale A scale for rating earthquake intensities. Devised in 1878 by de Rossi of Italy and Forel of Switzerland.

Rounding The degree to which the edges and corners of a grain (particle) become worn and rounded because of abrasion during transportation. Expressed as angular, subangular, subrounded, rounded, or well rounded.

Runoff Water that flows off the land.

Rupture A breaking apart or state of being broken apart.

S Symbol for secondary wave.

Salt (1) Any of a class of compounds derived from acids by replacement of part or all of the acid hydrogen by a metal or metallike radical; NaHSO and Na_2SO_4 are sodium salts of sulfuric acid (H_2SO_4). (2) Halite, common salt; sodium chloride, NaCl.

Saltation Mechanism by which a particle moves by jumping from one point to another.

Salt Dome A mass of NaCl generally of roughly cylindrical shape and with a diameter of about 1.6 km near the top. These masses have been pushed through surrounding sediments into their present positions, sometimes as far as 6,000 m. Reservoir rocks above and alongside salt domes sometimes trap oil and gas.

Sand Clastic particles of sand size, commonly but not always composed of the mineral quartz.

Sand Size A volume greater than that of a sphere with a diameter of $\frac{1}{16}$ mm (.0625 mm or .0025 in.), and less than that of a sphere with a diameter of 2 mm or $\frac{5}{64}$ in.

Sandstone A consolidated rock composed of sand grains cemented together. The size range and composition of the constituents are the same as for sand, and the particles may be rounded or angular. Although sandstones may vary widely in composition, they are usually made up of quartz; and if the term is used without qualification, a siliceous composition is implied.

Sapropel An aquatic ooze or sludge that is rich in organic matter. Believed to be the source material for petroleum and natural gas.

Schist A crystalline metamorphic rock that has closely spaced foliation and tends to split readily into thin flakes or slabs. Dominantly composed of fibrous or platy minerals.

Schistose Cleavage Rock cleavage in which grains and flakes are clearly visible and cleavage surfaces are rougher than in slaty or phyllitic cleavage.

Sea-floor Spreading The concept that there is a massive outward movement of the sea bottom from the crest of the midocean ridges. As the sides move laterally, new material rises from the mantle to fill the rift.

Seamount An isolated, steep-sloped peak rising from the deep ocean floor but submerged beneath the ocean surface. Most have sharp peaks, but some have flat tops and are called *guyots* or *tablemounts.* Seamounts are probably volcanic in origin.

Secondary Wave An earthquake body wave slower than the primary wave. A *shear, shake,* or *S wave.*

Secular Variation of the Magnetic Field A change in inclination, declination, or intensity of the earth's magnetic field. Detectable only from long historical records.

Sediment In the singular, the word is usually applied to material in suspension in water or recently deposited from suspension. In the plural, the word is applied to all kinds of deposits from the waters of streams, lakes, or seas, and in a more general sense to deposits of wind and ice. Such deposits that have been consolidated are generally called *sedimentary rocks.*

Sedimentary Facies An accumulation of deposits that exhibits specific characteristics and grades laterally into other sedimentary accumulations formed at the same time but exhibiting different characteristics.

Sedimentary Rocks Sedimentary rocks are composed of sediment: mechanical, chemical, or organic. They are formed through the agency of water, wind, glacial ice, or organisms and are deposited at the surface of the earth at ordinary temperatures. The materials from which they are made must originally have come from the disintegration and decomposition of older rocks, chiefly igneous. They cover about 75 percent of the land area of the globe.

Sedimentation Strictly, the act or process of depositing sediment from suspension in water. Broadly, all the processes whereby particles of rock material are accumulated to form sedimentary deposits. *Sedimentation,* as commonly used, involves not only aqueous but also glacial, aeolian, and organic agents.

Seif Dune A very large longitudinal dune. As high as 90 m and as long as 97 km.

Seismic Prospecting A method of determining the nature and structure of buried rock formations by generating waves in the ground (commonly by small charges of explosive) and measuring the length of time these waves require to travel different paths.

Seismic Sea Wave A large wave in the ocean generated at the time of an earthquake. Popularly, but incorrectly, known as a *tidal wave.* Sometimes called a *tsunami.*

Seismogram The record obtained on a seismograph.

Seismograph An instrument for recording vibrations, most commonly employed for recording earth vibrations.

Seismology The scientific study of earthquakes and other earth vibrations.

Serpentine A silicate of magnesium common among metamorphic minerals. Occurs in two crystal habits, one platy, known as *antigorite,* the other fibrous, known as *chrysotile.* Chrysotile is an asbestos. The name *serpentine* comes from mottled shades of green on massive varieties, suggestive of the markings of a serpent.

S.G. Symbol for specific gravity.

Shake Wave Wave that advances by causing particles in its path to move from side to side or up and down at right angles to the direction of the wave's advance, in a shake motion. Also called *shear wave* or *secondary wave.*

Shale A general term for lithified muds, clays, and silts that are fissile and break along planes parallel to the original bedding. A typical shale is so fine-grained as to appear homogeneous to the unaided eye, is easily scratched, and has a smooth feel. The lamination, or fissibility, is usually best displayed after weathering.

Shallow Focus Earthquake focus within 60 km (37 miles) or less of the earth's surface.

Shear Change of shape without change of volume.

Shear Modulus See *rigidity modulus.*

Shear Wave Wave that advances by shearing displacements (which change the shape without changing the volume) of a medium. This causes particles in its path to move from side to side or up and down at right angles to the direction of the wave's advance. Also called *shake wave* or *secondary wave.*

Sheeting Joints that are essentially parallel to the ground surface. They are more closely spaced near the surface and become progressively farther apart with depth. Particularly well-developed in granite rocks, but sometimes in other massive rocks as well.

Shell (1) The crust of the earth or some other continuous layer beneath the crust. (2) A thin layer of hard rock. (3) The hard outer covering of an organism or the petrified remains of the covering.

Shield The Precambrian nuclear mass of a continent around which and to some extent on which the younger sedimentary rocks have been deposited. The term was originally applied to the shield-shaped Precambrian area of Canada but is now used for the primitive areas of other continents, regardless of shape.

Shield Volcano A volcano built up almost entirely of lava, with slopes seldom as great as 10° at the summit and 2° at the base. Examples: the five volcanoes on the island of Hawaii.

Sial A term coined from the symbols for silicon and aluminum. Designates the composite of rocks dominated by granites, granodiorites, and their allies and derivatives, which underlie continental areas of the globe. Specific gravity considered to be about 2.7.

Sialic Rock An igneous rock composed predominantly of silicon and aluminum. The term is constructed from "si" for silicon and "al" for aluminum. Average specific gravity about 2.7.

Silica Silicon dioxide, SiO_2. Silica

forms the natural crystalline minerals quartz, cristobalite, and tridymite, and the noncrystalline mineral opal, which carries 2 to 13 percent water. Quartz is the most abundant mineral in the visible portions of the earth's crust and occurs in a great variety of igneous, metamorphic, and sedimentary rocks and as the filling in veins.

Silicate Minerals Minerals with crystal structure containing SiO_4 tetrahedra arranged as (1) isolated units, (2) single or double chains, (3) sheets, or (4) three-dimensional networks.

Silicic Pertaining to or derived from silica or silicon; specifically, designating compounds of silicon, as silicic acid.

Silicon-oxygen Tetrahedron A complex ion composed of a silicon ion surrounded by four oxygen ions. It has a negative charge of four units, is represented by the symbol $(SiO_4)^{4-}$, is the diagnostic unit of silicate minerals, and is the central building unit of nearly 90 percent of the materials of the earth's crust.

Silification The process of combining with or being impregnated with silica. A common method of fossilization.

Sill A tabular body of igneous rock that has been injected while molten between layers of sedimentary or igneous rocks or along the foliation planes of metamorphic rocks. Sills are relatively more extensive laterally than they are thick.

Silt Size A volume greater than that of a sphere with a diameter of $\frac{1}{256}$ mm (.0039 mm or .00015 in.), and less than that of a sphere with a diameter of $\frac{1}{16}$ mm (.0625 mm or .0025 in.).

Sima A term coined from "si" for silicon and "ma" for magnesium. Designates a worldwide shell of dark, heavy rocks. The sima is believed to be the outermost rock layer under deep, permanent ocean basins, such as the mid-Pacific. Originally, the sima was considered basaltic in composition, with a specific gravity of about 3.0. It has been suggested also, however, that it may be peridotitic in composition, with a specific gravity of about 3.3.

Simatic Rock An igneous rock composed predominantly of ferromagnesian minerals. The term is constructed from "si" for silicon and "ma" for magnesium. Average specific gravity 3.0 to 3.3.

Sink A sinkhole.

Sinkhole Depression in the surface of the ground caused by the collapse of the roof over a solution cavern.

Skeleton The hard structure that constitutes the framework supporting the soft parts of any organism. It may be internal as in the vertebrates, or external as in the invertebrates.

Slate A fine-grained metamorphic rock with well-developed slaty cleavage. Formed by the low-grade regional metamorphism of shale.

Slaty Cleavage Rock cleavage in which ease of breaking occurs along planes separated by microscopic distances.

Slickensides Parallel grooves or scratches on a fault plane; may show direction of movement.

Slip Face The steep face on the lee side of a dune.

Slope Failure See *slump*.

Slope Wash Soil and rock material that is being or has been moved down a slope predominantly by the action of gravity assisted by running water that is not concentrated into channels. The term applies to the process as well as the materials.

Slump The downward and outward movement of rock or unconsolidated material as a unit or as a series of units. Also called *slope failure*.

Snowfield A stretch of perennial snow existing in an area where winter snowfall exceeds the amount of snow that melts away during the summer.

Snow Line The lower limit of perennial snow.

Soapstone See *talc*.

Soft Water Water that is free of calcium and magnesium carbonates and other dissolved materials of hard water.

Soil The surficial material that forms at the earth's surface as a result of organic and inorganic processes. Soil varies with climate, plant and animal life, time, slope of the land, and parent material.

Soil Horizon A layer of soil approximately parallel to the land surface with observable characteristics that have been produced through the operation of soil-building processes.

Solar Nebula The cloud of gas and dust from which the solar system supposedly formed.

Solid Matter with a definite shape and volume and some fundamental strength. May be crystalline, glassy, or amorphous (*q.v.*).

Solid Solution Series A series of minerals of identical structure that can contain a mixture of two elements over a range of ratios (the plagioclase feldspars).

Solifluction Mass movement of soil affected by alternate freezing and thawing. Characteristic of saturated soils in high latitudes.

Space Lattice In the crystalline structure of a mineral, a three-dimensional array of points representing the pattern of locations of identical atoms or groups of atoms which constitute a mineral's *unit cell* (*q.v.*). There are 230 pattern types.

Species A group of plants or animals that normally interbreed producing fertile offspring and that resemble each other in structure, habits, and functions.

Specific Gravity A number that represents the ratio between the weight of a given volume of a material and the weight of an equal volume of water at 4°C (39.2°F).

Specific Heat The amount of heat necessary to raise the temperature of one gram of any material through one degree Centigrade.

Spheroidal Weathering The spalling off of concentric shells from rock masses of various sizes as a result of pressures built up during chemical weathering.

Spirifer A general name for brachiopods with wide, pointed, or winged shells.

Spit A sandy bar built by currents into a bay from a promontory.

Splay Deposit A small delta deposited on a flood plain when the stream breaches a levee during a flood.

Spontaneous Generation The appearance of life or living things from dead material without the intervention of outside or supernatural forces.

Spring A place where the water table crops out at the surface of the ground and where water flows out more or less continuously.

Stack A small island that stands as an isolated, steep-sided rock mass just off the end of a promontory. Has been isolated from the land by erosion and by weathering concentrated just behind the end of a headland.

Stage The time-stratigraphic unit next in rank below a series. It is the fundamental working unit in local time-stratigraphic correlation and therefore is employed most commonly to relate any of the various types of minor stratigraphic units in one geologic section or area to the rock column of another nearby section or area with respect to time of origin.

Stalactite Icicle-shaped accumulation of dripstone hanging from a cave roof.

Stalagmite Post of dripstone growing upward from the floor of a cave.

Staurolite A silicate mineral characteristic of middle-grade metamorphism. Its crystalline structure is based on independent tetrahedra with iron and aluminum. It has a unique crystal habit that makes it striking and easy to recognize: six-sided prisms intersecting at 90° to form a cross, or at 60° to form an X.

Stegocephalian Any large extinct amphibian with a broad, flat, bone-covered skull.

Stock A discordant pluton that increases in size downward, has no determinable floor, and shows an area of surface exposure less than 100 sq km. Compare with *batholith.*

Stoping A mechanism by which batholiths have moved into the crust by the breaking off and foundering of blocks of rock surrounding the magma chamber.

Strain Change of dimensions of matter in response to stress. Commonly, unit strain, such as change in length per unit length (total lengthening divided by original length), change in width per unit width, change in volume per unit volume. Contrast with *stress.*

Strata The plural of stratum.

Stratification The characteristic structural feature of sedimentary rocks produced by the deposition of sediments in beds, layers, strata, laminae, lenses, wedges, and other essentially tabular units. Stratification stems from many causes—differences of texture, hardness, cohesion or cementation, color, mineralogical or lithological composition, and internal structure.

Stratigrapher One who studies, or who has expert knowledge of, stratigraphy.

Stratigraphic Section The sequence or aggregation of stratified rock formations characteristic of a particular time or place.

Stratigraphic Trap A structure that traps petroleum or natural gas because of variation in permeability of the reservoir rock, or the termination of an inclined reservoir formation on the up-dip side.

Stratigraphy The branch of geology that deals with the definition and interpretation of the stratified rocks, the conditions of their formation, their character, arrangements, sequence, age, distribution, and especially their correlation, by the use of fossils and other means. The term is applied both to the sum of the characteristics listed above and to the study of these characteristics.

Stratum A single layer of homogeneous or gradational lithology deposited parallel to the original dip of the formation. It is separated from adjacent strata or cross strata by surfaces of erosion, nondeposition, or abrupt changes in character. Stratum is not synonymous with the terms *bed* or *lamination* but includes both. Bed and lamination carry definite connotations of thickness.

Streak The color of the fine powder of a mineral. May be different from the color of a hand specimen. Usually determined by rubbing the mineral on a piece of unglazed porcelain (hardness about 7), known as a *streak plate,* which is, of course, useless for minerals of greater hardness.

Stream Piracy or Stream Capture The process whereby a stream rapidly eroding headward cuts into the divide separating it from another drainage basin, and provides an outlet for a section of a stream in the adjoining valley. The lower portion of the partially diverted stream is called a *beheaded stream.*

Stream Terrace A surface representing remnants of a stream's channel or flood plain when the stream was flowing at a higher level. Subsequent downward cutting by the stream leaves remnants of the old channel or flood plain standing as a terrace above the present level of the stream.

Strength The stress at which rupture occurs or plastic deformation begins.

Stress Force applied to material that tends to change the material's dimensions. Commonly, unit stress, or total force divided by the area over which it is applied. Contrast with *strain.*

Striation A scratch or small channel gouged by glacial action. Bedrock, pebbles, and boulders may show striations produced when rocks trapped by the ice were ground against bedrock or other rocks. Striations along a bedrock surface are oriented in the direction of ice flow across that surface.

Striations (of a mineral) Parallel threadlike lines or narrow bands on the face of a mineral; reflect the internal atomic arrangement.

Strike The direction of the line formed by intersection of a rock surface with a horizontal plane. The strike is always perpendicular to the direction of the dip.

Strike-slip Fault A fault in which movement is almost in the direction of the fault's strike.

Stromatolite A stony structure of convex, columnar, or hemispherical shape with internal laminations built up by blue-green algae. The material is chiefly sediment trapped by the algae when alive.

Structural Geology The study of the architecture of the earth insofar as it is determined by earth movements. *Tectonics* and *tectonic geology* are terms that are synonymous with structural geology. The movements that affect solid rock result from forces within the earth and cause folds, joints, faults, and cleavage. The movement of magma, because it is often intimately associated with the displacement of solid rocks, is also a subject that lies within the domain of structural geology.

Subduction Zone A dipping planar zone descending away from a trench and defined by high seismicity; believed to be a shear zone between a sinking oceanic plate and an overriding plate.

Sublimation The process by which solid material passes into the gaseous state without first becoming a liquid.

Subsequent Stream A tributary stream flowing along beds of less erosional resistance, parallel to beds of greater resistance. Its course is determined subsequent to the uplift that brought the more resistant beds within its sphere of erosion.

Subsidence A sinking of a large area of the earth's crust.

Subsurface Pertaining to, formed, or occurring beneath the surface of the earth.

Subsurface Water Water below the surface of the ground. Also referred to as *underground water*, and *subterranean water*.

Sulfate Mineral (sulphate mineral) Mineral formed by the combination of the complex ion $(SO_4)^{2-}$ with a positive ion. Common example: gypsum, $CaSO_4 \cdot 2H_2O$.

Sulfide Mineral (sulphide mineral) Mineral formed by the direct union of an element with sulfur. Examples: argentite, chalcocite, galena, sphalerite, pyrite, and cinnabar.

Superimposed Stream A stream whose present course was established on young rocks burying an old surface. With uplift, this course was maintained as the stream cut down through the young rocks to and into the old surface.

Superposition The natural order in which rocks are accumulated in beds one above the other. See *law of superposition*.

Surface Wave Wave that travels along the free surface of a medium. Earthquake surface waves are sometimes represented by the symbol *L*.

Surficial Applied to material at or near the surface of the earth; usually unconsolidated soil or rubble.

Suspended Load The fine sediment suspended in a stream because the settling velocity is lower than the upward velocity of eddies.

Suspended Water Underground water held in the zone of aeration by molecular attraction exerted on the water by the rock and earth materials and by the attraction exerted by the water particles on one another.

Swash The landward rush of water from a breaking wave up the slope of the beach.

S wave The secondary seismic wave, traveling slower than the *P* wave, and consisting of elastic vibrations transverse to the direction of travel. Cannot penetrate liquids.

Symmetrical Fold A fold in which the axial plane is essentially vertical. The limbs dip at similar angles.

Syncline A configuration of folded stratified rocks in which the rocks dip downward from opposite directions to come together in a trough. The reverse of an *anticline*.

System A fundamental division or unit of rocks. It is of worldwide application and consists of the rocks formed during a period, as the Cambrian System.

Tablemount See *guyot*.

Tabular A shape with large area relative to thickness.

Taconite Unleached iron formation of the Lake Superior District. Consists of chert with hematite, magnetite, siderite, and hydrous iron silicates. An ore of iron. It averages 25 percent iron, but natural leaching turns it into an ore with 50 to 60 percent iron.

Talc A silicate of magnesium common among metamorphic minerals. Its crystalline structure is based on tetrahedra arranged in sheets. Greasy and extremely soft. Sometimes known as *soapstone*.

Talus A slope established by an accumulation of rock fragments at the foot of a cliff or ridge. The rock fragments that form the talus may be rock waste, sliderock, or pieces broken by frost action. Actually, however, the term "talus" is widely used to mean the rock debris itself.

Tarn A lake formed in the bottom of a cirque after glacier ice has disappeared.

Tar Sand A sandstone containing the densest asphaltic components of petroleum—the end product of evaporation of volatile components or of some thickening process (also, oil-impregnated sandstone).

Taxon A group of organisms making up any specific taxonomic category such as species, genus, family, and so on.

Taxonomy The science of classification, especially the classification of living things.

Tectonic Pertaining to rock structures or earth movements, especially those movements that are widespread.

Tektite A rounded object of glasslike appearance presumed to be of extraterrestrial origin, although none has been observed to fall. The chemical composition is different from obsidian, more like that of shale.

Temperature An arbitrary number that represents the activity of atoms. Degree of heat.

Temporary Base Level A base level that is not permanent, such as that formed by a lake.

Terminal Moraine A ridge or belt of till marking the farthest advance of a glacier. Sometimes called *end moraine*.

Terminal Velocity The constant rate of fall eventually attained by a grain when the acceleration caused by the influence of gravity is balanced by the resistance of the fluid through which the grain falls.

Terrace A nearly level surface, relatively narrow, bordering a stream or body of water, and terminating in a steep bank. Commonly the term is modified to indicate origin, as in *stream terrace* and *wave-cut terrace*.

Terrestrial Deposit A deposit that accumulated above sea level in lakes, streams, moraines, eolian dunes, and so forth.

Terrigenous Deposit Material derived from above sea level and deposited in deep ocean. Example: volcanic ash.

Test With regard to organisms, any

outer covering or supporting structure such as a shell.

Tethys A dominating seaway that lay between the northern landmass (Laurasia) and the southern landmass (Gondwanaland) during Paleozoic and Mesozoic time. The Alpine-Himalayan ranges reveal the deposits of this seaway.

Tetrahedron (pl. **tetrahedra**) A four-sided solid. Used commonly in describing silicate minerals as a shortened reference to the silicon-oxygen tetrahedron (*q.v.*).

Texture The general physical appearance of a rock, as shown by the size, shape, and arrangement of the particles that make up the rock.

Therapsid Any member of the extinct reptilian order *Therapsida.* The various species had many mammallike characteristics, and there were both herbivorous and carnivorous forms of varied size.

Thermal Gradient In the earth, the rate at which temperature increases with depth below the surface. A general average seems to be around 30°C increase per kilometer of depth or 150°F per mile.

Thermal Spring A spring that brings warm or hot water to the surface. Sometimes called *warm spring,* or *hot spring.* Temperature usually 15°F or more above mean air temperature.

Thermonuclear Reaction A reaction in which atomic nuclei fuse into new elements with a large release of heat, especially a reaction that is self-sustaining.

Thermo Remanent Magnetism Magnetism acquired by an igneous rock as it cools below the Curie temperatures of magnetic minerals in the rock. Abbreviation, TRM.

Thin Section A slice of rock ground so thin as to be translucent.

Thrust Fault A fault in which the upper or "hanging" wall appears to have moved upward at a relatively low angle. Also called *reverse fault.*

Tidal Current A water current generated by the tide-producing forces of the sun and the moon.

Tidal Inlet Waterway from open water into a lagoon.

Tidal Wave Popular but incorrect designation for *tsunami.*

Tide Alternate rising and falling of the surface of the ocean, other bodies of water, or the earth itself, in response to forces resulting from motion of the earth, moon, and sun relative to each other.

Till Unstratified and unsorted glacial drift deposited directly by glacier ice.

Tillite Indurated till. The term is reserved for pre-Pleistocene tills that have been indurated or consolidated by processes acting after deposition.

Time-rock Unit (time-stratigraphic unit) A mass of rock defined on the basis of arbitrary time limits and not physical characteristics. The Cambrian System is a time-rock unit including all rocks deposited during the Cambrian Period.

Tombolo A sand bar connecting an island to the mainland, or joining two islands.

Tongue A subdivision of a formation that passes in one direction into a thicker body of similar type of rock and dies out in the other direction.

Topographic Deserts Deserts deficient in rainfall either because they are located far from the oceans toward the center of continents, or because they are cut off from rain-bearing winds by high mountains.

Topset Bed Layer of sediment constituting the surface of a delta. Usually nearly horizontal, and covers the edges of inclined foreset beds.

Toreva Block A large-scale prehistoric slump characteristic of now arid and semiarid sections, as in New Mexico.

Tourmaline A silicate mineral of boron and aluminum, with sodium, calcium, fluorine, iron, lithium, or magnesium. Formed at high temperatures and pressures through the agency of fluids carrying boron and fluorine. Particularly associated with pegmatites.

Trace Element An element that appears in minerals in a concentration of less than 1 percent (frequently $<$ 0.001 percent).

Traction The process of carrying material along the bottom of a stream. Traction includes movement by saltation, rolling, or sliding.

Transcurrent Fault A fault of large displacement in which the movement is lateral and the surface of displacement is steeply inclined.

Transform Fault A large fault displacing the midoceanic ridges.

Transgression The gradual large-scale encroachment of water across a land surface. As a result, the water-laid deposits overlap progressively landward.

Transition Element An element in a series in which an inner shell is being filled with electrons after an outer shell has been started. All transition elements are metallic in the free state.

Transpiration The process by which water vapor escapes from a living plant and enters the atmosphere.

Transverse Dune A dune formed in areas of scanty vegetation and in which sand has moved in a ridge at right angles to the wind. It exhibits the gentle windward slope and the steep leeward slope characteristic of other dunes.

Trap (oil) A sedimentary or deformational structure that impedes the movement of oil and gas, permitting them to collect in the structure.

Travertine A form of calcium carbonate, $CaCO_3$, formed in stalactites, stalagmites, and other deposits in limestone caves, or as incrustations around the mouths of hot and cold calcareous springs. Sometimes known as *tufa,* or *dripstone.*

Trellis Pattern A roughly rectilinear arrangement of stream courses in a pattern reminiscent of a garden trellis, developed in a region where rocks of differing resistance to erosion have been folded, beveled, and uplifted.

Trench Generally a narrow elongate depression. Specifically such a depression found near the edge of a continent or where lithospheric plates converge. The oceanic trenches are about 2 km deeper than surrounding sea floors.

Trilobite A general name for an important group of extinct marine animals (phylum Arthropoda, class Crustacea) whose remains are found in rocks of Paleozoic age. They have a compressed trilobate body with numerous segments in the thoracic region. Some forms are especially valuable guide fossils for the Cambrian Period.

Triple Junction A point that is common to three plates, and which must also be the meeting place of three boundary features, such as divergence zones, convergence zones, or transform faults.

Tritium A radioactive isotope of hydrogen that is produced in the upper atmosphere as a result of cosmic-ray activity. It has a half-life of 12.5 years, and it can be used in tracing and dating water masses in the ocean or underground.

TRM See *thermo remanent magnetism.*

Truncated Spur The beveled end of a divide between two tributary valleys where they join a main valley that has been glaciated. The glacier of the main valley has worn off the end of the divide.

Tsunami (pl. **tsunami**) A large wave in the ocean generated at the time of an earthquake. Popularly, but incorrectly, known as a *tidal wave.* Sometimes called *seismic sea wave.*

Tufa Calcium carbonate, $CaCO_3$, formed in stalactites, stalagmites, and other deposits in limestone caves, as incrustations around the mouths of hot and cold calcareous springs, or along streams carrying large amounts of calcium carbonate in solution. Sometimes known as *travertine,* or *dripstone.*

Tuff Rock consolidated from volcanic ash.

Tundra A stretch of arctic swampland developed on top of permanently frozen ground. Extensive tundra regions have developed in parts of North America, Europe, and Asia.

Turbidite A deposit laid down by a turbidity (density) current.

Turbidity Current A current in which a limited volume of turbid or muddy water moves faster relative to surrounding water because of its greater density.

Turbulent Flow Mechanism by which a fluid such as water moves near a rough surface. Fluid not in contact with the irregular boundary outruns that which is slowed by friction or deflected by the uneven surface. Fluid particles move in a series of eddies or whirls. Most stream flow is turbulent, and turbulent flow is important in both erosion and transportation. Contrast with *laminar flow.*

Type Locality The place from which the name of a geologic formation is taken or from which the type specimen of an organism comes.

Type Section The particular section of a formation that is formally presented as being typical of the formation as a whole.

Ultimate Base Level Sea level, the lowest possible base level for a stream.

Ultramafic Rock An igneous rock mainly of mafic minerals, with less than 10 percent feldspar. Includes dunite, peridotite, amphibolite, and pyroxenite.

Unconformity A buried erosion surface separating two rock masses, the older of which was exposed to erosion for a long interval of time before deposition of the younger. If, in the process, the older rocks were deformed and were not horizontal at the time of subsequent deposition, the surface of separation is an *angular unconformity.* If the older rocks remained essentially horizontal during erosion, the surface separating them from the younger rocks is called a *disconformity.* An unconformity that develops between massive igneous rocks that are exposed to erosion and then covered by sedimentary rocks is called a *nonconformity.*

Underground Water Water below the surface of the ground. Also referred to as *subsurface water,* and *subterranean water.*

Uniformitarianism The belief or principle that the past history of the earth and its inhabitants is best interpreted in terms of what is known about the present. Uniformitarianism would explain the past by appealing to known laws and principles acting in a gradual, uniform way through past ages.

Unpaired Terrace A terrace formed when an eroding stream, swinging back and forth across a valley, encounters resistant rock beneath the unconsolidated alluvium and is deflected, leaving behind a single terrace with no corresponding terrace on the other side of the stream.

Upwarp A broad area uplifted by internal forces.

Upwelling Current The upward movement of cold bottom water in the sea, which takes place when wind or currents displace the lighter surface water.

Vadose Zone The region in the ground between the surface and the water table in which pore space is not filled with water.

Valley Glacier A glacier confined to a stream valley. Usually fed from a cirque. Sometimes called *alpine glacier* or *mountain glacier.*

Valley Train Gently sloping plain underlain by glacial outwash and confined by valley walls.

Valve One of the separate movable pieces that make up the shells of invertebrates such as pelecypods and brachiopods.

Van Allen Belts Belts composed mostly of energetic ionized nuclei of hydrogen atoms and electrons, trapped in the outer atmosphere by the earth's magnetic field.

Variety A subdivision of a species.

Varves The regular layers or alternations of material in sedimentary deposits that are caused by annual seasonal influences. Each varve represents the deposition during a year and consists ordinarily of a lower part deposited in summer and an upper, fine-grained part deposited in the winter. Varves of silt and claylike material occur abundantly in glacial-lake sediments, and varves are believed to have been recognized in certain marine shales and in slates. The counting of varves and correlation of sequences have been applied toward establishing both the absolute and relative ages of Pleistocene glacial deposits. Varved "clays" of glacial origin commonly consist largely of very finely divided quartz, feldspar, and micaceous minerals rather than mostly true clay minerals.

Vascular Plant Any plant with tissues specifically adapted to conduct liquids or gas.

Vein A deposit of foreign minerals within a rock fracture or joint.

Velocity of a Stream Rate of motion of a stream measured in terms of the distance its water travels in a unit of time, usually in meters per second.

Ventifact A pebble, cobble, or boulder that has had its shape or surface modified by wind-driven sand.

Vertebrate An animal with a backbone.

Virtual Geomagnetic Pole For any one locality, the pole consistent with the magnetic field as measured at that locality. The term refers to magnetic-field direction of a single point, in contrast to "geometric pole," which refers to the best fit of a geocentric dipole for the entire earth's field. Most paleomagnetic readings are expressed as virtual geomagnetic poles.

Viscosity An internal property of rocks that offers resistance to flow. The ratio of deforming force to rate at which changes in shape are produced.

Volatile Components Materials in a magma, such as water, carbon dioxide, and certain acids, whose vapor pressures are high enough to cause them to become concentrated in any gaseous phase that forms.

Volcanic Ash The unconsolidated, fine-grained material thrown out in volcanic eruptions. It consists of minute fragments of glass and other rock material, and in color and general appearance may resemble organic ashes. The term is generally restricted to deposits consisting mainly of fragments less than 4 mm in size. Very fine volcanic ash composed of particles less than 0.05 mm may be called *volcanic dust.* The indurated equivalent of volcanic ash is tuff.

Volcanic Bomb A rounded mass of newly congealed magma blown out in an eruption.

Volcanic Dust Pyroclastic detritus consisting of particles of dust size.

Volcanic Earthquakes Earthquakes caused by movements of magma or explosions of gases during volcanic activity.

Volcanic Eruption The explosive or quiet emission of lava, pyroclastics, or volcanic gases at the earth's surface, usually from a volcano but rarely from fissures.

Volcanic Neck The solidified material filling a vent or pipe of a dead volcano.

Volcanism The phenomena related to or resulting from the action or actions of a volcano.

Volcano A landform developed by the accumulation of magmatic products near a central vent.

Warm Spring A spring that brings warm water to the surface. A *thermal spring.* Temperature 15°F (−9°C) or more above mean air temperature.

Water Gap The gap cut through a resistant ridge by a superimposed or antecedent stream.

Water Table The upper surface of the zone of saturation for underground water. It is an irregular surface with a slope or shape determined by the quantity of ground water and the permeability of the earth materials. In general, it is highest beneath hills and lowest beneath valleys.

Wave-cut Terrace A level surface formed by wave erosion of coastal beds to the bottom of the turbulent breaker zone. May be seen above sea level if region is uplifted.

Weathering The response of materials that were once in equilibrium within the earth's crust to new conditions at or near contact with water, air, or living matter.

Wind Gap The general term for an abandoned water gap.

Xenolith A strange rock broken from the wall surrounding a magma chamber and frozen in the intrusion as it solidified.

X-ray Diffraction In mineralogy or petrology, the process of identifying mineral structures by exposing crystals to X rays and studying the resulting diffraction pattern.

Yardang A sharp-edged ridge between two troughs or furrows excavated by wind action.

Yazoo-type River A tributary that is unable to enter its main stream because of natural levees along the main stream. The Yazoo-type river flows along the back-swamp zone parallel to the main stream.

Yield Point The maximum stress that a solid can withstand without undergoing permanent deformation either by plastic flow or by rupture.

Zeolite A class of silicates containing H_2O in cavities within the crystal structure. Formed by alteration at low temperature and pressure of other silicates, frequently volcanic glass.

Zone A subdivision of stratified rock based primarily on fossil content. It may be named after the fossil or fossils it contains. Strictly speaking, the zone of paleontological stratigraphy is based not on one but on two or more designated fossils. No fixed thickness or lithology is implied by the term *zone.*

Zone of Aeration A zone immediately below the surface of the ground, in which the openings are partially filled with air, and partially with water trapped by molecular attraction. Subdivided into (a) belt of soil moisture, (b) intermediate belt, and (c) capillary fringe.

Zone of Flow The subsurface part of the earth in which the breaking of material is prevented by pressure and all deformation is by some type of flow. The term is also used in reference to the deeper parts of glaciers where fracturing of the ice does not take place.

Zone of Fracture The upper portion of the earth's crust in which rocks are deformed mainly by fracture. The term is also applied to the outer part of glaciers where fracturing occurs.

Zone of Saturation Underground region within which all openings are filled with water. The top of the zone of saturation is called the *water table.* The water that is contained within the zone of saturation is called *ground water.*

Index

Italicized page numbers indicate illustrations.

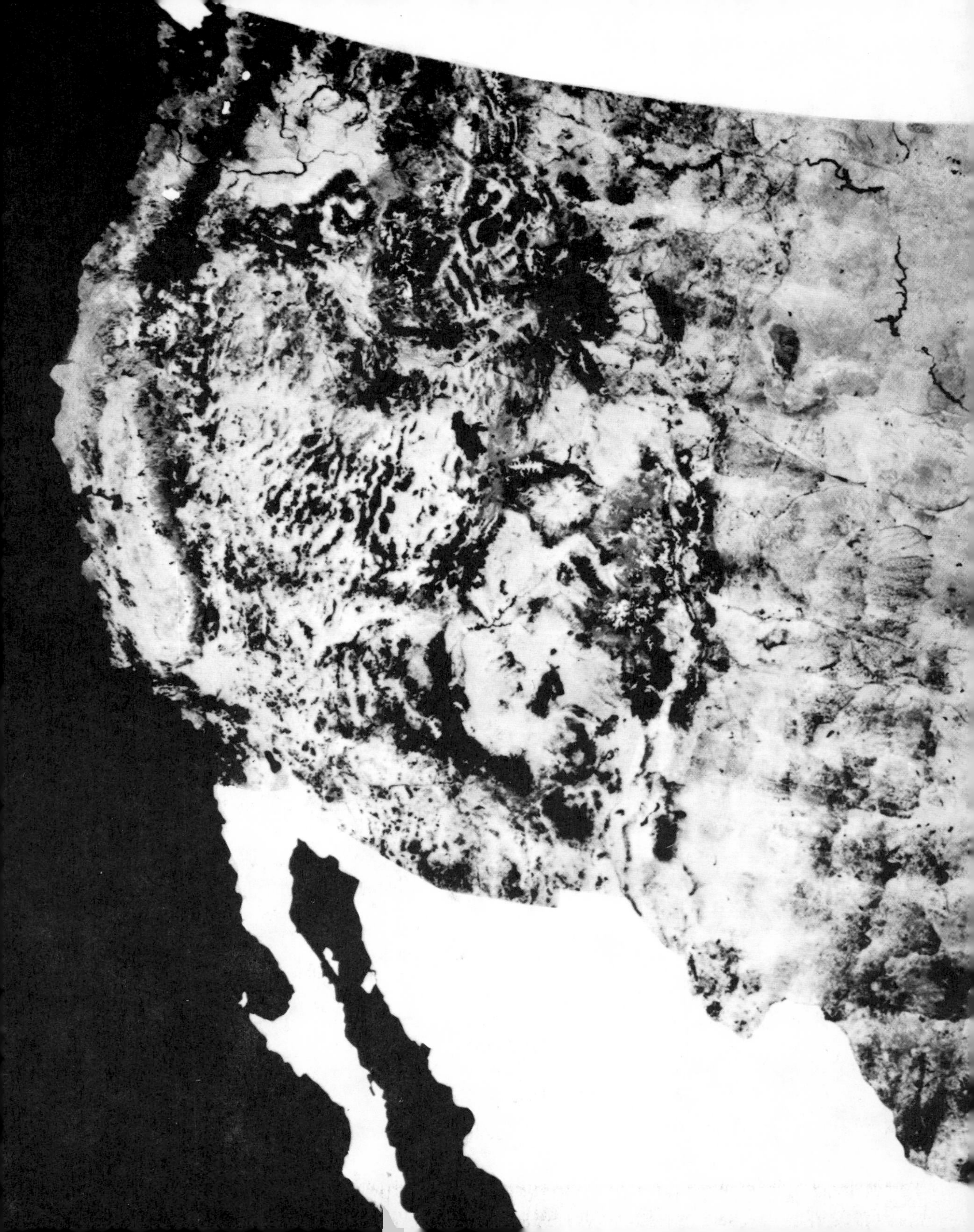